Calcium in Internal Medicine

Springer
London
Berlin
Heidelberg
New York
Barcelona
Hong Kong
Milan
Paris
Singapore
Tokyo

Hirotoshi Morii, Yoshiki Nishizawa and
Shaul G. Massry (Eds)

Co-editors: Masaaka Inaba and Miroslaw J. Smogorzewski

Calcium in Internal Medicine

 Springer

Hirotoshi Morii, MD, PhD
Professor Emeritus, 2nd Department of Internal Medicine
Osaka City University Medical School, 1-4-3 Asahi-machi, Abeno-ku
Osaka 545-8585, Japan

Yoshiki Nishizawa, MD, PhD
2nd Department of Internal Medicine
Osaka City University Medical School, 1-4-3 Asahi-machi, Abeno-ku
Osaka 545-8585, Japan

Shaul G. Massry, MD
Professor of Medicine, Division of Nephrology
Raulston Memorial Research Building, Room 104
Keck School of Medicine, University of Southern California
2025 Zonal Avenue, Los Angeles, CA 90033, USA

British Library Cataloguing in Publication Data
Calcium in internal medicine
  1. Calcium in the body 2. Calcium – Physiological effect
  I. Morii, H. (Hirotoshi) II. Nishizawa, Yoshiki III. Massry, Shaul G.
  612.3'924
  ISBN-13:978-1-4471-1173-3

Library of Congress Cataloging-in-Publication Data
Calcium in internal medicine / Hirotoshi Morii, Yoshiki Nishizawa and Shaul G. Massry (eds.).
    p. ; cm.
  Includes bibliographical references and index.
  ISBN-13:978-1-4471-1173-3   e-ISBN-13:978-1-4471-0667-8
  DOI:10.1007/978-1-4471-0667-8

    1. Calcium – Metabolism – Disorders. 2. Calcium in the body. 3. Calcium – Pathophysiology. I. Morii, H. (Hirotoshi) II. Nishizawa, Yoshiki, 1945– III. Massry, Shaul G.
  [DNLM: 1. Calcium Metabolism Disorders. 2. Minerals – metabolism. 3. Nutritional Requirements. WD 200.5.C2 C144 2002]
  RC632.C26 C35 2002
  616.3'99 – dc21   2001049594

ISBN-13:978-1-4471-1173-3
Springer-Verlag London Berlin Heidelberg
a member of BertelsmannSpringer Science+Business Media GmbH
http://www.springer.co.uk

Typeset by SNP Best-set Typesetter Ltd., Hong Kong

28/3830-543210   Printed on acid-free paper   SPIN 10786399

This book is dedicated to our wives:
Kumi Morii
Toshiko Nishizawa
Meira Massry

Foreword

Calcium in Internal Medicine was edited by Drs. Hirotoshi Morii, Yoshiki Nishizawa and Shaul G. Massry with contributions by many specialists in bone and calcium studies in the world especially from the standpoint of clinical medicine. Generally, it is very important for clinical medicine to be aware of the advances in the basic sciences to help solve the problems in diagnosis and treatment of disease. Basic sciences have continued to make great contributions to an understanding of many unsolved problems in clinical medicine, leading to the development of new methods of treatment. Calcium metabolism is linked to many disorders, thus the factors regulating calcium both at the systemic level, such as 1,25-dihydroxyvitamin D3, parathyroid hormone and calcitonin, or at the cellular level such as prostaglandins, cytokines and intracellular signaling systems are of great interest. This monograph attempts to link diseases through calcium metabolism, suggesting areas of needed investigation. One such issue presented is the relationship between bone and vascular tissues. Atherosclerosis and osteoporosis may be related and there may be factors common to metabolism and function of both bone and vascular tissues. The relationship between osteoporosis and colon cancer as well as breast cancer through calcium will be another issue presented. By linking disorders through calcium, this book may find a unique position among the many other volumes on bone and calcium.

Hector F. DeLuca, PhD
Steenbock Research Professor
Chairman
Department of Biochemistry
University of Wisconsin-Madison

Preface

Calcium in Internal Medicine was planned and turned into practice after many discussions among editors for almost ten years. They have backgrounds as internists and endocrine nephrologists or nephrologic endocrinologists. They have come to notice more and more the importance of calcium and its regulatory systems in clinical medicine.

One of the editors, Hirotoshi Morii, whose late teacher was Shigeo Okinaka, Professor Emeritus at the University of Tokyo, was advised to dedicate himself to the study of calcium in the early 1960s, even before discovery of calcitonin by Harold Copp and the mechanism of activation of vitamin D by Hector F. DeLuca. That was just before the time of tremendous progress in this field of basic and clinical studies on bone and calcium metabolism. The understanding of calcium physiology was deepened by studies by many investigators: Fuller Albright, William F. and Margaret W. Neumans, Franklin C. McLean, M. R. Urist, B. E. C. Nordin, L. G. Riggs, Lawrence G. Raisz, John Potts, Gerald Aurbach, Roy V. Talmage, A. M. Parfitt, Robert Lindsay, Pierre J. Meunier, Eduardo Slatopolsky, Etsuro Ogata, Tatsuo Suda, Yasuho Nishii and Takuo Fujita.

Clinically, haemodialysis treatment has become more popular for patients with chronic renal failure. Although longevity was the main concern in the initial stage of the treatment, complications associated with chronic renal failure or with the procedure of the treatment have raised serious problems. Secondary hyperparathyroidism was one of the most important issues. Shaul G. Massry was the first to systematise the importance of excess PTH in chronic renal failure. Renal bone disease is very important, but so are complications in other organs; nervous, haematopoietic, and immune systems, and so on. Metabolism of glucose and insulin, and of lipids was also demonstrated to have relations with PTH. Some of these topics are discussed in the chapters of this monograph.

On the other hand, with the increase of longevity worldwide, the importance of the problem of osteoporosis has come to be realised. Primary and secondary osteoporosis is a serious health problem in all countries in the world. There are many risk factors, which should be overcome in the early stage of the disease. Osteoporosis is not a simple bone disease; it has higher mortality and morbidity compared with control populations. Cardiovascular complications including atherosclerosis and hypertension, dyslipidaemia or cancer are those that should be considered. Yoshiki Nishizawa has been interested in the problems of calcification of the arteries; abnormal lipid metabolism may be linked to excess PTH in some situations.

In preparing this monograph, there were contributions from many investigators in this field of study. Readers can enjoy their own point of view in each title. The basic concept is to raise the clinical standpoint, to solve the problems in clinical medicine through basic medicine and to feedback the results to clinical medicine.

Written medicine is medicine in the past, and annoying patients in front of
you are teaching you the content of the medicine of tomorrow.
– Shigeo Okinaka, 1971 (translation by H. Morii)

There was also much help and advice in the process of this publication: Dr.
Shuzo Otani, Dr. Shoji Fukushima, Dr. Takashi Inoue, Dr. Masao Kim, Dr.
Kiyoshi Nakatsuka were especially appreciated. We also received help and
advice in the process of publication, and we would like to thank Dr. Yasuhisa
Okuno, Dr. Kazuko Iba, Dr. Kei Tsumura, Dr. Tsutomu Tabata, Dr. Senji
Okuno, Dr. Hiroshi Kishimoto, Dr. Hitoshi Goto, Dr. Masanori Emoto, Dr.
Yasuo Imanishi, Prof. Morii, Mr. Masamitsu Higashihata, Dr. Sayumi Higashi,
and Dr. Ayako Shiraishi.

We would also like to express our sincere gratitude for the support and
the contribution from Dr. Shunsaku Uchida and all members of the Alumni
Association of Second Department of Internal Medicine, Osaka City University, Graduate School of Medicine.

We are grateful to Miss Hiroko Inage and Miss Junko Ukita for their sec-
retarial assistance in arrangement of the manuscripts.

In the final analysis very little is known about anything, and much that seems
true today turns out to be only partly true tomorrow.
– Fuller Albright and Edward C Reifenstein Jr, 1948

January, 2001
Hirotoshi Morii
Yoshiki Nishizawa
Shaul G. Massry

Contents

List of Contributors

Tadao Akizawa, MD, PhD
Professor and Chief
Center of Blood Purification Therapy
Wakayama Medical University
Wakayama, Japan

Kazuo Chihara, MD
Professor and Chief
Department of
    Endocrinology/Metabolism,
Neurology and Hematology/Oncology
Kobe University Graduate School of
    Medicine
Kobe, Japan

Jack W. Coburn, MD
Nephrology Section, 111L
West Los Angeles VA Medical Center
Los Angeles, CA, USA

Hector F. DeLuca, PhD
Steenbock Research Professor
Chairman
Department of Biochemistry
University of Wisconsin-Madison
USA

Ginji Endo, MD, PhD
Department of Preventive Medicine
    and Environmental Health
Osaka City University Graduate School
    of Medicine
Osaka, Japan

João M. Frazão, MD, PhD
Department of Nephrology
Hospital S. João and
Porto School of Medicine
Porto, Portugal

Peter A Friedman, PhD
Professor of Pharmacology and
    Medicine
Departments of Pharmacology
School of Medicine
University of Pittsburgh
Pittsburgh, PA, USA

Takuo Fujita, MD
Professor Emeritus
Kobe University
Honorary Director, Katsuragi
    Hospital
Calcium Research Institute
Katsuragi Hospital
Osaka, Japan

Masafumi Fukagawa, MD, PhD FJSIM
Associate Professor and Chief
Division of Nephrology and Dialysis
    Center
Kobe University School of Medicine
Japan

Paul Glendenning MB.ChB, PhD
    FRACP
Department of Endocrinology and
    Metabolism
7th Floor, Services and Teaching Block
Royal Adelaide Hospital
Adelaide, South Australia, Australia

Naoyuki Hasebe, MD, PhD
First Department of Internal
    Medicine
Asahikawa Medical College
Asahikawa, Japan

Yasuo Imanishi, MD, PhD
Department of Metabolism,
    Endocrinology and Molecular
    Medicine
Osaka City University Graduate School
    of Medicine
Osaka, Japan

Masaaki Inaba, MD, PhD
Associate Professor
Department of Metabolism,
    Endocrinology and Molecular
    Medicine
Osaka City University Graduate School
    of Medicine
Osaka, Japan

Eiji Ishimura, MD, PhD, FJSIM
Associate Professor
Department of Metabolism,
  Endocrinology and Molecular
  Medicine
Osaka City University Graduate School
  of Medicine
Osaka, Japan

Masako Ito, MD, PhD
Associate Professor
Department of Radiology
Nagasaki University
1-7-1 Sakamoto
Nagasaki, Japan

Dirk Kerstan, PhD
Department of Medicine
University of British Columbia
Koerner Pavilion
University Hospital
Vancouver, BC, Canada

Kenjiro Kikuchi, MD, PhD
First Department of Internal Medicine
Asahikawa Medical College
Asahikawa, Japan

Eriko Kinugasa, MD, PhD
Associate Professor
Department of Internal Medicine
Showa University Northern Yokohama
  Hospital
Tsuzuki-ku
Yokohama, Japan

Toshiaki Konishi, MD, PhD
Department of Metabolism,
  Endocrinology and Molecular
  Medicine
Osaka City University Graduate School
  of Medicine
Osaka, Japan

Hidenori Koyama, MD, PhD
Dept Metabolism, Endocrinology &
  Molecular Medicine
Osaka City University Graduate School
  of Medicine
Osaka, Japan

Kiyoshi Kurokawa, MD, PhD
Dean and Professor of Internal
  Medicine
Tokai University School of Medicine
Kanagawa, Japan

M. Chris Langub, PhD
Assistant Professor of Medicine
Department of Internal Medicine
University of Kentucky
Lexington, KY, USA

Kiyoshi Maekawa, MD, PhD
Department of Metabolism,
  Endocrinology and Molecular
  Medicine
Osaka City University Graduate School
  of Medicine
Osaka, Japan

Hartmut H. Malluche, MD
Professor and Chief
G. "Robin" Luke Chair in Nephrology
Division of Nephrology
Department of Internal Medicine
University of Kentucky
Lexington KY, USA

Shaul G. Massry, MD
Professor of Medicine
Keck School of Medicine
University of Southern California
Los Angeles, CA, USA

Ken-ichi Matsumoto, MS
Sakai City Institute of Public Health
Sakai, Japan

Takami Miki, MD, PhD
Associate Professor
Department of Geriatrics
Osaka City University Graduate of
  School of Medicine
Osaka, Japan

Ken-ichi Miyamoto, PhD
Department of Nutritional Science
School of Medicine, The University of
  Tokushima
Tokushima, Japan

Marie-Claude Monier-Faugere, MD
Professor of Medicine
Department of Internal Medicine
University of Kentucky
Lexington, KY, USA

Hirotoshi Morii, MD, PhD
Professor Emeritus
Osaka City University
Osaka, Japan

Kyoko Morita, PhD
Department of Clinical Nutrition
School of Medicine, The University of
  Tokushima
Tokushima, Japan

Hiroki I Motoyama, MD, PhD
Research Associate
Department of Pediatrics and Child
  Health
Kurume University School of
  Medicine
Fukuoka, Japan

Kiyoshi Nakatsuka, MD, PhD
Dept. Metabolism
Endocrinology and Molecular Medicine
Osaka City University, Graduate School
  of Medicine
Osaka, Japan

Allan G. Need MBBS, MD, FRACP,
  FRCPA
Division of Clinical Biochemistry
Institute of Medicine and Veterinary
  Science
Adelaide, South Australia, Australia

Yoshiki Nishizawa, MD, PhD
Porfessor
Department of Metabolism,
  Endocrinology and Molecular
  Medicine
Osaka City University Graduate School
  of Medicine
Osaka, Japan

B.E. Christopher Nordin, MD, FRCP,
  FRACP, DSc
Division of Clinical Biochemistry
Institute of Medicine and Veterinary
  Science
Adelaide, South Australia, Australia

Toshio Okano, PhD
Department of Hygienic Sciences
Kobe Pharmaceutical University
Kobe, Japan

Gary A.Quamme, DVM, PhD
Department of Medicine
University of British Columbia
University Hospital
Vancouver, BC, Canada

B. Peter Sawaya, MD
Associate Professor of Medicine
Department of Internal Medicine
University of Kentucky
Lexington, KY, USA

Debra H. Schussheim, MD
Instructor in Clinical Medicine
College of Physicians & Surgeons of
  Columbia University
Department of Medicine, Division of
  Endocrinology
New York, NY, USA

Yoshiki Seino, MD, PhD
Department of Pediatrics,
Okayama University Graduate School
  of Medicine and Dentistry
Okayama, Japan

Gaurang M. Shah, MD
Professor of Medicine
University of California,
Irvine, CA
Chief, Nephrology Section
Long Beach VA Medical Center
Long Beach, CA, USA

Atsushi Shioi, MD, PhD
Department of Cardiovascular
  Medicine
Institute of Geriatrics and Medical
  Science
Osaka City University Graduate School
  of Medicine
Osaka, Japan

Tetsuo Shoji, MD, PhD
Department Metabolism,
  Endocrinology and Molecular
  Medicine
Osaka City University Graduate School
  of Medicine
Osaka, Japan

Shonni J. Silverberg, MD
Professor of Clinical Medicine
College of Physicians & Surgeons of
  Columbia University
Department of Medicine, Division of
  Endocrinology
New York, NY, USA

Miroslaw Smogorzewski, MD, PhD
Associate Professor of Medicine
Division of Nephrology
University of Southern California
Keck School of Medicine
Los Angeles, CA, USA

Toshitsugu Sugimoto, MD, PhD
Associate Professor
Department of
   Endocrinology/Metabolism,
   Neurology and
   Hematology/Oncology
Kobe University Postgraduate School of
   Medicine
Kobe, Japan

Hideki Tahara, MD, PhD
Department of Metabolism,
   Endocrinology and Molecular
   Medicine
Osaka City University Graduate School
   of Medicine
Osaka, Japan

Yoshitomo Takaishi DDS, PhD
Takaishi Dental Clinic
Himeji, Japan

Eiji Takeda, MD, PhD
Department of Clinical Nutrition
School of Medicine, The University of
   Tokushima
Tokushima, Japan

Yutaka Taketani, PhD
Department of Clinical Nutrition
School of Medicine, The University of
   Tokushima
Tokushima, Japan

Yasuhiro Uchida, PhD
Chugai Lilly Clinical Research Co., Ltd.
Tokyo, Japan

Satoshi Uematsu, MD
Department of Metabolism,
   Endocrinology and Molecular
   Medicine
Osaka City University Graduate School
   of Medicine
Osaka, Japan

Toru Yamaguchi, MD
Department of Endocrinology/
   Metabolism/Neurology and
   Hematology/Oncology
3rd Division of Internal Medicine
Kobe University Postgraduate School of
   Medicine
Kobe, Japan

Hironori Yamamoto, PhD
Department of Clinical Nutrition
School of Medicine, The University of
   Tokushima
Tokushima, Japan

Yoshitaka Yamanaka, MD, PhD
Department of Pediatrics
Okayama University Graduate School
   of Medicine and Dentistry
Okayama, Japan

Clinical Presentation of Derangements of Mineral Metabolism

H. Morii, K. Matsumoto, G. Endo, S. Uematsu, T. Konishi and K. Maekawa

1

Studies on the role of calcium in physiology and pathogenesis of various kinds of diseases have made tremendous progress in recent years. Basic and clinical medicines have close relationships to each other. Pathogenesis of symptoms and signs in clinical medicine has relationships with calcium in various ways. Clinicians ask what are the mechanism of diseases, how to diagnose them and how to treat them. Responses of basic scientists are accepted by clinicians and again new questions are raised. Abnormalities of calcium metabolism have been classified in various ways but the recent progress in the understanding of the role of calcium has necessitated repeated revisions of the method of classification. There have been many contributions to studies on calcium physiology and diseases. The first monumental achievement was done by Albright and Reifenstein [1], followed by Fourman [2], Danowski [3], Jackson [4] and others, especially from the clinical standpoint.

Albright and Ellsworth described calcium physiology as being regulated by parathyroid hormone (PTH). They described how a patient was diagnosed as having hyperparathyroidism and how he was treated [5]. Based on clinical observations, Albright proposed that PTH may act primarily on phosphate metabolism in some way which not only increases the excretion of phosphate in the urine but also produces certain bone changes directly [1]. Until his conclusion was issued, there were two schools of thought regarding the role of PTH: one proposed that PTH acts directly on bone and another believed that PTH acts on the electrolyte equilibria of the body, and the bone changes are secondary to the chemical changes [1]. Albright reconciled the two schools and integrated them into the one which prevailed in the study of bone and calcium physiology thereafter. Massry and Smogorzewski added some important roles of PTH from their observations on patients with chronic renal failure, including the effects on cardiovascular, nervous, haematopoietic and other systems [36].

In the present monograph, the principle should be based on clinical observations like those of Albright, who made precise observations of each patient and proposed many new syndromes. There may be many diseases undiagnosed even at this time. Concepts and technologies of genome analysis with the recent elucidation of the complete human genomic sequences will open a new era in the understanding of calcium physiology and disorders. Pochet [6] collected articles from such viewpoints: apoptosis, autoimmune disease, diabetes, drug dependence, cancer, neurodegeneration, allergy and so on. Using the classical approach will help us to understand many aspects of the role of calcium in the physiology and mechanisms of diseases related to it. The following categories are considered in many articles in this monograph:

1. Deficiency and excess of calcium as a nutrient,
2. Deficiency and excess of calcium in extracellular fluids,
3. Abnormalities of calcium in intracellular compartments,
4. Abnormalities in calcium-regulating systems.

Deficiency and Excess of Calcium as a Nutrient

Calcium deficiency is the most important risk factor in the pathogenesis of osteo-porosis, but it may also be important in the prevention of hypertension and colon cancer. The effect of calcium intake on bone mass was, in the first stage of this kind of study, reported by Matkovic et al. [7], who showed that lower intake of calcium was associated with lower bone mass in one area of Yugoslavia compared to another area where calcium intake was higher and bone mass was higher. It was also shown that calcium intake is at very low levels in Asian countries [8], and the difference in dietary calcium intake influences bone mineral density (BMD) in Chinese women [9]. A typical case of osteoporosis is described elsewhere in the monograph [10].

The correlation between stroke and calcium intake from drinking water was pro-posed by Kobayashi [11]. McCarron [12] stated that the intervention trials intended to assess prospectively the efficacy of increasing calcium intake on lowering blood pressure have in general been inconclusive in their results. If, however, they are evaluated in terms of what has been suggested by the epidemiological surveys as to the conditions that would have to be met, then a more consistent pattern emerges. Using a combination of adequate study size and adequate levels of blood pressure, the results of the effects of lifestyle, including diet, become more consistent. But there are so many other factors affecting blood pressure. Recommendations were issued in Canada in 1999 regarding lifestyle modifications. However, increasing the intake of dietary supplements with potassium, magnesium or calcium is not associated with prevention of hypertension, nor is it effective in reducing high blood pressure [13]. Yang et al. reported that there was a significant protective effect of magnesium intake from drinking water on the risk of hypertension, but no difference was noted among various levels of calcium intake [14].

The correlation between calcium intake and colon cancer was studied by Garland et al. [15] and the possibility of an association between dietary vitamin D and calcium with a 19-year risk of colorectal cancer was examined in 1,954 men during 1957–59. The risk of colorectal cancer was inversely correlated with dietary vitamin D and cancer [15]. Scalmati et al. [16] reviewed many articles and concluded that the age-adjusted incidence of colon cancer was inversely correlated with calcium intake. Barger-Lux and Heaney [17] summarised 12 intervention studies, demonstrating the benefit of increased calcium intake for bones and the protective effect against salt-sensitive and pregnancy-associated hypertension. High intakes of both dietary calcium and vitamin D are associated with reduced development of precancerous changes in colonic mucosa, and the protective effect of vitamin D against breast cancer [17]. There are some areas in the world known as longevity villages – Sang Sa Village in Korea is one of them – where spring water is used which has a higher calcium content [18] (Table 1.1). Calcium vs. magnesium ratio was shown to be closely related to cardiovascular mortality [19]. Such results should be investigated further.

Recommended Calcium Intake

Calcium in drinking water could contribute to dietary calcium intake in some areas of the world (Table 1.1). The contribution of calcium intake from drinking water may be in the region of 100 mg/day if we assume that daily water intake is 2,000 ml, but

Table 1.1. Calcium and magnesium contents in drinking water

Area	Country	City or Village	Source of Water	Components mg/l				Source of Data
				Calcium	Magnesium	Ca/Mg	Hardness	
Europe	Turkey	Pamukkale	Natural W	427	108	3.95	983	analyzed*
	Greece	Athens	City W	54.9	9.9	5.55	165	analyzed*
		Kos Island	City W	131.1	18.5	7.09	310.8	analyzed*
		ITI mountain	Natural W	64.4	24.2	2.7	238.2	analyzed*
	Belgium	Mountain Area	Natural W	5.3	1.9	2.79	54.2	analyzed*
	France	Contrexe-ville	Natural W	486	84	5.79	1,503	Impression Gold April '97
South America	Chili	Santiago	City W	53.4	8.7	6.14	160	analyzed*
North pole	Norway	Arctic Ocean	See W**	309.5	1,200	0.26	5,360	analyzed*
Africa	Egypt	Nile River	Natural W	25.7	8.6	3.90	102.6	analyzed*
Oceania	Australia	Sydney	City W	11.6	4.9	2.37	54.2	analyzed*
		Sydney s suburb	Natural W	nd	nd	nd	5.0	analyzed*
Asia	Korea	Kwang-ju	City W	38.6	6.0	6.43	85.7	analyzed*
		Sang Sa	Natural W	53.5	7.0	7.64	106.0	analyzed*
	Japan	Mountain Area of Gifu	Natural W	11	0.99	11.1	32	indicated in the label
		Osaka	City W	36.0	2.5	14.4	43.6	analyzed*

W: water nd: non detectable.
* Analysis was performed by Ken-ichi Matsumoto, Sakai Municipal Institute of Hygiene, Sakai, Japan 590-0953 in collaboration with Professor Ginji Endo, Department of Environmental Medicine, Osaka City University, Osaka, Japan 545-8585.
** Contents of calcium and magnesium of the sea in the temperate zone (calcium 360 mg/l and magnesium 1212 mg/l: Nishizawa Y. Mystery of Calcium in Nishizawa Y, Shiraka M, Ezawa I, Hirota T. eds Calcium, Dairy Promotion and Research Association of Japan, Tokyo, 1999) are similar to those of arctic ocean.

there are some areas where the concentration exceeds that in the ocean, as in Turkey and France. Such areas are also characterized by high magnesium content.

Calcium intake is a major factor in preventing hip fracture. Matkovic and co-workers demonstrated that cortical bone mass is greater in populations with a higher intake of calcium than in those with lower calcium intake [7]. Finland is known as having the highest calcium consumption and the lowest incidence of hip fracture among the Scandinavian countries [20].

One point to note is that calcium and magnesium contents are relatively high in Greece and Turkey compared to other areas in Europe and other parts of the world.

Table 1.2.   Main Causes of Death in Some Countries*

Countries	Average life span	Number of Death per 100,000 Populations		
		Malignancy	Ischemic Heart Disease	Cerebral Vascular Accident
Finland	77	220.5	267.7	121.3
France	78	223.0	80.1	79.9
Germany	77	264.3	221.3	121.2
Greece	78	197.5	124.9	177.8
Chile	75	132.1	56.9	49.2
United States	76	232.0	183.2	60.1
Japan	80	226.8	57.4	110.9
Turkey**	not available	12.54%	46.76%	7.04%

* Cited from "World Data Book on State of Countries", Kokuseisha, Tokyo, 1999.
** Based on "Death Statistics from Provincial and District Centers 1994, State Institute of Statistics Prime Ministry Republic of Turkey".

When we compare the mortality due to malignancy and ischaemic heart disease, a considerable difference is found; much less in France, Greece and Japan. However, as far as the mortality rate due to malignancy is concerned, the incidence is low in Chile and Turkey. The precise statistics have not been available until recently from Turkey and it is noteworthy that the mortality rate due to malignancy is less than 30% of that due to ischaemic heart disease (Table 1.2). It may be difficult reach definite conclusions regarding the correlation between mineral intake and mortality cause. However, such factors should be taken into consideration in further studies.

There are still some differences in recommended dietary allowances (RDA) among nations. National Institutes of Health (NIH) recommended 1,500 mg for individuals over 65 years of both sexes [21], but the Japanese government recommends 600 mg for both sexes for ages older than 50 years (Table 1.3a, 1.3b). There is thus a considerable difference between Caucasian and Japanese RDA. The NIH recommendation is based on data that between 25 and 50 years of age; women who are otherwise healthy should maintain a calcium intake of 1,000 mg/day to maintain calcium balance and stabilise bone mass [21]. For elderly women, an intake of more than 1,500 mg/day may reduce the rates of bone loss in selected sites of the skeleton such as the femoral neck [21]. One of the studies in Japan showed that a calcium intake of 842 mg/day is needed to maintain calcium balance in elderly women [22]. From such data the present RDA recommended by the Japanese government may be too little for Japanese women, even if the difference in body size between Caucasians and Japanese is taken into consideration.

Deficiency and Excess of Minerals in Extracellular Fluids

Abnormalities in minerals in extracellular fluids are discussed as the derangement of calcium, magnesium and phosphate. Hypocalcaemia, hypercalcaemia, hypomag-

Table 1.3. Optimal Calcium Requirement of NIH (a) and Revommended Dietary Allowance of Japan (b)

a. NIH

Group	Optimal Daily Intake (mg)
Infant	
Birth to six months	400
Six months to one year	600
Children	
One to five years	800
Six to 10 years	800–1,200
Adolescents/Young Adults	
11 to 24 years	1,200–1,500
Men	
25 to 65 years	1,000
Over 65 years	1,500
Women	
25 to 50 years	1,000
Over 50 years (postmenopausal)	1,500
On estrogens	1,000
Not on estrogens	1,500
Over 65 years	1,500
Pregnant and nursing	1,200

b. Japan

Age (years)	Recommended Dietary Allowance (mg)		Upper Limit (mg)
	Men	Women	
0 ~ (Months)		300	
6 ~ (Months)		500	
1 ~ 2		500	
3 ~ 5		500	
6 ~ 8	600	600	
9 ~ 11	700	700	
12 ~ 14	900	700	
15 ~ 17	800	700	
18 ~ 29	700	600	
30 ~ 49	600	600	2,500
50 ~ 69	600	600	2,500
Over 70	600	600	2,500
Pregnant		+300	2,500
Nursing		+300	2,500

nesaemia, hypermagnesaemia, hypophosphataemia and hyperphosphataemia are described in the monograph. Symptomatology has been described in detail for many years, but there are still possibilities of being able to add new findings especially when the mechanisms of the diseases are clarified.

A Proposal of Abnormal Dreams as a Manifestation of Hypocalcaemia

Clinical presentations of hypocalcaemia are characterised by tetany, cataract and calcification of basal ganglia, associated with or without extrapyramidal signs [23]. Tetany appears as hyperexcitability not only of sensory and somatic nerves but also of autonomic nerves and the central nervous system, including general seizure, Parkinsonism and other extrapyramidal disorders. Two cases of hypoparathyroidism with overt tetany combined with eccentric dreams have been described. Such symptoms have never been reported before.

Case Reports

Case Study One

Fifty-four-year-old male, chief complaint: Tetany

Past History: Appendicitis at 24 years of age.

Alcohol intake: Japanese sake, 900 ml/day.

Tobacco consumption: 30 cigarettes/day for 35 years.

Present illness: Since the age of 49, he has paraesthesia, like an electric current in his extremities. The attending physician noted elevated levels of serum LDH and CK.

While general malaise and paraesthesia were becoming worse, hypocalcaemia and low serum PTH were discovered and the patient was referred to Osaka City University Hospital.

Physical findings: height 164.4 cm, body weight 67.8 kg, BMI 25.2 kg/m², blood pressure 124/88 mm Hg, consciousness clear, general condition good, skin normal, conjunctiva: no anaemia and no jaundice, abdomen flat and soft.

Neurological findings: no paresthaesia or motor dysfunction, normal deep tendon reflex, no tremor or rigidity, Chvostek sign positive and Trousseau sign positive, no sensory or motor abnormalities.

Laboratory findings: TP 7.2 g/dl, alb 4.5 g/dl, LDH 1076 U/l, CK 1985 IU/l, cortisol 13.2 μg/dl (10~15), urinary 17 OHCS 7 mg/day, urinary 17 KS 6 mg/day, serum Ca 2.3 mEq/l, Ca2+ 0.59 mmol/l, P 5.6 mg/dl, Mg 1.2 mEq/l, intact PTH < 10 pg/ml, calcitonin 44.6 pg/ml (30.9–120.1), ALP 152 IU/l, BGP 2.2 ng/ml (3.1~12.7), 1,25 dihydroxyvitamin D 32.8 pg/ml (27.5~68.7), %TRP 96% (80~90), urinary Ca/Cr 0.008 mg/mg (0.012~0.371), GFR 119 ml/min, Ellsworth-Howard Test (U4 + U3) – (U2 + U3) = 62.11 mg (>35: difference in urinary phosphate excretion before (U2 + U3) and after (U4 + U5) administration of 100 U of h (1–34)PTH. Each U represents one-hour excretion of phosphate. U1 was excreted during hydration.). Urinary excretion of cAMP: U4 – U3 = 101 μmol/h (>1: positive), U4/U3 = 102 (>10: positive). Lamina dura of alveolar bones were normal. Calcification was noted in the basal ganglia.

Diagnosis and clinical course: Clinical signs, hypocalcaemia, hyperphosphataemia, low intact PTH, normal kidney function and a positive response of urinary phosphate and cAMP led to the diagnosis of idiopathic hypoparathyroidism. Polyglandular autoimmune disease was negated. Administration of calcitriol at a daily dose of 1.5 μg resulted in normalisation of serum calcium, phosphate, CK and LDH as well as clinical signs.

Table 1.4. Dreams of patients with hypoparathyroidism

a. 54-year old male	
Days	Story of Dream
July 3	Four big injection needles were stuck onto the breast
4	Divorced wife strangled the neck of a woman with a chain
5	Quarreled with friends regarding weight of parcels with which he is concerned.
6	Went back to the era of feudalism and fell in love with a young girl, and wife accused.
28	Whale sexual organ appeared while handling marine products for delivery (his business).
29	Lying down on the bed and being treated by physicians, but results were unpleasant.
August 1	A young nurse advised him to be admitted to the urology ward.
	There were many dreams of the same sort on other days. He dreamed almost every day.
b. 34-year old female	
Days	Story of Dream
3 to 4 days in a week	She just talked about one feature of dream at each occasion: black cat, thunder, blood, being raped, singing karaoke, mimicking famous artist and so on.
	Emotional instability was the main feature of her symptoms, crying very often remembering dead baby with concomitant shivering. Sleep disturbance was experienced in almost every night. Dreaming was another charateristic of this patient.

Diary of dreams: The patient dreamed almost every night. The dream content consisted of his divorced wife, sexual desire, unpleasant medical treatments, quarrels and so on (Table 1.4a). Dreams were related to sexual desire and assault, unsuccessful job experiences and unexpected results of medical treatment. Sleep disturbance was one of the chief complaints and he also complained of emotional instability.

Case Study Two

Thirty-four-year-old female, chief complaints: Paresthaesia and palpitation.

Past and family history: Nothing remarkable.

Present illness: Since childhood she has not been sociable and becomes tense in the presence of other people. On such occasions she shivers and experiences palpitations. At the age of 25 years, general seizure occurred but consciousness was not lost. At the age of 33 years, she delivered her second baby who died at the age of four months because of congenital heart disease: symptoms including paresthaesia worsened after the episode. She was found to show Trousseau sign and hypocalcaemia of 7.5 mg/dl and was referred to Osaka City University Hospital.

Physical findings: body height 161.0 cm, body weight 58.3 kg and BMI 22.5 kg/m². Consciousness was clear. Positive Trousseau sign was noted but Chvostek sign was negative. There were no abnormal findings in the neck, chest or abdomen. No abnormal neurological signs were shown.

Laboratory findings: Serum calcium 3.9 mEq/l, ionised calcium 1.00 mmol/l, phosphate 3.6 mg/dl, intact PTH 13 pg/ml, calcitonin pg/ml (17.1~58.7) with normal thyroid func-

tion. Ellsworth-Howard Test showed an increase of 44 mg (the normal response amounts to more than 35 mg) of urinary excretion of phosphate and an increase of 13.0 μmol of cAMP excretion after administration of 100 U of human (1–34) PTH. No calcification was noted in basal ganglia.

Diagnosis: idiopathic hypoparathyroidism with hyperventilation syndrome by which tetanic syndrome may have been worsened.

Diary of dreams: seemed to reflect her rather unhappy life so far: her childhood as an illegitimate child, unhappy marriage, death of her second child and disease of this time. The theme of her dreams seems to be predicting her fate (Table 4.1b).

It is assumed that people dream for 20% of their sleeping hours but most of them are forgotten or cannot be memorised [24]. Sigmund Freud stated that dreams are "not nonsense" but are a time to sort out one's experiences [25]. Rapid eye movement (REM), which is associated with dreaming, is now known to be important for the acquisition of some tasks [25]. While both of the patients with hypoparathyroidism dreamed mostly about unhappy events, Haigren [26] showed that direct electrical stimulation or heightened states of emotion usually produced dreams of fear or anxiety. Vividly formed dreams or memory-like hallucinations, or intense feelings of familiarity, may be evoked from the hippocampal formation and amygdala. Both the patients discussed experienced dreadful dreams of violence, or dreams related to sexual desire and a gloomy fate. Although Case 1 dreamed of being injected into his breast wall (Table 1.4a), there was no record made of whether or not he experienced pain. Nielsen et al. [27] discusses the experience of pain in dreams, stating that a total of 13 dreams had one or more references to pain and most often these references appeared to be direct, untransformed incorporations of real sensations produced by stimulation. Pain was the principal motivating agent in a majority of these dreams and was in many cases associated with strong emotion, typically anger.

Hypoparathyroidism is associated with tetany – characterised by hyperexcitability of the nervous system, including sensory and motor nerves, autonomous nervous system and central nervous system. Symptoms related to the central nervous system are called tetanic psychoses, which include "Dammerzustand und Transitorische Verwirrtheit" [28]. Abnormal dreams in hypoparathyroidism may be explained as a symptom of tetany related to hyperexcitability of the limbic system.

Abnormalities of Calcium in Intracellular Compartments

Abnormalities in intracellular calcium have not been classified adequately. Intracellular calcium varies under the influence of extracellular calcium or calcium released from intracellular stores. One method of classification of such abnormalities is "ion channels-related diseases", which include hyper- and hypokalaemic periodic paralysis, myotonias, long QT syndrome, Brugada syndrome, malignant hyperthermia and myasthenia [29].

The pathogenic factor of Alzheimer's disease is beta-amyloid protein, which may cause toxic damage to neurons. But another interesting aspect is the mitochondrial spiral, which consists of reduced brain metabolism, oxidative stress and calcium dysregulation [30]. For the treatment of the disease accompanying psychiatric symptoms like anxiety, depression, hallucinations and sleeplessness, drugs with minimal influence on cognitive processes are recommended. Attempts at causal therapy are focussed on searching for the substances which can prevent the formation and toxicity of beta-amyloid, among which calcium the channel blocker is included [30].

Another field of study is the cardiovascular system. Hypertension [31] and vascular calcification [32] are the main topics in this monograph.

Abnormalities in Calcium-Regulating Systems

Hypocalcaemia and primary hyperparathyroidism are described in one of the chapters in the monograph by Schussman and Silverberg [37]. The authors described the natural history of the disease and expressed interesting views regarding the difference in the incidence among different nations from the standpoint of the status of vitamin D.

Secondary hyperparathyroidism in chronic renal failure, especially in haemodialysed patients, has been a one of the most important complications and is the main topic in this monograph also [33, 34, 35].

References

1. Albright F, Reifenstein Jr EC. The parathyroid glands and metabolic bone disease. Baltimore: Williams & Wilkins, 1948.
2. Fourman P. Calcium metabolism and the bones. Oxford: Blackwell, 1860.
3. Danowski TS. Clinical endocrinology, vol III, calcium, phosphorus and bone. Baltimore: Williams & Wilkins, 1962.
4. Jackson WPU. Calcium metabolism and bone disease. London: Edward Arnold Ltd., 1967.
5. Albright F, Ellsworth R. Uncharted seas. Ed Loriaux DL. Portland: Kalmia Press, 1990.
6. Pochet R. Calcium: The molecular basis of calcium action in biology and medicine. Dordrecht: Kluwer, 2000.
7. Maktovicc V, Kostial K, Simonovic I, Buzina R, Broderac A, Nordin BEC. Bone mass and fracture rates in two regions of Yugoslavia. Am J Clin Nutr 1979;32:540–9.
8. Nordin BEC. Nutritional considerations. In: Nordin BEC, editor. Calcium, phosphate and magnesium metabolism, Edinburgh: Churchill Livingstone, 1976;1–35.
9. Hu JF, Zhao XH, Parpia B, Campbell TC. Dietary calcium and bone density of women in China. In: Lau EMC, editor. Osteoporosis in Asia. Singapore: World Scientific, 1997;57–79.
10. Morii H. Primary osteoporosis. In: Morii H, Nishizawa Y, Massry SG, editors. Calcium in internal medicine. Oxford: Springer-Verlag, 2001.
11. Kobayashi J. On the geographical relationship between the chemical nature of river water and death rate from apoplexy. Berichte Ohara Institute, 1957;12–21.
12. McCarron DA. Epidemiological evidence and clinical trials of dietary calcium's effect on blood pressure. In: Morii H, editor. Calcium-regulating hormones I. Role in disease and ageing. Munich: Karger, 1991;2–10.
13. Burgess E, Lewanczuk R, Bolli P, Chockalingam A, Autler H, Taylor G et al. Lifestyle modifications to prevent and control hypertension. Recommendations on potassium, magnesium and calcium. Canadian Hypertension Society.
14. Yang CY, Chiu HF. Calcium and magnesium in drinking water and the risk of death from hypertension. Am J Hypertens 1999;12(9 Pt 1):894–9.
15. Garland C, Shekelle RB, Barett-Connor E, Criqui MH, Rossof AH, Pau O. Dietary vitamin D and calcium and risk of colorectal cancer: A 19-year prospective study in men. Lancet 1985;I:307–9.
16. Scalmati A, Lipkin M, Newmark H. Calcium, vitamin D, and colon cancer. Clin Appl Nutr 1992;2:67–74.
17. Barger-Lux MJ, Heaney RP. The role of calcium intake in preventing bone fragility, hypertension, and certain cancers. J Nutri 1994;124:1406SA–11S.
18. Morii H, Rowe SM, Chung SS, Park CJ, Kang JC, Kee HS et al. Bone mineral in elderly people in Sangsa Village, Korea. Chonnam J Med Sci 1995:76–84.
19. Karppanen H, Pennanen R, Passinen II. Minerals and sudden coronary death. In: Manninen V, Halonen P, editors. Advances in cardiology, Basel 1978;25:9–24.

20. Lindsay R, Cosman F. Primary osteoporosis. In: Coe FL, Favus MJ, editors. Disorders of bone and mineral metabolism. New York: Raven Press, 1992, Chapter 37:831–88.
21. National Institutes of Health. NIH Consensus Statement Volume 12, Number 4, June 6–8, 1994 Optimal Calcium Intake.
22. De Souza AC, Nakamura T, Storgitpoulos K, Shiraki M, Ouchi Y, Orimo H. Calcium requirement in elderly Japanese women. Gerontology 1991;37:43–7.
23. Glenndenning P, Need AG, Nordin BEC. Hypocalcaemia. In: Morii H, Nishizawa Y, Massry SG, editors. Calcium in internal medicine. Oxford: Springer-Verlag, 2001.
24. Fontana D. The secret language of dreams. London: Duncan Baird, 1994.
25. Poe GR, Niez DA, McNaughton BL, Barnes CA. Experience-dependent phase reversal of hippocampal neuron firing during REM sleep. Brain Res 2000;855:176–80.
26. Haigren E. Mental phenomena induced by stimulation in the limbic system. Hum Neurobiol 1982;1:251–60.
27. Nielsen TA, McGregor DL, Zadra A, Ilnicki D, Ouelett L. Pain in dreams. Sleep 1993;16:490–8.
28. Jesserer H. Tetanie, Stuttgart: Georg Thieme Verlag, 1858.
29. Dworakowska B, Dolowy K. Ion channels-related diseases. Acta Biochim Pol 2000;47:685–703.
30. Blass JP. The mitochondrial spiral. An adequate cause of dementia in Alzheimer's syndrome. Ann NY Acad Sci 2000;924:170–83.
31. Hasebe N, Kikuchi K. Hypertension. In: Morii H, Nishizawa Y, Massry SG, editors. Calcium in internal medicine. Oxford: Springer-Verlag, 2001.
32. Shioi A. Vascular calcification. In: Morii H, Nishizawa Y, Massry SG, editors. Calcium in internal medicine. Oxford: Springer-Verlag, 2001.
33. Massry SG. Parathyroid hormone toxicity. In: Morii H, Nishizawa Y, Massry SG, editors. Calcium in internal medicine. Oxford: Springer-Verlag, 2001.
34. Coburn JW. Vitamin D. In: Morii H, Nishizawa Y, Massry SG, editors. Calcium in internal medicine. Oxford: Springer-Verlag, 2001.
35. Akizawa T. Bone disease in chronic renal failure. In: Morii H, Nishizawa Y, Massry SG, editors. Calcium in internal medicine. Oxford: Springer-Verlag, 2001.
36. Massry SG, Smogorzewski M. Parathyroid hormone. In: Morii H, Nishizawa Y, Massry SG, editors. Calcium in internal medicine. Oxford: Springer-Verlag, 2001.
37. Schussman DH, Silverberg SJ. Hypocalcaemia and primary hyperparathyroidism. In: Morii H, Nishizawa Y, Massry SG, editors. Calcium in internal medicine. Oxford: Springer-Verlag, 2001.

Nutritional Needs of Mineral Metabolism

T. Okano

2

Biologically Important Minerals

Of the 95 naturally occurring elements in the periodic table, no less than 29 perform essential functions in the human body, and these are termed "minerals" or "mineral elements". Minerals are arbitrarily divided into two groups in accordance with the elemental composition of the human body, diet, and daily needs. Seven of these, namely calcium, phosphorus, potassium, sulphur, chlorine, sodium and magnesium, are termed "macrominerals". These are required in amounts greater than 100 mg/day from the diet, and comprise 60–80% of all the inorganic material in the body. Macrominerals play important roles in many vital processes; for instance, calcium and phosphorus in bone formation; sodium and potassium in the regulation of nerve and muscle function; chlorine in the regulation of acid-base balance, water balance and osmotic pressure; magnesium in the activation of enzymes; sulphur as an integral part of many important physiologic compounds such as sulphur-containing amino acids, heparin, glutathione, insulin, and so forth. Nine minerals; namely iron, zinc, copper, manganese, iodine, selenium, molybdenum, cobalt, and chromium, have been demonstrated to be essential to humans and are termed "microminerals" or "trace elements". These are required in amounts less than 100 mg/day from the diet and each comprises less than 0.01% of total body weight. In the body, they have similar functions; most of them are found at the active site of enzymes or of physiologically active substances of the body. For instance, iron is found in haemoglobin, zinc in insulin, copper in oxidase enzymes, manganese in mitochondrial super-oxide dismutase, iodine in thyroxin, selenium in glutathione peroxidase, molybdenum in xanthine oxidase, cobalt in vitamin B_{12}, trivalent chromium in a glucose tolerance factor that binds to and potentiates insulin use, and so forth. At least 13 other minerals, namely fluoride, silicon, rubidium, bromine, lead, aluminium, cadmium, boron, vanadium, arsenic, nickel, tin and lithium, are present in trace amounts in the body, but their functions, if any, are not well defined in humans or other animals. Table 2.1 summarises the general characteristics of minerals.

Adequate Intake of Minerals

Nutrition is concerned with both the qualitative and quantitative requirements of the human body supplied by the diet. Minerals needed to maintain life appear to be clearly defined qualitatively, since it is possible to sustain humans or other animals on chemically defined diets. However, the quantitative requirement of each mineral from the diet is not easily defined as it varies with age, sex, and lifestyle of the individual. For the past several decades, Recommended Dietary Allowances (RDAs) for the daily intakes of nutrients have been widely used in the United States, Canada, European countries, Japan, and many other countries. RDAs are the levels of nutrient intake that, on the basis of scientific knowledge, are judged by nutritionists in each country to be adequate to meet the known nutritional needs of practically all

Table 2.1. General Characteristics of the Minerals

Minerals	Amounts in the human adults body (mg)	Functions	Estimated average intake (mg) adults, Japan	General signs of deficiency	General signs of toxicity	Dietary sources
Calcium	1,160,000	Constituent of bones and teeth; Regulation of muscle; Nerve function	400–700	Children : rickets; Adults : osteomalacia; Elderly and postmenopausal women : osteoporosis	Hypercalcaemia; Renal stones	Dairy products; Leafy green vegetables; Grains
Phosphorus	670,000	Constituent of bones and teeth; Phosphorylation of biomolecules	1,000–1,500	Children : rickets; Adults : osteomalacia	Secondary hyperparathyroidism	Meats; Poultry; Fish
Potassium	150,000	Muscle, nerve function; Na^+/K^+ ATPase	1,200–2,500	Muscle weakness; Anorexia; Paralysis	Cardiac arrest	Fruits, vegetables; Milk, fresh meats
Magnesium	25,000	Enzyme cofactor	150–300	Impaired calcium metabolism; Cardiac complications; Hypertension	Hypotension; Respiratory; Depression	Leafy green vegetables, grains and nuts
Iron	4,500	Constituent of haemoglobin and cytochromes	6–12	Anaemia	Siderosis	Offal; Egg yolk, spinach
Zinc	2,000	Enzyme cofactor; Constituent of insulin	6–15	Hypogonadism; Impaired wound healing; Growth failure	Gastrointestinal irritation; Vomiting	meat, liver, eggs, seafoods
Copper	80	Enzyme cofactor; Iron absorption	1.5–3.0	Anaemia; Menke's syndrome	Rare: secondary to Wilson's disease	Offal, seafoods, nuts, seeds
Manganese	15	Enzyme cofactor	3–6	Impaired bone and cartilage formation	Neurotoxicity	Liver, kidney, meat, spinach
Iodide	15	Constituent of thyroxine and triiodothyronine	0.5–1.5	Children : cretinism; Adults : goitre	Thyrotoxicous goitre	Seafoods; offal
Molybdenum	9	Constituent of oxidase enzymes	0.05–0.35	Amino acid intolerance; Irritability	Loss of copper; Goutlike syndrome	Milk, beans, breads, cereals
Chromium	2	Constituent of glucose tolerance factor	0.01–0.07	Impaired glucose tolerance	Rare	Meats, liver grains, nuts, cheese

healthy persons. In the tenth edition of the US Recommended Daily Allowances published by the Food and Nutrition Board of the US National Research Council in 1989 [1], RDAs were principally set on the basis of several sources of evidence: 1) studies of subjects maintained on diets containing low or deficient levels of a nutrient; 2) nutrient balance studies that measure nutrient status in relation to intake; 3) biochemical measurements of tissue saturation or adequacy of molecular function in relation to nutrient intake; 4) nutrient intakes of fully breastfed infants and of apparently healthy people from their food supply; 5) epidemiological observation of nutrient status in populations in relation to intake; and 6) in some cases, extrapolation of data from animal experiments. More recently, Dietary Reference Intakes (DRIs) for nutrients were set by the Standing Committee on the Scientific Evaluation of Dietary Reference Intakes of the Food and Nutrition Board, Institute of Medicine, US National Academy of Sciences, with the involvement of Health Canada in 1998 [2], and by the Ministry of Health, Labour, and Welfare of Japan in 1999 [3]. The term DRIs refers to a set of four nutrient-based reference values; namely the Estimated Average Requirement (EAR), the Recommended Dietary Allowance (RDA), the Adequate Intake (AI), and the Tolerable Upper Intake Level (UL) as shown in Figure 2.1 [4]. The EAR is the daily intake value that is estimated to meet the requirements, as defined by a specified indicator of adequacy, in half of the individuals in an age or gender group. At this level of intake, the other half of a specified group would not have its nutritional needs met. The RDA is the average dietary intake level that is adequate to meet the nutritional requirements of most (97–98%) healthy individuals. If the standard deviation (SD) of the EAR is available and the requirement for the nutrient follows a normal, or Gaussian, distribution pattern, the RDA is set by adding 2 SDs to the EAR ($RDA = EAR + 2SD_{EAR}$). If the data on the variability in requirements are insufficient to calculate an SD, a coefficient of variation (CV_{EAR}) of 10% is ordi-

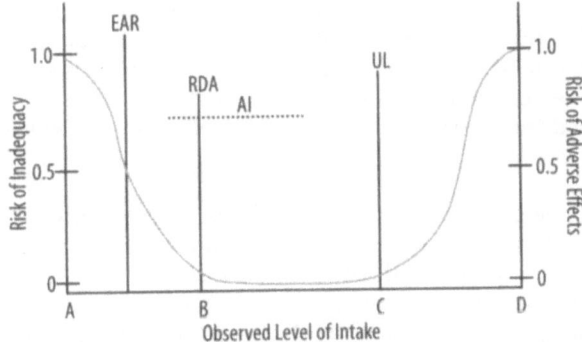

Figure 2.1.   Dietary reference intakes [2]. The Estimated Average Requirements (EAR) is the intake at which the risk of inadequacy is 0.5 (50%) for an individual. The Recommended Dietary Allowance (RDA) is the intake at which the risk of inadequacy is very small – only 0.02 to 0.03 (2–3%). The Adequate Intake (AI) does not bear a consistent relationship to the EAR or the RDA because it is set without being able to estimate the requirement. At intakes between the RDA and the Tolerable Upper Intake Level (UL), the risks of inadequacy and of adverse effects are both close to 0. The UL is the highest level of daily mineral intake that is likely to pose no risk of adverse health effects to almost all individuals in the general population. At intakes above the UL, the risk of adverse effects increases. A dashed line is used because the actual shape of the curve has not been determined experimentally. The distances between A and B, B and C, and C and D may differ much more than is depicted in this figure [4].

narily assumed. Since a CVEAR is calculated by dividing a SDEAR with an EAR (CVEAR = SDEAR/EAR or SDEAR = EAR × CVEAR), the equation for the RDA is: RDA = EAR + 2 (0.1 × EAR) or RDA = 1.2 × EAR.

If an EAR is not available and therefore an RDA is not set, an AI is used instead of an RDA. According to the definition of the US Institute of Medicine's Food and Nutrition Board [2], the AI is defined as a value based on experimentally derived intake levels or approximations of observed average nutrient intakes in apparently healthy populations. As with the RDA, the AI is also used to set the nutrient requirements; however, there is much less certainty with the AI value than with the RDA value. Because the AI depends on a greater degree of estimation than is used in determining the EAR and subsequently the RDA, the AI may deviate significantly from the RDA, when both are available, and may be numerically higher than the RDA. For this reason, AIs must be interpreted with greater caution than the RDAs. As data become available in the future, it should be possible to replace AI values with EARs and subsequently RDAs.

Table 2.2 represents the dietary reference intake values for minerals and vitamins related to mineral metabolism in an age or gender group set in Japan [3].

Safe Intake Levels of Minerals

With increasing evidence that certain minerals at intakes higher than the RDAs over an extended period of time may be helpful in protecting against health problems such as osteoporosis, heart disease, anaemia, and some cancers, there is much interest in the use of mineral supplements and food fortification to increase mineral intakes. The corollary is that with increasing consumption of mineral supplements and fortified foods, there is growing fear that some consumers will ingest excessive amounts of minerals. Thus, the safe intake level of minerals should be considered carefully. The safety of minerals is assessed on the basis of a scientific approach that selects the highest intake not associated with adverse effects (No Observed Adverse Effect Level, the NOAEL), or the lowest intake associated with adverse effects (Lowest Observed Adverse Effect Level, the LOAEL). If the NOAEL is used to set a safe limit for mineral intake (a Tolerable Upper Intake Level, the UL), no safety factor is needed. If the LOAEL is used for the same purpose as described in the NOAEL, a safety factor should be applied.

No Observed Adverse Effect Level (NOAEL). NOAEL is the highest level of mineral intake at which no significant adverse effects on human health have been identified. This is the intake level for which the human data are adequate to establish safety for longer-term use, and thus no additional safety factor is needed to define a safe intake level. If human safety data are insufficient, animal data may be used to supplement the human data. If a NOAEL from animal data were used alone, a safety factor would be required to calculate safe levels for humans.

Lowest Observed Adverse Effect Level (LOAEL). LOAEL is the lowest level of mineral intake at which significant adverse effects on human health have been observed. Use of this value to define safe intake levels requires reducing the value with a safety factor or an uncertainty factor (UF).

Uncertainty Factor (UF). According to the definition of the Vitamin Mineral Safety [5] published by the Council for Responsible Nutrition (CRN) of the US, an

Table 2.2. Dietary Reference Intake Values for Minerals, Vitamin D and Vitamin K *(continued next page)*

Life stage group	Calcium AI (mg) Male	Calcium AI (mg) Female	Calcium UL (mg)	Iron AI (mg) Male	Iron AI (mg) Female	Iron UL (mg)	Phosphorus AI (mg) Male	Phosphorus AI (mg) Female	Phosphorus UL (mg)	Magnesium AI (mg) Male	Magnesium AI (mg) Female	Magnesium UL (mg)	Potassium AI (mg) Male	Potassium AI (mg) Female	Copper AI (mg) Male	Copper AI (mg) Female	Copper UL (mg)	Iodine AI (µg) Male	Iodine AI (µg) Female	Iodine UL (mg)
0–5 months	200	200	–	6	6	10	130	130	–		25	–		500		0.3	–	40	40	–
6–11 months	500	500	–	6	6	15	280	280	–		30	–		700		0.7	–	50	50	–
1–2 years	500	500	–	7	7	20	600	600	–		60	130		900		0.8	–	70	70	–
3–5 years	500	500	–	8	8	25	700	700	–		80	200		1,100		1	–	80	80	–
6–8 years	600	600	–	9	9	30	900	900	–	120	120	250	1,350	1,200	1.3	1.2	–	100	100	3
9–11 years	700	700	–	10	10	35	1,200	1,200	–	170	170	500	1,550	1,400	1.4	1.4	–	120	120	3
12–14 years	900	700	–	12	12	35	1,200	1,200	–	240	220	600	1,750	1,650	1.8	1.6	–	150	150	3
15–17 years	800	700	–	12	12	40	1,200	1,200	–	290	250	650	2,000	2,000	1.8	1.6	–	150	150	3
18–29 years	700	600	2,500	10	12	40	700	700	4,000	310	250	700	2,000	2,000	1.8	1.6	9	150	150	3
30–49 years	600	600	2,500	10	12	40	700	700	4,000	320	260	700	2,000	2,000	1.8	1.6	9	150	150	3
50–69 years	600	600	2,500	10	12	40	700	700	4,000	300	260	650	2,000	2,000	1.8	1.6	9	150	150	3
>70 years	600	600	–	10	10	40	700	700	–	280	240	650	2,000	2,000	1.6	1.4	–	150	150	3
Pregnancy		+300			+8	40		+0			+35	700		+0		+0.4	9		+25	3
Lactation		+500			+8	40		+0			+0	700		+500		+0.6	9		+25	3

Table 2.2. *Continued.*

Life stage group	Manganese AI (mg) Male	Manganese AI (mg) Female	Manganese UL (mg)	Selenium AI (µg) Male	Selenium AI (µg) Female	Selenium UL (µg)	Zinc AI (mg) Male	Zinc AI (mg) Female	Zinc UL (mg)	Chromium AI (µg) Male	Chromium AI (µg) Female	Chromium UL (µg)	Molybdenum AI (µg) Male	Molybdenum AI (µg) Female	Molybdenum UL (µg)	Vitamin D AI (µg) Male	Vitamin D AI (µg) Female	Vitamin D UL (µg)	Vitamin K AI (µg) Male	Vitamin K AI (µg) Female	Vitamin K UL (µg)
0–5 months	0.003		–		15	10		1.2	10		–	–		–	–		10	25		5	5,000
6–11 months	1.2		–		20	15		4	15		–	–		–	–		10	25		10	5,000
1–2 years	1.8		–		25	20		5	20		16	60		6	60		10	50		15	10,000
3–5 years	2.5		–		35	25		6	25		20	80		8	80		10	50		20	14,000
6–8 years	3.0	3.0	–	40	40	30	6	6	30	25	25	120	12	12	120	2.5	2.5	50	25	25	17,000
9–11 years	3.5	3.0	–	50	45	35	7	7	35	30	30	150	15	15	150	2.5	2.5	50	35	35	22,000
12–14 years	3.5	3.0	–	55	50	35	8	8	35	35	30	200	20	20	200	2.5	2.5	50	50	50	27,000
15–17 years	4.0	3.0	–	60	45	40	10	9	40	35	30	250	30	25	250	2.5	2.5	50	60	55	28,000
18–29 years	4.0	3.0	10	60	45	40	11	9	40	35	30	250	30	25	250	2.5	2.5	50	65	55	30,000
30–49 years	4.0	3.5	10	55	45	40	12	10	40	35	30	250	30	25	250	2.5	2.5	50	65	55	30,000
50–69 years	4.0	3.5	10	50	45	40	11	10	40	30	25	250	30	25	250	2.5	2.5	50	65	55	30,000
>70 years	3.5	3.0	–	45	40	40	10	9	40	25	20	200	25	25	200	2.5	2.5	50	55	50	30,000
Pregnancy	+0	+0	10		+7	40		+3	40		+0	250		+0	250		+5	50		+0	30,000
Lactation	+0	+0	10		+20	40		+3	40		+0	250		+0	250		+5	50		+0	30,000

Association of the Dietary Supplement Industry, a UF is defined as a number by which the NOAEL or LOAEL is divided to obtain the UL. UFs are used in risk assessment to account for gaps in data (for example, data uncertainties) and knowledge (for example, model uncertainties). The size of the UF varies depending on the quality of the data and the nature of the adverse effect. In the Vitamin Mineral Safety, the UFs for calcium, phosphorus, magnesium, vitamin D, and fluoride are as follows: For fluoride and magnesium intakes, a UF of 1.0 was selected primarily based on the very mild nature of the adverse effects observed. For vitamin D intake in adults and other age groups except infants, a UF of 1.2 was selected. For vitamin D intake in infants, a UF of 1.8 was selected due to the insensitivity of the critical endpoint, the small sample size of the studies, and limited data on the susceptibility at the tails of the distribution. For calcium intake, a UF of 2.0 was selected due to the increased risk of high calcium intake by individuals who have renal stones, and due to the risk of mineral depletion in vulnerable populations due to the interference of calcium with mineral bioavailability, especially iron and zinc. For phosphorus intake, a UF of 2.5 was selected due to the lack of information on potential adverse effects in the range between normal phosphorus levels and levels associated with ectopic mineralisation.

Tolerable Upper Intake Level (UL). UL is the highest level of mineral intake at which no significant adverse effects on human health have been observed. As mineral intake increases above the UL, the risk of adverse effects increases. Thus, the UL is defined as the level of intake that can, with high probability, be tolerated biologically. The UL is not intended to be a recommended level of intake, and there is no established benefit for healthy individuals if they consume the minerals in amounts exceeding the RDA or the AI. The UL is based on an evaluation conducted using the methodology for risk assessment of minerals. The need for setting ULs grew out of the increased fortification of foods. The UL encompasses the total intake from food, fortified food, and mineral supplements. The UL applies to daily use. The NOAEL and LOAEL values for minerals and vitamins related to mineral metabolism are listed in Table 2.3 [5].

Nutritional Requirements for Minerals

Calcium Requirements

The adult human body contains approximately 1,200 g of calcium, over 99% of which is present as hydroxyapatite, which is combined with phosphorus in bones and teeth. The skeleton provides strength to the body and also serves as a reservoir for calcium. The remaining 1% of body calcium is present in the extracellular fluid, intracellular spaces, muscles and other soft tissues, where it plays an essential role in such vital processes as smooth muscle contraction, skeletal and cardiac muscle contraction, nerve conduction, glandular secretion, blood clotting, and membrane permeability. For optimal growth and function of the human body, an adequate intake of calcium for maintenance of normal calcium metabolism is required.

Bone is continuously undergoing osteoclastic resorption and osteoblastic formation. The rate of bone formation predominates over that of bone resorption in growing children and adolescents, whereas in later life, resorption predominates over formation, resulting in a gradual loss of bone. Each year, a portion of the skeleton is

Table 2.3. No Observed Adverse Effect Level (NOAEL) and Lowest Observed Adverse Effect Level (LOAEL) for Minerals and Vitamins

Minerals and Vitamins	The US (CRN)		Japan (DRIs)	
	NOAEL	LOAEL	NOAEL	LOAEL
Calcium	1,500 mg/day	over 2,500 mg/day		5,000 mg/day
Phosphorus	1,500 mg/day	over 2,500 mg/day		
Magnesium	700 mg/day			5 mg/day
Iron	65 mg/day	100 mg/day		60 mg/day
Zinc	30 mg/day	60 mg/day	30 mg/day	
Copper	9 mg/day		9 mg/day	
Manganese	10 mg/day			0.015 mg/day
Iodine	1 mg/day		3 mg/day	
Selenium	0.2 mg/day	0.91 mg/day		0.015 mg/kg/day
Chromium	1 mg/day			0.0147 mg/kg/day
Molybdenum	0.35 mg/day			0.14 mg/kg/day
Vitamin D	0.02 mg/day (800 IU)	0.05 mg/day (2000 IU)	0.05 mg/kg/day (2000 IU)	
Vitamin K (phylloquinone)	30 mg/day		30 mg/day	

remodelled. The rate of cortical bone remodelling can be as high as 50% per year in young children and is about 5% per year in adults. Trabecular bone remodelling is about five-fold higher than cortical bone remodelling in adults.

Although calcium is widely distributed in plant and animal tissues, it is not present in most foods in sufficiently high concentrations. According to the Survey of the Center for Nutrition Policy and Promotion, US Department of Agriculture (CNPP, USDA) in 1994, 73% of calcium in the US food supply is from milk products, 9% is from fruits and vegetables, 5% is from grain products and the remaining 12% is from all other sources [6]. Grains are not rich in calcium, but they contribute substantially to the daily intake of calcium because they are usually consumed in large quantities. Dairy products are the richest and most absorbable calcium sources. In general, vegetables contain more calcium than do animal foods, however, they have an appreciable amount of either oxalic acid (spinach, sweet potatoes, rhubarb, and beans) and/or phytic acid (unleavened bread, raw beans, seeds, nuts and grains, and soy isolates) which will form insoluble calcium complexes in the intestinal tract and thus reduce the absorption and utilisation of some of the calcium present. Water is a variable calcium source, depending on the geochemical environment. Other calcium-rich foods include such leafy green vegetables as the Chinese cabbage, kale, broccoli, collards, calcium-fortified orange juice, and calcium-precipitated tofu. A large number of supplements containing calcium are currently available in the market. In the US and Canada, the AIs for ages 0–6 months are 210 mg/day, for ages 7–12 months 270 mg/day, for ages 1–3 years 500 mg/day, for ages 4–8 years 800 mg/day, for ages 9–18 years 1,300 mg/day, for ages 19–50 years 1,000 mg/day, for ages 51–70 years and over 1,200 mg/day, and for pregnancy and lactation between 14–18 years

1,300 mg/day and 19–50 years 1,000 mg/day [7]. The ULs for ages 19–70 years are 2,500 mg/day, for ages 1–12 months is not established, for ages 1–70 years and over 2,500 mg/day, and for pregnancy and lactation between 14–50 years 2,500 mg/day [8]. In Japan, the AIs for men aged 0–5 months are 200 mg/day, for ages 6 months to 5 years 500 mg/day, for ages 6–8 years 600 mg/day, for ages 9–11 years 700 mg/day, for ages 12–14 years 900 mg/day, for ages 15–17 years 800 mg/day, for ages 18–29 years 700 mg/day, for ages 30–70 years and over 600 mg/day [3]. The AIs for women aged 0–5 months are 200 mg/day, for ages 6 months to 5 years 500 mg/day, for ages 6–8 years 600 mg/day, for ages 9–17 years 700 mg/day, for ages 18–70 years and over 600 mg/day, and an additional 300 mg/day and 500 mg/day during pregnancy and lactation, respectively [3]. The ULs for ages 18–70 years and over are 2,500 mg/day, and for pregnancy and lactation 2,500 mg/day [3].

Phosphorus Requirements

Phosphorus is present in foods and biological materials in its pentavalent form in combination with oxygen, as phosphate. Phosphorus is required for the formation of bones and teeth, the composition of intra- or extra-cellular fluids, the phosphorylation of carbohydrates, lipids, proteins, nucleotides, and nucleic acids, the maintenance of acid-base balance, the storage and transfer of energy derived from metabolism, and the activation of enzymes, water-soluble vitamins, and other important physiologic factors. Phosphorus is found in almost all foods. Meats, poultry, and fish contain 15 to 20 times more phosphorus than calcium. Eggs, grains, nuts, dry beans, peas, and lentils contain twice as much phosphorus as calcium. In contrast, milk, natural cheeses, and green leafy vegetables contain more calcium than phosphorus. Phosphorus is absorbed efficiently in the small intestine as free phosphate. Thus, phosphorus deficiency is rare in humans with a normal diet. Human milk provides calcium and phosphorus at a ratio of 2:1, whereas cow's milk has a Ca/P ratio of 1.2:1.

In Japan, the AIs of phosphorus range from 130 mg for children under 1 year of age to 900 mg for children under 8 years of age of both sexes, 1,200 mg for children ages 9–17 years, and 700 mg for youths and adults, with no additional phosphorus suggested during pregnancy and lactation [3]. The UL of phosphorus is 4,000 mg for adults, with no additional phosphorus during pregnancy and lactation [3].

Magnesium Requirements

Magnesium is required for numerous biological and physiological processes, including modulation of enzyme activities, glycolysis, mitochondrial oxidative phosphorylation, cyclic AMP formation, energy-dependent trans-membrane transport, cell replication, and transmission of the genetic code. Magnesium is widely distributed in foods. Green leafy vegetables are rich in magnesium because of the high chlorophyll content, a magnesium chelate of porphyrin. Foods such as unpolished grains and nuts are good sources of magnesium, whereas starches, meats, and milk are intermediate sources. Water, hard water in particular, contains a high concentration of magnesium. The dietary intake of magnesium is about 350 to 400 mg/day for adults. Magnesium deficiency induces hypocalcaemia, resulting in a variety of abnormalities of mineral homeostasis, including rickets, impaired release of parathyroid hormone (PTH), impaired peripheral responsiveness to PTH, low serum 1,25-dihydroxyvitamin D concentrations [9] and skeletal resistance to

1,25-dihydroxyvitamin D [10]. Magnesium deficiency is also associated with cardiac complications, including electrocardiographic changes, arrhythmias, and increased sensitivity to cardiac glycosides. Epidemiological studies suggest that populations that have a low dietary intake of magnesium have been reported to have an increased incidence of hypertension [11]. Magnesium depletion has been shown to induce insulin resistance as well as impaired insulin secretion, and thereby may impede the management of diabetes [12]. In Japan, the AIs of magnesium range from 25 mg for children under 1 year of age to 80 mg for children under 5 years of age of both sexes, from 120 to 320 mg for boys over 6 years of age, youths, and adult males, and from 120 to 260 mg for girls over 6 years of age, youths, and adult females, with an additional 35 mg daily during pregnancy, but not during lactation [3]. The UL of magnesium ranges from 130 to 700 mg for children, youths, and adults, with an additional 700 mg daily during pregnancy and lactation [3].

Iron Requirements

Iron is an essential component of haemoglobin, myoglobin, and a number of enzymes including cytochromes, peroxidases, and catalases. The need for iron in the diet varies greatly at different ages and under different circumstances. During growth, pregnancy, and lactation, an increased intake of iron is required because of the increased demand for haemoglobin synthesis. In patients with surgical resection of portions of the stomach and the small intestine, iron deficiency may occur as a result of malabsorption from the gastrointestinal tract. In women with excessive menstrual blood loss and a resultant chromic iron-deficiency anaemia, an increased intake of iron is also required. Traces of copper are required for intestinal iron absorption and utilisation of iron in haemoglobin formation [13]. Ascorbic acid enhances iron absorption from the intestine by conversion of ferric iron (Fe^{+++}, in the diet) to ferrous iron (Fe^{++}, in the gastric fluids) [14]. The richest sources of iron are offal, including liver, heart, kidney and spleen. Other good sources are egg yolk, fish, oysters, clams, whole wheat, beans, asparagus, spinach, molasses, and oatmeal. The dietary intake of iron is about 11 to 12 mg/day for adults. Iron deficiency has been linked with anaemia, immune dysfunction, and reduced physical performance.

In Japan, the AIs of iron range from 6 mg for children under 1 year of age to 12 mg for youths of both sexes [3]. The adult intake should be 10 mg per day for men and 12 mg for women, with an additional 8 mg daily during pregnancy and lactation [3]. These values are considerably higher than the amounts actually required. Allowance is thus made for the fact that only about 10% of food iron is absorbed by the normal adult. The ULs of iron range from 10 mg for children under 1 year of age to 40 mg for adults of both sexes [3].

Zinc Requirements

Zinc is required for certain enzymes, including carboxypeptidase, carbonic anhydrase, and alcohol dehydrogenase. Zinc is also well known to be a constituent of insulin. The dietary intake of zinc is about 12 to 20 mg/day for adults. Zinc is widely distributed in foods and deficiency is rare in humans with a normal diet. In malnourished humans, zinc deficiency causes stunted growth, anaemia due to concomitant iron deficiency, enlarged liver and spleen, and underdevelopment of the genitals and secondary sex characteristics. Zinc deficiency has also resulted in reduced bone size and strength, as well as deficient bone mineralisation [15]. A syndrome of zinc deficiency and radiologic rickets has been described in children with thalassaemia major treated with deferoxamine [16].

In Japan, the AIs of zinc range from 1.2 mg for children under 1 year of age to 6 mg for children under 6 years of age of both sexes, and from 6 to 12 mg for boys over 6 years of age, youths and adult males, and from 6 to 10 mg for girls over 6 years of age, youths and adult females, with an additional 3 mg daily during pregnancy and lactation [3]. The UL of zinc is 30 mg for adults, including women during pregnancy and lactation [3].

Copper Requirements

Copper is a constituent of certain enzymes, including cytochrome, cytochrome oxidase, catalase, tyrosinase, monoamine oxidase, ascorbic acid oxidase, and uricase. Copper is also necessary for the synthesis of haemoglobin along with iron. Copper is widely distributed in foods and the richest sources of it are offal, especially liver. The dietary intake of copper is about 1 mg/day for adults. Similar to other trace elements, copper deficiency is rare with a normal diet but has occurred under conditions of malnutrition. Copper deficiency has been associated with osteoporosis and metaphyseal abnormalities [17].

In Japan, the AIs of copper range from 0.3 mg for children under 1 year of age to 1.0 mg for children under 5 years of age of both sexes, and from 1.3 to 1.8 mg for boys over 6 years of age, youths and adult males, and from 1.2 to 1.6 mg for girls over 6 years of age, youths and adult females, with an additional 0.4 mg and 0.6 mg daily during pregnancy and lactation, respectively [3]. The UL of copper is 9 mg for adults, including women during pregnancy and lactation [3].

Manganese Requirements

Manganese is an activator of a number of different enzymes in general metabolism, including phosphatases, superoxide dismutase, carboxylases, arginase, and adenosine triphosphatase. Manganese is also required for normal formation of bone and cartilage because of its role as a cofactor for glycosyl transferase and other metalloproteins. It is found rather widely in plant and animal tissues. The richest sources are liver, kidney, muscle, spinach, whole-grain cereals, and lettuce. The dietary intake of manganese is about 2 to 3 mg/day for adults. Manganese deficiency is rare with a normal diet in humans and animals because of the abundant supply of manganese from foods compared to the relatively low requirements of mammals. Abnormally developed bones and cartilage have been observed in manganese-deficient animals [18].

In Japan, the AIs of manganese range from 0.003 mg for children under 1 year of age to 2.5 mg for children under 5 years of age of both sexes, and from 3.0 to 4.0 mg for boys over 6 years of age, youths and adult males and from 3.0 to 3.5 mg for girls over 6 years of age, youths and adult females, with no addition during pregnancy and lactation [3]. The UL of manganese is 10 mg for adults, including women during pregnancy and lactation [3].

Iodine Requirements

Iodine is a constituent of the thyroid hormones thyroxine and triiodothyronine, which are involved in a variety of general metabolic functions, development, and tissue differentiation. Iodine is a trace component of soil, and hence there is little in foods. Seafood and drinking water in coastal areas are important sources. Iodine deficiency is related to the occurrence of goitre, and in some cases, severe cretinism with mental retardation.

In Japan, the AIs of iodine range from 40 μg for children under 1 year of age to 120 μg for children under 11 years of age of both sexes, and 150 μg for children over 12 years of age, youths and adults of both sexes [3]. Women during pregnancy and lactation should take an additional 25 μg daily [3]. The UL of iodine is 3 mg for children over 6 years of age, youths and adults, including women during pregnancy and lactation [3].

Selenium Requirements

Selenium is an essential component of glutathione peroxidase, an enzyme responsible for the destruction of hydroperoxide and lipid peroxides by reduced glutathione, protecting membrane lipids and haemoglobin against oxidation by peroxides. Selenium is usually found in seafood, kidney, liver, and other meats. The dietary intake of selenium ranges from 80 to 130 mg/day and averages about 100 mg/day. Some cases of cardiomyopathy have been observed in selenium deficiency [19].

In Japan, the AIs of selenium range from 15 μg for children under 1 year of age to 35 μg for children under 5 years of age of both sexes, and from 40 to 60 μg for boys over 6 years of age, youths and adult males and from 40 to 50 μg for girls over 6 years of age, youths and adult females, with an additional 7 μg and 20 μg daily during pregnancy and lactation, respectively [3]. The UL of selenium is 250 μg for adults, including women during pregnancy and lactation [3].

Chromium Requirements

Trivalent chromium is required for normal glucose metabolism. Chromium may act together with insulin in promoting glucose use. Chromium is found mainly in dairy products, meats, and fish. Chromium content in tissues decreases with age. The dietary intake of chromium ranges from 25 to 200 μg/day and averages about 100 μg/day. In chromium deficiency, a relative insulin resistance and peripheral or central neuropathy have been observed [20].

In Japan, the AIs of chromium are 16 μg for children ages 1 through 2 years and 20 μg for children ages 3 through 5 years of both sexes, and range from 25 to 35 μg for boys over 6 years of age, youths and adult males and from 25 to 30 μg for girls over 6 years of age, youths and adult females, with no addition during pregnancy and lactation [3]. The UL of chromium ranges from 60 to 200 μg for children ages 1 through 14 years and 250 μg for children over 15 years of age, youths, and adults, including women during pregnancy and lactation [3].

Molybdenum Requirements

Molybdenum is a constituent of certain enzymes, including xanthine oxidase, aldehyde oxidase, and sulphite oxidase. The concentration of molybdenum in food varies depending on the environment in which the food was grown. The dietary intake of molybdenum in the US ranges from 120 to 240 μg/day and averages about 180 μg/day. Molybdenum deficiency is rare in humans, but it has been reported that deficiency leads to amino acid intolerance, irritability, and, ultimately, coma.

In Japan, the AIs of molybdenum are 6 μg for children ages 1 through 2 years and 8 μg for children ages 3 through 5 years of both sexes, and range from 12 to 30 μg for boys over 6 years of age, youths and adult males and from 12 to 25 μg for girls over 6 years of age, youths and adult females, with no addition during pregnancy and lactation [3]. The UL of molybdenum ranges from 60 to 200 μg for children ages 1 through 14 years and 250 μg for children over 15 years of age, youths, and adults, including women during pregnancy and lactation [3].

Fluoride Requirements

It is well recognised that fluoride is important for the development of bones and especially teeth, although a slight excess causes mottling of tooth enamel. Fluoride intake varies greatly depending on the fluoride content of drinking water. The average dietary intake of fluoride by US infants and children has been reported to be close to 0.05 mg/kg/day. In the US and Canada, the AIs of fluoride range from 0.01 mg for children under 1 year of age to 3 mg for children under 18 years of age of both sexes, and 4 mg for adult men and 3 mg for adult women, with no addition during pregnancy and lactation. The UL of fluoride ranges from 0.7 to 2.2 mg for children ages 1 through 8 years and 10 mg for children over 8 years and adults, with no addition during pregnancy and lactation.

Potassium Requirements

In the body, potassium is mainly intracellular, whereas sodium predominates in the plasma and extracellular fluid. Potassium and sodium have a reciprocal influence; that is, as one increases in a certain fluid and tissue, the other decreases. Since many foods are higher in potassium than sodium, the sodium content of the body tends to decrease following ingestion of food if sodium chloride is not added. Potassium is essential for muscle contraction, neuronal activity, and carbohydrate metabolism.

In Japan, the AIs of potassium range from 500 mg for children under 1 year of age to 1,100 mg for children under 5 years of age of both sexes, and from 1,350 to 1,750 mg for boys over 6 years of age and male youths, and from 1,200 to 1,650 mg for girls over 6 years of age and female youths, and 2,000 mg for adults, with an additional 500 mg daily during lactation, but not during pregnancy [3]. The UL of potassium has not been set for men and women [3].

Vitamin D Requirements

Vitamin D is a secosteroid, which is generated from provitamin D by ultraviolet irradiation in the skin. In nature, there are two forms of vitamin D: vitamin D_2 (known as ergocalciferol) and vitamin D_3 (known as cholecalciferol). These two forms are of equal potency in humans, but whereas vitamin D_2 arises from plant ergosterol, vitamin D_3 arises from the skin sterol, 7-dehydrocholesterol. Collectively, vitamin D_3 and vitamin D_2 can be simply termed vitamin D. Vitamin D is usually expressed in terms of international units (IU), which is the amount of activity contained in 0.025 μg of cholecalciferol. Under normal physiologic conditions, vitamin D is produced in adequate quantities by the skin through exposure to sunlight, but at higher latitudes and during winter, dietary supplementation becomes increasingly important. Vitamin D plays an essential role in the regulation of calcium and phosphorus homeostasis, and bone mineralisation. To become biologically active, vitamin D must undergo a 25-hydroxylation in the liver to produce 25-hydroxyvitamin D, and a subsequent 1α-hydroxylation in the kidney to produce 1,25-dihydroxyvitamin D, which is the most active form of vitamin D that is responsible for most, if not all, of its biologic functions. Vitamin D derived from photoproduction in the skin or from the diet is rapidly stored in fat and muscle or metabolised to 25-hydroxyvitamin D in the liver, since this form is much more stable in blood and its half-life in human circulation is approximately 10 to 20 days. Therefore, the circulating concentration of 25-hydroxyvitamin D represents a combination of skin vitamin D production and oral ingestion of vitamin D.

Elevations in alkaline phosphatase and PTH concentrations in the circulation have been found to correlate with serum (plasma) 25-hydroxyvitamin D levels. The production of 1,25-dihydroxyvitamin D in the kidney is strictly regulated by parathyroid hormone in response to serum calcium and phosphorus levels. The half-life of 1,25-dihydroxyvitamin D in human circulation is approximately 4 to 6 hours. Thus, 1,25-dihydroxyvitamin D is not an appropriate indicator for assessing vitamin D status. Serum 25-hydroxyvitamin D concentration is used instead as a primary indicator for determining the vitamin D status of an individual. The normal range of serum 25-hydroxyvitamin D is 10–35 ng/ml. According to the Dietary Reference Intakes, the lower limit of the normal range can be as low as 8 ng/ml and as high as 15 ng/ml. In the US and Canada, and Japan as well, a 25-hydroxyvitamin D concentration below 11 ng/ml is considered to be consistent with vitamin D deficiency in infants, neonates, and young children and is therefore used as the key indicator for determining the vitamin D reference value. Little information is available on the level of 25-hydroxyvitamin D that is essential for maintaining normal calcium metabolism and peak bone mass in older children or in young and middle-aged adults. For the elderly, there is growing evidence to support an increased requirement for dietary vitamin D to maintain normal calcium and phosphorus metabolism and bone health.

Regular exposure of the skin to sunlight is the best and simplest way to obtain vitamin D. Latitude, time of day, and season of the year have a dramatic influence on the production of vitamin D in the skin. According to the Dietary Reference Intakes, median vitamin D intakes from food by young women in the US were estimated to be 2.9 μg (114 IU)/day, with a range of 0 to 49 μg (0 to 1,960 IU)/day. A similar median vitamin D intake, 2.3 μg (90 IU)/day, was estimated for a sample of older women. Dietary intakes of vitamin D in Japanese have been investigated by three methods and the results are as follows [21].

1) Dietary vitamin D intakes in Japan, calculated from the findings of the National Nutrition Survey over 20 years from 1971, were 10.5 μg (421 IU)/day, with a range of 9.8 to 11.3 μg (392 to 452 IU)/day.
2) When dietary vitamin D intakes in Japanese were estimated by a model menu method, findings for 25-, 45-, and 65 year-old men were 4.7 μg (184 IU)/day with a range of 1.9 to 11.6 μg (77 to 463 IU)/day, 6.8 μg (272 IU)/day with a range of 1.8 to 19.0 μg (70 to 760 IU)/day, and 18.1 μg (723 IU)/day with a range of 6.5 to 54.2 μg (260 to 2,168 IU)/day, respectively. Findings for women were relatively lower than those of men.
3) When dietary vitamin D intakes in Japanese were estimated by questionnaire, findings for healthy men and women were 12.8 μg (513 IU)/day with a range of 3.1 to 29.5 μg (122 to 1,180 IU)/day and 9.3 μg (373 IU)/day with a range of 1.6 to

21.9 μg (65 to 874 IU)/day, respectively. Most of the dietary vitamin D intakes for Japanese originated from fish.

Vitamin D is especially abundant in fish, fungi, and eggs. Since significant amounts of vitamin D_3 in zooplankton and small fish have been found, the most probable origin of vitamin D_3 in fish is believed to result from the plankton food chain [22]. The high content in woody ear fungus (dried) and silver ear fungus (dried) is derived from the photoconversion of ergosterol to vitamin D_2 by exposure to sunlight during the drying process. The vitamin D content of meat and liver from cattle, swine, and chicken is very low, whereas that in domesticated duck meat is high because of high vitamin D content in their foods. No vitamin D has been found in shellfish, cereals, fruits, vegetables, or algae. In the US, dietary sources of vitamin D are largely fortified milk products and other fortified foods such as breakfast cereals. The concentrations of vitamin D in foods are listed in Table 2.4 [23].

Vitamin D deficiency results in rickets in growing children and osteomalacia in adults. In addition, secondary hyperparathyroidism associated with vitamin D deficiency enhances mobilisation of calcium from bone, and excretion of bone collagen degradation by-products including hydroxyproline, pyridinoline, deoxypyridinoline, and N-telopeptide into urine, resulting in porotic bone. Serum alkaline phosphatase usually increases in vitamin D deficiency. Epidemiological studies have suggested the possibility that vitamin D deficiency is associated with an increased risk of breast, colon, and prostate cancer; however, the pathophysiology of these cancers in relation to vitamin D deficiency remain largely unexplained.

In the US and Canada, the AIs for ages 0 to 50 years are 5 μg (200 IU)/day, for ages 51–70 years 10 μg (400 IU)/day, for ages 70 and older 15 μg (600 IU)/day, and for pregnancy and lactation 5 μg (200 IU)/day. Excessive vitamin D intake causes hypercalcaemia, hyperphosphataemia, and nephrocalcinosis with renal failure, calcification of soft tissues including large and small blood vessels, and kidney stones. The ULs are 25 μg (1,000 IU)/day for infants aged 0–12 months and 50 μg (2,000 IU)/day for children and adults, with no addition during pregnancy and lactation. In Japan, the AIs for ages 0 to 5 years are 10 μg (400 IU)/day and for ages 6 to 70 years and older, 2.5 μg (100 IU)/day, and for women an additional 5 μg (200 IU)/day during pregnancy and lactation [3]. The ULs are 25 μg (1,000 IU)/day for infants aged 0 to 12 months and 50 μg (2,000 IU)/day for children and adults, with no addition during pregnancy and lactation [3].

Vitamin K Requirements

Vitamin K is a cofactor for certain enzymes involved in the processes of blood coagulation and bone metabolism. Compounds with vitamin K activity are polyisoprenoid-substituted 2-methyl-1,4-naphthoquinones. Vitamin K_1, termed phylloquinone, is the major form of vitamin K found in plants. Vitamin K_2, termed menaquinones (MK-n), is most commonly found in animal tissues and is synthesised by bacteria in the intestine. MK-4 is the major form of menaquinones and can be produced from phylloquinone and menadione in animal tissues. Vitamin K_3, termed menadione, is the parent form of the vitamin K series, but it is not found naturally. This is converted to one of the menaquinones in the body after ingestion. Green leafy vegetables such as collards, spinach, salad greens, broccoli, sprouts, cabbage, and lettuce, and plant oils are good dietary sources containing between 100 and 400 μg of phylloquinone per 100 g.

Table 2.4. Concentration of Vitamin D in Foods

Food and description	μg/100 g (IU/100 g)	Food and description	μg/100 g (IU/100 g)	Food and description	μg/100 g (IU/100 g)
FISH		MEATS		EGGS	
Skipjack, viscera preserves	120 (4,800)	Beef, rib loin, separable lean	0 (0)	Duck's egg	18 (720)
Anglerfish, liver	110 (4,400)	Beef, rib loin, total edible	0 (0)	Japanese quail's egg	3 (100)
Indo-Pacific blue, raw	35 (1,400)	Beef, liver	0 (0)	Chicken egg, whole	3 (120*)
Chum salmon, raw	32.5 (1,300)	Swine, loin, separable lean	0.7 (28)	Chicken egg, yolk	6 (230*)
Herring, raw	27.5 (1,100)	Swine, loin, pork separable fat	1.4 (55)	Chicken egg, white	0 (0)
Flatfish, raw	23 (920)	Swine, liver	1.3 (50)	FUNGI	
Bastard halibut, cultured, raw	18 (720)	Chicken, breast, fresh only	0 (0)	Woody ear fungus, dried	400 (16,000)
Bluefin tuna, fatty meat	18 (720)	Chicken, liver	0.2 (8)	Silver ear fungus, dried	400 (16,000)
Sandeel, raw	15 (600)	Duck, domesticated, meat	32.5 (1,300)	Shiitake, dried	16 (640)
Grunt, raw	15 (600)	Turkey, meat, fresh only	0.1 (4)	Shiitake, raw	2 (90)
Rainbow trout, raw	15 (600)	MILK		Shimeji, raw	4 (160)
Eel, raw	14 (560)	Cow's milk	0.3 (13*)	Matsutake, raw	4 (140)
Red sea bream, cultured, raw	13 (520)	Human milk	0.3 (13*)	Common mushroom, raw	3 (100)
Mackerel, raw	11 (440)	Yogurt	0 (0)	Winter fungus, raw	1 (50)
Pacific saury, raw	11 (440)	Butter	0.6 (24)	Nameko, raw	0.4 (16)
Skipjack, raw	10 (400)	Cheese, Cheddar	0 (0)		

Notes: 1) The data shown in this table are picked up from the Standard Tables of Food Composition in Japan -Vitamin D, 1993 edited by Resources Council, Science and Technology Agency, Japan.
2) *These foods contain 25-OH-D and/or 24,25(OH)$_2$D and 1,25(OH)$_2$D besides vitamin D itself.

Margarine and butter contain small but significant amounts of vitamin K, whereas eggs, meats, fish, liver, and milk contain smaller amounts. The concentration of vitamin K in human milk is relatively low (1 to 2 μg/L) and thus, exclusively breast-fed infants are vulnerable to deficiency. According to the Dietary Reference Intakes by the Institute of Medicine, the dietary phylloquinone intake of the North American population is about 150 μg/day for older (over 55 years) and 80 μg/day for younger men and women. Gender differences have not been observed.

Vitamin K deficiency is rare in humans because vitamin K is widely distributed in foods, and is produced by the microflora in the intestine. However, vitamin K deficiency can be caused by fat malabsorption in biliary disease, steatorrhoea, and pancreatic dysfunction. Furthermore, sterilisation of the intestine by antibiotics results in vitamin K deficiency when dietary intake or supplementation of vitamin K is insufficient. Vitamin K deficiency is usually identified as a vitamin K-responsive hypoprothrombinaemia that is associated with an increase in prothrombin time.

Another indicator used to assess vitamin K status is the serum (plasma) concentration of vitamin K. Serum (plasma) phylloquinone concentration reflects recent intakes and has been shown to respond to changes in dietary intake within 24 hours. In healthy individuals, phylloquinone concentrations are higher in older subjects than in younger subjects, irrespective of dietary intake. According to the Dietary Reference Intakes, normal range for plasma phylloquinone concentration in healthy adults aged 20 to 49 years was 0.25 to 2.55 nmol/L and for those aged 65 to 92 years, 0.32 to 2.67 nmol/L [24].

There are some reports suggesting that vitamin K may play a role in osteoporosis. It is well established that vitamin K is essential for the post-translational conversion of specific glutamyl residues of bone protein osteocalcin (undercarboxylated osteocalcin) to γ-carboxyglutamyl residues (γ-carboxylated osteocalcin). Carboxylated osteocalcin is thought to be the active form of osteocalcin, involved in bone formation. Thus, serum (plasma) concentrations of vitamin K and undercarboxylated osteocalcin have been used as indicators to assess vitamin K status in humans. In a number of reports, lower circulating phylloquinone and menaquinone concentrations have been observed in subjects with reduced bone mineral density and osteoporotic patients with bone fractures [25]. Plasma undercarboxylated osteocalcin levels have been associated with increased age, bone status, and risk of hip fracture. These reports are of interest with respect to a potential role of vitamin K in bone health, however, other studies have not confirmed these findings. As the circulating vitamin K concentration can be altered through diet within a few days, and since an immunochemical assay specific for carboxylated osteocalcin is not yet available, the nutritional significance of vitamin K and osteocalcin in bone mineral metabolism remains to be established.

Because of the lack of data to estimate an average requirement, an AI is set based on representative dietary intake data from healthy individuals. In Japan, the AIs of vitamin K range from 5 μg for children under 1 year of age to 20 μg for children under 5 years of age of both sexes, and from 25 to 65 μg for boys over 6 years of age, youths and adult males, and from 25 to 55 μg for girls over 6 years of age, youths and adult females, with no addition during pregnancy and lactation [3]. The ULs of vitamin K range from 5,000 μg for children under 1 year of age to 30,000 μg for adults men and women with no addition during pregnancy and lactation [3].

References

1. National Research Council. Recommended dietary allowances, 10th Ed. Washington, DC: National Academic Press. 1989.
2. Standing Committee on the Scientific Evaluation of Dietary Reference Intakes, Food and Nutrition Board, Institute of Medicine, National Academy. Dietary reference intakes for thiamine, riboflavin, niacin, vitamin B_6, folate, vitamin B_{12}, pantothenic acid, biotin, and choline. Introduction to dietary reference intakes. Washington, DC: National Academic Press. 1998;1–8.
3. Ministry of Health, Labour, and Welfare of Japan. Recommended dietary allowances for the Japanese – dietary reference intakes. 1999;14–16.
4. Standing Committee on the Scientific Evaluation of Dietary Reference Intakes, Food and Nutrition Board, Institute of Medicine, National Academy. Dietary reference intakes for calcium, phosphorus, magnesium, vitamin D, and fluoride. Washington, DC: National Academic Press. 1999;21–7.

5. Hathcock JN. Vitamin and mineral safety. Washington, DC: Council for Responsible Nutrition. 1997.
6. Center for Nutrition Policy and Promotion, US Department of Agriculture. Nutrient content of the US food supply, 1990–1994. Preliminary data. Washington, DC: US Department of Agriculture. 1996.
7. Standing Committee on the Scientific Evaluation of Dietary Reference Intakes, Food and Nutrition Board, Institute of Medicine, National Academy. Dietary reference intakes for calcium, phosphorus, magnesium, vitamin D, and fluoride. Washington, DC: National Academic Press. 1999;91–134.
8. Standing Committee on the Scientific Evaluation of Dietary Reference Intakes, Food and Nutrition Board, Institute of Medicine, National Academy. Dietary reference intakes for calcium, phosphorus, magnesium, vitamin D, and fluoride. Washington, DC: National Academic Press. 1999;134–42.
9. Rude RK, Adams JS, Ryzen E, et al. Low serum concentrations of 1,25-dihydroxyvitamin D in human magnesium deficiency. J Clin. Endocrinol Metab 1985;61:933–40.
10. Carpenter TO, Carnes DL, Anast CS. Effect of magnesium depletion on metabolism of 25-hydroxyvitamin D in rats. Am J Physiol 1987;253:E106–13.
11. Ma J, Folsom AR, Melnick SL, Eckfeldt JH, Sharrrett AR, Nabulsi AA, et al. Associations of serum and dietary magnesium with cardiovascular disease, hypertension, diabetes, insulin, and carotid arterial wall thickness. The ARIC study. Atherosclerosis Risk in Community Study. J Clin Epidemiol 1995;48:927–40.
12. Paolisso G, Sgambato S, Gambardella A, Pizza G, Tesauro P, Varricchio M, et al. Daily magnesium supplements improve glucose handling in elderly subjects. Am J Clin Nutr 1992;55:1161–7.
13. Frieden E. Perspectives on copper biochemistry. Clin Physiol Biochem 1986;4:11–19.
14. Skikne B, Baynes RD. Iron absorption. In: Brock JH, Halliday JW, Pippard MJ, Powell LW, editors. Iron metabolism in health and disease. London: WB Saunders. 1994;151–87.
15. Yamaguchi M, Yamaguchi R. Action of zinc on bone metabolism in rats. Biochem Pharmacol 1986;35:773–7.
16. Tinker D, Rucker RB. Role of selected nutrients in synthesis, accumulation, and chemical modification of connective tissue proteins. Physiol Rev 1985;65:607–57.
17. Strause L, Saltman P, Glowacki J. The effect of deficiencies of manganese and copper on osteoinduction and on resorption of bone particles in Rats. Calcif Tissue Int 1987;41:145–50.
18. Strause L, Saltman P, Smith KT, Bracker M, Andon MB. Spinal bone loss in postmenopausal women supplemented with calcium and trace minerals. J Nutr 1994;124:1060–4.
19. Yang G, Ge K, Chen J, Chen X. Selenium-related endemic disease and the daily selenium requirement of humans. World Rev Nut. Diet 1988;55:98–152.
20. Jeejeebhoy KN, Chu RC, Marliss EB, Greenberg GR, Bruce-Robertson A. Chromium deficiency, glucose tolerance, and neuropathy reversed by chromium supplementation in a patient receiving long-term total parenteral nutrition. Am J Clin Nutr 1977;30:531–8.
21. Takeuchi A, Okano T, Hirahara F, Kobayashi T. Dietary intake of vitamin D in normal Japanese. Vitamins 1993;67(8):417–27.
22. Takeuchi A, Okano T, Tannda M, Kobayashi T. Possible origin of extremely high contents of vitamin D_3 in some kinds of fish liver. Comp Biochem Physiol 1991;100A:483–7.
23. Resources Council, Science and Technology Agency of Japan. Standard tables of food composition in Japan – vitamin D. 1993.
24. Standing Committee on the Scientific Evaluation of Dietary Reference Intakes, Food and Nutrition Board, Institute of Medicine, National Academy. Dietary reference intakes for vitamin A, vitamin K, arsenic, boron, chromium, copper, iodine, iron, manganese, molybdenum, nickel, silicon, vanadium, and zinc. Washington, DC: National Academic Press. 1999;5.1–5.27.
25. Hodges SJ, Akesson K, Vergnaud P, Obrant K, Delmas PD. Circulating levels of vitamins K_1 and K_2 decreased in elderly women with hip fracture. J Bone Miner Res 1993;8:1241–5.

Calcium Homeostasis

Intestinal Absorption of Calcium

T. Fujita

Introduction

Calcium (Ca), the most abundant inorganic element in the human body, is characterised by a rather limited and tightly regulated absorption from the intestine. In addition to being one of the most important constituents of the whole organism, especially of the skeleton, Ca is a vital regulator of metabolism. Serum Ca is the most strictly controlled and maintained biological constant and intracellular free Ca is a decisive factor in regulating cellular function; secretion, locomotion, differentiation and proliferation. Strict compartmentalisation with an extremely wide concentration gradient is an indispensable prerequisite for the action of Ca. A wide concentration gradient of approximately 10^4 is maintained between intracellular compartments with a low concentration level and extracellular fluid with a much higher and strictly constant level. Another concentration difference of similar magnitude is maintained between extracellular fluid and bone, where more than 99% of the Ca in the whole organism is found. It is no wonder, therefore, that the borderlines between each of these compartments should be strictly controlled. The intestinal canal is the only gateway for entrance of Ca to the organism. Intestinal absorption of Ca thus occupies the central position in the homeostatic mechanism of the organism. It therefore appears quite reasonable that Ca absorption should be accurately controlled so as not to allow unlimited entrance of Ca into the vital compartments of extracellular fluid. On the other hand, as much Ca as required must be permitted to enter these compartments whenever it is indicated. The control mechanism for Ca absorption should thus provide a limited but flexible entry of Ca into the organism.

Intestinal Ca absorption, an important and decisive factor for the Ca supply of the organism along with dietary Ca intake, may vary among individuals. Ca deficiency causes parathyroid hormone (PTH) hypersecretion or secondary hyperparathyroidism, with a loss of Ca from bone leading to osteoporosis and inadvertent entry of Ca into blood vessels, brain and intracellular compartments of other vital tissues with resulting dysfunction. Common diseases of old age such as hypertension, arteriosclerosis, Alzheimer's disease, diabetes mellitus and maligancy may thus be some of the consequences of Ca deficiency. An insufficient Ca absorption may therefore represent an important risk factor for these diseases. Measurement of intestinal Ca absorption in an imperative clinical test for this reason. Despite numerous attempts at measuring intestinal Ca absorption leading to the establishment of several sophisticated research methods, however, a simple and accurate test for practical clinical use has not been developed. The factors controlling intestinal calcium absorption and changes of these factors under various circumstances and their implications will be discussed in this Chapter.

Mechanism of Intestinal Ca Absorption

Intestinal calcium absorption mainly occurs in two distinct processes; active, saturable and cellular pathway utilising metabolic energy mainly in the duodenum and upper portion of the jejunum, and passive, unsaturable and possibly paracellular pathway following a concentration gradient throughout the intestinal canal and mainly in the lower part of the jejunum and ileum. The former takes place under precise control faster by $1,25(OH)_2$ vitamin D and other regulators to meet the acute need of the body, whereas the latter takes place more or less automatically and more slowly over a longer timespan, mainly for the maintenance of the need of Ca for the whole organism. The analysis of the mechanism of the active process has been clarified, but the impotance of the quiet passive process has been receiving increased attention. Currently, however, no reliable and practical tests are yet available to analyse these two mechanisms separately.

The passage of Ca through the intestinal wall is a bidirectional process. While the major pathway is the absorption or entry of Ca from the intestinal lumen into the bloodstream in the form of digestive juice, Ca secretion from blood into the lumen in an opposite direction cannot be ignored. Apparent absorption is the algebraic sum of these two. Absorption cannot be simply calculated by subtracting faecal Ca from oral Ca intake, but endogenous faecal Ca secreted from the intestine should be further subtracted. Some of the secreted calcium may be absorbed back again into the bloodstream (Figure 3.1). Measurement of true Ca absorption thus presents a complex problem [1].

Tests for Ca Absorption

Fractional Ca absorption measured by using two kinds of Ca isotope is the most widely used method (Figure 3.2). In order to overcome the complex passages of ingested Ca from plasma into bone and other tissues and back again, multiple measurements of two isotopes, one ingested with 250 mg stable Ca used as a carrier and the other intravenously injected, in urine and blood samples were carried out to calculate the specific activity. The ratio between the intravenously injected isotope to orally ingested isotope gives the rate of fractional absorption. The double isotope method gives results highly correlated with those obtained by the classical metabolic balance method with complete collection of food samples actually ingested, and stool and urine samples for analysis of calcium. In order to interpret the results or metabolic balance studies accurately, endogenous faecal calcium should be measured separately by injecting the Ca isotope intravenously and collecting stool samples. Techniques utilising stable isotopes for safety in children have also been developed.

Use of a single isotope was proposed for simplification and a wider application. Bandarkar et al. used 250 mg stable Ca as a carrier for ^{45}Ca, and Avioli et al. used 20 mg. According to the compartmental analysis of calcium absorbed from the intestine by Marshall and Nordin [2], plasma radioactivity 60 minutes after the tracer ingestion was highly correlated with the results of balance studies and the double

Figure 3.1. Main Ca compartments in the human body and their mutual relationship. Ca transition from the intestinal lumen to the blood pool, indicated by a strong arrow, indicates intestinal Ca absorption, which occupies a central and decisive position in the control of whole body Ca metabolism.

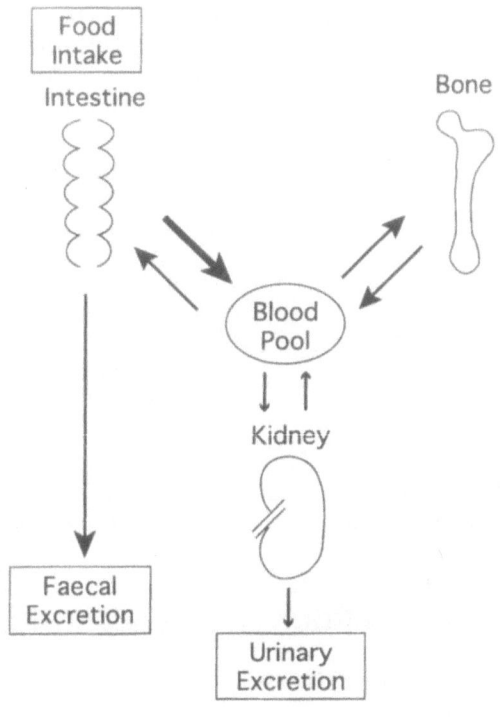

Double Isotope Method

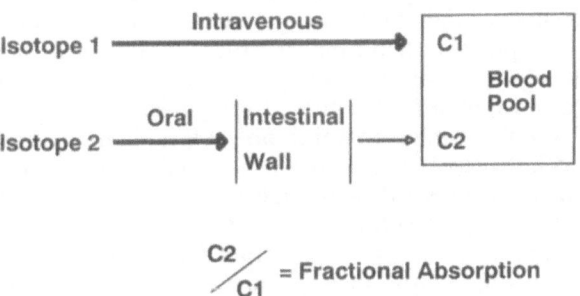

Figure 3.2. Principle of the double isotope method to measure intestinal Ca absorption. In essence, the method is based on a comparison of the fate of an orally administered calcium tracer with that of an intravenously injected tracer which bypasses the intestinal barrier.

isotope method. Assuming the initial dilution space of 15% of body weight in the calculation of the results of the single isotope method of intestinal Ca absorption, Ca absorption may be underestimated in underweight subjects such as osteoporotics. A correction formula using 1/body weight was devised accordingly. A low carrier dose of 20 mg is usually used for the single isotope method and a high dose of 250 mg for

Table 3.1.   Tests for Calcium Absorption in Humans. Currently available tests for humans are shown in this Table. Metabolic balance study is time-consuming and impractical. Indirect tests depend on factors other than Ca absorption

---

I. Direct Tests
1. Metabolic Balance Study
2. Double Isotope Tracer Test
3. Single Isotope Tracer Test
4. Blood Ca Measurement after Oral Ca load
5. Urine Ca Measurement after oral Ca Load

II. Indirect Tests
1. Parthyroid Hormone Measurement after Oral Ca Load
2. Bone Marker Measurement after Oral Ca Load
3. Bone Density Measurement after Long-Term Oral Ca Load

---

the double isotope method. The choice between the simple and complex test should depend on the purpose of the test and circumstances (Table 3.1).

Factors Influencing Ca Absorption

Anatomical Factors

Absorption Surface Area

Since Ca is absorbed from all intestinal segments, not only in the duodenum and upper jejunum where the active absorption mainly takes place, but also in the ileum and colon, resection of any length of the intestine is expected to reduce the absorptive surface and result in a fall of Ca absorption, but a compensatory mechanism was also reported to take place after bowel resection. A fall of fractional Ca absorption (FAC) from 60% to 34% was noted after extensive small bowel resection, and the postoperative FAC was proportional to the length of the remaining small bowel. In patients with the length of the remaining bowel less than 100 cm, a significant correlation was noted between FAC and the time after surgery, suggesting a compensatory mechanism in progess [3]. Although scarcely any Ca is absorbed from the stomach, total or subtotal gastrectomy markedly reduces Ca absorption and gastrectomy is known as one of the major risk factors for Ca deficiency and osteoporosis. This may be explained by a rapid passage of the ingested food over the segment of the intestine active in Ca absorption and failure of hydrolysis of some of the Ca compounds due to the lack of gastric acid secretion.

Absorptive Surface Quality

Even if the absorptive surface area does not change remarkably, function of the intestinal epithelium due to inflammation, circulatory disturbance, metabolic changes or neoplasma may influence Ca absorption. Ageing may cause some change in the absorptive surface quality to interfere with Ca absorption. Neomycin may also make the intestine inefficient for absorption of Ca and other nutrients.

Physiological Factors

Genetic Factors

Even if calcium intake is similar between individuals, the metabolic effect may be different, suggesting a genetic difference in Ca absorption. One of the rare, hereditary causes of hypercalciura and repeated kidney stones is the absorptive type of idiopathic hypercalciura. In absorptive-type hypercalciuria with intestinal hyperabsorption of Ca, a gene defect was recently pointed out, suggesting one of the gene loci related to Ca absorption [4]. Three kindreds were evaluated in a systematic autosomal genome-wide linkage analysis study which indicated 1q23.3–q24 as the chromosomal site responsible for hyperabsorption of Ca.

Ageing

Efficiency of Ca absorption changes markedly with age. Growing children with a large Ca requirement may absorb as much as 75% of the ingested Ca, adults usually absorb 20–40% depending on circumstances, and further ageing decreases this efficiency.

Intestinal Ca absorption decreases with advance in age. A longitudinal study on heathy nuns by Heaney et al. extending over a period of 17 years convincingly demonstrated a fall of fractional Ca absorption with age [5]. In 189 middle-aged women in good health, double isotope tests and balance studies were performed over a period of 17 years with approximately five-year intervals. A highly significant inverse correlation was noted between Ca intake and absorption fraction, varying between 0.45 on an intake of 200 mg/day and 0.15 on an intake of 2,000 mg/day. A highly significant fall of absorptive efficiency was associated with advance in age, at a rate of 0.0021/day.

The decrease of Ca absorption usually becomes evident after the age of 60 and parallels with the fall of serum $1,25(OH)_2$ vitamin D. Suboptimal vitamin D intake and its conversion in the skin in reponse to ultraviolet rays is quite common among elderly people, especially those in institutions. Ageing may also be associated with a decrease in the number of vitamin D receptors in the intestinal wall. Wood et al. [6] altered plasma $1,25(OH)_2$ vitamin D over a wide range in rats by continuous infusion. The slope of linear regression between plasma $1,25(OH)_2$ vitamin D and duodenal Ca transport in old rats was only 46% of that in young rats, indicating an age-bound resistance of the intestine to $1,25(OH)_2$ vitamin D. In enterocytes isolated from aged rats, $1,25(OH)_2$ vitamin D stimulation of Ca^{++} channels through the CAMP/PKC pathway is blunted. After $1,25(OH)_2$ vitamin D treatment in vivo, this change was restored to normal [7]. Increase of Ca absorption in response to low Ca diet was also diminished with advance in age. Multiple factors may be responsible for the age-bound fall of calcium absorption; 1) decreased vitamin D intake, solar exposure and biosynthesis of $1,25(OH)_2$ vitamin D in the kidney, 2) decreased oestrogen secretion or conversion of androgen to oestrogen by aromatase and 3) changes of the intestine itself such as simple atrophy, circulatory disturbance and decrease of receptors for vitamin D and other active substances. Age-bound increase of blood parathyroid hormone is mainly explained by the progressive decrease of intestinal Ca absorption. Calcium intake as high as 2,500 mg/day, shown to suppress the elevated blood PTH to the level of young normal subjects probably by overcoming the low intestinal Ca absorption associated with ageing, might be called a rejuvenation of Ca metabolism.

Acid-Base Balance

Alkalinity of the intestinal lumen enhances Ca absorption, as suggested by milk-alkali syndrome, an unfortunate complication of one-time peptic ulcer treatment with large amounts of milk and sodium bicarbonate, causing Ca hyperabsorption and hypercalcaemia. In addition to an oral Ca load as high as 5,000 mg/day, alkalinisation of the intestinal lumen is thought to have caused entrance of an enormous amount of Ca into the bloodstream, disrupting the constancy of serum calcium, leading to hypercalcaemia with metabolic alkalosis. Alkalosis may also utilise an indirect pathway to augment intestinal Ca absorption. About one-half of calcium in the blood is loosely bound to protein and another one-half remains free in an ionised form. Alkalosis facilitates Ca binding to protein, thereby decreasing the ionised fraction of Ca in blood. Although the total calcium concentration in the blood is unchanged, the fall of ionised Ca in blood gives a signal to increase secretion of PTH, which stimulates the biosynthesis of $1,25(OH)_2$ vitamin D, augmenting Ca absorption. Acidification of the intestinal lumen, on the contrary, decreases Ca absorption. Chronic acidosis, however, may enhance Ca absorption through enhanced $1,25(OH)_2$ vitamin D biosynthesis secondary to phosphorus depletion or urinary Ca loss.

Autonomic Nervous System

The parasympathetic nervous system stimulates gastrointestinal motility and function including absorbing activity, whereas they sympathetic nervous system suppresses these functions. Selective vagotomy decreased net intestinal Ca absorption in rats, being harmful to the calcium economy of the body [8]. Asian Africans were suggested to have higher parasympathetic and lower sympathetic activity than Caucasians. The pupillary size is smaller and intestinal Ca absorption higher in Asian Africans that in Caucasians. This may explain the lower risk for some diseases of old age such as osteoporosis, hypertension, arteriosclerosis and Alzheimer's disease in Asians [9].

Endocrine Factors

Hormonal Vitamin D

By far the most important regulator of intestinal Ca absorption is hormonal vitamin D. After conversion of the dietary precursors in the skin under ultraviolet solar exposure, vitamin D is hydroxylated in the liver to become 25(OH) vitamin D, the bulk of vitamin D derivative in blood, serving as an index of vitamin D intake, followed by the final conversion to $1,25(OH)_2$ vitamin D or calcitriol, which acts directly on the intestinal epithelial cells to facilitate Ca absorption. The mechanism of $1,25(OH)_2$ vitamin D action on the intestinal epithelial cells is complex, involving genomic action mediated by nuclear recoptors increasing biosynthesis of calbindin used in transcellular Ca movement or buffering under a possible participation of lysosomes, polyamine metabolism regulating differentiation of intestinal epithelial cells and more rapid non-genomic action possibly mediated by plasma membrane receptors.

Fluidity of plasma membranes may also undergo changes under vitamin D action, in favour of Ca absorption.

Vitamin D receptor polymorphism was suggested to be related to bone mineral density and the risk for osteoporosis [10–12]. Tolerance to a low-Ca diet may be impaired in subjects with vitamin D receptor gene polymorphism BB, thought to be exposed to higher risk of osteoporosis [13]. The actual relationship between vitamin D receptor gene polymorphism and receptor function as to Ca absorption still remains to be established, and no definite changes of the receptor molecule itself have been connected with the gene polymorphism except for Fok I polymorphism [14]. FF homozygotes absorbed 41.5% greater Ca from the intestine than the ff counterparts.

Oestrogen

Oestrogen deficiency is associated with a decrease of intestinal Ca absorption. Ovariectomy in rats decreased intestinal Ca absorption and oestrogen restored it [15]. Oestrogen enhances intestinal Ca absorption. Part of this effect may be explained by the increased biosynthesis of $1,25(OH)_2$ vitamin D, but a direct effect of oestrogen on the intestinal epithelium to augment Ca absorption was also suggested. Attempts were also made to demonstrate oestrogen receptors in the intestine to explain such direct action. In addition to the direct action of oestrogen deficiency to enhance bone resorption via its effect on cytokines, decrease of intestinal Ca absorption causing PTH hypersecretion and consequent augmentation of bone resorption should also be taken into account to reconstruct the complete picture of postmenopausal bone loss. In later years, negative Ca balance due to Ca malabsorption becomes more distinct as the cause of bone loss and so-called Type II or senile osteoporosis supervenes, but Ca malabsorption and secondary hyperparathyroidism must already be in action in the immediate postoperative period, only to become more manifest in later years. Strict distinction between Type I and II osteoporosis may be misleading, since the same and continuous trend of Ca malabsorption is operating throughout the ageing process in both males and females.

A sudden increase of Ca absorption in early menarche when the girl reaches Tanner stage 2 with the beginning of breast development compared to late premenarche six months earlier may also suggest the effect of the rising oestrogen secretion on intestinal Ca absorption [16]. In adult females, premenstrual symptoms are associated with a midcycle fall of $1,25(OH)_2$ vitamin D along with oestrogen with transient secondary hyperparathyroidism; this favourably responds to Ca and vitamin D supplementation [17]. Down-regulation of the vitamin D receptor in oestrogen receptors may explain vitamin D resistance in the intestine of the elderly [18].

Ipriflavone, a synthetic phyto-oestrogen used in the treatment of osteoporosis in Europe and Japan because of its inhibitory effect on bone resorption, was reported to increase intestinal Ca absorption. Ca uptake by the duodenal cells was significantly greater in ipriflavone- or oestrogen-treated animals than in the ovariectomised controls [19].

Testosterone also increases intestinal Ca absorption, and this is further enhanced by oestradiol. Conversion of some androgen to oestrogen by aromatase may also occur.

Glucocorticoid

In Cushing's sydrome, a prolonged endogenous corticosteroid excess, and under corticosteroid treatment, intestinal Ca absorption decreases, causing a negative Ca balance, secondary hyperparathyroidism and osteoporosis. Glucocorticoid was reported to suppress intestinal Ca transport [20]. The effect of glucocorticoid on intestinal Ca transport is apparently site-dependent, and distal segments of the intestine may respond to glucocorticoid with an increase in Ca transport [21].

Growth Hormone

Growth hormone administration increased intestinal Ca absorption in humans and animals deficient in growth hormone [22]. Growth hormone deficiency suppressed intestinal Ca transport and administration restored it promptly. This action of growth hormone is apparently not mediated by 1,25(OH)$_2$ vitamin D. Growth hormone may augment biosynthesis of 1,25(OH)$_2$ vitamin D, but the amount of 1,25(OH)$_2$ vitamin D required to explain the growth hormone effect on intestinal Ca absorption is much greater than that achieved by the stimulation of biosynthesis. Somatostatin, an inhibitor of growth hormone secretion, decreased intestinal Ca absorption.

Insulin

In patients with Type I insulinopaenic diabetes mellitus and rats with streptozotocin-induced diabetes mellitus, intestinal Ca absorption is reduced, as one of the possible causes of diabetic osteopaenia. In patients with idiopathic hypercalciuria and recurrent calcium urolithiasis, a pronounced post-prandial hyperinsulinaemia was associated with an increase of intestinal Ca absorption. Glucose-clamp studies have revealed that high plasma insulin levels seen after meals enhances intestinal Ca absorption independently of 1,25(OH)$_2$ vitamin D [23].

Thyroid Hormone

Intestinal Ca absorption is reduced and Ca balance may become negative in hyperthyroidism. A decrease of serum 1,25(OH)$_2$ vitamin D and intestinal hypermotility, shortening the food passage time, may be responsible. No consistent changes of Ca absorption or 1,25(OH)$_2$ vitamin D metabolism were found in hypothyroidism.

Other Hormones

Prolactin was reported to increase Ca absorption in addition to enhancing 1,25(OH)$_2$ vitamin D biosynthesis, apparently in response to the need for more Ca by the body during lactation [24]. Serotonin also increased Ca absorption. Bombesin infusion causes hypercalcaemia and PTH suppression, along with a rise of plasma insulin and glucagon. Increase of intestinal Ca absorption was suggested as the mechanism of hypercalcaemia.

Nutritional Factors

Wolf et al. [25] anlaysed the contributions of dietary and lifestyle factors to intestinal Ca absorption. In addition to the inverse relationship with Ca intake ($r = -0.18$, $p = 0.03$), dietary fibres ($r = -0.19, p = 0.028$), alcohol intake ($r = -0.14, p = 0.094$) and physical activity ($r = -0.22, p = 0.007$) also showed an inverse relationship or tendency towards it. Fat intake showed a positive relationship ($r = 0.29, p = 0.001$).

Minerals

Autoregulation by Ca

Ca is a unique nutrient for which intestinal absorption is autoregulated in a homeostatic mechanism. Low Ca intake increases and high intake decreases Ca absorption. This is in part explained by the well-known endocrinological homeostatic mechanism involving PTH and $1,25(OH)_2$ vitamin D. Low calcium intake, causing even a slight fall of serum calcium, prompts an increase of parathyroid hormone secretion, which augments 1α-hydroxylation of 25(OH) vitamin D in the kidney, leading to an increased production of $1,25(OH)_2$ vitamin D, enhancing intestinal Ca absorption. High Ca intake does exactly the opposite, suppressing Ca absorption.

Magnesium

Calcium and magnesium, divalent alkali earth metals of similar molecular weight, may be absorbed from the intestine by a similar mechanism, exerting some influence on each other. Actually, however, the mechanisms involved are quite different. Magnesium absorption is not as tightly controlled as calcium, without any participation of $1,25(OH)_2$ vitamin D. Although ingestion of a large amount of magnesium may suppress Ca absorption, no evident competition of the gate for absorption of these two elements was demonstrated.

Phosphorus

Since Ca and phosphorus readily bind to each other forming calcium phosphate, which is hardly absorbable, a large amount of phosphorus is expected to inhibit Ca absorption. Phosphorus intake of not more than twice the amount of Ca intake has, therefore, been generally recommended. In a separate action, phosphorus decrease urinary excretion of Ca and decreases renal $1,25(OH)_2$ vitamin D biosynthesis directly. The action of phosphorus on Ca metabolism is thus quite complex and the range of optimum phosphorus intake also appears to be wide. Alkalinity of the intestinal lumen may facilitate calcium phosphate precipitation. Therapeutically administered bisphosphonates may bind Ca and inhibit its absorption.

Protein and Amino Acids

Low protein intake of 0.7 g/kg body weight may decrease Ca absorption, giving fractional Ca absorption of 0.19 compared to high Ca intake of 2.1 g/kg body weight, which gives a corresponding value of 0.26. The low Ca intake apparently induces insufficient intestinal Ca and elevation of PTH levels in the blood [26]. In eight young normal subjects, 0.7 to 0.8 g/kg body weight protein intake caused blood PTH elevation in four days, but 0.9 to 1.0 g/kg did not. High-casein meals, 500 g/kg body weight, enhanced Ca absorption in rats, but the mechanism remains to be clarified [27]. According to a metabolic balance study by Heaney et al. [28], on the other hand, no influence of changes of intestinal Ca absorption were found in changes of dietary protein and phosphorus in 191 nuns with a mean age of 48 ± 7 years.

Amino acids, especially the basic varieties such as lysine and arginine, are known to increase Ca absorption. Dietary supplementation of 800 mg L-lysine, but not L-valine or L-tryptophane, increased intestinal Ca absorption.

Carbohydrate

Carbohydrates generally enhance intestinal Ca absorption. The effect of lactose apparently facilitating Ca absorption from milk is well known, but the mechanism involved is incompletely understood. Stable lactose reaching far into the distal segment of the intestine was suggested to be taken up by villous cells to facilitate Ca uptake. Glucose also increases Ca absorption. An increase of Ca absorption by 49% was reported on coadministration of 222 mmol glucose with the Ca load. Since no effect of lactose is found in lactase deficiency, hydrolysis of lactose into galactose and glucose may be necessary for the lactose facilitation of Ca absorption. Indigestible sugars such as inulin, lactulose and other oligosaccharides may increase Ca absorption by stablising and keeping the ingested food longer within the absorptive range of the intestinal canal. Production of volatile fatty acids to decrease the pH of the intestinal lumen may also help to increase free Ca ions to be absorbed. Mannitol increases Ca absorption by facilitating Ca access to the cell by forming complexes with Ca.

Fat

Fat and fatty acids may also bind to Ca and decrease Ca absorption. Ingestion of linoleic acid suppressed Ca absorption, but not its methyl ester or triglyceride, probably because micelle formation prevents them from binding to Ca. Short-chain fatty acids, on the other hand, may increase Ca ions through hydrolysis to facilitate absorption.

Fibres

Dietary fibres are usually thought to interfere with Ca absorption from the intestine by the simple physical principle of binding with Ca depending upon the uronic acid content, and also by shortening the passage time through the intestine. Breakdown products of fibres may acidify the luminal contents to hydrolyse Ca compounds, releasing free Ca ions to be absorbed.

Ca Binders

Oxalates and phytates, contained in many vegetables, are strong binders of Ca and potent inhibitors of Ca absorption. Spinach, containing abundant oxalate, is also rich in Ca, but resists intestinal absorption.

Lifestyle Factors

Smoking inhibits gastrointestinal motility and Ca absorption. Alcohol may also inhibit Ca absorption. Coffee intake exceeding 1,000 ml may cause negative Ca balance by inhibiting absorption, while two cups are assumed to be safe. Excessive exercise may also inhibit Ca absorption.

Pharmaceutical Factors

Diuretics

Thiazides and chlorthalidone may reduce intestinal Ca absorption, but the effect is not consistent and is unrelated to $1,25(OH)_2$ vitamin D. Since urinary Ca excretion

Table 3.2. Factors Influencing Intestinal Ca Absorption. Intestinal Ca absorption is controlled by many endogenous and exogenous factors, some of which are listed in this Table

1. Anatomical Factors
   Absorptive Surface Area
   Absorptive Surface Quality
 	
2. Physiological Factors
   Genetic Factors
   Ageing
   Acid-Base Balance
   Autonomic Nervous System
 	
3. Endocrine Factors
   Hormonal Vitamin D
   Oestrogen
   Glucocorticoid
   Growth Hormone
   Insulin
   Thyroid Hormone
   Other Hormones
	4. Nutritional Factors
	   Minerals
	      Autoregulation by Ca
	      Mg
	      P
	   Proteins and Amino Acids
	   Carbohydrate
	   Fat
	   Fibres
	   Ca Binders
	
	5. Lifestyle Factors
	   Smoking
	   Alcohol
	   Coffee
	
	6. Pharmaceutical Factors
	   Diuretics
	   Ca Channel Blockers

is also reduced by thiazides, the net result does not necessarily produce a negative Ca balance. Furosemide, on the other hand, increases both urinary Ca excretion and intestinal Ca absorption.

Ca Channel Blockers

Since Ca uptake by the intestinal epithelial cells through the Ca channel is involved in the process of Ca absorption, Ca channel blockers may be thought to interfere with Ca absorption. Each Ca channel antagonist compound apparently has cell-specific actions, and verapamil stimulates Ca absorption, while dilitiazem and nifedipine do not. The problem is still controversial and further studies are required (Table 3.2).

Intestinal Ca Absorption from Ca Supplements

Calcium is better absorbed from food sources than from supplements, making Ca ready to be absorbed probably because of gastric and other digestive juices. In order to provide optimum Ca intake larger than 1,000 mg ~1,500 mg/day, supplements may have to be used. Among various Ca compounds used as supplements, calcium carbonate was thought to be absorbed less efficiently than other calcium compounds such as calcium citrate, especially in subjects with gastric hypoacidity. Food ingestion increases gastric acid secretion and facilitates Ca absorption from Ca carbonate even in subjects with achlorhydria. Ca carbonate has the advantage of containing the largest amount of elementary Ca per unit weight, approximately 400 mg per gram compared to 90 mg per gram for Ca gluconate and other organic Ca salts. On comparing five commonly used Ca compounds used as supplements, absorption of Ca from Ca carbonate, gluconolactate and carbonate mixture, citrate, pidolate, Ca carbonate mixture and Ca hydroxyapatite complex were found to be absorbed similarly in the initial phase, since they raised serum Ca equally well. In vitro tests of availability of Ca salts is conducted by dissolving the tablets in vinegar. A good dissolution may indicate a high availability with a high correlation with the results of in vivo tests.

Active Absorbable Algal Calcium, known as AAACa or AdvaCal, is a widely used calcium preparation. Originally made from oyster shell heated to a moderately high temperature of 800°C under reduced pressure, with oxidation of Ca carbonate under persistence of the characteristic shell lattice structure [29], plus the addition of similarly treated seaweed, Heated Algal Ingredient (HAI), provided a high absorbability from the intestine [30]. In a crossover study in 10 normal males, urinary excretion of calcium after oral ingestion of 1,000 mg elementary calcium was significantly higher from AAACa than Ca carbonate. Plasma ionised Ca was measured after oral doses of 300 mg calcium as AAACa, Calcium Citrate Malate (CCM) and milk one, two and three hours later. As shown in Figure 3.3, rise of plasma Ca over the baseline level was similar in magnitude in the first and second hour after the oral load of all three Ca preparations, all significantly above the level obtained by placebo, and the plasma Ca level at three hours after the oral load of AAACa was significantly higher not only than the level obtained by placebo, but also from those achieved by CCM or milk. These results would suggest a high absorbability of AAACa comparable to CCM and milk.

HAI increased Ca absorption of calcium. Dried and powdered seaweed (Cystophyllum fusiforme) was heated at 800°C under reduced pressure, suspended in 6N HCl, mixed, kept for 24 hours and centrifuged, then neutralised with NaOH to pH 1.0. To 100 ml of this solution containing 75.67 g crude HAI was added a 500 ml butanol:125 ethanol mixture. The butanol–ethanol soluble fraction was further extracted with butanol to obtain 0.77 g, corresponding to approximately 1% of the original material used as the active fraction. Administration of 0.5 μg of this fraction

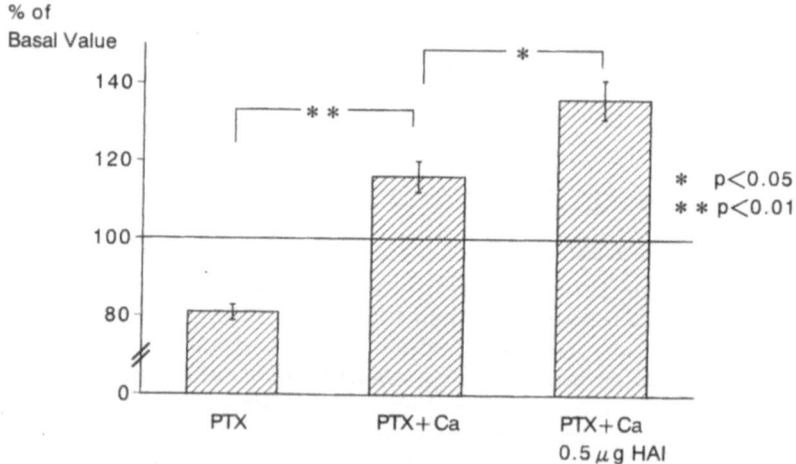

Figure 3.3. Effect of heated algal ingredient (HAI) on intestinal calcium absorption in rats. In order to detect an increased intestinal absorption as a rise of plasma calcium, rats were parathyroidectomised and maintained on a moderately low-calcium diet. Administration of 20 mg calcium raised plasma calcium, and HAI augmented it significantly.

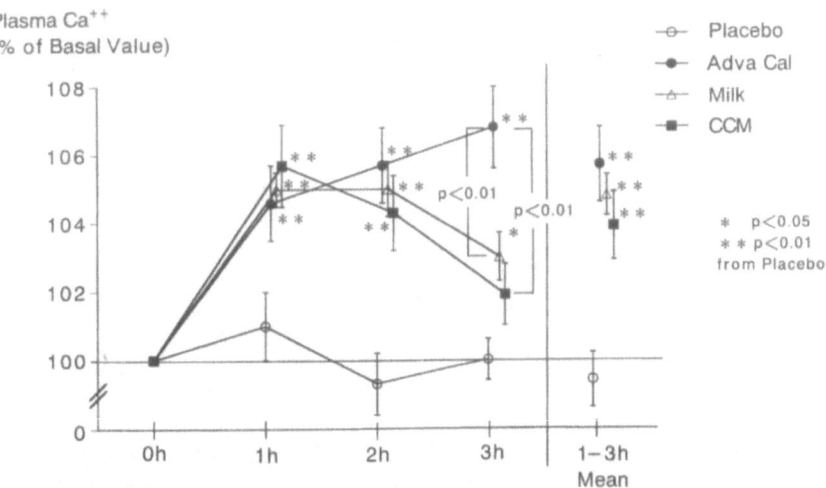

Figure 3.4. Plasma ionised calcium one to three hours after oral loading of 300 mg calcium as calcium citrate malate (CCM), milk and AAACa (AdvaCal). At one and two hours, the plasma ionised calcium level was similar, but AAACa gave a significantly higher value at three hours.

via a stomach tube in parathyroiroidectomised rats, along with 20 mg Ca, increased Ca absorption as indicated by a significantly greater rise of plasma Ca than that induced by Ca alone as shown in Figure 3.4. In parathyroidectomised rats maintained on a moderately low Ca intake of 0.3%, which does not interfere with their survival, serum Ca is maintained at a low level of 6–7 mg/dl, but may be raised towards the normocalcaemic range according to the amount of Ca contained in the diet and its absorbability. Plasma Ca in this model of rats was above the level achieved by Ca alone, therefore indicating the activity of HAI in stimulating Ca absorption (Table 3.3).

Summary

Intestinal Ca absorption is a central mechanism for the human body to control the Ca economy. As expected from the unique role of Ca as the body constituent and regulator in strict compartmentalisation with an extremely wide concentration difference between each compartment (bone, blood and cytosol), Ca absorption is limited and precisely regulated by many endogenous and exogenous factors. Multiple exchange between these compartments makes an accurate clinical test for the intestinal absorption of Ca quite difficult. In view of the universally deficient Ca intake and the many diseases provoked by it, understanding and assessment of Ca absorption is of the utmost importance. Based on such recognition, attempts should be made to overcome the problem of Ca deficiency by increasing Ca intake or augmenting Ca absorption. In addition to the nutritional approach of increasing the intake of Ca-containing food, Ca supplementation by using effective supplements should be employed. Since ageing is invariably associated with a decrease in intestinal Ca absorption with a parallel increase in osteoporosis and other diseases of Ca defi-

Table 3.3.   Available Ca Sources for Supplementation. Calcium carbonate is most
widely used. Organic Ca compounds contain less Ca per weight than inorganic Ca
salts

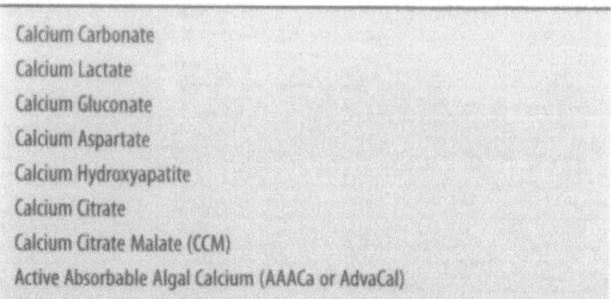

Calcium Carbonate

Calcium Lactate

Calcium Gluconate

Calcium Aspartate

Calcium Hydroxyapatite

Calcium Citrate

Calcium Citrate Malate (CCM)

Active Absorbable Algal Calcium (AAACa or AdvaCal)

ciency, and secondary hyperparathyroidism with increase of Ca in soft tissues such
as blood vessels, heart, brain and cartilage, improvement of the intestinal Ca absorption is a key to preventing these disease and providing a healthy life.

References

1. Heaney RP, Recker RR. Estimation of true calcium absorption. Ann Int Med 1985;103:516–21.
2. Marshall DH, Nordin BEC. Kinetic analysis of plasma radioactivity after oral ingestion of radiocal-
    cium. Nature 1969;222:797.
3. Colette C, Gouttebel MC, Monnier LH, Saint-Aubert B, Joyeux H. Calcium absorption following small
    bowel resection in man. Evidence for an adaptive response. Europ J Clin Invest 1986;16:271–6.
4. Reed BY, Heller HJ, Gitomer WL, Pak CY. Mapping a gene defect in absorptive hypercalciuria to chro-
    mosome 1q23.3–q24. J Clin Endocrinol Metab 2000;84:3907–13.
5. Heaney RP, Recker RR, Stegman MR, Moy AJ. Calcium absorption in women: Relationship to calcium
    intake, estrogen status and age. J Bone Miner Res 1989;4:469–75.
6. Wood RJ, Fleet JC, Cashman K, Bruns ME, DeLuca HF. Intestinal calcium absorption in the aged rat;
    Evidence of intestinal resistance to 1,25(OH)$_2$ vitamin D. Endocrinol 1998;139:3843–8.
7. Massheimer V, Boland R, Boland AR de. In vivo treatment with calcitriol (1,25(OH)$_2$D$_3$) reverses age-
    dependent alterations of intestinal calcium uptake in rat enterocytes. Calcif Tissue Int 1999;64:173–8.
8. Engelhardt W, Rumenapf G, Schwikke PO. Effects of highly selective vagotomy on small intestinal
    calcium transport in the rat. Miner Electrol Metab 1984;10:239–43.
9. Gilbert C. Low risk to certain diseases in aging: Role of the autonomic nervous system and calcium
    metabolism. Mechanism Aging Developm 1993;70:95–113.
10. Cooper GS, Umbach DM. Are vitamin D receptor polymorphisms associated with bone mineral
    density? A metaanalysis. J Bone Miner Res 1996;11:1841–9.
11. Eisman JA. Vitamin D receptor gene alleles and osteoporosis: An affirmative view. J Bone Miner Res
    1995;10:1289–93.
12. Peacock M. Vitamin D receptor gene alleles and osteoporosis: A contrasting view. J Bobe Miner Res
    1995;10:1294–7.
13. Dawson-Hughes B, Harris SS, Finneran S. Calcium absorption on high and low calcium intakes in
    relation to vitamin D receptor genotype. J Clin Endocrinol Metab 1995;80:3657–61.
14. Ames SK, Ellis KJ, Gunn SK, Copeland KC, Abrams SA. Vitamin D receptor gene Fok 1 polymorphism
    predicts calcium absorption and bone mineral density in children. J Bone Miner Res 1999;14:740–6.
15. O'Loughlin PD, Morris HA. Oestrogen deficiency impairs intestinal calcium absorption in the rat.
    J Physiol 1998;511:313–22.
16. Abrams SA, Copeland KC, Gunn SK, Gundberg CM, Klein KO, Ellis KJ. Calcium absorption, bone mass
    accumulation, and kinetics increase during early pubertal development in girls. J Clin Endocrinol
    Metab 2000;85:1805–9.
17. Thys-Jacobs S, Alvir MJ. Calcium-regulating hormones across the menstrual cycle: Evidence of a sec-
    ondary hyperparathyroidism in women with PMS. J Clin Endocrinol Metab 1995;80:2227–32.

18. Chen C, Noland KA, Kalu DN. Modulation of intestinal vitamin D receptor by ovariectomy, estrogen and growth hormone. Mechanism of Aging and Development 1997;99:109–22.
19. Arjmandi BH, Khalil DA, Hollis BW. Ipriflavone, a synthetic phytoestrogen, enhances intestinal calcium transport in vitro. Calcif Tissue Int 2000;67:225–9.
20. Favus MJ, Walling MW, Kimberg DV. Effects of 1,25-dihydroxy-cholecalciferol on intestinal calcium transport in cortisone-treated rats. J Clin Invest 1973;52:1680–5.
21. Yeh JK, Aloia JF. Influence of glucocorticoids on calcium absorption in different segments of the rat intestine. Calcif Tissue Int 1986;38:282–8.
22. Fleet JC, Bruns ME, Hock JM, Wood RJ. Growth hormone and parathyroid hormone stimulates intestinal calcium absorption in aged female rats. Endocrinol 1994;134:1755–60.
23. Rumenapf G, Schmidtler J, Schwille PO. Intestinal calcium absorption during hyperinsulinemic euglycemic glucose clamp in healthy humans. Calcif Tissue Int 1990;46:73–9.
24. Krishnamra N, Thumchai R, Limlomwongse L. Acute effect of prolactin on the intestinal calcium absorption in normal, pregnant and lactating rats. Bone Miner 1990;11:31–41.
25. Wolf RL, Cauley JA, Baker CE, Ferrell RE, Charron M, Caggiula AW et al. Factors associated with calcium absorption efficiency in pre- and postmenopausal women. Am J Clin Nutr 2000;72:466–71.
26. Kerstetter JE, Svastisalee CM, Caseria DM, Mitnick ME, Insogna KL. A threshold for low protein diet-induced elevation in parathyroid hormone. Am J Clin Nutr 1998;72:168–73.
27. Knowles JB, Wood RJ, Rosenberg IH. Response of fractional calcium absorption in women to various coadministered oral glucose doses. Am J Clin Nutr 1988;48:1471–4.
28. Deroisy R, Zartarian M, Meurmans L, Nelissenne N, Micheletti MC, Albert A, Reginster JY. Acute changes of serum calcium and parathyroid hormone circulating levels induced by the oral intake of five currently available calcium salts in healthy male volunteers. Clin Rheumatol 1997;16:249–53.
29. Fujita T, Fukase M, Nakada M, Koishi M. Intestinal absorption of oyster shell electrolysate. Bone Miner 1988;11:85–91.
30. Fujita T, Fujii Y, Goto B, Miyauchi A, Takagi Y, Kobayashi S et al. Increase of intestinal calcium absorption and bone mineral density by heated algal ingredient (HAI) in rats. J Bone Miner Metab 2000; 18:165–9.

Renal Handling of Calcium

H. Motoyama and P. A. Friedman

Introduction

The amount of calcium contained in the body of vertebrate mammals is prominently large when compared to other minerals. For example, total body calcium amounts to 600,000 mEq, whereas sodium amounts to 3,500 mEq [1]. Most of the calcium exists in bone in the form of hydroxyapatite and only 1% of the calcium is present in extracellular fluid [1]. The regulation of extracellular calcium is critical for many physiological processes and is complex. However, homeostasis is achieved by the cooperative interaction between the parathyroid gland, intestine, kidney, and bone. In this chapter, the mechanism and regulation of calcium handling in the kidney is discussed.

Overviews of Calcium Handling in a Body and the Kidney

In adult humans, the calcium concentration of extracellular fluid averages 10 mg/dl (5 mEq/l, 2.5 mM). The calcium in blood exists in three distinct chemical forms; protein-bound (non-diffusible), complexed (but diffusible), and as ionised calcium (diffusible). Forty per cent of the serum calcium is bound to plasma proteins, with albumin accounting for some 90% of this. Smaller percentages are bound, although with greater affinity, to globulins. Ten per cent of serum calcium is complexed with small polyvalent anions. The major complexes are formed by ion pairing with phosphate and citrate and, to a lesser extent, with bicarbonate and sulphate. Thus, 50% of serum calcium is free calcium in the normal state. The degree of complexation depends on the concentrations of ionised calcium and the complexing anion, and the ambient pH. The different moieties of calcium are important because diffusible calcium (i.e., the ionised calcium plus complexed calcium) is filtered at the glomerulus and crosses cell membranes.

The concentration of calcium in the extracellular fluid represents a dynamic balance between intestinal absorption, renal reabsorption and osseous resorption (Figure 3.5).

In the case of a 70-kg individual, net calcium absorption by the intestine is 200 mg on a daily dietary intake of 1,000 mg. In the steady state, this amount matches urinary calcium excretion.

Calcium is transported throughout the nephron. The major nephron sites involved in calcium reabsorption are proximal convoluted and straight tubules, thick ascending limbs, and distal tubules. Final adjustments of calcium excretion are achieved in

51

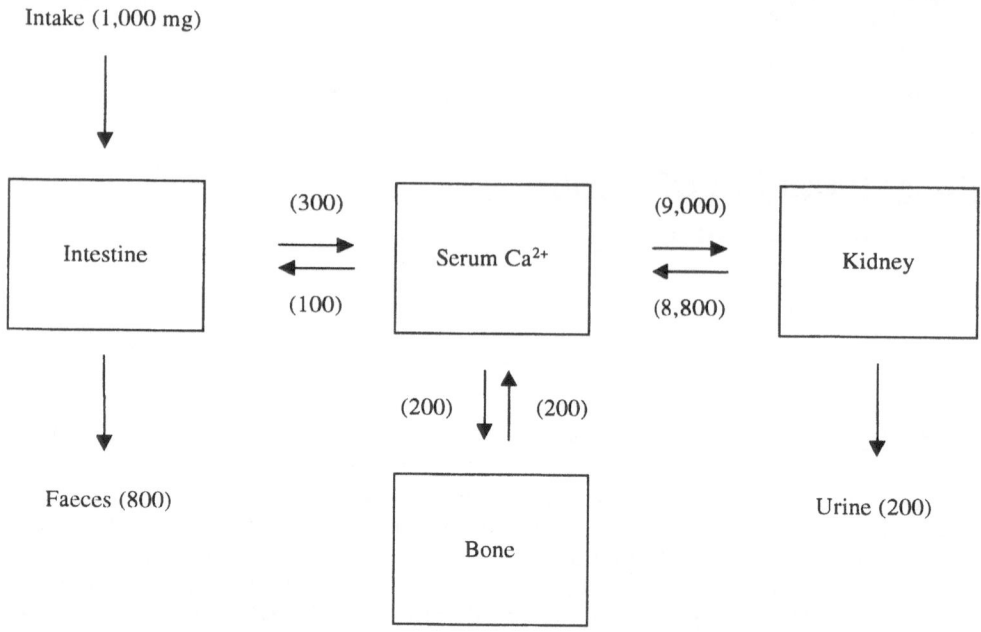

Intake (1,000 mg)

Figure 3.5. Extracellular calcium balance in the adult human. Numeric values for calcium intake, excretion, and fluxes are in milligrams per day, approximated for a 70-kg individual.

collecting ducts, where transport may be absorptive or secretory. As shown in Figure 3.6, 60–70% of the filtered calcium is reabsorbed by proximal tubules, an additional 20–25% by thick ascending limbs, and 8–10% by distal tubules. The net result of these processes is that only 0.5–1.5% of the filtered calcium is normally excreted in the voided urine.

Calcium reabsorption along the nephron, like that of other solutes, can be conceptually described by the relation:

$$\text{Calcium absorption} = \text{passive transport} + \text{active transport,}$$

where passive transport is the sum of diffusion and solvent drag through the paracellular pathway, and active transport is an energetically dependent, transcellular process.

The extent of calcium reabsorption is determined by the balance of driving forces for calcium movement and the permeability of the tight junction between adjacent cells to calcium.

The driving forces for passive paracellular movements involve reabsorptive water flow, chemical gradient and transepithelial voltage. In proximal tubules, calcium reabsorption is paralleled by sodium and water reabsorption at an insignificant transepithelial voltage [2] or chemical gradient. In the thin limbs of Henle's loop, because the permeability to calcium is very low [3], calcium reabsorption is negligibly small. In the thick ascending limbs, calcium ions are driven by the lumen positive transepithelial voltage [4]. In distal tubules, calcium reabsorption through paracellular pathways is negligibly small because of the markedly low permeability to calcium [5].

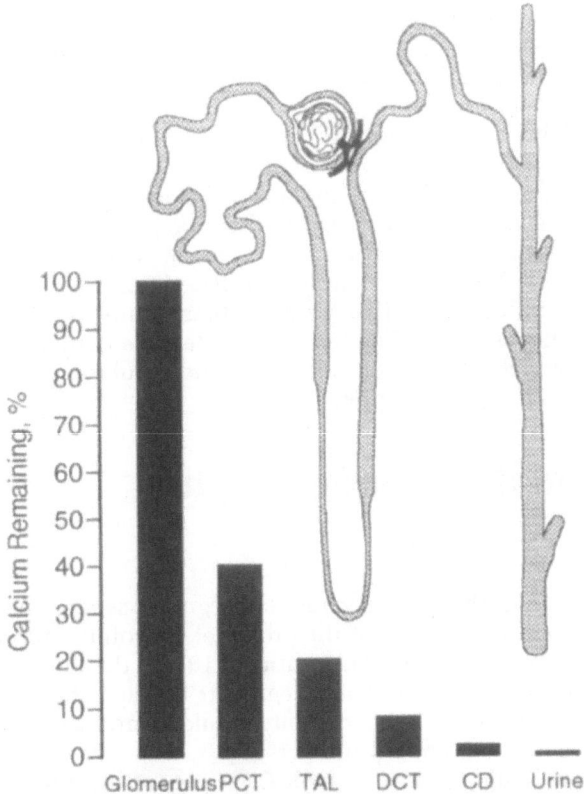

Figure 3.6. Renal calcium absorption. The figure represents a stylised view of the nephron, and the graph depicts the percentage of filtered calcium remaining at the glomerulus, proximal convoluted tubule (PCT), thick ascending limb (TAL) of Henle's loop, distal convoluted tubule (DCT), collecting duct (CD), or in the final urine.

In the collecting system, calcium transport through the paracellular route is thought to be absorptive or secretory. Though the mechanism of calcium absorption in collecting ducts is obscure, the lumen negative voltage may drive secretory calcium movement, whereas reabsorptive calcium movement may parallel with sodium reabsorption [6].

Transcellular calcium reabsorption is a three-step process; calcium entry from the lumen to cytosolic space through the apical membrane, diffusion to basolateral membranes through cytosol, and extrusion from cytosol across basolateral membranes. Calcium entry along the nephron is mediated by several types of calcium channel. The driving force for diffusion is favourable for calcium entry because the cytosolic calcium ion concentration is generally 50–100 nM, which is significantly lower than that in the tubular fluid or urine (i.e., 0.43 nM and 0.20 nM in urine, calculated from results of clearance studies on PTH-stimulated [7] and calcitonin-stimulated [8] rats). Furthermore, the inside of the cell is charged negatively with respect to the exterior, thereby favouring entry. Thus, ATP-activated transporters are not required for calcium entry. This step is also enhanced by calbindin D28k in distal tubules [9].

Calcium diffusion across the cytoplasm of distal tubule cells in some undefined fashion may involve calbindin D28k [10]. Calbindin D28k may act as a cytosolic calcium buffer to maintain low calcium during changes in transcellular transport. Because of its slow binding kinetics, calcium passing through the cytosol does not affect the intracellular calcium signalling system [10].

Basolateral extrusion is mediated by the Na^+/Ca^{2+} exchanger and the plasma membrane Ca^{2+}-ATPase. Because of the adverse electrochemical gradient, calcium extrusion is energetically supported. Na^+/Ca^{2+} exchange is driven by the sodium gradient, where the sodium concentration in the cytosol is maintained low by the basolateral Na^+/K^+-ATPase. Needless to say, Ca^{2+}-ATPase is energetically dependent.

There are some differences of isoforms of these transporter proteins and their expression along the nephron. Furthermore, differences among species may contribute to further variability. These differences and regulation of calcium transport are discussed in detail in this chapter.

The Mechanism of Calcium Reabsorption

Proximal Tubule

The proximal tubule is the primary tubular segment where most of the filtered calcium is reabsorbed. By the end of the proximal convoluted tubule, some 60–70% of the filtered calcium is absorbed. An additional 10% of the filtered calcium is recovered in proximal straight tubules. These segments are also characterised by the fact that calcium reabsorption is not hormonally regulated and is iso-osmotic, implicating closely linked sodium reabsorption [11–13].

The ratio of tubule fluid to ultrafilterable Ca^{2+} ranges from 1.02 to 1.1 [14, 15] along the proximal convoluted tubule. In other words, the concentration of calcium within the proximal tubular fluid remains essentially identical to that in glomerular fluid. This suggests that Ca^{2+} absorption in this segment is principally passive and secondary to sodium and water absorption through the paracellular pathway between adjacent cells. However, the slight rise in the ratio of ultrafilterable to luminal calcium may indicate some active transport. Between 80% and 85% of the reabsorption is passive and 15–20% is active [16–18]. The model of proximal tubular calcium absorption is shown in Figure 3.7.

Paracellular Pathway

As described above, calcium reabsorption in proximal tubules is mainly passive and occurs as a consequence of sodium and chloride transport, which establish an osmotic driving force for fluid abstraction and consequent solvent drag of calcium. The very high passive paracellular permeability of this segment to water and sodium permits water reabsorption in parallel with sodium reabsorption. The passive permeability to calcium equals that of sodium, leading to paracellular calcium reabsorption. As sodium and fluid are absorbed during their passage through the proximal tubule, the concentration of calcium in the tubular fluid increases [19], thereby enhancing the driving force for its passive absorption. Additionally, tubular fluid traversing late proximal tubules contains more chloride than bicarbonate [20]. Because proximal convoluted tubules are more permeable to chloride than to bicarbonate [21], the concentration differences for these anions result in a diffusion

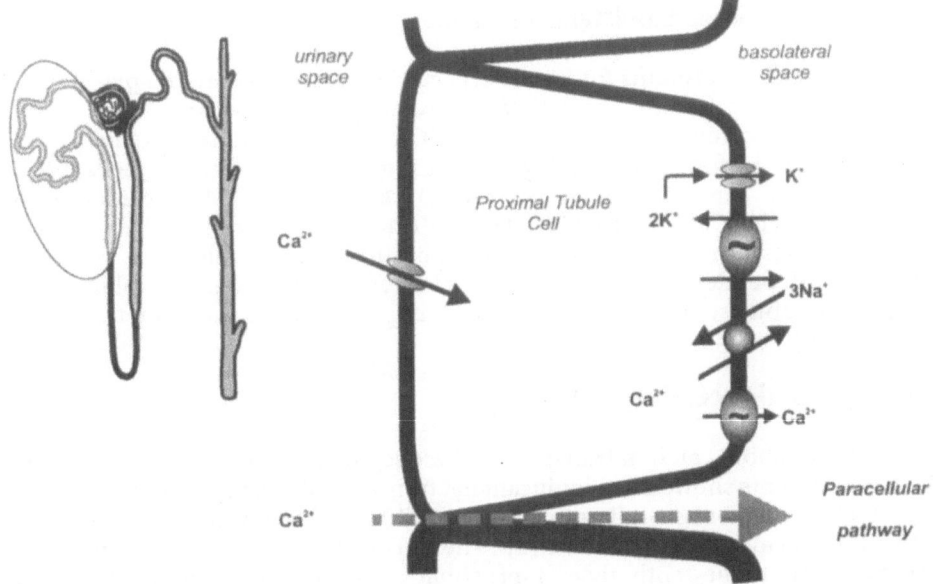

Figure 3.7. Model of proximal tubular calcium absorption. The inset shows the portion of the nephron referred to in the cell model. In proximal tubules, the majority of calcium is absorbed by passive mechanisms through the paracellular pathway. Evidence supports the presence of a small component of active, transcellular calcium transport. Calcium reabsorption in this segment is paralleled by sodium reabsorption and is not hormonally regulated.

voltage that is oriented electropositive in the lumen with respect to the peritubular fluid [2]. This positive voltage serves as an additional driving force for passive calcium absorption.

Transcellular Pathway

Calcium Entry Through Apical Membranes

Though calcium channels are most probably the mechanism of apical calcium entry, details of this process and the molecular structures involved are largely lacking. Pharmacological characterisation revealed one L-type calcium channel in rabbit proximal tubule [22]. A voltage-activated calcium channel was also identified in the apical membrane of cultured rabbit proximal tubule cells with patch-clamp technique [23]. This channel is mostly regulated by membrane voltage and appears to be an epithelial class of L-type calcium channel [23]. The same group further demonstrated that the calcium channel in rabbit proximal tubules was identical to the rabbit cardiac calcium channel (alpha-1) except for a 33 base deletion at the fourth S3–S4 linker in proximal tubule cells, suggesting that this channel is a splice variant of the cardiac calcium channel [24]. Furthermore, rat proximal tubules also express a voltage-dependent calcium channel [25]. The calcium channel in proximal tubules differs from that in distal segments. Immunohistochemistry for an epithelial calcium channel (ECaC) that has been cloned and expressed in rabbit distal segments failed to detect it in rabbit proximal tubules [26].

Calcium Extrusion Across Basolateral Membranes

Putative transporter proteins for basolateral calcium extrusion are the plasma membrane Ca^{2+}-ATPases (PMCA) and Na^+/Ca^{2+} exchanger. PMCAs are P-type ATPases [27], encoded by four discrete genes, PMCA1–PMCA4. PMCA gene products are homologous isoforms of ~140 kD [28]. In mouse proximal tubule cells, PMCA1 and PMCA4 are expressed in S1, S2 and S3 cells [29]. In rat proximal straight tubules, PMCA4b is expressed [30]. The Na^+/Ca^{2+} exchanger is also expressed in proximal tubules of rats [31] and dogs [32], but is absent in mice [29] and rabbits [33]. An ATP-dependent Ca^{2+}/H^+ antiporter also resides in the basolateral membrane of proximal tubules, though its functional importance for calcium transport is uncertain [34].

Descending and Ascending Thin Limbs

Little is known about calcium transport in these segments. Isolated single tubule perfusion experiments showed no significant net transport of calcium in descending thin limbs (DTL) [3, 17] or ascending thin limbs (ATL) [3]. Moreover, these segments are relatively impermeable to calcium [3]. The permeability of the DTL and ATL to calcium is about one-tenth that of proximal convoluted tubules. Micropuncture studies revealed calcium reabsorption in DTL under the influence of calcitonin [8]. DTL of rats expressed isoform 3 and 4 of plasma membrane calcium pump [30]. Though these suggest the possibility of calcium transport in the DTL, it is still uncertain whether DTL participates in renal calcium handling.

Thick Ascending Limbs

Approximately 20–25% of the calcium initially filtered at the glomerulus is absorbed by medullary (MTAL) and cortical (CTAL) thick ascending limbs of Henle's loop. The distinguishing feature of calcium absorption by thick ascending limbs is the capacity for parallel transcellular and paracellular transport. Basal calcium reabsorption is passive and proceeds through the paracellular pathway in both segments. While transcellular calcium reabsorption appears to be quiescent under resting conditions, PTH and calcitonin may stimulate calcium reabsorption by transcellular routes in both CTAL and MTAL in rabbits [35, 36], but in only CTAL in mice [36]. The model of calcium absorption in thick limbs is shown in Figure 3.8.

Paracellular Pathway

Basal calcium absorption proceeds through the paracellular pathway, where its rate of movement is governed by the prevailing electrochemical driving forces. Net calcium transport was measured on rabbit CTAL at various transepithelial voltages, which were established by various combinations of sodium in the lumen and the bath [4]. In the presence of a voltage oriented electropositive in the lumen with respect to the bath, absorptive calcium transport occurred. Conversely, at lumen negative voltages, calcium secretion occurred. In the absence of a voltage gradient, no calcium transport was detected. As the magnitude of this lumen positive voltage was elevated, calcium absorption increased in direct proportion. These findings suggest that basal calcium reabsorption through the paracellular pathway is strictly voltage dependent. The driving forces for passive calcium transport – in other words, lumen-

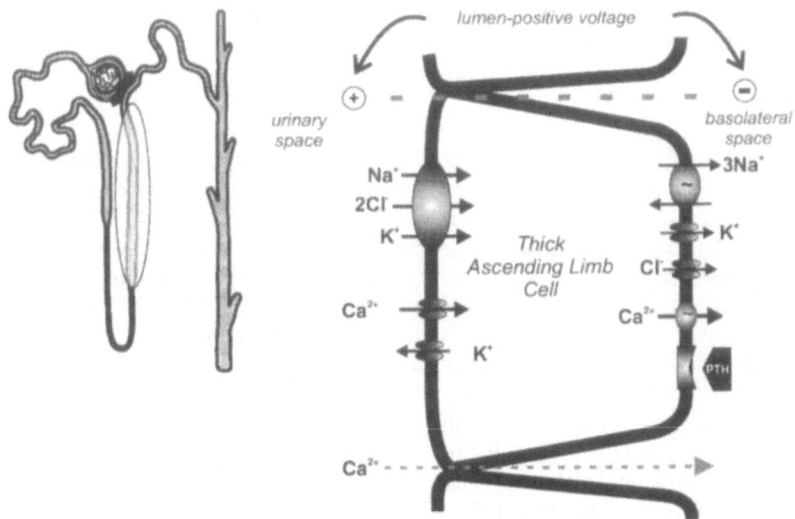

Figure 3.8. Model of calcium absorption in thick ascending limbs. The inset shows the portion of the nephron referred to in the cell model. Basal calcium absorption is mediated by the paracellular pathway and parallels sodium reabsorption. Active transcellular calcium absorption is quiescent in the absence of hormonal stimulation and is independent of sodium reabsorption.

positive transepithelial voltages – are generated by sodium reabsorption. Sodium reabsorption in thick ascending limbs is mediated by the basolateral Na^+/K^+-ATPase and the Na^+-$2Cl^-$-K^+ co-transporter in apical membranes. Because K^+ absorbed by the Na^+-$2Cl^-$-K^+ co-transporter is recycled through the ROMK apical membrane K^+ channel [37], a lumen-positive voltage is generated. Thus, calcium reabsorption through the paracellular pathway occurs in parallel with sodium reabsorption. Increases in sodium absorption are attended by parallel increases of calcium absorption, and decreases of sodium transport are accompanied by diminished calcium transport.

Transcellular Pathway

Little is known of the molecular structures responsible for hormone-mediated calcium entry and basolateral efflux.

Calcium entry through apical membranes is presumably mediated by calcium channels. Poorly selective or non-selective cation channels have been reported in apical membranes of thick limbs [38]. The epithelial calcium channel (ECaC), which has been identified in distal tubules, has not been found in thick ascending limbs [39]. RT-PCR on isolated tubules failed to detect ECaC in rabbit MTAL and CTAL [39], suggesting a different type of calcium channel from ECaC.

Calcium efflux across basolateral membranes is mediated by the PMCA and/or Na^+/Ca^{2+} exchanger (NCX1). The mRNA encoding plasma membrane Ca^{2+}-ATPase (PMCA) isoforms is expressed in rat MTAL [30] and rat CTAL and MTAL [40]. Although porcine CTAL cells express the NCX1 and exhibited transport activity [41, 42], NCX1 immunofluorescence and single nephron RT-PCR analyses failed to uncover the exchanger protein in CTALs of rabbits or other species [43–46]. Isolated single tubule perfusion experiments failed to support participation of NCX1 in calcium reabsorption in rabbit CTALs [47].

Distal Tubules

The distal nephron consists of three distinct nephron segments: the distal convoluted tubule (DCT), the connecting tubule (CNT), and the initial part of the cortical correcting duct (initial CCD), with distinct cell types.

The distal tubule is responsible for reabsorbing 8–10% of the filtered load [14, 48, 49]. The fine-tuning of calcium excretion in the kidney takes place in this segment, where calcium reabsorption is regulated by PTH, $1,25(OH)_2$ vitamin D_3, and calcitonin.

The distinguishing feature of this segment is that all calcium reabsorption is active and proceeds through a transcellular route. The absence of paracellular movement is due to the low calcium permeability of the tight junctions, which is only one-twentieth that of CTALs [5]. The lumen is negatively charged ($-20\,mV$), with respect to the serosal surface [50] and the calcium concentration within the lumen is uniformly below that of plasma [48]. Thus, calcium reabsorption occurs against an electrochemical gradient. Another characteristic feature is that calcium and sodium absorption are inversely related in this segment, whereas it is parallel in proximal tubules and thick ascending limbs. The reason for the inverse relation may be due to the fact that calcium extrusion through the basolateral membrane is mediated mainly by the Na^+/Ca^{2+} exchanger, or that inhibition of Na^+ entry through apical membranes hyperpolarises the apical membrane and increases calcium entry (see below). It was suggested that increased calcium absorption involved enhanced passive calcium entry across apical plasma membranes followed by proportionally elevated calcium extrusion [48]. It was also suggested that extrusion increased in response to the elevation of intracellular Ca^{2+} activity because the NCX1 has a Ca^{2+} binding site in its intracellular domain which activates the forward mode of NCX1 activity [51] (see below). Furthermore, "load dependence" is an additional feature of calcium absorption in distal convoluted tubules. This term describes the relations between the amount of solute delivered to a nephron site and the corresponding change in its transport. The model of distal tubular calcium absorption is shown in Figure 3.9.

Calcium Entry Through Apical Membranes

Calcium entry through the apical membrane in this segment is facilitated by the epithelial calcium channel (ECaC) [10]. ECaC is exclusively expressed in $1,25(OH)_2D_3$-responsive epithelia; namely, kidney, small intestine and placenta [26]. ECaC is localised in the apical membrane of distal tubule segments including DCT, CNT, and principal cells of the CCD, but is not located in proximal tubules or thick ascending limbs [26]. The ECaC cDNA contains an open reading frame of 2,190 nucleotides that encodes a protein of 730 amino acids with a predicted molecular mass of $83\,kDa$. ECaC has three structural domains: a large hydrophilic amino-terminal domain containing potential protein kinase C phosphorylation sites, suggesting an intracellular location: a six transmembrane-spanning domain with a hydrophobic stretch between transmembrane segments five and six, indicative of an ion pore region: and a hydrophobic carboxyl terminus containing potential protein kinase A and C phosphorylation sites [26]. The ECaC is also suspected to be a tetrameric ion channel [10].

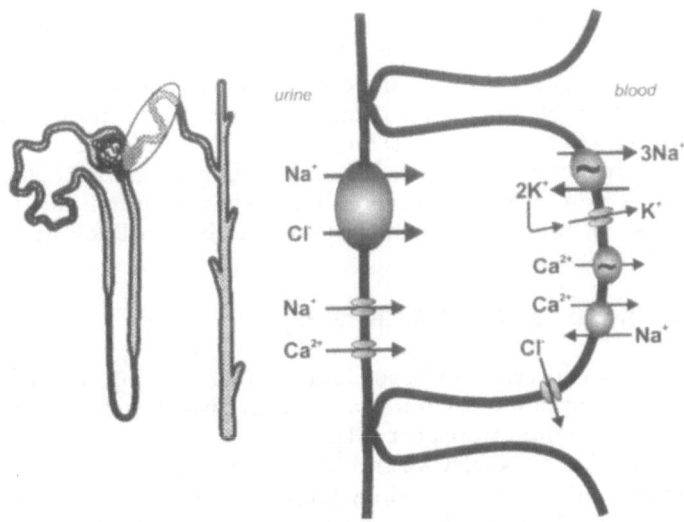

Figure 3.9. Model of calcium absorption in distal convoluted tubules. The inset shows the portion of the nephron referred to in the cell model. In this segment, paracellular calcium movement is absent because of its remarkably low permeability to calcium. Active transcellular calcium reabsorption exhibits an inverse relation to sodium reabsorption.

Several reports prior to cloning the ECaC indicated that the putative calcium entry channel was dihydropyridine-sensitive, a feature of L-type calcium channels, and activated at hyperpolarising membrane voltages [52, 53]. Recent studies demonstrated that ECaC was activated by hyperpolarisation in *Xenopus laevis* oocytes that expressed EcaC [54].

The primary ion transport proteins in the distal tubule include two apical membrane sodium entry pathways: the Na^+–Cl^- co-transporter and the amiloride-sensitive epithelial Na^+ channel (ENaC). Inhibition of either of these blocks Na^+ entry, hyperpolarises the membrane voltage and activates the apical membrane calcium channel [55]. This is one of the explanations for the inverse relation between Ca^{2+} and Na^+ reabsorption. Likewise, PTH [53, 56] and calcitonin [57], acting through their respective receptors, hyperpolarise DCT membrane voltage. Hyperpolarisation of membrane voltage by PTH is a direct effect since this action is present even when calcium entry is blocked by nifedipine [56].

However, kinetic analyses [58] and targeted antisense experiments [59] suggest that two distinct mechanisms are involved in calcium entry. According to Brunette et al. [58], a high-affinity, low-velocity system mediates PTH-stimulated calcium entry [58], whereas a low-affinity, high-velocity mechanism mediates thiazide [58] and calcitonin-induced [60] calcium uptake. It is not clear whether calcium entry in these segments involves ECaC or multiple calcium entry channels.

Cytosolic Calcium Transport

Calbindins are vitamin D-dependent calcium-binding proteins that belong to a family of intracellular proteins with high calcium affinities and are suspected to mediate calcium transport. Calbindin D28k is expressed in DCT, CNT, and principal cells in CCD. ECaC, calbindin D28k, NCX1 and plasma membrane Ca^{2+}-ATPase are completely co-localised in these segments [39]. Calbindin D28k is expressed throughout the cytoplasm, while ECaC is restricted to apical membranes, and NCX1 is expressed in basolateral membranes [39]. It has been suggested that calbindin D28k facilitates calcium transport through the cytoplasm and acts as a cytosolic calcium buffer to maintain low calcium during changes in transcellular transport. Because of its slow binding kinetics, calcium passing through the cytosol does not affect the intracellular calcium signalling system [10]. Preloading calbindin D28k into luminal membrane vesicles from distal tubules of rabbits enhanced initial calcium uptake by these membranes [9], suggesting participation of calbindin D28k in calcium entry through apical membranes.

Calcium Extrusion Across Basolateral Membranes

The NCX1, which is expressed in basolateral membranes [39, 46], is the primary calcium transporter protein in calcium transport in distal tubules and is a prerequisite for transcellular calcium transport [61].

Three genes for NCX: NCX1, NCX2 and NCX3, have been identified and the NCX1 gene products are expressed in the kidney [50, 62]. An immunohistological study of rabbit kidneys revealed that NCX1 was expressed in basolateral membranes of distal tubules, except for intercalated cells in CCD, and completely co-localised with ECaC in apical membranes [39]. A PCR cloning strategy in rat kidneys found the mRNA of the NCX1 only in DCT, but not in proximal tubules, thick limbs, or CCD [46]. In mouse DCT cells, the cDNAs encoding a conserved region and the variable regions of three alternatively spliced isoforms of the NCX1, NACA2, NACA3 and NACA6 were isolated in a ratio of $7:12:1$ [62].

The NCX1 may operate in several different modes: forward and reverse, as well as Ca^{2+}/Ca^{2+} and Na^+/Na^+ self exchange [63, 64]. In the forward mode, the exchange mediates calcium efflux that is coupled to sodium influx. The magnitude and direction of Na^+/Ca^{2+} exchange is dictated by the ambient electrochemical driving forces. Physiological and pharmacological manoeuvres that directly stimulate apical membrane calcium entry in distal tubule cells may inherently augment the rate of basolateral calcium extrusion [50]. Conversely, when the cells were treated with ouabain, intracellular Na^+ increased with concomitant membrane depolarisation: under these condition the driving forces favour reverse Na^+/Ca^{2+} exchange [50]. There is another explanation for regulation of the NCX. The NCX may exhibit intrinsic regulation. The predicted protein structure of the NCX1 gene product consists of a cleaved signal sequence followed by five transmembrane domains, a large intracellular loop, and six transmembrane domains near the carboxyl terminus [65]. The NCX1 is regulated by calcium at a high-affinity Ca^{2+}-binding site on the large intracellular loop, which is distinct from the calcium transporter site [51]. The intracellular loop also has a sodium-binding site located at the very beginning of the loop. According to this model, elevation of calcium in the cytosol activates the forward mode of NCX1 and, conversely, elevation of cytosolic sodium inactivates it [51].

A plasma membrane Ca^{2+}-ATPase (PMCA) that exists in proximal tubules is also expressed in distal tubules [29, 39, 40, 66] and may mediate calcium extrusion. As noted above, this ATP-dependent Ca^{2+} transporter is completely co-localised with other transporter proteins, ECaC, calbindin D28k, and NCX in these segments [39, 66]. PMCA exhibits a restricted localisation to lateral and occasionally the basal membranes of connecting tubules and cortical collecting ducts [66].

The PMCA is encoded by four genes, PMCA1, PMCA2, PMCA3 and PMCA4. PMCA1 and PMCA4 are expressed in human kidneys [28], whereas all four isoforms are expressed in rat kidneys [40]. It is not known which PMCA isoforms participate in calcium transport. Because of the wide expression of PMCA1 and PMCA4 throughout the mammalian body, these two isoforms are suspected to be the house-keeping isoforms [28].

It is not clear to what extent PMCA participates in cellular calcium extrusion and how or if these processes are hormonally regulated. In view of the earlier theoretical analysis of Na^+/Ca^{2+} exchange, the calculations are consistent with the conclusion that Na^+/Ca^{2+} exchange is minimally active under resting conditions [50]. Conversely, when cellular calcium transport is stimulated, Na^+/Ca^{2+} exchange may be more dominant than the PMCA in calcium efflux.

Calbindin D9k, which is another vitamin D-dependent Ca^{2+}-binding protein, is also expressed in distal tubules [67]. The protein is exclusively present in a thin layer along the basolateral membrane of the thick ascending limbs, the distal convoluted tubules, connecting tubules and intercalated cells of the cortical collecting ducts [67]. Preloading calbindin D9k in the basolateral, not luminal, membrane vesicles from distal tubules of rabbits increased ATP-dependent calcium uptake [68], suggesting that the protein enhances PMCA activity and promotes calcium efflux.

Collecting Ducts

The model of calcium absorption in collecting ducts is shown in Figure 3.10. Indirect assessment of calcium transport in the terminal nephron segments, derived from the difference between the amount of calcium present at the last accessible surface distal site and that in the final urine, suggested that calcium transport continued in the medullary collecting ducts [69, 70]. These indirect approaches imply that 3–7%

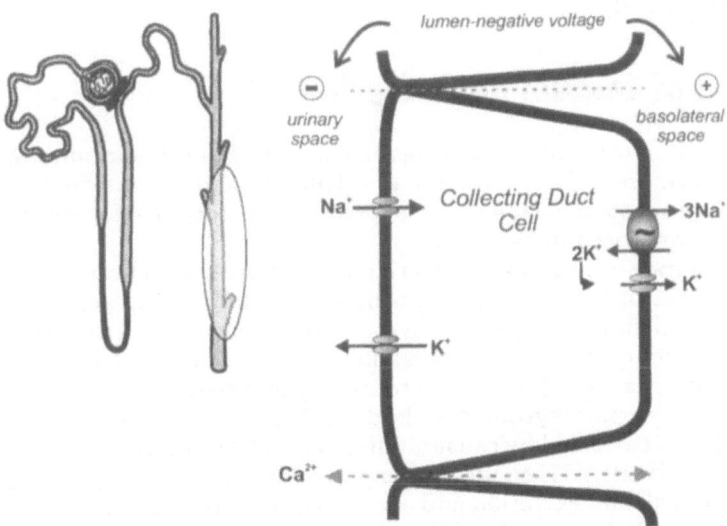

Figure 3.10. Model of calcium absorption in collecting ducts. The inset shows the portion of the nephron referred to in the cell model. Collecting ducts are not thought to be a primary site of calcium transport. Studies suggest that calcium movements are small in magnitude and thermodynamically passive. Transport may be absorptive or secretory, driven by the lumen-negative transepithelial voltage.

of the filtered load of calcium is absorbed along tubular segments beyond the last distal nephron.

Isolated perfused rabbit cortical collecting ducts exhibit only modest calcium reabsorption, which is about one-third that of connecting tubules, and PTH did not alter its magnitude [71]. However, another group reported no net calcium movement in the rabbit cortical collecting duct in the presence or absence of PTH [72]. Monolayers of primary cultured rabbit cortical collecting duct cells exhibit calcium transport from the apical side to basolateral side but PTH had no effect on this calcium movement [73]. The calcium permeability of this segment is very low (less than 2% of that of the cortical thick ascending limb of Henle's loop) [74]. There is no evidence of active calcium transport in perfused cortical collecting ducts [74]. However, as noted above, principal cells in cortical collecting ducts have calcium transporter proteins; epithelial calcium channel, calbindin D28k, Na^+/Ca^{2+} exchanger and plasma membrane Ca^{2+}-ATPase [26]. Because of the lumen-negative voltage and the low calcium permeability of cortical collecting ducts [74], relatively small active calcium transport may contribute to calcium transport in the kidney. However, at spontaneous lumen-negative transepithelial voltages, a small net secretory flux of calcium was observed [74]. This result suggests the possibility of calcium secretion in this segment.

Because only little information is available about calcium transport in inner medullary collecting ducts, it is not certain if or how much this segment facilitates calcium transport. Calcium transport was shown in rat inner medullary collecting ducts [6, 75]. PTH is not likely to alter the calcium reabsorption in this segment [6] and does not affect calcium permeability, which is low [76]. Calcium reabsorption was influenced by sodium reabsorption, suggesting the presence of a passive diffusion mechanism [6]. Taken together, it is supposed that passive calcium transport may be present in this segment, while the magnitude may be very small and not regulated by PTH directly.

Regulation of Calcium Reabsorption

There are at least four factors regulating renal handling of calcium, among which are parathyroid hormone (PTH), vitamin D calcitonin, and calcium itself through a calcium-sensing receptor (CaSR). The expression of receptors for these hormones and CaSR is summarised in Table 3.4.

In vivo clearance experiments reflecting integral renal function demonstrate the effect of these factors on renal calcium handling. It is widely accepted that PTH increases renal calcium absorption. Thyroparathyroidectomy was a frequently used technique to reduce circulating concentrations of PTH, calcitonin, and calcium. Fractional excretion of calcium (FE_{Ca}) was reduced by maximal hypocalciuric concentrations of PTH in thyroparathyroidectomised (TPTX) rats, in which the serum calcium concentration was corrected by calcium infusion [77]. Similarly, in PTH-, calcitonin- and ADH-deprived rats, PTH elicited the expected decrease in FE_{Ca} [7]. TPTX reduced the threshold of calcium excretion and addition of PTH returned it to normal in rats [78]. Furthermore, in humans, infusion of exogenous PTH decreased FE_{Ca} [79].

Calcitonin (CT), which decreases serum calcium concentration, also acts on renal calcium handling. Some clearance experiments demonstrated that CT increased urinary calcium excretion. In TPTX rats with normal serum calcium levels, hypocalciuric concentrations of CT reduced FE_{Ca}, and combined administration of maximal

Table 3.4. Expression of receptors along the nephron

	PTH/PTHrP		CTR			VDR		CaSR		
	mRNA	mRNA	mRNA	mRNA	protein	mRNA	protein	mRNA	mRNA	protein
Glom.	+	++	++	–	–	–	+	+	–	–
PCT	+	+++	+	–	–	+	+	+	–	+
PST	+	++	++	–	–	–	+	+	–	+
DTL			–	–	–	–				
ATL			–	–	–	–				
MTAL	–	+/–	–	++	+	–		+	++	+
CTAL	+	+	++	+++	++	–	+	+	++	++
DCT	+	+/–	+		+	+	+	+	+++	++
CNT					–	+	+			
CCD	+	–	–	++	–	+	+	+	+/–	–
OMCD	–	–	–	+/–	–	–		+	–	+
IMCD	–	–	–		–	–		+	+/–	
semiquantitative?	no	yes	yes	yes	no	no	no	no	yes	no
animal	rat	rat	rat	rat	rat	mouse, rat, pig	human	rat	rat	rat
Ref.	93	92	94	95	96	98	99	93	92	97

PTH1R, PTH/PTHrP receptor; CTR, calcitonin receptor; VDR, $1,25(OH)_2$ vitamin D_3 receptor; CaSR, calcium-sensing receptor; Glom., glomerulus; PCT, proximal convoluted tubule; PST, proximal straight tubule; DTL, descending thin limb; ATL, ascending thin limb; MTAL, medullary thick ascending limb; CTAL cortical thick ascending limb; DCT, distal convoluted tubule; CNT, connecting tubule; CCD cortical collecting duct; OMCD outer medullary collecting duct; IMCD inner medullary collecting duct.

PTH and CT did not show an additive effect [77]. Similar protocols also demonstrated decreased FE_{Ca} by CT in rats [80, 81]. In hormone-deprived rats with reduced circulating CT, antidiuretic hormone, PTH, and glucagon, administration of CT decreased FE_{Ca} [8]. Maximal calcium-conserving doses of CT inhibited urinary calcium excretion in acutely TPTX rats [82]. In rats, where increased serum calcium concentrations achieved by infusing calcium provoked PTH suppression and CT secretion, the presence of high CT levels was associated with a reduced FE_{Ca} in spite of a high serum calcium concentration, when compared to a TPTX group with low CT [35]. In the case of humans made acutely hypercalcaemic to inhibit endogenous PTH secretion, infusion of CT promptly inhibited the hypercalcaemia in a dose-dependent way and also reduced FE_{Ca} [83]. These results are consistent with the conclusion that CT is a calcium-conserving hormone and enhances calcium reabsorption in the kidney.

The effects of vitamin D on calcium absorption in the kidney are somewhat controversial. In some clearance experiments, vitamin D showed an increase [84], no change [78] or a decrease [79, 85] in urinary calcium excretion. In acutely TPTX hamsters, where PTH was infused continuously during clearance experiments in a low dose to reduce the hypocalcaemic effect of TPTX, an infusion of $1,25(OH)_2$ vitamin D_3 increased FE_{Ca} [84]. In vitamin D-replete rats, an inhibitor of $1,25(OH)_2$ vitamin D_3 formation did not alter the renal handling of calcium [78]. The threshold of urinary calcium excretion was lower in vitamin D-depleted rats than that in vitamin D-replete rats, indicating that vitamin D conserved calcium by the kidney [85]. Interestingly, the effect of low-dose PTH, which increased the threshold of calcium excretion, was abolished in the absence of vitamin D [85]. Furthermore, pharmacological doses of PTH did not require the presence of vitamin D for its effect [85], indicating that vitamin D may modulate PTH effects. In patients with vitamin D-resistant rickets, an autosomal recessive disease caused by mutated, non-functioning vitamin D receptors, PTH failed to decrease FE_{Ca} while PTH significantly decreased FE_{Ca} in normal controls, indicating that vitamin D facilitates PTH-induced calcium reabsorption [79].

Calcium ions also act as a regulating factor of calcium homeostasis by binding to a calcium-sensing receptor (CaSR). The CaSR was first cloned from bovine parathyroid gland [86]. The receptor is a member of the G-protein-coupled receptor superfamily and is characterised by three structural domains, including a large amino-terminal extracellular domain that is the ligand-binding site, a seven-membrane-spanning domain, and cytoplasmic carboxyl-terminal tail [86]. The receptor resides in the membrane as a dimer, which is dimerised within the extracellular domain [87]. Intermolecular interactions between dimeric CaSR monomers are important for its normal function [88]. Though the exact renal functions of the receptor are still ambiguous, a hypocalcaemic action is expected. Parathyroid glands expressing the CaSR [86] modulate PTH secretion in response to small changes of plasma calcium concentration [89]. Extracellular calcium suppressed PTH secretion by cultured bovine parathyroid cells and the calcium-mediated suppression was diminished due to decreased CaSR mRNA [90]. Moreover, expression of the CaSR enables parafollicular cells of the thyroid gland to release calcitonin [91]. The CaSR is also expressed in the kidney [92, 93]. Insufficient information is presently available to understand the role of the CaSR on renal calcium handling.

Proximal Tubules

This segment, including proximal convoluted tubule (PCT) and proximal straight tubule (PST), expresses the receptors for three of the four regulating factors of renal handling of calcium described above. The mRNA of PTH/PTHrP receptor (PTH1R) is expressed in rats [92–94] and the expression of its mRNA is most significant in the proximal tubule when compared with other segments [92, 94]. The RT-PCR technique could not detect the mRNA of the calcitonin receptor (CTR) in rats [95] and ^{125}I-labelled salmon calcitonin failed to bind to the proximal tubule in rats [96], indicating the absence of the CTR in this segment. The mRNA of CaSR is also expressed in the proximal tubule in rats [93] and a polyclonal anti-CaSR antibody bound to the apical membranes of the proximal tubule [97]. Furthermore, immunohistochemistry revealed that apical fluorescence intensity for the CaSR diminished along the proximal tubule from S1 to S3 segment and the fluorescence was absent in the S3 proximal straight tubule profiles in the outer stripe of the outer medulla [97]. However, semiquantitative analysis for the mRNA of CaSR did not detect it in the proximal tubules; strong expression of the mRNA was found in thick ascending limbs and distal tubules [92]. These results may suggest that the proximal tubule possesses the CaSR though its expression is relatively weak when compared with thick ascending limbs and distal tubules where the most important regulation of calcium homeostasis proceeds.

The existence of vitamin D receptor (VDR) for calcitriol (1,25(OH)$_2$ vitamin D$_3$) was shown in this segment by immunohistochemistry and in situ hybridisation in pigs, rats, mice [98] and humans [99]. The expression of VDR was detected in both nucleus and cytoplasm by a polyclonal antibody directed against the human VDR [98]. In situ hybridisation suggests that VDR gene expression is lower in proximal tubules than in distal tubules in mice, and immunohistochemistry supports this pattern of expression [98].

Despite the expression of the PTH1R, CaSR and VDR in proximal tubules, there is little evidence that these receptors participate in regulating calcium handling. PTH was reported to increase [100], decrease [101, 102] or not to change calcium entry through apical membranes. Neither PTH, the calcium channel blocker nitrendipine, nor the calcium channel agonist BAY K 8644 affected calcium entry in apical membrane vesicles prepared from proximal tubules [58]. PTH effects on basolateral calcium transport are also controversial. Parathyroidectomy decreased the V_{max} of calcium pump in basolateral membrane vesicles purified from rats [103], indicating that PTH accelerates the calcium pump. However, PTH did not influence the ATP-dependent calcium uptake by basolateral membrane vesicles in rabbits [104].

Consistent with the lack of CTR expression, CT was not reported to participate in calcium reabsorption in this segment. Isolated single tubules showed that CT had no effect on calcium transport in proximal convoluted and straight tubules in rats [105]. Similarly, in basolateral membrane vesicles purified from rabbit proximal tubules, preincubation with CT had no effect on calcium uptake [60].

Vitamin D was also reported not to have direct effects on calcium transport in this segment. Though decreased serum calcium concentrations evoked by parathyroidectomy were restored by a vitamin D supplement, vitamin D treatment did not affect calcium pump activity [103]. Likewise, vitamin D depletion did not affect calcium uptake by basolateral or luminal membrane vesicles purified from rabbit proximal tubules [9].

Taken together, calcium reabsorption by the proximal tubule may not be regulated by these hormones and, if any, the participation is not important to the fine tuning of calcium homeostasis.

The more important role of the proximal tubule is an indirect effect on calcium regulation; in other words, this segment is a major part of vitamin D activation and also PTH metabolism.

Proximal tubules are the site where biologically active $1,25(OH)_2$ vitamin D_3 is formed. Based on studies using proximal-like cell culture systems, where directly adding $25(OH)D_3$ resulted in $1,25(OH)_2D_3$ formation, it was surmised that in the kidney, $25(OH)D_3$ diffuses from the blood across basolateral cell membranes into proximal cells, where it is hydroxylated by the mitochondrial 1-25 vitamin D hydroxylase [106]. In the circulation, $25(OH)D_3$ is tightly ($5-10^{-8}$ M) and extensively bound (99.6%) to the vitamin D binding protein (DBP) [107]. DBP, which is comparable in size to albumin, is excreted in the urine [108]. $25(OH)D_3$ complexed to DBP is filtered at the glomerulus and taken up across apical cell membranes by megalin in proximal tubules [109]. Such uptake was demonstrated to be required for the formation of $1,25(OH)_2D_3$.

Megalin (gp 330) is a multifunctional clearance receptor that is expressed at the apical surface of proximal tubules and of several other epithelia, in the parathyroid glands, and placenta [110, 111]. It belongs to the low-density lipoprotein (LDL) receptor gene family and is probably the most important receptor for endocytosis of macromolecules filtered at the glomerulus [110]. Megalin is located in apical coated pits, small and large endocytic vacuoles, and in dense apical tubules of proximal tubule cells [112], with expression being greater in proximal segments one and two than in segment three. Megalin is also expressed in lysosomes of segments one and two [112]. Megalin exhibits broad ligand specificity that includes albumin, vitamin-binding protein complexes, various polybasic antibiotics, and calcium [113, 114]. Complexes of $25(OH)D_3$–DBP are filtered at the glomerulus and taken up in proximal tubules by megalin-mediated endocytosis [109]. The complex is targeted to lysosomes, where it is degraded, releasing $25(OH)D_3$ into the cytoplasm, where it is taken up by mitochondria and converted to $1,25(OH)_2D_3$ or released unchanged from the cells.

Proximal tubules contribute importantly to the clearance of PTH by glomerular filtration and metabolism [115, 116]. Megalin also mediates the endocytosis of PTH [117]. Approximately 50% of [^{125}I]PTH microinfused into proximal tubules was taken up by the tubules. Thus, megalin would appear to mediate about one-half of PTH catabolism. Megalin-mediated PTH endocytosis was specific; purified megalin specifically recognised full-length PTH(1–84) and synthetic amino terminal peptide fragments. Mid-region and carboxy-terminal fragments were not bound. Thus, a fair degree of primary if not secondary peptide sequence is specifically recognised by megalin. In the case of the PTH1R, the structural determinants of PTH (and PTHrP) binding have been identified as the 1–14 and 15–34 domains that interact when binding to the receptor, and that residues 5, 19, and 21 contribute either directly or indirectly to this interaction [118]. It will be of significant interest to identify the PTH binding components of megalin. In megalin-deficient mice, excretion of amino-terminal, but not of carboxy-terminal, PTH fragments increased substantially. Expression of the PTH1R in proximal tubules was not affected. Thus, these studies show that megalin contributes to renal PTH clearance. Therefore, proximal tubule has indirect effects on calcium homeostasis through its participation in PTH and vitamin D metabolism.

Thick Ascending Limbs

The thick ascending limb is comprised of medullary thick ascending limb (MTAL) and cortical thick ascending limb (CTAL). It is not known if changes in cellular architecture, which are gradual between medullary and cortical portions of the thick ascending limb, are related to differences in the route and magnitude of calcium transport. There are two distinct types of cells observed by scanning electron microscopy in this segment; smooth-surfaced cells and rough-surfaced cells [119]. Though these are mixed along the thick ascending limb, smooth-surfaced cells are dominant in MTAL and this relationship is reversed in CTAL. Though the functional difference between these two types of cells is not known, there are some functional differences between medullary and cortical thick ascending limbs.

The mRNA of PTH1R was reported to be expressed in CTAL, but not in MTAL in rats [93, 94]. Another report describes the expression of PTH1R mRNA in both CTAL and MTAL, with the expression weaker in MTAL than in CTAL [92].

The calcitonin receptor (CTR) mRNA was detected in both CTAL and MTAL in rats and the expression was somewhat greater in CTAL than that in MTAL [95]. The existence of the CTR was also confirmed by a histological techniques, where radio-labelled CT bound to both CTAL and MTAL [96].

Though the PTH1R may be expressed in the rat MTAL [92], isolated tubule perfusion experiments failed to show an effect of PTH on calcium reabsorption in rabbits [36, 120]. However, PTH simulated calcium reabsorption in CTAL of rabbits [36, 121, 122] and mice [123, 124], consistent with the expression of the PTH1R in rats [93, 94].

Although the CTR exists in both CTAL and MTAL in rats [95, 96], the CT effects of calcium transport are limited to MTAL in rats and rabbits [35, 105, 120]. Isolated single renal tubule perfusion experiments and micropuncture techniques revealed that CT stimulated calcium reabsorption only in MTAL in rats [105] and rabbits [35, 120] but not in CTAL in rabbits [35].

The opposite result was also reported that CT simulated calcium reabsorption in CTAL but not in MTAL in mice [124].

There appears to be some inconsistencies between the expression of these two receptors and their regulatory effects on calcium transport in CTAL and MTAL. As the opposite results between mice and other animals in CT effects, there are significant species differences that complicate drawing a comprehensive view of calcium absorption in these nephron segments. However, there are important differences in the responsiveness to PTH and calcitonin between CTAL and MTAL.

Limited information is available about the mechanism by which PTH and CT regulate calcium reabsorption in these segments. PTH does not affect paracellular reabsorption of calcium, which is voltage-dependent [4], because PTH did not alter the transepithelial voltage in single isolated perfused CTALs obtained from rabbits [121, 122] and mice [124], and MTALs from mice [124]. Thus, PTH stimulates transcellular calcium movement and activates latent calcium channels or promotes the recruitment of calcium channels to apical membranes. In single CTAL cells of mice, intracellular calcium concentration ($[Ca]_i$) was measured to show that the dihydropyridine-sensitive calcium channel agonist, Bay k 8644 and antagonist, nifedipine did not affect $[Ca]_i$, indicating the absence of calcium channel in the cell membrane in resting conditions. A subsequent challenge of PTH added to Bay k 8644 raised $[Ca]_i$, indicating recruitment of new dihydropyridine-sensitive calcium channels into cell membranes [125]. Similar results were obtained in patch-clamp studies [53].

CT may also stimulate transcellular calcium reabsorption because the transepithelial voltage did not change upon challenge with CT in single isolated perfused MTALs of rabbits, while it increased net calcium reabsorption [35]. It is unclear how CT regulates transcellular calcium transport.

The CaSR mRNA is expressed in both CTAL and MTAL in the rat [92, 93]. Relative levels of CaSR mRNA were 25% greater in CTAL than in MTAL [92]. An immunohistological study also supported the view that the CaSR exists in both CTAL and MTAL and is most abundant in CTAL compared to other nephron sites [97]. Moreover, confocal images showed that the expression of CaSR was restricted to the basolateral membrane in CTAL and MTAL [97]. Though the function of the CaSR on calcium absorption has not been fully elucidated, the receptor is expected to inhibit calcium reabsorption by thick ascending limbs (TAL). Single isolated CTAL perfusion revealed that increasing the calcium concentration in the bath from 1 to 5 mM inhibited calcium reabsorption [97]. One possible mechanism by which the CaSR inhibits calcium reabsorption by thick ascending limbs derives from patch-clamp experiments where gadolinium, which has been shown to stimulate the CaSR, reversibly decreased the activity of the 70 pS K^+ channel in apical membranes [126]. Calcium is absorbed through paracellular and transcellular routes in TAL and paracellular movement is driven by the lumen-positive transepithelial voltage. The transepithelial voltage is generated by a combination of $Na^+–K^+–2Cl^-$-co-transporter and K^+ channels in apical membranes. The CaSR may inhibit calcium reabsorption by reducing the transepithelial voltage and the driving force for passive calcium transport. In isolated perfused mouse cortical thick ascending limbs, the CaSR, stimulated by gadolinium, decreased transepithelial voltage (Ve) and net calcium transport even in the presence of relatively high lumen-positive Ve. Moreover, CaSR activation inhibited PTH-stimulated calcium transport without affecting Ve when Ve was low, indicating that the CaSR inhibits both transcellular and paracellular calcium reabsorption (unpublished data).

The effect of vitamin D in thick ascending limbs has not been well studied. Net calcium reabsorption and transepithelial voltage in basal and PTH stimulated-conditions of vitamin D-deficient rabbits were compared with those of vitamin D-replete rabbits. There were no differences of calcium reabsorption or of voltage in single isolated perfused CTALs of the two groups [127], indicating no participation of vitamin D in calcium handling in these segments.

Distal Tubule

As noted above, the distal nephron consists of the distal convoluted tubule, connecting tubule, and the initial portion of the cortical collecting duct. These segments are the most important sites for the fine-tuning of calcium homeostasis. Although there may be some functional and regulatory differences along these three segments, the technical and anatomical limitations make distinguishing transport by each cell type impossible at the present time. Thus, studies on DCT and CNT are described together.

Calcium absorption by the distal tubule is regulated by PTH, calcitonin (CT), vitamin D, and calcium itself directly or indirectly. The PTH/PTHrP receptor (PTH1R), calcitonin receptor (CTR), vitamin D receptor (VDR) and calcium-sensing receptor (CaSR) are expressed in this segment.

The expression of PTH1R mRNA in DCT was confirmed by RT-PCR on isolated microdissected tubules in rats [92, 93]. Expression in rat CCD was reported in one study [93] but not in another [92]. The microscopic localisation of PTH1R mRNA was also examined in rat kidney sections using emulsion autoradiography and showed that it is expressed in DCT but not in CCD [94]. CNT was not investigated in these papers.

CTR was found in DCT but not CCD in rats by autoradiography on kidney sections with radio-labelled calcitonin [96]. CNT was not investigated in this paper. RT-PCR for mRNA of CTR on microdissected tubules detected it in CCD in rats but DCT and CNT were not examined [95].

The VDR was demonstrated in DCT, CNT, and the initial portion of the CCD by immunocytochemistry in rats and pigs [98]. In situ hybridisation in mice showed that CCD beyond the initial portion did not express the VDR mRNA [98]. A polyclonal antibody against VDR was used to show that human distal tubules and CCD also expressed the VDR [99].

The mRNA of the CaSR was detected in DCT and CCD by RT-PCR on microdissected tubules of rats [92, 93]. Again, the CNT was not examined. The expression is greatest in DCT and weak in CCD [92]. Confocal fluorescence imaging with CaSR-specific antibody revealed that the CaSR is expressed in DCT and TAL in basolateral membranes [97]. The CaSR-specific fluorescence became abruptly less at the transition of the cortical TAL and DCT and its intensity diminished along the distal tubules, until it was absent in CNT [97].

PTH stimulates calcium absorption by distal tubules. Intracellular calcium concentration is increased in single mouse DCT cells [56]. PTH stimulated calcium uptake by luminal membrane vesicles from rabbit distal tubules, which include membranes originating from CTAL and CCD [58]. Net calcium reabsorption by isolated rabbit DCT and CNT perfused in vitro was increased by PTH but not in CCD [72]. However, another group reported that PTH stimulated net calcium reabsorption in the rabbit CNT but not in DCT or CCD [71]. The stimulatory effect of PTH on calcium reabsorption was confirmed by micropuncture studies of the distal tubule in rats [128, 129]. In PTH-depleted rats, TPTX caused a significant decrease in the calcium reabsorptive rate as compared to intact animals. PTH replacement in TPTX animals restored calcium transport to control levels [128, 129].

Little information is available about the mechanism by which PTH simulates calcium absorption in distal tubules. Calcium entry across apical membranes is mediated by the epithelial Ca^{2+} channel (ECaC) or other. In the absence of PTH, calcium entry and calcium channel activity are undetectable [53, 125]. Similar results were seen in luminal membrane vesicles of rabbit distal tubules [58]. These are indicative that the calcium channel is inactive until stimulated by PTH or, perhaps, does not reside in the cellular membrane in the absence of PTH, which elicits insertion of calcium channels. Pre-treatment with the microtubule depolymerising agent, colchicine, blocked exocytosis of sub-apical plasma membrane vesicles, and abolished the effect of PTH on calcium entry [125]. Taken together, PTH may mobilise calcium channels into apical membrane and increase open channel probability by hyperpolarising the apical membrane.

Calcium extrusion across basolateral membranes is achieved by both the plasma membrane Ca^{2+}-ATPase (PMCA) and Na^+/Ca^{2+} exchanger (NCX) in the distal tubule. PTH appears to activate NCX. Pre-treatment with PTH increased Na^+-dependent calcium uptake by basolateral membrane vesicles from rabbit distal tubule [104]. Though PTH increased net calcium reabsorption by single isolated perfused CNT, the absence of sodium on the basolateral side abolished the effect of PTH [71], indicating that the PTH effect is sodium-dependent. ATP-dependent calcium uptake by basolateral membrane vesicles from rabbit distal tubule was not influenced by pre-treatment of PTH [104]. In the isolated perfused rabbit CNT, PTH did not increase net calcium reabsorption without the participation of NCX [71], indicating no effect of PTH on the PMCA. In rat basolateral membrane vesicles, PTH stimulated calcium pump activity [103]. This inconsistency may depend on species differences.

CT may also enhance calcium reabsorption by distal tubules. CT increased net calcium reabsorption by single rabbit DCT but not by the CNT or CCD of rabbits [71]. Micropuncture experiments on the distal tubule in thyroparathyroidectomised rats showed that CT enhanced [81] or failed to affect [105] calcium reabsorption in the distal tubule. The reason for this discrepancy is unknown.

CT was reported to increase calcium uptake by luminal membrane vesicles from rabbit distal tubules [60], indicating that CT may stimulate the calcium channel. CT stimulated calcium entry by mouse DCT cells and the influx was abolished by the calcium channel blocker, nifedipine [57], consistent with the view that CT activates dihydropyridine-sensitive calcium channels that reside in apical membranes. As with PTH, CT also hyperpolarised mouse DCT cells, which may be necessary for activation of the calcium entry channel because blocking hyperpolarisation abolished calcium influx [57].

Though both PTH and CT stimulate calcium reabsorption in the distal tubule, these effects are not additive at maximal doses [57]. Because the magnitude of PTH and CT effects on DCT cells is similar [56, 57] and the effects of PTH and CT are not additive [57], it is probable that PTH and CT share a common mechanism of action despite the difference of their kinetics. Dual kinetics of calcium entry were also reported in luminal membranes of distal tubule [130]; a low affinity system with high velocity, and a high affinity system with low velocity [60, 130]. PTH and vitamin D-dependent calbindin D28k stimulate the high-affinity component by increasing its velocity, and CT targets the low affinity component [60].

The target transporter protein of calcitonin action in distal tubule basolateral membranes appears to be the Na^+/Ca^{2+} exchanger (NCX), but not Ca^{2+}-ATPase. CT increased Na^+-dependent calcium uptake, but not ATP-dependent calcium uptake by basolateral membrane vesicles prepared from rabbit distal tubules [60]. In single isolated perfused rabbit DCTs, CT had no effect on net calcium transport in the absence of basolateral sodium, but increased calcium absorption in the presence of sodium [71], indicating that CT stimulates Na^+-dependent calcium transport as PTH does on CNT [71]. Though neither PTH nor CT stimulated Ca^{2+}-ATPase, the transport could be regulated indirectly by vitamin D in the distal tubule [68, 131].

Vitamin D also stimulates calcium reabsorption in the distal tubule. Transcellular calcium absorption by monolayers of cultured cells originating from CNT and CCD of rabbits was increased by $1,25(OH)_2$ vitamin D_3 [132]. It is unclear whether vitamin D directly regulates calcium entry and extrusion in this segment. ATP-dependent calcium uptake by basolateral membrane vesicles from the distal tubule obtained from vitamin D-deficient rabbits was decreased when compared to vitamin D-replete rabbits, while there was no change in the activity of the Na^+/Ca^{2+} exchanger, indicating an effect of vitamin D on Ca^{2+}-ATPase but not on NCX [131]. Northern and Western blotting of cultured rabbit CNT and CCD cells incubated with $1,25(OH)_2$ vitamin D_3 revealed that neither NCX nor Ca^{2+}-ATPase mRNA, or protein content, respectively, was noticeably altered [66]. Though thyroparathyroidectomy decreased serum calcium concentration and calcium pump activity of basolateral membrane vesicles from the distal tubule of rats, vitamin D supplementation did not restore the calcium pump activity, while it did restore the serum calcium concentration [103], indicating no effect of vitamin D on Ca^{2+}-ATPase in the absence of PTH.

Vitamin D may enhance or modify the effect of PTH. PTH increased intracellular calcium concentration ($[Ca]_i$) in mouse DCT cells and $1,25(OH)_2$ vitamin D_3 reduced the latency before $[Ca]_i$ increased following addition of PTH. The magnitude of the PTH effect on $[Ca]_i$ was not altered [133]. Furthermore, $1,25(OH)_2$ vitamin D_3 increased PTH1R mRNA levels in mouse DCT cells [134].

Vitamin D enhances the expression of calbindins. Calbindin D28k is expressed exclusively in distal tubules, where it co-localised with the Ca^{2+}-ATPase and Na^+/Ca^{2+} exchanger [66]. The calbindin D28k contents of cultured CNT and CCD cells from rabbit increased by incubation with 1,25(OH)$_2$ vitamin D$_3$ [132]. Intraperitoneal injection of 1,25(OH)$_2$ vitamin D$_3$ increased the calbindin D28k protein in the rat kidney 48 hours after injection [135]. Forty-eight-hour treatment of 1,25(OH)$_2$ vitamin D$_3$ increased calbindin D28k mRNA in rabbit CNT and CCD [66]. These results suggest that vitamin D stimulates calbindin D28k synthesis [136]. Though it is hypothesised that vitamin D stimulates calbindin D9k production in the kidney, this has not been demonstrated. In this connection, calcitonin did not change the amount of calbindin D28k in the rat kidney [137]. Though thyroidectomy reduced serum calcitonin, calbindin D28k in the kidney determined by enzyme-linked immunoassay was not changed, and repletion of calcitonin by subcutaneous infusion for four days did not increase calbindin D28k [137], suggesting the absence of an effect of calcitonin on calbindin synthesis.

The 25-hydroxyvitamin D3 24-hydroxylase cytochrome P-450, which inactivates 1,25(OH)$_2$ vitamin D$_3$, was demonstrated in the human distal tubule and the proximal tubule [99], indicating that both the distal tubule and proximal tubule are sites of vitamin D metabolism [136].

The exact function of the calcium-sensing receptor (CaSR) is unclear despite its expression in the distal tubule. Activation of CaSR by neomycin inhibited the PTH- and calcitonin-stimulated cAMP release in mouse DCT cells [138]. Unidirectional calcium transport by MDCK cell monolayers was inhibited by stimulation of the CaSR [138]. Ca^{2+}-ATPase was also measured in MDCK cells and its activity was inhibited upon CaSR [138]. These results suggest that the CaSR may suppress calcium reabsorption in the distal tubule through the inhibition of Ca^{2+}-ATPase. The effect of the CaSR on PTH- and calcitonin-stimulated calcium reabsorption is unknown.

Collecting Ducts

Between 3% and 7% of the filtered load of calcium is absorbed by tubular segments beyond the last distal nephron [5]. However, there is no evidence available indicating that calcium transport in these segments is regulated by hormones or calcium itself, though the initial portion of the CCD could be regulated by hormones and calcium as described above.

PTH1R mRNA was detected by RT-PCR in rat CCD [93]. However, other groups failed to detect it in the same segment of the same species using similar protocols [92, 94]. The receptor mRNA was not found in rat outer medullary collecting duct (OMCD) and inner medullary collecting duct (IMCD) [92–94].

Consistent with these findings, PTH does not appear to regulate calcium absorption in these segments. Though monolayers of cultured rabbit CCD cells absorbed a certain amount of calcium from apical to basolateral compartments, the transport was not enhanced by PTH [73]. Isolated single rabbit CCD perfused in vitro did not absorb calcium and PTH did not elicit calcium absorption [72]. Microcatheterisation was performed on rats to collect samples along the IMCD [75]. Though the calcium concentration of collected fluid declined along the IMCD, indicating calcium reabsorption by this segment, the magnitude of transport was not changed by thyroparathyroidectomy [75]. This suggests that there is no effect of PTH in IMCD. According to an isolated tubule perfusion experiment, the permeability of the IMCD to calcium, which was low, was not altered by PTH [76].

The calcitonin receptor (CTR) mRNA was detected strongly in CCD and faintly in OMCD in rats [95]. However, radio-labelled calcitonin did not bind to CCD, OMCD, or IMCD on rat kidney slices [96]. It is unclear whether CT participates in calcium handling in these segments. CT could play a role in acid-base balance in intercalated cells [139, 140].

The CaSR was detected in rat CCD, OMCD, and IMCD [93]. The expression in these segments is weak when compared to that in thick ascending limbs and DCT [92]. Immunohistochemistry on kidney slices with co-staining for both CaSR and H^+-ATPase or anion exchanger-1 (Cl^-/HCO_3^- exchanger) showed that the CaSR was positive only in type A intercalated cells in rat CCD [97]. Further, the pattern of cellular localisation varied from the entire cytosol to the apical or basolateral surface with co-localisation with H^+-ATPase [97]. The role of the CaSR in calcium handling by these segments is unknown. Stimulating the CaSR by increasing basolateral calcium increased intracellular calcium concentration ($[Ca]_i$) of single microdissected CCDs and OMCDs but not IMCDs obtained from rats [141]. It is not known whether the increase of $[Ca]_i$ reflects a signalling pathway to inhibit calcium reabsorption in these segments. Stimulation of the CaSR by increasing calcium concentration did not change the calcium permeability of monolayers consisting of cultured rat CCD [76].

The vitamin D receptor mRNA is expressed in CCD but not in OMCD or IMCD in mice, rats, and pigs [98]. The role of vitamin D in these segments is unclear.

Diuretics and Related Diseases

Carbonic Anhydrase Inhibitors

These drugs, developed in the 1950s, inhibit sodium and HCO_3^- absorption by proximal tubules. As described in the section on the mechanism of calcium reabsorption in proximal tubules, calcium transport parallels Na^+ reabsorption. Thus, CA inhibitor blocks calcium reabsorption in this segment. Micropuncture studies in dogs show that acetazolamide, the prototype carbonic anhydrase inhibitor, elicits similar reductions of proximal tubule sodium and calcium reabsorption [142]. However, because sodium excretion increased appreciably, but final urinary calcium excretion was unchanged, compensatory calcium absorption by thick ascending limbs and distal tubules was enhanced. Short-term acetazolamide administration does not affect serum calcium concentrations. "Load dependence" of calcium reabsorption in distal tubules (as described in the section of the mechanism of calcium reabsorption) may explain the compensatory increases of distal calcium absorption. Chronic use of CA inhibitors for the management of glaucoma is associated with nephrocalcinosis.

Loop Diuretics

The major diuretics acting in the loop of Henle are furosemide, bumetanide, piretanide and torasemide. Collectively, the loop diuretics have been referred as "high ceiling" diuretics because of their pronounced natriuretic potency. Loop diuretics target Na^+–K^+–$2Cl^-$ co-transporters in apical membranes of thick ascending limbs. The mechanism of calcium absorption in this segment is shown in Figure 3.8. Calcium reabsorption in this segment involves both passive paracellular absorption, and active transcellular reabsorption, which is stimulated by PTH. The driving force of calcium reabsorption through the paracellular pathway is the lumen-positive transepithelial voltage, which is generated by sodium reabsorption. Inhibiting the Na^+–$2Cl^-$–K^+ co-transporter by loop diuretics abolishes Na^+ and Ca^{2+} reabsorption in parallel.

Bartter's Syndrome and Loop Diuretics

As described above, Ca^{2+} reabsorption is tightly coupled to Na^+ reabsorption in thick ascending limbs. Therefore, impaired Na^+ reabsorption diminishes Ca^{2+} reabsorption. Na^+ absorption is mediated by the apical Na^+–K^+–$2Cl^-$ co-transporter. Cl^- is extruded by basolateral Cl^- channels, and K^+ is recycled across apical membranes to generate the lumen-positive transepithelial voltage.

Patients with Bartter's syndrome exhibit hypokalaemic metabolic alkalosis with renal salt wasting and elevated urinary calcium excretion. Lifton et al. identified mutations in the Slc12a1 (Na^+–K^+–$2Cl^-$co-transporter) [143], the KCNJ1 (ROMK apical K^+ channel) [144], and CLCNKB (the human homologue of the CLC-K2 basolateral Cl^- channel) [145] genes in a number of families with Bartter's syndrome. Thus, loss of Na^+–K^+–$2Cl^-$ co-transporter function by inactivating gene mutations results in the same clinical presentation as does inhibition of co-transporter activity by loop diuretics (*viz.*, salt wasting, volume depletion, hypokalaemic metabolic alkalosis, and hypercalciuria). These findings underscore the parallel nature of Na^+ and Ca^{2+} absorption by thick ascending limbs.

Thiazide Diuretics

Thiazide diuretics such as chlorothiazide, chlorothalidone, polythiazide, and metolazone target the Na^+–Cl^- co-transporter in apical membranes of distal convoluted tubules. Thiazide diuretics have the unique characteristic of concomitantly decreasing calcium excretion while increasing sodium excretion. The mechanism of calcium reabsorption in distal convoluted tubule is shown in Figure 3.9.

Two sodium-entry mechanisms are thought to coexist in distal convoluted tubules; an electroneutral Na^+–Cl^- co-transporter and an amiloride-sensitive, epithelial Na^+ channel (ENaC). The transport of each Na^+ by the Na^+–Cl^- co-transporter is directly and tightly coupled in an electroneutral fashion to the influx of one Cl^- ion. The entry of Na^+ ions, by both the Na^+–Cl^- co-transporter and the ENaC, is a secondary active transport process that depends on the favourable electrochemical Na^+ gradient, which is sustained by the continuous extrusion of Na^+ across basolateral cell membranes by the Na^+/K^+-ATPase. Calcium entry in distal convoluted tubule is mediated by an epithelial calcium channel (ECaC) that is activated by membrane hyperpolarisation. Manoeuvres that inhibit Na^+ entry hyperpolarise the membrane and augment the driving force of calcium entry. Thus, thiazide diuretics increase calcium entry through apical membranes by their hyperpolarising effect on cell membrane voltage.

Calcium extrusion across basolateral membrane is mediated by the NCx1 Na^+/Ca^{2+} exchanger in distal convoluted tubules. Thiazide diuretics may indirectly stimulate the exchanger. Thiazide diuretics inhibit Na^+–Cl^- co-transporter, thereby decreasing intracellular Na^+ concentration, with an attendant increase of the driving force for basolateral Na^+ entry and an accompanying stimulation of the rate of Na^+/Ca^{2+} exchange. Thus, by decreasing Na^+ entry through apical membranes, thiazide diuretics increase calcium reabsorption.

With prolonged administration, thiazide diuretics cause a sustained reduction in calcium excretion that is accompanied by a modest but persistent increase of serum calcium [146–148].

Gitelman's Syndrome and Thiazide Diuretics

The variant of Bartter's syndrome, described by Gitelman et al. [149], also exhibits salt wasting, hypokalaemia, and hypermagnesuria but is associated with decreased calcium excretion [149]. Gitelman's syndrome is due to inactivating mutations of the apical membrane Na^+–Cl^- co-transporter gene, Slc12a3, which is located on human chromosome 16 [150]. The hypocalciuria of Gitelman's syndrome can be understood from the model of distal tubular calcium absorption and the mechanism of thiazide diuretics discussed above. Thiazide diuretics target the Na^+–Cl^- co-transporter in apical membrane in distal convoluted tubules, which is inactivated in Gitelman's syndrome. Because the direction and magnitude of sodium and calcium absorption are inversely related in distal convoluted tubules, impaired or diminished Na^+–Cl^- co-transport enhances calcium absorption. Thus, the disordered mineral ion metabolism accompanying Gitelman's syndrome reinforces the conclusions drawn regarding the cellular mechanisms of distal calcium absorption and the inverse relations between calcium and sodium transport by distal tubules.

Potassium-Sparing Diuretics

Na^+ reabsorption by distal convoluted tubules and principal cells of the cortical collecting duct is mediated by the epithelial Na^+ channel (ENaC) in apical membrane and Na^+/K^+ exchanger in basolateral membrane.

Potassium-sparing diuretics, amiloride and triamterene, inhibit ENaC in apical membranes. Consequently, cytosolic Na^+ concentration decreases, thereby hyperpolarising the cell membranes and stimulating Ca^{2+} absorption by ECaC in distal convoluted tubules.

The dissociation of Na^+ and Ca^{2+} excretion by amiloride has been shown in experimental animals but is less convincing in clinical settings. Triamterene blocks ENaC. However, whereas amiloride decreases fractional Ca^{2+} excretion in the dog, triamterene does not; i.e., the dissociation of Na^+ and Ca^{2+} clearance by triamterene was caused solely by its effect on Na^+ elimination. In humans, however, triamterene increased Ca^{2+} excretion in one study, while having no effect on Ca^{2+} excretion, but increasing Mg^{2+} excretion, in another [151, 152]. When administered in combination with furosemide, triamterene exhibited a biphasic action on urinary Ca^{2+} excretion. An initial decrease of Ca^{2+} excretion was followed by an increase thereafter. Over 12 hours, triamterene neither enhanced nor diminished Ca^{2+} excretion relative to the administration of furosemide alone.

References

1. Oh M. External balance of electrolytes and acid and alkali. In: Seldin DW, Giebisch G, editors. The kidney, 3rd edn. Philadelphia: Lippincott Williams and Wilkins, 2000;33–59.
2. Barratt LJ, et al. Factors governing the transepithelial potential difference across the proximal tubule of the rat kidney. Journal of Clinical Investigation 1974;53:454–64.
3. Rocha AS, Magaldi JB, Kokko JP. Calcium and phosphate transport in isolated segments of rabbit Henle's loop. Journal of Clinical Investigation 1977;59:975–83.

4. Bourdeau JE, Burg MB. Voltage dependence of calcium transport in the thick ascending limb of Henle's loop. American Journal of Physiology 1979;236:F357–64.
5. Friedman PA, Gesek FA. Cellular calcium transport in renal epithelia: measurement, mechanisms, and regulation. Physiological Reviews 1995;75:429–71.
6. Magaldi AJ, Van Baak AA, RochaAS. Calcium transport across rat inner medullary collecting duct perfused in vitro. American Journal of Physiology 1989;257:F738–45.
7. Elalouf JM, Roinel N, de Rouffignac C. Effects of glucagon and PTH on the loop of Henle of rat juxtamedullary nephrons. Kidney Int 1986;29(4):807–13.
8. Elalouf JM, Roinel N, de Rouffignac C. Effects of human calcitonin on water and electrolyte movements in rat juxtamedullary nephrons: inhibition of medullary K recycling. Pflugers Arch 1986; 406(5):502–8.
9. Bouhtiauy I et al. Two vitamin D_3-dependent calcium binding proteins increase calcium reabsorption by different mechanisms. I. Effect of CaBP 28K. Kidney Int 1994;45:461–8.
10. Hoenderop JG, Willems PH, Bindels RJ. Toward a comprehensive molecular model of active calcium reabsorption. Am.J Physiol Renal Physiol 2000;278:F352–60.
11. Duarte CG, Watson JF. Calcium reabsorption in proximal tubule of the dog nephron. American Journal of Physiology 1967;212:1355–60.
12. Murayama Y, Morel F, Le Grimellec C. Phosphate, calcium and magnesium transfers in proximal tubules and loops of Henle, as measured by single nephron microperfusion experiments in the rat. Pflugers Archiv European Journal of Physiology 1972;333:1–16.
13. Ng RCK, Rouse D, Suki WN. Calcium transport in the rabbit superficial proximal convoluted tubule. Journal of Clinical Investigation 1984;74:834–42.
14. Lassiter WE, Gottschalk CW, Mylle M. Micropuncture study of renal tubular reabsorption of calcium in normal rodents. American Journal of Physiology 1963;204:771–5.
15. Le Grimellec C, Roinel N, Morel F. Simultaneous Mg, Ca, P, K, Na and Cl analysis in rat tubular fluid. I. During perfusion of either inulin or ferrocyanide. Pflugers Archiv European Journal of Physiology 1973;340:181–96.
16. Ullrich KJ, Rumrich G, Kloss S. Active Ca^{2+} reabsorption in the proximal tubule of the rat kidney. Dependence on sodium and buffer transport. Pflugers Archiv European Journal of Physiology 1976;364:223–8.
17. Rouse D, Ng RCK, Suki WN. Calcium transport in the pars recta and thin descending limb of Henle of rabbit perfused in vitro. Journal of Clinical Investigation 1980;65:37–42.
18. Bomsztyk K, George JP, Wright FS. Effects of luminal fluid anions on calcium transport by proximal tubule. American Journal of Physiology 1984;246:F600–8.
19. Sutton RAL, Dirks JH. The renal excretion of calcium: a review of micropuncture data. Canadian Journal of Physiology and Pharmacology 1975;53:979–88.
20. Rector Jr FC. Sodium, bicarbonate, and chloride absorption by the proximal tubule. American Journal of Physiology 1983;244:F461–71.
21. Berry CA, Rector Jr FC. Relative sodium-to-chloride permeability in the proximal convoluted tubule. American Journal of Physiology 1978;235:F592–604.
22. Rose UM et al. Effects of Ca^{2+} channel blockers, low Ca^{2+} medium and glycine on cell Ca^{2+} and injury in anoxic rabbit proximal tubules. Kidney International 1994;46:223–9.
23. Zhang MIN, O'Neil RG. An L-type calcium channel in renal epithelial cells. Journal of Membrane Biology 1996;154:259–66.
24. Zhang MI, O'Neil RG. Molecular characterization of rabbit renal epithelial calcium channel. Biochem Biophys Res Commun 2001;280(2):435–9.
25. Tanaka H et al. Pathways involved in PTH-induced rise in cytosolic Ca^{2+} concentration of rat renal proximal tubule. American Journal of Physiology 1995;268:F330–7.
26. Hoenderop JGJ et al. Molecular identification of the apical Ca^{2+} channel in 1,25-dihydroxyvitamin D_3-responsive epithelia. Journal of Biological Chemistry 1999;274:8375–8.
27. Pedersen PL, Carafoli E. Ion motive ATPases. I. Ubiquity, properties, and significance to cell function. Trends in Biochemical Sciences 1987;12:146–50.
28. Stauffer TP, Guerini D, Carafoli E. Tissue distribution of the four gene products of the plasma membrane Ca^{2+} pump. A study using specific antibodies. Journal of Biological Chemistry 1995; 270:12184–90.
29. White KE et al. Molecular dissection of Ca^{2+} efflux in immortalized proximal tubule cells. Journal of General Physiology 1997;109:217–28.
30. Caride AJ, et al. Unique localization of mRNA encoding plasma membrane Ca^{2+} pump isoform 3 in rat thin descending loop of Henle. American Journal of Physiology 1995;269:F681–5.
31. Dominguez JH et al. Na^+ electrochemical gradient and Na^+–Ca^{2+} exchange in rat proximal tubule. American Journal of Physiology 1989;257:F1497–538.

32. Scoble JE, Cragoe Jr EJ, Hruska KA. Na^+–Ca^{2+} exchange and calcium permeability in canine basolateral membrane vesicles: the effects of dibutyryl cAMP and specific inhibitors. Biochimica et Biophysica Acta 1988;944:233–41.
33. Ramachandran C, Brunette MG. The renal Na^+/Ca^{2+} exchange system is located exclusively in the distal tubule. Biochemical Journal 1989;257:259–64.
34. Tsukamoto Y, Sugimura K, Suki WN. Role of Ca^{2+}/H^+ antiporter in the kidney. Kidney International 1991;40 Suppl 33:S90–4.
35. Suki WN, Rouse D. Hormonal regulation of calcium transport in thick ascending limb renal tubules. American Journal of Physiology 1981;241:F171–4.
36. Suki WN et al. Calcium transport in the thick ascending limb of Henle. Heterogeneity of function in the medullary and cortical segments. Journal of Clinical Investigation 1980;66:1004–9.
37. Greger R, Bleich M, Schlatter E. Ion channel regulation in the thick ascending limb of the loop of Henle. Kidney International 1991;40 Suppl 33:S119–24.
38. Lau K, Quamme G, Tan S. Patch-clamp evidence for a Ca channel in apical membrane of cortical thick ascending limb (cTAL) and distal tubule (DT) cells. Journal of the American Society of Nephrology 1991;2:775.
39. Hoenderop JGJ et al. Localization of the epithelial Ca^{2+} channel in rabbit kidney and intestine. Journal of the American Society of Nephrology 2000;11(7):1171–8.
40. Caride AJ et al. mRNA encoding four isoforms of the plasma membrane calcium pump and their variants in rat kidney and nephron segments. Journal of Laboratory and Clinical Medicine 1998;132:149–56.
41. Dai LJ et al. Na^+/Ca^{2+} exchanger in epithelial cells of the porcine cortical thick ascending limb. American Journal of Physiology 1996;270:F411–8.
42. Dai LJ et al. Modulation of Na^+/Ca^{2+} exchange in epithelial cells of porcine thick ascending limb. American Journal of Physiology 1996;270:F953–9.
43. Reilly RF et al. Immunolocalization of the Na^+/Ca^{2+} exchanger in rabbit kidney. American Journal of Physiology 1993;265:F327–32.
44. Lytton J et al. The kidney sodium–calcium exchanger. Annals of the New York Academy of Sciences 1996;779:58–72.
45. Bourdeau JE, Taylor AN, Iacopino AM. Immunocytochemical localization of sodium–calcium exchanger in canine nephron. Journal of the American Society of Nephrology 1993;4:105–10.
46. Yu ASL et al. Identification and localization of renal Na^+–Ca^{2+} exchanger by polymerase chain reaction. American Journal of Physiology 1992;263:F680–5.
47. Hanaoka K et al. Mechanisms of calcium transport across the basolateral membrane of the rabbit cortical thick ascending limb of Henle loop. Pflugers Archiv European Journal of Physiology 1993;422:339–46.
48. Costanzo LS, Windhager EE. Calcium and sodium transport by the distal convoluted tubule of the rat. American Journal of Physiology 1978;235:F492–506.
49. Greger R, Lang F, Oberleithner H. Distal site of calcium reabsorption in the rat nephron. Pflugers Archiv European Journal of Physiology 1978;374:153–7.
50. Friedman PA. Codependence of renal calcium and sodium transport. Annual Review of Physiology 1998;60:179–97.
51. Philipson KD et al. Molecular regulation of the Na^+–Ca^{2+} exchanger. Annals of the New York Academy of Sciences 1996;779:20–8.
52. Tan S, Lau K. Patch-clamp evidence for calcium channels in apical membranes of rabbit kidney connecting tubules. Journal of Clinical Investigation 1993;92:2731–6.
53. Matsunaga H et al. Epithelial Ca^{2+} channels sensitive to dihydropyridines and activated by hyperpolarizing voltages. American Journal of Physiology 1994;267:C157–65.
54. Hoenderop JGJ et al. The epithelial calcium channel, ECaC, is activated by hyperpolarization and regulated by cytosolic calcium. Biochemical and Biophysical Research Communications 1999;261:488–92.
55. Friedman PA. Mechanisms of renal calcium transport. Exp Nephrol 2000;8(6):343–50.
56. Gesek FA, Friedman PA. On the mechanism of parathyroid hormone stimulation of calcium uptake by mouse distal convoluted tubule cells. Journal of Clinical Investigation 1992;90:749–58.
57. Gesek FA, Friedman PA. Calcitonin stimulates calcium transport in distal convoluted tubule cells. American Journal of Physiology 1993;264:F744–51.
58. Lajeunesse D, Bouhtiauy I, Brunette MG. Parathyroid hormone and hydrochlorothiazide increase calcium transport by the luminal membrane of rabbit distal nephron segments through different pathways. Endocrinology 1994;134:35–41.
59. Barry ELR et al. Distinct calcium channel isoforms mediate parathyroid hormone and chlorothiazide-stimulated calcium entry in transporting epithelial cells. Journal of Membrane Biology 1998;161:55–64.

60. Zuo Q et al. Effect of calcitonin on calcium transport by the luminal and basolateral membranes of the rabbit nephron. Kidney International 1997;51:1991–9.
61. Bindels RJM et al. Role of Na^+/Ca^{2+} exchange in transcellular Ca^{2+} transport across primary cultures of rabbit kidney collecting system. Pflugers Archiv European Journal of Physiology 1992;420:566–72.
62. White KE, Gesek FA, Friedman PA. Structural and functional analysis of Na^+/Ca^{2+} exchange in distal convoluted tubule cells. American Journal of Physiology 1996;271:F560–70.
63. Philipson KD. Sodium–calcium exchange in plasma membrane vesicles. Annual Review of Physiology 1985;47:561–71.
64. Quabius ES, Murer H, Biber J. Expression of proximal tubular $Na-P_i$ and $Na-SO_4$ co-transporters in MDCK and LLC-PK$_1$ cells by transfection. American Journal of Physiology 1996;270:F220–8.
65. Nicoll DA, Longoni S, Philipson KD. Molecular cloning and functional expression of the cardiac sarcolemmal $Na^+–Ca^{2+}$ exchanger. Science 1990;250:562–5.
66. van Baal J et al. Localization and regulation by vitamin D of calcium transport proteins in rabbit cortical collecting system. American Journal of Physiology 1996;271:F985–93.
67. Bindels RJM et al. Calbindin-D_{9k} and parvalbumin are exclusively located along basolateral membranes in rat distal nephron. Journal of the American Society of Nephrology 1991;2:1122–9.
68. Bouhtiauy I et al. Two vitamin D3-dependent calcium binding proteins increase calcium reabsorption by different mechanisms. II. Effect of CaBP 9K. Kidney Int 1994;45(2):469–74.
69. Agus ZS, Chiu PJS, Goldberg M. Regulation of urinary calcium excretion in the rat. American Journal of Physiology 1977;232:F545–9.
70. Sutton RAL, Wong NLM, Dirks JH. Effects of parathyroid hormone on sodium and calcium transport in the dog nephron. Clinical Science and Molecular Medicine 1976;51:345–51.
71. Shimizu T et al. Effects of PTH, calcitonin, and cAMP on calcium transport in rabbit distal nephron segments. American Journal of Physiology 1990;259:F408–14.
72. Shareghi GR, Stoner LC. Calcium transport across segments of the rabbit distal nephron in vitro. American Journal of Physiology 1978;235:F367–75.
73. Rochelle LG et al. Active calcium absorption in primary cultures of cortical collecting duct cells. Canadian Journal of Physiology and Pharmacology 1993;71:491–6.
74. Bourdeau JE, Hellstrom-Stein RJ. Voltage-dependent calcium movement across the cortical collecting duct. American Journal of Physiology 1982;242:F285–92.
75. Bengele HH, Alexander EA, Lechene CP. Calcium and magnesium transport along the inner medullary collecting duct of the rat. American Journal of Physiology 1980;239:F24–9.
76. Carney SL, Dirks JH. Effect of parathyroid and antidiuretic hormone on water and calcium permeability in the rat collecting duct. Mineral and Electrolyte Metabolism 1988;14:142–5.
77. Carney SL. Comparison of parathyroid hormone and calcitonin on rat renal calcium and magnesium transport. Clinical and Experimental Pharmacology and Physiology 1992;19:433–8.
78. Hugi K, Bonjour J-P, Fleisch H. Renal handling of calcium: influence of parathyroid hormone and 1,25-dihydroxyvitamin D_3. American Journal of Physiology 1979;236:F349–56.
79. Even L et al. Selective modulation by vitamin D of renal response to parathyroid hormone: A study in calcitriol-resistant rickets. Journal of Clinical Endocrinology and Metabolism 1996;81:2836–40.
80. Carney S, Thompson L. Acute effect of calcitonin on rat renal electrolyte transport. American Journal of Physiology 1981;240:F12–16.
81. Elalouf JM, Roinel N, de Rouffignac C. Stimulation by human calcitonin of electrolyte transport in distal tubules of rat kidney. Pflugers Archiv European Journal of Physiology 1983;399:111–8.
82. Carney SL, Thompson L. Chronic calcitonin administration and renal calcium transport in the rat. Clinical and Experimental Pharmacology and Physiology 1998;25:236–9.
83. Carney SL. Calcitonin and human renal calcium and electrolyte transport. Mineral and Electrolyte Metabolism 1997;23:43–7.
84. Burnatowska MA et al. Effects of vitamin D on renal handling of calcium, magnesium, and phosphate in the hamster. Kidney International 1985;27:864–70.
85. Yamamoto M et al. Vitamin D deficiency and renal calcium transport in the rat. Journal of Clinical Investigation 1984;74:507–13.
86. Brown EM et al. Cloning and characterization of an extracellular Ca^{2+}-sensing receptor from bovine parathyroid. Nature (London) 1993;366:575–80.
87. Pace AJ, Gama L, Breitwieser GE. Dimerization of the calcium-sensing receptor occurs within the extracellular domain and is eliminated by Cys→Ser mutations at Cys[101] and Cys[236]. Journal of Biological Chemistry 1999;274:11629–34.
88. Bai M et al. Intermolecular interactions between dimeric calcium-sensing receptor monomers are important for its normal function. Proceedings of the National Academy of Science USA 1999; 96:2834–9.
89. Chattopadhyay N. Biochemistry, physiology and pathophysiology of the extracellular calcium-sensing receptor. Int J Biochem Cell Biol 2000;32(8):789–804.

90. Brown AJ et al. Loss of calcium responsiveness in cultured bovine parathyroid cells is associated with decreased calcium receptor expression. Biochem Biophys Res Commun 1995;212(3):861–7.
91. McGehee DS et al. Mechanism of extracellular Ca^{2+} receptor-stimulated hormone release from sheep thyroid parafollicular cells. J Physiol 1997;502(Pt 1):31–44.
92. Yang TX et al. Expression of PTHrP, PTH/PTHrP receptor, and Ca^{2+}-sensing receptor mRNAs along the rat nephron. American Journal of Physiology 1997;272:F751–8.
93. Riccardi D et al. Localization of the extracellular Ca^{2+}-sensing receptor and PTH/PTHrP receptor in rat kidney. American Journal of Physiology 1996;271:F951–6.
94. Lee KC et al. Localization of parathyroid hormone parathyroid hormone-related peptide receptor mRNA in kidney. American Journal of Physiology 1996270:F186–91.
95. Firsov D et al. Quantitative RT-PCR analysis of calcitonin receptor mRNAs in the rat nephron. American Journal of Physiology 1995;269:F702–9.
96. Sexton PM et al. Localization and characterization of renal calcitonin receptors by in vitro autoradiography. Kidney International 1987;32:862–8.
97. Riccardi D et al. Localization of the extracellular Ca^{2+} polyvalent cation-sensing protein in rat kidney. Am J Physiol Renal Physiol 1998;274:F611–22.
98. Liu L et al. Vitamin D receptor gene expression in mammalian kidney. Journal of the American Society of Nephrology 1994;5:1251–8.
99. Kumar R et al. Immunolocalization of calcitriol receptor, 24-hydroxylase cytochrome P-450, and calbindin D_{28k} in human kidney. American Journal of Physiology 1994;266:F477–85.
100. Khalifa S, Mills S, Hruska KA. Stimulation of calcium uptake by parathyroid hormone in renal brush-border membrane vesicles. Relationship to membrane phosphorylation. Journal of Biological Chemistry 1983;258:14400–6.
101. Agus ZS et al. Effects of parathyroid hormone on renal tubular reabsorption of calcium, sodium, and phosphate. American Journal of Physiology 1973;224:1143–8.
102. Dolson GM, Hise MK, Weinman EJ. Relationship among parathyroid hormone, cAMP, and calcium on proximal tubule sodium transport. American Journal of Physiology 1985;249:F409–16.
103. Tsukamoto Y, Saka S, Saitoh M. Parathyroid hormone stimulates ATP-dependent calcium pump activity by a different mode in proximal and distal tubules of the rat. Biochimica et Biophysica Acta 1992;1103:163–71.
104. Bouhtiauy I, Lajeunesse D, Brunette MG. The mechanism of parathyroid hormone action on calcium reabsorption by the distal tubule. Endocrinology 1991;128:251–8.
105. Quamme GA. Effect of calcitonin on calcium and magnesium absorption in rat nephron. American Journal of Physiology 1980;238:E573–8.
106. Takeyama K et al. 25-hydroxyvitamin D_3 1a-hydroxylase and vitamin D synthesis. Science 1997;277:1827–30.
107. Haddad JG. Plasma vitamin D-binding protein (Gc-globulin): multiple tasks. Journal of Steroid Biochemistry and Molecular Biology 1995;53:579–82.
108. Haddad JG, Fraser DR, Lawson DE. Vitamin D plasma binding protein. Turnover and fate in the rabbit. Journal of Clinical Investigation 1981;67:1550–60.
109. Nykjaer A et al. An endocytic pathway essential for renal uptake and activation of the steroid 25-(OH) vitamin D_3. Cell 1999;96:507–15.
110. Farquhar MG. The unfolding story of megalin (gp330): now recognized as a drug receptor. Journal of Clinical Investigation 1995;96:1184.
111. Lundgren S et al. Tissue distribution of human gp330/megalin, a putative Ca^{2+}-sensing protein. Journal of Histochemistry and Cytochemistry 1997;45:383–92.
112. Christensen EI et al. Segmental distribution of the endocytosis receptor gp330 in renal proximal tubules. European Journal of Cell Biology 1995;66:349–64.
113. Christensen EI et al. Megalin-mediated endocytosis in renal proximal tubule. Renal Failure 1998;20:191–9.
114. Cui S et al. Megalin/gp330 mediates uptake of albumin in renal proximal tubule. American Journal of Physiology 1996;271:F900–7.
115. Kau ST, Maack T. Transport and catabolism of parathyroid hormone in isolated rat kidney. American Journal of Physiology 1977;233:F445–54.
116. Brown RC, Silver AC, Woodhead JS. Binding and degradation of NH_2-terminal parathyroid hormone by opossum kidney cells. American Journal of Physiology 1991;260:E544–52.
117. Hilpert J et al. Megalin antagonizes activation of the parathyroid hormone receptor. Journal of Biological Chemistry 1999;274:5620–5.
118. Gardella TJ et al. Parathyroid hormone (PTH)-PTH-related peptide hybrid peptides reveal functional interactions between the 1–14 and 15–34 domains of the ligand. Journal of Biological Chemistry 1995;270:6584–8.

119. Allen F, Tisher CC. Morphology of the ascending thick limb of Henle. Kidney Int 1976;9(1):8–22.
120. Murphy E, Chamberlin ME, Mandel LJ. Effects of calcitonin on cytosolic Ca in a suspension of rabbit medullary thick ascending limb tubules. American Journal of Physiology 1986;251:C491–5.
121. Bourdeau JE, Burg MB. Effect of PTH on calcium transport across the cortical thick ascending limb of Henle's loop. American Journal of Physiology 1980;239:F121–6.
122. Friedman PA. Basal and hormone-activated calcium absorption in mouse renal thick ascending limbs. American Journal of Physiology 1988;254:F62–70.
123. Wittner M et al. Hormonal stimulation of Ca^{2+} and Mg^{2+} transport in the cortical thick ascending limb of Henle's loop of the mouse: Evidence for a change in the paracellular pathway permeability. Pflugers Archiv European Journal of Physiology 1993;423:387–96.
124. Di Stefano A et al. Effects of parathyroid hormone and calcitonin on Na^+, Cl^-, K^+, Mg^{2+} and Ca^{2+} transport in cortical and medullary thick ascending limbs of mouse kidney. Pflugers Archiv European Journal of Physiology 1990;417:161–7.
125. Bacskai BJ, Friedman PA. Activation of latent Ca^{2+} channels in renal epithelial cells by parathyroid hormone. Nature (London) 1990;347:388–91.
126. Wang W et al. Phospholipase A_2 is involved in mediating the effect of extracellular Ca^{2+} on apical K^+ channels in rat TAL. American Journal of Physiology 1997;273:F421–9.
127. Bourdeau JE, Langman CB, Bouillon R. Parathyroid hormone-stimulated calcium absorption in cTAL from vitamin D-deficient rabbits. Kidney International 1987;31:913–7.
128. Costanzo LS, Windhager EE. Effects of PTH, ADH, and cyclic AMP on distal tubular Ca and Na reabsorption. American Journal of Physiology 1980;239:F478–85.
129. Bailly C, Roinel N, Amiel C. Stimulation by glucagon and PTH of Ca and Mg reabsorption in the superficial distal tubule of the rat kidney. Pflugers Archiv European Journal of Physiology 1985;403:28–34.
130. Brunette MG, Mailloux J, Lajeunesse D. Calcium transport through the luminal membrane of the distal tubule. I. Interrelationship with sodium. Kidney International 1992;41:281–8.
131. Bouhtiauy I, Lajeunesse D, Brunette MG. Effect of vitamin D depletion on calcium transport by the luminal and basolateral membranes of the proximal and distal nephrons. Endocrinology 1993;132:115–20.
132. Bindels RJM et al. Active Ca^{2+} transport in primary cultures of rabbit kidney CCD: stimulation by 1,25-dihydroxyvitamin D_3 and PTH. American Journal of Physiology 1991;261:F799–807.
133. Friedman PA, Gesek FA. Vitamin D_3 accelerates PTH-dependent calcium transport in distal convoluted tubule cells. American Journal of Physiology 1993;265:F300–8.
134. Sneddon WB et al. Regulation of renal parathyroid hormone receptor expression by 1,25-dihydroxyvitamin D_3 and retinoic acid. Cellular Physiology and Biochemistry 1998;8:261–77.
135. Hemmingsen C, Staun M, Olgaard K. The effect of 1,25-vitamin D3 on calbindin-D and calcium-metabolic variables in the rat. Pharmacol Toxicol 1998;82(3):118–21.
136. Johnson JA, Kumar R. Renal and intestinal calcium transport: roles of vitamin D and vitamin D-dependent calcium binding proteins. Seminars in Nephrology 1994;14:119–28.
137. Hemmingsen C et al. Regulation of renal calbindin-D28K: The role of calcitonin. Calcified Tissue International 1995;56:372–5.
138. Blankenship KA et al. The calcium-sensing receptor regulates calcium absorption in MDCK cells by inhibition of PMCA. Am J Physiol Renal Physiol 2001;280(5):F815–22.
139. Colnot S et al. Transgenic analysis of the response of the rat calbindin-D 9k gene to vitamin D. Endocrinology 2000;141(7):2301–8.
140. Armbrecht HJ et al. Capacity of 1,25-dihydroxyvitamin D to stimulate expression of calbindin D changes with age in the rat. Archives of Biochemistry and Biophysics 1998;352:159–64.
141. Champigneulle A et al. Relationship between extra- and intracellular calcium in distal segments of the renal tubule. Role of the Ca^{2+} receptor RaKCaR. Journal of Membrane Biology 1997;156:117–29.
142. Beck LH, Goldberg M. Effects of acetazolamide and parathyroidectomy on renal transport of sodium, calcium, and phosphate. American Journal of Physiology 1973;224:1136–42.
143. Simon DB et al. Bartter's syndrome, hypokalaemic alkalosis with hypercalciuria, is caused by mutations in the Na-K-2Cl co-transporter NKCC2. Nature Genetics 1996;13:183–8.
144. Simon DB et al. Genetic heterogeneity of Bartter's syndrome revealed by mutations in the K^+ channel, ROMK. Nature Genetics 1996;14:152–6.
145. Simon DB et al. Mutations in the chloride channel gene, CLCNKB, cause Bartter's syndrome type III. Nature Genetics 1997;17:171–8.
146. Brickman AS, Massry SG, Coburn JW. Changes in serum and urinary calcium during treatment with hydrochlorothiazide: studies on mechanisms. Journal of Clinical Investigation 1972;51:945–54.
147. Parfitt AM. The interactions of thiazide diuretics with parathyroid hormone and vitamin D. Studies in patients with hypoparathyroidism. Journal of Clinical Investigation 1972;51:1879–88.

148. Seitz H, Jaworski ZF. Effect of hydrochlorothiazide on serum and urinary calcium and urinary citrate. Canadian Medical Association Journal 1964;90:414–20.
149. Gitelman HJ, Graham JB, Welt LG. A new familial disorder characterized by hypokalemia and hypomagnesemia. Transactions of the Association of American Physicians 1966;79:221–35.
150. Simon DB et al. Gitelman's variant of Bartter's syndrome, inherited hypokalaemic alkalosis, is caused by mutations in the thiazide-sensitive Na–Cl co-transporter. Nature Genetics 1996;12:24–30.
151. Pecchini F et al. Effeto del triamterene sulla calciuria. Giornale di Clinica Medica 1967;48:1140–8.
152. Walker BR, Hoppe RC, Alexander F. Effect of triamterene on the renal clearance of calcium, magnesium, phosphate, and uric acid in man. Clinical Pharmacology & Therapeutics 1972;13:245–50.

Hypercalcaemia and Primary Hyperparathyroidism

D. H. Schussheim and S. J. Silverberg

Introduction

The differential diagnosis of hypercalcaemia is vast, yet over 90% of afflicted patients carry one of two common diagnoses: either primary hyperparathyroidism or malignancy. The former is the most frequent cause of elevated blood calcium levels in a free-living population, while malignancy is the most common diagnosis in hospitalised patients. It is essential to develop a rational strategy for evaluating patients with hypercalcaemia: first, in order to ascertain the cause underlying the abnormality, and second, in order to determine the need for treatment. Not all patients with hypercalcaemia require intervention. In some cases, appropriate therapy is directed to controlling the calcium level per se; in other situations, it is targeted to the underlying disease; and in other clinical situations, no treatment at all is indicated. This chapter will emphasise primary hyperparathyroidism as the central disorder characterised by hypercalcaemia. Other causes of hypercalcaemia will be discussed in less detail thereafter (see Table 3.5).

Primary Hyperparathyroidism

Primary hyperparathyroidism is a common endocrine disorder, characterised by the excessive and incompletely regulated secretion of parathyroid hormone (PTH) from one or more parathyroid glands. The major actions of PTH, to mobilise calcium from bone, to conserve calcium in the kidney, and indirectly to increase gastrointestinal calcium absorption, lead to one of the major biochemical hallmarks of the disease, hypercalcaemia. Another major sign of the disorder is an elevated circulating concentration of PTH, now readily detected by accurate assays for the hormone. Advances in evaluation of metabolic bone diseases by sensitive, circulating and urinary markers of calcium metabolism have permitted a more detailed assessment of patients who do not appear to be suffering from overt clinical consequences of primary hyperparathyroidism. In addition, bone densitometry and analysis of bone by quantitative histomorphometry have provided direct insight into important current features of the disease. The result is a profile of primary hyperparathyroidism that not only is quite different from earlier historical descriptions but also requires consideration of a new set of issues insofar as the clinical management of the disease is concerned.

Table 3.5.   Differential Diagnosis of Hypercalcaemia

Hyperparathyroidism
  Primary
    Adenoma, hyperplasia, cancer (sporadic or familial)
  Tertiary

Malignancy-associated hypercalcaemia
  Humoral hypercalcaemia of malignancy (PTHrp)
  Tumour production of 1,25-dihydroxyvitamin D
  Local osteolytic bone metastases

Granulomatous disorders
  Sarcoidosis
  Tuberculosis
  Fungal disease
  Other (berylliosis, Crohn's disease, silicon-induced granulomas, leprosy)

Familial hypocalciuric hypercalcaemia

Drug induced
  Milk-alkali syndrome
  Vitamin D
  Vitamin A
  Thiazide diuretics
  Lithium
  Other (oestrogen, antioestrogen, aminophylline, growth hormone
    therapy)

Nonparathyroid endocrine disorders
  Hyperthyroidism
  Adrenal insufficiency
  Pheochromocytoma
  Vasoactive intestinal polypeptide-secreting tumour

Immobilization

Renal failure
  Acute and chronic

Total parenteral nutrition

Hypophosphataemia

Jansen's metaphyseal chondrodysplasia

Williams–Beuren syndrome

In the United States today, over 80% of cases are asymptomatic. The clinical presentation of the disease differs in other parts of the world, where primary hyperparathyroidism does seem to present more along classical lines [1–3]. A dramatic change in the incidence of primary hyperparathyroidism occurred in the late 1960s and early 1970s, due primarily to the introduction of the multichannel autoanalyser [4–9]. It is difficult to estimate true incidence figures for primary hyperparathyroidism [10], but the experience in Rochester, Minnesota, of 27.7 per 100,000 person-years is essentially identical to figures from Sweden [6] and from Birmingham, England [5]. On the basis of these figures, an estimate of approximately 100,000 new cases of primary hyperparathyroidism per year in the United States is likely to be accurate.

The prevalence of primary hyperparathyroidism (the proportion of the population affected with the disease at a given point in time) is higher than earlier estimates of incidence (the number of *new* cases diagnosed over a specified period of time). Prevalence estimates have been as high as 1 in 100 [11], but 1 per 1,000 would appear to be closer to the true prevalence rate in the early autoanalyser era. A recent report from Rochester, Minnesota suggests that newly diagnosed cases of primary hyperparathyroidism have been declining continuously since the mid 1970s. This experience has not been clearly repeated in other American centres. The incidence of primary hyperparathyroidism peaks in the middle years, with women predominating over men by a 2–3:1 margin. The disease is recognised most commonly in women who are in the first postmenopausal decade, between ages 50 and 60 years. It is perhaps because of the effects of oestrogens to oppose some of the skeletal actions of PTH that the disease may surface clinically when oestrogen levels fall. There do not appear to be any well-established predisposing factors for the development of primary hyperparathyroidism, but a history of irradiation to the neck and upper chest area in childhood is obtained in as many as 15–25% of patients with the disease [14, 15]. Most patients with primary hyperparathyroidism (80–85%) harbour a single adenoma; 15–20% of patients with primary hyperparathyroidism have a pathologic process involving all four parathyroid glands. The rarest form of primary hyperparathyroidism is parathyroid carcinoma [16]. Recent incidence figures place parathyroid carcinoma as a cause of primary hyperparathyroidism in well under 0.5% of all patients with primary hyperparathyroidism.

Clinical Manifestations of Primary Hyperparathyroidism

The clinical manifestations of primary hyperparathyroidism have changed dramatically over the past 40 years. In the United States today, hyperparathyroid bone disease is so rarely seen that most clinicians dispense with routine radiological assessment of primary hyperparathyroidism. Also noteworthy is the drop in the incidence of nephrolithiasis among these series, from approximately two-thirds of all cases to one-fifth. If potential manifestations of primary hyperparathyroidism (hypertension, neuropsychiatric abnormalities, etc.) are excluded from consideration, simply because we do not yet know whether they are specific for the hyperparathyroid process, one reaches the conclusion that the vast majority of patients seen with primary hyperparathyroidism today (approximately 80%) are asymptomatic. The definition of asymptomatic primary hyperparathyroidism was stated best by Heath et al. [4]: "well-documented primary hyperparathyroidism in which there are neither complications nor symptoms that are clearly and commonly attributable to either hypercalcaemia or parathyroid hormone excess".

Physical Findings

The most noteworthy aspect of the physical examination in primary hyperparathyroidism is that the examination is not noteworthy. Hypertension, which is frequently found, is not causally related to primary hyperparathyroidism (see below). Today, calcium phosphate deposition in the medial and lateral limbic margins of the cornea (band keratopathy) is rare, and is seen only by ophthalmologic slit-lamp examinations. Enlarged parathyroid tissue is usually palpable only when parathyroid carcinoma is present. The neurological examination, which used to be of interest in the days when the neuromuscular manifestations of primary hyperparathyroidism were often seen (see below), is similarly normal.

Symptoms and Signs

The extent to which patients may be symptomatic is related directly to whether the hypercalcaemia is in the range usually associated with specific symptomatology. It is important to note that the symptoms of hypercalcaemia are a function of the rate of rise of the serum calcium as well as the actual level. In addition, symptoms of hypercalcaemia vary from patient to patient. Most patients with mild primary hyperparathyroidism and calcium levels within 1 mg/dl of the upper limits of normal do not have features that can readily be attributed to the hypercalcaemia per se. Specific manifestations of primary hyperparathyroidism are those traditionally viewed as features of the disease itself. Along with the dramatic increase in the incidence of this disease, its clinical presentation has also undergone a major change.

Skeletal Manifestations of Primary Hyperparathyroidism

The classic bone disease of primary hyperparathyroidism is osteitis fibrosa cystica. When symptomatic, osteitis fibrosa cystica is experienced by patients as bone pain. Pathological fractures may occur. Overt bone resorption caused by excessive concentrations of PTH is associated with several typical radiological signs. Subperiosteal bone resorption of the distal phalanges is the most sensitive radiological sign of primary hyperparathyroidism. It is appreciated best on the radial side of the middle phalanges. Similar radiological changes may be present in the skull in the form of a moth-eaten or salt-and-pepper pattern. The distal one-third of the clavicles may appear to be tapered. Local destructive lesions, bone cysts, and "brown tumours" in the long bones and pelvis constitute other skeletal manifestations of the disease. Brown tumours are collections of osteoclasts intermixed with poorly mineralised woven bone. Non-specific generalised skeletal demineralisation is sometimes evident in the absence of these other features of hyperparathyroid bone disease. Both non-specific demineralisation and the specific radiological manifestations outlined above reflect the catabolic skeletal actions of PTH.

Although osteitis fibrosa cystica is distinctly unusual in patients who present with primary hyperparathyroidism in the United States, this does not imply that the skeleton is unaffected. There is now ample evidence of skeletal involvement in the hyperparathyroid process. Bone density has been measured at three sites to evaluate areas enriched in cortical bone (distal radius), cancellous bone (vertebral spine), and a mixture of both (the hip region) [17, 18]. At the distal radius, bone density was <80% of age- and sex-matched control values in 58% of our patients (see Figure 3.11). The mean value at this site was only 79% of expected. In contrast, at the lumbar spine, bone mineral density was relatively well preserved. Only 13% of patients had lumbar bone density <80% of age- and sex-matched control values, and the mean value was within 5% of expected. The values for the hip region were midway between data obtained for the spine and those for the distal radius (Figure 3.11). These results indicate that, in mild primary hyperparathyroidism, reductions in cortical bone density are seen regularly and that the cancellous bone is relatively well preserved. This preservation of cancellous bone is of particular importance in that primary hyperparathyroidism is a disease that disproportionately affects postmenopausal women. Women in their postmenopausal years are at risk for bone loss due to oestrogen deficiency, which is first seen in the cancellous bone of the spine. Thus, the effects of primary hyperparathyroidism on vertebral cancellous bone appear to be opposite to those of oestrogen deficiency, supporting the significant

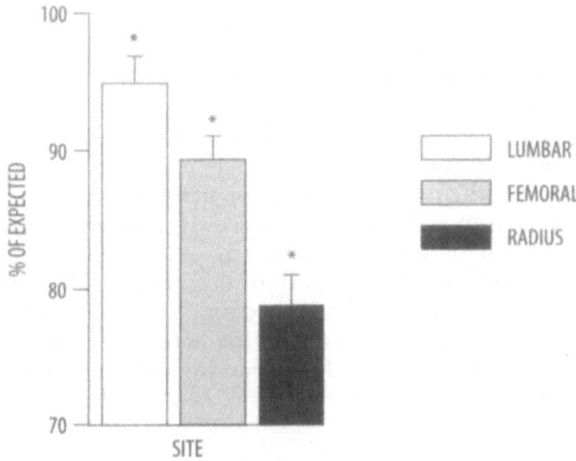

Figure 3.11.   Bone densitometry in primary hyperparathyroidism. Data are shown in comparison to age- and sex-matched normal subjects. Divergence from expected values is different at each site (p = 0.0001) (reprinted from ref. 18).

body of data advocating the use of parathyroid hormone as a therapeutic agent for postmenopausal osteoporosis.

Bone densitometry data have been supported by analysis of the bone biopsy by quantitative histomorphometry [19–24]. Bone biopsy results are consistent with a loss of cortical bone and preservation of cancellous bone. The measurements included both static and dynamic parameters of bone as well as a newer approach; strut analysis. The work of Parisien et al. has shown that static parameters of bone such as osteoid surface, osteoid volume, and eroded surface are all elevated in both men and women with primary hyperparathyroidism compared to normal values. Dynamic parameters of bone turnover (mineralising surface and bone formation rate) are also elevated. In summary, the data obtained from histomorphometric studies support the idea that patients with primary hyperparathyroidism have increased bone turnover, thinning of cortical bone, and increased connectivity of cancellous bone.

From the data presented, one might expect that, in primary hyperparathyroidism, fracture incidence would not be increased at the spine, whereas fracture of the distal radius or other cortical sites might depend, in part, on the degree to which there has been a reduction in bone mineral density at those site(s). Unfortunately, information in this area is limited, with studies arguing for or against an increased incidence of fractures [46–49]. Until a multi-centre trial with sufficient statistical power is performed, the question of fracture incidence in primary hyperparathyroidism in the United States will not be resolved.

Kidney Stone Disease

Kidney stones constitute another classic manifestation of primary hyperparathyroidism. Although the incidence of nephrolithiasis in primary hyperparathyroidism has diminished along with the incidence of bone disease, stones nevertheless are still seen. Most series now place the incidence of kidney stones at 15–20% of all patients with primary hyperparathyroidism. Because nephrolithiasis is such an important complication of primary hyperparathyroidism and because primary hyperparathyroidism is such a common disorder, it is still advisable to investigate the possibility of primary hyperparathyroidism in any patient who develops a kidney stone. Besides nephrolithiasis, the kidneys may be affected in other ways. Deposition of calcium phosphate crystals throughout the renal parenchyma, a process known as *nephrocalcinosis*, may occur. Nephrocalcinosis may or may not be associated with frank stones and/or a reduction in creatinine clearance. Hypercalciuria, defined as a total urinary calcium excretion of >250 mg (women) or >300 mg (men), occurs in up to 35–40% of patients with primary hyperparathyroidism [29]. The hypercalciuria is caused by the greater load of filtered calcium, which exceeds the capacity of the kidney to reabsorb it despite the conserving actions of PTH on renal calcium handling. Some patients with primary hyperparathyroidism will show reduced creatinine clearance without stones, hypercalciuria, nephrocalcinosis, or other predisposing factors.

Involvement of Other Organs

Primary hyperparathyroidism has the potential to involve organ systems besides the skeleton and the kidneys. The common complaints of weakness and easy fatigability used to be associated, in an earlier time, with a particular neuromuscular syndrome characterised histologically by atrophy of type II muscle fibres [30, 31]. Recent experience suggests that the weakness and easy fatigability sometimes ascribed to the hyperparathyroid syndrome are no longer associated with overt neurological findings [31]. However, in detailed electromyographical studies of muscles in primary hyperparathyroidism, abnormalities are still reported [32]. In one study, carpal tunnel syndrome was seen in a higher proportion of patients than expected in the normal population [31]. The significance of this finding remains unclear.

The gastrointestinal tract may also appear to be a target of the hyperparathyroid state. Historically, peptic ulcer disease was regarded as a frequent complication. It is now seen predominantly with the multiple endocrine neoplasia type 1 (MEN1) syndrome, in which primary hyperparathyroidism and peptic ulcer disease may coexist. Aside from this specific association, there is continuing debate over a pathophysiological link between these two relatively common disorders [33]. Similarly, the association between primary hyperparathyroidism and acute pancreatitis, apart from that related to hypercalcaemia per se, remains to be established [34].

Gout or pseudogout may affect the articular system of those with primary hyperparathyroidism. Older patients with asymptomatic chondrocalcinosis of the knees and bones of the wrist may be at risk for the development of pseudogout [35, 36]. Cardiovascular effects of primary hyperparathyroidism remain the subject of debate. Hypertension has been thought for many years to be a complicating feature of primary hyperparathyroidism. The experience of Heath et al. [4], in which there seems to be a greater incidence of hypertension among those with primary hyperparathyroidism in comparison to a control population, is in agreement with the experience of others [37, 38]. However, most patients do not experience a significant reduction in the blood pressure after successful parathyroid surgery [38, 39]. Although hypercalcaemia has been associated with numerous cardiac abnormalities (i.e., left ventricular hypertrophy, myocardial and valvular calcification), the cardiovascular effects of primary hyperparathyroidism remain unclear [40, 41]. Much of the available data in this regard has come from European centres, which describe patients with more severe disease than is usually seen in the United States. The applicability of these data to American patients who present with mild disease is controversial.

The neuropsychiatric manifestations of primary hyperparathyroidism are the subject of great uncertainty [42–47]. In part, this is because the symptomatology is exceedingly non-specific. Affective disorders, anxiety, cognitive difficulties, and somatisation exemplify the kinds of manifestations that have been ascribed to the hyperparathyroid syndrome. Certainly, most clinicians who care for patients with primary hyperparathyroidism agree that such symptoms are sometimes elicited from or volunteered by their patients. Moreover, some reports are intriguing for an apparent reversal of these symptoms after successful parathyroid surgery. Ljunghall and his colleagues [43–45] have investigated this problem in a series of impressive studies attempting to quantify this symptom complex using psychopathological rating scales before and after surgery in a population of European patients with primary hyperparathyroidism. More recently, Solomon et al. reported on the improvement of a series of neuropsychiatric measures after parathyroidectomy [47]. These data remain difficult to interpret, as the control group (patients who underwent thyroid surgery) also improved. It is unclear to what extent patients with primary hyperparathyroidism have manifestations that can be ascribed with certainty to a neuropsychiatric complex specifically due to primary hyperparathyroidism, and the extent to which such symptoms may be reversible after parathyroidectomy.

Longitudinal Course of Primary Hyperparathyroidism

Untreated Primary Hyperparathyroidism

There is ample documentation that the biochemical abnormalities so characteristic of primary hyperparathyroidism are stable during long-term follow up of mild, asymptomatic primary hyperparathyroidism patients [48]. Calcium and parathyroid hormone levels do not tend to rise over a decade of observation. The serum phosphorus tends to be in the lower range of normal while frank hypophosphataemia is present in less than one-quarter of patients. Average total urinary calcium excretion is at the upper end of the normal range, with less than half of all patients having hypercalciuria. Serum 25-hydroxyvitamin D levels tend to be in the lower end of the normal range. While mean values of 1,25-dihydroxyvitamin D_3 are in the high normal range, approximately one-third of patients have frankly elevated levels of this important hormone. There is no evidence that mild primary hyperparathyroidism is associated with progressive renal impairment, at least as measured by serum creatinine, blood urea nitrogen, or endogenous creatinine clearance. With regard to skeletal involvement, serum alkaline phosphatase, both total and bone specific, is also stable during several years of follow-up. This suggests lack of progression of PTH-mediated bone disease, which is discussed in detail below.

It must be noted that although groups of patients with primary hyperparathyroidism exhibit remarkable stability of biochemical indices, a minority of patients do have evidence of disease progression over time [48]. Although no clinically overt complication (i.e., fracture, kidney stone, etc.) occurred in a decade of observation, 4% of our asymptomatic patients developed substantial worsening of their hypercalcaemia (serum calcium >12 mg/dl), and 15% developed marked hypercalciuria (urinary calcium excretion >400 mg/day). There were no demographic or biochemical predictors of disease progression identified. Therefore, close follow-up of these parameters is necessary in all patients who do not undergo surgery.

Most patients with mild asymptomatic primary hyperparathyroidism do not appear to experience progressive loss of bone if the disease is left untreated [48]. Data from our group and others show remarkable stability in bone density over time in such patients (Figure 3.12). Only 12% of patients suffered a progressive decline in bone mineral density to the point that they met NIH Consensus Conference Guidelines for Surgery (radius Z-Score <−2).

Thus, the natural history of primary hyperparathyroidism, once the diagnosis has been established, is one of very mild clinical manifestations, with little objective or subjective evidence of progression if left untreated. A small group of patients do have laboratory evidence of worsening disease, with gradually increasing levels of serum or urinary calcium, or decreasing bone mineral density. Since there are no predictors for patients at risk for progressive disease, all patients followed without surgery must have regular monitoring. Surgical cure of primary hyperparathyroidism, on the other hand, leads to biochemical normalisation, and increased bone density at sites of cancellous bone.

Course of Primary Hyperparathyroidism Following Surgery

Surgery currently provides the only avenue to cure primary hyperparathyroidism. Following parathyroidectomy, there is a prompt normalisation of serum and urinary calcium levels, as well as a return of parathyroid hormone concentrations to normal. Surgery, additionally, leads to marked increases in bone mass. We have published data on a decade of postoperative follow-up [48]. Parathyroidectomy leads to a 10–12% rise in bone density at the lumbar spine and femoral neck (Figure 3.12). The increase at the lumbar spine and femoral neck is prompt, with the greatest increment in the first postoperative year. In our study, the trend towards a further increase after year one was significant only at the femoral neck, but the increases at both sites, which contain a significant amount of cancellous bone, is sustained over a decade after surgery, despite the tendency of advancing age to decrease bone density over time.

Lumbar spine and femoral neck bone mineral density increased to the same extent in the 28 postmenopausal women. Postmenopausal women showed a similar pattern of increased cancellous bone density. This is a curious observation in view of the fact that the lumbar region, enriched in cancellous bone, appears to be relatively well protected by PTH in primary hyperparathyroidism. The higher turnover rate of cancellous bone and the filling in, postoperatively, of the expanded remodelling space at this region could account for at least part for these observations. In the group as a whole, there was no significant change in bone mineral density at the distal radius. This is consistent with earlier data suggesting that, at least to some extent, PTH-mediated cortical bone loss may be irreversible.

Patients with primary hyperparathyroidism who have vertebral osteopenia or osteoporosis at the time of diagnosis experience an even greater postoperative increase in vertebral bone density, averaging 20% [49]. The marked improvement seen in patients with low vertebral bone density argues for surgery in those who present with cancellous as well as cortical bone loss.

Nephrolithiasis is a well-accepted indication for surgery in primary hyperparathyroidism. Urinary calcium excretion is reduced following surgery. Surgery is also of clear benefit in reducing the incidence of recurrent nephrolithiasis. Over 90% of patients with stone disease and hyperparathyroidism do not form additional

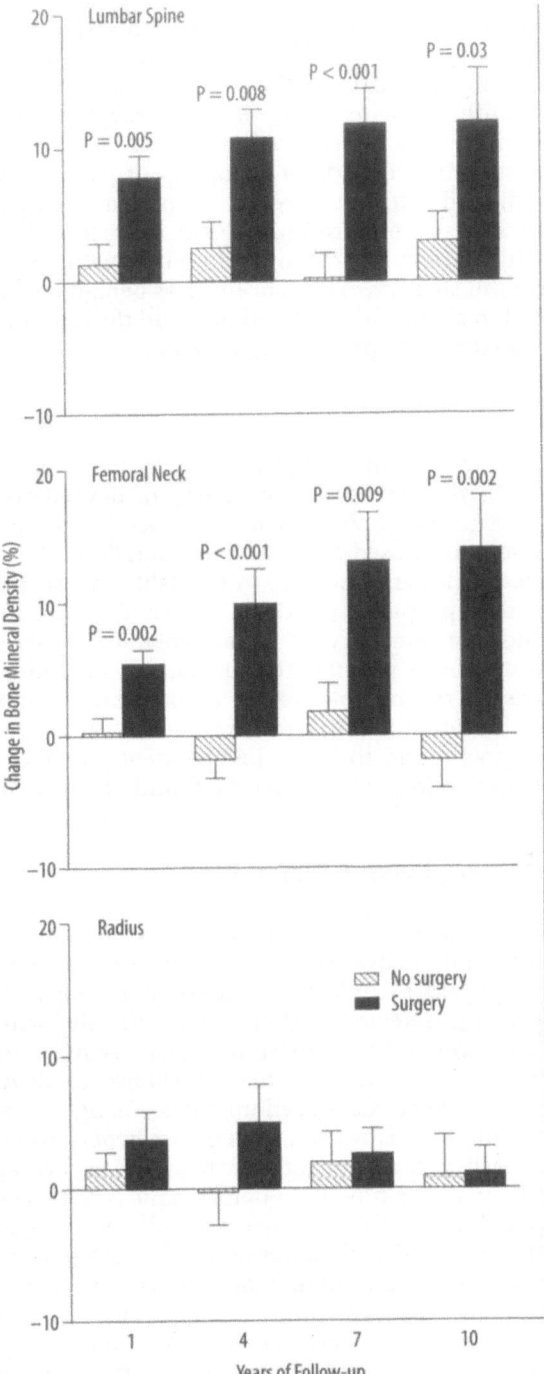

Figure 3.12.   Bone densitometry in primary hyperparathyroidism over ten years of observation with and without parathyroidectomy. Divergences among treated and untreated groups are shown at each site (p < 0.05; reprinted from ref. 48).

stones after parathyroid surgery. The 5–10% of patients who continue to form kidney stones after parathyroidectomy are thought to have a second, alternative cause for their stone disease, which persists despite cure of their hyperparathyroidism. Those patients with nephrolithiasis who do not undergo surgery are at risk for recurrent stone disease, and progressive hyperparathyroidism.

In our recent report on a decade of follow-up of patients with primary hyperparathyroidism with and without surgery, 20 patients with kidney stones were included. No recurrence of nephrolithiasis was reported in the 12 patients who underwent successful parathyroidectomy. The eight patients who refused, or had undergone unsuccessful surgery, experienced a less benign course. Recurrent kidney stones were reported in six of the eight patients, and the remaining two patients had other evidence of worsening hyperparathyroidism.

Mortality

Data supporting an increase in mortality rates in patients with primary hyperparathyroidism come from a retrospective study of limited size from Scandinavia [50]. These data have not been confirmed in a more recent study from the United States [51]. In a population-based study of hyperparathyroidism in Rochester, Minnesota over a 28-year period, no increase in mortality was seen in 435 patients with surgically confirmed hyperparathyroidism, hypercalcaemia with inappropriately elevated parathyroid hormone levels, or unexplained hypercalcaemia for more than one year's duration. It is possible that the change in clinical profile to a largely asymptomatic disease is responsible for the apparent reduction in mortality rates. Consistent with this notion, in the aforementioned survival study, higher maximal serum calcium level was found to be an independent predictor of mortality. Thus, the very mild elevations in serum calcium found today could account for the improved survival rates.

Primary Hyperparathyroidism: A Global View

While there is no question that this disease is most commonly asymptomatic in the United States and many other countries, this presentation is not seen worldwide. One explanation for the disparate clinical presentations is that in many countries the disorder is merely being detected earlier. Thus, with the widespread use of the autoanalyser, the old-fashioned presentation of primary hyperparathyroidism, with its specific symptomatology, does not have a chance to develop, and asymptomatic primary hyperparathyroidism predominates among the many other potential presentations of the disease. However, if primary hyperparathyroidism is simply being detected at an earlier stage, one might expect that the average age of the patient at diagnosis would be younger now than before. This is not clearly the case.

An alternative possibility is that the disease itself has changed. Perhaps the predominance of asymptomatic disease in our population is a reflection of different aetiologies or environmental or nutritional factors. One speculative example might suffice. Residents of many Western countries supplement their diet with vitamin D. Thus, vitamin D stores are more likely to be replete in those countries today than in the past. Reports of primary hyperparathyroidism from nations in which the disease has the phenotype of classical primary hyperparathyroidism include India, Saudi Arabia, Brazil and Vietnam [52–55]. In these countries, vitamin D deficiency is common, and it is possible that the symptomatic patients were afflicted by both primary hyperparathyroidism and vitamin D deficiency. The latter condition could increase the manifestations of the former. Moreover, sufficient vitamin D might protect against some of the skeletal effects of vitamin D deficiency in conjunction with elevated levels of PTH. In France, primary hyperparathyroidism is more likely to present with osteitis fibrosa cystica if levels of vitamin D metabolites are low [56]. This hypothesis thus suggests that the disease itself has changed along with development of a technology that has improved recognition.

Malignancy-Associated Hypercalcaemia

Malignancy-associated hypercalcaemia is relatively common, occurring in approximately 10–20% of patients with cancer [57]. In the vast majority of such patients, the presence of neoplastic disease is clinically evident by the time hypercalcaemia develops. Hypercalcaemia can occur in patients with solid tumours or leukaemia. The most common cancers associated with hypercalcaemia are breast, lung, kidney and multiple myeloma. Other malignancies include squamous cell carcinomas of the head and neck, lymphoma, leukaemia, and genitourinary cancer.

Hypercalcaemia in patients with malignancy is due to increased bone resorption. It may arise from three general mechanisms: tumour secretion of parathyroid hormone-related protein (PTHrp); tumour production of 1,25-dihydroxyvitamin D; and osteolytic metastases with local release of cytokines.

Humoral Hypercalcaemia of Malignancy

Humoral hypercalcaemia of malignancy (HHM), in broad terms, refers to a condition characterised by hypercalcaemia caused by the secretion by a cancer of a circulating calcaemic factor. The most common cause of malignancy-associated hypercalcaemia in patients with few or no bony metastases is secretion of PTHrp, and the term HHM is currently used to describe this specific clinical syndrome [58, 59]. With few notable exceptions, when HHM is identified, survival can be measured in weeks to a few months. Recognition of a syndrome of HHM was first described by Fuller Albright in 1941 and subsequently confirmed with analysis of biochemical indices in hypercalaemic cancer patients which showed suppressed PTH levels but elevated nephrogenous cAMP [60]. Extensive research efforts later identified and characterised the PTHrp peptide and gene [61]. Human PTHrp is encoded by a single-copy gene located on chromosome 12 [62]. It is a polypeptide that is homologous with PTH at the 13 N-terminal residues, the region believed to interact with the shared PTH/PTHrp G protein-coupled receptor [63]. PTHrp is smaller than PTH in length and differs at the C-terminal sequence, which prevents the two molecules from cross-reacting in the two-site antibody assays. Interestingly, the PTHrP gene is expressed not only in cancers but also in the vast majority of normal tissues. Under normal circumstances, PTHrP plays predominantly paracrine and/or autocrine roles in the regulation of smooth muscle tone, local calcium transport, and tissue development [64].

The mechanism by which PTHrp mediates hypercalcaemia is similar to PTH; it promotes bone resorption and reduces renal calcium clearance. However, unlike primary hyperparathyroidism, PTHrp-mediated HHM is associated with low 1,25-dihydroxyvitamin D, metabolic alkalosis, and reduced bone formation [60, 65]. Rarely, ectopic secretion of PTH itself mediates the malignancy-associated hypercalcaemia [66]. However, the association of elevated PTH and cancer is more likely to represent coincident primary hyperparathyroidism and malignancy.

In some cases, the mechanism of malignancy-associated hypercalcaemia cannot be easily identified [67]. Also, there are cancer types that are rarely associated with hypercalcaemia such as adenocarcinomas of the prostate, colon, stomach and non-small-cell lung carcinoma. The recognition of hypercalcaemia in these clinical scenarios should alert the physician to search for an alternate cause of the metabolic disturbance.

The diagnosis of HHM can be confirmed by identifying a high serum PTHrp concentration. Serum PTHrp levels in HHM can help predict the calcium-lowering effect of bisphosphonate therapy and give an indication of patient prognosis. Serum PTHrp concentrations greater than 12 pmol/L are associated with a lesser reduction in hypercalcaemia and with a greater likelihood of recurrent hypercalcaemia within two weeks of treatment [68]. In addition, those patients who respond to bisphosphonate therapy have a significantly better, although still short, survival prognosis compared to those who remain hypercalcaemic (53 versus 19 days in one study) [69].

Other humoral mechanisms of malignancy-associated hypercalcaemia exist, although less commonly. For example, hypercalcaemia can result from increased production of 1,25-dihydroxyvitamin D in patients with lymphoma [70, 71]. The elevation of plasma 1,25-dihydroxyvitamin D found in these cases is in contrast to findings in other types of malignancy-associated hypercalcaemia. PTH and PTHrp levels in these patients are not elevated. Treatment of the lymphoma usually reverses the elevations of 1,25-dihydroxyvitamin D and the hypercalcaemia. No unifying histological type of lymphoma has been identified. This syndrome appears to be due to excessive and unregulated conversion of circulating 25-hydroxyvitamin D to 1,25-dihydroxyvitamin D by the lymphoma cells. Hypercalcaemia ensues through 1,25-dihydroxyvitamin D-mediated intestinal calcium hyperabsorption and bone resorption.

Hypercalcaemia Associated with Localised Bone Destruction

This category comprises approximately 20% of patients with malignancy-associated hypercalcaemia. It appears that local factors elaborated by either the tumour cells or host cells reacting to the tumour cells cause bone resorption by stimulating osteoclast precursor cells to differentiate into mature osteoclasts. Such local factors include cytokines such as tumour necrosis factor and interleukin-1 [72, 73]. Multiple myeloma, breast cancer and non-small-cell lung cancer are the most common malignancies that cause hypercalcaemia by this mechanism.

Multiple myeloma with lytic bone lesions is complicated by hypercalcaemia in 20–40% of patients at some time during the course of the disease. Osteolytic bone lesions may occur as discrete local lesions or as diffuse disease throughout the axial skeleton. The bone destruction that occurs is due to increased osteoclast activity. Malignant cells in the marrow produce cytokines that stimulate adjacent endosteal osteoclasts to resorb bone. The identity of the specific cytokines responsible for stimulating osteoclasts is still unclear; however, myeloma cells in culture have been shown to elaborate lymphotoxin, tumour necrosis factor, IL-1, IL-6, PTHrp, macrophage colony-stimulating factor, macrophage inflammatory protein, vascular cell adhesion molecule-1, and hepatocyte growth factor (HGF) [73–79]. In addition, the nephropathic effects of the myeloma paraprotein, which can impair renal calcium excretion, can accelerate the development of hypercalcaemia in these patients.

The hypercalcaemia of breast cancer is associated with extensive bone metastases in the majority of patients. Recent findings suggest that PTHrp may play an important role in breast cancer that metastasises to bone. In these cases, PTHrp appears to be expressed in excessive amounts, but produced locally in the bone microenvironment [80]. It appears that a cycle is set up where bone destruction caused by tumour invasion leads to the production of bone-derived transforming growth factor B (TGF_B), which in turn stimulates tumour cells to produce PTHrp, which leads to more production by bone of TGF_B. It should be noted that among patients with breast cancer and skeletal metastases, the administration of oestrogen or an antioestrogen may lead to hypercalcaemia by causing the release of bone resorbing factors from the malignant cells [81]. Treatment of patients with malignancy-associated hypercalcaemia should include measures to decrease tumour burden, enhance renal calcium excretion and reduce osteoclastic bone resorption.

Hypercalcaemia Due to Granulomatous Disorders

Granulomatous diseases which have been associated with hypercalcaemia include sarcoidosis, tuberculosis, berylliosis, coccidioidomycosis, histoplasmosis, candidiasis, Langerhans-cell histiocytosis, silicone-induced granulomas, Wegener's granulomatosis and both Hodgkin's and non-Hodgkin's lymphoma [82–91]. Hypercalcaemia is the most commonly seen in sarcoidosis; studies indicate that hypercalcaemia is detected in 10% of patients and that up to 50% will become hypercalciuric at some time during the course of their disease [92]. This condition appears to be the benign counterpart to hypercalcaemia associated with malignant lymphoma. The mechanism responsible for abnormal calcium metabolism is the endogenous overproduction of an active metabolite of vitamin D causing increased intestinal calcium absorption and bone resorption [93]. Normally, renal conversion of 25-hydroxyvitamin D to 1,25-dihydroxyvitamin D by 1 alpha-hydroxylase is tightly regulated. However, in granulomatous diseases, the synthesis of 1,25-dihydroxyvitamin D in activated macrophages is not inhibited by its end product and appears substrate dependent. Therefore, even small amounts of dietary vitamin D and exposure to sunlight or ultraviolet radiation lead to an increase in 25-hydroxyvitamin D and subsequent conversion to 1,25-dihydroxyvitamin D. The first demonstration that an extrarenal source of 1 alpha-hydroxylation of 1,25-dihydroxyvitamin D existed occurred in an anephric patient with hypercalcaemia, elevated 1,25-dihydroxyvitamin D levels and active sarcoidosis [94]. It was subsequently shown that sarcoidotic lymphoid tissue and pulmonary alveolar macrophages from patients with sarcoidosis could metabolise 1,25-dihydroxyvitamin D from 25-hydroxyvitamin D [95]. PTHrp, the major cause of HHM, may also cause hypercalcaemia in some patients with sarcoidosis. PTHrp has been identified in sarcoid granulomatous tissue and some patients with sarcoid-associated hypercalcaemia had elevated serum PTHrp concentrations [96].

The macrophage hydroxylation reaction in granulomatous disease is sensitive to inhibition by glucocorticoids, chloroquine, and ketoconazole, and refractory to inhibition by 1,25-dihydroxyvitamin D [97–99]. This is in contrast to the renal enzymatic reaction that is relatively insensitive to inhibition by glucocorticoids and is downregulated by 1,25-dihydroxyvitamin D. These differences likely result from the differential regulation of the 1 alpha-hydroxylation gene in the two different cell types [100].

Treatment of hypercalcaemia is aimed at reducing 1,25-dihydroxyvitamin D and intestinal calcium absorption. Dietary calcium and vitamin D should be limited, as should ultraviolet light exposure. Glucocorticoids are highly effective in the treatment of granulomatous disease-associated hypercalcaemia and often hypercalcaemia will begin to correct within several days of treatment. Treatment with choloquine (and its analogues) and ketoconazole should be limited to patients in whom steroid therapy is unsuccessful or contraindicated. In patients with active sarcoid, hypercalciuria frequently precedes the development of hypercalcaemia. Therefore, all patients should be screened for hypercalciuria.

Familial Hypocalciuric Hypercalcaemia

Familial hypocalciuric hypercalcaemia (FHH) is an autosomal dominant disorder of high penetrance characterised by normal or elevated PTH levels in the face of elevated serum calcium and low urinary calcium excretion [101]. The phenotype is evident at birth and affected individuals usually present in childhood with the incidental discovery of hypercalcaemia, hypocalciuria, and frequently hypermagnesaemia. Patients with FHH have mild hypercalcaemia and inappropriately elevated serum PTH levels. In most kindreds with FHH, the pathophysiology is due to an inherited loss-of-function mutation in the calcium-sensing G protein-coupled receptor [102]. In the parathyroid gland, this results in relative insensitivity to serum calcium, and an elevation in the calcium set point to suppress hormone release. In the kidney, this defect leads to an increase in tubular calcium and magnesium reabsorption. Affected patients are FHH heterozygotes and have mild hypercalcaemia. Patients who are homozygous for the FHH gene have a more severe disease, presenting with neonatal hyperparathyroidism and usually marked hypercalcaemia (serum calcium concentration above 15 mg/dl) [103].

It is important to distinguish this syndrome from primary hyperparathyroidism because FHH is benign, and not cured by parathyroidectomy. A fractional excretion of calcium of less than 1% helps to support the diagnosis of FHH [101]. This indicates that greater than 99% of filtered calcium has been reabsorbed in the presence of hypercalcaemia. Patients also have normal urinary excretion of cyclic AMP versus increased excretion in other states of PTH excess. In FHH, nephrolithiasis and bone disease are absent, though there may be an association with pancreatitis and gallstones in some cases [104]. The parathyroid glands in FHH cases are slightly or moderately hyperplastic. Because the elevated serum calcium is refractory to subtotal parathyroidectomy and the course of the disease is benign, patients should not undergo surgery. Family screening is recommended.

Drug-Induced Hypercalcaemia

Milk-Alkali Syndrome

The classic features of this syndrome include hypercalcaemia, metabolic alkalosis and renal failure. It was first described in the 1920s, usually in patients treated for peptic ulcer disease with large quantities of milk together with systemically absorbable alkali, particularly calcium carbonate [105]. With the introduction of new effective medicines for the treatment of peptic ulcer disease, the incidence of this syndrome had been in decline. Recently, however, an increase in case reports of milk-alkali syndrome has been associated with the popularisation of calcium for the prevention and treatment of osteoporosis [106]. Most cases report an ingestion of calcium in the range of 3,000 to 15,000 mg daily. The syndrome is often completely reversible with discontinuation of the supplemental calcium ingestion, hydration and diuresis. Diagnosis is associated with suppressed PTH and 1,25-dihydroxyvitamin D measurements. Renal function is likely to recover in cases of brief duration, but may not in more chronic cases.

Vitamin D

Hypercalcaemia may occur in patients receiving vitamin D or its analogues. The recommended daily allowance of vitamin D is 400 to 800 IU. The amount of vitamin D required to produce hypercalcaemia is in excess of 50,000 IU per week. Toxicity can be severe and long lasting because of the fat solubility of this vitamin. When parent vitamin D is the offending agent, high levels of 25-hydroxyvitamin D are generated in the liver and bind and activate the vitamin D receptor, though with much weaker affinity than 1,25-dihydroxyvitamin D. Suppressed PTH and elevated phosphate levels, typical of vitamin D-induced hypercalcaemia, inhibit 1 alpha-hydroxylation of 25-hydroxyvitamin D in the kidney which accounts for the suppressed levels of 1,25-dihydroxyvitamin D in these patients. 1,25-dihydroxyvitamin D intoxication has a similar profile except for the elevation in the aetiologic 1,25-dihydroxyvitamin D and normal levels of serum 25-hydroxyvitamin D. The hypercalcaemia in vitamin D intoxication results from both increased intestinal absorption and bone resorption and responds to withdrawal of the vitamin D agent, volume expansion and calciuresis. Occasionally, glucocorticoid treatment is required as it inhibits vitamin D action on the intestine and skeleton. The contribution of bone resorption to vitamin D-induced hypercalcaemia can be great and treatment with anti-resorptives has been used successfully in this setting [107].

Vitamin A

Hypercalcaemia has been described in patients with vitamin A intoxication [108]. The recommended daily allowance of vitamin A is 5,000 IU. In adults, vitamin A-associated hypercalcaemia requires in excess of 50,000 IU daily for weeks to months. Recently, the widespread use of vitamin A analogues such as cis-retinoic acid and all-trans-retinoic acid for the treatment of dermatological disorders as well as for the treatment of neuroblastoma, haematologic and other malignancies has been associated with hypercalcaemia. The typical biochemical profile includes suppressed PTH, PTHrp and 1,25-dihydroxyvitamin D. Hypercalcaemia is believed to result from direct stimulation of osteoclast-mediated bone resorption, though the exact mechanism is unclear. Treatment involves termination of the offending retinoid and general antihypercalcaemic measures. Glucocorticoids are also effective in this condition.

Thiazide Diuretics

Thiazide diuretics act on the distal renal tubule to promote the reabsorption of calcium, a finding that has been useful in the treatment of patients with hypercalciuria. Decreased serum PTH levels should correct any hypercalcaemia induced by increased renal calcium reabsorption; nevertheless, a subset of patients on thiazides remain hypercalcaemic for weeks or longer. When the thiazide is discontinued, a subset of patients remains hypercalcaemic and the ultimate diagnosis is hyperparathyroidism [109]. Thus, it appears that thiazide use can unmask or precipitate hyperparathyroidism in patients with a propensity for the disease. The diagnosis of thiazide-associated hypercalcaemia is confirmed by the reversal of hypercalcaemia after the cessation of thiazide therapy. The diagnosis of primary hyperparathyroidism cannot be made with certainty while a patient remains on thiazide diuretics, and the medication should be stopped for several months prior to retesting.

Lithium

Lithium carbonate has been reported to cause hypercalcaemia in approximately 5% of patients receiving the drug [110]. There appears to be an alteration in PTH dynamics in patients on chronic therapy. PTH secretion is increased, most likely due to an increase in the set point at which calcium suppresses PTH release [111]. The hypercalcaemia usually, but not always, resolves when lithium therapy is discontinued. Patients with previously unrecognised mild hyperparathyroidism can have an unmasking of their disease. Evidence suggests an additional anticalciuric effect of lithium on the kidney. Such "hypocalciuric hypercalcaemia" resembles the profile of patients affected by familial hypocalciuric hypercalcaemia (FHH). Also, like FHH, patients with lithium-associated calcium elevations show no evidence of abnormal bone metabolism [112].

Other Drugs

Oestrogens and antioestrogens have been reported to cause hypercalcaemia in patients with breast cancer and extensive skeletal metastases [113]. This reaction may be indicative of tumour regression when the offending agent can be continued. Aminophylline has also been associated with hypercalcaemia [114], as has growth hormone therapy [115]. Typically, the hypercalcaemia induced by aminophylline and growth hormone is mild. Hypercalcaemia has also been reported in association with 8-Cloro-cAMP, an antineoplastic agent, and Foscarnet, an antiviral agent used in the treatment of AIDS [116, 117].

Nonparathyroid Endocrine Causes of Hypercalcaemia

Mild hypercalcaemia can occur in approximately 15–20% of patients with hyperthyroidism [118]. The hypercalcaemia is the result of increased bone resorption stimulated by thyroid hormone and is reversible upon correction of the hyperthyroid state. In this setting, phosphate may be slightly elevated and PTH and 1,25-dihydroxyvitamin D levels are suppressed. Hypercalcaemia occurs rarely in patients with adrenal insufficiency [119]. The cause is likely multifactorial and includes volume contraction, increased proximal tubular calcium reabsorption and withdrawal of glucocorticoid antagonism of vitamin D actions on intestine and bone. The hypercalcaemia corrects with glucocorticoid replacement. Pheochromocytoma is another endocrine tumour associated with hypercalcaemia. This may be due to concurrent multiple endocrine neoplasia type 2 disease, but PTHrp has also been implicated [120]. Hypercalcaemia is common in patients with vasoactive intestinal polypeptide-secreting tumours of the pancreatic islet cells [121]. Because of the association of this tumour with MEN1, it is possible that this may reflect unrecognised hyperparathyroidism, though in some instances the calcium disturbance is corrected after tumour resection.

Immobilisation

Immobilisation leads to increased bone resorption and hypercalcaemia in individuals with high rates of bone turnover such as growing children and patients with Paget's disease, hyperparathyroidism, skeletal fractures or osteolytic malignancies and thyrotoxicosis [122]. Hypercalciuria in this setting is more common than hypercalcaemia and is associated with both upper and lower tract nephrolithiasis. Serum levels of PTH and 1,25-dihydroxyvitamin D are both suppressed. The cellular mechanism responsible for the increased bone resorption and uncoupling of bone remodelling is incompletely understood. Antiresorptive drugs may ameliorate the hypercalcaemia [123] as can limited weight-bearing exercise, but passive range-of-motion exercises are ineffective.

References

1. Leite MOR, Correa PHS, Jorgetti V, Batalha JFR, Pereira RC, Mechica JB et al. Dynamic bone histomorphometry in hyperparathyroidism. J Bone Min Res 1990;5(Suppl 2):664a.
2. Meah FA, Tan TT, Taha A, Khalid BA. Primary hyperparathyroidism – a surgical review of 12 cases. Med J Malaya 1991;46:144–9.
3. Dotzenrath C, Goretzki PE, Roher HD. West Germany: still an underdeveloped country in the diagnosis and early treatment of primary hyperparathyroidism? World J Surg 1990;14:660–1.
4. Heath III H , Hodgson SF, Kennedy M. Primary hyperparathyroidism: incidence, morbidity and potential economic impact in a community. N Engl J Med 1980;302:189–93.
5. Mundy GR, Cove DH, Fisken R. Primary hyperparathyroidism: changes in the pattern of clinical presentation. Lancet 1980;1:1317–20.
6. Stenstrom G, Heedman P. Clinical findings in patients with hypercalcaemia: a final investigation based on biochemical screening. Acta Med Scand 1974;195:473–7.
7. Aitken RE, Bartley PC, Bryant SJ, Lloyd HM. The effect of multiphasic biochemical screening on the diagnosis of primary hyperparathyroidism. Aust NZ J Med 1975;5:224–6.
8. Trigonis C, Hamberger B, Farnebo LO, Abarca J, Granberg PO. Primary hyperparathyroidism. Changing trends over fifty years. Acta Chir Scand 1983;149:675–9.
9. Heath III H. Clinical spectrum of primary hyperparathyroidism: evolution with changes in medical practice and technology. J Bone Mineral Res 1991;6(Suppl 2):S63–S70.
10. Melton III LJ. Epidemiology of primary hyperparathyroidism. J Bone Mineral Res 1991;6(Suppl 2):S25–S30.
11. Palmer M, Jakobsson S, Akerstrom G, Ljunghall S. Prevalence of hypercalcaemia in a health survey: a 14-year follow up study of serum calcium values. Eur J Clin Invest 1988;18:39–46.
12. Boonstra CE, Jackson CE. Serum calcium survey for hyperparathyroidism: results in 50,000 clinic patients. Am J Clin Pathol 1971;55:523–6.
14. Beard CM, Heath III H, O'Fallon WM, Anderson JA, Earle JD, Melton III LJ. Therapeutic radiation and hyperparathyroidism: a case-control study in Rochester, Minnesota. Arch Intern Med 1989;149:1887–90.
15. Cohen J, Gierlowski TC, Schneider AB. A prospective study of hyperparathyroidism in individuals exposed to radiation in childhood. JAMA 1990;264:581–4.
16. Shane E, Bilezikian JP. Parathyroid carcinoma. In: Williams CJ, Krikorian JC, Green MR, Raghavan D, editors. Textbook of uncommon cancer. New York: Wiley, 1988;763–71.
17. Bilezikian JP, Silverberg SJ, Shane E, Parisien M, Dempster DW. Characterization and evaluation of asymptomatic primary hyperparathyroidism. J Bone Mineral Res 1991;6(Suppl I):585–9.
18. Silverberg SJ, Shane E, DeLaCruz L. Skeletal disease in primary hyperparathyroidism. J Bone Mineral Res 1989;4:283–91.

19. Parisien MV, Silverberg SJ, Shane E, de la Cruz L, Lindsay R, Bilezikian JP et al. The histomorphometry of bone in primary hyperparathyroidism: preservation of cancellous bone structure. J Clin Endocrinol Metab 1990;70:930–38.

20. van Doorn L, Lips P, Netelenbos JC, Hackengt WHL. Bone histomorphometry and serum intact PTH (I-84) in hyperparathyroid patients. Calcif Tissue Int 1989;44S:N36.

21. Parisien M, Mellish RWE, Silverberg SJ. Maintenance of cancellous bone connectivity in primary hyperparathyroidism: trabecular strut analysis. J Bone Mineral Res 1992;7:913–20.

22. Christiansen P, Steiniche T, Vesterby A, Mosekilde L, Hessov I, Melsen F. Primary hyperparathyroidism: iliac crest trabecular bone volume, structure, remodelling, and balance evaluated by histomorphometric methods. Bone 1992;13:144–9.

23. Parisien M, Cosman F, Mellish RWE, Schnitzer M, Nieves J, Silverberg SJ et al. Bone structure in postmenopausal hyperparathyroid, osteoporotic and normal women. J Bone Min Res 1995;10:1393–9.

24. Dempster DW, Parisien M, Silverberg SJ, Liang X-G, Schnitzer M, Shen V et al. On the mechanism of cancellous bone preservation in postmenopausal women with mild primary hyperparathyroidism. J Clin Endocrinol Metab 1999; 84:1562–6. 17 proven cases from one clinic. JAMA 1934;102: 1276–87.

25. Wilson RJ, Rao DS, Ellis B, Kleerekoper M, Parfitt AM. Mild asymptomatic primary hyperparathyroidism is not a risk factor for vertebral fractures. Ann Intern Med 1988;109:959–62.

26. Larsson K, Ljunghall S, Krusemo UB, Naessen T, Lindh E, Persson I. The risk of hip fractures in patients with primary hyperparathyroidism: a population-based cohort study with a follow-up of 19 years. J Int Med 1993;234:585–93.

27. Kenny AM, MacGillivray DC, Pilbeam CC, Crombie HD, Raisz LG. Fracture incidence in postmenopausal women with primary hyperparathyroidism. Surgery 1995;118:109–14.

28. Khosla S, Melton LJ, Wermers RA, Crowson CS, O'Fallon WM, Riggs BL. Primary hyperparathyroidism and the risk of fracture: a population-based study. J Bone Mineral Res 1999;14:1700–7.

29. Silverberg SJ, Shane E, Jacobs TP. Nephrolithiasis and bone involvement in primary hyperparathyroidism. Am J Med 1990;89:327–34.

30. Patten BM, Bilezikian JP, Mallette LE, Prince A, Engel WK, Aurbach GD. The neuromuscular disease of hyperparathyroidism. Ann Intern Med 1974;80:182–94.

31. Turken SA, Cafferty M, Silverberg SJ. Neuromuscular involvement in mild, asymptomatic primary hyperparathyroidism. Am J Med 1989;87:553–7.

32. Joborn C, Rastad J, Stalberg E, Akerstrom G, Ljunghall S. Muscle function in patients with primary hyperparathyroidism. Muscle Nerve 1989;12:87–94.

33. Linos DA, vanHeerdan JA, Abboud CF, Edis AJ. Primary hyperparathyroidism and peptic ulcer disease. Arch Surg 1978;113:384–6.

34. Bess MA, Edis AJ, vanHeerden JA. Hyperparathyroidism and pancreatitis. Chance or a causal association? JAMA 1980;243:246–7.

35. Bilezikian JP, Aurbach GD, Connor TB. Pseudogout following parathyroidectomy. Lancet 1973; 1:445–7.

36. Geelhoed GW, Kelly TR. Pseudogout as a clue and complication in primary hyperparathyroidism. Surgery 1989;106:1036–41.

37. Rapado A. Arterial hypertension and primary hyperparathyroidism. Am J Nephrol 1986;6(Suppl 1):49–50.

38. Diamond TW, Botha JR, Wing J, Meyers AM, Kalk WJ. Parathyroid hypertension. A reversible disorder. Arch Intern Med 1986;146:1709–12.

39. Sancho JJ, Rouco J, Riera-Vidal R, Sitges-Serra A. Long-term effects of parathyroidectomy for primary hyperparathyroidism on arterial hypertension. World J Surg 1992;16:732–5.

40. Lind L, Jacobsson S, Palmer M, Lithell H, Wengle B, Ljunghall S. Cardiovascular risk factors in primary hyperparathyroidism: a 15-year follow-up of operated and unoperated cases. J Intern Med 1991;230:29–35.

41. Stefenelli T, Mayr H, Bergler-Klein J, Globits S, Woloszczuk W, Niederle B. Primary hyperparathyroidism: incidence of cardiac abnormalities and partial reversibility after successful parathyroidectomy. Am J Medicine 1993;95:197–202.

42. Cogan MG, Covery CM, Arieff S, Wisniewski A, Clark OH. Central nervous system manifestations of primary hyperparathyroidism. Am J Med 1978;65:963–70.

43. Joborn C, Hetta J, Johansson H. Psychiatric morbidity in primary hyperparathyroidism. World J Surg 1988;12:476–81.

44. Joborn C, Hetta J, Frisk P, Palmer M, Akerstrom G, Ljunghall S. Primary hyperparathyroidism in patients with organic brain syndrome. Acta Med Scand 1986;219:91–8.

45. Ljunghall S, Jakobsson S, Joborn C, Palmer M, Rastad J, Akerstrom G. Longitudinal studies of mild primary hyperparathyroidism. J Bone Mineral Res 1991;6(Suppl 2):Sl I I–S116.

46. Brown GG, Preisman RC, Kleerekoper MD. Neurobehavioral symptoms in mild primary hyperparathyroidism: related to hypercalcaemia but not improved by parathyroidectomy. He my Ford Med J 1987;35:211–15.
47. Solomon BL, Schaaf M, Smallridge RC. Psychologic symptoms before and after parathyroid surgery. Am J Med 1994;96:101–6.
48. Silverberg SJ, Shane E, Jacobs TP, Siris E, Bilezikian JP. The natural history of treated and untreated asymptomatic primary hyperparathyroidism: a ten-year prospective study. N Engl J Med 1999; 341:1249–55.
49. Silverberg SJ, Locker FG, Bilezikian JP. Vertebral osteopenia: a new indication for surgery in primary hyperparathyroidism. J Clin Endocrinol Metab 1996;81:4007–12.
50. Palmer M, Adami H-O, Bergstrom R, Akerstroom G, Ljunghall S. Mortality after surgery for primary hyperparathyroidism: a follow-up of 441 patients operated on from 1956 to 1979. Surgery 1987;102:1–7.
51. Wermers RA, Khosla S, Atkinson EJ, Grant CS, Hodgson SF, O'Fallon M et al. Survival after the diagnosis of hyperparathyroidism. Am J Med 1998;104:115–22.
52. Harinarayan DV, Gupta N, Kochupillai N. Vitamin D status in primary hyperparathyroidism in India. Clin Endocrinol 1995;43:351–8.
53. Luong KVQ, Nguyen LTH. Co-existing hyperthyroidism and hyperparathyroidism with vitamin D-deficient osteomalacia in a Vietnamese immigrant. Endocr Practice 1996;2:250–4.
54. Meng XW, Xing XP, Liu SQ, Zhan ZW. The diagnosis of primary hyperparathyroidism – analysis of 134 cases. Acta Acad Med Sinicae 1994;16:13.
55. Bilezikian JP, Meng X, Shi Y, Silverberg SJ. Primary hyperparathyroidism in women: New York and Beijing (a tale of two cities). Int J Fertility 2000 (in press).
56. Patron P, Gardin J-P, Paillard M. Renal mass and reserve of vitamin D. Determinants in primary hyperparathyroidism. Kidney Int 1987;31:1174–80.
57. Rosol TJ, Capen CC. Mechanisms of cancer-induced hypercalcemia. Lab Invest 1992;67:680.
58. Ratcliffe WA, Hutchesson CJ, Bundred NJ, Ratcliffe JG. Role of assays of parathyroid hormone-related protein in investigation of hypercalcaemia. Lancet 1992;339:164.
59. Ikeda K, Ohno H, Hane M, Yokoi H, Okada M, Honma T et al. Development of a sensitive two-site immunoradiometric assay for parathyroid hormone-related peptide: Evidence for elevated levels in plasma from patients with adult T-cell leukaemia/lymphoma and B-cell lymphoma. J Clin Endocrinol Metab 1994;79:5.
60. Stewart AF, Horst R, Deftos LJ, Cadman EC, Lang R, Broadus AE. Biochemical evaluation of patients with cancer-associated hypercalcemia: evidence for humoral and nonhumoral groups. N Engl J Med 1980;303:1377–83.
61. Mangin M, Ikeda K, Dreyer BE, Broadus AE. Isolation and characterization of the human parathyroid hormone-like peptide gene. Proc Natl Acad Sci USA 1989;86(7):2408–12.
62. Dunbar ME, Wysolmerski JJ, Broadus AE. Parathyroid hormone-related protein: from hypercalcaemia of malignancy to developmental regulatory molecule. Am J Med Sci 1996;312:287–94.
63. Juppner H, Abou-Samra AB, Freeman M, Kong XF, Schipani E, Richards J et al. A G protein-linked receptor for parathyroid hormone and parathyroid hormone-related peptide. Science 1991; 254:1024–6.
64. Philbrick WM, Wysolmerski JJ, Galbraith S, Holt E, Orloff JJ, Yang KH et al. Defining roles of parathyroid hormone-related protein in normal physiology. Physiol Rev 1996;76:127–73.
65. Ikeda K, Ogata E. Humoral hypercalcaemia of malignancy: some enigmas on the clinical features. J Cell Biochem 1995;57:384–91.
66. Nussbaum SR, Gaz RD, Arnold A. Hypercalcemia and ectopic secretion of parathyroid hormone by an ovarian carcinoma with rearrangement of the gene for PTH. N Engl J Med 1990;323: 1324–8.
67. Stewart AF, Romero R, Schwartz PE, Kohorn EI, Broadus AE. Hypercalcemia associated with gynecologic malignancies: biochemical characterization. Cancer 1982;49(11):2389–94.
68. Wimalawansa SJ. Significance of plasma PTH-rp in patients with hypercalcemia of malignancy treated with bisphosphonate. Cancer 1994;73:2223.
69. Ling PJ, A'Hern RP, Hardy JR. Analysis of survival following treatment of tumour-induced hypercalcaemia with intravenous pamidronate (APD). Br J Cancer 1995;72:206–9.
70. Rosenthal NR, Insogna KL, Godshall JW, Smalldone L, Waldron JA, Stewart AF. Elevations in circulating 1,25(OH)2D in three patients with lymphoma-associated hypercalcemia. J Clin Endocrinol Metab 1985;60:29–33.
71. Seymour JF, Gagel RF, Hagemeister FB, Dimopoulos MA, Cabanillas F. Calcitriol production in hypercalcemia and normocalcemia patients with non-Hodgkin lymphoma. Ann Intern Med 1994; 121:633–40.

72. Thompson BM, Mundy GR, Chambers TJ. Tumor necrosis factors alpha and beta induce osteoblastic cells to stimulate osteoclastic bone resorption. J Immunol 1987;138:775.
73. Cozzolino F, Torcia M, Aldinucci D, Rubartelli A, Miliani A, Shaw AR et al. Production of interleukin-1 by bone marrow myeloma cells. Blood 1989;74:387–90.
74. Garrett IR, Durie BG, Nedwin GE, Gillespie A, Bringman T, Sabatini M et al. Production of lymphotoxin, a bone-resorbing cytokine, by cultured human myeloma cells. N Engl J Med 1987;317:526–32.
75. Bataille R, Jourdan M, Zhang Xue-Guang, Klein B. Serum levels of interleukin-6, a potent myeloma cell growth factor, as a reflection of disease severity in plasma cell dyscrasias. J Clin Invest 1989; 84:2008–11.
76. Firkin F, Seymour JF, Watson AM, Grill V, Martin TJ. Parathyroid hormone-related protein in hypercalcaemia associated with haematological malignancy. Br J Haematol 1996;94:486–92.
77. Choi SJ, Cruz JC, Craig F, Chung H, Devlin RD, Roodman GD et al. Macrophage inflammatory protein 1-alpha is a potential osteoclast stimulatory factor in multiple myeloma. Blood 2000;96:671.
78. Michigami T, Shimizu N, Williams PJ, Niewolna M, Dallas SL, Mundy GR et al. Cell–cell contact between marrow stromal cells and myeloma cells via VCAM-1 and alpha(4)beta(1)-integrin enhances production of osteoclast-stimulating activity. Blood 2000;96:1953.
79. Hjertner O, Torgersen ML, Seidel C, Hjorth-Hansen H, Waage A, Borset M et al. Hepatocyte growth factor (HGF) induces interleukin-11 secretion from osteoblasts: A possible role for HGF in myeloma-associated osteolytic bone disease. Blood 1999;94:3883.
80. Guise TA, Yin JJ, Taylor SD, Kumagai Y, Dallas M, Boyce BF et al. Evidence for a causal role of parathyroid hormone-related protein in the pathogenesis of human breast cancer-mediated osteolysis. J Clin Invest 1996;98:1544–9.
81. Valentin-Opran A, Eilon G, Saez S, Mundy GR. Estrogens and anti-estrogens stimulate release of bone resorbing activity in cultured human breast cancer cells. J Clin Invest 1985;75:726.
82. Winnacker JL, Becker KL, Katz S. Endocrine aspects of sarcoidosis. N Engl J Med 1968;278:483–92.
83. Abbasi AA, Chemplavil JK, Farah S, Muller BF, Arnstein AR. Hypercalcemia in active pulmonary tuberculosis. Ann Intern Med 1979;90:324–8.
84. Stoeckle JD, Hardy HL, Weber AL. Chronic beryllium disease. Long-term follow-up of sixty cases and selective review of the literature. Am J Med 1969;46:545–61.
85. Lee JC, Catanzaro A, Parthemore JG, Roach B, Deftos LJ. Hypercalcemia in disseminated coccidioidomycosis. N Engl J Med 1977;297:431–3.
86. Murray JJ, Heim CR. Hypercalcemia in disseminated histoplasmosis. Aggravation by vitamin D. Am J Med 1985;78(5):881–4.
87. Kantarjian HM, Saad MF, Estey EH, Sellin RV, Samaan NA. Hypercalcemia in disseminated candidiasis. Am J Med 1983;74(4):721–4.
88. Jurney TH. Hypercalcemia in a patient with eosinophilic granuloma. Am J Med 1984;76(3):527–8.
89. Kozeny GA, Barbato AL, Bansal VK, Vertuno LL, Hano JE. Hypercalcemia associated with silicone-induced granulomas. N Engl J Med 1984 Oct 25;311(17):1103–5.
90. Edelson GW, Talpos GB, Bone III HG. Hypercalcemia associated with Wegener's granulomatosis and hyperparathyroidism: etiology and management. Am J Nephrol 1993;13(4):275–7.
91. Seymour JF, Gagel RF. Calcitriol: the major humoral mediator of hypercalcemia in Hodgkin's disease and non-Hodgkin's lymphomas. Blood 1993;82(5):1383–94.
92. Studdy PR, Bird R, Neville E, James DG. Biochemical findings in sarcoidosis. J Clin Pathol 1980; 33:528–33.
93. Insogna KL, Dreyer BE, Mitnick M, Ellison AF, Broadus AE. Enhanced production rate of 1,25-dihydroxyvitamin D in sarcoidosis. J Clin Endocrinol Metab 1988;66(1):72–5.
94. Barbour GL, Coburn JW, Slatopolsky E, Norman AW, Horst RL. Hypercalcemia in an anephric patient with sarcoidosis: evidence for extrarenal generation of 1,25-dihydroxyvitamin D. N Engl J Med 1981;305(8):440–3.
95. Adams JS, Sharma OP, Gacad MA, Singer FR. Metabolism of 25-hydroxyvitamin D3 by cultured pulmonary alveolar macrophages in sarcoidosis. J Clin Invest 1983;72(5):1856–60.
96. Zeimer HJ, Greenaway TM, Slavin J, Hards DK, Zhou H, Doery JC et al. Parathyroid hormone-related protein in sarcoidosis. Am J Pathol 1998;152:17.
97. Adams JS, Gacad MA, Characterization of 1 hydroxylation of vitamin D3 sterols by cultured alveolar macrophages from patients with sarcoidosis. J Exp Med 1985;161:755–65.
98. Adams JS, Diz MM, Sharma OP. Effective reduction in the serum 1,25-dihydroxyvitamin D and calcium concentration in sarcoidosis-associated hypercalcemia with short-course chloroquine therapy. Ann Intern Med 1989;111:437–8.
99. Adams JS, Sharma OP, Diz MM, Endres DB. Ketoconazole decreases the serum 1,25-dihydroxyvitamin D and calcium concentration in sarcoidosis-associated hypercalcemia. J Clin Endocrinol Metab 1990;70:1090–5.

100. St-Arnaud R, Messerlian S, Moir JM, Omdahl JL, Glorieux FH. The 25-hydroxyvitamin D 1-alpha-hydroxylase gene maps to the pseudovitamin D-deficiency rickets (PDDR) diseases locus. J Bone Min Res 1997;12:1552–9.

101. Marx SJ, Attie MF, Levine MA, Speigel AM, Downs RW Jr, Lasker RD. The hypocalciuric or benign variant of familial hypecalcemia: clinical and biochemical features in fifteen kindreds. Medicine (Baltimore) 1981;60:397–412.

102. Brown EM. Physiology and pathophysiology of the extracellular calcium-sensing receptor. Am J Med 1999;106:238–53.

103. Cole DE, Janicic N, Salisbury SR, Hendy GN. Neonatal severe hyperparathyroidism, secondary hyperparathyroidism, and familial hypocalciuric hypercalcemia: Multiple different phenotypes associated with an inactivation ALU insertion mutation of the calcium-sensing receptor gene. Am J Med Genet 1997;71:202.

104. Davies M, Klimiuk PS, Adams PH, Lumb GA, Large DM, Anderson DC. Familial hypocalciuric hypercalcaemia and acute pancreatitis. Br Med J 1981;282:1023.

105. Orwoll ES. The milk-alkali syndrome: current concepts. Ann Intern Med 1982;97(2):242–8.

106. Muldowney WP, Mazbar SA. Rolaids-yogurt syndrome: a 1990s version of milk-alkali syndrome. Am J Kidney Dis 1996;27(2):270–2.

107. Selby PL, Davies M, Marks SJ, Mawer EB. Vitamin D intoxication causes hypercalcaemia by increased bone resorption which responds to pamidronate. Clin Endocrinol (Oxf) 1995;43:531–6.

108. Ragavan VV, Smith JE, Bilezikian JP. Vitamin A toxicity and hypercalcemia. Am J Med Sci 1982; 283:161–4.

109. Christensson T, Hellstrom K, Wengle B. Hypercalcemia and primary hyperparathyroidism. Prevalence in patients receiving thiazides as detected in a health screen. Arch Intern Med 1977;137:1138–42.

110. Haden ST, Stoll AL, McCormick S, Scott J, Fuleihan G el-H. Alterations in parathyroid dynamics in lithium-treated subjects. J Clin Endocrinol Metab 1997;82(9):2844–8.

111. Spiegel AM, Rudorfer MV, Marx SJ, Linnoila M. The effect of short-term lithium administration on suppressibility of parathyroid hormone secretion by calcium in vivo. J Clin Endocrinol Metab 1984;59(2):354–7.

112. Nordenstrom J, Elvius M, Bagedahl-Strindlund M, Zhao B, Torring O. Biochemical hyperparathyroidism and bone mineral status in patients treated long-term with lithium. Metabolism 1994;43(12):1563–7.

113. Legha SS, Powell K, Buzdar AU, Blumenschein GR. Tamoxifen-induced hypercalcemia in breast cancer. Cancer 1981;47(12):2803–6.

114. McPherson ML, Prince SR, Atamer ER, Maxwell DB, Ross-Clunis H, Estep HL. Theophylline-induced hypercalcemia. Ann Intern Med 1986;105(1):52–4.

115. Knox JB, Demling RH, Wilmore DW, Sarraf P, Santos AA. Hypercalcemia associated with the use of human growth hormone in an adult surgical intensive care unit. Arch Surg 1995;130(4):442–5.

116. Saunders MP, Salisbury AJ, O'Byrne KJ, Long L, Whitehouse RM, Talbot DC et al. A novel cyclic adenosine monophosphate analog induces hypercalcemia via production of 1,25-dihydroxyvitamin D in patients with solid tumors. J Clin Endocrinol Metab 1997;82(12):4044–8.

117. Gayet S, Ville E, Durand JM, Mars ME, Morange S, Kaplanski G et al. Foscarnet-induced hypercalcaemia in AIDS. AIDS 1997;11(8):1068–70.

118. Burman KD, Monchik JM, Earll JM, Wartofsky L. Ionized and total serum calcium and parathyroid hormone in hyperthyroidism. Ann Intern Med 1976;84(6):668–71.

119. Muls E, Bouillon R, Boelaert J, Lamberigts G, Van Imschoot S, Daneels R et al. Etiology of hypercalcemia in a patient with Addison's disease. Calcif Tissue Int 1982;34(6):523–6.

120. Stewart AF, Hoecker JL, Mallette LE, Segre GV, Amatruda Jr TT, Vignery A. Hypercalcemia in pheochromocytoma. Evidence for a novel mechanism. Ann Intern Med 1985;102(6):776–9.

121. Venkatesh S, Vassilopoulou-Sellin R, Samaan NA. Somatostatin analog: use in the treatment of vipoma with hypercalcemia. Am J Med 1989;87(3):356–7.

122. Stewart AF, Adler M, Byers CM, Segre GV, Broadus AE. Calcium homeostasis in immobilization: an example of resorptive hypercalciuria. N Engl J Med 1982;306(19):1136–40.

123. Meythaler JM, Tuel SM, Cross LL. Successful treatment of immobilization hypercalcemia using calcitonin and etidronate. Arch Phys Med Rehabil 1993;74(3):316–19.

Hypocalcaemia

P. Glendenning, A. G. Need and B. E. C. Nordin

Introduction

Calcium is critically important in a diverse number of physiological processes, necessitating close regulation of its intracellular and extracellular concentration. This demands the interplay of two principal hormones, parathyroid hormone (PTH) and 1,25 dihydroxyvitamin D_3 (calcitriol) and three body organs, the gut, the kidney and bone, which act in a coordinated fashion to maintain the circulating plasma-ionised calcium concentration within preset defined limits. This concept helps to explain the importance of the regulation of intracellular calcium homeostasis since it is this process that determines the regulation of the extracellular fluid (ECF) calcium concentration. Recent experimental work has further elucidated some of the molecular mechanisms of the action of PTH and calcitriol. These data have increased our knowledge of how the two principal calcitropic hormones may act to maintain ECF calcium concentration.

Plasma Calcium Homeostasis

Calcium circulates in the plasma as three fractions: protein-bound (35%), complexed but soluble (15%), and ionised (50%). The extravascular concentration is therefore about 65% of the plasma concentration and it is this fraction that passes through the renal glomeruli. The ionised calcium is the biologically important fraction and throughout this chapter, references to plasma calcium are generally to be taken as references to ionised calcium. Total calcium may, of course, change in the face of a constant ionised calcium level if there are changes in albumin or other ligands (e.g., at the menopause [1]) but unless otherwise stated, the term hypocalcaemia denotes a low ionised calcium concentration.

The plasma calcium concentration at any one time is the result of the relationship between the rate of input (net intestinal calcium absorption and net bone resorption), the glomerular filtration rate (GFR), and the renal tubular reabsorption of calcium, which is most elegantly expressed as a notional TmCa [2]. The mean total fasting plasma calcium concentration is about 2.4 mmol/L with a c.v. of some 5.0%, but even within this narrow range there is constancy within individuals such that those at the upper and lower ends of the range tend to remain in these relative positions over years, particularly in respect of the ionised fraction [3]. These intra- and inter-individual differences may have biological significance as

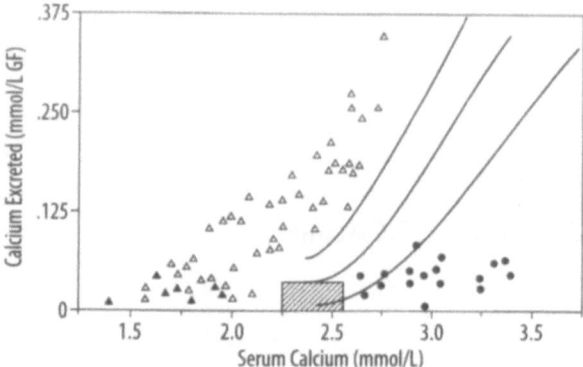

Figure 3.13. The relationship between serum calcium and calcium excretion before and during calcium infusions in normal subjects (solid lines) and hypoparathyroid individuals (triangles). The solid circles represent cases of primary hyperparathyroidism in the basal state. The shaded area represents the reference interval in the basal state.

exemplified by the positive correlation between filtered and excreted calcium in normal women [4].

Despite the intra- and inter-individual constancy of the fasting plasma-ionised calcium in the steady state, the equilibrium value can be quite easily perturbed by any change in the flow of calcium into or out of the immediate calcium pool, which amounts to about 25 mmol. Thus, there is a suggestion that an added calcium load as small as 0.5 mmol may raise the ionised calcium sufficiently to lower plasma PTH [5]. This may seem improbable but it is compatible with the 1.1% c.v. of intraindividual ionised calcium over a three-year period [3]. There is certainly no doubt that larger calcium loads produce easily measurable changes in plasma calcium and PTH concentrations [5]. However, hour-to-hour and day-to-day regulation is largely dependent on renal function. The kidney filters about 275 mmol of calcium daily through the glomeruli (170 litres at 1.6 mmol/L), reabsorbs some 98% of this and excretes 2%, or about 5 mmol. Thus, the plasma calcium can be modified by very small changes in renal tubular handling of calcium. The main regulator of the system is parathyroid hormone secreted in response to changes in ionised calcium as sensed by the calcium-sensing receptor (CaSR). This is clearly seen in cases of hyper- and hypoparathyroidism where the relationships between plasma and urine calcium are very different. For any given calcium flow relative to GFR, the plasma calcium is higher in hyperparathyroid cases and lower in hypoparathyroid cases than in normal controls (Figure 3.13).

Cell Biology of Calcium Homeostasis

Tubular Reabsorption of Calcium

Although filtered calcium is reabsorbed throughout the nephron, the mechanisms and relative proportional transport differ between kidney segments. Most calcium is reabsorbed in the proximal tubule, largely passively and by a paracellular route. At the end of the nephron, in the distal tubule, less calcium is reabsorbed but calcium

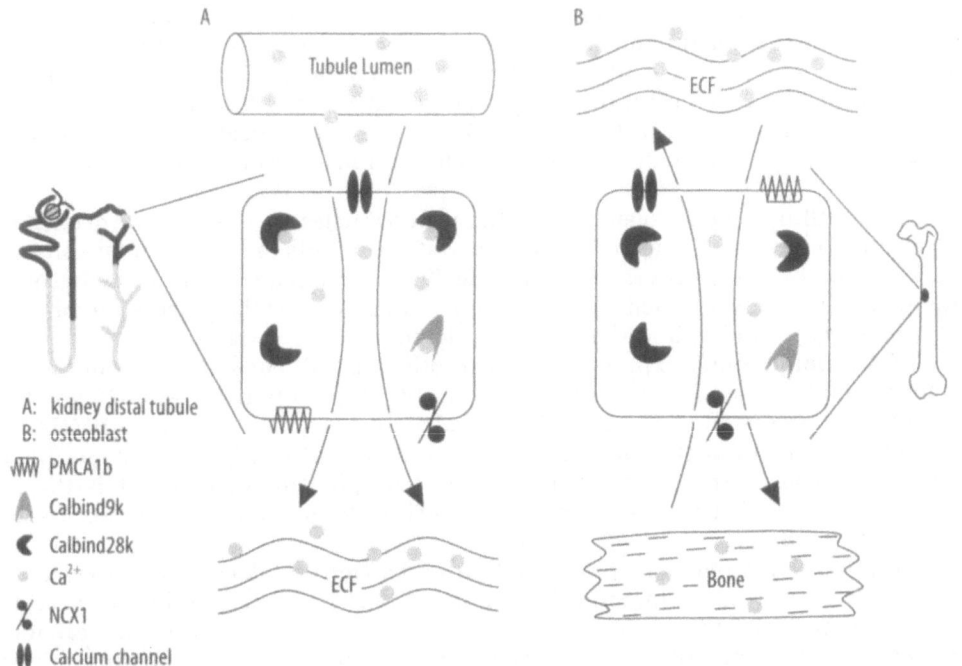

Figure 3.14.   Comparison of calcium transport in the kidney distal tubule and osteoblast. Calcium transport is unidirectional in the kidney distal tubule but bidirectional in the osteoblast. PMCA1b is expressed on the apical plasma membrane of osteoblasts, but on the basolateral membrane in the kidney distal tubule. The remaining features of transmembrane calcium transport in both cells are distinctly similar.

movement is largely mediated by active transport mechanisms which are under hormonal regulation [6].

Molecular Mechanisms Subserving Renal Calcium Reabsorption

Transepithelial calcium movement in the distal kidney tubule occurs by a three-step process involving calcium entry, transport from the apical to the basolateral membrane and extrusion of calcium across the basolateral membrane (Figure 3.14).

Apical calcium entry is largely passive. Calcium enters through at least two types of calcium channel. One such channel, the epithelial calcium channel (ECaC) has been recently cloned and is found in rabbits and humans [7]. ECaC belongs to a larger family of calcium-sensitive calcium channels and includes a calcium transport protein found in rat intestine and distal kidney tubule (CaT2) [8]. The second apical calcium channel family of proteins are voltage-dependent calcium channels (VDCC) and are sensitive to calcium channel blockers. In contrast to the VDCC family of proteins, the ECaC family of proteins are insensitive to calcium channel blockers.

Transcellular calcium movement is mediated by two cytosolic, calcium-binding proteins (calbindins), which may have more than one physiological role. Either of the calbindin proteins may act to expedite the transcellular movement of calcium from the apical site of entry to the basolateral site of calcium efflux. Calbindin D28k is localised in the cytosol and nucleus and is not associated with the plasma membrane [9]. Consequently, the role of calbindin D28k may be to facilitate intracellular Ca^{2+} diffusion from the apical membrane site of entry to the basolateral membrane and extrusion [10]. In contrast, calbindin D9k expression is localised to the basolateral membrane [10]. This difference in localisation may account for differences in function.

Calcium efflux at the basolateral membrane necessitates an active transport mechanism since the extracellular calcium concentration exceeds the intracellular calcium concentration by many orders of magnitude. Two mechanisms exist on the basolateral membrane of the epithelial cell to subserve this process: the plasma membrane calcium ATPase (PMCA) and the sodium–calcium exchanger (NCX).

PMCA is ubiquitously expressed in eukaryotic cells and has a much higher affinity for calcium than NCX. Thus, PMCA is probably more important than NCX in the physiological regulation of intracellular calcium concentration [11]. PMCA is encoded by four different genes and each gene transcript can be further modified, resulting in a vast array of different isoforms. Despite this diversity, it is the 1b isoform of PMCA that is expressed and is sensitive to hormonal regulation in the distal kidney tubule.

NCX is not an active transporter but an exchanger of calcium and relies on a sodium gradient, which is maintained by a membrane-expressed sodium ATPase, to move calcium across the plasma membrane. NCX has a lower affinity for calcium than PMCA but because of its higher calcium efflux capacity, plays an important role in organs where the rapid, large-volume movement of calcium is needed. Thus, NCX is particularly abundantly expressed in contractile and neuronal cells as well as in the distal tubule of the kidney [12].

Only three other genes and their products appear to have a role in calcium transport in the kidney. The vitamin D receptor is a nuclear receptor that binds calcitriol and probably mediates some of its hormonal effects. Early studies localised the nuclear vitamin D receptor (VDR) exclusively to the distal kidney tubule but more recent, sensitive methods have indicated that VDR is also expressed in the proximal tubule, although at a lower level than in the distal tubule [13]. Studies of the parathyroid hormone (PTH) receptor, using in situ hybridisation, have indicated that like VDR, the PTH receptor is also expressed in the proximal tubule and glomerulus as well as in the distal tubule [14]. The fundamental importance of CaSR in calcium homeostasis is highlighted by recent evidence showing that inactivating or activating mutations of the receptor result in hypercalcaemia or hypocalcaemia, respectively, in humans. CaSR is expressed on the apical membrane of the proximal tubule. Its polarity is reversed in the distal tubule where the CaSR protein is expressed on the basolateral membrane [14].

Hormonal Regulation of Calcium Transport in the Kidney

The two basolateral membrane-expressed calcium-transport mechanisms appear to be differentially regulated by calcitriol and PTH. In vivo experiments in the rabbit have demonstrated a specific effect of calcitriol on PMCA activity in the distal kidney tubule. Calcitriol did not exert any effect on PMCA activity in the proximal kidney tubule or on NCX activity in either nephron segment [15]. Recent evidence in vitro indicates that calcitriol mediates its effect through the 1b isoform of PMCA [16] and that this effect is probably post-transcriptionally mediated [17]. In contrast, PTH mediates its effect through the NCX protein and this effect is dependent on protein kinase A and C [18].

Both of the calbindin proteins have unique and independent effects on membrane calcium transport. Calbindin D9k increases PMCA activity in the basolateral membrane of renal cells. This effect is comparable to that seen within the intestinal absorptive cell. Calbindin D28k stimulates Ca^{2+} uptake across apical membrane vesicles. Consequently, both Ca^{2+} binding proteins may facilitate transcellular Ca^{2+} transport in the kidney distal tubule via an increase in Ca^{2+} diffusion and through direct effects on basolateral transporters. The expression of calbindin D28k is regulated by $1,25(OH)_2D_3$ in the kidney and these effects appear to be predominantly post-transcriptionally mediated [19].

One report has indicated that PTH increases calbindin D28k expression in rats independently of $1,25(OH)_2D_3$ [20]. Calbindin D9k mRNA expression is increased following $1,25(OH)_2D_3$ treatment of vitamin D-deficient mice but in contrast to calbindin D28k, the peak induction is delayed. Renal and intestinal calbindin D9k increase in late pregnancy in rats [21] at which time there is also an increase in VDR and plasma $1,25(OH)_2D_3$. While studies of the rat calbindin D9k promoter demonstrated the presence of a VDRE within the 5′ region of the gene, this site is not responsible for vitamin D_3 induction in vivo [22]. Recent studies of the promoter of the human calbindin D9k gene have demonstrated that in humans, the promoter is not responsive to $1,25(OH)_2D_3$ [23]. Thus, effects of $1,25(OH)_2D_3$ on calbindin D9k are probably also mostly mediated post-transcriptionally.

Intestinal Absorption of Calcium

Absorption of intestinal calcium may occur by three independent processes [24]. The first process employs direct transcellular transport of calcium across the enterocyte and involves apical calcium channels, intracellular calcium binding proteins and basolateral membrane active transport proteins similar to those expressed in the kidney distal tubule. It is this process which is energy-dependent and largely hormone-responsive. The second process is passive and proceeds by a paracellular route which is dependent on the intraluminal intestinal concentration gradient of calcium. The last process occurs by an endocytotic–exoxcytotic process termed transcaltachia.

Molecular Mechanisms Subserving Calcium Absorption

The active intestinal transport of calcium involves two distinct transport mechanisms, which are also expressed in the kidney distal tubule. Similar to the kidney distal tubule, both transport mechanisms PMCA1b and NCX are expressed on the basolateral membrane of the intestinal enterocyte. Of the two transport mechanisms described, PMCA1b is probably far more important than NCX in the intestine. This view is supported by functional studies and the lack of expression of demonstrable NCX protein on immunohistochemical studies of proximal intestinal epithelium [25]. Calbindin D9k but not calbindin D28k is expressed within the proximal mammalian small intestine. The recently described apical calcium channel in rats, CaT1, and in rabbits and humans (ECaC) are expressed in the proximal small intestine. Co-localisation studies have demonstrated that ECaC, PMCA and calbindin D9k are expressed in the duodenal villi and not within the crypts [25]. These findings indicate the importance of each of these transport proteins in the intestinal absorptive process [24].

Hormonal Regulation of Calcium Transport in the Intestine

As in the kidney distal tubule, calcitriol up-regulates activity and expression of PMCA1b and has similar effects on calbindin D9k [26]. More recent studies have not fully defined the molecular mechanism of action of this hormone but support the concept that the calcium transporting genes are not regulated transcriptionally but are probably regulated post-transcriptionally [23]. The lack of demonstrable NCX in the intestine has limited studies of its regulation in this tissue [25].

Although CaT1 and ECaC are expressed within the proximal small intestine, their regulation by hormonal agents has not been studied.

Calcium Movement in Bone

One of the most intriguing challenges is to understand how calcium enters and leaves the bone fluid compartment. Current evidence supports the concept that the basic muliticellular unit (BMU), which is composed of osteoclasts and osteoblasts, is primarily responsible for the regulation of calcium transport into and out of the bone compartment, but a concept that calcium may be transported into and out of the bone fluid compartment independently of the BMU has recently re-emerged [27]. In this concept, bone fluid is believed to be separated from the ECF by a bone-lining cell membrane. The bone-lining cells may be under hormonal regulation and thus calcium transport across this cell layer may occur independently of the BMU.

The skeleton is the major body storage site for calcium and contains approximately 99% of the total body calcium largely stored in a form of hydroxyapatite. An additional smaller component is present in the ECF and an even smaller fraction is present within the bone fluid compartment.

Molecular Mechanisms Subserving Calcium Movement in Bone

A description of the proteins involved in calcium flux in bone will immediately lead to the recognition that the same molecular mechanisms exist in this organ as in the kidney distal tubule and proximal small intestine (Figure 3.14). In bone, calcium transport proteins appear to be exclusively expressed within the osteoblast lineage of cells. The lack of demonstrable expression within the osteoclast cell lineage in bone lends support to the theory that the bone-lining cells, which are derived from the osteoblast lineage, may indeed play an important role in calcium transport within the bone compartment.

Hormonal Regulation of Calcium Transport in Bone

Calcitriol has been demonstrated to reduce NCX activity in osteoblast cells and this effect could be explained by a reduction in NCX protein [28]. In contrast, avian tibial calbindin D28k expression is up-regulated by calcitriol in vitamin D-deficient chickens in vivo and in human bone marrow stromal cells in vitro [29]. Calbindin D28k may act to expedite calcium transport across the osteoblast cell but may also act as an intracellular buffer and consequently prevent osteoblast apoptosis [30]. Finally, PMCA1b is up-regulated by calcitriol in vitro in bone cells [31] and this effect is probably mediated by post-transcriptional mechanisms [17].

An Overview of Molecular Mechanisms of Calcium Homeostasis

From the perspective of an overview of the entire area, several broad concepts emerge. First, there is the similarity between the existence and regulation of calcium transport mechanisms in the kidney, intestine and bone (Figure 3.14). The 1b isoform of PMCA is responsive to calcitriol although the exact molecular mechanism may differ between tissues. By changing the polarity of cellular distribution from the apical membrane in the osteoblast to the basolateral membrane in the kidney distal tubule and intestinal epithelium, opposing effects can occur.

Similar concepts apply to an understanding of the molecular mechanism of the action of PTH. However, although PTH acts to stimulate NCX activity in the kidney distal tubule, the opposing effect is demonstrable in bone. How the same protein can have opposing effects in different tissues is not understood.

Thus, we are left to contend with the emerging concept that each calcium transporter has a restricted repertoire of response to hormonal agents. Understanding that NCX is regulated by PTH and that PMCA1b is under the regulatory control of calcitriol is fundamental. The newly characterised apical calcium channels and their hormonal regulation should provide further interesting data in this field of work. Whether these newly described calcium channels are expressed in bone, and if so, which bone cells, as well as further work on calcium transport in this tissue, should prove to be a fruitful area of future evolving research.

Having explored some of the fundamental issues regarding the biology and physiology of calcium transport, now let us consider some of the clinical signs and symptoms that define the nature and aetiology of hypocalcaemia in humans.

The Clinical Picture and Classification of Hypocalcaemia

Many organs are dependent on normal plasma-ionised calcium levels for optimal function so it is not surprising that hypocalcaemia may have widespread deleterious effects, the most obvious of which are neuromuscular (Table 3.6).

Affected individuals may suffer overt tetany, they may find it hard to let go of objects in their grasp or may complain of paraesthesiae of the lips or fingers. There may be carpo-pedal spasm, aphonia, laryngeal stridor and even seizures. In severe chronic cases, dementia, choreoathetosis and Parkinsonism have been described, associated with calcification in the basal ganglia. Occasionally, delirium, stupor and coma may supervene. Most of the clinical effects are biochemical in origin and potentially reversible. A metabolic myopathy with high serum creatine kinase levels in children with hypoparathyroidism has been reported to abate rapidly after correction of plasma calcium [32].

Cardiovascular effects include arrhythmias associated with a prolonged ST interval, cardiomyopathy with congestive heart failure and hypotension [33].

Ocular effects include posterior lenticular cataract formation; the cataracts may decrease in size with successful medical treatment.

If hypocalcaemia occurs during development of the teeth, there may be delayed tooth eruption and formation of short or blunted tooth roots. These changes may be used to fix the time of onset of the hypocalcaemia [34].

Table 3.6. Clinical manifestations of hypocalcaemia

Neuromuscular	
Paraesthesiae	Seizures
Tetany	Laryngeal spasm
Chvostek's and Trousseau's signs	Confusion
Carpo-pedal spasm	Choreoathetosis
Cardiovascular	
Prolonged ST interval	Hypotension
Arrhythmia	Bradycardia
Other	
Cataract	
Delayed tooth eruption	

Table 3.7. Causes of hypocalcaemia

Spurious hypocalcaemia
Hypoalbuminaemia

Chronic hypocalcaemia
Hereditary hypoparathyroidism
Surgical hypoparathyroidism
Pseudohypoparathyroidism
Renal failure
Vitamin D deficiency
Vitamin D dependency
Autosomal dominant hypocalcaemia

Acute hypocalcaemia
Magnesium deficiency
Parathyroid surgery
Acute pancreatitis
Neonatal hypocalcaemia
Hyperphosphataemia
Acute alkalosis
Iatrogenic chelation
Cancer
Critical illness
Crush syndrome

Hypocalcaemia can be classified into a variety of chronic and acute forms, each of which has its own particular aetiology (Table 3.7). These will now be considered in turn.

Chronic Hypocalcaemia

The presence of chronic hypocalcaemia implies deficiency of or resistance to either PTH or calcitriol and may display any degree of severity.

Hypoparathyroidism

Low plasma-ionised calcium levels in the presence of decreased plasma PTH indicate hypoparathyroidism. The serum phosphate is high and the urine calcium may be paradoxically normal. The plasma calcitriol level is low and calcium absorption is probably low as well. The commonest cause used to be damage to the parathyroid glands during thyroid surgery but today hereditary hypoparathyroidism probably accounts for most of the cases. Hereditary hypoparathyroidism may be seen in isolation but is usually associated with abnormalities of the immune system such as defective development of the thymus (Di George syndrome) or failure of other endocrine organs such as the adrenal and ovary (autoimmune polyglandular syndrome type I (APS I)). In APS I, there is often deficient T cell function causing mucocutaneous candidiasis. Hereditary hypoparathyroidism often appears in the first decade of life but may present later.

Surgically induced hypoparathyroidism usually presents in the first few days after surgery but occasionally presents with a gradual onset many years later, presumably as a result of scarring and damage to the blood supply of the parathyroid glands.

Preventive measures against hypoparathyroidism include selection of experienced surgeons for thyroid and parathyroid surgery. In difficult cases there may be value in auto-transplantation of parathyroid tissue into peripheral sites at the time of surgery. Allo-transplantation has also been performed; patients with the Di George syndrome are most suitable because of their inherent immune deficiency.

Treatment with calcitriol or 1α-hydroxyvitamin D is the conventional treatment for most patients with hypocalcaemia. These, the most potent vitamin D metabolites, increase intestinal absorption of calcium, ensure calcium can be mobilised from bone and may enhance renal tubular reabsorption of calcium. They correct the low levels of serum calcitriol found in patients with hypoparathyroidism. Calcium supplements are usually required to provide adequate calcium for absorption. It is wise to aim at a plasma-ionised calcium level at the bottom of the normal range since hypercalciuria is the price to be paid for achieving normal plasma calcium levels; the 24-hour excretion of calcium should therefore be measured. It is sometimes necessary to accept a degree of hypocalcaemia in order to avoid excessive hypercalciuria and the risk of nephrocalcinosis and renal failure.

The doses of calcitriol and calcium required vary considerably, depending on tolerability, severity of disease, calcium absorption efficiency, renal function, dietary and other factors. Treatment therefore needs to be tailored to the individual and careful monitoring is required, at least initially. Dosage changes should be made either to calcium intake or to the vitamin D metabolite but not both at once. It is useful to remember that calcium is much cheaper than the vitamin D metabolites and can substitute for them to a degree. Treatment with the vitamin D metabolites allows rapid titration of dosage because of their short plasma and biological half lives, which are measured in days.

Vitamin D itself has been used to treat hypoparathyroidism but very large doses are required and over-dosage may lead to prolonged hypercalcaemia. Steady-state levels take months to achieve because of the volume of distribution of vitamin D in fat and muscle, but measurement of serum 25-hydroxyvitamin D levels may help in titrating the vitamin D dose. Hyperphosphataemia may occasionally be severe enough to require the use of aluminium or magnesium hydroxide orally as a phosphate binder to be taken with meals. This may cause a small rise in plasma calcium levels.

Thiazide diuretics help control the hypercalciuria, which threatens the successful treatment of hypocalcaemia. Thiazides increase renal tubular reabsorption of calcium and are most effective when the patient takes a low-sodium diet.

Nephrocalcinosis can be detected early by ultrasound examination and, if detected, may require adjustment to the therapy or more frequent monitoring.

Pseudohypoparathyroidism

In this condition, first described by Albright et al. [35], the signs and symptoms of hypoparathyroidism are present but serum PTH is normal or elevated. Two major types are now recognised. In pseudohypoparathyroidism (PHP) type I there is both insufficient production of 1,25-dihydroxyvitamin D and insensitivity of bone to the resorbing action of parathyroid hormone. The cause appears to be one of a number of point mutations in the gene coding for the G protein, $G_s\alpha$ (chromosome 20q13.3). Affected individuals are short and stocky, with round face and short neck and shortening of the fourth and fifth metacarpals. Intellectual development is delayed but does not respond to correction of the hypocalcaemia. Many develop basal ganglia calcification. Bone density is normal but occasionally there are changes consistent with hyperparathyroidism such as osteitis fibrosa cystica. Some subjects exhibit only the somatic phenotype of PHP type I and have normal plasma calcium levels (pseudo-pseudohypoparathyroidism).

In PHP type I, the mutation in the G protein-coupled receptor in the plasma membrane impairs its ability to respond to the PTH signal, resulting in reduced production of the second messenger, cyclic AMP. PTH is thus rendered ineffective. There is usually a decreased amount of G protein present but some cases have normal levels of a presumably defective protein. Other endocrine defects have been described. The prevalence of diabetes and hypothyroidism is increased. There may be impaired production of pituitary hormones such as thyroid stimulating hormone and prolactin [36]. Peripheral resistance to gonadotropins and vasopressin has been described.

In PHP type II, there are the biochemical changes of PHP but without the somatic changes. In PHP type Ib, there is a similar phenotype to PHP type I but there is increased urinary cyclic AMP excretion in response to PTH. The phosphaturic response to PTH is impaired. An uncoupling mutation in the carboxyl terminus of $G_s\alpha$ has been described [37].

Other variants of PHP have been described. Some subjects increase their cyclic AMP but not phosphate excretion in response to PTH.

The hypocalcaemia in patients with pseudohypoparathyroidism is usually not severe and there is some renal tubular responsiveness to PTH, enabling a higher plasma calcium to be reached without hypercalciuria. The treatment of the hypocalcaemia is similar to that for primary hypoparathyroidism but the doses required are less and phosphate binders are more strongly indicated.

Renal Failure

Several factors contribute to the hypocalcaemia of renal failure. Renal production of 1,25-dihydroxyvitamin D is diminished very early in the course of renal failure, primarily because of loss of renal tissue. Later, hyperphosphataemia lowers plasma ionised calcium levels both directly and through further suppression of 1,25-dihydroxyvitamin D production. Lack of circulating 1,25-dihydroxyvitamin D leads to a decrease in intestinal calcium absorption and a decrease in bone resorption. Lastly, the toxic effects of uraemia may also impair bone resorption.

The low levels of serum calcitriol in renal failure can easily be corrected with oral calcitriol. This, together with oral calcium, will help control the secondary hyperparathyroidism that develops in such patients. The calcium supplements also help bind phosphate in the gut and minimise hyperphosphataemia. This may help the plasma calcium level to rise. Magnesium and aluminium hydroxide have a similar effect.

Vitamin D Deficiency

Vitamin D insufficiency is thought to be present at plasma levels below 40 nmol/L. Severe vitamin D deficiency, with serum 25-hydroxyvitamin D levels below 20 nmol/L commonly causes mild hypocalcaemia leading to secondary hyperparathyroidism with a low plasma phosphate and raised serum alkaline phosphatase. The commonest cause is lack of adequate sunlight exposure but it may result from excessive gastrointestinal losses of $25(OH)D_3$ or certain types of anti-convulsant therapy.

Treatment is directed at the primary cause, e.g., coeliac disease, and then at replenishing vitamin D stores. This may require giving 10,000 IU per day of vitamin D orally for two weeks or more. There is a growing view that vitamin D requirements have been underestimated in the past and that the elderly may require 1,000 IU daily to maintain serum levels of 25(OH)D over 40 nmol/L, which is an accepted indicator of the adequacy of vitamin D stores. In young children and in adolescents, vitamin D deficiency may lead to clinical rickets with "flaring" of the metaphysis, the rickety rosary from enlarged costal cartilages, muscular weakness, bow legs, and so on.

Vitamin D Dependency

Affected patients present with the signs and symptoms of vitamin D-deficient rickets but have none of the recognised causes of vitamin D deficiency and normal plasma levels of 25-hydroxyvitamin D. Type I vitamin D-dependent rickets (VDDR) is characterised by a low serum 1,25-dihydroxyvitamin D due to poor conversion from its precursor, 25-hydroxvitamin D. In French Canadians it is an autosomal recessive disorder with a mutation in chromosome 12q14. Type II VDDR includes a number of defects affecting vitamin D responses at the receptor and post-receptor levels. Point mutations in the vitamin D receptor gene have been described.

Replacement doses of calcitriol are all that is required for treatment of vitamin D-dependent rickets type I. In vitamin D-dependent rickets type II, resistance to calcitriol makes therapy difficult and large doses of oral calcium may be required if large doses of calcitriol are ineffective.

Autosomal Dominant Hypocalcaemia

This condition is caused by an activating mutation of the calcium sensing receptor. There may be normal or increased urine calcium and the serum PTH is normal or low. It can be severe enough to cause seizures. Treatment is difficult since calcium and calcitriol can cause marked hypercalciuria, nephrocalcinosis and renal dysfunction.

Acute Hypocalcaemia

Acute hypocalcaemia is often encountered in hospital practice and is due to a variety of causes, of which the most common is magnesium deficiency. When caused by low plasma levels of its binding protein, albumin, it is of no clinical consequence since it is the ionised or free plasma fraction of calcium which exerts its physiological effects.

Magnesium Deficiency

Magnesium is a necessary constituent of many metallo-enzymes, and low plasma levels have many effects. Both the production and action of PTH are impaired, leading to a state of functional hypoparathyroidism. Cyclic AMP requires magnesium for normal function, offering an explanation for resistance to PTH in hypomagnesaemia.

Dietary magnesium deficiency is generally caused by malnutrition and/or alcoholism. Alcohol, cisplatinum, amphotericin and aminoglycosides all promote renal magnesium loss. There are also inherited conditions, including an autosomal recessive form segregating with chromosome 9q12–22, an autosomal dominant form mapping to chromosome 11q23, a recessive form caused by a mutation of the claudin 16 gene (3q27), and Gitelman's syndrome caused by mutations in the distal tubular NaCl co-transporter (16q13). An additional feature in Gitelman's syndrome is renal potassium wasting causing hypokalaemia.

Intestinal magnesium loss may be caused by vomiting, diarrhoea, nasogastric suction or fistula. Intestinal malabsorption due to coeliac disease, Crohn's disease, radiation or intestinal bypass may also cause magnesium malabsorption.

Other, rarer causes of hypomagnesaemia include parathyroidectomy, post-renal tubular necrosis diuresis and parenteral nutrition without magnesium supplementation.

Hypomagnesaemia may cause weakness, nausea, trembling, muscle fasciculation, vertigo and convulsions. Some of these symptoms are a result of the accompanying hypocalcaemia; on the other hand hypocalcaemic tetany is made worse by accompanying hypomagnesaemia.

Decreased PTH production during hypomagnesaemia may cause low levels of serum PTH while decreased tissue sensitivity to PTH may cause elevated levels. Plasma magnesium levels below 0.5 mmol/L will inhibit PTH secretion, with low serum PTH in the face of hypocalcaemia, whereas a plasma magnesium between 0.5 and 0.7 mmol/L can be associated with high levels of PTH [38]. On correction of hypomagnesaemia, serum PTH may rise transiently above the normal range as the brake on PTH production is removed. Magnesium depletion may also aggravate hypocalcaemia by decreasing bone resorption directly.

Treatment includes management of the cause of the hypomagnesaemia, restoring the deficit of magnesium and replacement of ongoing losses. The deficit may be as much as 0.5 mmol per kg body weight and may require intravenous replacement in severe cases. The aim should be to replace half the deficit within 24 hours. Oral magnesium is given as the aspartate, orotate or alginate salt to avoid the diarrhoea induced by magnesium sulphate.

Parathyroid and Thyroid Surgery

Removal of a parathyroid adenoma regularly induces a transient state of hypoparathyroidism as the remaining normal parathyroid glands take time to resume the functions which have been depressed during the preceding state of hyperparathyroidism. The nadir of plasma-ionised calcium occurs 48 hours after surgery and can be severe enough to cause symptoms. Its severity can be predicted from the level of bone turnover before surgery, e.g., from the serum alkaline phosphatase activity. Thyroid surgery for hyperthyroidism can cause similar changes but there may also be damage to or removal of parathyroid tissue causing either transient or permanent hypoparathyroidism. Occasionally, hypoparathyroidism develops years after thyroid surgery, presumably as a result of scarring and vascular deprivation of remaining parathyroid tissue. The effects can be subtle. Defects in parathyroid "reserve" have been reported after thyroid surgery [39].

Symptoms can usually be relieved with oral calcium supplements but calcitriol may be required in severe cases. If preoperative bone resorption is very high, calcitriol may be given the day before surgery to minimise postoperative hypocalcaemia.

Acute Pancreatitis

Plasma albumin levels are low in acute pancreatitis and total plasma calcium is therefore low. However, there is often a low ionised calcium level, which results from a number of observed abnormalities. First, there is deposition of calcium "soaps" in the peritoneum following the development of fat necrosis, and second, there is failure of PTH production, with inappropriately normal serum levels in the face of low plasma-ionised calcium levels. Whether there is increased degradation of PTH by pancreatic digestive enzymes in the circulation is not clear. Hypomagnesaemia may also play a role, particularly in alcoholic patients. There may also be intracellular accumulation of calcium in pancreas, liver and muscle [40].

Management may require intravenous calcium and magnesium in the acute phase. Long-term serum 25(OH)D should be measured and oral vitamin D given to maintain normal levels.

Neonatal Hypocalcaemia

This is frequently observed in premature infants with its severity being related to the degree of prematurity. It may cause symptoms such as "jitteriness" or even convulsions.

Neonatal hypocalcaemia occurs in the first few days of life and is related to complicated delivery, birth asphyxia, prematurity and maternal diabetes. The risk in infants of diabetic mothers seems to be higher if the mother has a low serum 25(OH)D level [41].

Late neonatal hypocalcaemia (occurring after the first few days) is often associated with feeding with undiluted cow's milk, which has three times the phosphate content of human milk. Hypocalcaemia persisting beyond the first few weeks of life suggests suppressed parathyroid activity, which may be caused by maternal hyperparathyroidism. It may also indicate the presence of agenesis of the parathyroid glands. Congenital absence of the thymus and parathyroids in association with cardiovascular abnormalities is linked and known as the Di George syndrome.

Severe cases of neonatal hypocalcaemia may require intravenous calcium infusion. Hypocalcaemia occurring after the first few days of life may require reduction in the amount of phosphate being fed to the infant. Congenital hypoparathyroidism is treated with calcitriol and calcium as described above.

Hyperphosphataemia

Hyperphosphataemia arises in such situations as renal failure, the tumour lysis syndrome, after phosphate enemas and from intravenous infusion. There may be precipitation of calcium phosphate salts in soft tissues. In addition, hyperphosphataemia reduces renal production of $1,25(OH)_2D$ which impairs intestinal calcium absorption and may diminish bone resorption. Treatment includes removal of the primary cause, if possible, and the administration of oral phosphate binders such as calcium carbonate or aluminium hydroxide if required.

Acute Alkalosis

The binding of plasma calcium to albumin is dependent on pH, with more binding at higher pH. Thus an acute rise in pH – seen, for example, in acute hyperventilation – will lead to increased calcium binding and a fall in ionised calcium, which may cause or aggravate the neuromuscular irritability which is a feature of alkalosis and which leads to muscle spasms. In severe cases, seizures may occur. Treatment requires management of the alkalosis.

Iatrogenic Chelation

During blood transfusion, large amounts of sodium citrate may be infused. The resulting calcium citrate complex increases the complexed fraction of the plasma calcium at the expense of the ionised fraction. A similar effect may be seen during platelet phoresis but it can be prevented by ingesting 200 ml of cow's milk 30 minutes before the procedure [42]. Rapid infusion of fresh frozen plasma to burns patients has also been reported to cause hypocalcaemia.

Foscarnet, a pyrophosphate analogue used to treat cytomegalovirus and herpes simplex virus infections in immuno-compromised patients, can cause a decrease in ionised calcium when given intravenously by increasing the bound fraction. Hydrofluoric acid skin burns may also cause severe hypocalcaemia. This may be due to a direct reaction of calcium with fluoride, precipitating calcium salts in the skin.

Cancer

Low plasma albumin is nearly always present in patients with cancer. There is, therefore, a low total plasma calcium but ionised levels are usually normal. However, bony metastases from prostate cancer may exhibit osteoblastic features with markedly increased bone formation. The requirement of calcium for new bone formation may be sufficient to lower the plasma-ionised calcium level. This is also occasionally seen in patients with lung and breast cancer.

Chemotherapy for severe leukaemia and lymphoma may cause the tumour lysis syndrome (see hyperphosphataemia above). Chemotherapy with mithramycin, methotrexate, bleomycin and busulphan may cause hypocalcaemia by inhibiting osteoclast function, while chemotherapy with cisplatinum and actinomycin D may cause hypocalcaemia by lowering plasma magnesium (see hypomagnesaemia above).

Aminoglycoside antibiotics, commonly used in cancer patients, may also cause hypomagnesaemia.

In cancer patients, both surgery and radiotherapy to the head and neck may damage the parathyroid glands. Occasionally, radio-iodine therapy for hyperthyroidism or thyroid cancer will cause hypoparathyroidism by directly damaging the parathyroid glands.

It is important to confirm that true, i.e., ionised hypocalcaemia is present and that hypocalcaemia is not simply due to the reduction in plasma albumin, which is common in cancer. In many cases, hypomagnesaemia will need treatment, as described above, while others will respond to oral calcium therapy. Those with long-term hypoparathyroidism will require calcitriol.

Critical Illness

Hypocalcaemia is common in critically ill patients, especially in those with sepsis. Serum PTH and $1,25(OH)_2D$ levels are often inappropriately low.

Levels of circulating free fatty acids have been related to the degree of reduction in plasma-ionised calcium; they appear to enhance binding of calcium to plasma proteins [43]. Hypocalcaemia is also common in critically ill children where it carries a grave prognosis. Appropriately high levels of PTH have been reported along with inappropriately elevated calcitonin levels. Hypocalcaemia is especially common in toxic shock syndrome. Both hypomagnesaemia and hypercalcitoninaemia are suspected to be major causes.

Treatment will often require intravenous calcium therapy but some will respond to magnesium replacement alone. Care should be taken to avoid hyperphosphataemia, which tends to exacerbate the hypocalcaemia.

Crush Syndrome

In the crush syndrome with renal failure, hypocalcaemia may be severe. There may be calcium and phosphate deposition in dystrophic muscle. Subsequent release of phosphate from muscle may decrease $1,25(OH)_2D$ production and so aggravate the effect of impaired renal function. Oral calcitriol is the mainstay of therapy but oral calcium is also useful.

Bisphosphonate-Induced Hypocalcaemia

Bishosphonates are such rapid and potent inhibitors of bone resorption that they can induce a marked imbalance between bone resorption and formation. This may be sufficient to cause hypocalcaemia. It is particularly noticeable in patients being treated for Paget's disease of bone and for the hypercalcaemia of malignancy where bone turnover is high. It is seldom symptomatic and recovers without treatment. Very mild hypocalcaemia, presumably from the same cause, may be found in the early stages of bisphosphonate treatment for osteoporosis.

Conclusions

True hypocalcaemia, while often asymptomatic, is not rare. Even in its milder forms, it should not be ignored but fully investigated, the cause uncovered and the appropriate treatment administered. It may frequently be a pointer to an underlying metabolic disorder (such as hypomagnesaemia), which might otherwise be over-looked. The differential diagnosis of hypocalcaemia elucidated in this chapter should enable the reader to determine the cause in most cases.

References

1. Young MM, Nordin BEC. Effects of natural and artificial menopause on plasma and urinary calcium and phosphorus. Lancet 1967;2:118–20.
2. Need AG, Guerin MD, Pain RW, Hartley TF, Nordin BEC. The tubular maximum for calcium reabsorption: normal range and correction for sodium excretion. Clinica Chimica Acta 1985;150:87–93.
3. Morris HA, Wishart JM, Horowitz M, Need AG, Nordin BEC. The reproducibility of bone-related biochemical variables in postmenopausal women. Annals Clin Biochem 1990;27:562–8.
4. Nordin BEC, Need AG, Morris HA, Horowitz M. Biochemical variables in pre- and postmenopausal women: reconciling the calcium and oestrogen hypotheses. Osteoporos Int 1999;9:494–8.
5. Horowitz M, Wishart JM, Goh D, Morris HA, Need AG, Nordin BEC. Oral calcium suppresses biochemical markers of bone resorption in men. Am J Clin Nutr 1994;60:965–8.
6. Reilly RF, Ellison DH. Mammalian distal tubule: physiology, pathophysiology, and molecular anatomy. Physiol Rev 2000;80:277–313.
7. Hoenderop JG, van der Kemp AW, Hartog A, van de Graaf, SF van Os CH, Willems PH et al. Molecular identification of the apical Ca2+ channel in 1, 25-dihydroxyvitamin D3-responsive epithelia. J Biol Chem 1999;274:8375–8.
8. Peng J-B, Chen X-Z, Berger UV, Vassilev PM, Brown EM, Hediger MA. A rat kidney-specific calcium transporter in the distal nephron. J Biol Chem 2000;275:28186–94.
9. Bronner F. Current concepts of calcium absorption: an overview. J Nutr 1992;122:641–3.
10. Bindels RJ, Timmermans JA, Hartog A, Coers W. van Os CH. Calbindin-D9k and parvalbumin are exclusively located along basolateral membranes in rat distal nephron [published erratum appears in J Am Soc Nephrol 1992 Mar;2(9):1461]. J Am Soc Nephrol 1991;21122–9.
11. Carafoli E. Biogenesis: plasma membrane calcium ATPase: 15 years of work on the purified enzyme. FASEB J 1994;8:993–1002.
12. Hoenderop JG, Willems PH, Bindels RJ. Toward a comprehensive molecular model of active calcium reabsorption. Am J Physiol 2000;278:F352–F360.
13. Kumar R, Schaefer J, Grande JP, Roche PC. Immunolocalization of calcitriol receptor, 24-hydroxylase cytochrome P-450, and calbindin D28k in human kidney. Am J Physiol 1994;266:F477–F485.
14. Riccardi D, Lee WS, Lee K, Segre GV, Brown EM, Hebert SC. Localization of the extracellular Ca(2+)-sensing receptor and PTH/PTHrP receptor in rat kidney. Am J Physiol 1996;271:F951–F956.
15. Bouhtiauy I, Lajeunesse D, Brunette MG. Effect of vitamin D depletion on calcium transport by the luminal and basolateral membranes of the proximal and distal nephrons. Endocrinology 1993;132:115–20.
16. Glendenning P, Ratajczak T, Dick IM, Prince RL. Calcitriol upregulates expression and activity of the 1b isoform of the plasma membrane calcium pump in immortalised distal kidney tubular cells. Arch Biochem Biophys 2000;380:126–32.
17. Glendenning P, Ratajczak T, Prince RL, Garamszegi N, Strehler EE. The promoter region of the human PMCA1 gene mediates transcriptional downregulation by 1,25-dihydroxyvitamin D₃. Biochemical Biophys Res Commun 2000;277:722–8.
18. Hilal G, Claveau D, LeClerc M, Brunette MG. Ca²⁺ transport by the luminal membrane of the distal nephron: action and interaction of protein kinase A and C. Biochem J 1997;328:371–5.
19. Enomoto H, Hendy GN, Andrews GK, Clemens TL. Regulation of avian calbindin-D28k gene expression in primary chick kidney cells: importance of post-transcriptional mechanisms and calcium ion concentration. Endocrinology 1992;130:3467–74.
20. Hemmingsen C, Staun M, Lewin E, Nielsen PK, Olgaard K. Effect of parathyroid hormone on renal calbindin-D28k. J Bone Miner Res 1996;11:1086–93.
21. Zhu Y, Goff JP, Reinhardt TA, Horst RL. Pregnancy and lactation increase vitamin D-dependent intestinal membrane calcium adenosine triphosphatase and calcium binding protein messenger ribonucleic acid expression. Endocrinology 1998;139:3520–4.
22. Colnot S, Ovejero C, Romagnolo B, Porteu A, Lacourte P, Thomasset M et al. Transgenic analysis of the response of the rat calbindin-D9k gene to vitamin D. Endocrinology 2000;141:2301–8.
23. Barley NF, Prathalingam SR, Zhi P, Legon S, Howard A, Walters JR. Factors involved in the duodenal expression of the human calbindin-D9k gene. Biochem J 1999;341:491–500.

24. Walters JR, Howard A, Lowery LJ, Mawer EB, Legon S. Expression of genes involved in calcium absorption in human duodenum. Eur J Clin Invest 1999;29:214–9.
25. Hoenderop JG, Hartog A, Stuiver M, Doucet A, Willems PH, Bindels RJ. Localisation of the epithelial Ca^{2+} channel in rabbit kidney and intestine. J Am Soc Nephrol 2000;11:1171–8.
26. Armbrecht HJ, Boltz MA, Christakos S, Bruns ME. Capacity of 1,25-dihydroxyvitamin D to stimulate expression of calbindin D changes with age in the rat. Arch Biochem Biophys 1998;352:159–64.
27. Mundy GR, Guise TA. Hormonal control of calcium homeostasis. Clin Chem 1999;45:1347–52.
28. Krieger NS. Parathyroid hormone, prostaglandin E_2, and 1,25-dihydroxyvitamin D_3 decrease the level of Na^+-Ca^{2+} exchange protein in osteoblastic cells. Calcif Tissue Int 1997;60:473–8.
29. Faucheux C, Bareille R, Amedee J. Synthesis of calbindin-D28k during mineralization in human bone marrow stromal cells. Biochem J 1998;333:817–23.
30. Bellido T, Huening M, Raval-Pandya M, Manolagas SC, Christakos S. Calbindin-D28k is expressed in osteoblastic cells and suppresses their apoptosis by inhibiting Caspase-3 activity. J Biol Chem 2000;34:26328–32.
31. Glendenning P, Ratajczak T, Dick IM, Prince RL. Regulation of the 1b isoform of the plasma membrane calcium pump by 1,25-dihydroxyvitamin D_3 in rat osteoblast-like cells. J Bone Miner Res 2001;16:525–34.
32. Ishikawa T, Inagaki H, Kanayama M, Manzai T. Hypocalcemic hyper-CK-emia in hypoparathyroidism. Brain Dev 1999;12:249–52.
33. Vered I, Vered Z, Perez J, Jaffe A, Whyte MP. Normal left ventricular performance documented by Doppler echocardiography in patients with longstanding hypocalcemia. Am J Med 1989;85:413–6.
34. Jensen SB, Illum F, Dupont E. Nature and frequency of dental changes in idiopathic hypoparathyroidism and pseudohypoparathyroidism. Scand J Dent Res 1981;89:26.
35. Albright F, Burnett CH, Smith PH, Parson W. Pseudohypoparathyroidism – an example of the "Seabright–Bantam's syndrome". Endocrinology 1942;30:922–32.
36. Breslau NA. Pseudohypoparathyroidism: current concepts. Am J Med Sci 1989;298:130–40.
37. Wu WI, Schwindinger WF, Aparicio LF, Levine M. Selective resistance to parathyroid hormone caused by a novel uncoupling mutation in the carboxyterminal of G αs: a cause of pseudohypoparathyroidism type Ib. J Biol Chem 2001;276:165–71.
38. Desai TK, Carlson RW, Geheb MA. Hypocalcaemia and hypophosphataemia in acutely ill patients. Crit Care Clin 1987;5:927–41.
39. Jones KH, Fourman P, Edetic acid test of parathyroid insufficiency. Lancet 1963;2:119–21.
40. Bhattacharya SK, Crawford AJ, Pate JW, Clemens MG, Chaudry IH. Mechanism of calcium and magnesium translocation in acute pancreatitis: a temporal correlation between hypocalcemia and membrane-mediated excessive intracellular calcium accumulation in soft tissue. Magnesium 1988;7:91–102.
41. Martinez ME, Catalan P, Balaguer G, Lisbona A, Quero J, Reque A, Pallardo F. 25(OH)D levels in diabetic pregnancies: relation with neonatal hypocalcemia. Horm Metab Res 1991;23:38–41.
42. Cassidy MJD, Wood L, Jacobs P. Hypocalcaemia during plateletpheresis. Transfus Sci 1990;11:217–21.
43. Zaloga GP, Willey S, Tomasic P, Chernow B. Free fatty acids alter calcium binding: a cause for misinterpretation of serum calcium values and hypocalcaemia in critical illness. J Clin Endocrinol Metab 1987;64:1010–14.

Phosphate Homeostasis

A: Intestinal Absorption of Phosphate
T. Yamaguchi, T. Sugimoto, K. Chihara

B: Renal Handling of Phosphate
E. Takeda, K. Morita, Y. Taketani, H. Yamamoto, K. Miyamoto

C: Hyper- and Hypophosphatemia
E. Ishimura, Y. Imanishi, M. Inaba

Intestinal Absorption of Phosphate

T. Yamaguchi, T. Sugimoto and K. Chihara

Introduction

Phosphate homeostasis in humans is controlled by the balance between dietary intake, intestinal absorption, bone deposition/resorption, and renal excretion. This control is partly achieved by interacting endocrine regulatory loops that mainly involves parathyroid hormone (PTH) and vitamin D in the form of the biologically most active compound $1,25(OH)_2D_3$. Phosphate homeostasis is primarily adjusted to provide sufficient amounts of inorganic phosphate (Pi) to a variety of body compartments ranging from the bone skeleton to cells in soft tissues. Pi is especially indispensable for both bone formation and cellular metabolism [1, 2].

Daily dietary Pi intake in humans is approximately 1,000 mg on average, and approximately 700 mg is absorbed from the small intestine. The rest of Pi that is not absorbed is excreted in the faeces. Absorbed Pi is delivered to various body compartments and is ultimately excreted in the urine. Approximately 150 mg Pi in the intestinal fluid undergoes circulation with excretion in and absorption from the intestine. Although dietary Pi intake fluctuates daily, Pi homeostasis is adjusted by its absorption from the intestine and excretion from the kidneys (Figure 4.1). The intestinal absorption step is carried out by two different pathways: passive Pi diffusion and Na^+-dependent active Pi transportation. The former is positively related to the luminal Pi concentration prevailing after a meal and is uncontrollable by hormonal actions, while the latter is altered in response to changes in dietary Pi intake and $1,25(OH)_2D_3$ levels. In contrast, the renal tubular reabsorption of Pi is carried out mainly by Na^+-dependent active Pi reabsorption, which is suppressed by PTH in response to changes in dietary Pi intake. This process is more important for the maintenance of Pi homeostasis than that of the intestine because of its swifter response and higher capacity for transportation (Table 4.1).

After Pi is absorbed from the intestine, it is distributed in body compartments with 85–90% in the skeleton and 10–15% in soft tissues. This indicates that the skeleton is the principal storage site for Pi, where Pi acts to maintain proper bone function by depositing hydroxyapatite in the calcification of the organic bone matrix. On the other hand, in metabolically active cells of soft tissue, adequate intracellular Pi is necessary for normal cell function through the synthesis of numerous organic compounds such as nucleotides, phospholipids, phosphorylated metabolic intermediates: membrane phospholipids consist of Pi, and the mitochondria utilise Pi to generate adenosine triphosphate. Pi is also involved in both glycogenolysis and glycolysis. Compared to the skeleton and soft tissues, less than 0.3% of total Pi in the body is distributed in the

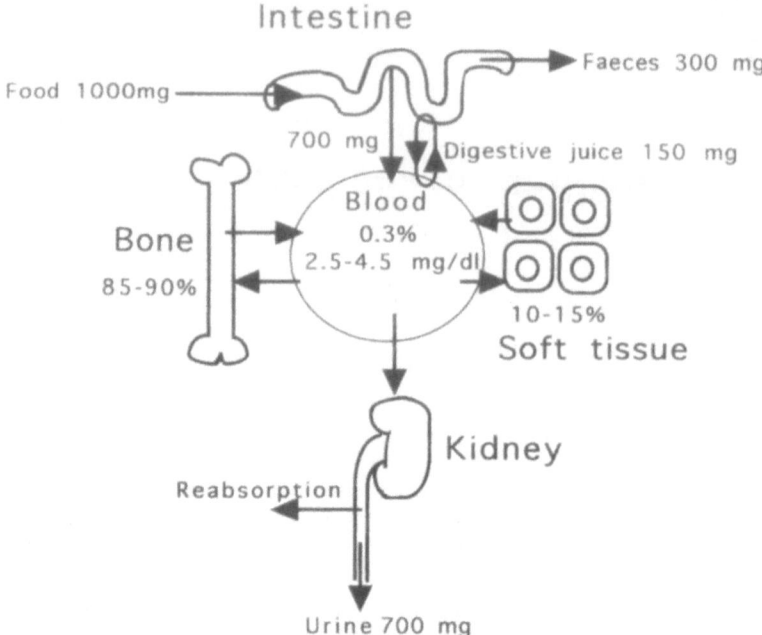

Figure 4.1. Pi homeostasis in humans. Although dietary Pi intake is fluctuating daily, Pi homeostasis is adjusted by its absorption from the intestine and excretion from the kidneys. After Pi is absorbed from the intestine, it is distributed in body compartments with 85–90% in the skeleton and 10–15% in soft tissues. Only less than 0.3% of total Pi in the body is distributed in the blood. In the skeleton, Pi acts to maintain proper bone function by depositing hydroxyapatite in the calcification of the organic bone matrix. In metabolically active cells of soft tissue, adequate intracellular Pi is necessary for normal cell function through the synthesis of numerous organic compounds such as nucleotides, phospholipids, phosphorylated metabolic intermediates.

Table 4.1. Comparison of Na^+-dependent active Pi transport between the kidney and the intestine. Na^+-dependent active Pi transportation in the intestine is altered in response to changes in dietary Pi intake and $1,25(OH)_2D_3$ levels, while that in the kidney is suppressed by PTH in response to changes in dietary Pi intake. This renal process is more important for the maintenance of Pi homeostasis than that of the intestine because of its more prompt response and higher capacity for transportation as shown in this table

	Kidney	Intestine
Capacity (V_{max}) (pmol/5 sec/mg protein)	1,000 to 1,500	60 to 100
Response to dietary Pi restriction (hour)	2 to 4	48
pH dependence	Neutral to alkaline	Acidic
Preferentially transported ion	HPO_4^{2-}	$H_2PO_4^-$
Main regulating hormone	PTH	$1,25(OH)_2D_3$
Molecular structure	Type II (NPT2)	Unknown

blood. The adaptive changes in the renal tubular Pi reabsorption control its serum concentration between 2.5 and 4.5 mg/dl in human adults (Figure 4.1).

Intestinal Pi Absorption Pathways

Although Pi absorption occurs throughout the small intestine, the intensity of absorption varies among intestinal segments as well as between species. In both humans and rats, Pi absorption is greatest in the jejunum, lower in the duodenum, and minimal in the ileum [3, 4]. In rabbits, Pi absorption is maximal in the duodenum and decreases along the length of the small intestine. Little or no active Pi absorption is observed in the colon regardless of species [3], although one study suggests that the large intestine may have the ability to absorb Pi [5].

Pi transport across the intestinal microvilli can occur in two different fashions. One is non-saturable Na^+-independent passive diffusion that is intensified with an increase in dietary Pi intake but is uncontrollable by hormonal actions, and the other is Na^+-dependent active Pi transport that is stimulated by $1,25(OH)_2D_3$ and low-Pi diet [1, 2] (Table 4.2).

Thirty per cent to fifty per cent of intestinal Pi absorption depends on non-saturable Na^+-independent passive diffusion through the paracellular pathway [6, 7], which is related to the luminal Pi concentration prevailing after a meal and readily occurs at high phosphate levels [8]. This component comprises a much higher portion of the intestinal Pi transport compared to the kidney, where the Na^+-independent passive Pi transport is generally less than 10% of the total Pi transport rate (Table 4.2). Passive Pi diffusion is significantly higher in the ileum than in the proximal parts of the intestine [3, 4]. Metabolic balance studies in healthy adults show that there is a strong positive correlation between intestinal Pi absorption and dietary Pi intake [8]. When dietary Pi intake varies over the normal range of 775 to 1,860 mg P/day, 60–80% of dietary Pi is absorbed. Because Pi is a major constituent of a variety of foods, dietary Pi intake rarely drops to less than 620 mg P/day among healthy adults. However, when synthetic diets are used to lessen dietary Pi intake to around 310 mg P/day, faecal Pi excretion surpasses dietary Pi intake.

Na^+-dependent active Pi transport occurs maximally at low phosphate concentrations. The driving force for the active Pi transport, Na^+ gradient across the brush

Table 4.2. Comparison between two pathways of intestinal Pi absorption. Thirty to fifty per cent of intestinal Pi absorption depends on non-saturable Na^+-independent passive diffusion through the paracellular pathway, which is related to the luminal Pi concentration prevailing after a meal and readily occurs at high phosphate levels. This component comprises a much higher portion of the intestinal Pi transport compared to the kidney, where the Na^+-independent passive Pi transport is generally less than 10% of the total Pi transport rate

	Passive diffusion	Active transport
Na^+-dependency	Independent	Dependent
Route	Paracellular	Transcellular
Site of maximal activity	Ileum	Duodenum, Jejunum
Regulator	None (uncontrollable)	$1,25(OH)_2D_3$, Dietary Pi

border membrane (BBM) of the enterocyte, is generated by the action of Na^+-K^+adenosine triphosphate at the basolateral membrane through pumping out intracellular Na^+[9]. This transport occurs mainly in the duodenum and jejunum. Studies in mouse, rat, and humans suggest the existence of high-affinity Na^+–Pi co-transporters, with Michaelis constants (K_m) of 14–93 µM, in the basolateral membrane [10–12]. Although the active intestinal Pi transport has properties similar to those of the renal Na^+–Pi co-transporter, several differences exist between the two transport systems (Table 4.1). The Na:Pi stoichiometry of the intestinal transporter is generally 1:1, as compared with 2:1 for the renal Na^+–Pi co-transporters [5, 13]. The two systems also differ with respect to their pH-dependence [5–7]. At physiological concentrations of sodium, the rate of proximal tubular Na^+–Pi co-transport is higher at alkaline pH values; a difference in rate of about threefold was observed between pH 6.5 and 8. In contrast, the Na^+-dependent Pi transport across intestinal brush border membrane vesicles (BBMVs) as well as intact rat jejunum is significantly higher at an acidic pH (pH 6.0) than at the physiologic pH of 7.4. Since the surface of intestinal microvilli is acidic at a pH of 5.3 to 6.1 [14], this property accordingly facilitates the intestinal Pi transport rate. Since $H_2PO_4^-$ is the predominant ionic species at slightly acidic pH, the Na^+-dependent Pi transport system in the intestine is thought to prefer monovalent $H_2PO_4^-$ to divalent HPO_4^{2-} [9], or at least there is agreement that the transport system accepts $H_2PO_4^-$ and HPO_4^{2-} alike. In contrast, HPO_4^{2-} is predominantly transported in the kidney. Peerce actually showed that Na^+–Pi co-transporter in the intestine occluded $H_2PO_4^-$, but not HPO_4^{2-}, and that phosphate occlusion had an absolute requirement for Na^+, with K^+ or Cs^+ unable to substitute [15]. Another difference between the two transport systems is that the maximal velocity (V_{max}) for intestinal Na^+–Pi co-transporters is about 10 times lower than that for their renal counterpart (Table 4.1). Although these differences between intestinal and renal active Pi transporters suggest that the two transport systems are molecularly distinct, the intestinal Na^+–Pi co-transporter DNA and protein structure is not clearly established at present. It has also been demonstrated that both electroneutral [16–18] and electrogenic [19, 20] Na^+–Pi co-transporters were present in the small intestinal epithelium, suggesting that multiple Na^+–Pi co-transporter isoforms may exist in the mammalian intestine. In contrast, the molecular structure of the renal Na^+–Pi co-transporter has been well documented and designated as NPT2 [21]. The importance of NPT2 in the maintenance of Pi homeostasis has been established by accumulating evidence.

Regulators of Na^+-Dependent Active Phosphate Absorption in the Intestine

1,25(OH)$_2$D$_3$

Two of the most important physiological and pathophysiological regulators of Na^+-dependent active Pi absorption in the small intestine are vitamin D, specifically 1,25(OH)$_2$D$_3$, and dietary Pi.

It is generally accepted that 1,25(OH)$_2$D$_3$ activates Na^+–Pi transport through increasing the number of transporters. The hormone was shown to enhance the V_{max} of the carrier system [22]. The inhibitory effect of cycloheximide shows that there is a requirement of intact protein synthesis for induction by 1,25(OH)$_2$D$_3$ of Na^+–Pi transport in the small intestine [23]. This strongly suggests that one of the many actions of $1,25(OH)_2D_3$ is hormone receptor-mediated activation of gene(s) that code for as yet unidentified Na^+–Pi transporter protein(s).

The response of Na^+-dependent active intestinal Pi transport to $1,25(OH)_2D_3$ is only apparent after weaning. In a mutant pig model that defects renal synthesis of $1,25(OH)_2D_3$, intestinal BBM Na^+–Pi co-transport is similar between normal and mutant littermates while they are newborns and are fed by lactation. However, the transport significantly reduces in the mutant group compared with the normal one after weaning [24]. Moreover, although exogenous $1,25(OH)_2D_3$ had no effect on apical Pi transport in the mutant group while fed by lactation, the correction of Pi transport in the group was achieved by the administration of $1,25(OH)_2D_3$ after weaning [24]. The same study also showed that intestinal Na^+–Pi co-transport significantly reduces over one week after birth in both normal and mutant groups.

A study on a vitamin D receptor knockout mouse showed that manifestations of rickets including hypocalcaemia, hypophosphataemia, and high serum alkaline phosphatase levels that are caused by reduced intestinal absorption of both calcium and Pi due to the defunct receptor became only conspicuous after weaning [25], providing additional evidence that $1,25(OH)_2D_3$ is not crucial for the control of intestinal Pi absorption before weaning.

Na^+-dependent active intestinal absorption of Pi is enhanced by $1,25(OH)_2D_3$, whereas the Na^+-independent passive diffusion of Pi was unaffected. The activities of Pi transport in the rat small intestine have been studied in vitamin D-deficient rats. The Pi uptake in BBMVs from vitamin D-deficient rat jejunum was markedly increased after the administration of $1,25(OH)_2D_3$ [26, 27]. Similar stimulation of Pi uptake by $1,25(OH)_2D_3$ was obtained in Xenopus oocytes microinjected with duodenum poly(A)$^+$RNA isolated from the rabbit intestine [28]. It has also been well documented that restriction of dietary calcium or phosphate increases circulating levels of $1,25(OH)_2D_3$ and, consequently, leads to increased absorption of Pi in a Na^+-dependent fashion [23].

In vitamin D-replete healthy subjects, however, experimental increases of serum $1,25(OH)_2D_3$ do not facilitate jejunal Pi absorption significantly. Studies using the balance technique show that fractional Pi absorption increases only slightly, as serum $1,25(OH)_2D_3$ concentrations vary from normal to high levels among adults taking diets that provide normal amounts of Pi [8]. On the other hand, intestinal Pi absorption more readily increases in response to an increase in dietary Pi intake [8]. Moreover, even in the presence of very low serum $1,25(OH)_2D_3$ levels, patients with chronic renal failure exhibit significant concentration-dependent jejunal Pi absorption through nonsaturable passive diffusion after taking Pi-containing meals [8]. Thus, the availability of Pi in the diet rather than the action of $1,25(OH)_2D_3$ appears to be the major determinant of net Pi input to the body from the intestine under vitamin D-replete circumstances.

The thyroid hormone (T_3) has a permissive effect on the enhancement of activation of genes that are modulated by $1,25(OH)_2D_3$. In cultured embryonic chick small intestine, the combination of T_3 and $1,25(OH)_2D_3$ synergistically induces Na^+-dependent Pi uptake [29]. Another study also provided evidence for synergistic interaction of the two hormones [30]: when BBMV was isolated from cultured embryonic chick jejunal segments and Na^+ gradient-driven Pi uptake was measured, exposure of cultures to $1,25(OH)_2D_3$ caused a significant increase in Na^+-dependent Pi accumulation in the isolated vesicles compared to the rate of uptake by BBMV from the thyroid hormone-free cultures. The effect of $1,25(OH)_2D_3$ was augmented when the gut preparations were treated with $1,25(OH)_2D_3$ plus T_3. In these experiments, although T_3 at the concentration of 10^{-8} M alone had no remarkable effect on Pi uptake by BBMV, addition of this dose of T_3 to the samples treated with $1,25(OH)_2D_3$ clearly potentiated the action of $1,25(OH)_2D_3$.

Dietary Phosphate

Dietary Pi is the other important regulator of Na^+-dependent active Pi absorption in the intestine as well as $1,25(OH)_2D_3$. It is well documented in several studies that the active intestinal Pi absorption responds in adaptation to changes in dietary Pi intake in intestinal preparations from different species. Na^+-dependent active intestinal Pi absorption is decreased when dietary Pi is increased and is enhanced in animals fed a low-Pi diet [31]. Similar to its renal counterpart, this adaptive response specifically occurs in the Na^+–Pi co-transporter, but not in the transport of D-glucose or L-proline as measured in the same BBMV preparations. The increased Pi transport in response to low-Pi diet is explained by an increased V_{max} for transport, whereas K_m for transport is either unchanged or decreased [32]. However, the adaptive increase in the intestinal Na^+-dependent Pi transport in response to dietary Pi restriction can be observed as early as 48 hours and continues for the duration of the Pi-restricted diet [32]. This intestinal adaptation is generally slower than that of the kidney, where adaptation occurs within two to four hours after the start of dietary Pi restriction. Studies on Na^+-dependent Pi transport in jejunal BBMVs isolated from vitamin D-replete rats have shown that there was a temporal relationship between the rise in plasma $1,25(OH)_2D_3$ and stimulation of the transport by Pi deprivation [32]. Other studies showed that dietary Pi restriction did not change Pi and Na^+-binding characteristics of the Na^+–Pi co-transporter but significantly augmented its intrinsic activity as indicated by an increase in the V_{max} of the translocation process [32, 33]. Since this kinetic change must be regarded as a distinguished feature of $1,25(OH)_2D_3$ action on the Na^+–Pi co-transporter [22], these studies provided confirmative evidence for the assumption that adaptation of intestinal Pi absorption to dietary requirements is mediated by $1,25(OH)_2D_3$. However, BBMVs from vitamin D-deficient rats have also shown adaptive changes in intestinal Pi transport to dietary Pi intake despite diminished activities of $1,25(OH)_2D_3$ [34], providing another possibility that the process is independent of $1,25(OH)_2D_3$. One study reported that the intestinal levels of type III Na^+–Pi co-transporter PiT-2 mRNA were increased after $1,25(OH)_2D_3$ administration to vitamin D-deficient animals, whereas this mRNA level was not elevated in the rats fed with the low-Pi diet [35], showing that the effects of $1,25(OH)_2D_3$ and Pi restriction on the up-regulation of the PiT-2 transporter are different and suggesting that each effect might be conveyed by different signals. Therefore, further studies are needed to determine whether or not the adaptation of intestinal Pi absorption to dietary Pi restriction is inseparable from $1,25(OH)_2D_3$ action.

Although the qualities of these adaptive reactions in intestinal Na^+–Pi co-transporter are partly similar to those described for the renal Na^+–Pi co-transporter, their overall effect on total body Pi is less prominent than that resulting from the renal adaptation. First, the intestinal adaptation by the active carrier-mediated component of Pi transport is confined primarily to and greatest in the proximal intestine. On the other hand, non-saturable Na^+-independent passive Pi diffusion is quantitatively dominant throughout the small intestine. Second, the capacity of the intestinal Na^+–Pi co-transport, as measured in jejunal BBMVs, is significantly lower than that measured in renal BBMVs [33]. In typical BBMV uptake experiments, the V_{max} for intestinal Na^+–Pi co-transporter is 60 to 100 pmol/5 sec/mg protein, whereas that for renal Na^+–Pi co-transporter is as high as 1,000 to 1,500 pmol/5 sec/mg protein [33] (Table 4.1).

A Putative Activator Protein for Na^+-Dependent Phosphate Transport (PiUS)

A putative activator protein for Na^+-dependent Pi transport (PiUS) has been detected in the small intestine. PiUS was cloned from rabbit small intestine by expression cloning [36]. The putative amino acid sequence of PiUS cDNA revealed a highly hydrophilic 425-amino acid protein with no membrane-spanning domain. The protein is also expressed in the kidney, liver, and heart. PiUS not only stimulated Na^+–Pi co-transporter activity but also Na^+-independent passive Pi transport when injected in Xenopus oocytes. Therefore, PiUS action may not be specific for Na^+-dependent Pi transport, but the molecule would rather function as a general regulator of cellular Pi metabolism.

Stanniocalcin

Intestinal Pi transport is also regulated by stanniocalcin [37], a peptide hormone that is produced by bony fish as well as humans. A recent study has identified the intestine of bony fish as a target organ for stanniocalcin: stanniocalcin is shown to decrease the absorption of calcium and to stimulate the absorption of Pi in both swine and rat duodenum primarily through an effect on the mucosal-to-serosal movement [38]. It has been documented that the primary function of stanniocalcin in fish is to prevent hypercalcaemia, and thus its secretion is stimulated by a rise in serum calcium levels. Release of stanniocalcin into the circulatory system causes a decrease in calcium uptake by the gills and intestine, which reduces calcium movement from the aquatic environment into the circulatory system and thereby reduces serum calcium levels [37, 39]. A second function of stanniocalcin in fish is the stimulation of phosphate reabsorption by the renal proximal tubules through a cAMP-dependent pathway [39]. This renal effect causes increased levels of serum Pi. The increased Pi combines with circulating calcium, thus promoting its deposition into bones and scales and reducing serum calcium levels. This combined effect of stanniocalcin on calcium and Pi movement results in a synergistic effect in lowering serum calcium levels.

Molecular Biology of Intestinal Na^+–Pi Co-transporters

Na^+–Pi Co-transporters Identified in Non-Intestinal Tissues

Three classes of Na^+–Pi co-transporter genes have been cloned and characterised. The type I (NPT1) and II (NPT2) transporters are expressed primarily in the kidney, whereas type III transporters are widely distributed and probably serve as housekeeping Na^+–Pi co-transporters. NPT1 can also function as a channel for chloride and organic anions, whereas NPT2 has the features of renal BBM Na^+–Pi co-transport, and is subject to hormonal and dietary Pi regulation [40]. Mice in which the NPT2 gene has been disrupted by targeted mutagenesis exhibit hypophosphataemia, renal Pi wasting and skeletal abnormalities [41], clearly showing the importance of renal NPT2 in the maintenance of Pi homeostasis. On the other hand, no hybridisation signal related to either NPT1 or NPT2 transporters was observed in mRNA isolated from the mucosa of the small intestine, suggesting that small intestinal apical Na$^+$–Pi co-transporter may represent another type of transporter. Type III transporters were isolated as receptors for gibbon ape leukaemia virus (GLVR1 or PiT-1) in mice and humans and amphotropic murine retrovirus (Ram-1 or PiT-2) in rats [42], and only later were functionally characterised as Na$^+$–Pi co-transporters [43]. The amino acid sequences of PiT-1 and PiT-2 transporters are approximately 60% identical [42], and exhibit no significant overall sequence similarity to the NPT1 or NPT2 transporters.

Putative Intestinal Na$^+$–Pi Co-transporters

Compared to the well-established role of renal NPT2 in the maintenance of Pi homeostasis, less is known about the molecular biology of the intestinal Na$^+$–Pi co-transporters. Recently, however, an isoform of the type II Na$^+$-Pi co-transporter (type IIb) cDNA was cloned from mouse small intestine [44]. Its mRNA expression, however, was not confined to the small intestine, but was found in a variety of tissues. Homology of this isoform to described mammalian and non-mammalian type II co-transporters is between 57% and 75%, with diversities found at the C terminus. The type IIb protein consists of 697 amino acids with putative eight transmembrane domains and was detected as a 108-kDa protein by Western blots using isolated small intestinal brush border membranes. By immunohistochemistry it was localised at the apical membrane of mucosal enterocytes. The injection of type IIb cRNA into oocytes of Xenopus laevis resulted in the expression of Na$^+$-dependent Pi transport with its acidic pH dependence and a K_m for Pi of approximately 50 μM. These pH dependence and kinetic properties are in agreement with the previously reported functional characteristics of Na$^+$–Pi co-transport in mouse small intestine [2], and support the notion that the type IIb co-transporter may represent a candidate for a small-intestine Na$^+$–Pi co-transporter. Since 1,25(OH)$_2$D$_3$ and Pi are two of the most important regulators of Na$^+$-dependent active Pi absorption in the intestine, further studies are needed to clarify the effects of these agents on the active absorption mediated by the type IIb co-transporter.

Xu et al. [45] have isolated a cDNA (SLC34A2) encoding a human homologue of the type IIb transporter. The cDNA is shown to be 4,135 bp in length with an open reading frame that predicts a 689-amino-acid polypeptide. The putative protein has 76% homology to mouse intestinal type IIb transporter and lower homologies with renal type II transporters. Northern blots showed a singular transcript of 5.0 kb in human lung, small intestine, and kidney. Computer analysis suggests a protein with 11 transmembrane domains and several potential post-translational modification sites. Functional characterisation in Xenopus laevis oocytes showed that this cDNA encodes a functional Na$^+$–Pi transporter. Furthermore, the gene encoding this cDNA was mapped to human chromosome 4p15.1–p15.3 by the FISH method.

More recently, Bai et al. [46] have isolated a cDNA that encodes a type III Na$^+$-dependent phosphate co-transporter from mouse small intestine (mPiT-2). The nucleotide sequence of mPiT-2 predicts a protein of 653 amino acids with at least 10 putative transmembrane domains. This protein has 94% homology to rat PiT-2 transporter. Kinetic studies, carried out in Xenopus oocytes, showed that mPiT-2 cRNA induces significant Na$^+$-dependent Pi uptake with an apparent K_m for phosphate of

38 μM. The transport of phosphate by mPiT-2 is inhibited at high pH. Northern blot analysis demonstrated the presence of mPiT-2 mRNA in various tissues, including intestine, kidney, heart, liver, brain, testis, and skin. The highest expression of mPiT-2 in the intestine was found in the jejunum. In situ hybridisation revealed that mPiT-2 mRNA is expressed throughout the vertical crypt–villus axis of the intestinal epithelium. Given the distinct structure of the predicted protein and epithelial expression of mPiT-2 mRNA, they suggest that mPiT-2 may be the second identified mammalian intestinal Na$^+$–Pi co-transporter.

Katai et al. reported that the intestinal PiUS mRNA levels showed an approximate doubling in the rats fed with the low-Pi diet compared with those fed with the normal-Pi diet. Moreover, the intestinal levels of type III Na$^+$–Pi co-transporter PiT-2 mRNA were increased 24 and 48 hours after 1,25(OH)$_2$D$_3$ administration to vitamin D-deficient animals, whereas PiUS or the type IIb Na$^+$–Pi co-transporter mRNA levels were unchanged [34]. These results suggest that PiT-2 might be one of the candidate high-affinity Na$^+$–Pi co-transporters in the vitamin D-responsive system. In contrast, they did not detect an elevation of the PiT-2 mRNA level in the rats fed with the low-Pi diet, and therefore speculated that 1,25(OH)$_2$D$_3$ and Pi restriction might give different signals in the up-regulation of intestinal Na$^+$–Pi co-transport mediated by either the PiUS or the PiT-2 co-transporter.

Shibui et al. reported that a cDNA clone (kaia2138), which shows a significant similarity with the renal Na$^+$–Pi co-transporter, was isolated from a human intestinal mucosa cDNA library [47]. The cDNA is 2,626 bases long, with one open reading frame encoding a protein of 497 amino acids. The deduced amino acid sequence shows an overall homology of 48% with the human renal NPT1 protein. This gene is expressed in the intestine, colon, liver, and pancreas. The chromosomal location of the gene was determined on the chromosome 6p21.3–p22 region by polymerase chain reaction-based analysis with both a human/rodent mono-chromosomal hybrid cell panel and a radiation hybrid mapping panel. Thus, this gene may code for intestinal type Na$^+$–Pi co-transporter or closely related protein. However, there were no data showing the physiological function of this protein in the report.

Taken together, these recent studies suggest that not one but multiple Na$^+$-dependent co-transporters are involved in controlling Pi absorption in the mammalian intestine. The putative intestinal Na$^+$–Pi co-transporters reported to date are summarised in Table 4.3.

Disorders Related to Intestinal Phosphate Absorption

In chronic renal failure, as the ability of the kidneys to secrete excess Pi into the urine reduces, intestinal absorption of Pi via uncontrolled passive diffusion has more influences on Pi homeostasis than in healthy subjects. Although the uraemic state itself does not alter the renal or intestinal adaptation to dietary Pi restriction, the overall impact of such adaptive responses is small once the renal function is severely impaired. At the intestinal level, the active Na$^+$-dependent and regulatable transport of Pi involves only the proximal intestine with its lower capacity and slower response to dietary Pi restriction than its renal counterpart [3] (Table 4.1). Uncontrolled passive diffusion of large amounts of Pi throughout the entire intestine readily surpasses the active Pi transport and facilitates concentration-dependent Pi absorption in uraemic patients. Furthermore, as the filtered load of Pi decreases, the renal tubular adaptive mechanisms become insufficient to maintain Pi homeostasis. Even

Table 4.3. Intestinal Na$^+$–Pi co-transporters cloned to date. The type IIb co-transporter is considered to be a possible candidate for a small intestine Na$^+$–Pi co-transporter because its pH dependence and kinetic properties are in agreement with the functional characteristics of Na$^+$–Pi co-transport in mouse small intestine. However, not one but multiple Na$^+$-dependent co-transporters might be involved in controlling Pi absorption in the mammalian intestine. All co-transporters shown here are not confined to the small intestine but are expressed in a variety of tissues

	Type IIb	SLC34A2	mPiT2	kaia2138
Cloned species	Mouse	Human	Mouse	Human
Location in the intestine	Apical	Apical	Basolateral?	Unknown
Amino acid number	697	689	653	497
Transmembrane domain	8	11	10	Unknown
Homology	57% to mNPT2	76% to mType IIb	94% to rPiT2	48% to hNPT1
K$_m$ (µM)	50	Unknown	38	Unknown
pH dependence	Acidic	Unknown	Acidic	Unknown
Reference number	44	45	46	47

Abbreviation: m, mouse; r, rat; h, human.

in the presence of very low serum 1,25(OH)$_2$D$_3$ levels, patients with chronic renal failure exhibit significant concentration-dependent Pi absorption at the distal intestine. Thus, in uraemic patients, net Pi input to the body from the intestine mainly depends on the amounts of Pi in the diet [8]. Indeed, net intestinal Pi absorption among such patients is nearly identical to that among normal subjects despite existing hyperphosphataemia, as dietary Pi intake varies over the range of 310 to 1,550 mg P/day. If absorption of dietary Pi persists, it is directly linked to the pathogenesis of secondary hyperparathyroidism in patients with progressive kidney disease. This pathological state intensifies hyperphosphataemia further, with its clinical complications such as stimulation of PTH secretion and ectopic calcification (Figure 4.2). Among patients with disorders associated with chronic hyperphosphataemia, such as hypoparathyroidism and tumoral calcinosis, intestinal Pi absorption is also not apparently significantly reduced, suggesting again that the role of adaptive responses of intestinal Pi absorption to hyperphosphataemia is minimal. In those patients, intestinal Pi absorption should therefore be suppressed by oral administration of Pi binders. Because of the ability of Ca^{2+} to form insoluble PO$_4$ salts and thus limit intestinal Pi absorption, calcium carbonate is routinely used in the care of patients with advanced renal failure [8].

On the other hand, hypophosphataemia is observed in about 1% of all hospitalised patients in industrialised countries. Major causative factors for hypophosphataemia are alcoholism, gross malnutrition, and gastrointestinal diseases that cause reduced absorption of Pi and vitamin D from the gut. Hypophosphataemia can also be seen in liver, renal or endocrine diseases, which are associated with disturbances of vitamin D metabolism. These diseases can result in low circulating levels of 1,25(OH)$_2$D$_3$ and thus in low activity of the 1,25(OH)$_2$D$_3$-mediated Na$^+$-dependent Pi transport system at the brush border of the enterocyte. Clinical symptoms of hypophosphataemia include osteomalacia (rickets), dysfunctions of the central nervous system or cardiac arrhythmias.

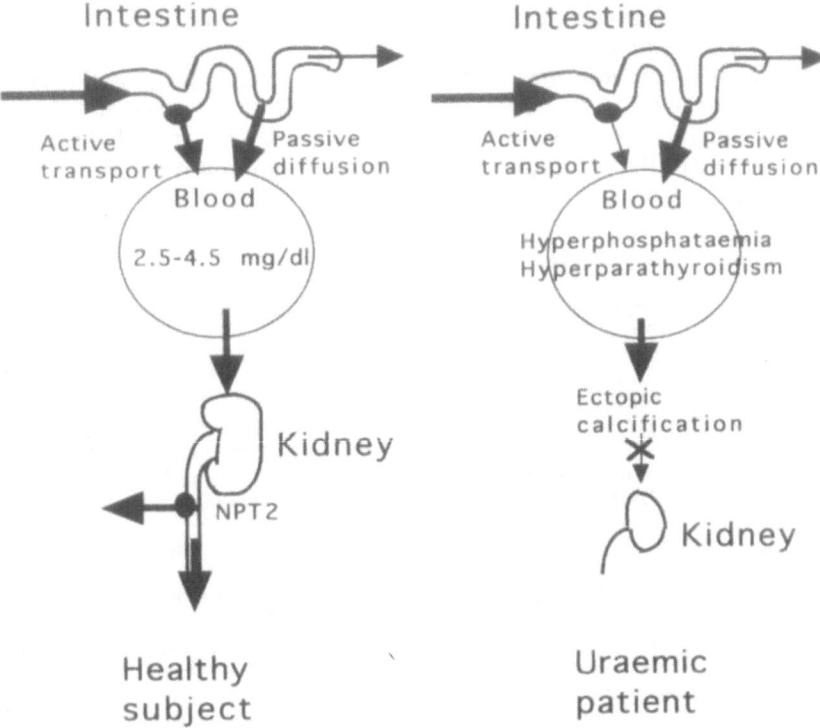

Figure 4.2. Comparison of Pi homeostasis between healthy subjects and uraemic patients. In chronic renal failure, Na$^+$-dependent active Pi transport in the proximal intestine cannot substitute for the role of its abolished renal counterpart because of lower capacity and slower adaptation. Uncontrolled passive diffusion of large amounts of Pi throughout the entire intestine readily surpasses the active Pi transport and facilitates concentration-dependent Pi absorption in uraemic patients. Thus, in uraemic patients, net Pi input to the body from the intestine mainly depends on the amounts of Pi in the diet. If absorption of dietary Pi persists, it is directly linked to the pathogenesis of secondary hyperparathyroidism, intensifying hyperphosphataemia with its clinical complications such as stimulation of PTH secretion and ectopic calcification.

Summary

Compared to the mechanisms of adaptation of the renal tubular reabsorption of Pi where the importance of the type II (NPT2) co-transporter was clearly established, those of intestinal Pi transport are less well understood. It is partly because of the lack of knowledge about the molecular structure of intestinal Na$^+$–Pi co-transporters, although recent studies present some candidate molecules such as the type IIb Na$^+$–Pi co-transporter. Passive Pi diffusion is quantitatively more important for Pi homeostasis in the intestine than in the kidney. This process depends on the luminal Pi concentration prevailing after a meal but undergoes no adaptive changes in response to either 1,25(OH)$_2$D$_3$ or low-Pi diet as seen in Na$^+$-dependent active Pi absorption. Although Na$^+$-dependent active Pi absorption is also observed in the intestine, its capacity and adaptation are much lower and slower than those in the kidney. Thus, the overall effect of the adaptive changes of intestinal Pi absorption on total body Pi homeostasis is minimal compared to that of their renal counterpart. This notion is supported by the clinical observation that intestinal Pi absorption is not significantly reduced among patients with chronic renal failure or other diseases associated with chronic hyperphosphataemia, despite the diminished activity of $1,25(OH)_2D_3$ and the retention of Pi in the body. In these patients, avoidance of excessive dietary Pi intake and reduction in intestinal Pi absorption by administering phosphate binders are important to prevent the symptoms of secondary hyperparathyroidism.

References

1. Loghman-Adham M. Adaptation to changes in dietary phosphorus intake in health and renal failure. J Lab Clin Med 1997;129:176–88.
2. Cross HS, Debiec H, Peterlik M. Mechanism and regulation of intestinal phosphate absorption. Miner Electrolyte Metab 1990;16:115–24.
3. Kayne LH, D'Argenio DZ, Meyer JH, Hu MS, Jamgotchian N, Lee DBN. Analysis of segmental phosphate absorption in intact rats. A compartmental analysis approach. J Clin Invest 1003;91:915–22.
4. Loghman-Adham M, Szczepanska-Konkel M, Yusufi ANK, VanScoy M, Dousa TP. Inhibition of Na^+-Pi co-transport in small gut brush border by phospho-carboxylic acids. Am J Physiol 252:G244–G249.
5. Lee DBN, Walling MW, Corry DB. Phosphate transport across rat jejunum: influence of sodium, pH, and 1,25-dihydroxyvitamin D_3. Am J Physiol 1986;251:G90–5.
6. Danisi G, Murer H, Straub RW. Effect of pH on phosphate transport into intestinal brush-border membrane vesicles. Am J Physiol 1984;246:G180–6.
7. Borowitz SM, Ghishan FK. Phosphate transport in human jejunal brush border membrane vesicles. Gastroenterology 1989;96:4–10
8. Lemann J, Favus MJ. The intestinal absorption of calcium, magnesium, and phosphate. In: Favus MJ, editor. Primer on the metabolic bone diseases and disorders of mineral metabolism. Philadelphia: Lippincott Williams & Wilkins, 1999;63–7.
9. Berner W, Kinne R, Murer H. Phosphate transport into brush border membrane vesicles isolated from rat small intestine. Biochem J 1976;160:467–74.
10. Ghishan FK, Kikuchi K, Arab N. Phosphate transport by rat intestinal basolateral-membrane vesicles. Biochem J 1987;243:641–6.
11. Kikuchi K, Ghishan FK. Phosphate transport by basolateral plasma membranes of human small intestine. Gastroenterology 1987;93:106–13.
12. Nakagawa N, Ghishan FK. Transport of phosphate by plasma membranes of the jejunum and kidney of the mouse model of hypophosphatemic vitamin D-resistant rickets. Proc Soc Exp Biol Med 1993;203:328–35.
13. Quamme GA. Phosphate transport in intestinal brush border membrane vesicles: effect of pH and dietary phosphate. Am J Physiol 1985;249:G168–76.
14. Shiau Y-F, Fernandez P, Jackson MJ, McMonagle S. Mechanism maintaining a low-pH microclimate in the intestine. Am J Physiol 1985;248:G608–17.
15. Peerce BE. Simultaneous occlusion of Na^+ and phosphate by the intestinal brush border membrane Na^+/phosphate co-transporter. Kidney Int1996;49:988–91.
16. Berner W, Kinne R, Murer H. Phosphate transport into brush-border membrane vesicles isolated from rat small intestine. Biochem J 1076;160:467–74.
17. Caverzasio J, Danisi G, Straub RW, Murer H, Bonjour JP. Adaptation of phosphate transport to low-phosphate diet in renal and intestinal brush border membrane vesicles: influence of sodium and pH. Pflugers Arch 1987;409:333–6.
18. Quamme GA. Phosphate transport in intestinal brush border membrane vesicles: effect of pH and dietary phosphate. Am J Physiol 1985;249:G168–76.
19. Danisi G, Murer H, Straub RW. Effect of pH on phosphate transport into intestinal brush border membrane vesicles. Am J Physiol 1984;246:G180–6.
20. Shirazi-Beechey SP, Gorvel JP, Beechey RB. Phosphate transport in intestinal brush border membrane. J Bioenerg Biomembr 1988;20:273–88.
21. Murer H, Biber J. Molecular mechanisms of renal apical Na phosphate co-transport. Annu Rev Physiol 1996;58:607–18.
22. Peterlik M, Wasserman RH. Effect of vitamin D on transepithelial phosphate transport in chick intestine. Am J Physiol 1978;234:E379–88.
23. Peterlik M, Wasserman RH. Regulation by vitamin D of intestinal phosphate absorption. Horm Metab Res 1980;12:216–9.

24. Schroder B, Hattenhauser O, Breves G. Phosphate transport in pig proximal small intestines during postnatal development. Lack of modulation by calcitriol. Endocrinology 1998;139:1500–7.
25. Yoshizawa T, Handa Y, Uematsu Y, Takeda S, Sekine K, Yoshihara Y et al. Mice lacking the vitamin D receptor exhibit impaired bone formation, uterine hypoplasia and growth retardation after weaning. Nat Genet 1997;16:391–6.
26. Matsumoto T, Fontaine O, Rasmussen H. Effect of 1,25-dihydroxyvitamin D_3 on phosphate uptake into chick intestinal brush border membrane vesicles. Biochim Biophys Acta 1980;599:13–23.
27. Fuchs R, Peterlik M. Vitamin D-induced phosphate transport in intestinal brush border membrane vesicles. Biochem Biophys Res Commun 1980;93:87–92.
28. Yagci A, Werner A, Murer H, Biber J. Effect of rabbit duodenal mRNA on phosphate transport in Xenopus laevis oocytes: dependence on 1,25-dihydroxy-vitamin-D_3. Pflugers Arch 1992;422:211–6.
29. Cross HS, Peterlik M. Calcium and inorganic phosphate transport in embryonic chick intestine: triiodothyronine enhances the genomic action of 1,25-dihydroxycholecalciferol. J Nutr 1988;118:1529–34.
30. Debiec H, Cross HS, Peterlik M. 1,25-Dihydroxycholecalciferol-related Na^+/D-glucose transport in brush border membrane vesicles from embryonic chick jejunum. Modulation by triiodothyronine. Eur J Biochem 1991;201:709–13.
31. Loghman-Adham M. Adaptation to changes in dietary phosphorus intake in health and in renal failure. J Lab Clin Med 1997;129:176–88.
32. Danisi G, Caverzasio J, Trechsel U, Bonjour JP, Straub RW. Phosphate transport adaptation in rat jejunum and plasma level of 1,25-dihydroxyvitamin D_3. Scand J Gastroenterol 1990;25:210–5.
33. Caverzasio J, Danisi G, Straub RW, Murer H, Bonjour JP. Adaptation of phosphate transport to low phosphate diet in renal and intestinal brush border membrane vesicles: influence of sodium and pH. Pflugers Arch 1987;409:333–6.
34. Cramer CF, McMillan J. Phosphorus adaptation in rats in absence of vitamin D or parathyroid glands. Am J Physiol 1980;239:G261–5.
35. Katai K, Miyamoto K, Kishida S, Segawa H, Nii T, Tanaka H et al. Regulation of intestinal Na^+-dependent phosphate co-transporters by a low-phosphate diet and 1,25-dihydroxyvitamin D_3. Biochem J 1999;343:705–12.
36. Norbis F, Boll M, Stange G, Markovich D, Verrey F, Biber J et al. Identification of a cDNA protein leading to an increased Pi-uptake in Xenopus laevis oocytes. J Membr Biol 1997;156:19–24.
37. Wagner GF, Hampong M, Park CM, Copp DH. Purification, characterization, and bioassay of teleocalcin, a glycoprotein from salmon corpuscles of Stannius. Gen Comp Endocrinol 1986;63:481–91.
38. Madsen KL, Tavernini MM, Yachimec C, Mendrick DL, Alfonso PJ, Buergin M et al. Stanniocalcin: a novel protein regulating calcium and phosphate transport across mammalian intestine. Am J Physiol 1998;274:G96–102.
39. Lu M, Wagner GF, Renfro JL. Stanniocalcin stimulates phosphate reabsorption by flounder renal proximal tubule in primary culture. Am J Physiol 1994;267:R1356–62.
40. Tenenhouse HS. Recent advances in epithelial sodium-coupled phosphate transport. Curr Opin Nephrol Hypertens 1999;8:407–14.
41. Beck L, Karaplis AC, Amizuka N, Hewson AS, Ozawa H, Tenenhouse HS. Targeted inactivation of Npt2 in mice leads to severe renal phosphate wasting, hypercalciuria, and skeletal abnormalities. Proc Natl Acad Sci USA 1998;95:5372–7.
42. Kavanaugh MP, Kabat D. Identification and characterization of a widely expressed phosphate transporter/retrovirus receptor family. Kidney Int 1996;49:959–63.
43. Olah Z, Lehel C, Anderson WB, Eiden MV, Wilson CA. The cellular receptor for gibbon ape leukemia virus is a novel high affinity sodium-dependent phosphate transporter. J Biol Chem 1994;269:25426–31.
44. Hilfiker H, Hattenhauer O, Traebert M, Forster I, Murer H, Biber J. Characterization of a murine type II sodium-phosphate co-transporter expressed in mammalian small intestine. Proc Natl Acad Sci USA 1998;95:14564–9.
45. Xu H, Bai L, Collins JF, Ghishan FK. Molecular cloning, functional characterization, tissue distribution, and chromosomal localization of a human, small intestinal sodium-phosphate (Na^+-Pi) transporter (SLC34A2). Genomics 1999;62:281–4.
46. Bai L, Collins JF, Ghishan FK. Cloning and characterization of a type III Na-dependent phosphate co-transporter from mouse intestine. Am J Physiol 2000;279:C1135–43.
47. Shibui A, Tsunoda T, Seki N, Suzuki Y, Sugane K, Sugano S. Isolation and chromosomal mapping of a novel human gene showing homology to Na^+/PO4 co-transporter. J Hum Genet 1999;44:190–2.

4B

Renal Handling of Phosphate

E. Takeda, K. Morita, Y. Taketani, H. Yamamoto and K. Miyamoto

Introduction

Dietary phosphorus is converted in the body to phosphates, in which form it exerts its widespread physiological functions, as an essential component of phospholipids, ATP, DNA, phosphorylated proteins, metabolic intermediates, body buffers, and bone. Phosphate homeostasis is dependent on the interaction of three organ systems – the gastrointestinal tract, the bone, and the kidneys – through the coordinated functions primarily of two hormones; parathyroid hormone (PTH) and vitamin D. Under conditions of low dietary phosphate intake, the intestine increases its absorptive efficiency to maximise phosphate absorption and the kidney increases renal phosphate transport to minimise urinary phosphate losses. The ability of the kidney to respond to acute increases or decreases in the filtered load of phosphate by altering excretion is well described. More intriguing is the intrinsic ability of the renal tubule cell to respond to alterations in body phosphate balance as a result of dietary availability or metabolic need.

Physiologic conditions associated with lifecycle changes, such as growth, pregnancy and lactation, are associated with increased phosphate need and a corresponding increase in phosphate absorption. It is well known that the rate of phosphate excretion and adaptation to dietary phosphate changes with age. Plasma phosphate concentration varies as a function of age in humans: in the range of 3.8 to 5.5 mg dl^{-1} (1.23 to 1.77 mM) in children and 3.0 to 4.5 mg dl^{-1} (0.97 to 1.45 mM) in adults. The kidney is the major organ of extracellular phosphate homeostasis and plays a key role in bone mineralisation and growth [1]. Retention of phosphate and associated hyperphosphataemia are important in the development of hyperparathyroidism secondary to renal failure [2]. High extracellular concentrations of phosphate have a direct effect on PTH secretion leading to renal osteodystrophy [3, 4].

Renal Handling of Phosphate

Most (93–98%) of the phosphate in plasma is ultrafilterable at the level of the glomerulus. Normally, the renal tubules reabsorb 80–97% of the filtered load, so only 3–20% of filtered phosphate appears in the urine. As plasma phosphate increases, filtered phosphate and phosphate reabsorption increase. However, the reabsorptive mechanism is quickly saturated, and excretion then increases in proportion to the filtered load (Figure 4.3). The plasma concentration at which maximal reabsorption

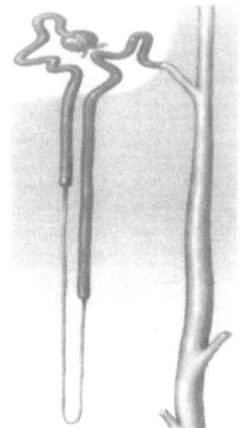

glomerular filtration
9 to 83% of phosphate in plasma

tubular reabsorption
80 to 97% of the filtered load
proximal convoluted tubule   60%
proximal straight tubule   15 to 20%
distal convoluted tubule   10 to 15%
collecting duct   5 to 10%

urine
3 to 20% of the filtered load

Figure 4.3. Renal handling of phosphate.

of phosphate (TmP) occurs lies close to the concentration of normal fasting plasma phosphate concentration, indicating that the kidneys regulate phosphate within a narrow range. There is a direct correlation between Tm phosphate values and glomerular filtration rate (GFR) even when the latter is varied over a broad range. Thus, the ratio TmP GFR^{-1} is a better index of tubule phosphate reabsorptive capacity. In humans, with a GFR above 40 mL min^{-1}, TmP varies proportionately with the GFR to maintain plasma phosphate and the TmP GFR^{-1} constant [5]. However, when the GFR falls below 40 mL min^{-1} or when nephrons hypertrophy because of a reduction in renal mass, TmP GFR^{-1} and plasma phosphate increase.

Most of the filtered phosphate is reabsorbed in the proximal tubule, with approximately 60% of the filtered load reclaimed in the proximal convoluted tubule and 15–20% in the proximal straight tubule. A significant amount of phosphate, perhaps on the order of 20–30%, is reabsorbed beyond the portion of the proximal tubule that is accessible to micropuncture. There is little phosphate transport within the loop of Henle, with most distal transport occurring in the distal convoluted tubule. In this segment, approximately 15% of filtered phosphate is reabsorbed under baseline conditions in animals subjected to parathyroidectomy, but the value falls to about 6% after administration of large doses of PTH. The collecting duct is a potential site for distal nephron reabsorption of phosphate.

It is also evident from data obtained from various micropuncture and microinjection studies that juxtamedullary and superficial nephrons have different capacities for phosphate transport. The increased responsiveness of the deep nephrons to phosphate intake suggests a key regulatory role for this system in phosphate homeostasis.

In contrast to other tubule transport processes, phosphate reabsorption in infancy is higher than that in adulthood and declines with age, despite a lower GFR in neonates. Studies in isolated perfused guinea pig kidneys demonstrated that proximal tubule phosphate reabsorption was 77% in newborns, compared to 67% in adults, and that distal tubule phosphate reabsorption was 16% in newborns and 11% in adults. Thus, immature animals have a higher rate of renal phosphate transport than adults. This increase is seen in proximal and distal tubule and is accompanied by an increased phosphate transport activity.

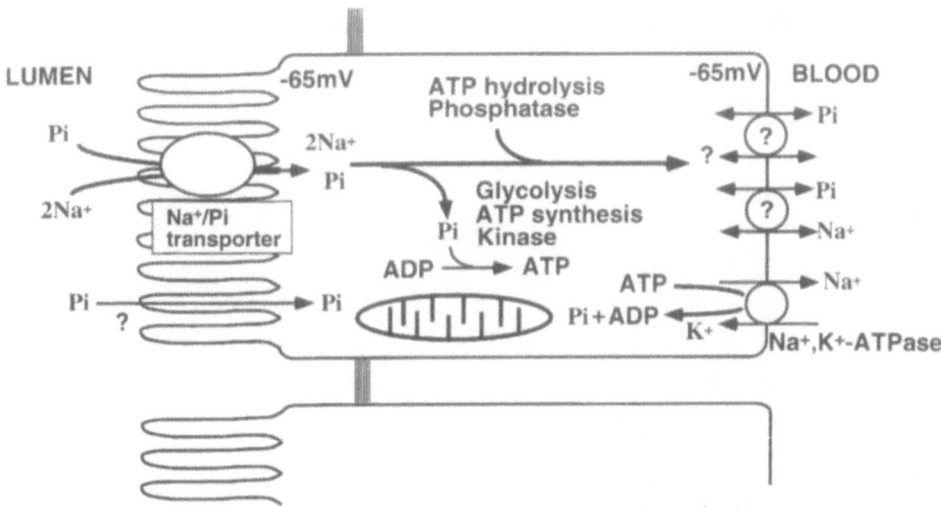

Figure 4.4. Schematic model of inorganic phosphate transport across the renal proximal tubular cell.

Mechanism of Renal Tubular Phosphate Reabsorption

Transcellular Phosphate Reabsorption

Transepithelial phosphate transport is effectively unidirectional and includes uptake at the brush border membrane (BBM) of the renal tubule cell, translocation across the cell, and efflux at the basolateral membrane (BLM) (Figure 4.4). Because the electrical charge of the cell interior is negative compared to the exterior, and phosphate concentrations are higher in the cytosol, phosphate must move against an electrochemical gradient into the cell interior, whereas, at the antiluminal membrane, the transport of phosphate into the peritubular capillary is favoured by the high intracellular phosphate concentration and the electronegativity of the cell interior. Studies with BBM vesicles (BBMV) have demonstrated co-transport of Na$^+$ with phosphate across the BBM, whereas the transport of phosphate across the BLM is independent of that of Na$^+$ [6]. Thus, it is probable that an active transport process is necessary to allow phosphate into the cell, but phosphate may exit the cell by diffusion.

Phosphate uptake at the BBM is the rate-limiting step in this process and the major site of regulation [7]. The uptake at the BBM is mediated by sodium-dependent phosphate (Na/Pi) transporters. Several hormonal and non-hormonal factors influence phosphate reabsorption via a change of apical Na/Pi transport activity (Table 4.4). This function is regulated by dietary phosphate and a variety of hormones, such as PTH, 1,25-dihydroxyvitamin D$_3$ (1,25(OH)$_2$D$_3$) and others. Thyroid hormones, 1,25(OH)$_2$D$_3$, all-transretinoic acid, insulin, insulin-like growth factor-1 and growth hormone increase transport rate, whereas glucocorticoids, oestradiol, calcitonin, atrial natriuretic peptide, epidermal growth factor, transforming growth factors, PTH and PTH-related peptide reduce transport activity [1].

Table 4.4.   Factors influencing renal phosphate reabsorption

Stimulating factors	Inhibiting factors
Low phosphate	High phosphate
Vitamin D_3	Parathyroid hormone
Hypercalcaemia	Calcitonin
Insulin	Dopamine
Growth hormone	Diuretics
Thyroid hormone	Glucocorticoid
All-transretinoic acid	Oestradiol
Insulin like growth factor	Atrial natriuretic peptide
Alkalosis	Acidosis

Table 4.5.   Sodium dependent phosphate transporters

Transporter type	Distribution	Affinity for phosphate (Km)	Characteristics
Type 1 (NPT1)	Kidney Proximal tubule brush border membrane	5–10 nM	Not specific for phosphate transport Functions as anion transporter Not regulated by PTH or phosphate
Type 2 (NPT2)	Kidney Proximal tubule brush border membrane	100–200 nM	Specific for phosphate transport Regulated by PTH and phosphate Inhibited by acidosis
Type 3 (NPT3)	Ubiquitous	25 nM	"Housekeeping" function Regulated by phosphate Inhibited by alkalosis

PTH: parathyroid hormone.

Na/Pi Transporters

The renal Na/Pi transporters are the primary regulators of the plasma phosphate level by altering the rate of phosphate reabsorption of the filtered phosphate load. The renal apical Na/Pi transporters have been cloned and divided into two types based on their predicted amino acid sequences: type I (NPT1), represented by NaPi-1, RNaPi-1, NPT-1 and Npt-1; and type II (NPT2), represented by NaPi-2, NaPi-3, NaPi-4, NaPi-6 and NaPi-7 [8, 9]. The cell surface receptors for the gibbon ape leukaemia virus (Glvr-1, PiT-1) and amphotropic murine retrovirus (Ram-1, PiT-2) have been demonstrated to be type III Na/Pi transporters (NPT3) [10, 11]. The NPT3 are expressed in a wide variety of tissues [12], and are regulated by changes in extracellular phosphate concentration (Table 4.5). The proteins of all three families use the favourable electrochemical gradient for Na^+ entry into the cell to drive the simultaneous movement of phosphate intracellularly against its gradient. Ultimately, the energy for this process derives from the Na^+,K^+-ATPase which pumps Na^+ extracellularly in exchange for the inward movement of K^+, maintaining the conditions favourable for Na^+ entry into the cell.

The oocyte expressing NPT1 increases the transmembrane permeability of different organic anions such as probenecid, phenol red and penicillin, which are all inhibited by chloride channel blockers [13, 14]. Glucose and insulin up-regulate the expression of NPT1 in cultured hepatocytes, whereas glucagon and elevated intra-cellular adenosine 3',5'-cyclic monophosphate (cAMP) levels down-regulate its expression [15].

In contrast, NPT2 primarily transports phosphate and does not interact with sulphate. The deduced amino acid sequences predict proteins of approximately 65 to 75 kDa with eight transmembrane regions, several potential sites for N-glycosylation and for phosphorylation by protein kinase C, and a characteristic leucine zipper motif. Immunohistochemical studies confirm proximal renal tubule apical membrane localisation with the highest concentrations in the early S1 segment, declining progressively to the S3 segment. NPT2 transports monovalent and divalent phosphate. Transport is electrogenic and inhibited by acidic pH owing to a decreased affinity of the transporter for Na^+. Immunoblot analysis of proximal renal tubule apical membranes and radiation inactivation studies show proteins ranging in size from 90 to 215 kDa and suggest that the functional transporter is glycosylated and may exist in multimeric states. Glycosylation is important for the function of type II transporters, possibly through targeting to the apical membrane. The expression of NPT2 mRNA was found exclusively in the human kidney cortex, with the exception of a related transcript found in human lung tissue. Immunohistochemical analyses have also shown that NPT1 is expressed in the straight part of proximal tubules (S3 segments) and in convoluted proximal tubules. Dietary phosphate depletion leads to an increase in the expression of NPT2 and a subsequent increase in the V_{max} for phosphate transport [7]. Conversely, PTH, which inhibits phosphate transport, decreases the expression of NPT2. This family of Na/Pi transporters, thus, is felt to be the target protein for regulation of phosphate homeostasis by the kidney.

NPT3 (PiT-1 and PiT-2) also transports sodium and phosphate with a stoichiometry that results in a net influx of positive charge. Northern blot analysis demonstrated the expression of NPT1 mRNA in the human kidney cortex and liver [16]. NPT3 is regulated by changes in the medium phosphate concentration in vitro, and shows a high affinity for phosphate. However, phosphate depletion and parathyroidectomy do not affect the amount of NPT3 (Glvr-1) mRNA in the rat kidney. The ubiquitous distribution of NPT3 suggests it serves a "housekeeping" function.

The regulation of phosphate reabsorption by PTH and adaptation of dietary phosphate occurs mainly in proximal tubules (S1 segments). The expression of the NPT2 is stronger in the proximal tubules of juxtamedullary nephrons. Both NPT3 (PiT-1 and PiT-2) transcripts were found in relatively high abundance in total mRNA prepared from the kidney, although the functional role in renal phosphate transport has not yet been determined [17]. These observations suggest that the NPT2, rather than the NPT1 or NPT3 mediate most Na/Pi transport in the kidney and serve to physiologically and pathophysiologically regulate the proximal tubular reabsorption of phosphate [17–19].

Factors Affecting Renal Handling of Phosphate

Regulatory control of phosphate transport is crucial because of the diverse roles of intracellular phosphate in cell biology, and the availability of NPT2 clones will now allow detailed molecular studies of the structural and functional bases of this regu-

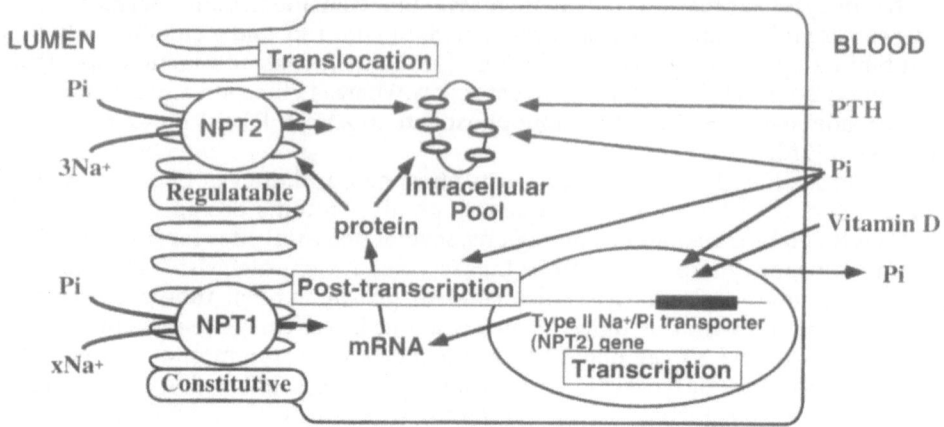

Figure 4.5. Regulatory mechanisms of Na+/Pi transport activity in renal proximal tubular cell.

lation (Figure 4.5). 1,25(OH)$_2$D$_3$ and dietary phosphate are important factors for the gene expression of NPT2 (Table 4.4) [20–22].

Phosphate

Regulation by Intracellular Translocation of NPT2

Alteration of the dietary intake of phosphate also leads to rapid changes in renal NPT2 protein. Acute adaptation is not associated with a change in NPT2 mRNA and is not blocked by actinomycin D, suggesting that acute phosphate depletion results in the insertion of transporters into the apical membrane from a subapical pool [23]. Treatment with colchicine, which disrupts microtubules, abolished the effect of high-phosphate diet on the endocytosis of NPT2. The effect of dietary high phosphate on rapid endocytosis was in a PTH-independent manner (Figure 4.5) [20, 24]. However, the signals mediating the effect of dietary high phosphate remain unknown.

Reshkin et al. demonstrated in opossum kidney (OK) cells that a low phosphate medium provoked up-regulation of transport only when bathing the apical surface, not the basolateral surface, of the cells [25]. Chronic phosphate deprivation has been shown to increase cytosolic Ca^{2+} in OK cells, which leads to increased transcription and translation of NPT2 in them [26]. This is in contrast to what is seen in MDCK cells, where chronic phosphate depletion results in a decrease in cytosolic Ca^{2+}. Apparently, the sensing mechanism is localised to the apical surface of proximal tubule cells and leads to an increase in intracellular Ca^{2+}.

Transcriptional Regulation of NPT2 Expression

NPT2 is up-regulated by chronic feeding of low-phosphate diets [20]. Chronic adaptation to alterations in phosphate intake emerges over two to three days and persists for the duration of phosphate deprivation or excess. Both the level of mRNA and the amount of transporter protein in the apical membrane of renal epithelial cells are increased in those animals [20]. The elevation of the NPT2 mRNA level is due, at least in part, to an increase in the transcription rate. To elucidate the molecular mechanisms by which dietary phosphate regulates gene transcription, a candidate phosphate-responsive element and its binding protein on the human NPT2 (NaPi-3) gene promoter have been identified [27].

In mice fed a phosphate-restriction diet, DNA footprinting analysis and gel mobility shift analysis showed that the position of −1,010 to −985 upstream of the transcription start site was important to the gene transcription. The yeast one-hybrid system was used to clone a transcription factor that binds to the sequences in the proximal promoter of the NPT2 gene, utilising as bait the sequence "CACGTG". Two cDNA clones were isolated that encoded protein of TFE3 transcription factor, which is a DNA-binding protein that activates transcription through the μE3 site of the immunoglobulin heavy-chain enhancer. Recombinant TFE3 was shown to bind to the phosphate-responsive element of the NPT2 gene promoter. Also, the mutation of the core motif completely abolished the DNA binding of TFE3. Coexpression of TFE3 in cloned OK-B cells transfected the NPT2 gene promoter markedly stimulated the luciferase activity. These observations suggest that TFE3 may participate in the transcriptional regulation of the NPT2 gene by dietary phosphate restriction [27].

Post-Transcriptional Regulation in NPT2 Expression

Hypophosphataemia also increases NPT2 gene expression post-transcriptionally, which correlates with a more stable transcript mediated by the 3'-untranslated region (UTR), and an increase in NPT2 translation, which involves protein binding to the 5'-UTR. The additive result is an increase in renal phosphate reabsorption. Renal hypophosphataemic proteins stabilise the NPT2 mRNA. Hypophosphataemia leads to an increase in a renal translation factor(s). Hypophosphataemic proteins led to an increased binding of renal cytosolic proteins, but not hepatic proteins, to the 5'-UTR of NPT2 mRNA and not the 3'-UTR. Kidney cytosolic proteins from hypophosphataemic rats led to increased synthesis of NPT2 in an in vitro translation assay, suggesting an increase in translation in addition to the increased mRNA levels [28].

Vitamin D

Long-term administration of $1,25(OH)_2D_3$ results in phosphaturia, independent of the effect of PTH. The tubule site of its action is thought to be the proximal tubule, because the effect can be demonstrated in renal cortical BBMV. The mechanism of action is probably secondary to increased intestinal absorption, which causes the same response in the proximal tubule as a high dietary phosphate intake.

In vitamin D-deficient rats, the amounts of NPT2 protein and mRNA were decreased in the juxtamedullary kidney cortex, but not in the superficial cortex, compared with control rats. The administration of $1,25(OH)_2D_3$ to vitamin D-deficient rats increased the initial rate of phosphate uptake as well as the amounts of NPT2 mRNA and protein in the juxtamedullary cortex.

To further investigate the regulation of the NPT2 gene expression using $1,25(OH)_2D_3$, the human NPT2 gene was cloned from the human placenta genomic library [21, 22]. The human NPT2 gene comprises 13 exons and spans approximately 14 kb. In the 5' flanking region of the NPT2 gene, a typical TATA box and various cis-acting elements, including a cAMP-responsive element, AP-1, AP-2 and SP-1 sites have been identified. The transcriptional activity of a luciferase reporter plasmid containing the promoter region of the human NPT2 gene was markedly increased by

1,25(OH)$_2$D$_3$ in COS-7 cells expressing the human vitamin D receptor [21, 22]. The human NPT2 gene promoter also contains three direct repeat-like sequences that resemble the consensus binding sequence for members of the steroid–thyroid hormone receptor superfamily. A deletion and mutation analysis of the NPT2 gene promoter identified the vitamin D-responsive element as the sequence 5'-GGGGCAGCAAGGGCA-3' nucleotides −1,977 to −1,963 relative to the transcription start site. This element bound a heterodimer of the vitamin D receptor and retinoid X receptor, and it enhanced the basal transcriptional activity of the promoter of the herpes simplex virus thymidine kinase gene in an orientation-independent manner [22].

Parathyroid Hormone (PTH)

PTH produces an increase in urinary phosphate excretion by the inhibition of BBM Na/Pi transport [29]. Renal Na/Pi transport activity was significantly increased in parathyroidectomised rats (TPTX rats). Western blots of isolated denatured BBM have shown the elevation of the abundance of NPT2 protein in TPTX rats. The level of NPT2 protein was decreased after a two-hour treatment with PTH [29]. The loss of NPT2 from the BBM was observed in proximal tubules throughout the cortex, both superficial and juxtamedullary. These findings strongly suggest that one mechanism by which PTH inhibits renal Na/Pi transport at the cellular level involves selective endocytic removal of Na/Pi transporters from the BBM of the renal proximal tubule [8, 9, 20]. The endocytosed NPT2 are redistributed to the cytoplasm but they appear to be targeted for rapid lysosomal degradation rather than recycling to the BBM if PTH treatment is prolonged [29]. The inhibition of phosphate uptake does not require protein synthesis but is blocked by colchicine, an inhibitor of microtubules. However, after withdrawal of PTH, recovery of phosphate transport does require protein synthesis, suggesting that PTH causes degradation of the transporters. Kempson et al. and others have demonstrated endocytotic withdrawal of Na/Pi transporters from the apical membrane, followed by disappearance of the protein [30, 31]. Na/Pi transporters have been demonstrated in endosomal vesicles, supporting the concept that there are pools of transporters that may be subject to shuffling in and out of the plasma membrane in response to different stimuli.

PTH stimulates two signalling pathways in proximal tubule cells – adenylate cyclase and phospholipase C – resulting in activation of protein kinase A and protein kinase C. Immunoblot analysis of membranes derived from OK cells shows that protein kinase A, but not protein kinase C, activation decreases the expression of NPT2 [32]. This finding suggests that PTH-stimulated protein kinase A triggers the pathway responsible for internalisation and degradation of the transporters. Le Goas et al. have demonstrated that protein kinase C activation inhibits a cAMP phosphodiesterase, the enzyme responsible for breakdown of intracellular cAMP [33]. Siegfried et al. have shown that PTH activates 5'-ectonucleotidase, the enzyme responsible for breakdown of extracellular cAMP to adenosine [34]. Thus, protein kinase C activation could prolong the half-life and the effectiveness of intracellular cAMP, and potentiate the effect of extracellular cAMP, which would enhance the effect of PTH-stimulated protein kinase A.

Calcitonin

Infusion of large concentrations of calcitonin is transiently hyperphosphaturic, even in the absence of PTH [35]. The effect appears to be at the level of the Na/Pi transporter, as calcitonin decreases phosphate uptake in rat BBMV [36]. In contrast to PTH, calcitonin-sensitive adenylate cyclase is localised in the medullary and cortical thick ascending limbs and in the distal tubule, but the calcitonin-sensitive 25-hydroxyvitamin D-1α-hydroxylase (1α-hydroxylase) is localised in the proximal straight tubule. Calcitonin acts in the proximal convoluted tubule and possibly in proximal straight tubule to inhibit phosphate reabsorption. From the facts that calcitonin acts on both phosphate reabsorption and 1a-hydroxylase activity in nephron segments that do not possess calcitonin-sensitive adenylate cyclase, it has been concluded that calcitonin regulates these processes by a cAMP-independent mechanism. The physiologic importance of calcitonin on phosphate homeostasis in general and on renal handling of phosphate is uncertain.

Role of Renal Phosphate Reabsorption

Defect of NPT2 in the Kidney

Mice deficient in the NPT2 gene exhibit increased urinary phosphate excretion, hypophosphataemia, an appropriate elevation in the serum concentration of $1,25(OH)_2D_3$ with attendant hypercalcaemia, hypercalciuria and decreased serum parathyroid hormone levels, and increased serum alkaline phosphatase activity [37]. NPT2-/- mice do not have rickets or osteomalacia. At weaning, NPT2-/- mice have poorly developed trabecular bone and retarded secondary ossification, but with increasing age there is a dramatic reversal and eventual overcompensation of the skeletal phenotype. These findings indicate that NPT2 is a major regulator of phosphate homeostasis and necessary for normal skeletal development.

X-Linked Hypophosphataemia

X-linked hypophosphataemia (XLH) is the most common form of familial hypophosphataemic rickets, which results from impaired renal phosphate reabsorption, renal vitamin D metabolism and skeletal mineralisation. It is well established that hypophosphataemia stimulates 1α-hydroxylase to elevate circulating $1,25(OH)_2D_3$ levels. However, circulating $1,25(OH)_2D_3$ is inappropriately low for the prevailing phosphate concentrations in patients with XLH [38]. Importantly, such patients can be cured with a therapeutic combination of oral phosphate and $1,25(OH)_2D_3$.

The murine Hyp and Gy homologues are useful models for human X-linked hypophosphataemia and have been used in recent studies on renal phosphate transport [39, 40]. We have shown that the BBM NPT2 protein content and the related mRNA content are reduced in affected animals, showing also reduced BBM Na/Pi transport activity. In contrast, the NPT1 appears to be unaffected. The NPT2 may thus represent a final target in the regulatory cascades affected in X-linked disorders. Recently, the gene causing X-linked hypophosphataemia was identified using positional cloning [41]. This gene was labelled PHEX (phosphate regulating gene with homologies to endopeptidase on the X-chromosome). However, the role that the PHEX gene plays in phosphate transport is not yet known. Expression of the PHEX gene has been detected using Northern blot analysis in mouse osteoblasts and in human lung and ovary [42]. The manner in which a functional loss of the putative PHEX enzyme in these tissues leads to the anatomically remote renal tubular defects of phosphate transport remains to be elucidated. Several reports suggest that a humoral factor (phosphatonin) may be responsible for the renal phosphate loss observed in XLH [43, 44].

Cai et al. detected a factor secreted by the cells of a recurrent haemangiopericytoma in a patient with tumour-associated osteomalacia that inhibited phosphate uptake in OK cells [45]. This factor was not PTH or PTHrP and was specific for inhibition of phosphate uptake. Screening of a tumour library revealed the presence of a peptide that contained a protein kinase C activation domain similar to the protein kinase C domain of PTH, but expression of the gene failed to produce the synthesis of the phosphate inhibitory peptide [46]. A similar but not identical factor has been isolated from the dialysate of patients with chronic renal failure [46].

Novel Factors Influencing Renal Handling of Phosphate

Stanniocalcin (STC) is a calcium- and phosphate-regulating hormone produced by the corpuscles of Stannius in bony fishes. The mammalian homologue of STC has recently been reported (STC 1), which stimulates the phosphate uptake of the kidney [47]. The cloning of a second mammalian stanniocalcin (STC2) from the human osteosarcoma cDNA library was also reported [48]. The effect of STC2 on the promoter activity of renal NPT2 was examined, using the culture medium of STC2-transfected CHO cells. The finding suggested that STC2 would inhibit the expression of NPT2 at the transcriptional level. Therefore, STC2 may inhibit phosphate transport in the long term through the transcriptional control of a phosphate transporter. Indeed, STC2 inhibited the phosphate uptake of OK cells after two-day incubation, suggesting that the expression of the endogenous NPT2 in OK cells was reduced.

Novel candidate genes for phosphatonin were reported as matrix extracellular phosphoglycoprotein (MEPE) in tumour tissue from oncogenic hypophosphataemic osteomalacia [49], osteoregulin in osteoblasts and osteocytes [50] and fibrogrowth growth factor 23 as a causal gene of autosomal dominant hypophosphataemic rickets [51], although their function remains to be clarified.

Summary

Phosphate homeostasis is primarily dependent on renal tubular phosphate transporters. NPT2 serves to physiologically and pathophysiologically regulate the proximal tubular phosphate reabsorption. Phosphate transport activity mediated by NPT2 is regulated by transcriptional level, intracellular translocation of NPT2, and post-transcriptional level. In addition, novel humoral factors may regulate the expression of NPT2 in renal tubular cells.

References

1. Drezner MK. Clinical disorders of phosphate homeostasis. In: Feldman D, Glorieux, Pike JW, editors. Vitamin D. San Diego: Academic Press, 1997;733–53.
2. Bover J, Rodriguez M, Trinidad P, Jara A, Machado L, Liach F, et al. Factors in the development of secondary hyperparathyroidism during graded renal failure in the rat. Kidney Int 1994;41: 953–61.
3. Silver J, Russell J, Sherwood LM. Regulation by vitamin D metabolism of messenger ribonucleic acid for preproparathyroid hormone in isolated bovine parathyroid cells. Proc Natl Acad Sci USA 1985; 84:1349–55.

4. Almaden Y, Canelejo A, Hernandez A, Ballesteros E, Garcia Navarro S, et al. Direct effect of phosphorus on PTH secretion from rat parathyroid glands in vitro. J Bone Miner Res 1996;11:970-6.
5. Bijvoet OLM. Relation of plasma phosphate concentration to renal tubular reabsorption of phosphate. Clin Sci 1969;37:23-36.
6. Hoffman N, Thees M, Kinne F. Phosphate transport by isolated renal brush border vesicles. Pflugers Arch 1976;362:147-56.
7. Suki WN, Lederer ED, Rouse D. Renal transport of calcium, magnesium, and phosphate. In: Brenner BM, editor. The Kidney. Philadelphia: WB Saunders Company, 2000;520-74.
8. Biber J, Custer M, Magagnin S, Hayes G, Werner A, Lotscher M, et al. Renal Na/Pi-cotransporters. Kidney Int 1996;49:981-5.
9. Murer H, Biber J. A molecular view of proximal tubular inorganic phosphate (Pi) reabsorption and its regulation. Pflugers Arch 1997;433:379-89.
10. Kavanaugh MP, Miller DG, Zhang W, Law W, Kozak SL, Kabat D, et al. Cell-surface receptors for gibbon ape leukemia virus and amphotropic murine retroviruses are inducible sodium-dependent phosphate symporters. Proc Natl Acad Sci USA 1994;91:7071-5.
11. Caverzasio J, Bonjour JP, Fleisch H. Tubular handling of Pi in young growing and adult rats. Am J Physiol 1982;242:F705-10.
12. Kavanaugh MP, Kabat D. Identification and characterization of a widely expressed phosphate transporter/retrovirus receptor family. Kidney Int 1996;49:959-63.
13. Busch AE, Biber J, Murer H, Lang F. Electrophysiological insights of type I and II Na/Pi transporters. Kidney Int 1996;49:986-7.
14. Yabuuchi H, Tanai I, Morita K, Kouda T, Miyamorto K, Takeda E, et al. Hepatic sinusoidal membrane transport of anionic drugs mediated by anion transporter Npt 1. J Pharmacol Exp Ther 1998; 268:1391-6.
15. Li H, Ren P, Onwochei M, Ruch RJ, Xie Z. Regulation of rat Na$^+$/Pi cotransporter-1 gene expression: the roles of glucose and insulin. Am J Physiol 1996;271:EI021-8.
16. Miyamorto K, Tatsumi S, Sonoda T, Yamamoto H, Minami H, Taketani Y, et al. Cloning and functional expression of a Na$^+$-dependent phosphate cotransporter from human kidney cDNA cloning and functional expression. Biochem J 1995;305:81-5.
17. Miyamorto K, Segawa H, Morita K, Nii T, Tatsumi S, Taketani Y, et al. Relative contributions of Na$^+$-dependent phosphate co-transporters to phosphate transport in mouse kidney: RNase H-mediated hybrid depletion analysis. Biochem J 1997;327:735-9.
18. Murer H, Markovich D, Biber J. Renal and small intestinal sodium-dependent tranporters of phosphate and sulphate. J Exp Biol 1994;196:167-81.
19. Takeda E, Taketani Y, Morita K, Tatsumi S, Katai K, Nii T, et al. Molecular mechanisms of mammalian inorganic phosphate homeostasis. Advan Enzyme Regul 2000;40:285-302.
20. Katai K, Segawa H, Haga H, Morita K, Arai H, S. Tatsumi S, et al. Acute regulation by dietary phosphate of the sodium-dependent phosphate transporter (NaPi-2) in rat kidney. J Biochem 1997; 121:50-5.
21. Taketani Y, Miyamorto K, Tanaka K, Chikamori M, Tatsumi S, Segawa H, et al. Gene structure and functional analysis of the human Na$^+$/phosphate co-transporter (NaPi-3). Biochem J 1997;324:927-34.
22. Taketani Y, Segawa H, Chikamori M, Morita K, Tanaka K, Kido S, et al. Regulation of type ll renal Na$^+$-dependent inorganic phosphate transporters by 1,25-dihydroxyvitamin D$_3$. J Biol Chem 1998;273:14575-81.
23. Loghman-Adham M. Adaptation to changes in dietary phosphorus intake in health and in renal failure. J Lab Clin Med 1997;129:176-88.
24. Takahashi F, Morita K, Katai K, Segawa H, Fujioka A, Kouda T, et al. Effects of dietary inorganic phosphate on the renal Na$^+$-dependent inorganic-phosphate transporter NaPi-2 in thyroparathyroidectomized rats. Biochem J 1998;333:175-81.
25. Reshkin SJ, Forgo J, Biber J, Murer H. Functional asymmetry of phosphate transport and its regulation in opossum kidney cells: Phosphate "adaptation". Pflugers Arch 1991;419:256-62.
26. Saxena S, Allon M. The role of cytosolic calcium in chronic adaptation to phosphate depletion in opossum kidney cells. J Biol Chem 1996;271:3902-6.
27. Kido S, Miyamorto K, Mizobuchi H, Taketani Y, Ohokido I, Ogawa N, et al. Identification of regulatory sequences and binding proteins in the type II sodium/phosphate cotransporter NPT2 gene responsive to dietary phosphate. J Biol Chem 1999;274:28256-63.
28. Moz Y, Silver J, Naveh-Many T. Protein-RNA interactions determine the stability of the renal NaPi-2 cotransporter mRNA and its translatiion in hypophosphatemic rats. J Biol Chem 1999;274:25266-72.
29. Kempson SA, Lotscher M, Kaissling B, Biber J, Murer H, Levi M. Parathyroid hormone action on phosphate transporter mRNA and protein in rat renal proximal tubules. Am J Physiol 1995;268: F784-91.

30. Kempson SA, Latscher M, Kaissling B, Biber J, Murer H, Levi M. Parathyroid hormone action on phosphate transporter mRNA and protein in rat renal proximal tubules. Am J Physiol 1995;268:F784–91.
31. Paraiso MS, McAteer JA, Kempson SA. Parathyroid hormone inhibits plasma membrane Pi transport without changing endocytic activity in opossum kidney cells. Biochim Biophys Acta 1995;1266:143–7.
32. Lederer ED, Sohi SS, Mathiesen JM, Klein JB. Regulation of expression of type II sodium-phosphate cotransporters by protein kinases A and C. Am J Physiol 1998;275:F270–7.
33. Le Goas F, Amiel C, Friedlander G. Protein kinase C modulates cAMP content in proximal tubular cells: Role of phosphodiesterase inhibition. Am J Physiol 1991;261:F587–92.
34. Siegfried G, Vrtovsnik F, Prie D, Amiel C, Friedlander G. Parathyroid hormone stimulates ecto-5′-nucleotidase activity in renal epithelial cells: Role of protein kinase C. Endocrinology 1995;136:1267–75.
35. Carney SL. Calcitonin and human renal calcium and electrolyte transport. Miner Electrolyte Metab 1997;23:43–7.
36. Yusufi AN, Szczepanska-Konkel M, Hoppe A, Dousa TP. Different mechanisms of adaptive increase in Na–Pi cotransport across renal brush-border membrane. Am J Physiol 1989;256:F852–61.
37. Beck L, Karaplis AC, Amizuka N, Hewson AS, Ozawa H, Tenenhouse HS. Targeted inactivation of Npt2 in mice leads to severe renal phosphate wasting, hypercalciuria, and skeletal abnormalities. Proc Natl Acad Sci USA 1998;95:5372–7.
38. Drezner MK, Lyles KW, Haussler MR, Harrelson JM. Evaluation of a role for 1α,25-dihydroxyvitamin D₃ in the pathogenesis and treatment of X-linked hypophosphatemic rickets and osteomalacia. J Clin Invest 1980;66:1020–32.
39. Tenenhouse HS, Beck L. Renal Na⁺-phosphate cotransporter gene expression in X-linked Hyp and Gy mice. Kidney Int 1996;49:1027–32.
40. Morita K, Fujioka A, Haga H, Nii T, Segawa H, Kouda T, et al. Dietary regulation of renal phosphate transporters in hypophosphatemic mice. J Bone Miner Metab 1998;16:234–40.
41. The HYP Consortium. Positional cloning of PEX: A gene with homologies to endopeptidases is mutated in patients with X-linked hypophosphatemic rickets. Nature Genet 1995;11:130–6.
42. Grieff M, Mumm S, Waeltz P, Mazzarella R, Whyie MP, Thakker RV, et al. Expression and cloning of the human X-linked hypophosphatemia gene cDNA. Biochem Biophys Res Commun 1997;231:635–9.
43. Nesbitt T, Econs ML, Byun JK, Martel J, Tenenhouse HS, Drezner MK. Phosphate transport in immortalized cell cultures from the renal proximal tubules of normal and Hyp mice: evidence that the HYP gene locus product is an extrarenal factor. J Bone Miner Res 1995;10:1827–83.
44. Rowe PS. The PEX gene: its role in X-linked rickets, osteomalacia, and bone mineral metabolism. Exp Nephrol 1997;5:355–63.
45. Cai Q, Hodsgon SF, Kao PC, Lennon VA, Klee GG, Zinsmiester AR, et al. Inhibition of renal phosphate transport by a tumor product in a patient with oncogenic osteomalacia. N Engl J Med 1994;330:1645–9.
46. Kumar R, Haugen JD, Wieben ED, Londowski JM. Inhibitors of renal epithelial phosphate transport in tumor-induced osteomalacia and uremia. Proc Assoc Am Physicians 1995;107:296–305.
47. Wagner GF, Vozzolo BL, Jaworski EM, Haddad M, Kline RL, Olsen HS, et al. Human stanniocalcin inhibits renal phosphate excretion in the rat. J Bone Miren Res 1997;12:165–71.
48. Ishibashi K, Miyamoto K, Taketani Y, Morita K, Takeda E, Sasaki S, et al. Molecular cloning of a second human stanniocalcin homologue (STC2). Biochem Biophys Res Commun 1998;250:252–8.
49. Rowe PS, de Zoysa PA, Dong R, Wang HR, White KE, Econs MJ, et al. MEPE, a new gene expressed in bone marrow and tumors causing osteomalacia. Genomics 2000;67:54–68.
50. Brown TA, Petersen DN, Tkalcevic GT, Vail AL, Rivera-Gonzalez R. Identification of "osteoregulin", a bone-specific cDNA encoding an RGD-containing protein that is highly expressed in osteoblasts and osteocytes. J Bone Miner Res 2000;15:S184.
51. ADHR Consortium. Autosomal dominant hypophosphatemic rickets is associated with mutations in FGF23. Nat Genet 2000;26:345–8.

Hyper- and Hypophosphataemia

E. Ishimura, Y. Imanishi and M. Inaba

Introduction

Abnormalities in serum concentrations of phosphate, whether hyperphosphataemia or hypophosphataemia, are caused by abnormalities in the following conditions: 1) abnormalities in oral intake and in intestinal absorption of phosphate, 2) abnormalities in renal excretion of phosphate, and 3) abnormal shift of phosphate from cells to extracellular fluid or abnormal shift of phosphate into cells. When considering the differential diagnosis of abnormalities in serum phosphate levels, these three mechanisms should be considered to enable correct treatment. In this chapter, homeostasis of phosphate metabolism in adults is described first, followed by hypophosphataemia and hyperphosphataemia.

The normal range of serum phosphate levels is 3.0 mg/dl to 4.5 mg/dl, namely a range of 1.0 mmol to 1.5 mmol. Hypophosphataemia is defined as serum concentrations less than 2.5 mg/dl, and hyperphosphataemia is defined as serum levels greater than 5.0 mg/dl. In Tables 4.6 and 4.7, the causes of hypophosphataemia and hyperphosphataemia, respectively, are shown.

Homeostasis of Phosphorus in Adults

In adults, the total phosphorus amount is about 1% of total body weight, approximately 700–800 g. Of that, 85% of phosphorus is present in the bone as hydroxyapatite complex $(Ca_2(PO_4)6(OH)_2)$, 14% is present in the intracellular compartment, and the remaining small amount (less than 1%) is present in extracellular fluid, including plasma. Although the net concentration of phosphorus is approximately 14 mg/dl in plasma, more than half is present as organic phosphorus (constituents of protein and phospholipid compounds), and the inorganic phosphorus concentration is approximately 4 mg/dl. Inorganic phosphorus is present as phosphate in biological systems, and the serum concentration of phosphorus is clinically measured and reported as a phosphate concentration. The plasma (serum) concentration of phosphate is similar to the extracellular fluid concentration. Approximately 85% of plasma phosphate is present as free ions, 10% is present in protein-bound forms, and the remaining 5% is present as complexes of $CaHPO_4$ and $MgHPO_4$. Ionised phosphate is present as an equilibrium condition of divalent (HPO_4^{2-}) and monovalent $(H_2PO_4^-)$ forms, and has a buffering function regulating the changes in plasma pH. The pK for these forms is 6.8, and at the normal plasma condition of pH 7.4, the ratio

149

Table 4.6. Causes of Hypophosphataemia

Decreased phosphate intake and/or decreased intestinal absorption of phosphate
    Antacids
        magnesium hydroxide, calcium carbonate, aluminium hydroxide
    Malabsorption syndrome
    Starvation
    Vitamin D deficiency
    Disturbance in the synthesis of the active form of vitamin D
    Alcoholism
    Intravenous hyperalimentation with low phosphate concentration

Increased urinary phosphate excretion
    Increased parathyroid hormone
        primary hyperparathyroidism
    Increased PTHrP (PTH related peptide)
        Humoral hypercalcaemia of malignancy
    Renal tubular acidosis (RTA)
        Particularly, distal type RTA
    Diuretics
    Excessive extracellular fluid accumulation
    Recovery phase of acute renal failure – polyuric phase
    Post-transplantations
    Oncogenic osteomalacia (tumour-induced osteomalacia)
    Fanconi Syndrome
    Familial hypophosphataemic rickets
    Diabetes mellitus with poor control and ketoacidosis
    Excessive glucocorticoids, such as Cushing's syndrome
    Hypomagnesaemia
    Hypokalaemia

Shift of phosphate into cells
    Acute respiratory alkalosis
    Recovery phase of diabetic ketoacidosis
    Intravenous hyperalimentation
    Insulin therapy, particularly glucose—insulin therapy
    Nutritional recovery from alcohol abstention
    Recovery phase from starvation
    Rapid proliferation of malignant cells, such as leukaemia

Phosphate absorbed by bone
    Hungry bone syndrome (after parathyroidectomy of secondary hyperparathyroidism)

Others
    Excessive haemodialysis
    Bisphosphonate

of HPO_4^{2-} and $H_2PO_4^-$ is approximately $1:4$. In this condition, one mmol phosphate in plasma is equal to one mEq or 3.10 mg/dl.

The serum phosphate concentration is affected by dietary intake, leading to diurnal changes of approximately one mg/dl. For clinical practice, serum phosphate is measured under overnight fasting conditions. The serum phosphate concentration is higher in neonates and infants than in adults, being 4.8 mg/dl to 7.4 mg/dl in neonates. The higher concentration of serum phosphate is advantageous to the growth of the skeleton.

Table 4.7.   Causes of Hyperphosphataemia

Excessive phosphate intake and/or increased absorption of phosphate in the intestine
   Phosphate-containing medicine
   Excessive administration of phosphate
   Vitamin D intoxication
   Excessive intravenous phosphate administration

Decrease in phosphate excretion from the kidney
   Renal failure
      Acute renal failure
      Chronic renal failure
   Decrease in phosphate excretion (decreased Tmp/GFR)
   Hypoparathyroidism
      Surgical parathyroidectomy (at the time of neck surgery)
      Idiopathic hypoparathyroidism
   Pseudohypoparathyroidism
      Type 1a (Gs protein mutation)
      Type 1b (functional abnormalities of PTH/PTHrP receptor/Gs protein)
      Type 2 (abnormalities in the catalytic unit of adenyl cyclase)
   Pseudo-pseudohypoparathyroidism
   Hyperthyroidism
   Increased growth hormone (GH)
      Gigantism and acromegaly
      Excessive GH administration
   Tumoral calcinosis
   Neonates and infants

Shift of phosphate into extracellular fluid from cells
   Acidosis, particularly acute respiratory acidosis
   Rhabdomyolysis
   Crush syndrome
   Haemolytic anaemia
   Massive necrosis of malignant tumours
      Chemotherapy, as in leukaemia
   Malignant hypertension
   Excessive catabolism

In Figure 4.6, homeostasis of phosphate is shown in a 70-kg adult. Daily oral intake of food contains 800 to 1,500 mg of phosphate, which is strongly positively correlated with protein intake. Eighty grams of protein contains approximately 1,000 mg of phosphate. In the intestine, 1,100 mg phosphate per day is absorbed, and 200 mg phosphate per day is excreted as digestive juice such as pancreatic fluid. The net balance of 900 mg phosphate per day is absorbed, and this amount is excreted in urine under normal conditions [1, 2].

Phosphate Absorption and Excretion

Phosphate Absorption in the Intestine

Dietary phosphate is mostly absorbed in the duodenum and proximal part of the jejunum. In normal conditions, phosphate absorption in the intestine is mostly

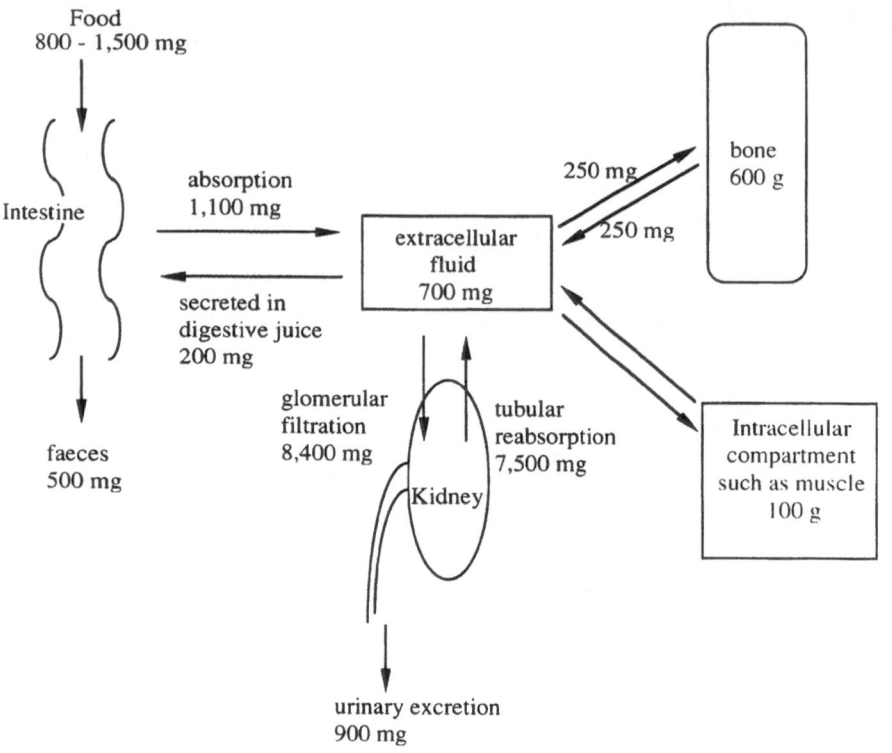

Figure 4.6.   Phosphate homeostasis in a 70-kg adult.

via passive transport (diffusion) and, in small parts, via active transport. Active transport of phosphate is via a sodium-dependent phosphate co-transporter whose activity is regulated by the active form of vitamin D (1,25-dihydroxivitamin D). In the phosphate depletion condition of bizarre diet or shortage of oral phosphate intake, the active transport of phosphate via the sodium-dependent phosphate co-transporter is activated. The detailed regulation of intestinal phosphate absorption is described in another chapter.

Phosphate Excretion from the Kidney

Total glomerular filtration of phosphate is approximately 6,000–9,000 mg/day, and 90% of filtered phosphate is reabsorbed in tubules, mostly proximal tubules (convoluted proximal tubules). In phosphate reabsorption, one molecule of phosphorus and two molecules of sodium are co-transported. Recently, two types of transporters (Na–Pi co-transporter type I and type II) of the kidney were cloned, and the functional differences between them were explored [3, 4]. (The details are described in another chapter).

To evaluate the tubular reabsorption, it is most useful to assess the maximal tubular reabsorptive rate of phosphorus (TmP/GFR) [5]. This theoretical value means that phosphate is excreted in urine if the serum phosphate concentration is beyond the value of TmP/GFR; in other words, phosphate is completely reabsorbed by the kidney if the serum phosphate concentration is below the TmP/GFR value. TmP/GFR can be calculated using the following formula:

$$TmP/GFR = Serum\ Pi - (Upi \times UV)/GFR$$
$$= Serum\ Pi\ (1 - Cpi/GFR)$$
$$= Serum\ Pi\ \{1 - (Upi \times Pcr)/(Ppi \times Ucr)\}$$

(Pi: phosphate, Upi: urinary phosphate concentration, UV: urinary volume, GFR: glomerular filtration rate, Cpi: phosphate clearance, Pcr: plasma creatinine concentration, Ppi: plasma phosphate concentration, Ucr: urinary creatinine concentration).

The normal range of TmP/GFR is similar to the normal serum concentration (2.5–4.2 mg/dl). The two main factors that regulate TmP/GFR levels are parathyroid hormone (PTH) and phosphate intake. When PTH and phosphate intake increase, TmP/GFR is decreased, leading to increased urinary phosphate excretion and decreased serum phosphate concentration. When PTH and phosphate intake decrease, TmP/GFR is increased, leading to decreased urinary phosphate excretion and increased serum phosphate concentration. Glucocorticosteroid decreases TmP/GFR, leading to phosphaturia. Growth hormone, insulin and insulin-like growth factor 1 increase TmP/GFR. Advanced ageing decreases TmP/GFR through the decrease of the expression of the Na–Pi co-transporter in proximal tubules [6, 7].

Percent tubular reabsorption of phosphate (%TRP) has a similar meaning to TmP/GFR, and can be calculated using the following formula:

$$\%TRP = (1 - Cpi/Ccr) \times 100 = \{1 - (Upi \times Pcr)/(Ppi \times Ucr)\} \times 100$$

The normal range of %TRP is 85–95%. This means that 85–95% of phosphate filtered from glomeruli is reabsorbed by the kidney. Similar to TmP/GFR, %TRP is decreased by increased PTH and increased phosphate intake. Since %TRP is greatly affected by the decreased glomerular filtration rate, TmP/GFR is more useful in clinical examinations.

Regulation of Serum Phosphate Concentration

Calcium Regulating Hormones

The most important hormone that regulates serum phosphate concentration is PTH and $1,25(OH)_2D$. In Figure 4.7, the relationship between serum phosphate, PTH and $1,25(OH)_2D$ is shown.

In bone, PTH receptor is present in osteoblasts and stromal cells, but not in osteo-clasts [8]. Osteoblasts and stromal cells stimulated by PTH activate osteoclast activity through secretion of osteoclast differentiation factor [9], which binds to RANK (receptor activator of NF-κB) [10, 11]. Thus, PTH stimulates bone absorption by osteoclasts, leading to mobilisation of calcium and phosphate into serum. In the kidney, PTH translocates the Na–Pi co-transporter type II from brush border into cytoplasm [12], and inhibits phosphate reabsorption, leading to a decrease in TmP/GFR. PTH also stimulates 24-hydroxy-vitamin D-1α-hydroxylase activity in the proximal tubuli and increases the synthesis of $1,25(OH)_2D$. Although PTH-induced mobilization of phosphate from bone is enhanced in primary hyper-

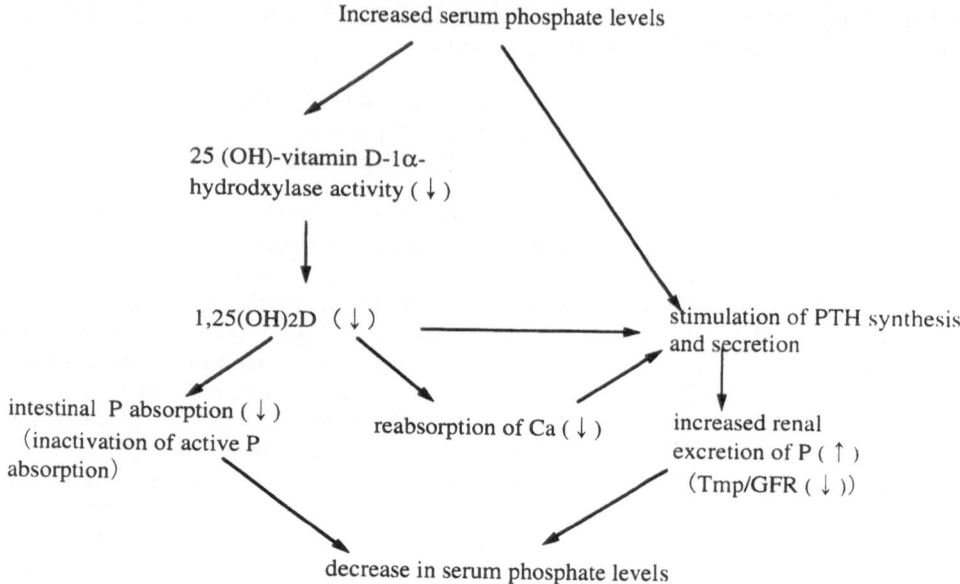

Figure 4.7. Regulation of calcium–phosphate-regulating hormones, 1,25-dihydroxy-vitamin D (1,25(OH)$_2$D) and parathyroid hormone (PTH), in hyperphosphataemia. In the case of hypophosphataemia, an opposite regulation of the hormones is seen.

parathyroidism, increased renal excretion of phosphate is usually greater than the phosphate mobilization from bone, leading to hypophosphataemia in normal renal function.

The active form of vitamin D, 1,25(OH)$_2$D, stimulates intestinal absorption of both calcium and phosphate. In high concentrations, 1,25(OH)$_2$D stimulates bone absorption, and inhibits PTH synthesis and secretion from the parathyroid gland. Thus, 1,25(OH)$_2$D increases the serum phosphate concentration.

Shift of Phosphate into Cells

The shift of phosphate into cells is an important factor that regulates the serum phosphate concentration. The largest organ of the intracellular phosphate pool is the muscle and soft tissue of the body. The intracellular shift of phosphate leading to hypophosphataemia occurs in such conditions as rapid insulin therapy, the recovery phase of malnutrition (nutritional recovery syndrome), acute respiratory alkalosis, rapid proliferation of malignant cells (such as leukaemia and malignant lymphoma) and so on. Inversely, the extracellular shift of phosphate leading to hyperphosphataemia occurs in such conditions as rapid necrosis of massive cells (such as crush syndrome, rhabdomyolysis, necrosis of malignant tumours) and acute respiratory acidosis.

Hypophosphataemia

Causes of Hypophosphataemia

Hypophosphataemia is defined as serum phosphate levels less than 2.5 mg/dl. The causes of hypophosphataemia are summarised in Table 4.6. Severe hypophosphataemia with phosphate levels less than 1.0 mg/dl is a particularly serious clinical condition. Severe hypophosphataemia is caused by 1) excessive use of antacids that contain phosphate binders such as aluminium hydroxide and calcium carbonate, 2) intravenous hyperalimentation with low concentrations or no phosphate, 3) severe, acute respiratory alkalosis, 4) polyuric phase of acute renal failure, 5) rapid intensive insulin therapy, 6) rapid proliferation of malignant cells, 7) nutritional recovery syndrome, 8) excessive carbohydrate intake at abstention of alcohol, and so on.

Clinical Manifestation of Hypophosphataemia

In moderate hypophosphataemia (serum phosphate of 1.0–2.5 mg/dl) and in relatively short-term hypophosphataemia, there are few clinical symptoms or signs. Clinical signs and symptoms are seen when severe hypophosphataemia (serum phosphate less than 1.0 mg/dl) and/or long-term hypophosphataemia are present. Under such conditions, the disease causes intracellular depletion of organic phosphate, which causes cellular dysfunction of several organs. In red blood cells, concentration of 2,3-diphosphoglycerate decreases to cause higher affinity of haemoglobin to oxygen [13]. This causes disturbance of the oxygen supply to tissues, and induces functional failure of several organs. In the central nervous system, a lower supply of oxygen causes intracellular depletion of ATP, and results in neuronal dysfunction, such as insomnia, uneasiness, loss of consciousness, convulsions, and even coma. Intracellular depletion of ATP in the muscles causes weakness. In severe cases of hypophosphataemia, rhabdomyolysis is seen [14]. In the heart, decreased oxygenation and intracellular decrease in ATP cause a decrease in the contractility of the heart muscles, leading to intractable heart failure [15]. Long-term hypophosphataemia brings about demineralization of the bone, a cause of osteomalacia and osteoporosis.

Differential Diagnosis of Hypophosphataemia

In Figure 4.8, a flowchart for the differential diagnosis of hypophosphataemia is shown. Here, the causes of hypophosphataemia, as shown in Table 4.6, should be considered. The use of phosphate-lowering drugs should be checked first, such as diuretics, insulin, phosphate-binding antacids, steroids, bisphosphonate [16], and hyperalimentation fluid with low phosphate concentration. Next, the clinical background of possible causes should be considered, such as the recovery phase of diabetic ketoacidosis, the polyuric phase of acute renal failure, leukaemia of acute proliferation, the recovery phase of malnutrition, acute respiratory alkalosis, malabsorption syndrome, and excessive carbohydrate intake at abstention of alcohol. The causes of hypophosphataemia can normally be found when a careful check of drugs used and the clinical background is assessed.

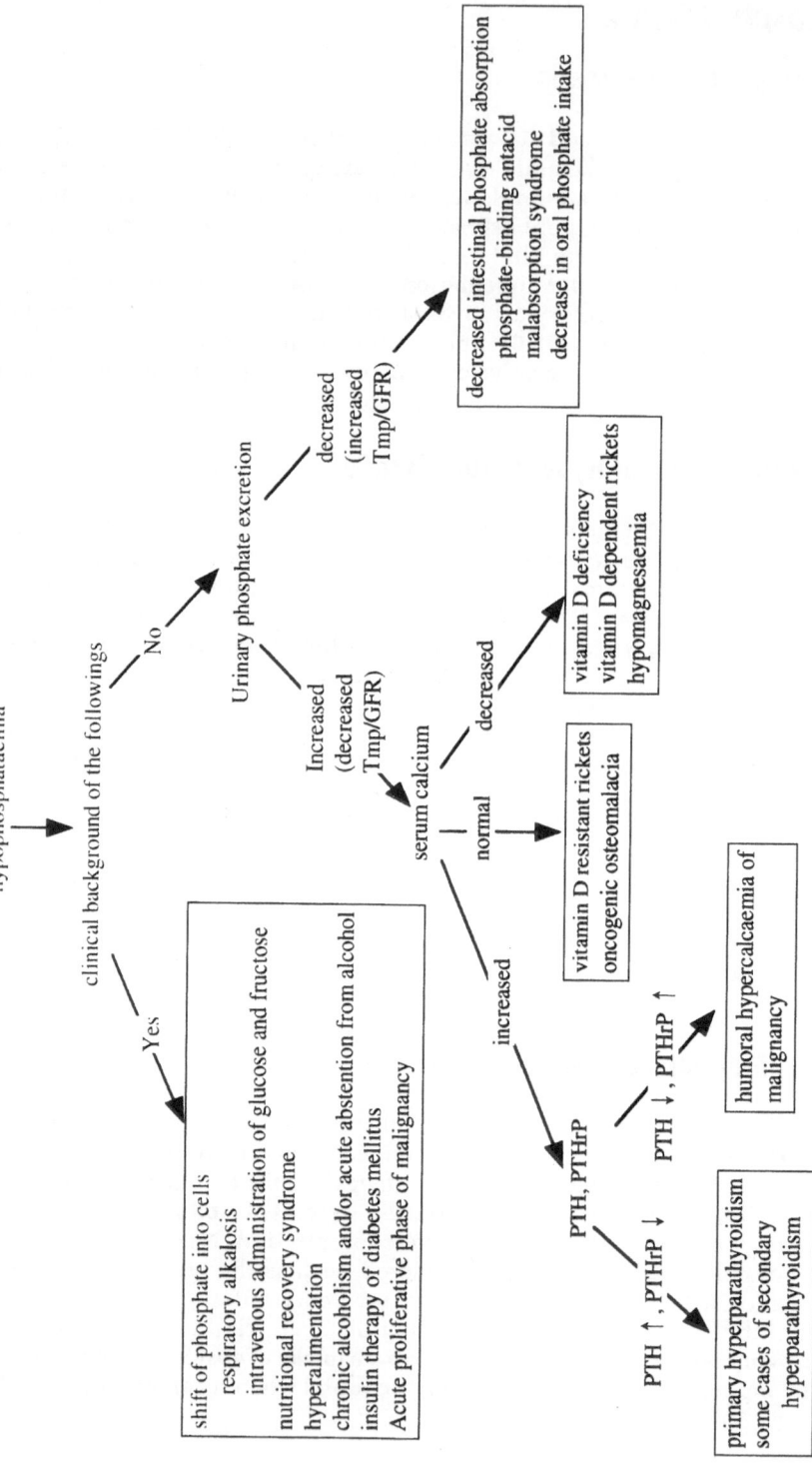

Figure 4.8. A flowchart of the differential diagnosis of hypophosphataemia. PTH: parathyroid hormone, PTHrP: PTH-related peptide, TmP/GFR: maximal tubular reabsorptive rate of phosphorus.

When the cause of hypophosphataemia is not clear even after checking the drugs and clinical background, examination of urinary excretion of phosphate is most useful [5]. Although urinary excretion of phosphate should be estimated by calculating TmP/GFR, %TRP is feasible when renal function is normal. As shown in Figure 4.8, primary hyperparathyroidism and humoral hypercalcaemia associated with malignancy should be considered when both hyperphosphaturia and hypercalcaemia are observed. In such a condition, elevated PTH levels confirm the former, and elevated PTHrP (PTH-related peptide) levels suggest the latter. Under the condition of phosphaturia and normal or decreased serum calcium, familial hypophosphataemic rickets and oncogenic osteomalacia, vitamin D deficiency, Fanconi syndrome and Cushing syndrome can be considered. Although most of these diseases that cause hypophosphataemia can be relatively easily diagnosed by measuring serum calcium, PTH and PTHrP, oncogenic osteomalacia is usually difficult to diagnose correctly. This is because the causative humoral agent has not yet been identified and also because severe hypophosphataemia is caused even by very small mesenchymal tumours, which are difficult to detect clinically [17]. In oncogenic osteomalacia, plasma 1,25-dihydroxyvitamin D is usually decreased [18, 19].

Treatment of Hypophosphataemia

Mild cases improve easily with treatment of the causative diseases. It is necessary, however, to treat moderate to severe cases (phosphate less than 1.0 mg/dl) and long-term hypophosphataemia with supplementation of oral or intravenous phosphate.

In oral phosphate supplementation, the safest method is an intake of skimmed low-fat milk. If the intake of milk is insufficient to correct the serum phosphate level, a neutrally adjusted (pH 7.4) phosphate salt is given orally ($NaH_2PO_4 : Na_2HPO_4 = 4 : 1$). Usually, the initial dose of phosphate will be less than 2,000 mg/day, and the amount should be gradually increased by monitoring the serum phosphate concentration and urinary phosphate excretion. When serum levels of calcium are low, administration of the active forms of vitamin D (alphacalcidol and calcitriol) is also effective in correcting both serum calcium and phosphate levels.

When oral supplementation of phosphate is unsuitable or serum phosphate levels are severely low (less than 1 mg/dl), continuous intravenous phosphate administration may be considered, using a solution of potassium phosphate salt (K_2HPO_4). In the administration of potassium phosphate, which contains high concentrations of potassium, slow intravenous administration over a few hours is needed to avoid rapid increase in serum potassium.

Hyperphosphataemia

Causes

Hyperphosphataemia is defined as serum phosphate levels greater than 5 mg/dl. Causes are summarized in Table 4.7.

If the renal function is normal, hyperphosphataemia is relatively uncommon. Increased serum phosphate levels caused by excessive oral intake of phosphate or increased phosphate shift from the intracellular compartments can be promptly corrected by renal excretion of phosphate. Rapid phosphate overload induces a decrease in TmP/GFR in the kidneys, and phosphate can be rapidly excreted in urine. However, when several causes that induce hyperphosphataemia exist simultaneously, such as tumour necrosis under the condition of severe acidosis, hyperphosphataemia is observed.

When TmP/GFR is increased, hyperphosphataemia is seen. Several types of hypoparathyroidism, listed in Table 4.7, cause increased TmP/GFR. The most common cause of hypoparathyroidism is parathyroidectomy at the time of neck surgery. In idiopathic hypoparathyroidism, PTH synthesis and/or secretion is impaired in the parathyroid gland, and serum PTH is decreased or absent. As a cause of idiopathic hypoparathyroidism, auto-antibodies to extracellular domains of the calcium-sensing receptor were reported in patients with acquired hypoparathyroidism [20]. A familial syndrome with mutations in the calcium-sensing receptor was also reported [21]. Pseudohypoparathyroidism is a rare disease [22] and is characterised by end-organ resistance to the action of PTH, leading to elevated serum PTH levels. Among the three types of pseudohypoparathyroidism, type 1a is caused by abnormalities in Gs protein, type 1b is caused by functional (not structural) [23, 24] abnormalities in PTH/PTHrP receptor coupled to Gs protein, and type 2 is caused by the abnormal catalytic unit of adenylate cyclase [25]. In type 1a, but not in type 1b, there are several specific somatic characteristics (Albright's hereditary osteodystrophy), such as short stature, round face, short metacarpal bones and phalanges, and some degree of mental retardation. Although no phosphaturic response to intravenous infusion of PTH is seen (Ellsworth–Howard test); urinary cAMP is increased after PTH infusion in patients with type 2, but not in type 1a or 1b. Recently, several mutations of the G(s) α protein gene (GNAS1) were identified as the cause of type 1a [26, 27]. As a cause of type 1b, abnormal unmethylation of alleles of a region upstream of the G(s) α promoter is posited [28]. Increased TmP/GFR is also caused by increased growth hormone (gigantism, acromegaly, and excessive use of growth hormone) and hyperthyroidism. In neonates and infants, TmP/GFR is usually increased, compared with adults, and relative hyperphosphataemia is seen.

The most frequent cause of hyperphosphataemia is renal failure, either acute or chronic. When GFR decreases to less than 30 ml/min, urinary excretion of phosphate is rapidly decreased and hyperphosphataemia is commonly seen [29]. In Figure 4.9, serum phosphate levels in non-dialysed patients with chronic renal failure are shown. Hyperphosphataemia is frequently seen in patients with maintenance haemodialysis, although dietary instructions to restrict excessive phosphate intake is given and oral phosphate binders are prescribed (Figure 4.10).

Clinical Manifestation of Hyperphosphataemia

There are few clinical symptoms and signs that are directly associated with hyperphosphataemia. However, the condition usually causes hypocalcaemia, which induces several clinical manifestations. Acute hyperphosphataemia causes rapid hypocalcaemia, which induces tetany and/or convulsions. In hypocalcaemia, Q–T elongation and arrhythmia are seen in ECG. In patients with various types of hypoparathyroidism, chronic hyperphosphataemia and hypocalcaemia are seen, frequently causing tetany and convulsion, cataracts, dental abnormalities and calcification of basal ganglia of the cerebrum.

In hyperphosphataemia associated with chronic renal failure, there are few clinical symptoms directly associated with hyperphosphataemia and hypocalcaemia. However, in chronic hyperphosphataemia in patients with end-stage renal disease, ectopic calcification is not infrequently observed, such as in soft tissue around large

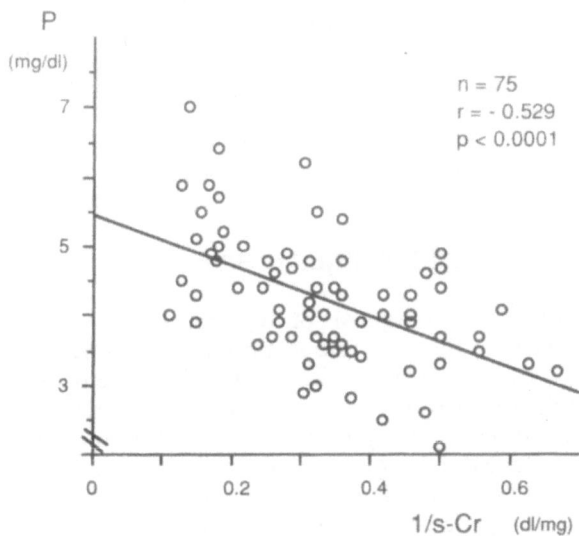

Figure 4.9.   Serum phosphate levels as a function of 1/serum creatinine (s-Cr) in non-dialysed patients with chronic renal failure. As renal function declines, serum phosphate increases (modified from the findings of reference number 29).

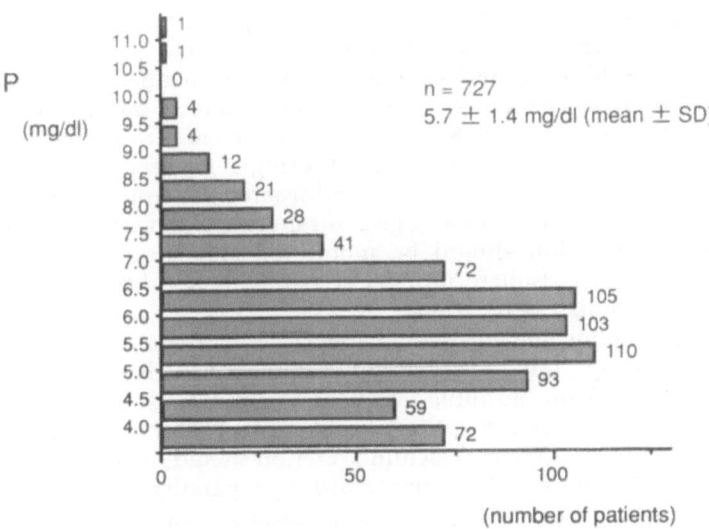

Figure 4.10.   Serum phosphate concentrations in patients with maintenance haemodialysis. Although serum phosphate levels less than 5.5 mg/dl are recommended [32], it is still difficult to achieve these levels even with dietary instructions and intake of oral phosphate binders in many patients. (Courtesy of Dr. Tsutomu Tabata and Shigeichi Shoji, Inoue Hospital, Suita, Japan.)

joints, in the skin, and in small-to-medium-sized arteries. Ectopic calcification around large joints causes intractable arthralgia. Under conditions of increased calcium–phosphate product (Ca [mg/dl] × P [mg/dl]) greater than 70, various ectopic calcifications are often seen, causing several organ dysfunction; pruritus, myalgia, corneal calcification (red eyes), conduction disturbance in the myocardium, heart failure due to valvular calcification, diffusion disturbance in the lung, and so on.

Differential Diagnosis of Hyperphosphataemia

A flowchart of the differential diagnosis of hyperphosphataemia is shown in Figure 4.11. To diagnose the cause, the clinical background should be considered first, such as excessive intake of vitamin D, tissue destruction (massive necrosis of malignancy, crush syndrome, and presence of acidosis). Next, renal function should be considered. The most frequent cause of hyperphosphataemia is renal failure, either acute or chronic. If renal failure does not exist, TmP/GFR or %TRP should be calculated. Increased TmP/GFR or %TRP in hyperphosphataemia is seen in idiopathic hypoparathyroidism, various types of pseudohypoparathyroidism, hyperthyroidism, and gigantism/acromegaly. Normal or decreased TmP/GFR or %TRP is seen in exogenous or endogenous phosphate overload. The former includes excessive vitamin D and oral phosphate intake, and the latter tissue destruction and acidosis.

Treatment of Hyperphosphataemia

Acute hyperphosphataemia without renal failure often causes symptomatic hypocalcaemia. In this situation, treatment of the causative diseases of hyperphosphataemia and an avoidance of phosphate overload usually lead to the disappearance of symptomatic hypocalcaemia. However, when the symptoms associated with hypocalcaemia are strong and protracted, such as severe tetany, intravenous calcium administration should be started, with use of calcium gluconate or calcium chloride. Ten millilitres of 8.5% calcium gluconate is intravenously administered over 5–10 minutes, and then 50–100 ml of 8.5% calcium gluconate is administered over a few hours by drip infusion with continuous monitoring of the serum calcium level. Since calcium chloride causes vasulitis if it is administered via small veins, calcium chloride should be administered via a large central vein. When administering calcium, urinary calcium excretion should be monitored and should be less than 200–250 mg/day. Calcium–phosphate product (Ca [mg/dl] × P [mg/dl]) should be less than 55 to avoid ectopic calcification.

In chronic hyperphosphataemia and hypocalcaemia associated with various types of hypoparathyroidism, renal excretion of calcium is usually increased. In patients with hypoparathyroidism, administration of vitamin D is the first choice, since vitamin D increases both reabsorption of calcium at renal tubules and absorption of calcium in the intestine. Urinary calcium excretion should be monitored and should be less than 200–250 mg/day. In patients with hypoparathyroidism, serum calcium should be in the lower normal range, since increased serum calcium induces hypercalciuria even when using vitamin D.

In hyperphosphataemia in patients with chronic renal failure, associated hypocalcaemia rarely causes clinical symptoms, because metabolic acidosis is usually present, which leads to a reduced frequency of the decrease in ionised calcium. The main therapeutic target of hyperphosphataemia in end-stage renal diseases is to

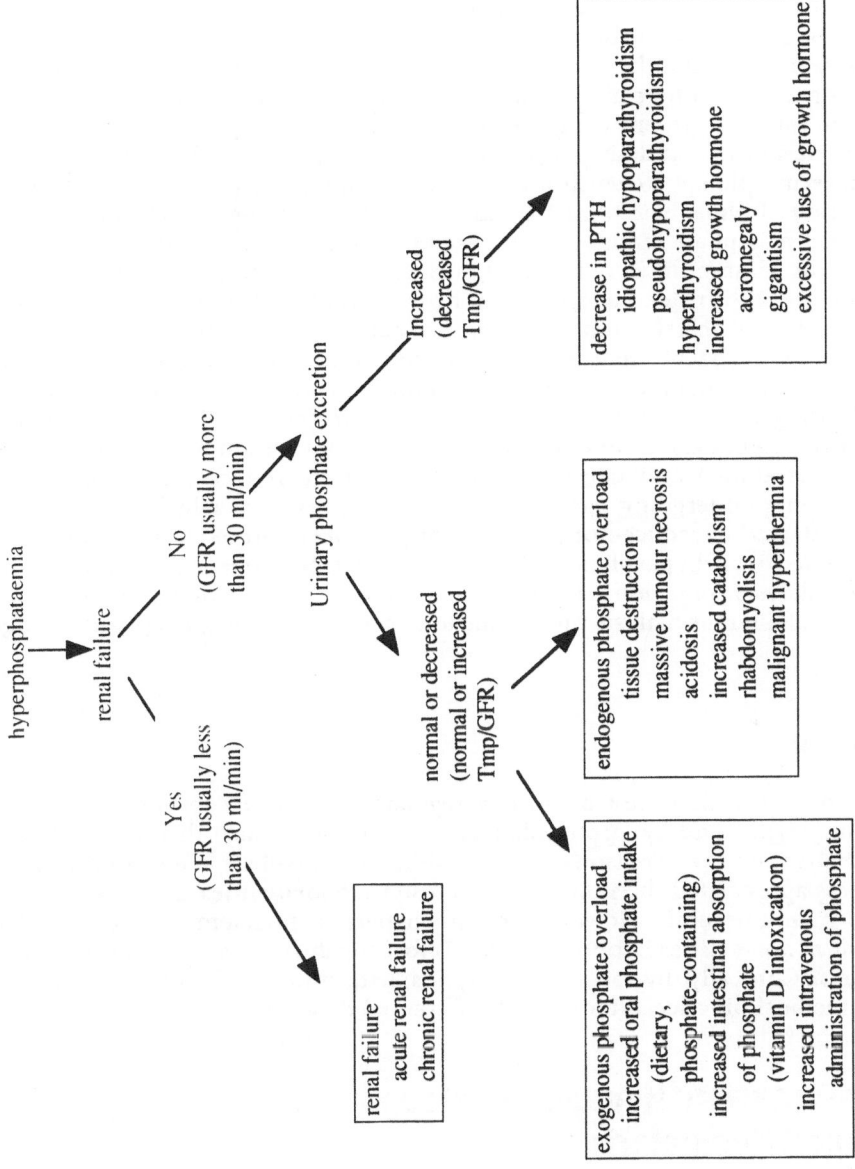

Figure 4.11. A flowchart of the differential diagnosis of hyperphosphataemia. PTH: parathyroid hormone, PTHrP: PTH-related peptide, TmP/GFR: maximal tubular reabsorptive rate of phosphorus.

prevent ectopic calcification and to prevent the progression of secondary hyper-parathyroidism [30]. Hyperphosphataemia has been demonstrated to directly stim-ulate parathyroid cell growth and the progression of secondary hyperparathyroidism [30, 31]. In patients with pre-terminal chronic renal failure or CAPD, a serum phos-phate level of 4.0–5.0 mg/dl is desirable, and in patients with maintenance haemodial-ysis, a level of 4.5–5.5 mg/dl should be achieved [32].

The most fundamental therapy of hyperphosphataemia in chronic renal failure is to restrict dietary phosphate intake. Since protein intake is correlated well with phos-phate intake (80 g of protein contains approximately 1,000 mg phosphate), dietary protein restriction is mandatory. Dietary phosphate intake should be 600–800 mg/day. When serum phosphate levels are high even with dietary phosphate restriction, administration of phosphate binder should be started. Although aluminum hydrox-ide is a potent phosphate binder and was used until the 1980s, serious complications associated with aluminum were reported, such as encephalopathy, osteomalacia and anaemia [33, 34, 35]. Instead of aluminum hydroxide, phosphate binders of calcium carbonate, calcium acetate and calcium succinate are currently widely used in patients undergoing haemodialysis. These calcium-containing phosphate binders not infrequently cause hypercalcaemia and induce increased calcium–phosphate product, leading to the risk of increased ectopic calcification. Recently, sevelamer hydrochloride (RenaGel), a non-absorbed calcium- and aluminum-free phosphate binder, was developed and demonstrated to effectively reduce serum phosphate levels in patients undergoing haemodialysis [36, 37]. Sevelamer hydrochloride has been demonstrated to reduce not only calcium–phosphate products, but also serum LDL levels [36, 38]. If hyperphosphataemia is still present after applying the above therapy, the dialysis conditions should be changed, such as an elongation of dialysis time and use of a larger dialysis membrane to effectively remove phosphate through dialysis.

Conclusions

Phosphate homeostasis is not as strictly regulated as that of sodium, potassium and calcium. Hypo- and hyperphosphataemia do not have such distinctive clinical manifestations. For these reasons, abnormalities in phosphate concentrations are clinically less appreciated than the other electrolyte abnormalities, except for hyper-phosphataemia in uraemia. However, not only hyperphosphataemia in uraemia, but also chronic abnormalities in phosphate levels for long durations cause several organ dysfunctions. Clinically, hypo- and hyperphosphataemia should be appropriately treated in medical practice, after correct differential diagnosis of the abnormalities.

Case Presentation: Hypophosphataemia Due to Oncogenic Osteomalacia

Oncogenic osteomalacia, or tumour-induced osteomalacia, is a relatively rare disease in which vitamin D-resistant osteomalacia, or rickets, occurs in association with tumours of soft tissue or bone. A case of oncogenic osteomalacia caused by small bone tumour of the femoral neck is presented.

Table 4.8. Laboratory findings of a case with hypophosphataemia due to oncogenic osteomalacia

Ca 8.6 mg/dl
P 1.5 mg/dl
Urinary Ca 112 mg/day
Urinary P 600 mg/day
%TRP (tubular reabsorption of phosphate) 79.0% (85–95)
Tmp/GFR 1.185 mg/dl (2.3–4.3)
ALP 772 IU/l (80–230)
Bone ALP 166.0 U/l (13–33)
PTH mid-lesion 174.0 pg/ml (74–276)
Intact PTH (1-82) 36.0 (10.0–65.0)
c-PTHrP (c-terminal PTH-related peptide) 24.8 pmol/l (13.8–55.3)
Calcitonin 13.1 pg/ml (30.9–120.1)
1,25(OH)$_2$D 7.0 pg/ml (27.5–68.7)
Osteocalcin 8.0 ng/ml (2.3–9.9)
Bone mineral density (BMD) of lumbar vertebra (L2–L4) 0.422 g/cm^2, Z-score − 4.31, 41% of the age-adjusted BMD

Parentheses show the normal range.

A 35-year-old male was admitted to Osaka City University Hospital with lumbago and arthralgia. He had been well until 32 years old, when he developed lumbago and pain in the lower extremities. At the age of 33, he noticed muscle weakness of the bilateral lower extremities. He presented a waddling gait on a visit to an orthopaedist, who found hypophosphataemia (P 1.9 mg/dl) and increased serum alkaline phosphatase. He was referred to the hospital for further examination and treatment.

The patient was 174 cm tall and weighed 82 kg. He complained of pain in both knee joints and ankle joints. Mild muscle weakness of the bilateral thigh was noted (4/5 by manual muscle test). Urinalysis and blood count were normal. Laboratory findings concerning calcium and phosphate metabolism are shown in Table 4.8. Serum phosphate was decreased to 1.5 mg/dl. Serum alkaline phosphatase was markedly increased to 772 IU/l (normal range: 80–230). TmP/GFR and %TRP were decreased (1.19 mg/dl and 79%, respectively). Serum calcium and intact PTH were normal, and serum 1,25(OH)$_2$D was decreased (7.0 pg/ml). X-ray of the lumbar vertebrae showed diminished bone mineral and compression fractures (Figure 4.12). Dual X-ray absorptiometry revealed a marked decrease in bone mineral density (BMD) at the lumbar spine (L2–4 BMD, 0.422 g/cm^2, 41% of the age-adjusted BMD). Histological examination of bone biopsy of the iliac crest showed typical osteomalacia (Figure 4.13).

Since oncogenic osteomalacia was suspected from the above findings, several examinations to detect mesenchymal tumours were performed, such as magnetic resonance image (MRI) of the whole skeleton and computed tomography (CT) scan. After several examinations of his skeleton, finally, a small bone tumour (2 cm) at the left femoral neck was found. After resection of this tumour, serum phosphate was rapidly increased to within the normal range in two weeks, and TmP/GFR and %TRP were normalised. Serum 1,25(OH)$_2$D was increased (40.0 pg/ml) at one month after resection of the

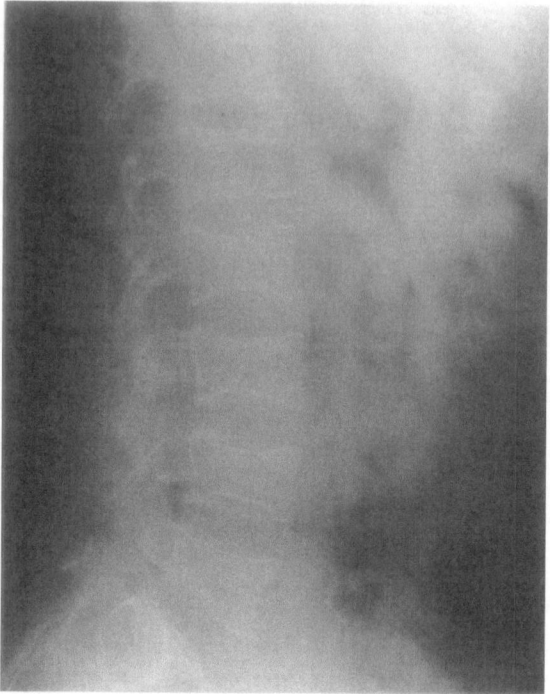

Figure 4.12.   Roentogenography of the lumber vertebral bones. Extreme loss of bone mineral and compression fractures of the vertebral bodies are seen.

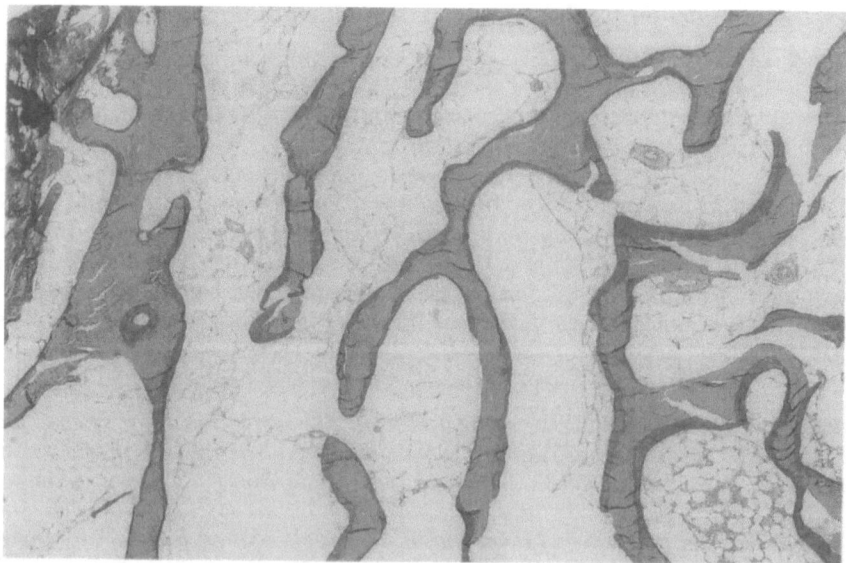

Figure 4.13.   Biopsy of the ileac bone shows the histology of osteomalacia with markedly increased uncalcified osteoid tissue, which is stained red (Goldner stain, ×38).

tumour. Pathological diagnosis of the mesenchymal tumour of the mixed connective tissue variant was made.

Oncogenic osteomalacia has a clinical similarity with a hereditary disease of X-linked hypophosphataemic rickets, in which severe hypophosphataemia, resistant to vitamin D therapy, is seen in infants. Oncogenic osteomalacia is caused by mesenchymal tumours, such as fibroma and haemangioma. The most frequent site of the tumour is at the rib and at head and neck [39]. Moderate-to-severe hypophosphataemia is caused by the tumour, even though it is small. Muscle weakness, compression fractures of vertebral bones and fractures of the rib bones are frequently observed. It is suggested that a humoral factor, phosphatonin [40], is secreted from the tumour, causing hypophosphataemia through phosphate secretion at the proximal tubules of the kidney. This factor is also thought to inhibit 24-hydroxyvitamin D-1α-hydroxylase activity [18]. Recently, the gene for phosphaturic factor secreted from the tumour was reported to be cloned and named MEPE (matrix extracellular phosphoglycoprotein) [41]. MEPE protein contains RGD motifs that are essential for integrin-receptor interactions. Furthermore, another gene, FGF-23 (fibroblast growth factor-23), is recently claimed to be the most responsible gene for the oncogenic osteomalacia [42]. Continuous production of FGF-23 in nude mouse is reported to reproduce clinical, biochemical, and histological features of oncogenic osteomalacia in vivo, such as hypophosphatemia with increased renal phosphate clearance, a high level of serum alkaline phosphatase, and deformity of the bone, in which histological examination showed marked increase of osteoid and widening of growth plate [42]. Although an MRI survey of the whole skeleton is useful in detecting the causative tumour [17], not infrequently the tumour is difficult to locate until it is large enough to clinically detect.

References

1. Nordin BEC Nutritional considerations. In: Nordin BEC, editor. Calcium, phosphate, and magnesium metabolism. Edinburgh: Churchill Livingstone, 1976;1–35.
2. Suki WN, Rouse D. Renal transport of calcium, magnesium, and phosphate. Philadelphia: W.B. Saunders; 1996;472–515.
3. Taketani Y, Miyamoto K, Chikamori M, Tanaka K, Yamamoto H, Tatsumi S, et al. Characterization of the 5′ flanking region of the human NPT-1 Na$^+$/phosphate co-transporter gene. Biochim Biophys Acta 1998;1396:267–72.
4. Tatsumi S, Miyamoto K, Kouda T, Motonaga K, Katai K, Ohkido I, et al. Identification of three isoforms for the Na$^+$-dependent phosphate co-transporter (NaPi-2) in rat kidney. J Biol Chem 1998;273: 28568–75.
5. Yanagawa N, Nakhoul F, Kurokawa K. Physiology of phosphorus metabolism, in clinical disorders of fluids and electrolytes, 5th ed, Narins RG, editor. New York: McGraw-Hill, 1995;345–50.
6. Portale AA, Halloran BP, Morris Jr RC, Lonergan ET. Effect of aging on the metabolism of phosphorus and 1,25-dihydroxyvitamin D in healthy men. Am J Physiol 1996;270:E483–90.
7. Sorribas V, Lotscher M, Loffing J, Biber J, Kaissling B, Murer H, et al. Cellular mechanisms of the age-related decrease in renal phosphate reabsorption. Kidney Int 1996;50:855–63.
8. Suda T, Takahashi N, Martin TJ. Modulation of osteoclast differentiation. Endocr Rev 1992;13:66–80.
9. Yasuda H, Shima N, Nakagawa N, Yamaguchi K, Kinosaki M, Mochizuki S, et al. Osteoclast differentiation factor is a ligand for osteoprotegerin/osteoclastogenesis-inhibitory factor and is identical to TRANCE/RANKL. Proc Natl Acad Sci USA 1998;95:3597–602.
10. Anderson DM, Maraskovsky E, Billingsley WL, Dougall WC, Tometsko ME, Roux ER, et al. A homologue of the TNF receptor and its ligand enhance T-cell growth and dendritic cell function. Nature 1997;390:175–9.
11. Yasuda H, Shima N, Nakagawa N, Yamaguchi K, Kinosaki M, Goto M, et al. A novel molecular mechanism modulating osteoclast differentiation and function. Bone 1999;25:109–13.

12. Katai K, Segawa H, Haga H, Morita K, Arai H, Tatsumi S, et al. Acute regulation by dietary phosphate of the sodium-dependent phosphate transporter (NaP(i)-2) in rat kidney. J Biochem (Tokyo) 1997;121:50–5.

13. Travis SF, Sugerman HJ, Ruberg RL, Dudrick SJ, Delivoria-Papadopoulos M, Miller LD, et al. Alterations of red-cell glycolytic intermediates and oxygen transport as a consequence of hypophosphatemia in patients receiving intravenous hyperalimentation. N Engl J Med 1971;285:763–8.

14. Knochel JP, Barcenas C, Cotton JR, Fuller TJ, Haller R, Carter NW. Hypophosphatemia and rhabdomyolysis. J Clin Invest 1978;62:1240–6.

15. O'Connor LR, Wheeler WS, Bethune JE. Effect of hypophosphatemia on myocardial performance in man. N Engl J Med 1977;297:901–3.

16. Ishimura E, Miki T, Koyama H, Harada K, Nakatsuka K, Inaba M, et al. Effect of aminohydroxypropylidene diphosphonate on the bone metabolism of patients with parathyroid adenoma. Horm Metab Res 1993;25:493–7.

17. Fukumoto S, Takeuchi Y, Nagano A, Fujita T. Diagnostic utility of magnetic resonance imaging skeletal survey in a patient with oncogenic osteomalacia. Bone 1999;25:375–7.

18. Huang QL, Feig DS, Blackstein ME. Development of tertiary hyperparathyroidism after phosphate supplementation in oncogenic osteomalacia. J Endocrinol Invest 2000;23:263–7.

19. Ohashi K, Ohnishi T, Ishikawa T, Tani H, Uesugi K, Takagi M. Oncogenic osteomalacia presenting as bilateral stress fractures of the tibia. Skeletal Radiol 1999;28:46–8.

20. Li Y, Song YH, Rais N, Connor E, Schatz D, Muir A, et al. Autoantibodies to the extracellular domain of the calcium-sensing receptor in patients with acquired hypoparathyroidism. J Clin Invest 1996; 97:910–14.

21. Pearce SH, Williamson C, Kifor O, Bai M, Coulthard MG, Davies M, et al. A familial syndrome of hypocalcemia with hypercalciuria due to mutations in the calcium-sensing receptor. N Engl J Med 1996;335:1115–22.

22. Nakamura Y, Matsumoto T, Tamakoshi A, Kawamura T, Seino Y, Kasuga M, et al. Prevalence of idiopathic hypoparathyroidism and pseudohypoparathyroidism in Japan. J Epidemiol 2000;10:29–33.

23. Schipani E, Weinstein LS, Bergwitz C, Iida-Klein A, Kong XF, Stuhrmann M, et al. Pseudohypoparathyroidism type Ib is not caused by mutations in the coding exons of the human parathyroid hormone (PTH)/PTH-related peptide receptor gene. J Clin Endocrinol Metab 1995;80:1611–21.

24. Fukumoto S, Suzawa M, Takeuchi Y, Kodama Y, Nakayama K, Ogata E, et al. Absence of mutations in parathyroid hormone (PTH)/PTH-related protein receptor complementary deoxyribonucleic acid in patients with pseudohypoparathyroidism type Ib. J Clin Endocrinol Metab 1996;81:2554–8.

25. Barrett D, Breslau NA, Wax MB, Molinoff PB, Downs Jr RW. New form of pseudohypoparathyroidism with abnormal catalytic adenylate cyclase. Am J Physiol 1989;257:E277–83.

26. Bastepe M, Juppner H. Pseudohypoparathyroidism. New insights into an old disease. Endocrinol Metab Clin North Am 2000;29:569–89.

27. Mantovani G, Romoli R, Weber G, Brunelli V, De Menis E, Beccio S, et al. Mutational analysis of GNAS1 in patients with pseudohypoparathyroidism: identification of two novel mutations [In Process Citation]. J Clin Endocrinol Metab 2000;85:4243–8.

28. Liu J, Litman D, Rosenberg MJ, Yu S, Biesecker LG, Weinstein LS. A GNAS1 imprinting defect in pseudohypoparathyroidism type IB [In Process Citation]. J Clin Invest 2000;106:1167–74.

29. Ishimura E, Nishizawa Y, Inaba M, Matsumoto N, Emoto M, Kawagishi T, et al. Serum levels of 1,25-dihydroxyvitamin D, 24,25-dihydroxyvitamin D, and 25-hydroxyvitamin D in nondialysed patients with chronic renal failure. Kidney Int 1999;55:1019–27.

30. Slatopolsky EA, Burke SK, Dillon MA. RenaGel, a non-absorbed calcium- and aluminium-free phosphate binder, lowers serum phosphorus and parathyroid hormone. The RenaGel Study Group. Kidney Int 1999;55:299–307.

31. Slatopolsky E, Finch J, Denda M, Ritter C, Zhong M, Dusso A, et al. Phosphorus restriction prevents parathyroid gland growth. High phosphorus directly stimulates PTH secretion in vitro. J Clin Invest 1996;97:2534–40.

32. Block GA, Port FK. Re-evaluation of risks associated with hyperphosphataemia and hyperparathyroidism in dialysis patients: recommendations for a change in management. Am J Kidney Dis 2000;35:1226–37.

33. Alfrey AC, LeGendre GR, Kaehny WD. The dialysis encephalopathy syndrome. Possible aluminum intoxication. N Engl J Med 1976;294:184–8.

34. Parkinson IS, Ward MK, Feest TG, Fawcett RW, Kerr DN. Fracturing dialysis osteodystrophy and dialysis encephalopathy. An epidemiological survey. Lancet 1979;1:406–9.

35. Pei Y, Hercz G, Greenwood C, Sherrard D, Segre G, Manuel A, et al. Non-invasive prediction of aluminium bone disease in haemo- and peritoneal dialysis patients. Kidney Int 1992;41:1374–82.

36. Fournier A, Oprisiu R, Albu AT, Dungaciu M, El Esper N, Morniere P. The crossover comparative trial of calcium acetate versus sevelamer hydrochloride (Renagel) as phosphate binders in dialysis patients. Am J Kidney Dis 2000;35:1248–50.
37. Collins AJ, St. Peter WL, Dalleska FW, Ebben JP, Ma JZ. Hospitalization risks between Renagel phosphate binder treated and non-Renagel treated patients. Clin Nephrol 2000;54:334–41.
38. Chertow GM, Burke SK, Dillon MA, Slatopolsky E. Long-term effects of sevelamer hydrochloride on the calcium × phosphate product and lipid profile of haemodialysis patients. Nephrol Dial Transplant 1999;14:2907–14.
39. Gonzalez-Compta X, Manos-Pujol M, Foglia-Fernandez M, Peral E, Condom E, Claveguera T, et al. Oncogenic osteomalacia: case report and review of head and neck associated tumours. J Laryngol Otol 1998;112:389–92.
40. Drezner MK. PHEX gene and hypophosphatemia. Kidney Int 2000;57:9–18.
41. Rowe PS, de Zoysa PA, Dong R, Wang HR, White KE, Econs MJ, et al. MEPE, a new gene expressed in bone marrow and tumors causing osteomalacia. Genomics 2000;67:54–68.
42. Shimada T, Mizutani S, Muto T, Yoneya T, Hino R, Takeda S, Takeuchi Y, Fujita T, Fukumoto S, Yamashita T. Cloning and characterization of FGF23 as a causative factor of tumor-induced osteomalacia. Proc Natl Acad Sci USA 2001;98:6500–5.

Magnesium Homeostasis

A: Intestinal Absorption of Magnesium
D. Kerstan, G. A. Quamme

B: Renal Handling of Magnesium
G. M. Shah

C: Hyper- and Hypomagnesemia
K. Nakatsuka, M. Inaba, E. Ishimura

Intestinal Absorption of Magnesium

D. Kerstan and G. A. Quamme

Dietary Magnesium and Magnesium Balance

The average daily diet in North America contains approximately 228 (adult females) to 323 mg (adult males) of elemental magnesium [1]. The recommended daily estimated average requirement for magnesium is 265 and 350 mg for adult females and males, respectively [2, 3]. This suggests that the average Western country diet is only adequate for maintenance of body magnesium levels. Moreover, magnesium requirements increase with rapid growth in infancy and adolescence and during pregnancy and lactation. Despite the suggestion of borderline dietary insufficiency, overt magnesium deficiency from dietary causes alone is uncommon. This is probably because of the ubiquity of magnesium in most diets, although recent dietary surveys have shown that the average magnesium intake in Western countries has declined over the last 100 years [1].

Magnesium balance is controlled by availability though gastrointestinal absorption and renal excretion (Figure 5.1). Obligatory secretory losses from the gastrointestinal tract play only a small role in magnesium balance. Cellular redistribution may be important, although this is indefinable at this time. Intracellular magnesium concentration and, more importantly, its availability to essential metabolic processes appear to be a major determining factor in magnesium homeostasis. Recent evidence indicates that only 1–2% of the total cellular magnesium is in a free form, Mg^{2+} [4]. The free Mg^{2+} concentration is carefully controlled within the cell and the total cellular magnesium content is maintained at the expense of extracellular fluid and bone magnesium levels. Bone reservoirs are more important over the long term because they contain about half the total body magnesium [5]. Bone turnover rates are slow, and the hourly or daily requirements are met by a balance between intestinal absorption and renal excretion.

Physiology of Intestinal Magnesium Absorption

About 30–40% of ingested magnesium is absorbed when dietary intake of magnesium is in the normal range, 300–350 mg/day (Figure 5.2). Fractional magnesium absorption may increase to as high as 80% when dietary intake is restricted and decrease to 20% of intake on high magnesium diets [6–8]. Accordingly, intestinal absorption is controlled by factors that respond appropriately to dietary magnesium content and magnesium balance. The factors that control intestinal absorption in

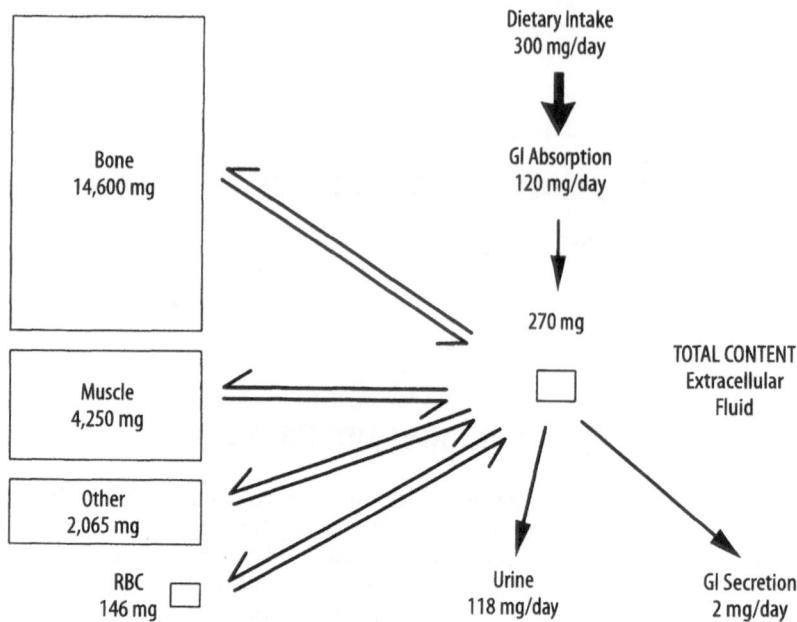

Figure 5.1. Distribution of magnesium in the body, including dietary intake, intestinal absorption, and urinary excretion. From Quamme [4].

response to magnesium requirements are poorly understood at the present time (see below).

Magnesium absorption in humans and animals occurs primarily in the distal portion of the small intestine, including the jejunum and ileum [9, 10]. This explains the high incidence of magnesium deficiency following surgical resection or bypass of the ileum for either obesity or diseases of the distal small bowel [11]. The colon absorbs small amounts of magnesium, which may be significant under circumstances in which intestinal magnesium is compromised or severe dietary restriction [12–15].

Passive Paracellular Magnesium Absorption

The evidence is that intestinal magnesium absorption occurs through paracellular and transcellular pathways [9, 10, 14–16]. Fine et al. determined net magnesium absorption in normal subjects consuming various magnesium concentrations. Net magnesium absorption increased with increasing magnesium intake, but fractional magnesium absorption fell progressively (from 65% to 11%) as magnesium intake was increased from 1.5 to 40 mg. Thus, magnesium absorption was not a linear function of magnesium intake. Their results were compatible with two simultaneously functioning absorptive processes; a mechanism that reaches an absorptive maximum beginning at an intake of 5–6 mg and a mechanism that absorbs 7.1% of ingested magnesium (Figure 5.3). This generally describes transcellular and paracellular magnesium absorption, respectively.

The majority, about 90%, of normal magnesium absorption occurs passively through the paracellular pathway between the enterocytes (Figure 5.2). The rate of

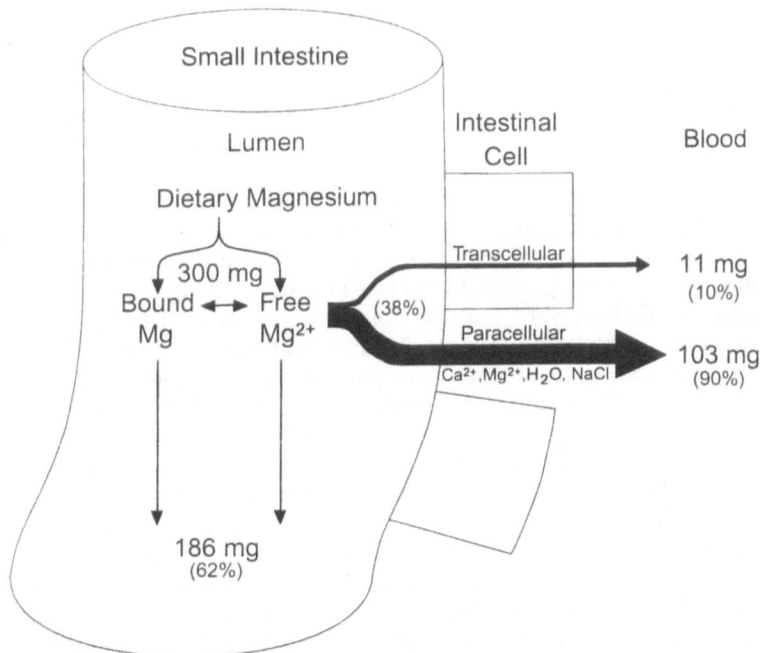

Figure 5.2. Schematic model of intestinal magnesium absorption. Normally about 38% of dietary magnesium is absorbed [6]. The majority of magnesium absorption occurs passively through the paracellular pathway entrained with salt and water. The driving force for passive Mg^{2+} movement comprises the transepithelial voltage and the Mg^{2+} concentration gradient. A smaller amount of Mg^{2+} absorption occurs through transcellular pathways. The numbers approximate absolute net magnesium absorption, and fractional absorption is indicated in brackets.

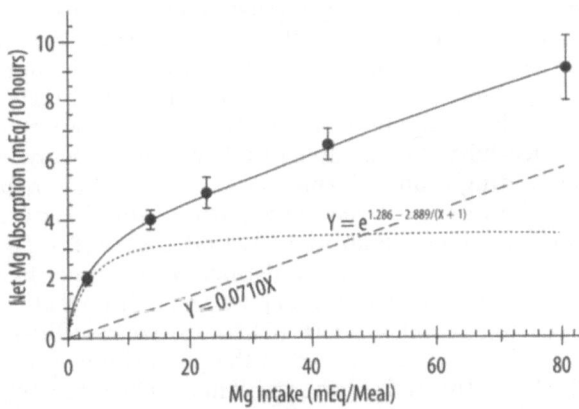

Figure 5.3. Net magnesium absorption versus intake from normal subjects. Fine et al. measured net magnesium absorption in eight normal subjects after they ingested a standard meal supplemented with 0, 5, 10, 20, and 40 mg of magnesium acetate [7]. Although absorption increased with each increment in intake, fractional magnesium absorption fell progressively from 65% at the lowest to 11% at the highest intake. Absorption is represented by a linear function (nonsaturable) and a hyperbolic function (saturable). These separate components and their respective equations are shown. From Fine et al. [7].

magnesium absorption across the intestinal epithelium is dependent on the trans-epithelial electrical voltage (which is normally about +5 millivolts, lumen positive with respect to blood) and the transepithelial concentration gradient. The luminal voltage is generated by Na^+ transport across the epithelium. The luminal magnesium concentration may be of the order of 1.0–5.0 mM depending on the dietary magnesium content and the presence of anionic chelators. Only the free Mg^{2+} moves through the paracellular pathway so that bound magnesium does not contribute to the transepithelial gradient. Serum Mg^{2+} concentration is 0.5–0.7 mM, so that there is normally a concentration gradient from lumen to the blood side. The intestinal paracellular pathway is readily permeable to NaCl divalent cations (Ca^{2+}, Mg^{2+}) and water. In addition to diffusion, Mg^{2+} may be absorbed passively through the paracellular pathway entrained with salt and water; i.e., solvent drag. Accordingly, any influence that changes water absorption may alter paracellular salt and water absorption and passive Ca^{2+} and Mg^{2+} transport. In summary, passive Mg^{2+} absorption is determined by Na^+ transport-induced transepithelial voltage and water movement, by the luminal Mg^{2+} concentration, and by the permeability of the paracellular pathway.

The permeability of the paracellular pathway is influenced by proteins comprising the tight junction. Tight junctions are the most apical structures of the junctional complex in epithelium cells. Tight junctions serve as a permeability barrier in regulating the passage of ions and small molecules through the paracellular pathway [17]. A number of the tight junction-associated proteins have been identified, including ZO-1, cingulin, 7H6 antigen, occludin, and claudin families [18]. A member of the claudin family, paracellin-1 or claudin 16, has been shown to be critical for passive Mg^{2+} reabsorption in the thick ascending limb of the kidney [19]. Tight junction assembly and function can be modulated by a number of signalling molecules, including G proteins, phospholipase C, cAMP, protein kinase C, intracellular Ca^{2+}, diacylglycerol, and MAP and MEK1 kinases [18]. Receptor-mediated activation of these signals may alter the phosphorylation state of the tight junctional proteins and the ionic permeability of the paracellular pathway [18]. Accordingly, passive paracellular absorption of Mg^{2+} may be sensitively regulated by hormonal and non-hormonal factors acting through a variety of intracellular signals. It is likely that many of the same hormones and factors that affect cellular transport also control passive Mg^{2+} absorption across the paracellular pathway. This has been illustrated in Figure 5.4.

The controls that modulate the permeability of the paracellular pathway may also possess some ionic selectivity. An example of this property has recently been elucidated in the thick ascending limb of Henle's loop of the kidney nephron [20]. Over 60% of the filtered Ca^{2+} and Mg^{2+} is reabsorbed within the thick ascending limb by passive means through the paracellular pathway [20]. A subset of patients with renal magnesium wasting disease present with a distinct syndrome of hypomagnesaemia with hypercalciuria and nephrocalcinosis [21]. Using positional cloning, Simon et al. have identified a human gene, claudin 16 (CLDN16) [paracellin-1 (PCLN-1)] that codes for a tight junctional protein located in the paracellular pathway of the thick ascending limb [19]. Mutations in this gene result in the absence of claudin 16 expression, abnormal permeability of the paracellular pathway, and diminished Mg^{2+} and Ca^{2+} reabsorption. Interestingly, the permeability to Na^+ is not changed in this disorder, so claudin 16 controls only divalent cations. Claudin 16 is not expressed in the small intestine or colon [19, personal unpublished observations]. Because of the significance of the paracellular pathway in passive Mg^{2+} and Ca^{2+} absorption, it may be speculated that other unidentified tight junction proteins may perform a similar role in the intestine.

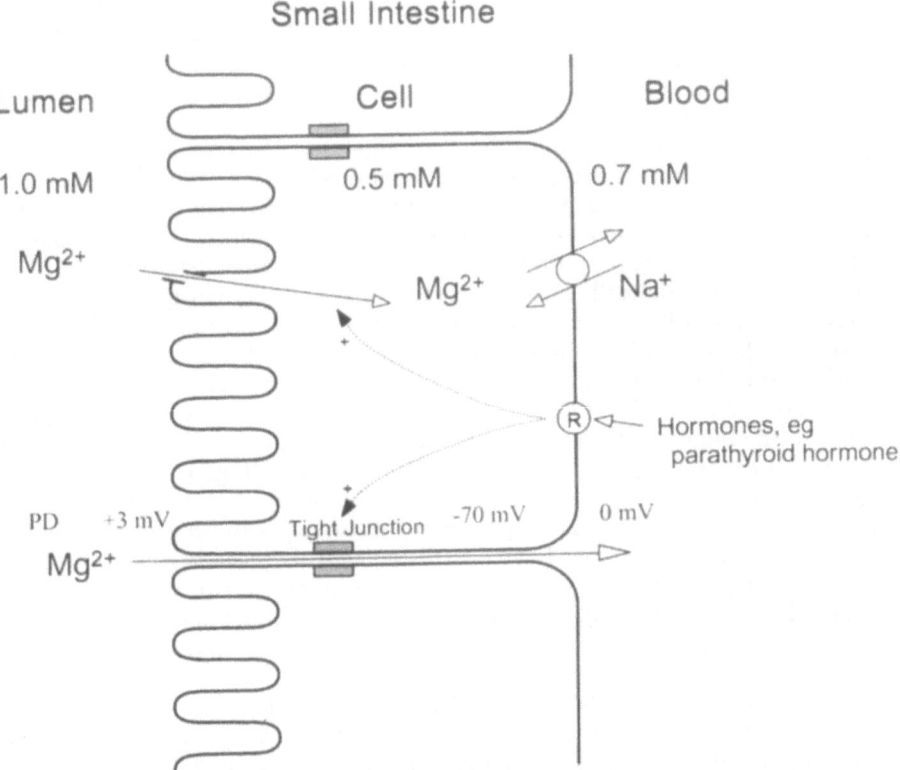

Figure 5.4. Schematic model of magnesium absorption in the small intestine. Passive magnesium absorption occurs through the paracellular pathway and active Mg^{2+} transport through the intestinal cell. The role of tight junctional proteins in passive magnesium absorption and the identity of Mg^{2+} transporters in active absorption remains to be identified.

Active Transcellular Magnesium Absorption

Transcellular magnesium absorption is active in nature and controlled by a number of hormones and factors. The evidence for active intestinal magnesium absorption is persuasive. Karbach and colleagues demonstrated active cellular Mg^{2+} transport across rat jejunum and ileum segments mounted in an Ussing chamber [12, 16]. Mg^{2+} transport was enhanced by vitamin D metabolites [13]. Hayashi and Hoshi measured steady-state net flux of Mg^{2+} in everted preparations of guinea pig jejunum and ileum and showed that active transcellular transport involves a verapamil-sensitive entry step and Na$^+$-dependent or ouabain-sensitive exit processes which were different from Ca^{2+} extrusion mechanisms [10]. Brannan et al. reported vitamin D metabolite-sensitive magnesium absorption in human subjects [6]. In support of these experimental studies, inherited disorders have been described with genetic linkage studies to chromosome 9q [22]. It was postulated that mutations present in these patients may locate to a gene coding unique Mg^{2+} transporters involved in active transcellular absorption. Active transport is important in absorption of small amounts of magnesium against large concentration gradients as observed with dietary magnesium restriction. The cellular mechanisms by which active magnesium absorption occurs remain to be defined. Based on a number of experimental observations, we are able to speculate on the mechanisms involved in transcellular magnesium absorption across the enterocyte (Figure 5.4). Magnesium may move passively into the cell across the luminal membrane driven by a favourable transmembrane voltage. The entry pathways are unknown but there is evidence for selective Mg^{2+} channel(s) present in the apical membrane of epithelial cells. Schweigel and Martens demonstrated electrodiffusive Mg^{2+} uptake in sheep rumen epithelium [23]. In these studies, unidirectional Mg^{2+} flux, determined by $^{28}Mg^{2+}$ and fluorescence, was correlated with the transmembrane voltage, measured by microelectrodes. These results support the assumption that the membrane potential acts as a principal driving force for Mg^{2+} entry in rumen epithelial cells and suggest that this pathway comprises a channel or a carrier. Using microfluorescence, we characterised Mg^{2+} entry in distal convoluted tubule cells [24]. Transcellular Mg^{2+} transport in the enterocyte is probably similar to that of the renal distal tubule cell. Our observations support the notion that selective (not shared with Ca^{2+}) Mg^{2+} channels provide the basis for Mg^{2+} entry into epithelial cells. We speculate that Mg^{2+} entry is through a unique, unidentified channel(s) and this transport is dependent on the transmembrane voltage (Figure 5.4). Magnesium moves across the cell, perhaps involving Mg^{2+} binding proteins such as calbindin D9k [25]. Experimental evidence supports the notion that these proteins bind magnesium, and some investigators have postulated that they are involved in epithelial magnesium transport [24, 26]. The active step in transcellular movement is predicted to be at the basolateral membrane where Mg^{2+} leaves the cell against both electrical and concentration gradients. The means by which Mg^{2+} actively moves across the basolateral membrane is unknown. Evidence taken from studies using non-epithelial cells suggests that Na^+/Mg^{2+} exchange may occur; Na^+ moving back into the cell coupled with Mg^{2+} exit from the cell into the interstitium [27]. In our view, the factors which influence transcellular magnesium absorption include alterations of Mg^{2+} entry across the luminal membrane and changes in Mg^{2+} exit across the basolateral membrane; both of these steps may be modulated by the transmembrane voltage and concentration gradients across the respective membrane (Figure 5.4).

Regulation of Intestinal Paracellular and Transcellular Mg^{2+} Absorption

Hormonal Control of Intestinal Magnesium Absorption

Vitamin D3 metabolites stimulate magnesium absorption in both the small and large intestine [6, 13]. Schmulen et al. reported that jejunal magnesium absorption was fourfold greater in chronic renal failure patients receiving $1,25(OH)_2D_3$ compared to those subjects not treated [28]. Chronic renal disease leads to diminished 1α-hydroxylation of $25(OH)D_3$ leading to deficiency in circulating vitamin D metabolite levels. Patients with renal failure have severely diminished jejunal and ileal calcium and magnesium absorption due to decreased $1,25(OH)_2D_3$ levels [6, 28]. Norman et al. studied normal human subjects consuming high (1,900 mg/day) or low (20 mg/day) calcium diets [29]. They perfused 30-cm segments of jejunum and ileum and measured calcium and magnesium absorption. Calcium absorption rate was higher when subjects had been on a low-calcium diet than when they had been consuming high levels of calcium; the ileum responded more rapidly and more completely than the jejunum. Similar results were obtained with magnesium absorption. The serum concentrations of parathyroid hormone (PTH) and 1,25(OH)$_2$D$_3$ were higher on the low diet compared to the high calcium intake. Their data are compatible with the concept that adaptation in calcium and magnesium absorption due to dietary calcium intake is mediated by changes in serum concentrations of PTH and 1,25(OH)$_2$D$_3$. Using similar techniques, Krejis et al. observed a doubling of jejunal calcium and magnesium absorption following 1,25(OH)$_2$D$_3$ administration to healthy subjects [30]. Levine et al. observed an augmentation of intestinal absorption of magnesium of about 50% in vitamin D-deficient rats given physiological amount of vitamin D sterols, particularly 1,25(OH)$_2$D$_3$ [31]. Hardwick et al. reported similar findings [32]. Karbach and Ewe showed greater cellular-mediated uptake in perfused rat colon [13]. We have shown that 1,25(OH)$_2$D$_3$ stimulates Mg^{2+} entry in distal tubule cells that demonstrate active Mg^{2+} transport not unlike enterocytes [26]. The response involves de novo protein synthesis as it was sensitive to actinomycin D and cyclohexamide inhibitors of transcription and translation. The evidence supports the notion that 1,25(OH)$_2$D stimulates synthesis of proteins that are involved with Mg^{2+} entry, perhaps as yet unidentified Mg^{2+} channels, that lead to increased Mg^{2+} transport [23]. These observations may have relevance to the intestine.

It is likely that intestinal magnesium absorption is influenced by a wide variety of hormones but these have not been extensively studied [9]. Hormonal control of renal magnesium transport has been more extensively investigated [20]. A large number of hormones stimulate passive magnesium absorption in Henle's loop and active transport in the distal tubule [24]. They include peptide hormones such as PTH, calcitonin, glucagon and arginine vasopressin (AVP). Adrenergic agonists, isoproterenol, insulin and prostaglandins also modulate magnesium transport [24]. Mineralocorticoids stimulate NaCl absorption and, in turn, the transepithelial voltage in the loop, and by inference increase passive magnesium absorption in the loop [20]. Aldosterone also potentiates hormone-mediated Mg^{2+} entry in the distal tubule [24]. All of these hormonal responses are mediated by changes in both transepithelial transport and paracellular permeability [20]. By analogy, intestinal magnesium absorption is probably controlled by many hormones acting through the coordinate changes in active transcellular and passive paracellular transport.

Intrinsic Control of Epithelial Magnesium Transport

The rate of intestinal transepithelial magnesium absorption is dependent on the concurrent magnesium status. Accordingly, fractional magnesium absorption is higher in magnesium-deficient subjects then normal individuals for any given luminal magnesium concentration. Fractional magnesium absorption may also be markedly greater, in excess of 80% in rapidly developing neonates [8] The adaptation of the intestine to a chronic high or low magnesium intake is independent of net water or NaCl movement. Levine et al. observed that increasing dietary magnesium from 0.03% to 0.2% decreased the fractional net magnesium absorption in vitamin D-deficient rats after three days [31]. The cellular basis for the adaptation of magnesium absorption to magnesium balance is unknown; it may involve 1,25(OH)$_2$D$_3$ or hormone feedback responses. We have shown that the rate of Mg^{2+} entry in kidney and intestinal cells is dependent on the magnesium status of the cell [24, unpublished personnel observations]. This response is rapid (within one to two hours), and specific for magnesium as there was no effect on sodium or calcium transport. The "adaptation" of magnesium transport rates is intrinsic as there were no hormones in the culture media. Furthermore, the adaptation was dependent on the concentration of media magnesium and the length of time in the culture media. Pre-treatment of the distal tubule cells with cyclohexamide inhibited this adaptation by about 50% [23]. Accordingly, it was concluded that magnesium transport is controlled by gene(s) that somehow respond to extracellular magnesium. We postulate that this intrinsic adaptation provides the discriminatory control of magnesium transport independent of sodium and calcium. Intrinsic adaptation provides the selective control that hormonal regulation does not have.

The paracellular pathway may also be modulated by prior magnesium status. Dietary magnesium restriction and hypomagnesaemia have recently been reported to stimulate increased passive magnesium absorption within the renal thick ascending limb [20]. Wittner et al. have shown that the increase in magnesium transport is due to an increase in the permeability of the paracellular pathway to magnesium, in keeping with the passive nature of magnesium transport within the loop. Interestingly, calcium absorption in the loop is also increased by magnesium restriction, so the response is not selective to magnesium [33]. As the paracellular pathway is not selective for individual divalent cations, a change in tight junction permeability would be expected to affect Ca^{2+} and Mg^{2+} equally.

In summary, intestinal magnesium absorption may be visualised as concerted changes of passive paracellular and active transcellular magnesium transport (Figure 5.5). Normally, about 40% of dietary magnesium is absorbed so that with a normal dietary magnesium intake of 300 mg, for example, a total of 114 mg is absorbed of which about 103 mg is through the paracellular pathway and 11 mg across the cell (Figure 5.5). Following chronic consumption of low dietary magnesium, for example 100 mg, fractional magnesium absorption increases to 80%. Absolute net transcellular Mg^{2+} increases to 20 mg in the face of diminished paracellular Mg^{2+} transport. Active transcellular Mg^{2+} transport increases because of cellular adaptation and an increase in apical Mg^{2+} entry and basolateral Na^+/Mg^{2+} exchange. Passive Mg^{2+} movement decreases because of diminished transepithelial, lumen-to-blood concentration gradient even though Mg^{2+} permeability of the paracellular pathway may be increased (Figure 5.5). Conversely, chronic consumption of a high-magnesium diet, for instance 1,000 mg, leads to diminished transcellular Mg^{2+} transport, 6 mg, and to an increase in passive paracellular, 194 mg, absorption (Figure 5.5). Again, the Mg^{2+} permeability of the paracellular pathway may be decreased but the concentration gradient is sufficiently large to increase overall transport. The fractional magnesium reabsorption would be expected to be about 20%, considerably less in subjects consuming diets with lower magnesium content. All of these changes occur without changes in salt and water transport.

Interactions of Intestinal Calcium and Magnesium Absorption and the Ca^{2+}-Sensing Receptor

The effect of luminal and serum calcium on magnesium absorption is a controversial issue (see Hardwick et al, for a review [9]). Calcium has been reported to inhibit or have no effect on intestinal magnesium absorption. On balance, those studies that report no effect of calcium on magnesium absorption were performed using intact or isolated intestinal perfusions where luminal calcium concentration was altered [9, 13]. Those reports that showed an inhibition of magnesium absorption with calcium were performed by manipulating dietary consumption [9]. To illustrate this view, we

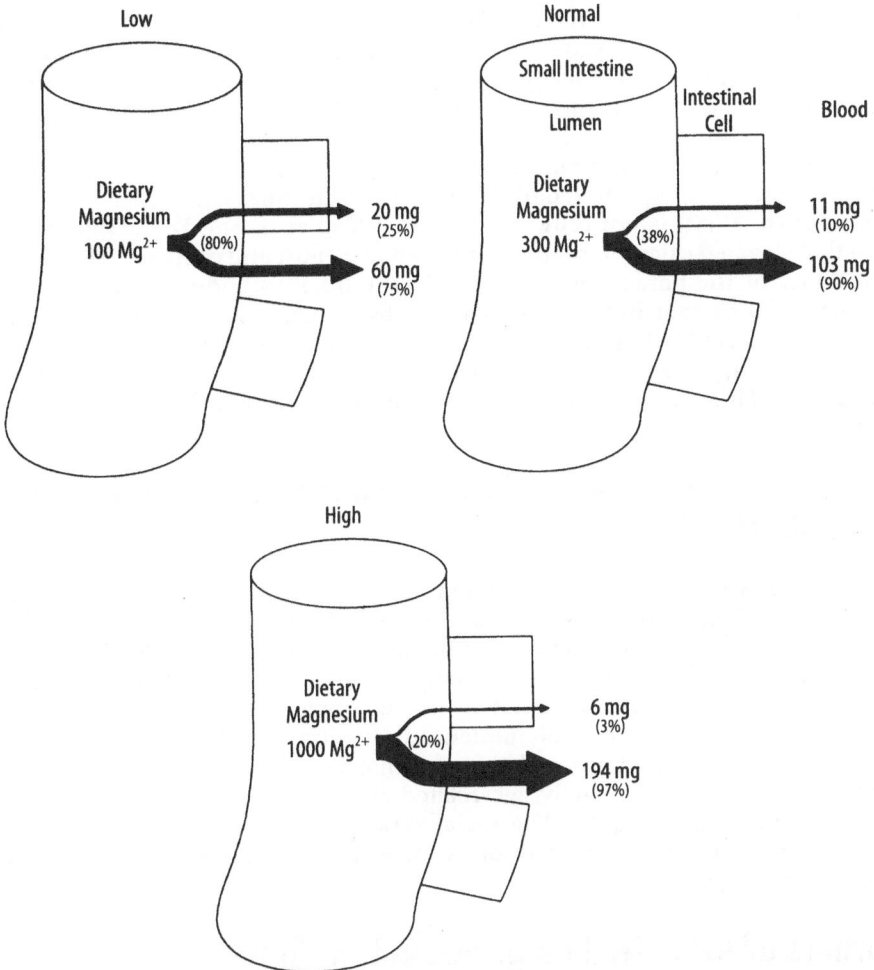

Figure 5.5. Schematic illustration of passive and active magnesium absorption in the intestine. The intestine clearly responds appropriately to dietary magnesium and body requirements, however, the cellular mechanisms for these changes are not fully understood.

use data from the same laboratory. Brannan et al. perfused segments of the small intestine of subjects with solutions containing variable magnesium and calcium concentrations; they showed that magnesium absorption was not affected with changes in luminal calcium [6, 7]. Norman et al. performed perfusions on normal subjects after four or eight weeks on high- or low-calcium diets [29]. They found that magnesium absorption was higher on a low-calcium compared to calcium-rich diets. These authors explained the different results by evoking changes in circulating vitamin D metabolites; the subjects of the first study would be expected to have constant $1,25(OH)_2D_3$ levels, whereas individuals of the latter report have concentrations appropriate for the calcium intake. We speculate that the interaction of calcium and magnesium absorption may involve the polyvalent cation-sensing receptor (CaSR).

This receptor is the same as the one expressed in the parathyroid gland and is sensitive to both extracellular Ca^{2+} and Mg^{2+} concentrations [34]. Chattopadhyay et al. showed by immunohistochemistry that the CaSR is expressed on the apical and basal membrane of the microvilli and crypts of the small intestine and of the colonic crypt epithelial cells [35]. They speculate that the CaSR may have diverse functions including modulation of salt, water and divalent cation absorption in these segments. The CaSR is expressed along the length of the gastrointestinal tract [34]. Although the function of the CaSR is not fully understood, Wang et al. and DiStefano et al. have shown that elevated calcium inhibits active salt transport and decreases Ca^{2+} and Mg^{2+} permeability of the paracellular pathway in the thick ascending limb [36, 37]. Dai et al. have shown that hypercalcaemia and hypermagnesaemia inhibit hormone-mediated transcellular Mg^{2+} transport in the distal tubule [24]. Accordingly, serum Ca^{2+} and Mg^{2+} may modulate both transcellular and paracellular Mg^{2+} absorption in the intestine. This would explain the diversity of experimental observations concerning the interaction of calcium and magnesium absorption.

Segmental and Maturational Regulation of Intestinal Paracellular and Transcellular Mg^{2+} Absorption

Although we have discussed regulation of magnesium absorption as a uniform entity along the intestine, there are significant segmental differences [6, 9, 12, 16]. There may be quantitative differences between the jejunum and ileum [6, 9]. The colon has a higher transepithelial electrical resistance, i.e., lower paracellular permeability and higher transcellular absorption relative to the small intestine. In contrast to the ileum, colonic Na^+ transport is influenced by mineralocorticoids that increase transepithelial voltage and water absorption that would influence passive and transcellular Mg^{2+} absorption. Finally, age-related changes occur along the length of the small and large intestine [38]. The serial arrangement of the intestinal segments allows for compensation of absorption if magnesium transport is compromised in any one segment [11].

Disorders of Intestinal Magnesium Handling

Primary Disorders of Intestinal Magnesium Handling

Hypomagnesaemia with secondary hypocalcaemia (HSH) is an autosomal recessive disorder that manifests in the newborn period and is characterised by very low serum magnesium and low calcium concentrations [21]. Patients usually present before six months of age with neurologic symptoms of hypomagnesaemic hypocalcaemia, including tetany, muscle spasms, and seizures. In older children with inadequate control, clouded sensorium and disturbed speech are often seen and choreoathetoid movements have been described. The hypocalcaemia is secondary to parathyroid failure and peripheral PTH resistance as a result of magnesium deficiency. Hypokalaemia is occasionally present that is only corrected with normalisation of plasma magnesium. The disease is primarily due to defective intestinal magnesium absorption and may be fatal unless treated with high oral intakes. Walder et al. reported that HSH is an autosomal recessive disease and showed by genetic linkage studies that the gene segregates to chromosome 9 (9q12–9q22.2) [22]. They suggest that the candidate gene codes a receptor or an ion channel involved in transcellular magnesium absorption. As passive transport is normal, the disease can be treated by high oral magnesium supplements, although the acute presentation may be more rapidly corrected with intramuscular or intravenous magnesium therapy. Renal magnesium conservation has been reported to be normal in most studies, suggesting that the kidney responds appropriately to low circulating magnesium levels by reabsorbing fractionally greater amounts of filtered magnesium. In some cases, however, a renal leak may also be present that manifests primarily when oral supplements are insufficient to normalise the serum magnesium [21]. Whether this condition is genetic heterogeneous remains to be seen, but further studies to address renal tubular magnesium absorption as a function of plasma concentration and filtered magnesium in patients with well-delineated intestinal defects are clearly warranted [21].

We have stressed the importance of the paracellular pathway in the control of intestinal magnesium absorption. Accordingly, it may be predicted that mutational changes in proteins comprising the tight junction may lead to intestinal calcium and magnesium malabsorption [17, 39]. No doubt, there are disorders in the general population that have not yet been identified or characterised. The continued use of molecular techniques to probe the constitutive and congenital disturbances of magnesium metabolism will increase our understanding of intestinal magnesium transport and provide new insights into the way in which these diseases are diagnosed and managed.

Acquired Disorders of Intestinal Magnesium Handling

A decrease in the absorptive surface area by any means would be expected to diminish intestinal magnesium transport. Patients undergoing intestinal resection or those with short bowel syndrome should be carefully followed for hypomagnesaemia [11]. The presence of non-absorbable anions, such as oxalate, or magnesium binding such as occurs in steatorrhoea diminish magnesium absorption [7, 9]. These anionic ligands act by decreasing the transepithelial concentration gradient; again, the free Mg^{2+} is the principle form transported through paracellular and transcellular pathways. Acidic luminal pH values increase Mg^{2+} solubility so that any dietary influence that acidifies the ingesta may result in greater magnesium absorption. Salt and water absorption affects paracellular Mg^{2+} absorption by solvent drag. Accordingly, any influence on paracellular salt and water absorption, such as the presence of luminal osmotic agents or diseases including acute and chronic diarrhoea affecting active Na^+ transport, would be expected to lead to a concomitant decrease in magnesium absorption [40].

References

1. Cleveland LE, Goldman JD, Borrud LG. Data tables: Results from USDA's 1994 continuing survey of food intakes by individuals and 1994 diet and health knowledge survey. Beltsville, MD: Agriculture Research Service, US Department of Agriculture, 1994.
2. Institute of Medicine. Dietary reference intakes: Calcium, phosphorus, magnesium, vitamin D, and fluoride. Washington, DC: National Academy Press, 1997;6-1-6-45.
3. Standing Committee on the Scientific Evaluation of Dietary Reference Intakes. Food and Nutrition Board, Institute of Medicine. Washington, DC: National Academy Press, 1997;190–249.
4. Quamme GA. Laboratory evaluation of Mg^{2+} status: renal function and free intracellular Mg^{2+} concentration. In: Preuss HG, editor. Clinics in laboratory medicine. Philadelphia: WB Saunders, 1993;-vol 13, 209–23.

5. Rude RK. Magnesium and bone. In: Rayssiguier Y, Mazur A, Durlach J, editors. Advances in magnesium research: nutrition and health. London: John Libbey, 2001.

6. Brannan PG, Vergne-Marini P, Pak CY, Hull AR, Fordtran JS. Magnesium absorption in the human small intestine. Results in normal subjects, patients with chronic renal disease, and patients with absorptive hypercalciuria. J Clin Invest 1976;57:1412–8.

7. Fine KD, Santa Ana CA, Porter JL, Fordtran JS. Intestinal absorption of magnesium from food and supplements. J Clin Invest 1991;88:396–402.

8. Lönnerdal B. Effects of milk and milk components on calcium, magnesium, and trace element absorption during infancy. Physiol Rev 1997;77:643–69.

9. Hardwick LL, Jones MR, Brautbar N, Lee DB. Site and mechanism of intestinal magnesium absorption. Mineral Electrolyte Metabol 1990;16:174–80.

10. Hayashi H, Hoshi T. Properties of active magnesium flux across the small intestine of the guinea pig. Jap J Physiol 1992;42:561–75.

11. Lisbona F, Alferez MJ, Barrionuevo M, Lopez-Aliaga I, Pallares I, Hartiti S et al. Effects of type of dietary fat and cholecalciferol on magnesium absorption in rats with intestinal resection. Int J Vit Nutr Res 1994;64:135–43.

12. Karbach U. Magnesium transport across colon ascendens of the rat. Dig Dis Sci 1989;34:1825–31.

13. Karbach U, Ewe K. Calcium and magnesium transport and influence of 1,25-dihydroxyvitamin D3. In vivo perfusion study at the colon of the rat. Digestion 1987;37:35–42.

14. Karbach U. Cellular-mediated and diffusive magnesium transport across the descending colon of the rat. Gastroenterology 1989;96:1282–9.

15. Ohta A, Baba S, Ohtsuki M, Takizawa T, Adachi T, Hara H. In vivo absorption of calcium carbonate and magnesium oxide from the large intestine in rats. J Nutr Sci Vitaminology 1997;43:35–46.

16. Karbach U, Rummel W. Cellular and paracellular magnesium transport across the terminal ileum of the rat and its interaction with the calcium transport. Gastroenterology 1990;98:985–92.

17. Anderson JM, Van Itallie CM. Tight junctions and the molecular basis for the regulation of paracellular permeability. Am J Physiol 1995;269:G467–75.

18. Muresan Z, Paul DL, Goodnough DA. Occludin 1B, a variant of the tight junction protein occludin. Mol. Biol. Cell 2000;11:627–34.

19. Simon DB, Lu Y, Choate KA, Velazquez H, Al-Sabban E, Praga M et al. Paracellin-1, a tight junction protein required for paracellular Mg^{2+} resorption. Science 1999;285:103–6.

20. Quamme GA, De Rouffignac C. Renal magnesium handling. In: Seldin DW, Giebisch G, editors. The Kidney: Physiology and pathophysiology. Third Edition. New York: Raven Press, 2000.

21. Cole DEC, Quamme GA. Inherited disorders of renal magnesium handling. J Am Soc Nephrol 2000;11:1937–47.

22. Walder RY, Shalev H, Brennan TM, Carmi R, Elbedour K, Scott DA et al. Familial hypomagnesemia maps to chromosome 9q, not to the X chromosome: genetic linkage mapping and analysis of a balanced translocation breakpoint. Human Molecular Genetics 1997;6:1491–7.

23. Schweigel M, Lang I, Martens H. Mg^{2+} transport in sheep rumen epithelium: evidence for an electrodiffusive uptake mechanism. Am J Physiol 1999;277:G976–82.

24. Dai, L-J, Ritchie G, Kerstan D, Kang HS, Cole DEC, Quamme GA. Magnesium transport in the distal nephron plays an important role in determining normal and abnormal renal magnesium balance. Physiol Rev 2001;81:51–84.

25. Van Os CH. Transcellular calcium transport in intestinal and renal epithelial cells. Biochim Biophys Acta 1987;906:195–222.

26. Ritchie G, Kerstan D, Kang HS, Dai L-J, Canaff L, Hendy GN et al. 1,25-Dihydroxyvitamin D3 stimulates Mg^{2+} uptake into mouse distal convoluted tubule cells: modulation by extracellular polyvalent cations. Am J Physiol 2001(in press).

27. Günther T. Mechanisms and regulation of Mg^{2+} efflux and Mg^{2+} influx. J Miner Electrolyte Metab 1993;19:259–65.

28. Schmulen AC, Lerman M, Pak CY, Zerwekh J, Morawski S, Fordtran JS et al. Effect of 1,25(OH)2D3 on jejunal absorption of magnesium in patients with chronic renal disease. Am J Physiol 1980;238: G349–52.

29. Norman DA, Fordtran JS, Brinkley LJ, Zerwekh JE, Nicar MJ, Strowig SM et al. Jejunal and ileal adaptation to alterations in dietary calcium: changes in calcium and magnesium absorption and pathogenetic role of parathyroid hormone and 1,25-dihydroxyvitamin D J Clin Invest 1981;67:1599–1603.

30. Krejs GJ, Nicar MJ, Zerwekh JE, Norman DA, Kane MG, Pak CY. Effect of 1,25-dihydroxyvitamin D3 on calcium and magnesium absorption in the healthy human jejunum and ileum. Am J Med 1983;75:973–6.

31. Levine BS, Brautbar N, Walling MW, Lee DBN, Coburn J. Effects of vitamin D and diet magnesium on magnesium metabolism. Am J Physiol 1980;239:E515–23.

32. Hardwick LL, Jones MR, Brautbar N, Lee DB. Magnesium absorption: mechanisms and the influence of vitamin D, calcium and phosphate. J Nutr 1991;121:13–23.
33. Wittner M, Joiner S, Deschênes G, De Roufignac C, Di Stefano A. Cellular adaptation of the mouse cortical thick ascending limb of Henle's loop (CTAL) to dietary magnesium restriction: enhanced transepithelial Mg^{2+} and Ca^{2+} transport. Pflügers Arch 2000;439:765–71.
34. Brown EM. Physiology and pathophysiology of the extracellular calcium-sensing receptor. Am J Med 1999;106:238–53.
35. Chattopadhyay N, Cheng I, Rogers K, Riccardi D, Hall A, Diaz R et al. Identification and localization of extracellular Ca^{2+}-sensing receptor in rat intestine. Am J Physiol 1998;274:G122–30.
36. Wang W-H, La M, Hebert SC. Cytochrome P-450 metabolites mediate extracellular Ca^{2+}-induced inhibition of apical K^+ channels in the TAL. Am J Physiol 1996;271:C103–11.
37. Di Stefano A, Desfleurs A, Simeone S, Nitschke R, Wittner M. Ca^{2+} and Mg^{2+} sensor in the thick ascending limb of the loop of Henle. Kidney & Blood Pressure Research 1997;20:190–3.
38. Ghishan FK, Meneely RL. Intestinal maturation: the effect of glucocorticoids on in vivo net magnesium and calcium transport in the rat. Life Sci 1982;31:133–8.
39. Bamforth SD, Kniesel U, Wolburg H, Engelhardt B, Risau W. A dominant mutant of occluding disrupts tight junction structure and function. J Cell Sci 1999;112:1879–88.
40. Agus ZS. Hypomagnesemia. J Am Soc Nephrol 1999;10:1616–22.

5B

Renal Handling of Magnesium

G. M. Shah

Magnesium (Mg) is the fourth most abundant cation in the body, following calcium, potassium and sodium. It follows potassium as the second most prevalent intracellular ion. This ion plays an essential role in the regulation of a variety of cellular processes, including activation of enzymes, regulation of channel activities and modulation of the nuclear transcriptional and translational processes.

The principal organ regulating magnesium homeostasis is the kidney. In adult humans, the filtered load of Mg is approximately 2.38 gm/day, given a normal GFR of 170 L/day, and ultrafilterable Mg of 1.4 mg/L. However, only 3–5% of the filtered load is excreted in urine, indicating that greater than 95% of the load is reabsorbed in the tubular segments. In dogs, there appears to be a tubular maximum (Tm) for Mg. Massry et al. [1] found a Tm of 140 ug/min/kg at a filtered load of 280 ug/min in parathyroid intact dogs. Studies in humans also support the existence of a Tm for magnesium [2, 3]. Next, we will briefly describe the renal handling of Mg by various segments (Table 5.1).

Proximal Tubule

Mg is unique among the cations in that only 10% of the filtered magnesium is absorbed in the proximal tubules, whereas the fractional reabsorption of sodium, potassium and calcium by the adult proximal tubule is in excess of 60%. Free-flow micropuncture studies have shown that TF/P Mg increases along proximal convoluted tubules in parallel with sodium and water. Further, these studies suggest that proximal tubular Mg permeability is low relative to that of sodium and water [4]. There appears to be a change in Mg permeability in the proximal segment during development. In the neonate, the rat proximal tubule reabsorbs approximately 70% of the filtered magnesium [5]. The mechanism of decreasing Mg reabsorption in mature adult kidney is unclear at the present time, but may involve a paracellular route, similar to developmental changes in the transcellular chloride gradient [5]. Interestingly, studies in proximal tubules of the glomerular and aglomerular fish show that Mg is secreted actively, presumably through basolateral Mg channels and/or a Mg/Cl co-transporter [6].

Mg reabsorption appears to be a passive and unsaturable process in this segment. FE Mg in proximal tubule is unchanged by a variety of conditions, including volume expansion, acid-base status, plasma calcium level or diuretics. In addition, the Mg transport increases with increasing luminal Mg levels [4]. These data suggest that the reabsorption of Mg occurs presumably by the paracellular route. Mg han-

Table 5.1. Characteristic of renal tubular magnesium transport. A = apical, BL = basolateral. *Mutations associated with idiopathic magnesium wasting syndromes. #Mutations associated with Bartter's syndrome. @Mutations associated with Gitelman's syndrome. ** Mutations may cause hypo- or hypermagnesaemia

	Proximal	cTAL	DCT
Voltage	Electroneutral	Electrogenic Luminal positive	Electrogenic Luminal negative
Nature of flux	Passive	Passive	Active
Route	Paracellular	Paracellular	Cellular Paracellular
Transport proteins/Channels	? Claudin family	Paracellin-1* NKCC2# ROMK# (A) CCLNKB# (BL) CaR** (BL)	Paracellin-1* NCC@ Na-channel (A) ? Na–Mg exchanger CaR** (BL)
Energy pump	? Mg ATPase Na-K ATPase	Na-K ATPase	Na-K ATPase*

dling appears to be essentially similar in the proximal convoluted and straight segments.

Loop of Henle

This is the most critical segment of the nephron, contributing to overall Mg absorption by the kidney. Direct studies of the thin descending or ascending segments are not available. The thin descending loop segment may add Mg in the lumen, and Mg may be recycled (via backflux) in the medulla as shown by microperfusion studies in the sand rat [4, 7]. Further, up to 20% of Mg may be reabsorbed by the thin descending loop under dehydrated conditions in the rat [7]. However, there is no evidence for significant Mg transport in this loop segment under normal physiological conditions.

The cortical segment of the thick ascending limb (cTAL) reabsorbs approximately 70% of the filtered magnesium, whereas the medullary segment (mTAL) does not participate in magnesium conservation. Further evidence indicates that transepithelial magnesium absorption is passive, moving from lumen to the interstitial space through the paracellular pathway [7]. There are three potential mechanisms which may govern Mg transport in cTAL.

(1) The driving force for magnesium movement is the luminal transepithelial voltage, which is positive [8]. The voltage in the loop is determined by the rate of activity of the apical Na-K-2Cl co-transporter (NKCC2) and active sodium absorption by the activity of the basolateral Na-K ATPase. In addition, apical potassium channel (ROMK) and basolateral Cl (CCLNKB) channel activities will affect overall sodium absorption in the cells of the thick segment. Changes in their transport rates will affect the transepithelial voltage and thus magnesium

absorption [4, 7]. Mutations in NKCC2, ROMK or CCLNKB are associated with Bartter's syndrome and, paradoxically, only mild Mg wasting [9].

(2) Recently, a paracellular protein, "paracellin-1" or "claudin 16", has been identified in the tight junction of the thick ascending loop [10]. The claudin 16 gene encodes a protein of 305 amino acids that is a component of the tight junction. Simon and colleagues have proposed that paracellin-1 forms an intercellular pore, which allows the conductance of divalent cations in the cTAL. Alternatively, it may act as a Mg sensor that alters the paracellular permeability affecting other mechanisms. Mutations in paracellin-1 causes dominant isolated renal Mg wasting [10].

(3) In addition to these mechanisms, an extracellular Ca^{2+}/Mg^{2+}-sensing receptor (CaR) has been discovered in the basolateral membrane of the TAL. This polycationic, G-protein-coupled receptor, which is also present in the parathyroid cells, intestinal cells and osteoblasts, plays a major role in calcium homeostasis [11, 12]. Studies have shown that CaR modulates magnesium transport in the cTAL. CaR promotes formation of 20-HETE, a metabolite of arachidonic acid, which inhibits the apical potassium channel (ROMK) and possibly the Na-K-2Cl co-transporter [11]. CaR also inhibits adenylate cyclase, thereby reducing hormone-mediated divalent ion transport. Inactivating mutations of CaR cause familial hypocalciuric hypercalcaemia with mild hypermagnesaemia, whereas activating mutations leads to autosomal dominant hypoparathyroidism and mild hypomagnesaemia [4].

Distal Tubule

Distal convoluted tubule (DCT) reabsorbs 10–15% of filtered Mg load, representing 70–80% of the amount delivered from the loop segment. Thus, this segment of the nephron plays an important role in determining final urinary Mg excretion. In vitro experiments have been performed using canine kidney cell lines and mouse distal convoluted tubule (MDCT) cell lines. These studies indicate that Mg transport in these cells depends on an electrochemical gradient generated by the activities of NCC transporter and apical Na channel, as well as putative apical Mg channels and basolateral Na–Mg exchange mechanism [4, 8]. Furthermore, Meij et al. [13] have recently discovered a mutation in the gamma subunit of the basolateral NA-K ATPase located in the distal convoluted tubule. This mutation causes misrouting of the subunit, entrapping the enzyme in the cytosol rather than translocating into the plasma membrane, presumably reducing the membrane ATPase pump activity and resulting in renal Mg wasting syndrome [13]. In addition, mutations in NCC transporter cause Gitelman syndrome [9]. Ellison has proposed that normal electroneutral NaCl reabsorption is changed to an electronegative Na reabsorption by increased activity of the apical Na channel. Accompanying hyperaldosteronism favours electrogenic Na absorption, thus increasing transepithelial voltage and Mg secretion [9]. Thus, these rare mutations underscore the crucial role of sodium reabsorption in modulating Mg transport in the distal tubule. In addition, like the cTAL cells, CaR is also present on the basolateral surface [12], and paracellin-1 is located in the tight junction of the early distal tubular cells [9]. Recent identification of these crucial proteins has enhanced our understanding of renal Mg transport and inherited disorders of renal Mg transport defects.

Table 5.2. Control of magnesium transport by physiological and pathophysiological process and therapeutic agents. cTAL = cortical thick ascending loop of Henle. DCT = distal convoluted tubule. MDCT = mouse distal convoluted tubule

	cTAL	DCT	MDCT cells
Hormones			
PTH	Increase	Increase	Increase
Calcitonin	Increase	Increase	Increase
Glucagon	Increase	Increase	Increase
Vasopressin	Increase	Increase	Increase
Insulin	Increase		Increase
Aldosterone			Increase
Other factors			
Isopreteronol	Increase		Increase
PGE2			Increase
1.25(OH)$_2$D$_3$			Increase
Acid base			
Metabolic acidosis		Decrease	Decrease
Metabolic alkalosis	Increase	Increase	Increase
Ionic excess/deficit			
Hypermagnesaemia	Decrease	Decrease	Decrease
Hypercalcaemia	Decrease	Decrease	Decrease
Mg restriction	Increase	Increase	Increase
Phosphate depletion		Decrease	Decrease
Potassium depletion			Decrease
Diuretics			
Furosemide	Decrease	No effect	No effect
Bumetanide	Decrease		
Chlorthiazide	No effect	Decrease	Decrease
Amiloride	No effect		Increase
Drugs			
Aminoglycosides	? Decrease		
Capreomycin	? Decrease		
Viomycin	? Decrease		
Amphotericin	? Decrease		
Pentamidine	? Decrease		
Cis-platinum	? Decrease		

Control of Renal Mg Transport

Mg is called an orphan ion, since no major hormonal control mechanism has yet been identified. A large number of factors, including hormones, transport of other ions and other substances, regulate magnesium transport within the cTAL and DCT (Table 5.2). The change in magnesium transport is rapid, sensitive, and selective for magnesium. This adaptive response of transport rates in the DCT provides for the selective renal magnesium conservation. In addition, several drugs have clinically important effects on renal magnesium transport, resulting in renal magnesium wasting or conservation (e.g., amiloride) including diuretics, anti-infective, anti-neoplastic and immunosuppressive agents [14] (Table 5.2). Several excellent reviews have been published discussing the control of renal magnesium transport under various physiological and pathological conditions [15, 16].

References

1. Massry SG, Coburn JW, Kleeman CR. Renal handling of magnesium in the dog. Am J Physiol 1969; 216:1460–7.
2. Rude RK, Bethune JE, Singer FR. Renal tubular maximum for magnesium in normal, hyperparathyroid and hypoparathyroid man. J Clin Metab 1980;51:1425–31.
3. Rude RK, Ryzen E. TmMg and renal Mg threshold in normal man and in certain pathophysiologic conditions. Magnesium1986;5:273–81.
4. Suki WN, Lederer ED, Rouse D. Renal transport of calcium, magnesium, and phosphate. In: Brenner BM, editor. The Kidney, 6th edn. Philadelphia: W.B. Saunders, 2000;520–74.
5. Leliévre-Pegorier M, Merlet-Bénichou C, Roinel N, de Rouffignac C. Developmental pattern of water and electrolyte transport in the superficial nephron. Am J Physiol 1983;244:F15–F21.
6. Beyenbach KW. Renal handling of magnesium in fish: From whole animal to brush border membrane vesicles. Front Bios 2000;5:D712–19.
7. Quamme GA, de Rouffignac C. Epithelial magnesium transport and regulation by the kidney. Front Bios 2000;5:d694–711.
8. Mandon BE, Siga E, Roinel N, de Rouffignac C. Ca^{2+}, Mg^{2+} and K^+ transport in the cortical and medullary thick ascending limb of the rat nephron: Influence of transepithelial voltage. Pflügers Arch 1993;424:558–60.
9. Ellison DH. Divalent cation transport by the distal nephron: insights from Bartter's and Gitelman's syndromes. Am J Physiol 2000;279:F616–25.
10. Simon DB, Lu Y, Choate KA, Velazquez H, Al-Sabban E, Praga M et al. Paracellin-1, a tight junction protein required for paracellular Mg^{2+} resorption. Science 1999;285:103–6.
11. Brown EM. Physiology and pathophysiology of the extracellular calcium-sensing receptor. Am J Med 1999;106:238–53.
12. Hebert SC. Extracellular calcium-sensing receptor: Implications for calcium and magnesium handling in the kidney. Kidney Int 1996;50:2129–39.
13. Meij IC, Koenderink JB, van Bokhoven H, Assink KFH, Groenestege WT, de Pont Jj. Dominant isolated renal magnesium loss is caused by misrouting of the Na-K ATPase gamma-subunit. Nat Genet 2000; 26:265–6.
14. Shah GM, Kirschenbaum MAK. Renal magnesium wasting associated with therapeutic agents. 1991;17:58–64.
15. Quamme GA. Renal magnesium handling: New insights in understanding old problems. Kidney Int 1997;52:1180–95.
16. Cole DE, Quamme GA. Inherited disorders of renal magnesium handling. J Am Soc Nephrol 2000; 11:1937–47.

Hyper- and Hypomagnesaemia

K. Nakatsuka, M. Inaba and E. Ishimura

Distribution, Functions, and Homeostasis of Magnesium

Magnesium (Mg) is the second most abundant ion followed by potassium (K) in intracellular fluid. The recommended daily requirement of Mg from food was reported to be 40 mg/kg, that is 216–396 mg in the US [1] and 240–350 mg in Japan. A negative Mg balance is considered to be induced when the Mg intake in individuals is less than 3 mg/kg. The main dietary sources of Mg are foods such as meat, nuts, green vegetables, grains, and seafood. In an average adult male, total Mg content in the body is 30 mg/kg. Most Mg in the body is in the bones (67%), followed by the intracellular fluid (31%) and extracellular fluid (slightly more than 1%). The serum level of Mg is normally maintained at a concentration of 1.4–2.0 mEq/L (1.7–2.4 mg/dL), about 20% of which is protein-bound. In normal adults, excretion of Mg in urine is estimated to be 4–16 mEq (4.8–19.2 mg) per day [2].

The important biological function of Mg is as a co-factor of several enzymes which catalyse proteins, nucleic acids and phospholipids. The amount of Mg in the cells of most tissues is 6–9 mM/kg wet weight, and most of this mineral is localised in membrane structures, such as in microsomes, mitochondria, and plasma membranes [3]. Significantly small amounts of free Mg in the cells lie in an exchanging equilibrium in a protein-bound form in the membranes. The buffering system regulates the intracellular free Mg at the optimal concentration of about 1 mM, suitable for most intracellular enzyme systems [3]. The role of free Mg on cellular physiology is related to enzymatic processes, which link to the transfer, storage, and utilisation of energy, such as in the reaction of ATPase [4]. Mg activates phosphatases that hydrolyse and transfer organic phosphate involved in the catalysis of ATP. Since Mg shares an important role with ATP in the metabolism of carbohydrates, lipids, nucleic acids, and proteins as well as many ATP-dependent enzymatic processes, several cellular functions are impaired in the event of Mg deficiency.

A change in the serum levels of one ion generally causes the other to deviate in the same direction. In the absence of K replacement, Mg replacement alone is sufficient to correct hypokalaemia in some conditions [5]. This association between these two ions is also supported by the observations that a deficient state of either ion produces clinical symptoms with much similarity and that the replacement of one ion without the other often induces symptoms specific to a deficiency of the ion not replaced [6]. Cellular levels of Mg and K concentrations tend to rise and fall together during periods of Mg and K deficiency. Since dietary Mg restriction and supplementation have little effect on urinary K excretion, decrease in the cellular K content during Mg deficiency must be a consequence of some effects at the tissue level, rather than a defect in renal K excretion [7]. Mg deficiency syndromes primarily affect membrane-bound Mg, which exerts important effects on membrane permeability to sodium (Na), K, and calcium (Ca). These resulting ionic changes, e.g., reduced intra-cellular K and elevated intracellular Na and Ca, lead to the clinical and metabolic effects of Mg deficiency.

In humans and most other species, however, the characteristic response to Mg depletion is hypocalcaemia as described below. Although there is little experimental and clinical evidence that Mg deficiency leads to Ca deficiency, the reverse may be true. In a Mg-deficiency state, the regulatory mechanisms for maintaining serum Ca in the normal range are impaired. Previous experimental observations are completely consistent with the clinical observations that hypocalcaemic and hypomagnesaemic patients have a reduced urinary Ca excretion, normal Ca content in bone, positive Ca balance, and normalisation of serum Ca following correction of Mg stores alone.

Hypermagnesaemia

Causes of Hypermagnesaemia

Since Mg is not routinely measured, hypermagnesaemia is not frequently en-countered in clinical settings (Table 5.3). However, clinicians should pay attention to severe hypermagnesaemia in patients taking agents containing Mg at therapeutic or conventional doses in the presence of impaired renal functions. Symptomatic hypermagnesaemia is usually due to excessive intake or administration of Mg salts. In patients with impaired renal function, injudicious use of antacids containing Mg may result in excessive and symptomatic levels of Mg in the circulation [8]. In addi-tion, Mg-containing cathartics, which are used in the treatment of drug intoxication, such as salicylate and theophylline intoxication [9, 10], may also result in significant hypermagnesaemia. A chronic excess of total Mg is seen in chronic renal failure [11]. It is unclear whether this excess of Mg is responsible for any of the complications of uraemia, although Mg is incorporated into the soft tissue calcification of uraemic patients. Hypermagnesaemia is usually seen in patients with renal failure who are

Table 5.3.   Causes of hypermagnesaemia

Increased intake
- Mg-containing antacids
- Mg-containing laxatives
- Parenteral excessive Mg administration

Decreased renal excretion
- Acute and chronic renal failure
- Volume depletion

Familial hypocalciuric hypercalcaemia (FHH)

Cellular redistribution
- Respiratory acidosis
- Metabolic acidosis
- Cell lysis

receiving Mg as an antacid [12], enema, or intravenous infusion. Hypermagnesaemia is also sometimes observed in acute renal failure accompanied by rhabdomyolysis. Oral intake of large amounts of Mg rarely causes symptomatic hypermagnesaemia in individuals with normal renal function. The rectal administration of Mg for purgation may result in hypermagnesaemia. An administration of Mg is established as a standard therapy for pregnancy-induced hypertension during pre-eclampsia and eclampsia, and may cause Mg intoxication in the mother as well as in the neonate. Modest elevations in the serum Mg concentration may also be seen in patients with familial hypocalciuric hypercalcaemia and patients with lithium ingestion.

Signs and Symptoms of Hypermagnesaemia

Serial symptoms and signs induced by hypermagnesaemia develop as serum Mg increases (Table 5.4). Neuromuscular symptoms are the most common problems associated with Mg excess. One of the earliest signs of hypermagnesaemia is the disappearance of deep tendon reflexes. This occurs at serum Mg concentrations of 4–7 mEq/L. Depressed respiration and apnea due to paralysis of the voluntary muscles may be seen at serum Mg concentrations in excess of 8 mEq/L [13]. Somnolence may be observed at levels as low as approximately 13 mEq/L.

A moderate elevation in the serum Mg concentration of 3–5 mEq/L results in a mild reduction in blood pressure. High concentrations of Mg may be cardiotoxic and result in severe symptomatic hypotension. At serum Mg concentrations greater than 5 mEq/L, electrocardiographic abnormalities such as prolonged PR intervals as well as increased QRS duration and QT intervals are reported to be observed. A complete heart block of a conduction system as well as cardiac arrest may occur at concentrations greater than 15 mEq/L.

Hypermagnesaemia causes a fall in the serum Ca concentration. The hypocalcaemia may be related to the suppressive effect of hypermagnesaemia on parathyroid hormone (PTH) secretion or to hypermagnesaemia-induced PTH resistance of target organs [14]. A direct effect of Mg on decreasing the serum Ca is suggested by the observation that hypermagnesaemia also causes hypocalcaemia in hypoparathyroid subjects.

Other non-specific manifestations of Mg excess include nausea, vomiting, and cutaneous flushing at serum levels of 3–9 mEq/L.

Table 5.4.   Serum levels of magnesium and symptoms/signs

Serum Mg (mEq/L)	Symptoms/signs
3–5	Nausea and vomiting
4–7	Sedation
4–7	Decreased deep tendon reflexes
4–7	Muscle weakness
5–10	Hypotension, bradycardia
10–15	Absent deep tendon reflexes, coma, respiratory paralysis
>15	Cardiac arrest

Treatment of Hypermagnesaemia

The possibility of Mg excess should be taken into account in any patient receiving Mg, especially in patients with renal insufficiency. Mg-containing drugs should be discontinued in patients with mild to moderate increase in the serum Mg level. The treatment of hypermagnesaemia is directed first at removal of the source of Mg and then it is aimed at lowering serum levels of Mg if the serum level is high enough to pose a threat to survival. The vital signs are affected usually at serum levels greater than 7 mEq/L. The acute intravenous infusion of 100–200 mg of Ca (Ca gluconate) over a period of 5 to 10 minutes will be effective at inducing rapid but transient reduction in the serum Mg and lead to a marked improvement in clinical signs and/or symptoms. This should be followed by continuous intravenous infusion of Ca and the use of loop diuretics to enhance urinary Mg excretion. High serum levels of Mg in the presence of impaired renal function may require peritoneal dialysis or haemodialysis using Mg-free dialysate or dialysate with low magnesium concentration. Since Mg is easily dialysed, a rapid and sustained response to this treatment can be expected [8].

Hypomagnesaemia

Causes of Hypomagnesaemia

Hypomagnesaemia is more common than previously suggested in clinical practice. Hypomagnesaemia is usually due to the loss of Mg from either the gastrointestinal tract or kidney (Table 5.5). The Mg concentration of fluids of the upper intestinal tract is approximately 1 mEq/L [4]. Vomiting and nasogastric drainage, therefore, may contribute to Mg depletion. The Mg concentration of diarrhoeal fluids and fistulous drainage are much higher (up to 15 mEq/L), and consequently hypomagnesaemia is common in acute and chronic diarrhoea, regional enteritis, ulcerative colitis, and intestinal and biliary fistulas [4]. Malabsorption syndromes due to nontropical sprue, or radiation enteritis also may result in Mg deficiency [15]. Steatorrhoea and resection or bypass of the small bowel, particularly the ileum, often results

Table 5.5.   Causes of hypomagnesaemia

Decreased gastrointestinal (GI) absorption	• Acquired tubular disorder
• Dietary deficiency	– Loop diuretics
• GI disease – malabsorption	– Hypoparathyroidism
• Alcohol excess	– Hypercalcaemia
• Vomiting/diarrhoea	– Alcohol
Increased renal excretion	• Diabetes mellitus
• Primary tubular disorder	• Tubulo-interstitial disease
– Bartter's syndrome	Cellular redistribution
– Renal tubular acidosis (RTA)	• Respiratory alkalosis
– Gitelman's syndrome	• Hungry bones
	Familial hypomagnesaemia

in intestinal Mg loss or malabsorption. Hypomagnesaemia is also associated with factors causing pancreatitis, such as alcohol excess or with saponification of Mg in a necrotic omentum.

Excessive excretion of Mg into the urine is a cause of Mg deficiency [16]. Renal Mg reabsorption in tubules is adversely proportional to tubular fluid flow as well as to Na and Ca excretion [17]. Therefore, chronic parenteral fluid therapy, particularly with saline, and volume-expansion states, such as primary aldosteronism, occasionally result in Mg deficiency. Hypercalcaemia and hypercalciuria have been reported to decrease renal Mg reabsorption and probably are a cause of increased urinary Mg excretion. Hypomagnesaemia is observed in hypercalcaemic states as well as in patients with vitamin D intoxication.

Diabetes mellitus is probably the most common clinical disorder associated with Mg deficiency, since osmotic diuresis induced by glucosuria results in increased urinary Mg excretion. Several pharmaceutical agents are also recognised as causing increased urinary Mg excretion, leading to hypomagnesaemia. The major site of renal Mg reabsorption is at the loop of Henle. Therefore, excessive use of diuretics such as furosemide and ethacrynic acid has been reported to result in marked hypomagnesaemia. Hypomagnesaemia is often seen in chronic alcoholism, and high levels of blood alcohol are a contributing factor to Mg deficiency [18]. Metabolic acidosis due to untreated diabetic ketoacidosis, starvation, or alcoholism also results in increased urinary Mg excretion. Hypomagnesaemia is also not infrequently accompanied with the "hungry bone" syndrome, a phase of rapid bone mineral apposition in subjects with hyperparathyroidism or hyperthyroidism following surgical parathyroidectomy or thyroidectomy, respectively. Finally, chronic tubulo-interstitial kidney disease is occasionally associated with increased urinary Mg excretion.

Signs and Symptoms of Hypomagnesaemia

The serum Mg concentration is generally measured to assess the Mg status in the body. The normal serum Mg concentration ranges from 1.5 to 1.9 mEq/L (1.8 to 2.2 mg/dL) and a value less than 1.5 mEq/L usually indicates Mg deficiency. However, the serum Mg concentration may not reflect the intracellular Mg content since Mg is principally an intracellular cation and only approximately 1% of the body Mg content exists in the extracellular fluid.

Since Mg depletion is usually secondary to other disease processes or therapeutic agents, the features of the primary disease process itself may complicate and mask the clinical manifestation of hypomagnesaemia. Signs and symptoms induced by hypomagnesaemia are listed in Table 5.6. Neuromuscular hyperexcitability is a presenting complaint. Latent tetany, as elicited by positive Chvostek's and Trousseau's signs, or spontaneous carpal-pedal spasms may be present as a sign of Mg depletion. Grand mal seizures may also occur in severe Mg deficiency. Although hypocalcaemia often contributes to the neurological signs in most patients, hypomagnesaemia without hypocalcaemia has also been reported to result in neuromuscular hyperexcitability. Other signs include vertigo, ataxia, nystagmus, and athetoid and choreiform movements, as well as muscular tremor, fasciculation, wasting, and weakness. Electrocardiographic abnormalities of Mg depletion in humans include prolonged PR and QT intervals. Mg depletion may also result in cardiac arrhythmias. Supraventricular arrhythmias including premature atrial complexes, atrial tachycardia, atrial fibrillation, and junctional arrhythmias have been described [19]. Ventricular pre-

Table 5.6. Symptoms and signs of hypomagnesaemia

Symptoms	Signs
Apathy	Trousseau's sign
Weakness	Chvostek's sign
Anorexia	Muscle fasciculation
Nausea	Tremors
Vomiting	Muscular spasticity
	Hyporeflexia
	Grand mal seizures
	Cardiac conduction disturbances
	Sudden death

mature complexes, ventricular tachycardia, and ventricular fibrillation are serious complications in Mg deficiency.

A laboratory feature frequently seen in Mg deficiency is hypokalaemia. During Mg deficiency, there is a loss of K from the cells leading to intracellular K deficiency. Decreased intracellular K causes an inability of the kidney to conserve K. Attempts to replete the K deficit with K supplementation alone are not successful without simultaneous Mg supplementation. This may be a cause of the electrocardiologic findings and cardiac arrhythmias mentioned above.

Hypocalcaemia is a common manifestation of moderate to severe Mg deficiency and may be a major contributing factor to the increased neuromuscular excitability often present in patients with Mg deficiency. The pathogenesis of hypocalcaemia is multifactorial. In normal subjects, acute changes in the serum Mg concentration will influence parathyroid hormone (PTH) secretion in a manner similar to changes in Ca concentration. An acute fall in serum Mg stimulates PTH secretion, whereas hypermagnesaemia inhibits PTH secretion. During chronic and severe Mg deficiency, however, PTH secretion is impaired [20]. The majority of patients had serum PTH concentrations that were undetectable or inappropriately normal for the degree of hypomagnesaemia-induced hypocalcaemia [21]. Some patients, however, may show serum PTH levels above the normal range that may reflect early Mg depletion. Regardless of the basal circulating PTH concentration, an acute injection of Mg stimulates PTH secretion. Impaired PTH secretion, therefore, appears to be a major factor in hypomagnesaemia-induced hypocalcaemia. Hypocalcaemia in the presence of normal or elevated serum PTH concentrations also suggests end-organ resistance to PTH [22]. Clinically, patients with hypocalcaemia due to Mg deficiency are resistant not only to PTH, but also to calcium and vitamin D therapy. The vitamin D resistance may be due to impaired metabolism of vitamin D, as serum concentrations of an active form of vitamin D, 1,25-dihydroxyvitamin D are low [23].

Treatment of Hypomagnesaemia

Patients who present with signs and symptoms of Mg depletion should be treated with Mg supplementation. The serum levels of Mg in these patients are usually low. Under these circumstances, oral Mg administration is first indicated. In severe Mg deficiency, however, an effective treatment regimen is the administration of 8–12 g of

$MgSO_4 \cdot 7H_2O$ (16.2 mEq Mg) intravenously or intramuscularly [24]. As intramuscular injections may be painful, a continuous intravenous infusion of 48 mEq of Mg over 24 hours is, therefore, preferred and is better tolerated. Either regimen usually results in a normal to slightly increased serum Mg concentration. Mg administration to patients with acute myocardial infarction was shown to decrease the incidence of arrhythmia and the mortality rate [25].

Despite the finding that PTH secretion increases within minutes after beginning Mg administration, the serum calcium concentration may not return to a normal range for three to seven days. This may reflect slow restoration of intracellular Mg. During this period of therapy the total body deficit may not yet be corrected, although the serum concentration of Mg may be normal. Administration of Mg should be continued until the clinical manifestations associated with hypocalcaemia and hypokalaemia in Mg depletion are resolved. Patients who are hypomagnesaemic and have seizures or acute arrhythmias may be administered 8 to 16 mEq Mg as an intravenous injection over a 5- to 10-minute period, followed by 48 mEq of Mg intravenously each day. The degree of Mg loss should be estimated during therapy by monitoring serum Mg concentration. If the patient continues to lose Mg from the intestine or kidney, therapy may have to be continued for a longer duration.

Once Mg deficiency of the body disappears, patients can usually maintain a normal Mg status on a regular diet. If repletion is not accomplished and the patient cannot eat, a maintenance dose of 100 mg of Mg should be added daily. Patients who have chronic Mg loss from the intestine or kidney may require continued oral Mg supplementation. A daily oral dose of 300–600 mg of elemental Mg per day should be administered in divided doses to avoid the cathartic effect of Mg. However, in patients with any degree of renal failure, attention should be paid during Mg supplementation. If there is a decrease in the glomerular filtration rate, the dose of Mg should be halved, and the serum Mg concentration must be monitored daily. If hypermagnesaemia is improved, therapy should be discontinued.

References

1. Jones JE, Manalo R, Flink EB. Magnesium requirements in adults. Am J Clin Nutr 1967;20:632–5.
2. Evans RA, Watson L. Urinary excretion of magnesium in man. Lancet 1966;1:522–3.
3. Gunther T. Biochemistry and pathobiochemistry of magnesium. Artery 1981;9:1967–81.
4. Wacker WEC, Parisi AF. Magnesium metabolism. New Engl J Med1968;278:658–776.
5. Shils ME. Experimental human magnesium depletion. Medicine 1969;48:61–82.
6. Flink EB, McCollister RJ, Prasad AS, et al. Evidence for clinical magnesium deficiency. Ann Intern Med 1957;47:956–68.
7. Carney SL, Wong NLM, Dirks HJ. Effect of magnesium deficiency and excess on renal tubular potassium transport in the rat. Clin Sci 1981;60:549–54.
8. Ferdinandus J, Pederson JA, Whatg R. Hypermagnesemia as a cause of refractory hypotension, respiratory depression, and coma. Arch Intern Med 1981;141:669–70.
9. Gren J, Woolf A. Hypermagnesemia associated with catharsis in a salicylate-intoxicated patient with anorexia nervosa. Ann Emerg Med 1989;18:200–3.
10. Weber CA, Santiago RM. Hypermagnesemia. A potential complication during treatment of theophylline intoxication with oral activated charcoal and magnesium-containing cathartics. Chest 1989; 95:56–9.
11. Contiguglia SR, Alfrey AC, Miller NL, et al. Total body magnesium excess in chronic renal failure. Lancet 1972;1:1300–2.
12. McLaughlin SA, Mckinney. Antacid-induced hypermagnesemia in a patient with normal renal fuction bowel obstruction. Ann Pharmacother 1998;32:312–5.
13. Cao Z, Bideau R, Valdes Jr R, Elin RJ. Acute hypermagnesemia and respiratory arrest following infusion of $MgSO_4$ for tocolysis. Clin Chem Acta 1999;285:191–3.

14. Cholst IN, Steinberg SF, Trooper PJ, et al. The influence of hypermagnesemia on serum calcium and parathyroid hormone in human subjects. N Engl J Med 1984;310:1221–5.
15. Booth CC, Babouris N, Hanna S, et al. Incidence of hypomagnesaemia in intestinal malabsorption. Br Med J 1963;2:141–4.
16. Shah GM, Hirschenbaum MA. Renal magnesium wasting associated with therapeutic agents. Miner Electrolyte Metab 1991;17:58–64.
17. Morger ID, Truttmann AC, von Vigier RO, Bettinelli A, Ramelli GP, Bianchetti MG. Plasma ionized magnesium in tubular disorder with and without total hypomegnesemia. Pediatr Nephrol 1999;13:50–3.
18. Flink EB. Magnesium deficiency in alcoholism. Alcoholism 1986;10:590–4.
19. Iseri LT, Freed J, Bures AR. Magnesium deficiency and cardiac disorders. Am J Med 1975;58:837–46.
20. Nadler JL, Rude RK. Disorder of magnesium metabolism. Endocrinol Metab Clin North Am 1995;24:224–8.
21. Zofkova I, Kancheva RL. The relationship between magnesium and calciotropic hormones. Magnesium Res 1995;8:77–84.
22. Rude RK. Magnesium deficiency in parathyroid function. In: Bilezikian JP, editor. The parathyroids. New York: Raven Press, 1994;829–42.
23. Rude RK, Oldham SB. Hypocalcemia of Mg deficiency: altered modulation of adenylate cyclase by Mg^{++} and Ca^{++} may result in impaired PTH secretion and PTH end-organ resistance. In: Altura BM, Aubach J, Seelig JS, editors. Magnesium in cellular processes and medicine. Basel: Karger, 1987;183–95.
24. Oster TW, Epstain M. Management of magnesium depletion. Am J Nephrol 1988;8:349–54.
25. Woods KL, Fletcher S. Long-term outcome after intravenous magnesium sulphate in suspected acute myocardial infarction: the second Leicester intravenous magnesium intervention trial (Limit-2). Lancet 1994;343:816–9.

The Parathyroid Gland

M. Fukagawa and K. Kurokawa

6

The Role of the Parathyroid Gland in Calcium Homeostasis

The parathyroid gland is a very small organ, but it is the key organ for the maintenance of extracellular calcium ion concentration within a narrow physiological range [1].

Calcium ion concentration always falls without regulating hormones, i.e., parathyroid hormone (PTH) and vitamin D. Parathyroid cells sense a small fall in plasma calcium ion from the normal value and secrete PTH. PTH primarily acts on the bone to release calcium. PTH then acts on the kidney to activate the vitamin D system by increasing production of 1,25-dihydroxyvitamin D and also enhances distal calcium reabsorption to maintain the set-point for plasma calcium ion at normal; meanwhile, PTH inhibits phosphate reabsorption by proximal tubule cells so that plasma calcium and phosphate products will not increase. PTH also acts in concert with vitamin D, to enhance bone resorption, allowing more calcium ions to appear in plasma. This sequence will again normalise plasma calcium ion (Figure 6.1).

Our ability to maintain plasma calcium ion concentration requires the constant renewal of bone, parathyroid cells, and PTH and its regulation of plasma calcium ion concentration at the kidney and bone levels. It seems that this ability has been necessary in vertebrates since the transition from aquatic life to terrestrial life at the appearance of amphibian some 300 million years ago (Table 6.1) [2, 2a]. Parathyroidectomy almost inevitably causes hypocalcaemia in reptiles, suggesting the critical role of the parathyroid gland in calcium metabolism.

Table 6.1. Aquatic vs. Terrestrial Life: Evolution of the Ca-Regulating System

High ambient [Ca^{++}] (10 mM) of seawater
Zero ambient [Ca^{++}] of terrestrial life

Development and Structure of the Parathyroid

The development of the parathyroid gland during embryogenesis is a complex process and takes place in the 8–10 mm embryo [3]. Interestingly, the third branchial pouch gives rise to the "lower" parathyroid gland together with the thymus, and the fourth branchial pouch, which is caudal to the third branchial pouch, gives rise to the "upper" parathyroid gland. Because of their longer migration process, the location of the lower parathyroid gland is more variable than the upper and may be distributed from the superior thyroid pole to the thymus and superior mediastinum [4]. DiGeorge syndrome is characterised by cardiovascular, thymus and parathyroid defects. Recently, the gene responsible for this syndrome, Tbx1, was identified in

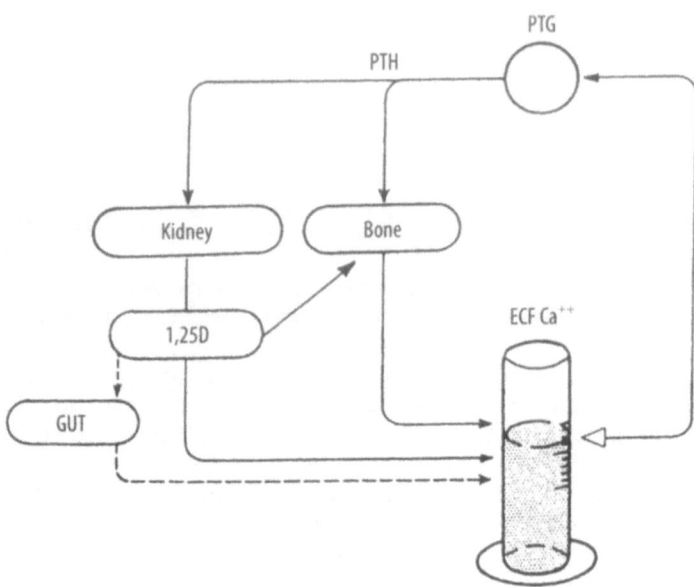

Figure 6.1. Regulatory system for maintaining extracellular calcium ion concentration. Reproduced from [1] with permission.

chromosome 22q11 [5, 6]. Deletion of this gene has been shown to affect the fourth pharyngeal arch arteries.

A recent paper suggests that the thymus may become a source of PTH in certain circumstances [7]. Glial cells missing2 (Gcm2) is a transcription factor whose expression is restricted to parathyroid glands. Gcm2-deficient mice lack parathyroid glands, however, their serum PTH levels are comparable with those of normal mice. The source of PTH in these mice was the thymus, where another Gcm homologue, Gcm1, was expressed, possibly as an additional down-regulatable source of PTH.

The number and location of the parathyroid glands are highly variable. About 80–90% of people have four glands, 5–10% have five, and 5% have three. Less than 5% of people have five or more glands, and up to twelve has been reported in one case [8].

The weight of parathyroid glands depends on age, sex, race, and nutritional status. It also reflects the functional status and fat content of the glands. Lower parathyroid glands tend to be larger than the upper glands. Weights of individual glands range between 35 mg and 55 mg, and seldom exceed 60 mg. The total weight of all glands in a normal adult is 120–150 mg.

In normal parathyroid glands, small clusters of cells and adipose tissue are dispersed unevenly. Normal parathyroid cells are chief cells, which are further classified into oxyphil cells and clear cells, depending upon their morphological and functional status. In clear cells, large amounts of glycogen are seen in the cytoplasm. Chief cells undergo a cyclic process depending upon the activity of PTH synthesis and secretion. The length of the resting phase is determined by calcium ion concentration. Thus, the resting phase is long with hypercalcaemia and short with hypocalcaemia. In normal parathyroid glands, 70–80% of chief cells are in the resting phase.

Parathyroid glands secrete several hormones other than PTH, including PTH-related peptide (PTHrP) [9], chromogranin A [10, 11], and so on, the roles of which are less well understood. Removal of the parathyroid glands of foetuses results in hypocalcaemia of the foetus, which is corrected by infusion of PTHrP (1–86, 1–108, 1–141), not of PTH, into the foetal circulation. In the PTHrP null foetus, placental calcium transport is impaired and serum calcium levels are low, which can be restored by intraperitoneal PTHrP (1–86, 67–86) injection in foetal mice. These data suggest that, during foetal life, PT cells primarily produce and secrete PTHrP, not PTH, and that PTHrP is responsible for maintaining foetal plasma calcium ion concentration through its action on the placenta. Immediately following birth, haematopoiesis moves to bone marrow, thus bone begins to turnover, and parathyroid cells begin to produce and secrete PTH instead of PTHrP. It has been also reported that PTHrP is synthesized in hyperplastic parathyroid glands [12], especially from oxyphil cells, the roles of which still remain unclear.

Control of Parathyroid Function

The main function of the parathyroid is the secretion of PTH. In addition to PTH secretion, synthesis of PTH and proliferation of parathyroid cells reflect the activity of parathyroid glands. All of these functions are regulated mainly by calcium, vitamin D and phosphate. Contribution of local factors, such as endothelin [13], FGF [14], TGFα [15], has also been suspected.

PTH is a peptide hormone with 84 amino acids in mammalians. The amino terminal portion of this hormone shows high homology among the different vertebrate species, and this portion has been shown to be responsible for most of the biologic actions [16]. However, it has been suggested that other portions, i.e., the middle and carboxyl-terminal portions, may have distinct biological properties [17].

PTH is initially synthesised as part of a larger precursor molecule, pre-pro PTH. The "pre" or signal sequence is cleaved from the amino terminus in the cisternae of the endoplasmic reticulum. In Golgi, the short amino-terminal "pro" sequence is removed and then concentrated in dense core secretory vesicles.

The human PTH gene is located on 11p15 and is composed of three exons. Exon 1 (85 bps) encodes of 5′-untranslated region. Exon 2 (90 bps) encodes the translation start site, signal peptide, and a part of the pro-hormone. Exon 3 encodes the rest of the prohormone, PTH peptide, and 3′-untranslated region [18].

Control of PTH Secretion

Secretion of PTH is tightly regulated by the extracellular calcium ion concentration. Although hypercalcaemia does not completely suppress PTH secretion, a negative correlation between the extracellular calcium and PTH has been clinically and experimentally demonstrated by many studies. These two parameters show a sigmoidal curve, which allows very sensitive secretion of PTH in response to the small changes of calcium ion concentration (Figure 6.2) [19].

The PTH level is maximally stimulated by hypocalcaemic challenge (maximal PTH) and minimally suppressed by hypercalcaemic stimulation (minimal PTH), representing the absolute or functional PTH secreting tissue mass. Maximal PTH correlates with the basal PTH level and parathyroid tissue volume, and minimal PTH correlates with the tonic PTH secreting level which can not be suppressed further

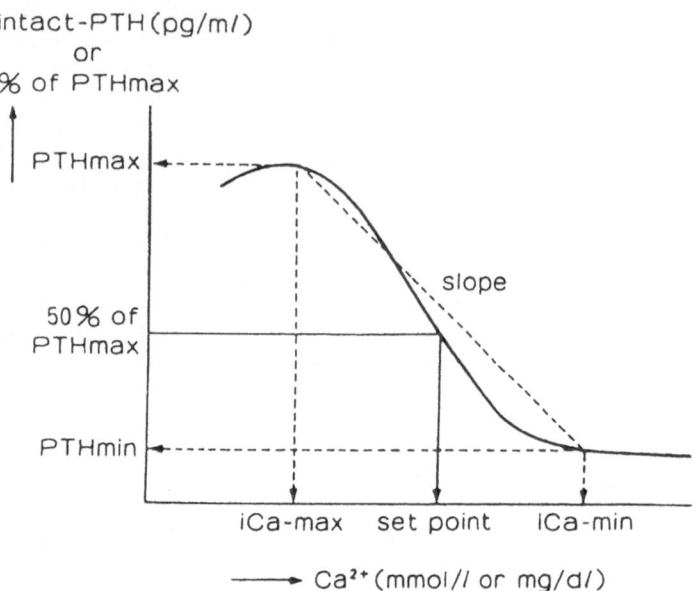

intact-PTH(pg/m/)
or
% of PTHmax

Figure 6.2.   PTH–calcium curve (sigmoidal curve). Reproduced from Fukagawa M, Akizawa T. Modulation of parathyroid cell function by calcium ion in health and uraemia. Am J Med Sci 1999;317:358–62.

even by increasing the extracellular calcium level. Basal PTH is defined as the PTH level at the baseline serum calcium concentration before inducing the change in serum calcium. The slope of the curve and the calcium concentration producing half-maximal change in PTH release (midpoint between the maximal PTH and minimal PTH) represent the sensitivity of parathyroid cells to serum calcium; in other words, the function of the calcium sensor of the parathyroid (set point of calcium for PTH secretion). In many clinical studies, the set point is usually defined as the calcium level that reduces the maximal PTH by 50% [20–22].

Hypocalcaemic stimulation of PTH secretion is explained by the following three stages; 1) direct action of calcium on parathyroid cells to release PTH stored for secretion within seconds, 2) new synthesis of PTH by hypocalcaemic stimulation for several hours, and 3) elevated PTH secretion by the increase in parathyroid cell numbers and growth after prolonged hypocalcaemia for several days. Such information is transmitted through calcium-sensing receptors on the surface of parathyroid cells [23].

The calcium-sensing receptor is a member of the superfamily of G-protein-coupled cell surface receptors, which activates phospholipase C and inhibits adenylate cyclase in the cells [24, 25]. This receptor mediates suppression of PTH secretion in parathyroid cells and the stimulation of calcitonin in thyroidal C-cells by high calcium ion concentration. In the kidney, the receptor is expressed in the thick ascending limb and in the collecting duct, and plays roles in the control of calcium ion and magnesium ion reabsorption, and also in the inhibition of urinary concentrating activity by hypercalcaemia [26]. In addition, the calcium-sensing receptor is also expressed in the pituitary, brain, intestine, and lung, suggesting the unknown action of calcium ions on the cells in these organs.

The deduced amino acid sequence of the calcium-sensing receptor predicts large amino-terminal extracellular domains, followed by seven membrane-spanning helices and cytoplasmic, carboxyterminal tail. The calcium-sensing receptor has significant homology only with the metabotropic glutamate receptor. Experiments with the chimeric receptor of these two receptors reveals that the determinant of ligand specificity resides within extracellular domains, where highly acidic regions are suspected to be the potential binding sites.

Demonstrations of the mutations of the calcium-sensing receptor gene in several inherited disorders strongly support the role of this receptor for the sensing of calcium ions by parathyroid cells. As can be expected, the mutations were mainly identified in the extracellular and membrane-spanning domains.

There are two kinds of mutations; inactivating mutations and activating mutations, which decrease or increase the sensitivity to calcium ions [27].

Familial hypocalciuric hypercalcaemia (FHH) is a disorder with autosomal dominant inheritance [28]. Despite mild hypercalcaemia, patients with this disorder exhibit slightly high PTH levels and low urinary calcium excretion (urinary calcium/creatinine ratio <1%). The functions of mutant receptors have been examined in Xenopus oocytes or HEK cell systems and the decreased sensitivity to calcium was actually confirmed in some mutants [29]. Neonatal severe hyperthyroidism (NSHPT) is a fatal disease, which shows marked elevation of serum calcium and PTH [30]. Pollak et al. demonstrated that patients with NSHPT were homozygous for the mutation responsible for FHH [31]. Such differences between heterozygous and homozygous mutation was also observed in knockout mice of the calcium-sensing receptor gene [32]. However, it has recently been reported that de novo heterozygous mutation of the gene may cause NSHPT [33].

In contrast, the parathyroid in patients with autosomal dominant hypocalcaemia (ADH) senses hypocalcaemia as normal, suggesting the activating mutation of the calcium-sensing receptor gene [34].

Administration of 1,25-dihydroxyvitamin D also suppresses PTH secretion both in vitro and in vivo [35, 36]. Suppression of PTH secretion may be achieved through suppression of PTH synthesis. In addition, 1,25-dihydroxyvitamin D may modulate the sensitivity of parathyroid cells to extracellular calcium. It has been reported that the set point for PTH secretion moves toward normal by 1,25-dihydroxyvitamin D3 pulse therapy in chronic dialysis patients [37]. Increase of calcium-sensing receptors by 1,25-dihydroxyvitamin D has been also reported [38]. However, these observations are still controversial and remain to be elucidated in future.

The effect of hyperphosphataemia on PTH secretion has been attributed to hypocalcaemia and decreased production of 1,25-dihydroxyvitamin D in the kidney. However, available data suggest that phosphate may directly modulate PTH secretion [39, 40]. By using organ cultures of parathyroid tissue, two groups revealed that high phosphate per se stimulated PTH secretion in vitro. Although the existence of phosphate transporters in parathyroid cells has been reported [41], no molecule for phosphate sensor has yet been identified. Another possible mechanism is that the phosphate load may decrease the number of calcium-sensing receptors as recently suggested [42].

PTHrP is also secreted from parathyroid cells. A recent paper suggests that PTHrP enhances the secretory response of PTH to hypocalcaemic stimulus [43]. Thus, PTHrP may act on parathyroid cells as a paracrine factor [44]. By contrast, chromogranin A, another auto- and paracrine factor, inhibits the secretory response of PTH to hypocalcaemic stimulus [45].

Control of PTH Synthesis

Synthesis of PTH is also regulated by calcium. Hypocalcaemia increases PTH mRNA. The small decrease in serum calcium from 2.6 to 2.1 mmol/l produces a maximally threefold increase in PTH mRNA within six hours [46], and also hypocalcaemia induced by 48 hours of EGTA infusion induced a sevenfold increase in PTH mRNA [47].

In contrast to such prompt responses to hypocalcaemia, it has been reported that acute [46] and even chronic hypercalcaemia failed to decrease PTH mRNA. Although several groups of investigators supported the reduced expression of pre-pro-PTH mRNA in high extracellular calcium concentration [47, 48, 49], these findings suggest that the parathyroid gland is geared to respond well to hypocalcaemia rather than to hypercalcaemia.

Even though the mechanism that transmits the message of the change of the extracellular calcium to the regulation of PTH mRNA has not been fully elucidated, a negative calcium regulatory element (nCaRE), identified in the atrial natriuretic peptide gene as well as in the PTH gene, appears to contribute to this mechanism [50]. Hypercalcaemia possibly suppresses the mRNA expression by the transcriptional repression through the negative regulatory element in the PTH gene [51].

In addition to this mechanism, recent papers suggest that PTH mRNA levels are also regulated by modulating the stability of mRNA. Silver and associates demonstrated that the 3'-uncoding region of PTH mRNA was important for such control of mRNA stability [52], and they have identified a protein that binds to this region and protects it from endonuclease [53].

1,25-dihydroxyvitamin D suppresses PTH gene transcription [54, 55]. Although less clearly demonstrated than other vitamin D-regulated genes, a negative responsive element has been identified in the 5'-flanking sequence of the human PTH gene [56]. Furthermore, 1,25-dihydroxyvitamin D also up-regulates vitamin D receptor expression in the parathyroid [57], which may further potentiate the inhibitory action of 1,25-dihydroxyvitamin D on PTH synthesis and secretion. Other steroid hormones have been reported to modulate PTH synthesis. According to recent papers, retinoic acid [58] suppresses, and glucocorticoid [59] and oestrogen [60] enhance, PTH synthesis.

Recent data also suggest that phosphate independently modulates PTH synthesis [40]. The underlying mechanisms have not been fully elucidated yet, however, Silver et al. demonstrated that a low-phosphate diet decreased the amount of protective protein for the 3'-uncoding region of PTH mRNA and enhanced degradation [53].

Control of Parathyroid Cell Proliferation

In normal conditions, turnover of parathyroid cells is extremely low [59]. However, they do have the potential to replicate by appropriate stimulation. As discussed later, several abnormal factors can serve as stimuli for parathyroid cell proliferation in uraemia. These include decreased plasma concentrations of ionised calcium and 1,25-dihydroxyvitamin D, and the accumulation of phosphate. The contribution of each factor has been evaluated experimentally both in vivo and in vitro, but there still remain some points of controversy. This is probably due to the absence of established parathyroid cells.

In weaning rats, hypocalcaemia induced by a low-calcium diet leads to vigorous proliferation of parathyroid cells [61]. Such effects of calcium concentration on cell proliferation were also demonstrated in vitro with primary culture cells of parathyroid by one group [63], but were not confirmed by another group [64]. The physiological relevance of these data still remains a matter of dispute because a number of calcium-sensing receptors on these cells have been shown to decrease rapidly during culture [65].

Suppression of parathyroid cell proliferation by 1,25-dihydroxyvitamin D has been demonstrated by several groups. Although thymidine incorporation into parathyroid cells excised from uraemic rats was decreased by 1,25-dihydroxyvitamin D administration as shown by Szabo et al. [66], Naveh-Many et al. failed to demonstrate the decrease of PCNA-positive cells by 1,25-dihydroxyvitamin D injection in vivo [62].

The effect of phosphate on parathyroid cell proliferation has recently been examined intensively. Parathyroid cell growth was stimulated by phosphate load and was prevented by phosphate restriction in uraemic rats without any changes of calcium and 1,25-dihydroxyvitamin D concentration [67, 68]. Thus, phosphate may regulate parathyroid cell proliferation directly as well as on PTH secretion and synthesis as recently shown [39, 40].

Although calcium ion and 1,25-dihydroxyvitamin D obviously play roles in the several steps of the apoptotic process, the role of apoptosis in the development and regression of parathyroid hyperplasia is confusing at present [69]. This is partly due to the sensitivity of methods used to demonstrate apoptosis directly or indirectly and also due to the difference between the animal model and human model. In most reports, it has been impossible to demonstrate apoptosis in animal models of parathyroid hyperplasia, even with treatment or diet modifications [62, 70]. In contrast, several groups did demonstrate apoptosis in normal and abnormal human parathyroid [71, 72]. They also examined the expression of Bcl-2 and/or Bax, key proteins that prevent or promote apoptosis. Uda et al. demonstrated that the number of apoptotic cells was lower with increased expression of Bcl-2 in secondary hyperparathyroidism as opposed to primary hyperparathyroidism [72]. They suggested that the remarkable proliferation of parathyroid cells in secondary hyperparathyroidism may be due to the reduction of apoptosis of parathyroid cells by Bcl-2 expression, while another group reported the increase of apoptosis which may counteract the elevated proliferation rates of parathyroid cells [73].

Actions and Receptors of Parathyroid Hormone

Receptors of Parathyroid Hormone

The action of PTH on mineral metabolism is mainly mediated through PTH/PTHrP receptors (Type 1 receptor) coupled with adenylate cyclase and phospholipase C [74]. This receptor is abundantly expressed in classic target organs of PTH, such as bone and kidney.

Recent studies suggest that there is more than one type of PTH receptor. Type 1 receptors bind both PTH and PTHrP, while PTH2 receptor binds only to PTH [75]. Recently, type 3 receptors have been cloned in zebra fish, which bind only to PTHrP [76].

In addition to bone and kidneys, type 1 receptors distribute in many organs such as the brain and cardiovascular systems. On the other hand, type 2 receptors have been confirmed in the central nervous system, the cardiovascular system, and the lung, but not in the kidney or bone (Table 6.2). Although the roles and the effects of

Table 6.2. Distribution of PTH Receptor Subtypes

PTH/PTHrP Receptor	PTH2 Receptor
Bone, Kidney	Brain
Heart, Blood Vessels	Heart, Blood Vessels
Brain, Lung, Muscle, etc.	Brain, Lung, etc.

these two types of receptor remain to be clarified, high PTH, especially in uraemia, may affect the function of many organs other than bone and the kidney as discussed in the next chapter.

Strangely, no mutation of the PTH/PTHrP receptor gene has been confirmed yet in pseudohypoparathyroidism type Ib [77]. Activating mutations of PTH/PTHrP receptors, which lead to uncontrolled cAMP accumulation, were identified in Jansen's metaphyseal chondrodysplasia [78]. This disease is characterised by short limbs, severe hypercalcaemia and hypophosphataemia, with normal or lower levels of PTH and PTHrP. Inactivating mutations of PTH/PTHrP receptors were identified in Bloomstrand's chondrodysplasia [79]. This lethal disease is characterised by advanced bone formation and accelerated chondrocyte differentiation with severe abnormalities in mineral homeostasis. Strangely, no mutation of the PTH/PTHrP receptor gene has been confirmed in pseudohypoparathyroidism.

Actions of PTH

PTH administration leads to acute release of calcium from bone. Prolonged administration of PTH leads to an increase in the number and activity of osteoclasts. Recently, mechanisms of osteoclastogenesis and osteoclast activation have been revealed at the molecular level [80]. PTH binds to PTH/PTHrP receptors on osteoblasts and induces a membrane-bound glycoprotein called osteoclast differentiation factor (ODF) [81], which is identical to TRANCE [82], OPGL [83], or RANKL [84]. ODF binds to its receptor expressed on the surface of osteoclast precursor cells [85] and promotes their differentiation into mature osteoclasts. ODF also acts on mature osteoclasts to induce their activation [85, 86]. These actions of ODF on osteoclastic lineage result in increased bone resorption. Interestingly, intermittent administration of PTH exerts anabolic effects on bone; i.e., it increases the formation of trabecular bone.

The major actions of PTH on the kidney are stimulation of calcium reabsorption, inhibition of phosphate reabsorption and stimulation of 1,25-dihydroxyvitamin D synthesis. PTH/PTHrP receptor distributes in the kidney along proximal convoluted tubule, proximal straight tubule, thick ascending limb of Henle, and distal tubule [87].

Reabsorption of calcium occurs both in proximal and distal tubules. Control of calcium reabsorption in the distal tubule determines the final urinary excretion of calcium. PTH binds to PTH/PTHrP receptor on the basolateral membrane of the distal connecting tubule cells. PTH stimulates calcium reabsorption by opening Ca channels on apical membranes through cAMP and membrane hyperpolarisation [88, 89].

Reabsorption of phosphate occurs mainly in the proximal tubules. PTH suppresses phosphate reabsorption by reducing the amount of a sodium–phosphate cotransporter, Npt2, in the apical membrane of proximal tubular cells. Such a reduction of apical Npt2 by PTH has been achieved by internalisation of Npt2 [90] and by suppression of Npt2 synthesis.

PTH activates vitamin D 1α-hydroxylase in the proximal tubule cells [91, 92]. Recently, the 1α-hydroxylase gene has been cloned. Promoter analysis of this gene localised a positive responsive element for PTH at 4 kb upstream in the 5'-flanking sequence [93].

Abnormal Structure and Function

Hyperparathyroidism

Primary hyperparathyroidism is one of the two most common causes of hypercalcaemia and is due to excess secretion of PTH.

This disease is mainly caused by a benign, solitary adenoma (80%) [4]. Hyperplasia of four glands is seen in 20% of patients, which may be associated with MEN. Parathyroid carcinoma is rare, occurring in less that 5% of patients with hyperparahyroidism [94]. The pathogenesis of primary hyperparathyroidism is not known except in a limited number of patients in whom genetic abnormalities discussed later were confirmed.

Patients are becoming less symptomatic than before. Bone lesions are seen in less than 5% of patients. Nephrolithiasis is seen in 20% of patients and is still the most common complication of hyperparathyroidism [95].

Diagnosis is mostly straightforward with elevated PTH concentration. In addition to the widely used intact PTH assay, whole PTH assay has been recently established. Whole PTH assay detects 1–84 PTH only, while traditional intact PTH assay may detect 7–84 PTH as well due to the recognition site of the antibody [96]. Thus, this whole PTH assay may be more useful than intact PTH assay.

Familial hypocalciuric hypercalcaemia (FHH), due to inactivating mutation of the calcium-sensing receptor, must be ruled out during the diagnostic process because this is a benign disease with little risk of renal dysfunction and because surgical parathyroidectomy is only rarely indicated. Screening for MEN needs to be performed during the diagnostic process before the indication of surgical parathyroidectomy, especially in patients with multiple parathyroid hyperplasia.

Patients with symptoms are indicated for surgical parathyroidectomy. For those without symptoms, guidelines originally recommended by the Consensus Development Conference on the Management of Asymptomatic Primary Hyperparathyroidism may be used [97].

1. Serum calcium >1 mg/dl above the upper limit of normal;
2. Any complications of primary hyperparathyroidism (e.g., overt bone disease, nephrolithiasis);
3. An episode of acute primary hyperparathyroidism with life-threatening hypercalcaemia;
4. Marked hypercalciuria (>400 mg daily excretion);
5. Reduction in bone mass at the distal radius, as determined by bone densitometry >2 standard deviations below age- and sex-matched control subjects;
6. Age younger than 50.

For the medical management of primary hyperparathyroidism, calcium restriction, oral phosphate, bisphosphonate and oestrogen therapy in postmenopausal women have been tried, but their effects and risks are still controversial. By contrast, recently developed calcimimetics, which suppress PTH secretion by acting on calcium-sensing receptors, may be useful for such patients [98].

Parathyroid Tumour

Parathyroid tumours are histologically divided into hyperplasia, adenoma, and cancer. Borderlines between these histological classes are less clear compared with those of other organs. In this section, parathyroid tumours associated with primary hyperparathyroidism or multiple endocrine neoplasia syndromes are discussed.

Ever since Arnold demonstrated monoclonal growth in sporadic cases of primary hyperparathyroidism [99], several mutations or translocations of specific genes and loss or inactivation of tumour-suppressor genes have been investigated [100].

In sporadic cases of primary hyperparathyroidism, PRAD1/cyclin D is a well-established oncogene [101] in which the 5′ promoter sequence of the PTH gene is translocated immediately upstream of the cyclin D gene (Figure 6.3) [35]. Thus, the stimulation of PTH gene transcription evokes the overexpression of cyclin D, leading to tumorous proliferation of parathyroid cells. In a recent paper, parathyroid-targeted overexpression of cyclin D1 in transgenic mice caused parathyroid hyperplasia and hyperparathyroidism [102], further supporting this pathophysiological model.

Loss of the tumour-suppressor gene, such as the retinoblastoma gene [103] and p53 [104], has also been reported. Although parathyroid hyperplasia is seen in inactivating mutation of the calcium-sensing receptor, such a mutation has not been demonstrated in primary hyperparathyroidism yet [105]. Mutation of the vitamin D-receptor gene has not been confirmed either [106]. In addition, several reports have been published on the loss or amplification of a specific part of chromosomes, which should be further investigated in future.

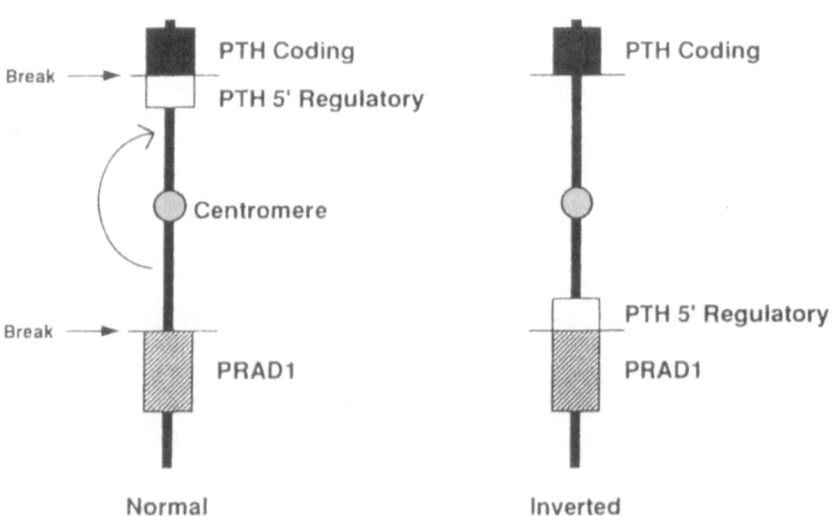

Figure 6.3.   DNA rearrangement involving PTH gene and PRAD1 gene. Chromosomal inversion resulted in PRAD1 being located under 5′-flanking sequence of the PTH gene. Reproduced from [100] with permission.

Multiple endocrine neoplasia syndrome (MEN) is a familial disorder with hyperparathyroidism. Parathyroid tumours develop in MEN type I with tumours of the pancreatic islets, anterior pituitary, and certain other tissues, and in MEN type IIA with medullary thyroid carcinoma and pheochromocytoma. The genes responsible have been recently cloned for these two types of MEN.

The MEN type I gene (menin) is located on chromosome 11q and encodes a tumour-suppressor gene [107]. Menin has been shown to suppress transcriptional activation by JunD through the AP-1 site [108]. The MEN type IIA gene was localised by linkage analysis on chromosome 10q and was then identified as the RET protooncogene [109].

Hypoparathyroidism

Hypoparathyroidism is characterised by hypocalcaemia and hyperphosphataemia due to the insufficient action of PTH. In this disorder, PTH secretion is impaired either due to developmental defects or destruction of the parathyroid glands, or altered regulation of parathyroid function.

Congenital agenesis or hypoplasia of the parathyroid glands causes hypoparathyroidism during the newborn period. DiGeorge syndrome results from the development of the third and fourth branchial pouches, thus hypoparathyroidism, thymic aplasia with immunodeficiency, and congenital conotruncal cardiac abnormalities. In patients with this syndrome, microdeletion of chromosome band 22q11.21–q11.23 has been detected. Recently, a genetic defect of Tbx1, a member of the T-box transcription factor family, has been shown to be responsible for this syndrome. Congenital aplasia of the parathyroid glands are also encountered in Kenny–Caffey syndrome, HDR (hypoparathyroidism, nerve deafness, renal dysplasia) syndrome, and so on, the molecular mechanisms of which are less well understood.

Destruction of the parathyroid glands occurs either by surgery, radiation, metal overload, granulomatous disease, or neoplastic invasion. In addition, autoimmune mechanisms are suspected to play an important role. In APECID (Autoimmune Polyglandular Endocrinopathy Candidiasis Ectodermal Dystrophy syndrome), hypoparathyroidism is manifested alone or in combination with other endocrine deficiency states, such as Addison's disease, insulin-dependent diabetes mellitus, primary hypogonadism, autoimmune thyroiditis and so on. Mutation of the autoimmune regulator gene (AIRE) on chromosome 21q22.3 has neen identified in this syndrome [110]. This protein has been shown to have transcriptional transactivating properties [111].

Secretion of PTH is impaired by hypomagnesaemia and maternal hyperparathyroidism. In addition, several mutations of the PTH gene have been reported. In a family with autosomal dominant isolated hypoparathyroidism, a single base substitution in exon 2 has been identified, which may result in the block of conversion of prepro PTH to pro PTH [112]. In another family with autosomal recessive isolated hypoparathyroidism, a mutation of the donor splice site at exon 2–intron 2 was identified. Such a mutation results in abnormal PTH mRNA without entire exon 2 which encodes the initiation codon and a portion of the signal sequence required for translocation at the endoplasmic reticulum [113].

Hypoparathyroidism can be also caused by activating mutation of the calcium-sensing receptor as already reported in patients with autosomal dominant hypocalcaemia (ADH) and familial syndrome of hypocalcaemia with hypercalciuria. In these syndromes, PTH secretion is impaired even in the presence of hypocalcaemia.

On the other hand, end-organ resistance to PTH also results in biochemical hypoparathyroidism. It is well known that severe hypomagnesaemia may cause such a resistance. Other abnormalities are known as pseudohypoparathyroidism, as first described by Albright. Pseudohypoparathyroidism has been classified into several types depending upon the responsible molecular defect [114]. Nephrogenous cyclic AMP production and phosphaturic response to exogenous PTH are blunted in Type I pseudohypoparathyroidism, which is further divided into subtypes. In pseudo-hypoparathyroidism type Ia – usually with typical developmental and somatic defects of Albright's hereditary osteodystrophy (AHO) such as short stature, round face, brachydactyly, and subcutaneous ossifications – a 50% reduction of Gsα protein is observed. By contrast, patients with pseudohypoparathyroidism type Ib lack typical features of AHO, have hormone resistance that is limited to PTH target organs, and have normal Gsα activity. Patients with multiple hormone resistance in the absence of any defect in Gs or Gi are classified as pseudohypoparathyroidism type Ic. Patients with type II pseudohypoparathyroidism show reduced phosphaturic response to exogenous PTH despite a normal increase in nephrogenous cAMP. Thus, abnormalities downstream of cAMP production has been suspected.

Renal Failure

Secondary hyperparathyroidism is one of the main features of bone diseases seen in chronic dialysis patients. This abnormality is characterised by the high-turnover bone disease caused by excess PTH secreted from hyperplastic parathyroid glands [115]. A number of stimuli for PTH secretion have been suggested in renal failure (Table 6.3).

In patients in the advanced phase of renal failure, circulating concentrations of ionised calcium and 1,25-dihydroxyvitamin D decrease without any treatment, partially due to phosphate accumulation. Both of these abnormalities stimulate PTH secretion. Thus, the treatment of hyperparathyroidism in chronic dialysis patients has been mainly aimed at ameliorating hypocalcaemia and at maintaining physiological concentrations of 1,25-dihydroxyvitamin D. Accordingly, phosphate binders, such as calcium carbonate and oral active vitamin D sterols are routinely used in these patients [116, 117]. In spite of these therapies, it is sometimes still difficult to control PTH secretion in chronic dialysis patients.

As reported in 1984 by Slatopolsky et al. [118], some of such patients respond to supraphysiological concentrations of 1,25-dihydroxyvitamin D achieved by bolus administration. These data suggest that parathyroid glands in chronic dialysis patients are resistant to physiological concentration of 1,25-dihydroxyvitamin D, but

Table 6.3. Stimuli of PTH Secretion in Uraemia

Decreased Calcium Ion Concentration
Decreased 1,25-dihydroxyvitamin D Production
Abnormal Sensitivity to Calcium Ion
Resistance to 1,25-dihydroxyvitamin D
Direct Effect of Phosphate
Skeletal Resistance to PTH

may be responsive to pharmacological concentration of 1,25-dihydroxyvitamin D. Thus, resistance of parathyroid cells to 1,25-dihydroxyvitamin D may serve as another stimulus for PTH secretion in chronic renal failure. In an animal model of mild chronic renal failure, PTH secretion, synthesis and parathyroid cell proliferation were all enhanced despite the presence of normal plasma concentration of calcium and 1,25-dihydroxyvitamin D. Since such parathyroid hyperfunction returned to normal with pharmacological doses of 1,25-dihydroxyvitamin D, resistance of parathyroid to physiological concentration of 1,25-dihydroxyvitamin D is clearly present even from the early phase of chronic renal failure [119].

As demonstrated in the parathyroid glands of chronic dialysis patients [120] as well as in those of uraemic animals [121], decrease in 1,25-dihydroxyvitamin D receptor density is currently considered the main abnormality responsible for such resistance to 1,25-dihydroxyvitamin D. Disturbance in the up-regulation of 1,25-dihydroxyvitamin D receptor by 1,25-dihydroxyvitamin D itself at various steps is thought to be the main mechanism leading to the decrease of 1,25-dihydroxyvitamin D receptor density. Uraemic toxins may play roles in this disturbance by blocking the interaction of the 1,25-dihydroxyvitamin D-receptor complex with target genes as reported by Hsu's group [122]. This abnormality may exist from the early phase of chronic renal failure. As renal failure progresses, this disturbance may form a vicious cycle of further reducing 1,25-dihydroxyvitamin D receptor density leading to progressive resistance to 1,25-dihydroxyvitamin D.

In uraemic patients, it is known that the PTH–calcium curve shifts to the right. This implies that a higher concentration of calcium is needed to suppress PTH secretion in uraemia. Recent data suggest that such a resistance to calcium is due to the decreased density of calcium-sensing receptors, at least in part [123].

The role of phosphate in the pathogenesis of secondary hyperparathyroidism has long been recognised as indirect effects through the modulation of serum calcium and 1,25-dihydroxyvitamin D production. Recent data suggest the direct action of phosphate on the parathyroid [39, 40].

Marked hyperplasia of parathyroid glands is a unique feature of hyperparathyroidism in chronic dialysis patients. Observations of the patients treated with 1,25-dihydroxyvitamin D3 pulse therapy revealed that the size of the parathyroid glands is the critical marker for the long-term prognosis of vitamin D therapy [124]. In patients with at least one gland enlarged more than one centimetre in diameter or $0.5 \, cm^3$ in volume, it is usually difficult to control PTH secretion. In such patients, hyperparathyroidism always persists or relapses even if it initially responds to 1,25-dihydroxyvitamin D3 pulse therapy. By contrast, patients with gland enlargement less than $0.5 \, cm^3$ not only respond to 1,25-dihydroxyvitamin D3 pulse therapy well, but also can be controlled with active vitamin D sterols in the long term. As demonstrated by Tominaga et al., autoimplantation of tissue fragments from glands heavier than $0.5 \, g$ resulted in frequent relapse of parathyroid hyperfunction [125]. Since $0.5 \, g$ in weight corresponds to $0.5 \, cm^3$ in volume, this seems the critical size of parathyroid for the successful control by 1,25-dihydroxyvitamin D3 pulse therapy.

Such differences in the response to 1,25-dihydroxyvitamin D seem to reflect the difference in the degree of decrease of 1,25-dihydroxyvitamin D receptor density. As we have shown, 1,25-dihydroxyvitamin D receptor density inversely correlates with the weight of enlarged glands (Figure 6.4) [126].

Parathyroid hyperplasia is divided into two types, nodular hyperplasia and diffuse hyperplasia (Figure 6.5). Nodular hyperplasia is usually seen in larger glands (90% of glands heavier than $0.5 \, g$) and is considered the more severe type of parathyroid hyper-

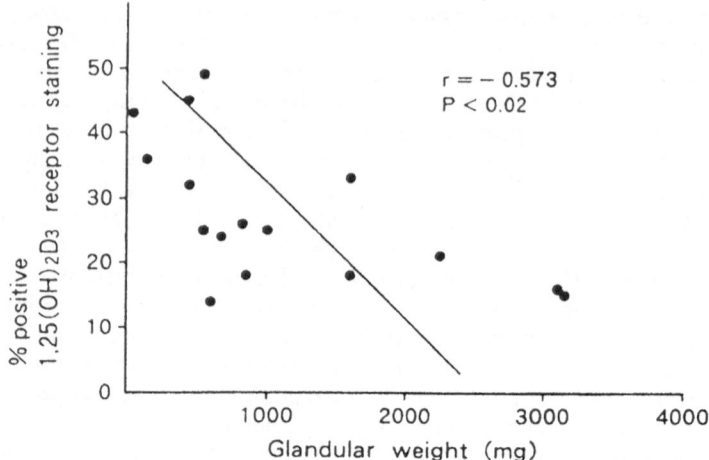

Figure 6.4. Vitamin D receptor and gland weight. They show a significant negative correlation. Thus, larger parathyroid glands have a lower density of vitamin D receptor and are thus resistant to medical therapy. Reproduced from [126] with permission.

plasia. Our data also demonstrated that 1,25-dihydroxyvitamin D receptor density is more decreased in nodular hyperplasia than in diffuse hyperplasia [126]. Thus, difference of the response to 1,25-dihydroxyvitamin D dependent upon gland size can be explained by that in histology and in 1,25-dihydroxyvitamin D receptor density.

In vitro study suggests that a higher concentration of calcium ion was needed to suppress PTH secretion from nodular hyperplasia than from diffuse hyperplasia. It has recently been demonstrated that the density of the calcium-sensing receptor was lower in nodular hyperplasia than in diffuse hyperplasia [127].

Interestingly, stimulation of PTH synthesis by high phosphate concentration was more evident with diffuse hyperplasia than nodular hyperplasia. Thus, undiscovered phosphate sensors may also decrease with the progression of parathyroid hyperplasia.

In uraemia, parathyroid hyperplasia develops initially from diffuse hyperplasia (Figure 6.6). Some cells with more severe reduction of 1,25-dihydroxyvitamin D receptor and calcium-sensing receptor within diffuse hyperplasia then begin to proliferate vigorously to nodule formation, and finally to nodular hyperplasia. In very severe cases, one of the nodules grows very rapidly to occupy the whole gland, forming a single nodule. Monoclonal cell growth was believed to occur only in the severest cases of nodular hyperplasia [128]; however, a recent study elegantly revealed that cells in nodules always show monoclonal proliferation irrespective of their size [129].

In uraemic hyperparathyroidism, Falchetti et al. demonstrated the allelic loss of 11q13 [130] and other groups also reported abnormalities of chromosome 11 [131] on which the MEN-1 gene is located. Recently, Inagaki et al. analysed genomic DNA in the parathyroid of uraemic patients for possible rearrangement or allelic losses of several gene markers located on chromosome 11p near the PTH gene, and found loss of heterozygosity of Ha–ras locus (9%), and allelic loss of the WT-1 gene [132].

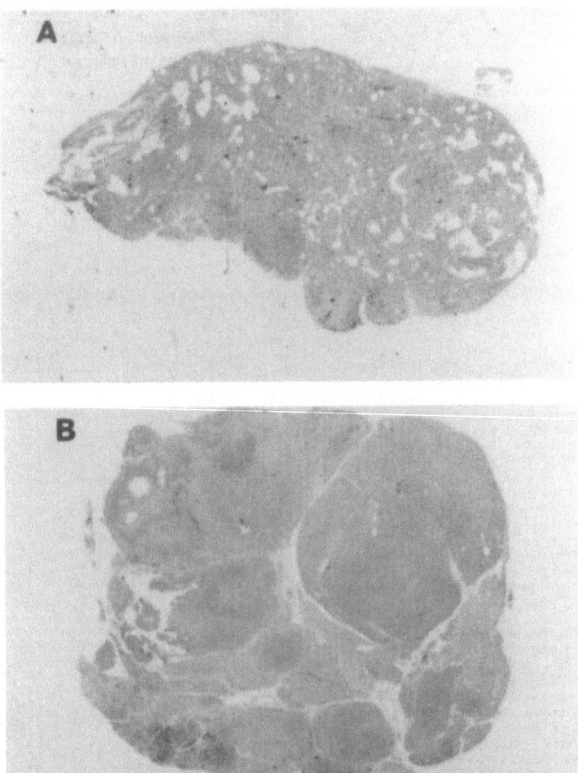

Figure 6.5. Type of parathyroid hyperplasia in uraemia. A: Diffuse hyperplasia. B: Nodular hyperplasia. Reproduced from [126] with permission.

Mutation of the calcium-sensing receptor gene is responsible for familial hypocalciuric hypercalcaemia and neonatal severe hyperparathyroidism with marked parathyroid hyperplasia. Although the decrease of the calcium-sensing receptor was confirmed in the parathyroid glands in uraemic patients, Hosokawa et al. could not find mutation of this gene at least within the coding region [105].

Based on pathophysiology, we propose three important principles for the general management of hyperparathyroidism in chronic dialysis patients. First, parathyroid hyperplasia should be prevented from the early phase of renal failure. This can be achieved by dietary phosphate restriction, and the early use of phosphate binders and active vitamin D sterols with caution. Second, parathyroid size should be routinely evaluated for the selection of therapy. As can be seen, it is of no use, and is sometimes even dangerous, to continue 1,25-dihydroxyvitamin D3 pulse therapy in patients with glands enlarged more than $0.5\,cm^3$. Third, PTH should not be suppressed too much [133]. This is especially important to avoid the risk of adynamic bone disease, recently recognised as a serious problem [134, 135]. Although the PTH levels for the start and the termination of 1,25-dihydroxyvitamin D3 pulse therapy are still controversial, recent consensus for the goal of intact PTH level is around

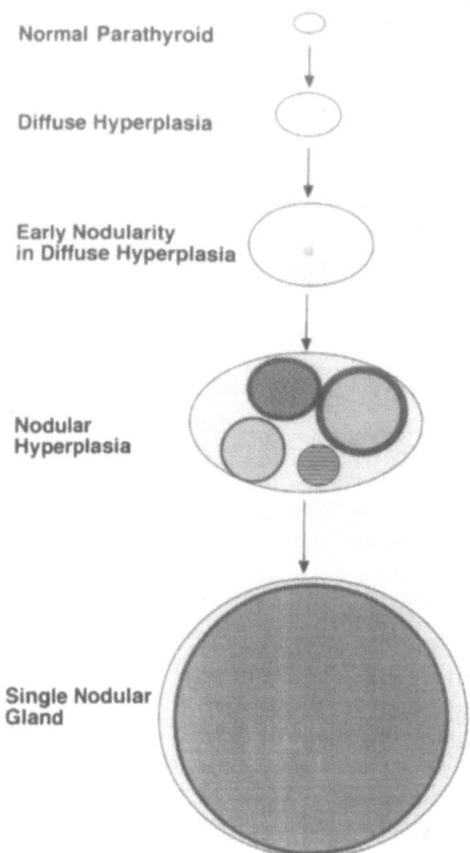

Normal Parathyroid

Diffuse Hyperplasia

Early Nodularity
in Diffuse Hyperplasia

Nodular
Hyperplasia

Single Nodular
Gland

Figure 6.6. Progression of parathyroid hyperplasia in uraemia. Parathyroid hyperplasia progresses initially from diffuse hyperplasia to nodular hyperplasia.

200 pg/ml. In addition, markers for bone turnover, such as serum alkaline phosphatase activity, should also be monitored carefully.

If secondary hyperparathyroidism is controllable by conventional oral active vitamin D therapy, the time course of these patients should be observed carefully. If it is not controllable by conventional therapy, the size of the parathyroid glands should be checked by ultrasonography. If the largest gland is smaller than 0.5 cm³, PTH usually returns to a normal level by 1,25-dihydroxyvitamin D3 pulse therapy in almost all cases and will likely remain well controlled by conventional oral active vitamin D therapy or low-dose intravenous 1,25-dihydroxyvitamin D3 pulse therapy. These patients should be followed up carefully for the relapse of hyperparathyroidism.

If the largest gland is larger than 0.5 cm³, 1,25-dihydroxyvitamin D3 pulse therapy may be tried initially, but should not be continued for too long. In such patients, PTH hypersecretion is usually uncontrollable in the long term and the risk of hypercalcaemia is very high. New vitamin D analogues which are more selective to the parathyroid and with low systemic effects, such as 22-oxa-1,25-dihydroxyvitamin D3, may be attractive [136]. Calcimimetic drugs, which modify the sensitivity of calcium-sensing receptors, may be useful for suppressing PTH secretion and will be available in the near future [98].

How, then, should we manage the patients with marked parathyroid hyperplasia? As for the medical management of such patients, we have developed two new techniques as adjuncts to 1,25-dihydroxyvitamin D3 pulse therapy.

The first technique is selective percutaneous ethanol injection therapy (PEIT). We selectively destroyed the large glands resistant to 1,25-dihydroxyvitamin D pulse therapy, thus leaving only small glands responsive to 1,25-dihydroxyvitamin D3. After successful destruction of the large glands confirmed by the disappearance of blood supply with Doppler ultrasonography, PTH secretion became controllable within a desirable range by the 1,25-dihydroxyvitamin D therapy and conventional active vitamin D therapy [137, 138].

The second technique is direct 1,25-dihydroxyvitamin D3 injection. Theoretically, higher concentration of 1,25-dihydroxyvitamin D3 should be more effective in suppressing the function of the parathyroid glands with a lower density of 1,25-dihydroxyvitamin D receptor. However, higher doses of 1,25-dihydroxyvitamin D3 usually lead to marked hypercalcaemia. To achieve very high concentrations of 1,25-dihydroxyvitamin D3 only locally within the parathyroid glands, we repeatedly injected 1,25-dihydroxyvitamin D3 solution (1 μg/ml) directly into glands enlarged more than $0.5\,cm^3$. Direct 1,25-dihydroxyvitamin D3 injections suppressed PTH hypersecretion and restored the responsiveness to oral vitamin D therapy in chronic dialysis patients [139]. Thus, very high local concentrations of 1,25-dihydroxyvitamin D3 not only suppressed the function of the parathyroid cells with a lower density of 1,25-dihydroxyvitamin D receptor, but may have also up-regulated the 1,25-dihydroxyvitamin D receptor density in parathyroid cells.

References

1. Kurokawa K. Calcium-regulating hormones and the kidney. Kidney Int 1987;32:760–71.
2. Kurokawa K. The kidney and calcium homeostasis. Kidney Int 1994;45 (Suppl. 44):S97–S105.
2a. Kurokawa K. How is plasma calcium held constant? Milieu interieur of calcium. Kidney Int 1996; 49:1760–4.
3. Moore KL, Persaud TVN. The developing human: clinically oriented embryology, 5th ed. Philadelphia: WB Saunders, 1993.
4. Gilmour JR. The embrology of the parathyroid glands, the thymus and certain associated remnants. J Pathol Bacteriol 1937;45:507–22.
5. Mersher S, Funke B, Epstein JA, Heyer J, Puech A, Lu MM, et al. TBX1 is responsible for cardiovascular defects in velo-cardio-facial/DiGeorge syndrome. Cell 2001;104:619–29.
6. Lindsay EA, Vitelli F, Su H, Morishima M, Huynh T, Promparo T, et al. Tbx1 haplp insufficiency in the DiGeorge syndrome region causes aortic arch defects in mice. Nature 2001;410:97–101.
7. Gunther T, Chen ZF, Kim J, Priemel M, Rueger JM, Amling M, et al. Genetic ablation of parathyroid glands reveals another source of parathyroid hormone. Nature 2000;406:199–203.
8. DeLellis RA. Atlas of tumor pathology: Tumors of the parathyroid gland. Washington, DC, AFIP, 1993.
9. Kovacs CS, Kronenberg KM. Maternal-fetal calcium and bone metabolism during pregnancy, puerperium, and lactation. Endocr Rev 1997;18:832–72.
10. Kemper B, Habener JF, Rich A, Potts Jr JT. Parathyroid secretion: discovery of a major calcium-dependent protein. Science 1974;184:167–9.
11. Cohn DV, Zangerle R, Fischer-Colbie R, Chu LLH, Elting JJ, Hamilton JW, et al. Similarity of secretory protein-I from parathyroid gland to chromograni A from adrenal medulla. Proc Natl Acad Sci USA 1982;79:6056–9.
12. Matsushita H, Usui M, Hara M, Shishiba Y, Nakazawa H, Honda K, et al. Co-secretion of parathyroid hormone and parathyroid hormone-related protein via a regulated pathway in human parathyroid adenoma cells. Am J Pathol 1997;150:861–71.
13. Fujii Y, Moreira JE, Orlando C, Maggi M, Aurbach GD, Brandi ML, et al. Endohelin as an autocrine factor in the regulation of parathyroid cells. Proc Natl Acad Sci USA 1991;88:4235–9.

14. Sakaguchi K. Acidic fibroblast growth factor autocrine system as a mediator of calcium-regulated parathyroid cell growth. J Biol Chem 1992;267:24554–62.
15. Gogusev J, Duchambon P, Stoermann-Chopard C, Giovannini M, Sarfati E, Drueke TB. De novo expression of transforming growth factor-α in parathyroid gland tissue of patients with primary and secondary uraemic hyperparathyroidism. Nephrol Dial Transplant 1996;11:2155–62.
16. Hendy GN, Kronenberg HM, Potts Jr JT, Rich A. Nucleotide sequence of cloned cDNAs encoding human preproparathyroid hormone. Proc Natl Acad Sci USA 1981;78:7365–9.
17. Inomata N, Akiyama M, Kubota N, Juppner H. Characterization of a novel PTH-receptor with specificity for the carboxyl-terminal region of PTH (1–84). Endocrinology 1995;136:4732–40.
18. Reis A, Hecht W, Groger R, Bohm I, Cooper DN, Lindenmaier W, et al. Cloning and sequence analysis of the human parathyroid hormone gene region. Hum Genet 1990;84:119–24.
19. Brown EM. Four-parameter model of the sigmoidal relationship between PTH release and extracellular calcium concentration in normal and abnormal parathyroid tissue. J Clin Endocrinol Metab 1983;56:572–81.
20. Felsenfeld AJ, Llach F. Parathyroid gland function in chronic renal failure. Kidney Int 1993;43:771–9.
21. Dumlay R, Rodriguez M, Felsenfeld A, Llach F. Direct inhibitory effect of calcitriol on parathyroid function (sigmoidal curve) in dialysis patients. Kidney Int 1989;36:1093–8.
22. Felsenfeld AJ, Rodriguez M, Dunlay R, Llach F. A comparison of parathyroid gland function in haemodialysis patients with different forms of renal osteodystrophy. Nephrol Dial Transplant 1991;6:244–51.
23. Brown EM, Gamna G, Riccardi D, Lombardi M, Butters R, Kifor O, et al. Cloning and characterization of an extracellular Ca^{2+}-sensing receptor from bovine parathyroid. Nature 1993;366:575–80.
24. Chattopadhyay N, Mithal A, Brown EM. The calcium-sensing receptor: a window into the physiology and pathophysiology of mineral ion metabolism. Endocrine Rev 1996;17:289–307.
25. Brown EM, MacLead RJ. Extracellular calcium sensing and extracellular calcium signaling. Physiol Rev 2001;81:239–97.
26. Brown EM, Hebert SC. Ca^{2+}-receptor-mediated regulation of parathyroid and renal function. Am J Med Sci 1996;312:99–109.
27. Heath III H, Odelberg S, Jackson C, The BT, Hyward N, Lasson C, et al. Clustered inactivating and benign polymorphism of the calcium receptor gene in familial benign hypocalciuric hypercalcemia suggest receptor functional domains. J Clin Endocrinol Metab 1996;81:1312–7.
28. Law Jr WM, Heath III H. Familial benign hypercalcemia (hypocalciuric hypercalcemia). Clinical and pathologic studies in fifteen kindred. Ann Int Med 1995;102:511–19.
29. Pearce SH, Bai M, Quinn SJ, Kifor O, Brown EM, Thakker RV. Functional characterization of calcium-sensing receptor mutations expressed in human embryonic kidney cells. J Clin Invest 1996;98:1860–6.
30. Spiegel AM, Harrison HE, Marx SJ, Brown EM, Auerbach GD. Neonatal primary hyperparathyroidism with autosomal dominant inheritance. J Pediatr 1997;90:269–72.
31. Pollak M, Brown EM, Chou Y-HW, Hebert SC, Marx SJ, Steinnmann B, et al. Mutation in the human Ca^{2+}-sensing receptor gene causes familial hypocalciuric hypercalcemia and neonatal severe hyperparathyroidism. Cell 1993;75:1297–303.
32. Ho C, Corner DA, Pollak M, Ladd DJ, Kifor O, Warren H, et al. A mouse model for familial hypocalciuric hypercalcemia and neonatal severe hyperparathyroidism. Nature Genet 1995;11:389–94.
33. Pearce SH, Trump D, Wooding C, Besser GM, Chew SL, Grant DB, et al. Calcium-sensing receptor mutations in familial benign hypercalcemia and neonatal severe hyperparathyroidism. J Clin Invest 1995;96:2683–92.
34. Pollak M, Brown EM, Kifor O, Estep H, Seidman C, Seidman JG. Autosomal dominant hypocalcemia due to an activating mutation in the human extracellular Ca^{2+}-sensing receptor gene. Nature Genet 1994;8:303–8.
35. Chertow BS, Baker GR, Henry HL, Norman AW. Effects of vitamin D metabolites on bovine parathyroid hormone release in vitro. Am J Physiol 1980;238:E384–8.
36. Silver J, Russell J, Sherwood LM. Regulation by vitamin D metabolites of messenger ribonucleic acid for preproparathyroid hormone in isolated bovine parathyroid cells. Proc Natl Acad Sci USA 1985;82:4270–3.
37. Delmez JA, Tindira C, Grooms P, Dusso A, Windus DW, Slatopolsky E. Parathyroid hormone suppression by intravenous 1,25-Dihydroxyvitamin D. A role for increased sensitivity to calcium. J Clin Invest 1989;83:1349–55.
38. Brown AJ, Zhong M, Finch J, Ritter C, McCracken R, Morrissey J, et al. Rat calcium-sensing receptor is regulated by vitamin D but not by calcium. Am J Physiol 1996;270:F454–60.
39. Slatopolsky E, Finch J, Denda M, Ritter C, Zhong M, Dusso A, et al. Phosphorus restriction prevents parathyroid gland growth. High phosphorus directly stimulates PTH secretion in vitro. J Clin Invest 1996;97:2534–40.

40. Almaden Y, Hernandez A, Torregrosa V, Canalejo A, Sabate L, Fernandez Cruz L, et al. Direct effect of phosphorus on parathyroid hormone secretion from whole rat parathyroid glands in vitro. J Bone Miner Res 1996;11:970–6.
41. Tatsumi S, Segawa H, Morita K, Haga H, Kouda T, Yamamoto H, et al. Molecular cloning and hormonal regulation of PiT-1, a sodium-dependent phosphate cotransporter from rat parathyroid glands. Endocrinology 1998;139:1692–9.
42. Brown AJ, Ritter CS, Finch JL, Slatopolsky E. Decreased calcium-sensing receptor expression in hyperplastic parathyroid glands of uremic rats: role of phosphate. Kidney Int 1999;55:1284–92.
43. Lewin E, Almaden Y, Rodriguez M, Olgaard K. PTHrP enhances the secretory response of PTH to a hypocalcemic stimulus in rat parathyroid glands. Kidney Int 2000;58:71–81.
44. Matsushia H, Hara M, Endo Y, Shishiba Y, Hara S, Ubara Y, et al. Proliferation of parathyroid cells negatively correlates with expression of parathyroid hormone-related protein in secondary parathyroid hyperplasia. Kidney Int 1999;55:130–8.
45. Fasciotto BH, Gorr SU, Bourdeau AM, Cohn DV. Autocrine regulation of parathyroid secretion: inhibition of secretion by chromogranin-A (secretory protein-I) and potentiation of secretion by chromogranin-A and pancreastatin antibodies. Endocrinology 1990;127:1329–35.
46. Naveh-Many T, Friedlander MM, Mayer H, Silver J. Calcium regulates parathyroid hormone messenger ribonucleic acid (mRNA), but not calcitonin mRNA in vivo in the rat. Dominant role of 1,25-dihydroxyvitamin D3. Endocrinology 1989;125:275–80.
47. Yamamoto M, Igarashi T, Muramatsu M, Fukagawa M, Motokura T, Ogata E. Hypocalcemia increases and hypercalcemia decreases the steady-state level of parathyroid hormone messenger RNA in the rat. J Clin Invest 1989;83:1053–6.
48. Naveh-Many T, Raue F, Grauer A, Silver J. Regulation of calcitonin gene expression by hypocalcemia, hypercalcemia, and vitamin D in the rat. J Bone Miner Res 1992;7:1233–7.
49. Russell J, Lettieri D, Sherwood LM. Direct regulation by calcium of cytoplasmic messenger ribonucleic acid coding for preproparathyroid hormone in isolated bovine parathyroid cells. J Clin Invest 1983;72:1851–5.
50. Okazaki T, Ando K, Igarashi T, Ogata E, Fujita T. Conserved mechanism of negative gene regulation by extracellular calcium. Parathyroid hormone gene versus atrial natriuretic polypeptide gene. J Clin Invest 1992;89:1268–73.
51. Okazaki T, Chung U, Nishishita T, Igarashi T, Ogata E, Fujita T, et al. A redox factor protein, ref 1, is involved in negative gene regulation by extracellular calcium. J Biol Chem 269:27855–62.
52. Moallem E, Kilav R, Silver J, Naveh-Many T. RNA-protein binding and post-transcriptional regulation of parathyroid hormone gene expression by calcium and phosphate. J Biol Chem 1998;27:5253–9.
53. Sela-Brown A, Silver J, Brewer G, Naveh-Many T. Identification of AUF1 as a parathyroid hormone mRNA 3'-untranslated region-binding protein that determines parathyroid hormone mRNA stability. J Biol Chem 2000;275:7424–9.
54. Silver J, Russell J, Sherwood LM. Regulation by vitamin D metabolites of messenger ribonucleic acid for preproparathyroid hormone in isolated bovine parathyroid cells. Proc Natl Acad Sci USA 1985;82:4270–3.
55. Silver J, Naveh-Many T, Mayer H, Schmelzer HJ, Popovtzer MM. Regulation by vitamin D metabolites of parathyroid hormone gene transcription in vivo in the rat. J Clin Invest 1986;78:1296–301.
56. Demay MB, Kiernan MS, DeLuca HF, Kronenberg HM. Sequences in the human parathyroid hormone gene that bind the 1,25-dihydroxyvitamin D3 receptor and mediates transcriptional repression in response to 1,25-dihydroxyvitamin D3. Proc Natl Acad Sci USA 1992;89:8097–101.
57. Strom M, Sandgren ME, Brown TA, DeLuca HF. 1,25-Dihydroxyvitamin D3 up-regulates the 1,25-dihydroxyvitamin D3 receptor in vivo. Proc Natl Acad Sci USA 1989;86:9770–3.
58. MacDonald PN, Ritter C, Brown AJ, Slatopolsky E. Retinoic acid suppresses parathyroid hormone (PTH) secretion and prepro PTH mRNA levels in bovine parathyroid cell culture. J Clin Invest 1994;93:725–30.
59. Peraldi MN, Rondeau E, Jousset V, el M'Selmi A, Lacave R, Delarue F, et al. Dexamethasone increased preproparathyroid hormone messenger RNA in human hyperplastic parathyroid cell in vitro. Eur J Clin Invest 1990;20:392–7.
60. Naveh-Many T, Almogi G, Livni N, Silver J. Estrogen receptors and biologic response in rat parathyroid tissue and C cells. J Clin Invest 1992;90:2434–8.
61. Parfitt AM. Parathyroid growth: Normal and abnormal. In: Bilezikian JP, Marcus R, Levine MA, aditors. The Parathyroids: Basic & Clinical Concepts. New York: Raven Press, 1994;373–406.
62. Naveh-Many T, Rachmaninov R, Livni N, Silver J. Parathyroid cell proliferation in normal and chronic renal failure rats: the effects of calcium, phosphate and vitamin D. J Clin Invest 1995;96:1786–93.
63. Bianchi S, Fabiani S, Muratori M, Arnold A, Sakaguchi K, Miki T, et al. Calcium modulates the cyclin D1 expression in a rat parathyroid cell line. Biochem Biophys Res Commun 1994;204:691–700.

64. Kremer R, Bolivar I, Goltzman D, Hendy GN. Influence of calcium and 1,25-dihydroxycholecalciferol on proliferation and proto-oncogene expression in primary cultures of bovine parathyroid cells. Endocrinology 1989;125:935–41.
65. Brown AJ, Zhong M, Ritter C, Brown EM, Slatopolsky E. Loss of calcium responsiveness in cultured bovine parathyroid cells is associated with decreased calcium receptor expression. Biochem Biophys Res Commun 1995;212:861–7.
66. Szabo A, Merke J, Beier E, Mall G, Ritz E. 1,25(OH)$_2$ vitamin D$_3$ inhibits parathyroid cell proliferation in experimental uremia. Kidney Int 1989;35:1045–56.
67. Lopez-Hilker S, Dusso AS, Rapp NS, Martin KJ, Slatopolsky E. Phosphorus restriction reverses secondary hyperparathyroidism in chronic renal failure independent of changes in calcium and 1,25 dihydroxycholecalciferol. Am J Physiol 1990;259:F432–7.
68. Yi H, Fukagawa M, Yamato H, Kumagai M, Watanabe T, Kurokawa K. Prevention of enhanced parathyroid hormone secretion, synthesis and hyperplasia by mild dietary phosphorus restriction in early chronic renal failure in rats: possible direct role of phosphorus. Nephron 1995.
69. Drüeke TB, Zhang P, Gogusev J. Apoptosis: background and possible role in secondary hyperparathyroidism. Nephrol Dial Transplant 1997;12:2228–33.
70. Wang Q, Palnitkar S, Parfitt AM. Parathyroid cell proliferation in the rat: effect of age and of phosphate administration and recovery. Endocrinology 1996;137:4558–62.
71. Wang W, Johannson H, Kvasnicka T, Farnebo L-O, Grimelius L. Detection of apoptotic cells and expression of Ki-67 antigen, bcl-2, p53 oncoproteins in human parathyroid adenoma. APMIS 1996;104:789–96.
72. Uda S, Yoshimura A, Sugaya Y, Inui S, Iwasaki S, Taira T, et al. Role of apoptosis in the progression of secondary hyperparathyroidism. Jpn J Nephrol 1996;38:323–8.
73. Zhang P, Duchambon P, Gogusev J, Nabarra B, Sarfati E, Bourdeau A, et al. Apoptosis in parathyroid hyperplasia of patients with primary or secondary uremic hyperparathyroidism. Kidney Int 2000; 57:437–45.
74. Jüppner H, Abou-Samra AB, Freeman M, Kong XF, Schipani E, Richards J, et al. A G-protein-linked receptor for parathyroid hormone and parathyroid hormone-related peptide. Science 1991;254: 1024–6.
75. Usdin TB, Gruber C, Bonner TI. Identification and functional expression of a receptor selectively recognizing parathyroid hormone, the PTH2 receptor. J Biol Chem 1995;270:15455–8.
76. Kronenberg HM, Lanske B, Kovacs CS, Chung UI, Lee K, Segre GV, et al. Functional analysis of PTH/PTHrP network of ligands and receptors. Rec Prog Hormone Res 1998;53:283–301.
77. Schipani E, Weinstein LS, Bergwitz C, Iida-Klein A, Kong XF, Stuhrmann M, et al. Psudohypoparathyroidism type Ib is not caused by mutations in the coding exons of the human parathyroid hormone (PTH)/PTH-related peptide receptor gene. J Clin Endocrinol Metab 1995;80:1611–21.
78. Schipani E, Langman CB, Parfitt AM, Jensen GS, Kikuchi S, Kooh SW, et al. Constitutively activated receptors for parathyroid hormone and parathyroid hormone-related peptide in Jansen's metaphyseal chondroplasia. N Engl J Med 1996;335:708–14.
79. Jobert AS, Zhang P, Couvineau A, Bonaventure J, Roume J, Le Merrer M, et al. Absence of functional receptors of parathyroid hormone and parathyroid hormone-related peptide in Bloomstrand chondrodysplasia. J Clin Invest 1998;102:34–40.
80. Suda T, Takahashi N, Udagawa N, Jimi E, Gillespie MT, Martin TJ. Modulation of osteoclast differentiation and function by the new members of the tumor necrosis factor receptor and ligand families. Endocrine Rev 1999;20:345–57.
81. Matsuzaki K, Udagawa N, Takahashi N, Yamaguchi K, Yasuda H, Shima N, et al. Osteoclast differentiation factor (ODF) induces osteoclast-like cell formation in human peripheral blood mononuclear cell cultures. Biochem Biophys Res Commun 1998;246:199–204.
82. Wong BR, Rho J, Arron J, Robinson E, Orlinick J, Chao M, et al. TRANCE is a novel ligand of the tumor necrosis factor receptor family that activates c-Jun N-terminal kinase in T cells. J Biol Chem 1997;272:25190–4.
83. Lacey DL, Timms E, Tan HL, Kelley MJ, Dunstan CR, Burgess T, et al. Osteoprotegerin ligand is a cytokine that regulates osteoclast differentiation and activation. Cell 1998;93:165–76.
84. Anderson DM, Maraskovsky E, Billingsley WL, Dougall WC, Tometsko ME, Roux ER, et al. A homologue of the TNF receptor and its ligand enhance T-cell growth and dendritic-cell function. Nature 1997;390:175–9.
85. Fuller K, Wong B, Fox S, Choi Y, Chambers TJ. TRANCE is necessary and sufficient for osteoblast-mediated activation of bone resorption in osteoclasts. J Exp Med 1998;188:997–1001.
86. Burgess TL, Qian Y, Kaufman S, Ring BD, Van G, Capparelli C, et al. The ligand for osteoprotegerin (OPGL) directly activates mature osteoclasts. J Cell Biol 1999;145:527–38.
87. Lee K, Brown D, Urena P, Ardaillou N, Ardaillou R, Deeds J, et al. Localization of parathyroid/parathyroid hormone-related peptide receptor mRNA in the kidney. Am J Physiol 1996;270:F186–91.

88. Shimizu T, Yoshitomi K, Nakayama M, Imai M. Effect of PTH, calcitonin and cyclic AMP on calcium transport in the rabbit distal nephron segments. Am J Physiol 1990;259:F408–14.
89. Lau K, Bordeau JE. Parathyroid hormone action in calcium transport in the distal nephron.. Curr Opin Nephrol Hypertens 1995;4:55–63.
90. Pfister MF, Ruf I, Stange G, Ziegler U, Lederer E, Biber J, et al. Parathyroid hormone leads to the lysosomal degradation of the renal type II Na/Pi cotransporter. Proc Natl Acad Sci USA 1998;95:1909–14.
91. Fraser DR, Kodicek E. Regulation of 25-hydroxycholecalciferol metabolism by parathyroid hormone. Nature 1973;241:163–6.
92. Kawashima H, Torikai S, Kurokawa K. Localization of 25-hydroxyvitamin D3–1-alpha-hydroxylase and 24-hydroxylase along the rat nephron. Proc Natl Acad Sci USA 1981;78:1199–203.
93. Murayama A, Takeyama K, Kitanaka S, Kodera Y, Hosoya T, Kato S. The promoter of the human 25-hydroxyvitamin D3 1a-hydroxylase gene confers positive and negative responsiveness to PTH, calcitonin and 1a, 25(OH)2D3. Biochem Biophys res Commun 1998;249:11–16.
94. Wynne AG, Van Heerden J, Carney JA, Fitzpatrick LA. Parathyroid carcinoma: clinical and pathological features in 43 patients. Medicine 1992;71:197–205.
95. Silverberg SJ, Shane E, Jacobs TP, Siris ES, Gartenberg F, Seldin D, et al. Nephrolithiasis and bone involvement in primary hyperparathyroidism. Am J Med 1990;89:327–34.
96. Gao P, Scheibel S, D'Amour P, John MR, Rao SD, Schmidt-Gayk H, et al. Development of a novel immunoradiometric assay exclusively for biologically active whole parathyroid hormone: implications for improvement of accurate assessment of parathyroid function. J Bone Mineral Res 2001;16:605–14.
97. Consensus Development Conference Panel. Diagnosis and management of asymptomatic primary hyperparathyroidism: Concensus Development Conference Statement. Ann Intern Med 1991;114:593–7.
98. Silverberg SJ, Bone III HG, Marriott TB, Locker FG, Thys-Jacobs S, Dziem G, et al. Short-term inhibition of parathyroid hormone secretion by a calcium-receptor agonist in patients with primary hyperparathyroidism. N Engl J Med 1997;337:1506–10.
99. Arnold A, Staunton CE, Kim HG, Gaz RD, Kronenberg HM. Monoclonality and abnormal parathyroid hormone gene in parathyroid adenomas. N Engl J Med 1988;318:658–62.
100. Arnold A. Molecular genetics of parathyroid gland neoplasia. J Clin Endocrinol Metab 1993;77:1108–12.
101. Motokura T, Bloom T, Kim HG, Juppner H, Ruderman JV, Kronenberg HM, et al. A novel cyclin encoded by bcl1-linked candidate oncogene. Nature 1991;350:512–5.
102. Imanishi Y, Hosokawa Y, Yoshimoto K, Schipani E, Mallya S, Papanikolaou A, et al. Primary hyperparathyroidism caused by parathyroid-targeted overexpression of cyclin D1 in transgenic mice. J Clin Invest 2001;107:1093–102.
103. Cryns VL, Thor A, Xu IIJ, Hu SX, Wierman ME, Vickery Jr AL, et al. Loss of the retinoblastoma tumor-suppressor gene in parathyroid carcinoma. N Engl J Med 330:757–61.
104. Cryns VL, Rubio MP, Thor AD, Louis DN, Arnold A. p53 abnormalities in human parathyroid carcinoma. J Clin Endocrinol Metab 1994;78:1320–4.
105. Hosokawa Y, Pollak MR, Brown EM, Arnold A. Mutational analysis of the extracellular Ca2+-sensing receptor gene in human parathyroid tumor. J Clin Endocrinol Metab 1995;80:3107–10.
106. Brown SB, Brierley TT, Palanisamy N, Salusky IB, Goodman W, Brandi ML, et al. Vitamin D receptor as a candidate tumor-suppressor gene in severe hyperparathyroidism of uremia. 2000;85:868–72.
107. Chandrasekharappa SC, Guru SC, Manickam P, Olufemi SE, Collins FS, Emmert-Buck MR, et al. Positional cloning of the gene for multiple endocrine neoplasia-type 1. Science 1997;276:404–7.
108. Agarwal SK, Guru SC, Heppner C, Erdos MR, Collins RM, Park SY, et al. Menin interacts with the AP1 transcription factor Jun D and represses JunD-activated transcription. Cell 1999;96:143–52.
109. Mulligan LM, Kwok JB, Healey CS, Elsdon MJ, Eng C, Gardner E, et al. Germ-line mutations of RET proto-oncogene in multiple endocrine neoplasia type 2A. Nature Year?;363:458–60.
110. Nagamine K, Peterson P, Scott HS, Kudoh J, Minoshima S, Heino M, et al. Positional cloning of the APECED gene. Nat Genet 11997;7:393–8.
111. Pitkanen J, Doucas V, Sternsdorf T, Nakajima T, Aratani S, Jensen K, et al. The autoimmune regulator protein has transcriptional transactivating properties and interacts with the common coactivator CREB-binding protein. J Biol Chem 2000;275:16802–9.
112. Arnold A, Horst SA, Gardella TJ, Baba H, Levine MA, Kronenberg HM. Mutation of the signal peptide-encoding region of the preproparathyroid hormone gene in familial isolated hypoparathyroidism. J Clin Invest 1990;86:1084–7.
113. Parkinson DB, Thakker RV. A donor splice site mutation in the parathyroid hormone gene is associated with autosomal recessive hypoparathyroidism. Nature Genet 1992;1:149–52.
114. Bestepe M, Juppner H. Pseudohypoparathyroidism. New insights into an old disease. Endocrinol Metab Clin North Am 2000;29:569–89.

115. Malluche H, Faugere MC. Renal bone disease 1990: An unmet challenge for the nephrologist. Kidney Int 1990;38:193–211.
116. Akizawa T, Fukagawa M, Koshikawa S, Kurokawa K. Recent progress in management of secondary hyperparathyroidism of chronic renal failure. Current Opinion Nephrol Hypertens 19932:558–65.
117. Coburn JW, Llach F. Renal osteodystrophy. In Narins, editor. Clinical disorders of fluid and electrolyte metabolism, 5th edn. New York: McGraw Hill, 1994;1299–377.
118. Slatopolsky EA, Weerts C, Thielan J, Horst R, Harter H, Martin KJ. Marked suppression of secondary hyperparathyroidism by intravenous administration of 1,25-dihydroxycholecalciferol in uremic patients. J Clin Invest 1984;74:2136–43.
119. Fukagawa M, Kaname S, Igarashi T, Ogata E, Kurokawa K. Regulation of parathyroid hormone synthesis in chronic renal failure in rats. Kidney Int 199239:874–81.
120. Korkor AB. Reduced binding of [^{3}H]1,25-dihydroxyvitamin D$_3$ in the parathyroid glands of patients with renal failure. N Engl J Med 1987;316:1573–7.
121. Brown A, Dusso A, Lopez-Hilker S, et al. 1,25(OH)$_2$D receptors are decreased in parathyroid glands from chronically uremic dogs. Kidney Int 1989;35:19–23.
122. Hsu CH, Patel SR, Young EW, Vanholder R. The biological action of calcitriol in renal failure. Kidney Int 1994;46:605–12.
123. Gogusev J, Duchambon P, Hory B, Giovannini M, Goureau Y, Sarfati E, et al. Depressed expression of calcium receptor in parathyroid gland tissue of patients with hyperparathyroidism. Kidney Int 1997; 51:328–36.
124. Fukagawa M, Kitaoka M, Yi H, et al. Serial evaluation of parathyroid size by ultrasonography is another useful marker for the long-term prognosis of calcitriol pulse therapy in chronic dialysis patients. Nephron 1994;68:221–8.
125. Tominaga Y, Takagi H. Molecular genetics of hyperparathyroid disease. Current Opinion Nephrol Hypertens 1996;5:336–41.
126. Fukuda N, Tanaka H, Tominaga Y, et al. Decreased 1,25-dihydroxyvitamin D$_3$ receptor density is associated with a more severe form of parathyroid hyperplasia in chronic uremic patients. J Clin Invest 1993;92:1436–43.
127. Yano S, Sugimoto T, Tsukamoto T, Chihara K, Kobayashi A, Kitazawa S, et al. Association of decreased calcium-sensing receptor expression with proliferation of parathyroid cells in secondary hyperparathyroidism. Kidney Int 2000;58:1980–6.
128. Drueke TB. The pathogenesis of parathyroid gland hyperplasia in chronic renal failure. Kidney Int 1995;48:259–72.
129. Tominaga Y, Kohara S, Namii Y, Nagasaka T, Haba T, Uchida K, et al. Clonal analysis of nodular parathyroid hyperplasia in renal hyperparathyroidism. World J Surg 1996;20:744–50.
130. Falchetti A, Bale AE, Amorosi A, Bordi C, Cicchi P, Bandini S, et al. Progression of uremic hyperparathyroidism involves alleic loss on chromosome 11. J Clin Endocrinol Metab 1993;76:139–44.
131. Farnebo F, Teh BT, Dotzenrach C, Wassif WS, Svensson A, White I, et al. Differential loss of heterozygosity in familial, sporadic, and uremic hyperparathyroidism. Hum Genet 1997;99:342–9.
132. Inagaki C, Dousseau M, Pacher N, Sarfati E, Drueke TB, Gogusev J. Structural analysis of gene marker loci on chromosome 10 and 11 in primary and secondary uremic hyperparathyroidism. Nephrol Dial Transplant 1998;13:350–7.
133. Quarles LD, Lobaugh B, Murphy G. Intact parathyroid hormone overestimates the presence and severity of parathyroid-mediated osseous abnormalities in uremia. J Clin Endocrinol Metab 1992;75: 145–50.
134. Hercz G, Pei Y, Greenwood C, Manuel A, Saiphoo C, Goodman et al. Aplastic osteodystrophy without aluminum: The role of "suppressed" parathyroid function, Kidney Int 1993;44:860–6.
135. Goodman WG, Ramirez JA, Belin TR, Chon Y, Gales B, Segre GV, et al. Development of adynamic bone in patients with secondary hyperparathyroidism after intermittent calcitriol therapy. Kidney Int 1994; 46:1160–6.
136. Hirata M, Katsumata K, Masaki T, Koike N, Endo K, Tsunemi K, et al. 22-oxacalcitriol (OCT) ameliorates high turnover bone and marked osteitis fibrosa in rats with slowly progressive nephritis. Kidney Int 1999;56:2040–7.
137. Kitaoka M, Fukagawa M, Ogata E, Kurokawa K. Reduction of functioning parathyroid cell mass by ethanol injection in chronic dialysis patients. Kidney Int 1994;46:1110–7.
138. Kakuta T, Fukagawa M, Fujisaki T, Hida M, Suzuki H, Sakai H, et al. Prognosis of parathyroid function after successful percutaneous ethanol injection therapy (PEIT) guided by color Doppler flow mapping in chronic dialysis patients. Am J Kidney Dis 1999;33:1091–9.
139. Kitaoka M, Fukagawa M, Kurokawa K. Direct injection of calciriol into parathyroid hyperplasia in chronic dialysis patients with severe parathyroid hyperfunction. Nephrology 1995;1:563–8.

7

Parathyroid Hormone Toxicity in Chronic Renal Failure

S. G. Massry and M. Smogorzewski

One of the major hormonal disturbances in uraemia is the state of secondary hyperparathyroidism and the elevation in blood levels of parathyroid hormone (PTH) [1]. In an editorial published in 1977, Massry suggested that the state of chronic excess of PTH in patients with chronic renal failure (CRF) may exert a widespread adverse effect on many organs, and that PTH may be a uraemic toxin [2]. Indeed, the available evidence shows that PTH satisfies the strict criteria of a uraemic toxin in that its nature and structure is known, its blood levels are elevated, a relationship between it and the uraemic manifestation(s) exists, an improvement in many signs and symptoms of uraemia follows a reduction in its blood levels, and the administration of PTH to experimental animals with normal renal function produces derangements similar to those seen in uraemic patients.

Acute Effects of Parathyroid Hormone on Cytosolic Calcium

Earlier studies showed that PTH enhances the entry of calcium into cells [3], and newer data demonstrated that PTH causes an acute rise in cytosolic calcium ($[Ca^{2+}]i$) in many cells including pancreatic islets, thymocytes, cardiac myocytes, hepatic cells, adipocytes, kidney cells, and osteoblasts; this effect of PTH is receptor mediated and uses several cellular pathways [4], but these pathways are not uniform among these cells (Table 7.1) and are depicted in Figure 7.1 [4]. In all these cells, PTH activates voltage-dependent and other calcium channels, and hence the hormone-mediated calcium influx into cells is blunted or prevented by calcium channel blockers such as verapamil, nifedipine or amlodipine.

It is of interest that some of these cells respond only to the intact molecule of PTH [PTH-(1-84)] [5], whereas in other cells the response to PTH-(1-84) is significantly greater than equimolar amounts of its aminoterminal fragment [PTH-(1-34)] [6]. These observations suggest that the biological activity of PTH assessed by calcium movement into cells resides not only in its aminoterminal fragment but also in a bigger moiety containing the 1-34 amino acid sequence or in both the 1-34 fragment and another part of the carboxyterminal sequence of the hormone [4]. Figure 7.2 shows a dose–response relationship between the rise in $[Ca^{2+}]i$ of rat hepatocytes and equimolar concentrations of PTH-(1-34) or PTH-(1-84), demonstrating the greater effect of the latter moiety of the hormone.

The aforementioned data indicate that the traditional (kidney and bone) and the non-traditional cells for PTH action must have the molecular machinery for the production of PTH receptors. Indeed, the mRNA for the PTH–PTH related protein (PTH-PTHrP) receptor is present in the kidney, bone, heart, brain, spleen, aorta, ileum, skeletal muscle, lung, and testis [7, 8]. Thus, both physiologic and molecular evidence exist supporting the notion that almost all body organs are targets for PTH action; therefore, it is not surprising that chronic excess of PTH in CRF may exert a widespread deleterious effect on body function in uraemia.

Table 7.1. The Effects of PTH on [Ca²⁺]i of Various Cells, the Cellular Pathways That Are Involved in this Action of PTH, and the Source of the Rise in [Ca²⁺]i

Cell Type	PTH		Generation of cAMP	Inhibition by PTH Antagonist	Cellular Pathways				Source of Calcium	
	1–84	1–34			Voltage-Dependent Calcium Channels	cAMP	G Protein(s)	Protein Kinase C	Extracellular	Intracellular
Cardiac myocytes	++	+	+	+	++	−	+	−		+
Pancreatic islets	++	+	+	+	+	+	+	+	+	−*
Thymocytes	++	−	+	+	+	+		+	+	−*
Hepatocytes	++	+	+	+	+	++	+	+	+	−*
Adipocytes	++	−	−	+	+	−	+	+	+	+
Renal cells	++	+	+	+	+	−*	+	+	+	+
Osteoblast‡		+	+	+	+	+	+	+	+	+

* cAMP pathway participates in the PTH-induced rise in [Ca²⁺]i of rabbit connecting tubular cells.

† PTH failed to increase [Ca²⁺]i in these cells when they were incubated in a free calcium media. It should be mentioned that cells incubated in free calcium media are calcium-depleted (low basal levels of [Ca²⁺]i) and, therefore, mobilisation of calcium from intracellular stores may be limited. However, other agonists were able to cause a rise in [Ca²⁺]i in cells incubated in calcium-free media when PTH failed to do so; examples include arginine vasopressin (AVP) and angiotensin II in hepatic cells, and AVP in adipocytes. It seems, therefore, that if PTH raises [Ca²⁺]i by mobilising intracellular calcium in these cells, such an action would depend on PTH-induced calcium influx, which is important in intracellular calcium mobilisation (calcium-calcium release phenomenon); absence of such influx, when cells are incubated in a free calcium media, may impede intracellular calcium mobilisation.

‡ The data on osteoblasts are derived from studies on rat osteosarcoma cell line UMR-106 and rat osteoblast-like cells ROS 17/2.8.

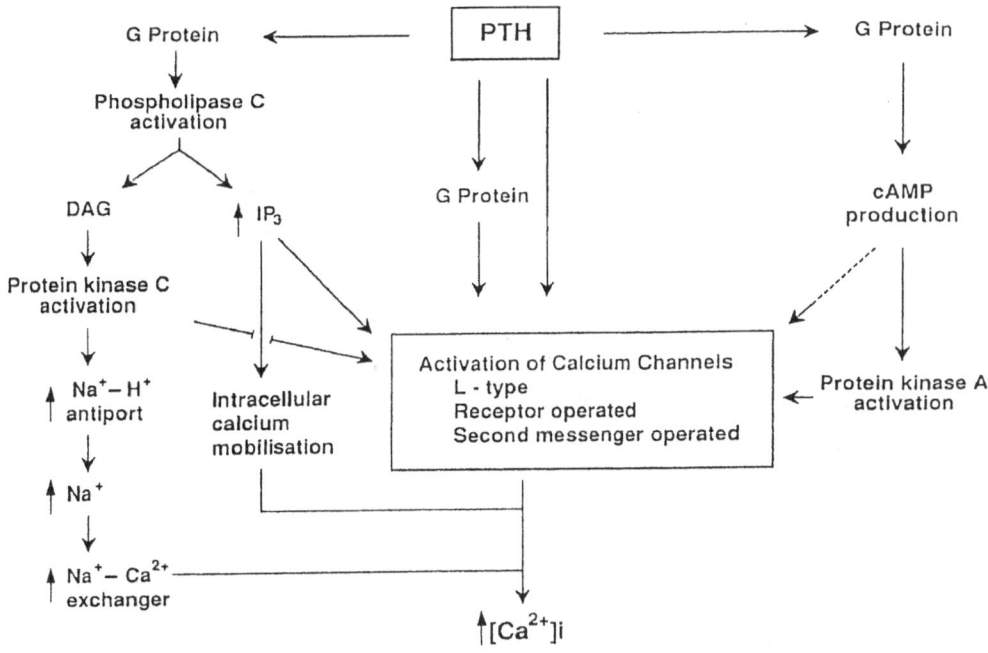

Figure 7.1.   The potential cellular pathways through which PTH may mediate its action on [Ca²⁺]i of cells. DAG = Diaclyglycerol (reproduced with permission [4]).

Chronic Effect of PTH on Tissue Calcium Content and Basal Levels of [Ca^{2+}]i

Soft tissue calcification is a common finding in uraemic patients, and increased calcium content was found in the cornea, skin, blood vessels, brain, peripheral nerves, heart, lungs, pancreas, liver, epididymal fat, and testis of patients and animals with renal failure [10]. These abnormalities have been attributed to the state of secondary hyperparathyroidism of CRF, because parathyroidectomy prevented the calcium accumulation in these tissues. These observations are consistent with the ability of PTH to augment entry of calcium into cells.

Newer data have shown that the increased calcium burden of the tissues in CRF is also associated with significant elevation of the basal levels of [Ca^{2+}]i of various cells including brain synaptosomes, pancreatic islets, cardiac myocytes hepatocytes, adipocytes, thymocytes, B cells, T cells, leukocytes, and platelets [11]. An example of the changes in the basal levels of [Ca^{2+}]i of the cardiac myocytes of uraemic rats is shown in Figure 7.3. This chronic and sustained elevation in [Ca^{2+}]i of these cells is prevented or reversed by parathyroidectomy and/or by treatment with the calcium channel blocker, verapamil. Again, these observations are in agreement with the property of PTH to augment entry of calcium into cells and with the finding that this action of the hormone uses voltage-dependent calcium channels that are blocked by verapamil.

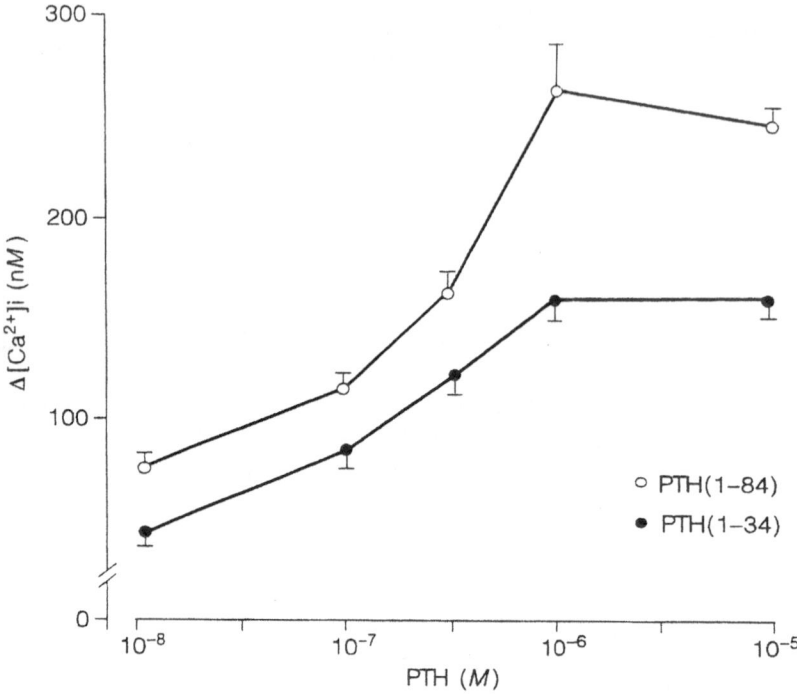

Figure 7.2. Dose-response relationship between the rise in [Ca^{2+}]i of rat hepatocytes and the concentration of PTH. Each datum point is the mean of 6–12 studies and brackets denote 1 SE. The changes in [Ca^{2+}]i were examined at five minutes after the exposure of hepatocytes to PTH (reproduced with permission [6]).

Effect of Parathyroid Hormone on Phospholipids of Cell Membrane

PTH enhances phospholipid turnover in the kidney [13], affects phospholipid content of human red blood cells [14], and reduces total content of phospholipids, phosphatidylserine, phosphatidylethanolamine, and phosphatidylinositol in brain synaptosomes from CRF rats [15]. This latter effect is prevented by parathyroidectomy of the CRF rats or by their treatment with verapamil. The action of PTH on phospholipid content of cell membranes may affect membrane fluidity, its permeability to ions and the agonist/receptor interaction.

Mechanism of Sustained Elevation of Basal Levels of Cytosolic Calcium

Although PTH augments entry of calcium into cells, this action is not adequate to induce a sustained rise in [Ca^{2+}]i because cells are endowed with powerful mechanisms that allow them to pump out excess calcium and/or buffer it by intracellular

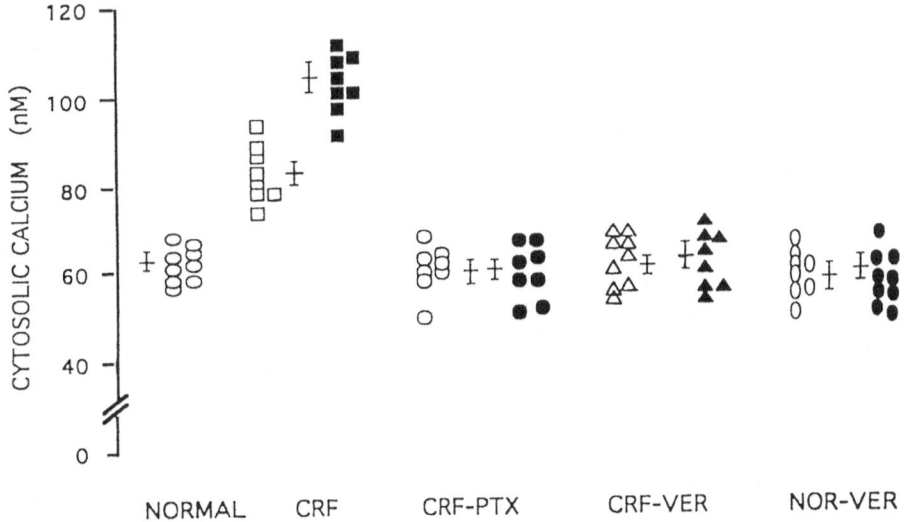

Figure 7.3. Cytosolic calcium in cardiac myocytes of rats. Each datum point represents one animal. Brackets denote mean ± SE. Open symbols are data after three weeks of study and closed symbols after six weeks of study. CRF = chronic renal failure, PTX = parathyroidectomy, V = verapamil (reproduced with permission [12]).

organelles [16]. Therefore, the finding that the basal levels of $[Ca^{2+}]i$ of many cells are elevated in CRF indicates that the balance between calcium entry into and its extrusion out of cells or its buffering within the cells is impaired.

Calcium extrusion from cells is regulated, in major part, either directly or indirectly, by calcium adenosine triphosphatase (Ca^{2+}-ATPase), Na^+–Ca^{2+} exchanger, and Na^+-K^+-ATPase [16]. The activities of Ca^{2+}-ATPase, Na^+-K^+-ATPase and Na^+–Ca^{2+} exchanger of many cells are impaired in CRF [12].

It appears that the initial event leading to the sustained elevation of the basal level of $[Ca^{2+}]i$ is a PTH-induced augmentation of calcium entry into the cells, which would impair mitochondrial production of ATP. The subsequent decrease in ATP content of cells would impair the activities of Ca^{2+}-ATPase and Na^+-K^+-ATPase. The decrease in the phospholipid contents of cell membrane would contribute to the inhibition of the activity of Ca^{2+}-ATPase and Na^+-K^+-ATPase. The reductions in the activity of these two enzymes and of Na^+–Ca^{2+} exchanger lead to a decrease in calcium extrusion out of the cells. Such an effect in the face of a continued PTH-induced increase in calcium entry would result in calcium accumulation in the cells and to an increase in their basal levels of $[Ca^{2+}]i$. The latter would further inhibit mitochondrial oxidation and ATP production. Thus, a vicious circle develops until a new steady state is achieved with a low ATP content, a reduced V_{max} of Ca^{2+}-ATPase and Na^+-K^+-ATPase, a decreased activity of Na^+–Ca^{2+} exchanger, and elevation in the basal levels of $[Ca^{2+}]i$.

The formulation presented in Figure 7.4 implies a sequence of events that occurs over time during the progression of CRF. The chronology of the events depicted in Figure 7.4 was studied in pancreatic islets during the evolution of CRF over a period of six weeks [17]. The data of this study are shown in Figure 7.5. The serum levels of

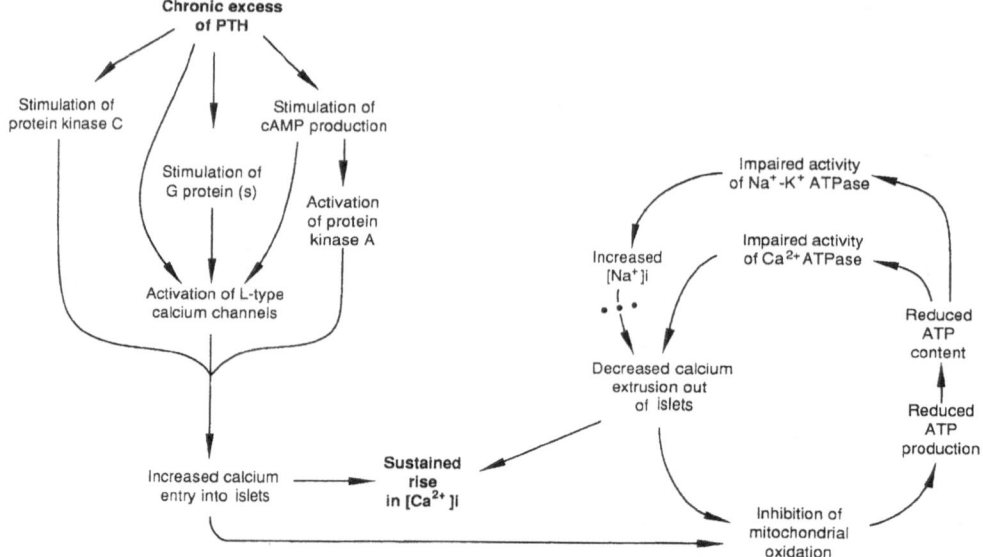

Figure 7.4. A schematic presentation of the sequence of events that leads to the elevation in the basal levels of $[Ca^{2+}]i$ in cells exposed to chronic excess of PTH (reproduced with permission [4]).

PTH begin to rise during the first week of CRF. The V_{max} of Ca^{2+}-ATPase was higher during weeks one to two, and of Na^+-K^+-ATPase during weeks one to three of CRF, but both activities fell to low levels thereafter. At week three of CRF, the ATP content started to fall, and the basal level of $[Ca^{2+}]i$ began to rise. Thus, these data support the formulation shown in Figure 7.4 and indicate that as serum levels of PTH begin to rise, calcium entry into islets is augmented; this in turn will stimulate the activity of Ca^{2+}-ATPase and the Na^+–Ca^{2+} exchanger, and hence, calcium extrusion out of the islets is increased. As a result, $[Ca^{2+}]i$ remains normal during the first two weeks of CRF. An activation of Na^+–Ca^{2+} exchanger may result in accumulation of sodium into islets, an event that would activate the Na^+-K^+-ATPase. As calcium entry is further augmented by the progressive rise in serum PTH levels, mitochondrial oxidation and ATP production would be reduced, resulting in lower ATP content. This fall in ATP causes a reduction in the V_{max} of Ca^{2+}-ATPase and Na^+-K^+-ATPase, and therefore calcium extrusion out of the islets is reduced; consequently, $[Ca^{2+}]i$ rises. This latter change was associated with the impairment in the function of the islets as manifested by a decrease in insulin secretion.

It is of interest that the sequence of events described in Figure 7.5 does not continue, but stabilises at a new steady state. This may be due to down-regulation of the PTH receptors in CRF, an adaptive process that would protect the cells from further actions of the high blood levels of PTH and consequent continued accumulation of calcium, which may eventually lead to cell death. Indeed, studies have demonstrated that the PTH-PTHrP receptor mRNA is down-regulated in CRF [18, 19], an event that would lead to reduced receptor synthesis and hence a decrease in the number of receptors. The reduction in the amount of mRNA of the PTH-PTHrP receptors was found in both the traditional (kidney) [18, 19], and the non-traditional organs (liver [19], and heart [20]) for PTH action.

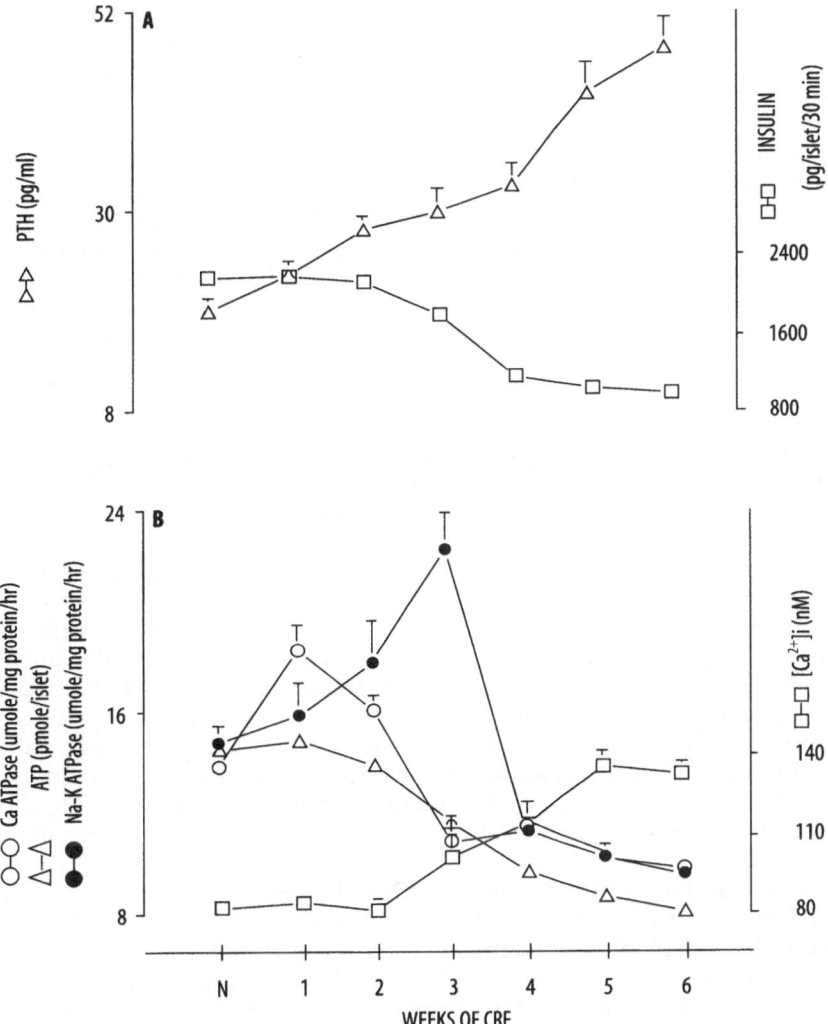

Figure 7.5.   The chronological relationship between the various parameters studied during the evolution of CRF over a period of six weeks (reproduced with permission [17]).

The down-regulation of the mRNA of the PTH-PTHrP receptor in various organs in uraemia is prevented by parathyroidectomy or by treatment with a calcium channel blocker. Because these latter procedures normalise the basal levels of $[Ca^{2+}]i$ in CRF, one may suggest that the elevation of $[Ca^{2+}]i$ in CRF plays a major role in the down-regulation of the PTH-PTHrP receptor. Such an effect may be mediated by the interference of the high levels of $[Ca^{2+}]i$ with the molecular machinery of the cells. The chronic increase in $[Ca^{2+}]i$ may impair the transcription or processing or turn-over of the mRNA of the receptor.

The available data are consistent with the proposition that the PTH-mediated rise in the basal levels of $[Ca^{2+}]i$ in CRF provides a negative feedback mechanism through which the mRNA of the PTH-PTHrP receptor is down-regulated to protect the cells against continued increase in blood levels of PTH, which further augments the elevation in $[Ca^{2+}]i$ to reach a level that would result in cell death. However, this useful adaptive process is not without an adverse effect. The trade-off may be a down-regulation of other hormone receptors if their molecular machinery is affected in a similar manner by the elevated $[Ca^{2+}]i$ of CRF. Indeed, the mRNA of angiotensin II (AT_1) and of vasopressin (AV1a) are down-regulated in CRF [21]. Again, the down-regulation of these two receptors is prevented by parathyroidectomy or by treatment with a calcium channel blocker. This potential generalised effect of the elevation in the basal levels $[Ca^{2+}]i$ in CRF may provide an explanation for the resistance to the action of many hormones commonly encountered in CRF. Indeed, the calcium signals induced by PTH, angiotensin II, vasopressin, or glucagon in hepatocytes of CRF rats are reduced [21].

Role of PTH-Mediated Elevation in $[Ca^{2+}]i$ in the Genesis of the Uraemic Syndrome (Tables 7.2, 7.3)

Available data indicate that the elevation of the basal levels of $[Ca^{2+}]i$ in various cells in uraemia plays a paramount role in the genesis of the manifestations of the uraemic syndrome. First, the prevention of the rise in $[Ca^{2+}]i$ in animals with CRF by prior parathyroidectomy or by their treatment with verapamil from day one of CRF was associated with correction of the uraemic derangements [22–40]. Second, reduction of blood levels of PTH by subtotal parathyroidectomy or by the medical suppression of the parathyroid gland activity with 1,25-dihydroxyvitamin D_3 ($1,25(OH)_2D_3$) in humans was associated with improvement of the uraemic abnormalities [41–52]. Third, treatment of humans or animals with pre-existing CRF with calcium channel blockers (verapamil or nifedipine) reversed the uraemic manifestations [29, 49, 53, 54].

Additional support for the proposition that elevated basal levels of $[Ca^{2+}]i$ is the culprit in the genesis of organ dysfunction in uraemia is provided by other observations. Phosphate depletion, which is a state with elevated basal levels of $[Ca^{2+}]i$, hypoparathyroidism, and normal renal function, is also associated with a wide range of organ dysfunction similar to those seen in CRF; and the normalisation of $[Ca^{2+}]i$ in phosphate-depleted animals by therapy with a calcium channel blocker that normalises $[Ca^{2+}]i$ is also followed by correction of the organ dysfunction [55]. Also, diabetes mellitus with even normal renal function is a state in which $[Ca^{2+}]i$ is elevated in many cells [56], and this abnormality is associated with cell dysfunction similar to those seen in uraemia. Calcium channel blockers can prevent the elevation in $[Ca^{2+}]i$ and this correction is followed by normalisation of the diabetic complication that was studied (i.e., polymorphonuclear dysfunction) [57]. In addition, chronic treatment of normal animals with PTH was also associated with a rise in $[Ca^{2+}]i$ and organ dysfunction despite the absence of CRF [58–61].

It appears, therefore, that a large body of evidence indicates that the secondary hyperparathyroidism of CRF generates a state of cell calcium toxicity that is responsible, in major part, for many of the uraemic manifestations. The clinical implication of this formulation is that the signs and symptoms of uraemic syndrome could be ameliorated, prevented, or reversed either by reduction of blood levels of PTH or by

Table 7.2. Uraemic Manifestations Attributed to PTH-Mediated Rise in $[Ca^{2+}]i$ in Animals With Chronic Renal Failure

Organ	Derangement	Prevented or Reversed by		Induced by PTH Administration in Normal Animals
		Total Parathyroidectomy	Calcium Channel Blocker	
Brain	Abnormal EEG	+		+
Brain synaptosomes	Decreased NE content	+	+	+
	Decreased NE release	+	+	+
	Decreased NE uptake	+	+	+
	Reduced V_{max} of tyrosine hydroxylase	+		
	Increased Km of monoamine oxidase	+		
	Reduced total phospholipid content, phosphoinositol phosphatidylserine and phosphatidylethanolamine	+	+	+
	Increased Ach content and release	+	+	
	Reduce choline kinase activity	+	+	
	Decreased choline content	+	+	
	Increased choline uptake	−	−	
	Increased choline release	−	−	
Peripheral nerves	Prolonged motor nerve conduction velocity	+		+
Heart	Impaired mitochondrial oxidation	+	+	+
	Impaired energy shuttle*	−		+
	Impaired energy use*	−		+
	Impaired fatty acid oxidation	+	+	+
	Decreased cardiac output*	−		+
	Right ventricular hypertrophy	+		
Lungs	Impaired diffusion capacity	+		
	Increased mean pulmonary wedge pressure	+		
Pancreas	Impaired glucose-induced insulin secretion	+	+	+
	Reduced leucine-induced insulin secretion	+	+	
	Decreased-potassium-induced insulin secretion	+	+	
Lipid metabolism	Triglyceridaemia	+	+	
	Impaired fat tolerance	+	+	
	Reduced postheparine lipolytic activity	+	+	
	Decreased hepatic lipase activity	+	+	
Polymorphonuclear leukocytes	Impaired phagocytosis	+	+	
	Reduced oxygen consumption	+	+	
Erythrocytes	Shortened survival	+		
Testes	Reduced blood levels of testosterone	+		+
B cells	Impaired immunoglobulin production	+		

* Total parathyroidectomy did not correct these abnormalities because normal amounts of PTH are required for the synthesis of creatinine phosphokinase.
NE, norepinephrine; Ach, acetylcholine.

Table 7.3.   Uraemic Manifestations Attributed to Excess PTH in Patients With Chronic Renal Failure

Organ	Derangements	Improved or Normalised by		
		Parathyroidectomy	Suppression of Parathyroid Glands Activity by $1.25(OH)_2D_3$ or Vitamin D_3	Calcium Channel Blocker
Brain	Abnormal EEG	+	+	
Peripheral nerve	Prolonged motor nerve conduction velocity*			
Heart	Decreased cardiac index	+	+	
	Reduced left ventricular ejection fraction	+		
	Decreased percent fibre shortening	+	+	
Pancreas	Impaired insulin secretion	+	+	
Polymorphonuclear	Impaired phagocytosis			+
	Impaired glycogen metabolism		+	+
Itching		+		
Tissue necrosis		+		
Soft-tissue calcification		+	+	
Bone resorption		+	+	
Sexual function impotence		+	+	

* A relationship between the derangement in motor nerve conduction velocity and blood levels of PTH was documented.[63]

blocking the action of the hormone on $[Ca^{2+}li$ by therapy with a calcium channel blocker.

Other Pathways Through which PTH May Exert Toxic Actions in Uraemia

PTH affects phospholipids of cell membrane, and alterations in these phospholipids may affect agonist-receptor interactions and cell membrane fluidity. These latter alterations may be associated with adverse effects on cell function.

PTH could be a catabolic agent and, as such, play a role in the wasting syndrome of uraemia and contribute in a modest manner to the accumulation of nitrogenous compounds in blood. Indeed, administration of PTH to humans is associated with a negative nitrogen balance. Nitrogen balance is usually negative in patients with primary hyperparathyroidism and this balance becomes significantly more positive after removal of the adenoma. Thus, it is possible that the elevated blood levels of PTH in CRF not only are directly toxic by affecting cell function but may also contribute to uraemic toxicity by enhancing catabolism and by augmenting the accumulation of nitrogenous compounds in the blood.

Specific Organ Dysfunction in Uraemia and the Role of PTH in Their Genesis

Neurotoxicity of PTH in Uraemia

Studies in animals and humans show that PTH satisfies all the criteria of the uraemic neurotoxin. The calcium content of the brain and peripheral nerves of dogs with intact parathyroid glands and CRF are elevated and these animals display abnormal EEG, characteristic of uraemia, as well as prolonged motor nerve conduction velocity (MNCV) [22, 23]. Parathyroidectomised dogs with comparable degree and duration of CRF did not display these abnormalities [22, 23]. Furthermore, the chronic administration of PTH to animals with normal renal function resulted in similar derangements seen in dogs with intact parathyroid glands and CRF [61]. In rats with CRF or normal rats chronically treated with PTH, the basal levels of $[Ca^{2+}]i$ of their brain synaptosomes were significantly elevated [62] and the metabolism of neurotransmitters (norepinephrine [31] and acetylcholine [39]) were markedly deranged. These abnormalities were prevented by parathyroidectomy of these rats or by their treatment with verapamil [31, 36, 39].

Studies in humans [41, 48, 63, 64] show that (a) patients with CRF, who usually have elevated blood levels of PTH, have increased calcium content of the brain; (b) the EEG abnormalities in patients with CRF were found to be reversed three months after parathyroidectomy; (c) a direct relationship exists between the EEG abnormalities in patients with CRF and their blood levels of the N-terminal fragment of PTH, and amelioration of secondary hyperparathyroidism by treatment with $1,25(OH)_2D_3$ is associated with improvement of the EEG; (d) patients with primary hyperparathyroidism and normal renal function display EEG abnormalities similar to those observed in uraemia, and these derangements progressively improve following parathyroidectomy, with the EEG recordings being normal within two to three weeks after surgery; and (e) MNCV is prolonged in dialysis patients with elevated blood levels of PTH, but it is within the normal range in age-matched dialysis patients who have markedly lower blood levels of PTH.

Finally, certain data also point toward a possible role of PTH in the pathogenesis of dialysis encephalopathy. Aluminium toxicity resulting in increased aluminium content of the brain has been implicated in the cause of dialysis encephalopathy. It has been reported that the accumulation of aluminium in the brain is increased by PTH [65], suggesting that the hormone may play a role in the overall process underlying the development of this syndrome. Interestingly, a patient with dialysis encephalopathy improved markedly after parathyroidectomy [66].

Role of PTH in the Genesis of the Anaemia of Uraemia (Figure 7.6)

The factors responsible for the anaemia of uraemia are multiple, including reduced erythropoiesis due to erythropoietin deficiency, decreased availability of red marrow secondary to bone marrow fibrosis, and shortened RBC survival. Clinical data suggest that PTH may play a role in the genesis of the anaemia in patients with CRF. It has been reported that the haematocrit and reticulocyte count increased and the blood transfusion requirements decreased after subtotal parathyroidectomy in dialysis patients with secondary hyperparathyroidism [67, 68]. Also, a significant rise in

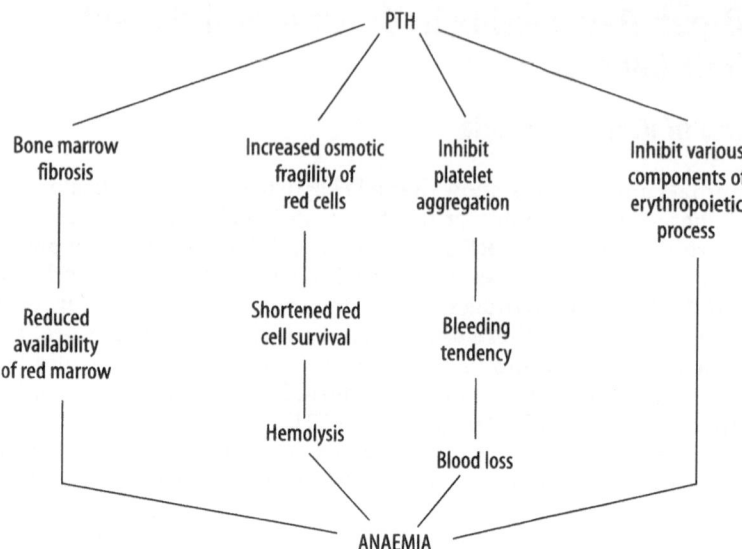

Figure 7.6. A schematic presentation of the pathways through which parathyroid hormone may participate in the genesis of the anaemia of chronic renal failure.

haemoglobin may follow suppression of parathyroid gland activity by the treatment of these patients with $1,25(OH)_2D_3$ [68].

Effect on Erythropoiesis

Intact PTH (1-84 PTH) produced marked inhibition of human peripheral blood and mouse bone marrow erythroid burst-forming units (BFU-E) but had no effect on mouse bone marrow erythroid colony-forming units [69]. Inactivation of 1-84 PTH abolished its action on erythropoiesis. Increasing the concentration of erythropoietin in the media from 0.67 to 1.7 U/ml overcame the inhibitory effect of 1-84 PTH on BFU-E. The N-terminal fragment of PTH (1-34 PTH) and the C-terminal fragment of PTH (53-84 human PTH) had no effect on BFU-E. These data demonstrate that (a) either the intact PTH molecule or a C-terminal fragment(s) bigger than the 53-84 moiety exerts the inhibitory effect on erythropoiesis and (b) adequate amounts of erythropoietin can overcome this action of PTH. Also, red blood cell production as measured by ^{59}Fe incorporation was reduced in rats with excess PTH compared with control rats [70].

If PTH interferes with erythropoiesis, one would expect patients with primary hyperparathyroidism and normal renal function to have anaemia. Indeed, 5–21% of patients with primary hyperparathyroidism have a moderate degree of chronic normocytic anaemia, which improves or normalises after removal of the parathyroid adenoma [71, 72]. The lower incidence and the milder degree of anaemia in patients with primary hyperparathyroidism than in patients with uraemia should not be construed as evidence against a role for PTH in the pathogenesis of anaemia. The presence of normal kidneys and hence normal production of erythropoietin in patients with parathyroid adenoma counterbalances the action of PTH on erythropoiesis. In contrast, patients with CRF are in double jeopardy: the excess blood levels of PTH inhibit erythropoiesis, and the inability to generate adequate erythropoietin permits the inhibitory action of PTH to proceed unchecked.

Effect on Osmotic Fragility of Human Red Blood Cells

Both the N-terminal (1-34) PTH and the intact (1-84) PTH but not the C-terminal (53-84) PTH have been shown to produce significant increases in osmotic fragility of human red blood cells (RBCs) [73]. This effect was abolished by prior inactivation of the hormone. There was a dose-response relationship between both moieties of PTH and the increase in osmotic fragility. This action of PTH required calcium, was mimicked by calcium ionophore, and was partially blocked by verapamil, and PTH caused significant influx of ^{45}Ca into RBCs. Scanning electron microscopy revealed that the incubation of RBCs with PTH was associated with the appearance of membrane filamentous extensions that anchor RBCs together.

These observations indicate that (a) the human RBC is a target organ for PTH; (b) the hormone increases osmotic fragility of RBCs; and (c) this effect of PTH is due to enhanced calcium entry into RBCs. It is possible that the increased calcium influx may affect the spectrin-actin of the cytoskeletal network of the RBCs and alter the stability and integrity of the cell membrane. Also, the survival of RBC was assessed with ^{51}Cr in an in vivo setting. RBC survival was shortened in dogs with intact parathyroid glands and CRF but was normal in normocalcaemic parathyroidectomised dogs with comparable degree and duration of CRF [74]. Furthermore, a significant inverse relationship was found between red cell survival and blood levels of PTH in 27 haemodialysis patients [75].

Role of PTH-Induced Marrow Fibrosis

Excess PTH induces moderate to marked degrees of bone marrow fibrosis [76] and therefore limits the availability of red marrow, hence reducing the number of red cell-forming units. Myelofibrosis in dialysis patients has been noted, and its relation to anaemia has been documented [77].

Effect of PTH on Polymorphonuclear Leukocyte Function

Uraemic patients have an increased susceptibility to infection. This may be partly due to defective leukocyte function. Indeed, polymorphonuclear leukocytes (PMNLs) from uraemic patients display impaired migration and defective phagocytic and bactericidal activities [78-80]. Certain clinical observations suggest that PTH affects leukocyte function. It has been reported that random migration and chemotaxis of PMNLs were impaired in patients with primary hyperparathyroidism and normal renal function, and these defects disappeared after removal of the parathyroid adenoma [81].

Studies from our laboratory have demonstrated that PMNLs are targets for PTH action. Indeed, acute exposure of leukocytes to 1-84 PTH increased elastase release [82] from these cells and impaired their random migration [78]. The aminoterminal fragment of the hormone was inert in regard to PMNLs function, and the 19-84 aminosequence fragment of PTH increased elastase release from PMNLs, as did the 1-84 PTH [82].

A large body of evidence exists implicating the state of secondary hyperparathyroidism in the genesis of PMNLs dysfunction in uraemic patients. Random migration of PMNLs is impaired in CRF patients, and an inverse relationship exists between random migration of PMNLs and blood levels of PTH in these patients [82]. PMNLs of CRF patients and those treated with haemodialysis have elevated basal levels of $[Ca^{2+}]i$, reduced ATP content, and impaired phagocytosis [79]. These derangements are due to the state of secondary hyperparathyroidism of CRF. Studies in CRF rats support this conclusion and further demonstrate that these derangements are prevented by prior parathyroidectomy of CRF animals or by their treatment with verapamil [28]. Also, glucose uptake by and glycogen content of PMNLs are reduced and the activity of their glycogen synthesis is impaired in patients with CRF [83]. The treatment of these patients with verapamil or $1,25(OH_2D_3)$ (which suppresses the activity of the parathyroid glands) reversed these abnormalities [49, 83]. Finally, oxygen consumption by PMNLs from humans and rats with CRF is decreased [29]; this abnormality is due to the state of secondary hyperparathyroidism of CRF, and is prevented by prior parathyroidectomy of the CRF animals or by their treatment with verapamil [29]. The treatment of rats with pre-existing CRF with verapamil reversed the derangement in the oxygen consumption by these cells [29]. Also, treatment with a calcium channel blocker of haemodialysis patients reversed the abnormalities in metabolism and function of their PMNLs [84, 85].

Role of PTH in the Genesis of the Abnormalities of the Immune System in Uraemia

Abnormalities in the immune system are encountered in clinical and experimental CRF, and both cellular and humoral immunity are impaired [86]. Metabolic and toxic consequences of CRF and/or the compounding effects of malnutrition, vitamin deficiency, drug therapy, and dialysis treatment may each alone, or in any combination, contribute to the genesis of the deranged immune system in CRF.

Lymphocytes have receptors for PTH [87]. Indeed, 1-34 PTH activated mononuclear leukocytes, most likely T cells, and caused them to produce a substance(s) that enhances bone resorption [88]. Also, PTH stimulates proliferation of thymic lymphocytes. It is plausible that the state of excess PTH in patients with CRF adversely affects the function of T and B cells and as such contributes to the impaired cellular and humoral immunity in these patients.

PTH and T Cell Function

Studies from our laboratory show that lymphocytes are targets for both 1-84 PTH and 1-34 PTH and that both moieties of the hormone augment PHA-induced lymphocyte proliferation [89]. This effect of the hormone is related to its biologic activity because inactivation of the hormone abolished its stimulatory effect. This action of PTH is most likely due to the ability of the hormone to enhance entry of calcium into cells and/or stimulation of protein kinase C, but is independent of cAMP production.

We must emphasise that these studies dealt with the effect of the acute exposure of the T cells to PTH. However, in patients with advanced CRF, there is chronic exposure to markedly elevated blood levels of PTH. Such sustained exposure may result in a calcium overload of the lymphocytes, down-regulation of their PTH receptors, and altered phospholipid metabolism of their cell membrane. Either or any combination of such events may have adverse effects on lymphocyte function. Indeed, the chronic exposure of T cells to high levels of PTH in patients with CRF is associated with elevated basal levels of $[Ca^{2+}]i$ of these cells and with derangements in T cell function. PHA stimulated proliferation of T cells from normal subjects and dialysis patients, but the magnitude of the increment in the dialysis patients was smaller than that observed in the normal subjects [90]. Although both 1-84 PTH and 1-34 PTH in a dose of 4×10^{-7} M stimulated PHA-induced proliferation of T cells from normal subjects, they failed to stimulate PHA-induced proliferation of T cells from dialysis patients.

T cells from dialysis patients failed to produce a significant increment in IL-2 after 48 hours of culture with PHA as compared to the significant increment in IL-2 by T cells from normal subjects [90]. Also, 1-84 PTH produced further augmentation in PHA-induced IL-2 production by T cells from normal subjects, but failed to increase PHA-induced IL-2 production by T cells from dialysis patients [90].

PTH and B Cell Function

The antibody response to viral [91] but not bacterial [92] antigens may be reduced, and the number of total B cells has been reported to be normal [93] or decreased [94]. Furthermore, it was found that in dialysis patients, both the T cell-dependent and T cell-independent B cell proliferation were reduced.

B Cell Proliferation

Available data indicate that Staphylococcus aureus Cowan I (SAC)-induced lymphocyte proliferation represents T cell-independent B cell proliferation. We showed that SAC induced a significant increase in proliferation of B cells from both normal subjects and dialysis patients, but the increment in the dialysis patients was significantly lower than in normal subjects [95]. In normal subjects, both 1-34 PTH and 1-84 PTH produced a dose-dependent inhibition of SAC-induced B cell proliferation, but the effect of 1-84 PTH was significantly greater than that produced by an equimolar dose of 1-34 PTH. 1-84 PTH also inhibited SAC-induced proliferation of B cells from dialysis patients, but the decrement in the dialysis patients was significantly smaller than in the B cells of normal subjects [95]. In 12 dialysis patients, the percentage inhibition of SAC-induced B cell proliferation with 1-84 PTH was inversely and significantly correlated with their blood levels of PTH.

1-84 PTH also produced an increase in the production of cAMP by B cells from normal subjects. It is of interest that agents that increased cAMP without receptor interaction (forskolin and cholera toxin) also inhibited SAC-induced proliferation of B cells from normal subjects and dialysis patients, and this effect was not different between the two groups [95].

These observations indicate that B cells are targets for PTH. It is worthwhile to mention again that the effect of the intact hormone is greater than that of its aminoterminal fragment. This phenomenon is similar to the effects of these two moieties of PTH on T cell function. As we suggested earlier, this difference between the magnitude of the effects of 1-84 PTH and 1-34 PTH is consistent with the notion that the intact hormone may attach more tightly to its receptor or that other parts of PTH, in addition to its aminoterminal fragment, may possess biologic activity. Certain data exist supporting such possibilities. For example, binding sites with specificity for the middle region or the carboxy-terminal fragment of PTH have been described in canine [96] and chicken renal membranes [97]. Also, we have found that the carboxy-terminal fragment (19-84 PTH) exerts a biologic activity on human PMNLs in that it stimulates elastase release from these cells [82].

Several observations suggest that cAMP may inhibit B cell function and interfere with its response to mitogens. First, forskolin inhibited SAC-induced B cell proliferation [98]. Second, cAMP inhibited mouse B cell proliferation, which is facilitated by B cell-stimulating factor-1 [99]. Third, cAMP caused a significant inhibition of immunoglobulin secretion by human lymphoblastoid-B cell line [100]. Fourth, our own data also showed that cAMP elevating agents, such as forskolin and cholera toxin, inhibited proliferation of B cells from normal subjects and dialysis patients.

It is possible, therefore, that PTH stimulates cAMP production through its interaction with its receptors on B cells: such an event could, at least in part, be responsible for the inhibitory effect of PTH on SAC-induced B cell proliferation. Indeed, our studies showed that 1-84 PTH stimulated cAMP production by B cells and both forskolin and cholera toxin, agents known to elevate intracellular cAMP, produced inhibition of the proliferation of normal B cells compared in magnitude with that induced by PTH.

It is of interest that the effect of PTH was similar to that of forskolin and cholera toxin on B cells from normal subjects but significantly less on B cells from dialysis patients. It should be mentioned that the increase in cAMP production by forskolin and cholera toxin does not require receptor interaction, but that induced by PTH does. It could be argued, therefore, that a down-regulation or desensitisation of the PTH receptors on B cells in the dialysis patients caused by prolonged exposure to high levels of PTH in blood is associated with less production of cAMP by PTH than by forskolin and cholera toxin. Such a phenomenon may explain the differences in the action of PTH and forskolin and cholera toxin on the proliferation of B cells of the dialysis patients.

Antibody Production by B Cells

The inhibition of B cell proliferation by PTH may also lead to impaired antibody production by these cells. We examined the production of IgG, IgM and IgA by B cells stimulated with SAC or with PWM after eight days of culture and evaluated the effect of PTH on this process in 34 haemodialysis patients and 44 normal subjects [101]. IgG, IgM and IgA production by B cells from patients was significantly lower than by B cells from normal subjects. Both 1-34 and 1-84 PTH inhibited immunoglobulin production by B cells from normal subjects and dialysis patients. However, this inhibitory effect was evident in dialysis patients only with the high dosage of PTH. The inhibition of immunoglobulin production by PTH occurred only when the hormone was added in the initiation of the B cell culture. Inactivation of PTH abolished its inhibitory effect on immunoglobulin production. Agents that stimulate cAMP production (forskolin, cholera toxin) and the cAMP analogue, 8-bromoadenosine 3′,5′cyclic monophosphate, inhibited immunoglobulin production by B cells from both normal subjects and dialysis patients, and the degree of inhibition was not different between the two groups. The calcium ionophore A23187 also inhibited IgG, IgA and IgM production by B cells from normal subjects and dialysis patients; there was no difference in the degree between the two groups. As mentioned previously, the resting levels of $[Ca^{2+}]i$ of B cells from dialysis patients is significantly higher than in normal subjects. These observations show that: (a) immunoglobulin production by B cells from dialysis patients is impaired; (b) PTH inhibits IgG, IgA and IgM production and this effect is, at least partly, mediated by PTH-induced cAMP generation and by alterations in $[Ca^{2+}]i$ of B cells; (c) this inhibitory effect is mediated by events that affect the initial stages of B cell proliferation and maturation; and (d) the requirement for high doses of PTH for its inhibitory effect on B cells from dialysis patients is probably due to desensitisation and/or down-regulation of PTH receptors on B cells. The results are consistent with the proposition that impaired immunoglobulin production by B cells from dialysis patients is at least partly due to the state of secondary hyperparathyroidism in these patients.

The observations that PTH inhibits immunoglobulin production in vitro do not provide definite proof for a role of excess PTH in the genesis of the abnormality in the in vivo immunoglobulin production. It is possible that other consequences of the uraemic state and/or another, as yet unidentified, factor accumulates in the blood of uraemic patients and underlies the impaired humoral immunity. To definitely incriminate excess PTH in the genesis of impaired humoral immunity in CRF, one must document that reduced antibody production in response to antigens is normal in a CRF state without excess PTH.

We examined in vivo antibody production in response to sheep red blood cells (SRBCs), bovine serum albumin (BSA) and influenza vaccine in normal rats, CRF fats and parathyroidectomised (PTX)-CRF rats maintained normocalcaemic [102]. The antibody responses to all three antigens in CRF rats were significantly and markedly lower than in normal or CRF-PTX rats. The response to SRBC, the IgG anti-BSA and the IgG and IgM anti-influenza vaccine in CRF-PTX rats were not different from normal, whereas the IgM anti-BSA was lower than in normal rats but higher than in CRF rats. These observations demonstrate that the state of secondary hyperparathyroidism of CRF plays a paramount role in the genesis of impaired humoral immunity of CRF.

Treatment of haemodialysis patients with the calcium channel blocker nifedipine produced marked and significant improvement in the basal levels of $[Ca^{2+}]i$ and ATP content of B cells and in their proliferation in response to SAC as compared with haemodialysis patients who were not receiving treatment with nifedipine [54]. The values of all these parameters of B cell metabolism and function in the group of patients treated with nifedipine approached the normal values. Furthermore, the blood levels of IgG were significantly higher in patients receiving nifedipine than those without treatment.

A prospective study further demonstrated that the treatment of haemodialysis patients with nifedipine reversed the abnormalities in B cells cited previously within two months, and that these derangements re-emerged after discontinuation of therapy with the drug [103]. It appears, therefore, that the treatment of haemodialysis patients with a calcium channel blocker is beneficial in combating the adverse effects of uraemia on the function of B cells. The treatment must be maintained in order to maintain its benefit.

Effect of PTH on Platelet Function

Bleeding tendencies are not infrequent in uraemic patients, and this abnormality is thought to be due to many factors, among which is impaired platelet aggregation. We have found that 1-84 PTH, and not 1-34 PTH, inhibits platelet aggregation induced by adenosine 5'-diphosphate (ADP), fibrin, and collagen. Similar observations were reported by others. The state of secondary hyperparathyroidism of uraemia may contribute to the pathogenesis of the bleeding tendencies and blood loss in these patients. This phenomenon may provide an additional pathway through which PTH participates in the aetiology of the anaemia of uraemia (Figure 7.6).

PTH and Uraemic Myopathy

Uraemic patients may experience muscle dysfunction [104], and patients with primary hyperparathyroidism may display diffuse neuromuscular defects and prominent myopathy [105]. These clinical observations suggest that the state of secondary hyperparathyroidism of renal failure could play an important role in the aetiology of uraemic myopathy. Indeed, available data indicate that the skeletal muscle is a target organ for PTH action, and the hormone may affect its protein metabolism, bioenergetics, and fatty-acid oxidation. Both intact PTH and its aminoterminal fragment exert marked effects on the metabolism of skeletal muscle protein, amino acids, and cyclic nucleotides. PTH stimulates the production of both cAMP and cyclic guanosine 3',5'-monophosphate (cGMP), enhances the release of alanine and glutamine from muscle without altering their intracellular levels, and decreases the incorporation of [^{3}H] leucine into muscle protein [106].

Studies in our laboratory [59] showed that the administration of PTH for four days to normal rats caused a significant increase in ^{45}Ca uptake by the skeletal muscle and was associated with a significant decrease in the muscle content of inorganic phosphorus, creatine phosphate, and ATP. The hormone also significantly reduced mitochondrial oxygen consumption and impaired the activity of mitochondrial and myofibrillar creatine phosphokinase; verapamil abolished all these adverse effects of PTH on muscle bioenergetics. CRF of 21-days duration produced derangements in skeletal muscle bioenergetics similar to those induced by PTH administration. These derangements were absent in normocalcaemic, parathyroidectomised rats with CRF or in CRF rats treated with verapamil. The findings demonstrate that excess PTH in animals with normal renal function or with CRF is associated with impaired energy production, transfer, and use by skeletal muscle.

Fatty acids are an important source of skeletal muscle energy, and certain data suggest that oxidation of long-chain fatty acids (LCFA) may be impaired in uraemia. Such an abnormality may contribute to uraemic myopathy. Studies in our laboratory showed that the administration to rats of either 1-84 or 1-34 PTH for four days was associated with impaired LCFA oxidation [32]. Parathyroidectomy of CRF animals or their treatment with verapamil prevent the derangement in the activity of carnitine-palntitoyl transferase and in the oxidation of LCFA [32, 33].

Effect of PTH on Myocardium

Patients with advanced renal failure may have myocardiopathy of unexplained origin. A large body of evidence obtained from experimental data suggests that the excess blood levels of PTH in uraemia adversely affect the heart. Uraemia is associated with cardiac lesions and accumulation of calcium in the heart. Administration of PTH enhanced the progression of these lesions, and parathyroidectomy prevented their evolution as well as the accumulation of calcium [107, 108].

An extensive body of evidence indicates that the heart is a target organ for PTH action. Myocardium contains mRNA for PTH-PTHrP receptor [7, 8]. Both N-terminal (1-34) PTH and intact (1-84) PTH produced an immediate and sustained significant rise in beats per minute, and the cells after exposure to PTH died earlier than control cells [109]. The effect was reversed if PTH was removed from the medium, and the action was abolished by inactivation of the hormone. There was a dose-response relationship between both moieties of PTH and the rise in heartbeats, but the effect of 1-84 PTH was significantly greater than that of the 1-34 moiety. PTH stimulated cAMP production within one minute and cAMP remained elevated there-after. The effect of PTH required calcium, was mimicked by calcium ionophore, was prevented by verapamil, and was not abolished by α- or β-adrenergic blockers. PTH action was additive to an α-adrenergic agonist and synergistic with a β-adrenergic agent. Sera from uraemic PTX rats did not affect heartbeats, but sera from uraemic rats with intact parathyroid glands or from uraemic PTX rats treated with PTH had effects similar to PTH [109]. PTH also has an inotropic effect on the heart. The hormone increased the contraction of the papillary muscle [110] and affected the isometric contractile force of isolated guinea pig auricles [111].

In vitro studies showed that PTH inhibited mitochondrial respiration and phosphorylation and uncoupled oxidative phosphorylation [112]. These effects of PTH were dose dependent and occurred only in the presence of calcium. The inhibitory effect of PTH on mitochondrial respiration and on oxidative phosphorylation would result in decreased ATP synthesis and hence reduced availability of ATP. Such a sequence of events may provide an explanation for potential long-term adverse effects of the hormone on the myocardium. Support for such a notion came from additional studies carried out in our laboratory evaluating the in vivo effect of PTH on myocardial metabolism [58].

Administration of 1-84 and 1-34 PTH to rats for four days significantly reduced mitochondrial oxygen consumption without altering the ADP:oxygen ratio, indicating reduced phosphorylation. Myocardial contents of inorganic phosphorus, ATP and creatine phosphate were significantly lower in 1-84 PTH-treated animals. Both moieties of the hormone significantly reduced mitochondrial and myofibrillar creatine phosphokinase. There were also significant increments in myocardial ^{45}Ca uptake and total calcium content. Inactivation of PTH and administration of verapamil, a calcium channel blocker, abolished these adverse effects of the hormone on myocardial bioenergetics. Treatment of rats with 1-84 PTH for 10 days was associated with a significant decrease in cardiac index and mean arterial pressure. The data demonstrate that both 1-84 PTH and 1-34 PTH impair myocardial energy production, transfer, and utilisation. These biochemical derangements, if maintained, produce a decrease in cardiac index.

We also examined myocardial energetics and cardiac index in normal rats, in rats with CRF and intact parathyroid glands, and in PTX, normocalcaemic rats with CRF [24]. There were significant decrements in myocardial ATP and creatine phosphate, mitochondrial oxygen consumption, and the activity of mitochondrial and myofibrillar creatine phosphokinase in rats with CRF and intact parathyroid glands compared with normal animals. Myocardial content and ^{45}Ca uptake in the animals with CRF were significantly higher than in normal rats. This increased calcium content in the myocardium is associated with significant elevation of the basal levels of $[Ca^{2+}]i$ in cardiac myocytes [12]; both prior parathyroidectomy of the CRF animals or their treatment with verapamil prevented the elevation in $[Ca^{2+}]i$. In the PTX rats with CRF, ATP content, mitochondrial oxygen consumption, ^{45}Ca uptake, and calcium content were normal, but parathyroidectomy did not normalise the activity of mitochondrial and myofibrillar creatine phosphokinase. Parathyroidectomy in rats with normal renal function caused significant reduction in myofibrillar and mitochondrial creatine phosphokinase activity. There was a significant decrease in the cardiac index in rats with CRF and intact parathyroid glands, and the cardiac index did not normalise in PTX rats with CRF.

These data allow the following conclusions: (a) in CRF with excess PTH there is impaired myocardial energy production, transfer, and utilisation, and a fall in cardiac index; (b) the decrease in energy production is due to excess PTH and not CRF; (c) the impaired energy transfer and utilisation is most likely due to excess PTH, but absence of the hormone did not correct these abnormalities, since parathyroidectomy even in rats with normal renal function was followed by reduced activity of mitochondrial and myofibrillar creatine phosphokinase; (d) the effect of PTH on myocardium is partially mediated by enhanced entry of calcium into the heart; (e) the optimal activity of creatine phosphokinase of heart mitochondria and myofibrils requires PTH, and either excess or lack of the hormone is deleterious; and (f) total parathyroidectomy is not the procedure of choice for the prevention of uraemic myocardiopathy.

Fatty acids constitute an important substrate used by the myocardium as a major fuel for energy production; impaired oxidation of fatty acids may result in myocardial dysfunction. As in the case of skeletal muscle, excess PTH in normal rats or in rats with CRF is associated with impaired oxidation of LCFA by the myocardium [34], and this effect is due to reduction in the activity of carnitine-palmitoyl transferase and inhibition in the mitochondrial β-oxidation sequence, respectively. No changes in carnitine content of the myocardium were observed, and parathyroidectomy in rats with CRF normalised these derangements in fatty-acid oxidation. Verapamil treatment of these rats prevented the adverse effects of PTH and of CRF on LCFA oxidation [35].

Several clinical studies examined the relationship between primary hyperparathyroidism or secondary hyperparathyroidism of CRF on myocardial function. It was found that reductions in blood levels of PTH in dialysis patients by parathyroidectomy was followed by significant improvement in cardiac function [46]. Indeed, cardiac index, left ventricular ejection fraction, and per cent fibre shortening increased 6–14 days after parathyroidectomy. Others examined the effects of reduction of blood PTH levels on left ventricular function in dialysis patients, either after six weeks of treatment with $1,25(OH)_2D_3$ or after parathyroidectomy [47]. They found that these medical or surgical procedures were followed by improvements in fractional fibre shortening and mean velocity of fibre shortening; there was also a significant inverse relationship between the ratio of pre-ejection period/left ventricular ejection time (an index of left ventricular function) and blood levels of PTH. We have previously provided experimental data indicating that total parathyroidectomy in CRF may not be beneficial for cardiac function. These experimental data appear to contradict the results of the aforementioned study; however, it seems that the total parathyroidectomy performed in the patients was not complete, because the blood levels of PTH after the surgical procedure were normal. Other investigators studied 37 haemodialysis patients and found that 62% of them had abnormal left ventricular function, including enlargement of left ventricular cavity, a reduction of myocardial contractility, and thickening of left ventricular posterior wall [113]. In seven of these 37 patients, congestive myocardiopathy was present, and these patients had severe secondary hyperparathyroidism. These authors concluded that secondary hyperparathyroidism plays an important role in the genesis of congestive myocardiopathy.

Symons et al. [114] reported that 15 of 16 patients with primary hyperparathyroidism and without renal insufficiency or hypertension had hypertrophic cardiomyopathy (five patients), asymmetric septal hypertrophy (six patients), and symmetric left ventricular hypertrophy (four patients). This was not due to the hypercalcaemia of the primary hyperparathyroidism because six patients with hypercalcaemia related to other causes did not have cardiac hypertrophy. These authors also found that in five of 18 patients (28%) who had hypertrophic myocardiopathy not caused by CRF, the blood levels of PTH were elevated. These data clearly demonstrate an effect of excess PTH on the myocardium even in the absence of CRF.

Others concluded that the major derangement in cardiac function in uraemia is an inappropriate change in the left ventricular wall in response to alterations in chamber size resulting in higher ventricular radius:wall thickness ratio [115–117]. They found a direct and significant correlation between this ratio and the blood levels of PTH, and an inverse relationship between left ventricular posterior wall thickness and blood levels of PTH. These observations incriminate secondary hyperparathyroidism of CRF patients in the genesis of myocardial dysfunction. It should be mentioned that these observations are at variance with those of others, demonstrating that dialysis patients frequently display left ventricular thickening. The reasons for these differences are not clear. Also, these observations differ from those demonstrating left ventricular hypertrophy in primary hyperparathyroidism. It is possible that the response of the myocardium to excess PTH in the presence or absence of CRF is different.

In contrast to the aforementioned data, others reported that parathyroidectomy in 10 dialysis patients did not modify the various aspects of left ventricular function [118]. However, it should be indicated that all the parameters of left ventricular function were normal before parathyroidectomy, and therefore improvement should not be expected.

Interstitial cardiac fibrosis has been recognized in CRF for quite some time [119]. It was brought to our attention lately by studies of Ritz and co-workers [120, 121]. They found that interstitial cardiac fibrosis develops in CFR rats and that PTX of CRF animals abolished the fibrous deposition, while the infusion of PTH (1-34) to the PTX-CRF animals restore the excess interstitial volume. They concluded that PTH plays a permissive role for fibroblast activation and for the genesis of cardiac fibrosis in uraemia [122]. The increment of the interstitial volume could add to the diastolic dysfunction of the heart in uraemia.

Taken together, the available experimental and clinical data clearly demonstrate that the heart is a target for PTH and that chronic excess of the hormone in the presence or absence of CRF exerts adverse effects on the metabolism, structure, and function of the myocardium. PTH satisfies all the criteria described earlier in this chapter for a cardiotoxin in CRF. Its structure is identified, its blood levels are elevated in CRF, a relationship between its blood level and various parameters of cardiac dysfunction is documented, the reduction of its blood levels is followed by improvement in cardiac function, and its administration to experimental animals with normal renal function produces cardiac alterations similar to those seen in CRF. Also, excess PTH in humans with normal renal function causes myocardial disease. In addition, it was found that left ventricular hypertrophy among patients with essential hypertension and normal renal function is independently affected by blood PTH levels, and there is a strong positive correlation between left ventricular mass index and blood level of PTH. We would like to emphasise, however, that this conclusion in no way means that PTH is the only cardiac toxin or the only factor underlying the myocardiopathy of uraemia.

Finally, it should be mentioned that it may not always be possible to demonstrate a direct relationship between the various components of left ventricular function and the blood levels of PTH, for several reasons. First, in CRF, there is a multitude of factors besides PTH that could be harmful to the myocardium. Second, the derangements in cardiac metabolism, structure, and function in CRF are due to chronic and progressive processes, and it may not always be possible to correlate the outcome of such chronic processes with a single measurement of blood PTH. Third, it is also possible that chronic excess of PTH may cause in certain CRF patients irreversible myocardial damage; therefore a reduction in blood levels of PTH may not be followed by improvement in cardiac function.

PTH and Vascular Reactivity in Uraemia

PTH is known to be a vasodilator and to exert a hypotensive action. We evaluated the possible mechanisms involved in this action, with special emphasis on the role of vasodilatory prostaglandins. The effects of intact 1-84 PTH and of its N-terminal (1-34 PTH) fragment on mean arterial pressure and on the vascular response to norepinephrine or angiotensin II were examined in rats before and after pre-treatment with indomethacin [123]. Bolus injections of 1-84 and 1-34 PTH produced a significant decrease in mean arterial pressure; however, the hypotensive response to 1-34 PTH was more marked than the response to 1-84 PTH. Both 1-84 and 1-34 PTH antagonised the pressor effects produced by bolus injection of norepinephrine or angiotensin II. Pre-treatment with indomethacin abolished the inhibitory effect of 1-84 and 1-34 PTH on the response of mean arterial pressure to norepinephrine or angiotensin II. The infusion of 1-84 and 1-34 PTH produced a significant rise in urinary excretion of 6-keto-PGFα2. These data are consistent with the hypothesis that the effect of PTH on blood vessels is mediated by increased prostaglandin production.

Patients with acute or chronic renal failure manifest reduced pressor response to norepinephrine. This abnormality is at least partly responsible for some of the manifestations of autonomic nervous system dysfunction in these patients. Because uraemia is associated with increased blood levels of PTH, and because PTH blunts the pressor effect of norepinephrine, most likely by the activation of prostaglandins, we studied the relationship between blood levels of PTH and the magnitude of the reduction in the pressor response to norepinephrine in uraemic patients and the effect of treatment with indomethacin on the response to norepinephrine in these patients [124]. There was a significant negative correlation between the changes in blood pressure and the blood levels of PTH in the uraemic patients. Treatment with indomethacin was followed by significant improvement or normalisation of the pressor response to norepinephrine. These data are consistent with the notion that the decreased pressor response to norepinephrine in uraemia is due to increased production of prostaglandins induced by excess PTH and provide a therapeutic tool for the treatment of some of the manifestations of autonomic nervous system dysfunction in uraemia.

To evaluate further the role of PTH in the genesis of this abnormality in uraemia, we examined the pressor responses to norepinephrine in rats with experimental CRF with and without (CRF-PTX) parathyroid glands [125]. The pressor response to norepinephrine was significantly reduced in CRF rats compared with normal rats and was significantly less than in CRF-PTX rats in which the pressor responses were the same as those in control rats. Treatment of CRF rats with indomethacin normalised the pressor response to norepinephrine. We also examined the dose-response curves between norepinephrine and perfusion pressure of the hind-limb preparation [125]. There was a significant shift to the right in the dose-response curve in CRF rats compared with controls and CRF-PTX rats, indicating a decreased sensitivity to norepinephrine in CRF rats. No significant difference was observed between the dose-response curves in control and CRF-PTX rats. Thus, excess PTH and not other consequences of CRF plays a paramount pathogenetic role in the reduced pressor response to norepinephrine, and this effect of PTH is due to direct action on the blood vessels. Furthermore, this action of the hormone is most likely to be mediated by increased production of vasodilating prostaglandins. It is also possible that down-regulation of norepinephrine and/or angiotensin II receptors in CRF, as discussed earlier, play an important role in the reduced pressor response to those vasoconstrictor agonists.

Role of PTH in Carbohydrate Intolerance in Uraemia

Chronic renal failure is associated with many disturbances in carbohydrate metabolism. The characteristic of glucose and insulin metabolism in uraemia is presented in Table 7.4. Studies in humans have demonstrated that both insulin resistance and impaired insulin secretion contribute to the pathogenesis of carbohydrate intolerance. The insulin resistance is almost always present in patients with uraemia [126].

Insulin secretion, as evaluated by the blood levels of insulin in response to hyperglycaemia, may be normal, increased, or decreased; these variations appear difficult to reconcile but may reflect different responses of β-cells to hyperglycaemia and/or different degrees of impairment in insulin degradation.

The normal response of β cells to the presence of insulin resistance is to enhance their secretion of insulin. If for any reason the β cells were unable to augment their secretion of insulin appropriately, an impaired glucose tolerance would ensue. Indeed, glucose intolerance is usually encountered in uraemic patients in whom both impaired tissue sensitivity to insulin and impaired β-cell secretion of insulin coexist.

Peripheral Resistance to the Action of Insulin

The first indirect and convincing evidence for the presence of impaired peripheral action of insulin was provided in 1962 by studies showing that glucose uptake by the forearm of uraemic patients is reduced [127]. This phenomenon was confirmed by studies using a euglycemic clamp technique, which demonstrated that the amount of glucose metabolised per unit of insulin in dialysis patients is reduced [128].

The two tissues that are involved in most of the peripheral uptake of glucose are the liver and the skeletal muscle. Therefore, resistance to insulin action may result from (a) impaired uptake of glucose by these tissues and/or (b) increased production of glucose by the liver. However, glucose uptake by the liver is small compared with that of skeletal muscle, and it is not impaired in uraemia. Also, glucose pro-

Table 7.4. Characteristics of Glucose and Insulin Metabolism in Uraemia

Normal fasting blood glucose

Spontaneous hypoglycaemia

Fasting hyperinsulinaemia[a]

Normal, elevated, or decreased blood insulin levels in response to hyperglycaemia induced by oral or intravenous glucose administration

Elevated blood levels of proinsulin and C-peptide

Elevated blood levels of immunoreactive glucagon

Impaired insulin secretion by pancreatic islets[b]

Multiple derangements in metabolism and function of pancreatic islets

Impaired glycolytic pathways

Reduced basal and glucose-stimulated ATP content

Elevated basal levels of cytosolic calcium

Decreased Vmax of Ca^{2+}-ATPase and Na^+-K^+-ATPase

Reduced calcium signal and response to glucose and potassium

Normal hepatic glucose production

Normal suppression of hepatic glucose production by insulin

Decreased peripheral sensitivity to insulin action

Impaired glucose tolerance

Decreased requirement for insulin by diabetic patients with diabetic nephropathy and uraemia

[a] Normal blood insulin levels may be encountered.
[b] This is observed only in the presence of established secondary hyperparathyroidism.
[c] This is present only when insulin secretion is impaired in the presence of the commonly encountered resistance to the peripheral action of insulin.

duction by the liver and its suppression by insulin are not affected by uraemia. Therefore, it is evident that the major site for the decreased sensitivity to insulin action is the skeletal muscle.

The insulin action on glucose transport by muscle is mediated through a complex process, which includes insulin binding to its receptor, autophosphorylation of the tyrosine residues and activation of tyrosine kinase, tyrosine phosphorylation of insulin receptor substrate I (IRS1). Subsequently, phosphatidylinositol 3-kinase pathway is activated resulting in the modulation of glucose transport. Available data indicate that the number and the affinity of the insulin receptors in uraemia are unchanged and may even be increased [129]. Furthermore, the insulin receptor kinase activity in the skeletal muscle of patients with CRF or in the skeletal muscle of uraemic rats is unchanged [129]. It is evident, therefore, that a post-receptor defect due to derangement in the downstream signalling most likely leads to impaired insulin-mediated glucose transport. Such a defect is responsible for the resistance to the peripheral action of insulin in uraemia. This notion would imply that even very high levels of insulin may not correct the defect in glucose uptake by skeletal muscle, and available data show that this indeed may be the case [130].

Skeletal muscles and adipose tissue possess a glucose transporter (GLUT-4), which is regulated by insulin. A defect in this system, including a reduction in the abundance of the GLUT-4, its translocation to cell surface, and/or its intrinsic activity, may contribute to the reduced glucose uptake by skeletal muscle in response to insulin. However, the abundance of GLUT-4 in the muscle of uraemic patients is not different from that of normal individuals [129].

The peripheral resistance to insulin action occurs early in the course of renal failure and before the onset of signs and symptoms of the uraemic syndrome become apparent [131]. It is observed in the majority of patients with advanced renal failure and those treated with haemodialysis [128, 132]. This defect is markedly improved after 10 weeks of haemodialysis [128, 132] therapy and following treatment with continuous ambulatory peritoneal dialysis (CAPD) [133]. Furthermore, treatment of uraemic patients with dietary protein restriction and supplementation with keto and amino acids for six months was followed by marked amelioration of the peripheral resistance to insulin action [134]. These observations are consistent with the notion that the defect in peripheral action of insulin is at least partly due to a dialyzable compound(s) produced by protein breakdown. It has been reported that the sera of uraemic patients contain a compound with a molecular weight of 1,000 to 2,000 D, which inhibits glucose metabolism by normal rat adipocytes [135]. This compound appears to be specific for uraemia because it is absent in the blood of patients with insulin resistance but without uraemia. Others proposed that hippurate, which accumulates in the blood of patients with renal failure, contributes to the insulin resistance in these patients; indeed, hippurate inhibits glucose utilisation by the diaphragm, brain, kidney cortex, and erythrocytes in rats [136].

Uraemic patients are weak, lead a sedentary lifestyle, and engage in little, if any, exercise. The sedentary lifestyle could contribute to the resistance of insulin action, and certain studies have demonstrated that uraemic patients displayed improvement in insulin sensitivity after exercise training.

Insulin Secretion and Pancreatic Islet Metabolism

Insulin secretion by β cells of the pancreatic islets is a complex process [137]. The β cells are stimulated by nutrients (glucose, amino acids, or fatty acids) or non-nutrient agents such as hormones and neurotransmitters. The scope of this chapter does not permit a detailed presentation of this process. Although, the mechanisms of recognition by the β cells of nutrient and non-nutrient stimuli may be different, the secretory process of insulin utilises the same intracellular processes when the β cells are activated by nutrient or non-nutrient secretagogues.

In the case of glucose-induced insulin secretion, the process begins by the uptake of glucose by the β cells. Glucose is then metabolised to produce adenosine triphosphate (ATP). The latter facilitates the closure of ATP-dependent potassium channels, which is followed by cell depolarisation and subsequent activation of voltage-sensitive calcium channels. As a consequence, calcium enters the islets, causing a rise in cytosolic calcium concentration that triggers cellular events that lead to insulin secretion. Others proposed that the ATP:adenosine diphosphate (ADP) ratio is an important factor in the sequence of events just described, and a rise in the ATP:ADP ratio initiates the closure of the ATP-sensitive potassium channel and the depolarisation of islets. Thus, a lower ATP content and/or a lower ATP:ADP ratio in islets in CRF, both in the resting state or after the stimulation with glucose, may contribute to the reduced insulin secretion.

As indicated, reduced insulin sensitivity is common in uraemic patients. Hence, glucose intolerance is found only in those who display reduced insulin secretion, and one must conclude that insulin secretion is impaired in a substantial segment of patients with renal failure.

In Table 7.4, it is stated that the blood levels of insulin in patients with renal failure may be normal, elevated, or decreased in response to hyperglycaemia. These observations may suggest that insulin secretion in these patients may not be reduced. However, it must be remembered that the blood levels of insulin are determined by insulin secretion and its metabolic clearance rate. Because the latter is impaired in patients with CRF, changes in the blood levels of insulin in response to hyperglycaemia are not a reliable indicator of insulin secretion. Evaluation of insulin secretion is better estimated by the hyperglycaemic clamp technique. Studies in humans or animals with CRF using this technique reported a normal initial and exaggerated late response [132], an exaggerated early and late response [138], or decreased initial and late responses [139]. In all these studies, glucose utilisation was lower than normal, indicating that insulin secretion is inappropriate relative to the insulin resistant state. Direct evidence for impaired insulin secretion in CRF was provided by in vitro dynamic perfusion studies of pancreatic islets obtained from rats with CRF. Both the initial and the late phases of glucose-induced insulin secretion were markedly reduced [27, 140].

Mechanisms of Impaired Insulin Secretion

Available data indicate that in renal failure, there is a generalised failure of the pancreatic islets to respond to glucose [27], amino acids [141], or potassium [142]. The mechanisms for this phenomenon are not fully understood. However, a substantial body of evidence has accumulated incriminating the chronic excess of parathyroid hormone (PTH) and/or the reduced blood levels of $1,25(OH)_2D_3$ in the genesis of the impaired insulin secretion in chronic renal failure.

Role of Secondary Hyperparathyroidism

Hyperglycaemic clamp studies in dogs with CRF with elevated blood levels of PTH showed that both the initial and late plasma insulin levels were reduced; however, insulin responses were normal in normocalcaemic parathyroidectomised CRF animals [139]. These observations were confirmed in uraemic children treated with haemodialysis. These studies showed marked improvement in insulin secretion after normalisation of the blood levels of PTH by the medical suppression of the parathyroid glands by treatment with $1,25(OH)_2D_3$ or after the surgical removal of the glands [50]. Furthermore, others showed that glucose-induced insulin secretion by islets from CRF rats is impaired but is normal by islets obtained from normocalcaemic parathyroidectomised rats with similar degree and duration of renal failure [25, 27]. It has also been reported that the daily administration of PTH for six weeks to rats with normal renal function caused a marked impairment in insulin secretion by their pancreatic islets. Also, the impairment in L-leucine and potassium-induced insulin secretion in CRF is due to chronic excess of PTH [141, 142]. All these observations clearly demonstrate that states with chronic excess of PTH inhibit insulin secretion whether renal failure is present or not.

Patients with CRF display variability in regard to glucose intolerance. This may be due in part to the variations in the severity of secondary hyperparathyroidism among these patients. It is reasonable to propose that patients with moderate elevations in blood levels of PTH may secrete adequate insulin in response to hyperglycaemia and hence have normal glucose tolerance. On the other hand, in patients with marked secondary hyperparathyroidism, insulin secretion is significantly reduced and glucose intolerance is present.

Chronic excess of PTH in renal failure is associated with a rise in basal levels of cytosolic calcium $[Ca^{2+}]i$ in many cells, including the pancreatic islets; furthermore, chronic excess of PTH without renal failure also causes a rise in the basal levels of $[Ca^{2+}]i$ of pancreatic islets. This is due to both a PTH-mediated increase in calcium entry into the islets and a decrease calcium exit out of the islets because of impaired activity of Ca^{2+}-ATPase, Na^+–K^+-ATPase and Na^+–Ca^{2+}exchange.

The elevation in the basal levels of $[Ca^{2+}]i$ of pancreatic islets in CRF appears to be responsible, in major part, for the impairment in their insulin secretion by glucose, potassium or L-leucine. Indeed, normalisation of $[Ca^{2+}]i$ of pancreatic islets of CRF rats by prior parathyroidectomy or by their treatment with verapamil, an agent that blocks the PTH-mediated entry of calcium into the islets, prevented the impairment in insulin secretion. Also, treatment of rats with established CRF with verapamil reversed the elevation in the basal levels of $[Ca^{2+}]i$ of their pancreatic islets and normalised the glucose-induced insulin secretion by these islets [37]. Finally, in rats with phosphate depletion that have normal renal function and low blood levels of PTH, the basal levels of $[Ca^{2+}]i$ of the islets are elevated, and both glucose or L-leucine-induced insulin secretion are impaired [143, 144]. Thus, an elevation in basal levels of $[Ca^{2+}]i$ of pancreatic islets is associated with impaired insulin secretion in the presence or absence of renal failure and whether blood levels of PTH are elevated or not.

PTH and the Hyperlipidaemia of Uraemia

Hyperlipidaemia is usually present in patients with chronic uraemia, and it has been reported in the oliguric phase of acute renal failure. A rise in blood lipids also occurs in animals rendered uraemic by either bilateral nephrectomy or bilateral ureteral ligation. An increase in production of triglyceride-rich lipoproteins by the liver, reduction in their removal by peripheral tissues, or a combination of both may be operative. In 1965, Cantin provided experimental evidence suggesting that PTH is involved in the genesis of the lipidaemia of uraemia [145]. He found that removal of the parathyroid glands partially inhibited the rise in blood lipids observed after bilateral nephrectomy, and the administration of parathyroid extract to PTX rats restored the hyperlipidaemia.

Both clinical and experimental data indicate that PTH can affect lipid metabolism: (a) PTH stimulates adipose tissue lipase in animals and in humans [146, 147], (b) the N-terminal fragment of PTH stimulates basal and GMP (PNP)-liganded adenylate cyclase of human fat cells [148], (c) a state of endogenous hyperparathyroidism, induced in rats by feeding them a calcium-poor diet, caused a significant increase in serum cholesterol and triglyceride concentrations, a significant decrease in the rate of serum clearance of intravenously infused lipids, and a significant increase in hepatic tissue triglyceride content, compared with the values obtained in control animals [149]. These authors also found that the administration of parathyroid extract to normal rats was associated with a modest increase in serum cholesterol and triglyceride concentrations. In the aparathyroid state, a significant decrease in serum levels of cholesterol and triglycerides and a significant increase in plasma post-heparin lipolytic activity were observed. These studies showed that (a) hyperparathyroidism induces increased serum cholesterol and triglyceride levels, the latter being due to decreased peripheral removal, and (b) the aparathyroid state induces opposite changes; the authors concluded that normal parathyroid function is required for normal lipid metabolism in the rat. Type IV hyperlipoproteinaemia was found in a large number of patients with primary hyperparathyroidism [150] and other studies showed that parathyroidectomy ameliorated the hyperlipidaemia of chronic uraemia, and the presence of excess PTH enhanced the magnitude of the hyperlipidaemia of CRF [151]. The authors of these studies suggested that PTH plays a permissive role in the genesis of the hyperlipidaemia of uraemia. Indirect evidence from these studies also suggested that PTH may affect serum lipids in experimental uraemia by modulating hepatic synthesis of lipoproteins.

Despite the impressive information implicating PTH in the genesis of abnormal lipid metabolism, certain clinical data argue against such a notion. Patients with primary hyperparathyroidism may have low serum levels of cholesterol and triglycerides. In some patients, blood levels of both cholesterol and triglycerides rose after surgical removal of the parathyroid adenoma, despite the lack of change in body weight or serum thyroxine.

We examined the interaction between PTH and adipocytes and investigated the role of excess PTH in CRF on lipid metabolism. Acute exposure of adipocytes to 1-84 but not 1-34 PTH causes a rise in their $[Ca^{2+}]i$ [152], and in vivo exposure to chronic excess of PTH was followed by a significant elevation in the basal levels of $[Ca^{2+}]i$ of adipocytes [153]. This latter effect is prevented by prior parathyroidectomy of the CRF rats or by their treatment with verapamil [153]. Both dogs [40] and rats [26] with CRF have fasting triglyceridaemia, impaired fat tolerance test, and reduced post-heparin lipolytic. These derangements were prevented by parathyroidectomy of the dogs or the rats [40, 26] and treatment of the CRF rats with verapamil [26].

The available data indicate that the hyperlipidaemia of CRF is due to impaired removal of lipids from the circulation. This derangement is secondary to abnormalities in the metabolism of both hepatic lipase (HL) and lipoprotein lipase (LPL). We have found that CRF is associated with (a) reduced V_{max} and increased Km of HL in liver homogenate, and (b) lower-than-normal activity of HL after eight hours of culture of hepatic cells, (c) down-regulation of the expression of mRNA of HL, (d) marked reduction of the heparin-induced release of HL in in vitro hepatic perfusion system [154]. All these derangements in HL were prevented by parathyroidectomy of the CRF animals or by their treatment with verapamil. Furthermore, perfusion of the liver with PTH 1-84 or PTH 1-34 inhibited the heparin-induced release of HL. These observations show that the molecular machinery of HL, its production, its activity, and its release are impaired in CRE These abnormalities are most likely caused by the PTH-mediated elevation in the basal levels of $[Ca^{2+}]i$ of hepatic cells.

We [155] and others [156] also found that CRF is associated with alterations in the molecular machinery of LPL in both adipose tissue and myocardium; both the expression of the mRNA and the protein mass of LPL are reduced. In addition, the activity of LPL and its release by heparin are impaired. Again, these abnormalities are prevented by parathyroidectomy of the CRF animals or by their treatment with verapamil, indicating that the impairment in the metabolism and function of LPL are mediated by the PTH-induced rise in $[Ca^{2+}]i$ of adipocytes and cardiac myocytes.

Effect of PTH on Nitric Oxide Synthase in CRF

CRF in rats is associated with reduced activity of nitric oxide synthase (NOS) [157]. This abnormality was accompanied by a significant reduction in both endothelial

NOS (eNOS) and inducible NOS (iNOS) proteins. These derangements were prevented by parathyroidectomy of the CRF animals or by their treatment with the calcium channel blocker felodipine. These observations again assign to a PTH-mediated rise in $[Ca^{2+}]i$ a critical role in the genesis of the abnormalities of NOS.

Role of Parathyroid Hormone in the Pathogenesis of Other Uraemic Manifestations

PTH may play a role in the genesis of the impotence of male uraemic patients [158]. Necrosis of soft tissue is a serious and life-threatening complication of uraemia. It was first seen in renal transplant recipients with secondary hyperparathyroidism and later reported in patients with chronic uraemia [42, 45]. These lesions are progressive, and in most cases, healing occurs only after parathyroidectomy. It has been suggested that this lesion is a manifestation of calciphylaxis induced by PTH [42]. Pruritus is a common manifestation among patients with advanced renal failure and has been attributed to the toxic effect of uraemia on the skin [43, 44]. The amelioration of uraemia by dialysis is not always associated with alleviation of pruritus, and the latter may even worsen despite dialysis therapy. It has been demonstrated that in patients with persistent pruritus, the itching may rapidly disappear after subtotal parathyroidectomy, assigning a possible role for PTH in the aetiology of uraemic pruritus [43, 44].

PTH has been shown to promote the deposition of calcium in the myocardium or in and around the conduction system. Such an event may cause cardiac block and arrhythmias. Pulmonary calcification may reduce the vital capacity and diffusion functions of the lung and may even lead to pulmonary fibrosis [159] and this may cause an increase in pulmonary artery pressure and right ventricular hypertrophy [159]. Thus, by promoting soft-tissue calcification, PTH participates in the genesis of the cardiac and pulmonary dysfunction of uraemic patients.

Osteitis fibrosa caused by excess PTH is an important component of renal osteodystrophy. Enhanced bone resorption and variable degrees of fibrosis of the bone marrow cavity are almost always observed in bone biopsy specimens from uraemic patients. Removal of the parathyroid glands or medical suppression of their activity is usually associated with healing of the osteitis fibrosa. Another bone disorder that may be related to excess PTH in the blood is aseptic necrosis. In a study of 60 renal transplant recipients, 7 of 16 patients who had persistent secondary hyperparathyroidism and elevated blood levels of PTH developed aseptic necrosis of the head of the femur, whereas only 2 of the remaining 46 patients had this complication, despite no significant differences in glucocorticoid regimen between the two groups [60]. Furthermore, aseptic necrosis of the femoral head has been reported in 22 dialysis patients who had secondary hyperparathyroidism and who had not been treated with glucocorticoid [61].

Clinical Implications

The interactions of PTH with the function of so many organ systems assign to the hormone a deleterious role when it accumulates in pharmacological quantities in the blood of uraemic patients. The documentation of the uraemic toxicity of PTH has tremendous clinical implications because we have means to prevent the development of secondary hyperparathyroidism and the accumulation of PTH in blood as renal failure progresses. This could be done by appropriate dietary phosphate restriction or by supplementation of small dosages of $1,25(OH)_2D_3$. Thus, it is theoretically possible to have patients with renal disease reach end-stage renal failure without secondary hyperparathyroidism. Such patients may not display or may have an ameliorated version of the uraemic syndrome, would feel better, and could easily be rehabilitated, with the quality of their lives as well as their contribution to society greatly improved. In addition, medical and surgical therapeutic modalities are available to manage an already existing state of severe secondary hyperparathyroidism in patients with advanced renal failure and those treated with haemodialysis. Furthermore, the use of calcium channel blockers can prevent or reverse the action of PTH on cells. Thus it is also possible to ameliorate the uraemic manifestations in those who already have the uraemic syndrome. The concept that PTH is a uraemic toxin is unique because it is the first time in the history of the search for uraemic toxins that we can identify a toxin whose source could be controlled, allowing us to prevent its accumulation and to reduce its level or its action if it is already elevated.

References

1. Arnaud CD. Hyperparathyroidism and renal failure. Kidney Int 1973;4:89–95.
2. Massry SG. Is parathyroid hormone a uremic toxin? Nephron 1977;19:125–30.
3. Borle AB. Effect of purified parathyroid hormone on the calcium metabolism of monkey kidney cells. Endocrinology 1968;1316–22.
4. Massry SG, Smogorzewski M. The mechanisms responsible for the PTH-induced rise in cytosolic calcium in various cells are not uniform. Miner Electrolyte Metab 1995;21:13–28.
5. Stojceva-Taneva O, Fadda GZ, Smogorzewski M, Massry SG. Parathyroid hormone increases cytosolic calcium of thymocytes. Nephron 1993;64:592–9.
6. Klin M, Smogorzewski M, Khilnani H,Michnowska M, Massry SG. Mechanisms of PTH- induced rise in cytosolic calcium in adult rat hepatocytes. Am J Physiol 1994;267:G754–63.
7. Urenia P, Kong X-F, Abu-Samra A-B, et al. Parathyroid hormone (PTH)/PTH-related peptide receptor messenger ribonucleic acids are widely distributed in rat tissue. Endocrinology 1993;133:617–23.
8. Tian J, Smogorzewski M, Kedes L, Massry SG. Parathyroid hormone-related protein receptor messenger RNA is present in many tissues besides the kidney. Am J Nephrol 1993;13:210–3.
9. Parfitt AM. Soft-tissue calcification in uremia. Arch Intern Med 1969;124:544–56.
10. Massry SG. Divalent ion metabolism and renal osteodystrophy. In Massry SG, Glassock, RJ, editors. Textbook of nephrology 3rd Edition. Baltimore: RJ Williams and Willkins, 1995;1441–73.
11. Massry SG, Smogorzewski M. Mechanisms through which parathyroid hormone mediates its deleterious effects on organ function in uremia. Seminars in Nephrol 1994;14:219–31.
12. Zhang Y-B, Ni Z, Smogorzewski M, et al. Altered cytosolic calcium homeostasis in rat cardiac myocytes in CRF. Kidney Int 1994;45:1113–19.
13. Hruska KA, Moskowitz D, Esbirt P, et al. Stimulation of inositol triphosphate and diacylglycerol production in renal tubular cells by parathyroid hormone. J Clin Invest 1987;79:230–9.
14. Brautbar N, Chakraborty J, Coates J, Massry SG. Calcium, parathyroid hormone and phospholipid turnover of human red blood cells. Miner Elect Metab 1985;11:111–16.
15. Islam A, Smogorzewski M, Massry SG. Effect of chronic renal failure and parathyroid hormone on phospholipid content of brain synaptosomes. Am J Physiol 1989;256:F705–10.
16. Carafoli E. Intracellular calcium homeostasis. Ann Rev Biochem 1987;56:395–433.
17. Levi E, Fadda GZ, Thanakitcharu P, Massry SG. Chronology of cellular events leading to derangements in function of pancreatic islets in chronic renal failure. J Am Soc Nephrol 1992;3:1139–46.
18. Urena P, Kubrusky M, Mannstadt M, et al. The renal PTH/PTHrP receptor is down-regulated in rats with chronic renal failure. Kidney Int 1994;45:605–11.
19. Tian J, Smogorzewski M, Kedes L, Massry SG. PTH/PTHrP receptor is down-regulated in chronic renal failure. Am J Nephrol 1994;14:41–6.
20. Smogorzewski M, Tian J, Massry SG. Down-regulation of PTH-PTHrP receptor of heart in CRF: role of $[Ca^{2+}]i$. Kidney Int 1995;47:1182–6.

21. Massry SG, Klin M, Ni Z, Kedes L, Smogorzewski. Impaired agonist-induced calcium signaling in hepatocytes from chronic renal failure rats. Kidney Int 1995;48:1324–31.
22. Akmal M, Goldstein DA, Multani S, Massry SG. Role of uremia, brain calcium, and parathyroid hormone on changes in electroencephalogram in chronic renal failure. Am J Physiol 1984;246:F575–6.
23. Akmal M, Massry SG. Role of parathyroid hormone in the decreased motor nerve conduction velocity of chronic renal failure. Proc Soc Exp Biol Med 1990;195:202–7.
24. El-Balbessi S, Brautbar N, Anderson K, et al. Effect of chronic renal failure on heart: Role of secondary hyperparathyroidism. Am J Nephrol 1986;6:369–75.
25. Fadda GZ, Akmal M, Prendas H, et al. Insulin release from pancreatic islets: Effect of CRF and excess PTH. Kidney Int 1988;33:1066–72.
26. Akmal M, Perkins S, Kasim SE, et al. Verapamil prevents chronic renal failure-induced abnormalities in lipid metabolism. Am J Kidney Dis 1993;22:158–63.
27. Fadda GZ, Hajjar SM, Perna AF, et al. On the mechanisms of impaired insulin secretion in chronic renal failure. J Clin Invest 1991;87:255–61.
28. Chervu I, Kiersztejn M, Alexiewicz JM, et al. Impaired phagocytosis in chronic renal failure is mediated by secondary hyperparathyroidism. Kidney Int 1992;41:1501–5.
29. Kiersztejn M, Smogorzewski M, Thanakitcharu P, et al. Decreased O_2 consumption by polymorphonuclear leukocytes from humans and rats with CRF. Role of secondary hyperparathyroidism. Kidney Int 1992;42:602–9.
30. Islam A, Smogorzewski M, Massry SG. Effect of verapamil on CRF-induced abnormalities in phospholipid contents of brain synaptosomes. Proc Soc Exp Biol Med 1990;194:1620.
31. Smogorzewski M, Campese VM, Massry SG. Abnormal norepinephrine uptake and release in brain synaptosomes in chronic renal failure. Kidney Int 1989;36:458–65.
32. Smogorzewski M, Piskorska G, Borum PR, Massry SG. Chronic renal failure, parathyroid hormone and fatty acids oxidation in skeletal muscle. Kidney Int 1988;31:555–60.
33. Perna AF, Smogorzewski M, Massry SG. Verapamil reverses PTH- or CRF-induced abnormal fatty acid oxidation in muscle. Kidney Int 1988;34:774–8.
34. Smogorzewski M, Perna AF, Borum PR, Massry SG. Fatty acid oxidation in the myocardium: Effects of parathyroid hormone and chronic renal failure. Kidney Int 1988;34:797–803.
35. Perna AF, Smogorzewski M, Massry SG. Effects of verapamil on the abnormalities in fatty acid oxidation of myocardium. Kidney Int 1989;36:453–7.
36. Smogorzewski M, Islam A, Minasian R, Massry SG. Verapamil corrects abnormalities in norepinephrine metabolism of brain synaptosomes in CRF. Am J Physiol 1990;258:F1036–41.
37. Thanakitcharu P, Fadda GZ, Hajjar SM, Massry SG. Verapamil prevents the metabolic and functional derangements in pancreatic islets in chronic renal failure rats. Endocrinology 1991;129:1749–54.
38. Fadda GZ, Akmal M, Soliman AR, et al. Correction of glucose intolerance and the impaired insulin release of chronic renal failure by verapamil. Kidney Int 1989;36:773–9.
39. Ni Z, Smogorzewski M, Massry SG. Derangements in acetylcholine metabolism in brain synaptosomes in chronic renal failure. Kidney Int 1993;44:630–7.
40. Akmal M, Kasim SE, Soliman AR, Massry SG. Excess parathyroid hormone adversely affects lipid metabolism in chronic renal failure. Kidney Int 1990;37:854–8.
41. Cogan MG, Covey CM, Arieff AI, et al. Central nervous system manifestations of hyperparathyroidism. Am J Med 1978;65:963–70.
42. Massry SG, Gordon A, Coburn JW, Kaplan L, Franklin SS, Maxwell MH, et al. Vascular calcification and peripheral necrosis in a renal transplant patient: Reversal of lesions following subtotal parathyroidectomy. Am J Med 1970;49:416–22.
43. Massry SG, Popovtzer MM, Coburn JW, et al. Pruritus as a manifestation of secondary hyperparathyroidism in uremia. Disappearance of itching following subtotal parathyroidectomy. N Engl J Med 1968;279:697–700.
44. Hampers CL, Katz Al, Wilson RE, Merrill JP. Disappearance of "uremic" itching after subtotal parathyroidectomy. N Engl J Med 1968;279:695–7.
45. Gipstein RH, Coburn JW, Adams DA, et al. Calciphylaxis in man. A syndrome of tissue necrosis and vascular calcification in 11 patients with chronic renal disease. Arch Intern Med 1976;13:1273–80.
46. Drueke T, Fleury I, Toure Y, et al. Effect of parathyroidectomy on left ventricular function in haemodialysis patients. Lancet 1980;1:112–14.
47. McGonigle RJS, Fowler MB, Timmis AB, et al. Uremic cardiomyopathy: Potential role of vitamin D and parathyroid hormone. Nephron 1989;36:94–100.
48. Goldstein DA, Feinstein El, Chui LA, et al. The relationship between the abnormalities in electroencephalogram and blood levels of PTH in dialysis patients. J Clin Endocrinol Metab 1980;51:130–4.

49. Haag-Weber M, Mai B, Horl WH. Normalization of enhanced neutrophil cytosolic calcium of hemodialysis patients with 1,25-dihydroxyvitamin D₃ and calcium channel blocker. Am J Nephrol 1993;13:467–72.

50. Mak RHK, Bettinelli A, Turner C, et al. The influence of hyperparathyroidism on glucose metabolism in uremia. J Clin Endocrinol Metab 1985;60:229–33.

51. Massry SG, Goldstein DA, Procci WR. Impotence in patients with uremia: A possible role for parathyroid hormone. Nephron 1977;19:305–10.

52. Malluche HH, Goldstein DA, Massry SG. Management of renal osteodystrophy with 1,25(OH)₂D₃ II. Effects on histopathology of bone: Evidence for healing of osteomalacia. Miner Elect Metab 1979;2:48–55.

53. Thanakitcharu P, Fadda GZ, Hajjar SM, et al. Verapamil reverses glucose intolerance in pre-existing chronic renal failure: Studies on mechanisms. Am J Nephrol 1992;12:179–87.

54. Alexiewicz JM, Smogorzewski M, Akmal M, Klin M, Massry SG. Nifedipine reverses the abnormalities in [Ca²⁺]i and proliferation of B cells from dialysis patients. Kidney Int 1996;50:1249–54.

55. Massry SG, Fadda GZ, Perna AF, et al. Mechanism of organ dysfunction in phosphate depletion: A critical role for a rise in cytosolic calcium. Miner Elect Metab 1992;18:133–40.

56. Damerdash TM, Syrek N, Smogorzewski M, Marcinkowski W, Nasser-Moaddeli S, Massry SG. The pathways through which glucose induces a rise in [Ca²⁺]i of PMNL of rats. Kidney Int 1996;50:2032–40.

57. Syrek N, Marcinkowski W, Smogorzewski M, Demerdash TM, Massry SG. Amlodipine prevents and reverses in [Ca²⁺]i and the impaired phagocytosis of PMNL of diabetic rats. Nephrol Dial Transplant 1997;12:265–72.

58. Baczynski R, Massry SG, Kohan R, et al. Effect of parathyroid hormone on myocardial energy metabolism in the rat. Kidney Int 1985;27:718–25.

59. Baczynski R, Massry SG, Magott M, et al. Effect of parathyroid hormone on energy metabolism of skeletal muscle. Kidney Int 1985;28:722–7.

60. Goldstein DA, Massry SG. Effect of parathyroid hormone administration and its withdrawal on brain calcium and electroencephalogram. Miner Elect Metab 1978;1:84–92.

61. Goldstein DA, Chiu LA, Massry SG. Effect of parathyroid hormone and uremia on peripheral nerve calcium and motor nerve conduction velocity. J Clin Invest 1978;62:88–93.

62. Smogorzewski M, Koureta P, Fadda GZ, et al. Chronic parathyroid hormone excess in vivo increases resting levels of cytosolic calcium in brain synaptosomes: Studies in the presence and absence of chronic renal failure. J Am Soc Nephrol 1991;1:1162–8.

63. Avram MD, Feinfeld DA, Huatuco AH. Search for the uremic toxin: Decreased motor nerve conduction velocity and elevated parathyroid hormone in uremia. N Engl J Med 1973;298:1000–4.

64. Allen EM, Singer FR, Melamed D. Electroencephalographic abnormalities in hypercalcemia. Neurology 1970;20:15–20.

65. Mayor GH, Keiser JA, Makdani D, Ku PK. Aluminum absorption and distribution: effect of parathyroid hormone. Science 1977;197:1187–9.

66. Ball JH, Butkus DE, Madison DS. Effect of subtotal parathyroidectomy on dialysis dementia. Nephron 1977;18:151–6.

67. Avram MM, Alexis H, Raham M, Son B, Iancu M. Decreased transfusional requirement following parathyroidectomy in long-term hemodialysis. Proc Am Soc Nephrol 1971;5:5.

68. Better OS, Shasha SM, Windver J, Chaimovitz C. Improvement in the anemia of hemodialysis patients following parathyroidectomy (PTX). Proc Am Soc Nephrol 1976;9:1.

69. Meytes D, Bogin E, Ma A, Dukes PP, Massry SG. Effect of parathyroid hormone on erythropoiesis. J Clin Invest 1981;67:1263–9.

70. Lacour B, Basile C, Drueke T. Hyperparathyroidism and red cell production Kidney Int 1980;18:137.

71. Boxer M, Ellman L, Geller R, et al. Anemia in primary hyperparathyroidism. Arch Int Med 1977;137:588.

72. Mallette LE, Bilezikian JP, Heath DA, et al. Primary hyperparathyroidism: Clinical and biochemical features. Medicine 1979;53:127.

73. Bogin E, Massry SG, Levi J, et al. Effect of parathyroid hormone on osmotic fragility of human erythrocytes. J Clin Invest 1982;69:1017–25.

74. Akmal M, Telfer N, Ansari AN, Massry SG. Erythrocyte survival in chronic renal failure: Role of secondary hyperparathyroidism. J Clin Invest 1985;76:1695–8.

75. Saltissi D, Carter GD. Association of secondary hyperparathyroidism with red cell survival in chronic hemodialysis patients. Clin Sci 1985;68:29–33.

76. Malluche HH, Goldstein DA, Massry SG. Management of renal osteodystrophy with 1,25(OH)₂D₃ II. Effects on histopathology of bone: Evidence for healing of osteomalacia. Min Elect Metab 1979;2:48–55.

77. Weinberg SG, Lubin A, Weiner SN, et al. Myelofibrosis and renal osteodystrophy. Am J Med 1977; 63:755.
78. Doherty CC, Labelle P, Collins JF, Brautbar N, Massry SG. Effect of parathyroid hormone on random migration of human polymorphonuclear leukocytes. Am J Nephrol 1988;8:212–19.
79. Alexiewicz JM, Smogorzewski M, Fadda GZ, Massry SG. Impaired phagocytosis in dialysis patients: Studies on mechanisms. Am J Nephrol 1991;11:102–11.
80. Salant DJ, Glover AM, Anderson R, Meyers AM, Rabbin R, Myburgh JA, et al. Polymorphonuclear leukocyte function in chronic renal failure and after renal transplantation. Proc Eur Dial Transplant Assoc 1975;12:370–9.
81. Khan F, Khan AJ, Papagaroufalls C, Warman J, Khan P, Evans HE. Reversible defect of neutrophil chemotaxis and random migration in primary hyperparathyroidism. J Clin Endocrinol Metab 1979;48:582–4.
82. Massry SG, Schaefer RM, Teschner M, Roeder M, Zull JF, Heidland A. Effect of parathyroid hormone on elastase release from human polymorphonuclear leukocytes. Kidney Int 1989;36:883–90.
83. Horl WH, Haag-Weber M, Mai B, Massry SG. Verapamil reverses abnormal $[Ca^{2+}]i$ and carbohydrate metabolism of PMNL of dialysis patients. Kidney Int 1995;47:1741–5.
84. Alexiewicz JM, Smogorzewski M, Klin M, Akmal M, Massry SG. Effect of treatment of hemodialysis patients with nifedipine on metabolism and function of polymorphonuclear leukocytes. Am J Kidney Dis 1995;25:440–4.
85. Alexiewicz JM, Smogorzewski M, Gill SK, Akmal M, Massry SG. Time course of the effect of nifedipine therapy and its discontinuation on $[Ca^{2+}]i$ and phagocytosis of polymorphonuclear leukocytes from hemodialysis patients. Am J Nephrol 1997;17:12–16.
86. Descamps-Latscha B, Chatenoud L. T cells and B cells in chronic renal failure. Semin Nephrol 1996;16:182–91.
87. Yamamoto I, Potts Jr JT, Segre GV. Circulating bovine lymphocytes contain receptors for parathyroid hormone. J Clin Invest 1983;71:404–7.
88. Perry HM. Parathyroid hormone-lymphocyte interactions modulate bone resorption. Endocrinology 1986;19:2333–9.
89. Klinger M, Alexiewicz JM, Linker-Israeli M, et al. Effect of parathyroid hormone on human T cell activation. Kidney Int 1990;37:1543–51.
90. Alexiewicz JM, Gaciong Z, Klinger M, et al. Evidence of impaired T cell function in hemodialysis patients: Potential role for secondary hyperparathyroidism. Am J Nephrol 1990;10:495–501.
91. Pabico RC, Douglas RG, Belts RF, et al. Influenza vaccination of patients with glomerular diseases. Effect on creatinine clearance, urinary protein excretion, and antibody response. Ann Intern Med 1974;4:81–171.
92. Goldblum SE, Reed WP. Host defenses and immunologic alterations associated with chronic hemodialysis. Ann Intern Med 1980;93:597–613.
93. Raskova J, Ghobrial I, Czerwinski DK, Shea SM, Eisinger RP, Raska K. B cell activation and immunoregulation in end-stage renal disease patients receiving hemodialysis. Arch Intern Med 1987;147:89–93.
94. Quadracci LI, Ringden O, Krzymanski M. The effect of uremia and transplantation on lymphocyte subpopulation. Kidney Int 1976;10:179–84.
95. Alexiewicz JM, Klinger M, Pitts TO, Gaciong Z, Linker-Isreali M, Massry SG. Parathyroid hormone inhibits B cell proliferation: Implications in chronic renal failure. J Am Soc Nephrol 1990;1:236–44.
96. Rizzoli RE, Murray TM, Marx SJ, Aurbach GD. Binding of radioiodinated bovine parathyroid hormone (1–84) to canine renal cortical membrane. Endocrinology 1983;112:1303–12.
97. McKee MD, Murray TM. Binding of intact parathyroid hormone to chicken renal plasma membranes. Evidence for a second binding site with carboxyl-terminal specificity. Endocrinology 1985;117: 1930–9.
98. Shenker BJ, Matt WC. Suppression of human lymphocyte responsiveness by forskolin: Reversal by 12-O-tetradecanoylphorbol-13-acetate, diacylglycerol and ionomycin. Immunopharmacology 1987;13:73–86.
99. Hoffmann MK. The requirement for high intracellular cyclic adenosine monophosphate concentrations distinguishes two pathways of B-cell activation induced with lymphokinese and antibody to immunoglobulin. J Immunol 1988;140:580–2.
100. Shearer WT, Patke CL, Gilliam EB, et al. Modulation of a human lymphoblastoid B-cell line by cyclic AMP. J Immunol 1988;141:1678–86.
101. Gaciong Z, Alexiewicz JM, Linker-lsraeli M, et al. Inhibition of immunoglobulin production by parathyroid hormone: Implications in chronic renal failure. Kidney Int 1991;40:96–106.
102. Gaciong Z, Alexiewicz JM, Massry SG. Impaired in vivo antibody production in CRF rats: Role of secondary hyperparathyroidism. Kidney Int 1991;40:862–7.

103. Alexiewicz JM, Smogorzewski M, Akmal M, Massry SG. A longitudinal study of the effect of nifedipine therapy and its discontinuation on [Ca^{2+}]i and proliferation of B lymphocytes of dialysis patients. Am J Kidney Dis 1997;29:233–8.
104. Floyd MA, Ayyar DR, Barnick DD, Hudgson P, Weightman D. Myopathy in chronic renal failure. Q J Med 1974;53:509–24.
105. Patten MB, Bilezikian JP, Mallatte LE, Prince A, Engel WK, Aurbach GD. Neuromuscular disease in primary hyperparathyroidism. Ann Intern Med 1974;80:182–93.
106. Garber AJ. Effect of parathyroid hormone on skeletal muscle protein and amino acid metabolism in the rat. J Clin Invest 1983;71:1806–21.
107. Seyle J. The Pleuricausal cardiomyopathies. Springfield: Thomas, 1961.
108. Lehr D. The role of certain electrolytes and hormones in disseminated myocardial necrosis. In: Bejusz E, editor. Electrolyte and cardiovascular disease, Basel: Karger, 1966;208.
109. Bogin E, Massry SG, Harary I. Effect of parathyroid hormone on rat heart cells. J Clin Invest 1981;67:1215–27.
110. Katoh Y, Klein KL, Kaplan RA, Sanborn WG, Kurokawa K. Parathyroid hormone has a positive inotropic action in the rat. Endocrinology 1981;109:2253.
111. Lhoste F, Drueke T, Larus S, Boissier J. Cardiac interaction between parathyroid, β-adrenoreceptor, and verapamil in the guinea-pig in vitro. Clin Exp Pharmacol Physiol 1980;7:377–85.
112. Bogin E, Levi J, Harary I, Massry SG. Effects of parathyroid hormone on oxidative phosphorylation of heart mitochondria. Min Electrol Metab 1982;7:151–6.
113. Lai KN, Ng J, Whitford J, Buttfield I, Fasset RG, Mathew TH. Left ventricular function in uremia: Echocardiographic and radionuclide assessment in patients on maintenance hemodialysis. Clin Nephrol 1985;23:125–33.
114. Symons C, Fortune F, Greenbaum RA, Dandona P. Cardiac hypertrophy, hypertrophic cardiomyopathy and hyperparathyroidism – an association. Br Heart J 1985;54:539–42.
115. London GM, Fabiani F, Marchais SJ, DeVernejoul MC, Guerin A, Safar ME, et al. Uremic cardiomyopathy: an inadequate left ventricular hypertrophy. Kidney Int 1987;31:973–80.
116. London GM, DeVernejoul MC, Fabiani F, Marchais SJ, Guerin AP, Metivier F, et al. Secondary hyperparathyroidism and cardiac hypertrophy in hemodialysis patients. Kidney Int 1987;32:900–7.
117. London GM, Marchais SJ, Guerin AP, Metivier F. Contributive factors to cardiovascular hypertrophy in renal failure. Am J Hypertension 1989;2:261S–5S.
118. Zucchelli P, Santoro A, Zucchelli A, Spongano M, Ferrari G. Long-term effects of parathyroidectomy on cardiac and autonomic nervous system function in hemodialysis patients. Nephrol Dial Transplant 1988;3:45–50.
119. Langendorf R, Piran CL. The heart in uremia. Am Heart J 1947;33:282–307.
120. Mall G, Rambausek M, Neumeister A, Kollmer U, Vetterle NF, Ritz E. Myocardial interstitial fibrosis in experiment uremia: Implication for cardiac compliance. Kidney Int 1988;33:804–11.
121. Mall G, Huther W, Schneider J, Lundin P, Ritz E. Diffuse intermyocardiocytic fibrosis in uremic patients. Nephrol Dial Transplant 1990;5:39–44.
122. Amann K, Ritz E, Wiest G, Klaus G, Mall G. A role of parathyroid hormone for activation of cardiac fibroblasts in uremia. J Am Soc Nephrol 1994;4:1814–19.
123. Saglikes Y, Massry SG, Iseki K, Nadler JL, Campese VM. Effect of PTH on blood pressure and response to vasoconstrictor agonists. Am J Physiol 1985;248:F674–81.
124. Collins J, Massry SG, Campese VM. Parathyroid hormone and the altered vascular response to norepinephrine in uremia. Am J Nephrol 1985;5:110–13.
125. Iseki K, Massry SG, Campese VM. Evidence for the role of PTH in the reduced pressor response to norepinephrine in chronic renal failure. Kidney Int 1985;28:11–15.
126. DeFronzo RA, Alverstrand A, Smith D, Hender R, Hendler E, Wahren J. Insulin resistance in uremia. J Clin Invest 1981;67:563–8.
127. Westervelt Jr FB, Schreiner GE. The carbohydrate intolerance of uremic patients. Ann Intern Med 1962;57:266–76.
128. DeFronzo RA, Alverstrand A. Glucose intolerance in uremia: Site and mechanism. Am J Clin Nutr 1980;33:1638–45.
129. Massry SG, Smogorzewski M. Carbohydrate metabolism in renal failure. In: Kopple JD, Massry SG, editors. Nutritional management of renal disease. Baltimore: Williams and Wilkins, 1997;63–76.
130. Smith D, DeFronzo RA. Insulin resistance in uremia mediated by postbinding defects. Kidney Int 1982;22:54–62.
131. Castellino P, Solini A, Luzi L, Barr JG, Smith DJ, Petrides A, et al. Glucose and amino acid metabolism in chronic renal failure: Effect of insulin and amino acids. Am J Physiol 1992;262:F168–76.
132. DeFronzo RA, Tobin JD, Rowe JW, Andres R. Glucose intolerance in uremia. Quantification of pancreatic beta cell sensitivity to glucose and tissue sensitivity to insulin. J Clin Invest 1978;62:425–35.

133. Heaton A, Taylor R, Johnston D, Ward WK, Wilkinson R, Alberti KGMM. Hepatic and peripheral insulin action in chronic renal failure before and during continuous ambulatory peritoneal dialysis. Clin Sci 1989;77:383–8.
134. Mak R, Turner C, Thompson T, Haycock G, Chantler C. The effect of a low protein diet with amino acid/keto acid supplements on glucose metabolism in children with uremia. J Clin Endocrinol Metab 1986;63:985–9.
135. McCaleb ML, Wish JB, Lockwood DH. Insulin resistance in chronic renal failure. Endocrinol Res 1985;11:113–25.
136. Dzurik R, Hupkova V, Cernacek P. The isolation of an inhibitor of glucose utilization from the serum of uremic subjects. Clin Chim Acta 1983;46:77–83.
137. Hedeskov CJ. Mechanisms of glucose-induced insulin secretion. Physiol Rev 1980;60:442–509.
138. Schmitz O. Effects of psychological and supraphysiologic hyperglycemia on early and late-phase insulin secretion in chronically dialyzed uremic patients. Acta Endocrinol (Copenh) 1989;121:251–8.
139. Akmal M, Massry SG, Goldstein DA, Fanti P, Weisz A, De Fronzo RA. Role of parathyroid hormone in glucose intolerance of chronic renal failure. J Clin Invest 1985;75:1037–44.
140. Nakamura Y, Yoshida T, Kajiyama S, Kajiyama S, Kitagawa Y, Kanatsuna T, et al. Insulin release from column-perfused isolated islets of uremic rats. Nephron 1985;40:467–9.
141. Oh HY, Fadda GZ, Smogorzewski M, Liou H-H, Massry SG. Abnormal leucine-induced insulin secretion in chronic renal failure. Am J Physiol 1994;267:F853–60.
142. Fadda GZ, Thanakitcharu P, Communale R, Lipson LG, Massry SG. Impaired potassium-induced insulin secretion in chronic renal failure. Kidney Int 1991;40:413–17.
143. Fadda GZ, Hajjar SM, Zhou X-J, Massry SG. Verapamil corrects abnormal metabolism of pancreatic islets and insulin secretion in phosphate depletion. Endocrinology 1992;130:193–202.
144. Oh H-Y, Fadda GZ, Smogorzewski M, Liou H-H, Massry SG. Phosphate depletion impairs leucine-induced insulin secretion. J Am Soc Nephrol 1994;5:1259–65.
145. Cantin M. Kidney, parathyroid and lipemia. Lab Invest 1965;14:1691–8.
146. Werner S, Low H. Stimulation of lipolysis and calcium accumulation by parathyroid hormone in rat adipose tissue in vitro after adrenalectomy and administration of high doses of cortisone acetate. Horm Metab Res 1973;5:292–6.
147. Cozariu L, Forster K, Faulhaber JD, Minne H, Ziegler R. Parathyroid hormone and calcitonin influences upon lipolysis of human adipose tissue. Horm Metab Res 1974;6:243–5.
148. Kather H, Simon B. Human fat cells adenylate cyclase: Responsiveness towards catecholamines, peptide hormones and prostaglandins. In: Hessel LW, Krauss HMJ, editors. Lipoprotein metabolism and endocrine regulation. Amsterdam: Elsevier, 1979;189.
149. Lacour B, Basile C, Drueke T, Funck-Brentano JL. Parathyroid function and lipid metabolism in rat. Miner Electrolyte Metab 1982;7:157.
150. Ljunghall S, Lithell H, Wide L, Vessley B. Glucose and lipoprotein metabolism in primary hyperparathyroidism: Effects of parathyroidectomy. Acta Endocrinol (Copenhagen) 1978;89:580–9.
151. Ritz E, Heuck CC, Boland R. Phosphate, calcium and lipid metabolism. In: Massry SG, Ritz E, Jahn H, editors. Phosphate and minerals in health disease. New York: Plenum Press, 1980;197–208.
152. Ni Z, Smogorzewski M, Massry SG. Effects of parathyroid hormone on cytosolic calcium of rat adipocytes. Endocrinology 1994;135:1837–44.
153. Ni Z, Smogorzewski M, Massry SG. Elevated cytosolic calcium of adipocytes in chronic renal failure. Kidney Int 1995;47:1624–9.
154. Klin M, Smogorzewski M, Ni Z, Zhang G, Massry SG. Abnormalities in hepatic lipase in chronic renal failure. Role of excess parathyroid hormone. J Clin Invest 1996;97:2167–73.
155. Yanru W, Smogorzewsi M, Varuzhan G, Massry SG. Abnormalities in lipoprotein lipase (LPL) metabolism in chronic renal failure (CRF) (abstract). J Am Soc Nephrol 1998:626A.
156. Vaziri ND, Wang XQ, Liang K. Secondary hyperparathyroidism downregulates lipoprotein lipase expression in chronic renal failure. Am J Physiol 1997;273:F925–30.
157. Vaziri ND, Ni Z, Wang XQ, Oveisi F, Zhou HJ. Downregulation of nitric oxide synthase in chronic renal insufficiency: role of excess PTH. Am J Physiol 1998;274:F642–9.
158. Massry SG, Goldstein DA, Procci WR, Kletzky. Impotence in patients with uremia: possible role of parathyroid hormone. Nephron 1977;305–10.
159. Akmal M, Barndt RR, Ansari AN, Mohler JG, Massry SG. Excess PTH in CRF induces pulmonary calcification, pulmonary hypertension and right ventricular hypertrophy. Kidney Int 1995;47:158–63.
160. Chatterjee SN, Friedler RM, Berne TV, Oldham SB, Singer FR, Massry SG. Persistent hypercalcemia after successful renal transplantation. Nephron 1976;17:1–7.
161. Bailey GL, Griffiths HJ, Mocelin AJ, Gundy DH, Hampers CL, Merrill JP. Avascular necrosis of the femoral head in patients on chronic hemodialysis. Transaction Am Soc Artificial Organs 1972;18:401–10.

8

Vitamin D: Normal Physiology and Vitamin D Therapeutics in Normal Nutrition and Various Disease States

J. W. Coburn and J. M. Frazão

8

Introduction

There has been a recent explosion of new information regarding the science of vitamin D; these findings include the discovery and synthesis of new analogues or derivatives of vitamin D that have actual or potential therapeutic value in several disorders related to calcium metabolism and other conditions, such as psoriasis or malignancies, that are unrelated to calcium metabolism.

This chapter reviews the chemistry and physiology of vitamin D and includes the disorders of calcium metabolism that involve vitamin D in their pathogenesis, and other conditions where vitamin D may be important in therapy. The review of biochemistry and physiology will be limited to processes and metabolic conversions that may be clinically relevant. For the interested reader, several outstanding reviews [1–4] and a recent multi-authored scholarly textbook [5] are available; the reader is referred to such sources for details not included in this chapter.

With regard to the nomenclature employed, the "generic" term, vitamin D, without a subscript, is used when vitamin D_3 and vitamin D_2 are interchangeable. For assays of plasma vitamin D and its analogues, the use of "D" without a subscript implies that the assay does not differentiate between the "D_3" and "D_2" forms and it may include both sterols. Table 8.1 lists the vitamin D metabolites and synthetic sterols that are considered, with their synonyms and abbreviations.

History

The events and technologic advances that brought about the industrial revolution during the twentieth century created an "epidemic" of childhood rickets, a disorder later shown to arise as a result of vitamin D deficiency. This occurred as a sizeable fraction of the population abandoned the agrarian lifestyle, moved to cities, and become urbanised. Moreover, smoke from numerous large factories polluted the atmosphere and led to a further reduction of exposure to sunlight. Sir Edward Mellanby reasoned that rickets might represent a dietary deficiency [6]. Rickets was very common in Scotland, where oats provided a major source of calories, and Mellanby thought that such a diet might produce a nutritional deficiency. He then raised dogs on a diet made up largely of oats, and, by accident, he maintained these animals indoors without exposure to ultraviolet light. As a result, the animals developed severe rickets; moreover, he was able to either cure or prevent the disorder by giving cod liver oil to the animals. At that time it was uncertain whether vitamin A, known to be present in cod liver oil, or some other nutrient was responsible, and this healing quality of cod liver oil was initially attributed to vitamin A. A few years later, McCollum et al. [7] demonstrated that cod liver oil was able to promote healing of rickets even after inactivation of the vitamin A that is present in cod liver oil. He proposed that cod liver oil contained a new essential nutrient; this was termed, "vitamin D"; subsequently, vitamin D became known as an essential nutrient.

Table 8.1. Vitamin D sterols: Abbreviations, synonyms, generic names, and trade names

Vitamin D Sterol	Abbreviations	Generic Name	Trade Name
Cholecalciferol	Vitamin D_3 D_3	cholecalciferol	
Ergocalciferol	Vitamin D_2 D_2	ergocalciferol	Drisdol™ Calciferol™
25-Hydroxyvitamin D	$25(OH)D_3$	calcifediol calcidol	Calderol™
24,25-Dihydroxyvitamin D	$24,25(OH)_2D$		
1,25-Dihydroxyvitamin D	$1,25(OH)_2D_3$	calcitriol	Rocaltrol™ Calcijex™
1α-Hydroxyvitamin D_3	$1α(OH)D_3, 1αD_3$	alphacalcidol	Et-Alpha™
1-Nor,1,25-Dihydroxyvitamin D_2	$19-nor,1,25(OH)_2D_2$, 19-Nor	paricalcitol	Zemplar™
1α-Hydroxyvitamin D_2	$1α(OH)D_2, 1αD_2$	doxercalciferol	Hectorol™
$26,27-F_6-1,25$-Dihydroxyvitamin D_3	$26,27-F_6-1,25(OH)_2D_3$	falecalcitriol	
22-Oxa-1,25-Dihydroxyvitamin D_3	22-oxa-calcitriol, OCT 22-oxa-$1,25(OH)_2D_3$	maxacalcitriol	
Dihydrotachysterol	dihydrotachysterol(D_3)	dihydrotachysterol	Intensol™ DHT™
Calcipotriene	calcipotriol		Dovonex™

At about the same time that Mellanby was finding an essential vitamin in cod liver oil, other studies documented that rachitic children could be cured by exposure to sunlight or following treatment with artificial ultraviolet light. This apparent paradox, with rickets reversed by either a nutrient or by sunlight, was ultimately resolved when it was discovered that irradiation of either the experimental animals themselves or feeding them with certain foods that had been irradiated could lead to healing of experimental as well as clinical rickets [8].

The ability to generate vitamin D from the irradiation of food led to the generation of clinically useful quantities of the antirachitic vitamin D and permitted the characterisation of the chemistry of several vitamin D compounds. The initial vitamin D, termed vitamin D_1, was later found to be a combination of at least two sterols: vitamin D_2 or ergocalciferol, which was produced by the irradiation of plant sterols, and vitamin D_3 or cholecalciferol, which is the sterol produced in the skin by the irradiation of its precursor, 7-dehydrocholesterol, a natural skin constituent. The structural identification of these compounds resulted in the award in 1928 of the Nobel Prize for Chemistry to Adolf Windaus.

Research during the last half of the twentieth century resulted in the identification of the mode of action of vitamin D. It was found that vitamins D_2 and D_3, themselves, were actually inactive but needed to undergo bioconversion, first, by hydroxylation at carbon (C) 25 to 25-hydroxyvitamin D [9], and subsequently by a second hydroxylation at C-1, to produce the highly potent, naturally active hormonal form, 1,25-dihydroxyvitamin D_3, or calcitriol [10]. These bioconversions and their control are considered in more detail below.

With the availability of useful quantities of vitamin D, the characterisation of its physiologic roles followed. The role of vitamin D to stimulate the intestinal absorption of both calcium and phosphate was established [11]. The effects of vitamin D on bone are more complex, with the mineralisation of bone and the healing of rickets being the major and obvious end-result of vitamin D treatment. However, the effect of vitamin D on promoting the normal mineralisation of bone was found to be indirect and to arise as a consequence of the normalisation of the levels of calcium and phosphate in the blood [12, 13].

Vitamin D Nutrition, Physiology and Biochemistry

Nutrition of Vitamin D and Its Photosynthesis

Absorption of Ingested Vitamin D_3 and Vitamin D_2

In a consideration of nutritional sources of vitamin D, the diet can include vitamin D obtained from plant sources, as vitamin D_2 or ergocalciferol, and from animal sources, as vitamin D_3 or cholecalciferol. The intestine absorbs about half of the vitamin D that is ingested; once absorbed, it is transported in chylomicrons via the lymphatics to the liver. The best dietary sources of vitamin D_3 are fatty fish or the oil from their livers; much smaller amounts are found in cream, butter and egg yolks. The content of vitamin D or 25(OH)-vitamin D in human colostrum or in human or cow milk is quite low; with levels of only 15–40 IU/L in the last two sources [14]. It is, thus, difficult to obtain adequate vitamin D from dietary sources alone; moreover, the problem of vitamin D nutrition is magnified in individuals who eat a vegetarian diet [15]. In North America, milk, butter, certain margarines and many cereals are required, by law, to be fortified with vitamin D_2; however, the actual quantity can differ substantially from that listed on the label [16]. The NHANES (National Health and Nutritional Examination Survey) II found the median dietary intake of vitamin D to be only 3 µg (120 IU) in normal adults and the amount was 20% lower in women over 75 years [15].

Dermal Biosynthesis of Vitamin D_3

The dermal generation of vitamin D_3 is known to be a complex process with a certain degree of control so that it is impossible to develop vitamin D intoxication as a result of excessive exposure to sunlight alone. The biochemical steps involved in the dermal synthesis of vitamin D_3 are illustrated in Figure 8.1. First, the exposure of the dermis to sunlight containing the appropriate ultraviolet spectrum leads to the conversion of 7-dehydrocholesterol (provitamin D_3), an abundant constituent in the dermis, to previtamin D_3 [17]. In the dermis, this labile intermediate sterol is converted slowly, over two to three days, to vitamin D_3. The latter becomes bound to the vitamin D-binding protein (DBP) present within the plasma of the skin capillaries; this provides the means for "extraction" of vitamin D_3 from the skin and the mechanism for its entry into the systemic circulation. The role of DBP in vitamin D metabolism and conservation is reviewed below. When a person has prolonged skin exposure to sunlight, the quantity of previtamin D_3 generated is limited: the stores of 7-dehydrocholesterol are temporarily depleted and the concentration of previtamin D_3 ceases to rise; instead, there is the generation and accumulation in the skin of two inactive sterols, lumisterol and tachysterol. There are important racial effects on vitamin D_3 synthesis; thus, the presence of increased melanin in the skin markedly delays the

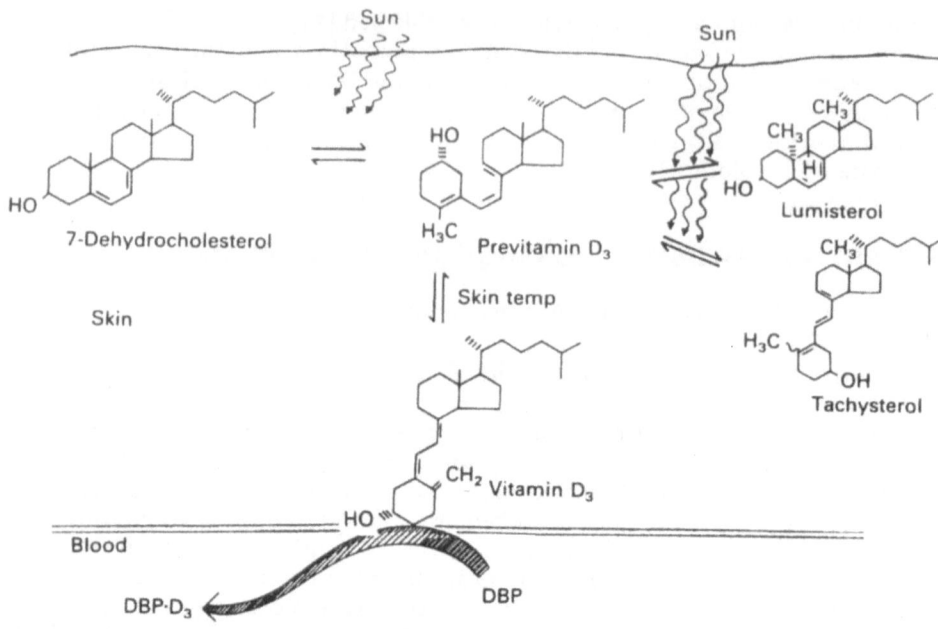

Figure 8.1. Schema showing the formation and metabolism of vitamin D_3 in the skin. First, there is the formation of previtamin D_3 from 7-dehydrocholesterol during exposure to ultraviolet light. This is followed by the slow thermal conversion of previtamin D_3 to vitamin D_3; the latter is then translocated into the circulation as it becomes bound to the plasma vitamin D-binding protein (DBP). With continued exposure to the sun, previtamin D_3 is converted to lumisterol$_3$ and tachysterol$_3$, two biologically inert sterols. The latter have little or no affinity for DPB and these will be lost with sloughing of the skin cells. When the stores of previtamin D_3 become depleted, sunlight exposure leads to isomerisation of lumisterol and tachysterol to previtamin D_3. Reproduced from Holick, MF, Vitamin D and the kidney, Kidney Int. 32:912–29, with permission of the publishers and the author.

conversion of 7-dehydrocholesterol to previtamin D_3, with the process slowed by as much as 80–90% [18].

With regard to quantification of vitamin D_3 generated through the process of photosynthesis, it has been shown that exposure of the body of a young, healthy white adult to sunlight long enough to produce mild erythema leads to an increase of serum levels of vitamin D_3 to values similar to those produced by the ingestion of 10,000 IU (250 μg) of vitamin D_3 [1]. It should be noted that the generation of vitamin D_3 produced after the identical exposure of the skin to sunlight in older individuals or in those with renal insufficiency is considerably lower than that occurring in a young white adult, as shown in Figure 8.2 [19]. Also, the application of a sunscreen with a rating of only eight largely prevents the photosynthesis of previtamin D_3 in human skin [20]. The season of the year is also very important with there being much more UV light during summer than winter; for example, several hours of sunlight exposure in Boston (Latitude, 42° N) in November and February failed to produce any increase of serum levels of vitamin D_3 [21].

The quantity of 7-dehydrocholesterol that is converted to previtamin D_3, the precursor to vitamin D_3 present in the skin, depends on several factors: 1) the surface of skin exposed, 2) the quantity of melanin present in skin, since melanin absorbs the

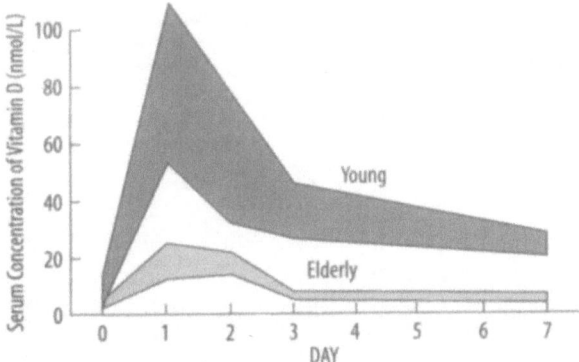

Figure 8.2. Circulating concentrations of vitamin D_3 in response to a whole-body exposure to one minimal erythema dose in healthy young individuals compared to elderly subjects. Reproduced with permission of the publishers and author. From Holick, MF. Vitamin D: Photobiology, metabolism, mechanism of action, and clinical applications. In: Favus, MJ, editor. Primer on the metabolic bone diseases and disorders of mineral metabolism, Third Edition. Philadelphia: Lippincott-Raven Publishers, 1996;74–81.

radiation responsible for the synthesis of previtamin D_3, 3) the time of day, with solar UV radiation being most intense between 1,100 and 1,400 hours, 4) the latitude, with solar UV radiation increasing as one nears the equator, and 5) the season, with little effective radiation penetrating the atmosphere during winter at latitudes more distant from the equator. Also, cultural habits with avoidance of sunshine or the wearing of garments that totally prevent exposure to the sun lead to greatly reduced dermal generation of vitamin D_3, even in sunny climates and at latitudes near the equator [22].

From these considerations, it is apparent that adequate total intake of vitamin D and dermal synthesis can arise via a proper diet combined with the appropriate exposure to sunlight; however, many situations lead to the inadequate intake or reduced generation of vitamin D. The latest update of the recommended dietary allowances and reasonable daily allowances for vitamin D from the US Food and Nutrition Board of the National Research Council are given in Table 8.2. In particular, extra supplements of vitamin D are recommended during pregnancy and lactation, for newborns and young children, and for the elderly, particularly those not exposed to adequate sunlight. As noted below under Vitamin D Deficiency, the requirement for vitamin D is greater when the dietary intake of calcium is low [23].

Bioconversion and Metabolism of Vitamin D

Generation of 25-Hydroxyvitamin D (25-Hydroxylase)

After vitamin D_3 is synthesised in the skin and vitamin D_2 and vitamin D_3 are ingested, they enter the circulation and are transported to the liver, where they are metabolised to 25(OH)-vitamin D via the vitamin D-25-hydroxylase. Two distinct but active vitamin D-25-hydroxylases are found in two separate intrahepatic organelles, a microsomal form (in the smooth endoplasmic reticulum), and the second in mitochondria [24]. Of these, the microsomal form is believed to be more relevant physiologically than that in the mitochondria of hepatocytes; however, the latter has been

Table 8.2.   Recommended dietary allowance (RDA); reasonable daily allowance; and tolerable upper limit of daily intake

Population Groups and Age in Years	RDA (IU/Day)	Reasonable Daily Allowance (IU/Day)	Tolerable Upper Limit (IU/Day)
Infants: 0–1.0	300	200–400	1,000
Children: 1–10	400	200–400	2,000
Adults: 11–24	400	200–400	2,000
Adults: 25–50	200	200–400	2,000
Adults: 51–70	200	400–600	2,000
Adults: >70	200	600–800	2,000
Women: Pregnant or lactating	400	200–400	2,000

Note: For vitamin D, 400 IU = 10 mcg.
From the Food and Nutrition Board, National Research Council, National Academy of Sciences: Recommended Dietary Allowances, 10th Ed., Washington DC, National Academy Press, 1989; and Dietary Reference Intakes for Calcium, Phosphorus, Magnesium, Vitamin D and Fluoride, Washington DC, National Academy Press, 1997.

the form identified in humans. The activation of vitamin D, either endogenous or exogenous, by the liver is highly efficient and does not seem to be under much regulation. There is substantial reserve capacity of the liver to generate 25(OH)-vitamin D, and low levels of 25(OH)D occur only with very severe cholestatic or parenchymal liver disease [25]. When the concentrations of 25(OH)D are found to be low, this can also arise as a consequence of reduced absorption of vitamin D due to any condition causing malabsorption of fat. There is evidence in various animals for the existence of extrahepatic vitamin D-25-hydroxylase, but such activity has not been identified in humans.

Generation of 1α,25-Dihydroxyvitamin D (25-Hydroxyvitamin D-1α-Hydroxylase)

25-hydroxyvitamin D, itself, is biologically inactive and requires further hydroxylation in the kidney to 1,25(OH)$_2$D, the highly active hormonal form of vitamin D. This conversion occurs through the mitochondrial enzyme, 25-hydroxyvitamin D-1α-hydroxylase [26]. Unlike the situation with little or no regulation of the generation of 25-hydroxyvitamin D, the latter's 1α-hydroxylation to 1,25(OH)$_2$-vitamin D is controlled by several factors noted below. The proximal renal tubule is the primary site of this 1α-hydroxylation that controls the circulating levels of 1,25(OH)$_2$D. The renal 1α-hydroxylase activity is suppressed by 1,25-dihydroxyvitamin D, itself, and by increased calcium and/or phosphate; it is stimulated by PTH, hypophosphataemia, hypocalcaemia, calcitonin, and insulin-like growth factor I [26, 27]. The inhibition produced by 1,25-dihydroxyvitamin D is probably indirect and involves several mechanisms; the stimulatory effect of PTH is also indirect and probably related to changes in the concentrations of both phosphate and calcium [26, 27]. The levels of calcium, inorganic phosphate, PTH, 1,25-dihydroxyvitamin D, and the more recently recognised calcium-sensing receptor [28, 29], interact together to account for the very tight regulation of ionised calcium (Ca^{++}) in blood and the extracellular fluid.

There are alterations in the plasma levels of 1,25-dihydroxyvitamin D that arise in other conditions, although the mechanism of activation of the renal 1α-Hydroxylase is not well clarified. Thus, the serum levels of 1,25-dihydroxyvitamin D, including the free or unbound portion, are elevated during human pregnancy [30]; oestrogens may be responsible for this effect but the mechanism is uncertain [31]. Lactation and treatment with prolactin can raise the plasma levels of 1,25-dihydroxyvitamin [32]. Also, growth hormone can acutely raise the serum levels of 1,25-dihydroxyvitamin [33]; however, the role of growth hormone and prolactin on vitamin D metabolism in normal human physiology is less certain [34].

Tissues other than the kidney have the capability of generating 1,25-dihydroxyvitamin D from 25(OH)-vitamin D. Thus, high levels of 1α-hydroxylase messenger RNA (mRNA) have been found in human keratinocytes and in murine macrophages. Thus, $1,25(OH)_2D_3$ can be produced in skin, both in isolated keratinocytes [35] and in a skin perfusion system [36]. Moreover, there is clinical evidence of excessive production of calcitriol by the macrophages from patients with various granulomatous disorders [37, 38] and by lymphocytes in various lymphoproliferative disorders [27, 37]. The function of 1,25-dihydroxyvitamin D generated in these tissues is poorly understood; it is unusual for serum levels of 1,25-dihydroxyvitamin D to be affected except under pathologic conditions, and it has been speculated 1,25-dihydroxyvitamin D may have an autocrine function in these tissues [26].

24-Hydroxylation of Vitamin D (24-Hydroxylase)

The enzyme 24-hydroxylase, also termed CPY, is a mixed-function hydroxylase that acts to hydroxylate both 25(OH)D and $1,25(OH)_2D$ on C-24, steps that are important in the catabolism of these sterols; it also catalyses the dehydrogenation of the 24-hydroxyl group and hydroxylates C-23 leading to an immediate precursor to the generation of calcitroic acid, which is a major end product of $1,25(OH)_2$-vitamin D [39]. Evidence for the in vivo role of 24-hydroxylase in the catabolism of 1,25-dihydroxyvitamin D is provided by studies with 1,25-dihydroxyvitamin D treatment of mice lacking the 24-hydroxylase gene: the results included changes in the kidney consistent with vitamin D intoxication [40]. The 24-hydroxylase gene is found in almost every tissue that contains the vitamin D receptor (VDR), and the induction of CPY is one of the most sensitive markers for the responsiveness of a specific tissue to 1,25-dihydroxyvitamin D.

There has been major disagreement about whether or not 24,25-dihydroxyvitamin D is essential or can exert unique actions that are separate from the effects of 1,25-dihydroxyvitamin D. Developmental abnormalities of collagen synthesis and defective mineralisation have been demonstrated in animals devoid of 24-hydroxylase [41, 42]. Also, studies reported an essential role of 24,25-dihydroxyvitamin D in combination with 1,25-dihydroxyvitamin D in being essential for producing normal "hatchability" of eggs from vitamin D-deficient chickens [43]. Other studies suggested that 24,25-dihydroxyvitamin D was essential for the healing of tibial fractures in chicks [44]. Genetic studies in which mice devoid of the 24-hydroxylase were bred with mice having a mutation in the vitamin D receptor gene have provided some clarification. The abnormalities of calcification of intramembranous bone observed in animals with absent 24-hydroxylase were associated with very high levels of 1,25-dihydroxyvitamin D. However, animals devoid of both the VDR and the 24-hydroxylase failed to exhibit the impaired mineralisation during their developmental period. From these observations, it was suggested that the very high levels of 1,25-dihydroxyvitamin D may have accounted for the skeletal abnormalities observed in animals devoid of the 24-hydroxylase gene [40].

There have been uncontrolled clinical trials with the administration of combinations of 1,25-dihydroxyvitamin D and 24,25-dihydroxyvitamin D in patients with end-stage renal disease and osteomalacia resistant to vitamin D treatment [45, 46]. The initial promising results [45] could not be confirmed with subsequent experience [46]. Also, data in dogs with experimental uraemia have shown no beneficial effect of treatment with 24,25-dihydroxyvitamin D [47]. In some dialysis patients, treatment with 24,25-dihydroxyvitamin D was associated with tolerance of treatment with larger doses of 1,25-dihydroxyvitamin D without the occurrence of hypercalcaemia [45]. Finally, it was shown subsequent to these reports that the osteomalacia of dialysis patients that was resistant to treatment with 1,25-dihydroxyvitamin D arose as a consequence of the marked accumulation of aluminium [48]; this syndrome has largely disappeared as the exposure to aluminium was either abandoned or greatly reduced.

Actions of Vitamin D

Genomic Actions of Vitamin D

The mechanisms whereby vitamin D produces its actions have been largely clarified: Thus, vitamin D produces its effects, in part, by binding to a nuclear receptor, the VDR, and thereby inducing translational changes that result in the synthesis of several new proteins that account for many of the actions of vitamin D. Thus, vitamin D acts in a manner similar to various steroid hormones. The VDR is a member of a large superfamily of nuclear receptors that form dimers with the retinoid X receptor (RXR); this group includes the thyroid hormone receptor, retinoic acid receptor, and several others [49].

The human VDR gene has been identified; there are several promoters and alternate splicing that give rise to several less abundant transcripts that encode the same 427-amino acid protein [50] and include two VDR proteins containing an additional 23 or 50 amino acids at the amino-terminus. The presence of multiple, functionally distinct isoforms is a feature similar to most nuclear receptors; this property may contribute to the tissue-specific effects of various steroid hormones [50]. The most distal VDR promoter exhibits exclusive activity in vitamin D target cells of organs with calciotropic activity, including the kidney, gut, and parathyroid glands [50]; varying stability or translation from these mRNAs may underlie the distinct action of vitamin D on the classic calciotropic tissues.

Before $1,25(OH)_2D_3$ binds to the VDR, there is the heterodimerisation with RXR combined with a series of events that ultimately lead to the transcription of $1,25(OH)_2D_3$-regulated genes. This complex process [51] involves co-activators or transcriptional intermediary factors that interact with the activating portions of the receptor. A large complex of proteins, termed DRIP, facilitates the conformational changes of the ligand-binding domain of the VDR [52]; there is dissociation of various repressor compounds and recruitment of co-activators that ultimately lead to the increase in transcriptional activity [49]. Regulation of various transcriptional co-regulators may provide for the regulation of function of the nuclear receptor; a schema of the proposed genomic action of vitamin D is shown in Figure 8.3.

Other critical factors in control of the response to vitamin D in target tissues for vitamin D include the concentration of $1,25(OH)_2D_3$ and the abundance of the VDR; the latter is regulated by a variety of signals [53], including the significant up-regulation by calcitriol, itself.

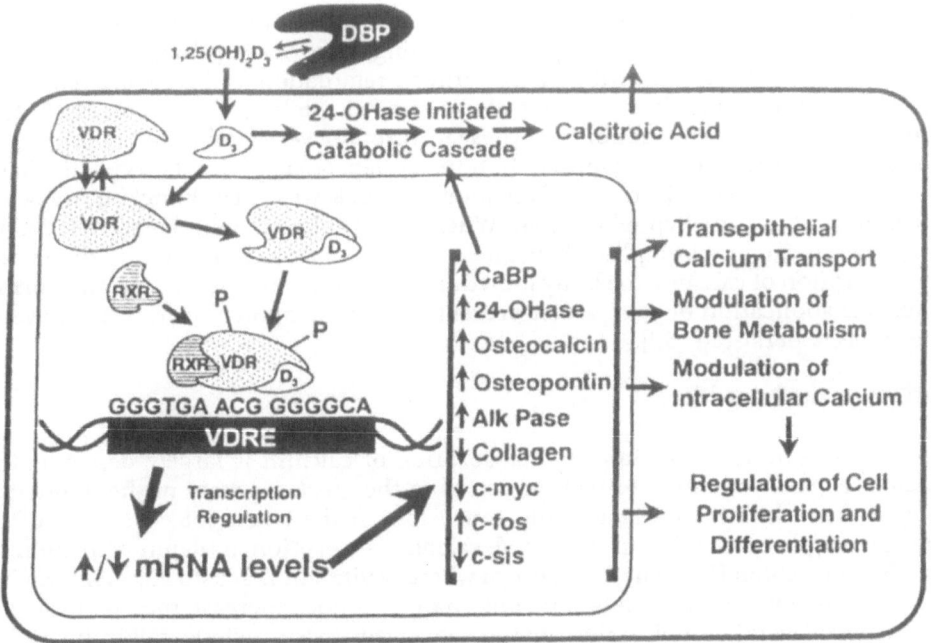

Figure 8.3. Cartoon showing the presumed mechanism of genomic action of 1,25-dihydroxyvitamin D on various target cells with the stimulation of different biologic responses. From the circulation, the unbound or "free" 1,25-dihydroxyvitamin D enters the target cell, interacts with nuclear vitamin D receptor (VDR) and is then is phosphorylated. The 1,25-dihydroxyvitamin D-VDR complex combines with the retinoic acid X-receptor (RXR) to form a heterodimer; the latter then interacts with the vitamin D response element (VDRE). There is consequent enhancement or inhibition of transcription of vitamin D response genes, such as 24-hydroxylase (24-OHase). Reproduced with permission of the publishers and the author; from Holick, MF. Vitamin D: Photobiology, metabolism, mechanism of action, and clinical applications. In: Favus MJ, editor. Primer on the metabolic bone diseases and disorders of mineral metabolism, Third Edition. Philadelphia: Lippincott-Raven Publishers, 1996;74–81.

Distribution of the Nuclear Vitamin D Receptor (VDR)

It is now apparent that a large number of tissues contain the VDR and are potential target tissues for the action of vitamin D. Foremost among these are the "classic" target tissues that control calcium and phosphate homeostases; these include the intestine, kidneys, parathyroid glands and bone. In addition, there are numerous tissues that contain receptors and in which various end-organ actions of calcitriol or other vitamin D analogues can be identified.

Other tissues that are target tissues or potential target tissues include 1) the dermis of the skin, 2) the immune system, 3) cardiovascular tissues, including the heart and vascular smooth muscle, 4) skeletal muscle, 5) both the male and female reproductive organs, 6) various endocrine tissues, including the pancreas, adrenal medulla, thyroid, and pituitary, 7) the liver, 8) lung, and 9) even the brain. In certain of these tissues, a role of vitamin D in enhancing growth and maturation has been demonstrated under various experimental conditions [2].

The development of genetic strains of mice that lack the vitamin D receptor, the so called "VDR knockout" strains, in two independent laboratories have contributed knowledge about vitamin D actions in various tissues. In these models, the animals appear normal at birth, but, soon after weaning, they develop hypocalcaemia and rising levels of PTH, along with rickets, growth retardation and alopecia. The levels of 1,25-dihydroxyvitamin D rise far above normal, while the levels of 24,25-dihydroxyvitamin D remain subnormal. These features are very similar to the findings in patients with the genetic disorder, vitamin D dependency rickets Type II, that is described below [54, 55]. In one of the mouse models with VDR knockout there was the finding of uterine hypoplasia [56]. When the VDR knockout mice were fed a diet very high in calcium and phosphate and with lactose added to facilitate the intestinal absorption of calcium, the body growth and serum calcium levels became normal. Thus, the application of such a diet prevented the development of rickets; however, the alopecia persisted [57].

Actions on the Intestine

The active component of intestinal absorption of calcium is largely dependent on vitamin D. Such calcium absorption occurs to the greatest extent in the duodenum and jejunum, less in the ileum, and much less in the colon [58]. Because of the greater length of the ileum, the total calcium absorption is ileum > jejunum > duodenum > colon [59]. The mechanism whereby vitamin D acts to augment calcium transport is not totally understood, but vitamin D does increase the synthesis and activity of several proteins that may be involved; these include calbindin-D9k, a low-affinity Ca-adenosine triphosphatase (Ca^{++}ATPase), and alkaline phosphatase [60]. The entry of calcium into the enterocyte is believed to occur by diffusion and its stimulation by vitamin D may represent a non-genomic action (see below). The calcium-binding protein, calbindin-D9k, is believed to facilitate the transcellular transfer of calcium, and the extrusion of calcium from the cell into the extracellular fluid involves the membrane calcium pump, Ca^{++}ATPase, and a calcium–sodium exchanger [61].

Phosphate absorption by the gut is also stimulated by vitamin D; the mechanism of its transport differs strikingly from that of calcium. Its entry into the cell involves a transport process that is dependent on sodium and a mucosal Na^{+},K^{+}-ATPase. Several specific sodium/phosphate (Na^{+}/P$_i$) co-transporters have been identified, and data suggest that 1,25-dihydroxyvitamin D increases the expression of the Na^{+}/P$_i$ co-transporter gene [62]. There is also evidence of an effect of vitamin D to enhance the extrusion of phosphate from the enterocyte into the extracellular fluid [60]. Although older studies suggested an interdependence between the vitamin D-stimulated absorption of calcium and that of phosphate [11], there is strong evidence for the independence of the processes of their absorption in the gut [58, 63]. It is well known that there is intestinal cell proliferation and thickening of the intestinal mucosa after the administration of 1,25-dihydroxyvitamin D to vitamin D-deficient animals [64] and to uraemic humans [65]. Such proliferation of the mucosa could enhance the absorption of many compounds in a non-specific manner.

Actions on the Kidney

The kidney plays a major role in the metabolism of calcium, phosphorus, and vitamin D. It is thus the major site of biosynthesis of the hormone 1,25-dihydroxyvitamin D. It is a site of 1,25-dihydroxyvitamin D action, with marked stimulation of the renal 24-hydroxylase by 1,25-dihydroxyvitamin D [66]. With the likely role of the 24-

hydroxylase as an initial step in the catabolism of 1,25-dihydroxyvitamin D, it is probable that the prolonged circulating half life of 1,25-dihydroxyvitamin D in uraemic patients compared to normal subjects [67, 68] arises, in part, as a consequence of the reduced renal 24-hydroxylase in renal tissue of patients with advanced renal failure. There is also evidence that 1,25-dihydroxyvitamin D has an effect to enhance the distal tubular reabsorption of calcium [69], although this effect may be subtle. The mechanisms for this are poorly understood. The distal renal tubular cells are known to contain various elements of the calcium transport system, including the calbindin-D28k and the membrane Ca^{++}-ATPase, in the same areas where VDR is found [69]. A 1,25-dihydroxyvitamin D-dependent epithelial calcium channel, termed ECaC1, is present in the renal tubule, and it may play a role in calcium homeostasis [70].

Actions on the Parathyroid Glands

The gene for the synthesis of the precursor to PTH, pre-pro-PTH, is regulated by a number of factors, including 1,25-dihydroxyvitamin D. Thus, 1,25-dihydroxyvitamin D decreases this gene transcription and reduces the synthesis of PTH and thereby directly suppresses PTH levels [71, 72]. At present, the mechanism for such a transcriptional down-regulation is not well understood. However, this effect of 1,25-dihydroxyvitamin D accounts for the major beneficial effects of treatment with calcitriol or one of its analogues in the management of uraemic secondary hyperparathyroidism. With studies done in vitro, this effect of 1,25-dihydroxyvitamin D may be offset by changes in calcium or phosphate concentrations; thus, high levels of phosphate can act to enhance the proliferation of parathyroid cells through mechanisms that are totally different. Whether 1,25-dihydroxyvitamin D has an effect to regulate the calcium-sensing receptor, which regulates the control of PTH secretion by Ca^{++}, is controversial; one study reported evidence for a role of 1,25-dihydroxyvitamin D to regulate the calcium sensing receptor [73], while another study failed to find any effect [74].

Actions on Bone

The effects of vitamin D on bone are complex and the global effect on bone is poorly understood. The application of documented in vitro actions of 1,25-dihydroxyvitamin D on osteoblasts to in vivo conditions is difficult. The actions of vitamin D on osteoclasts appear to arise indirectly as a consequence of its actions on the osteoblast. It seems apparent that the actions of vitamin D to bring about the mineralisation of bone occur as a consequence of the normalisation of levels of calcium and phosphate in the blood and extracellular fluid [75].

The effect of vitamin D on the osteoblasts in culture depends on whether they are in a proliferating [76] or differentiating phase [77]. During the proliferative phase, there seems to be an inhibitory effect [12, 76, 78], while the effect of vitamin D on mature osteoblasts may lead to up-regulation with more calcium accumulation [77], and osteoblasts may become even more "mature". There are clinical data suggesting that 1,25-dihydroxyvitamin D treatment can strikingly reduce bone formation [79, 80] and can even reduce linear bone growth in children [80]. Histomorphometric observations based on tetracycline labelling indicate that treatment of secondary hyperparathyroidism with 1,25-dihydroxyvitamin D leads to reduced numbers of active osteoblasts; however, the activity of the remaining osteoblasts is largely unchanged [82].

The effect of vitamin D to aid in calcium homeostasis by inducing mineral loss from bone may be applicable to the vitamin D-deficient state when there is accompanying secondary hyperparathyroidism. However, the applicability of such an effect in the vitamin D-replete state and in the absence of pharmacological amounts of vitamin D is less certain.

Vitamin D Actions on Other Tissues

The role of vitamin D in tissues that are not related to calcium and phosphate homeostasis is poorly understood. The finding of hair loss and alopecia in humans with defective VDR and animals devoid of the VDR suggest a major role of vitamin D in the normal differentiation of skin [55, 56, 83]. In cultures of keratinocytes, 1,25-dihydroxyvitamin D has been shown to lead to growth arrest and differentiation [84], with identification of specific effects on various cell cycle regulatory genes [85].

The in vitro effects of 1,25-dihydroxyvitamin D and other vitamin D analogues to produce substantial antiproliferative and differentiating actions in cell lines of various cancers has led to the study of anticancer effects of vitamin D sterols. A long list of animal models of malignancies with antitumoral effects of 1,25-dihydroxyvitamin D and its analogues are described in detail elsewhere [2]. In particular, there has been interest in the application of various vitamin D analogues with reduced calcaemic actions to various malignancies, including breast cancer, prostate cancer, and colonic cancer [86]. Epidemiological data showing relationships between conditions, such as sunlight exposure, latitude, and/or dietary calcium intake, and the incidence of certain malignancies (including prostate cancer [87, 88], colon cancer [89], and breast cancer [90]) provide some support for a potential role of natural vitamin D-insufficiency in the pathogenesis of certain of these disorders.

With the finding of VDR in almost all cells of the immune system, the immunomodulating effect of various vitamin D analogues has been viewed with interest. Also, activated macrophages are known to synthesise 1,25-dihydroxyvitamin D [91]. The 1α-Hydroxylase identified in such cells seems similar to that in the renal tubule, but its regulation is totally different, with no down-regulation by calcium or by 1,25-dihydroxyvitamin D, itself. In the immune system, 1,25-dihydroxyvitamin D may act as a negative signal, leading to down-regulation of antigen presentation, cytokine production by macrophages, and T cell proliferation [91]; it also may directly and indirectly inhibit immunoglobulin production by B lymphocytes [92]. In a number of animal models of autoimmune disease, 1,25-dihydroxyvitamin D or other vitamin D analogues have modulated the course of the disease [92]; also, the combination of 1,25-dihydroxyvitamin D or other vitamin D analogues have shown combined efficacy when added to cyclosporin [92]. In VDR knockout mice, there are major abnormalities of immune function; however, the role of hypocalcaemia, as a major factor producing or contributing to this effect, has yet to be totally excluded [93].

Non-Genomic Actions of 1,25-Dihydroxyvitamin D

In addition to the VDR-mediated genomic or transcriptional effects, there have been very rapid effects produced by 1,25-dihydroxyvitamin D that appear to be mediated by the action of 1,25-dihydroxyvitamin D on the cell membrane [94]. These effects may include the opening of calcium or chloride channels and activation of second messenger signalling pathways [95]. To date, these effects have only been studied in the intestine, where rapid calcium absorption (transcaltachia) has been described

[96], and in a leukaemia cell line, where the rapid induction of cellular differentiation has been reported [97]. It is believed that the vitamin D receptor responsible for these effects on the cell surface is totally different from the nuclear VDR [2].

Transport of Vitamin D in Plasma

The vitamin D absorbed in the intestine is transported in chylomicrons within the lymphatics; both vitamin D_3 and D_2 can be stored in several tissues, including fat, the liver, and muscle. The vitamin D_3 generated in the skin and the vitamin D_2 entering the circulation bind to a specific α-globulin, known as vitamin D binding protein or DBP. This glycoprotein, initially termed group-specific component before its role as a transport protein for vitamin D was recognised, has 458 amino acids and a molecular weight of 51,300 daltons. This α_2-globulin is synthesised by the liver and is known to have many polymorphisms that have been used in genetic testing and forensic medicine [98]. To date, a genetic deficiency of DBP has not been identified in man.

The same DBP is the plasma carrier for all the known metabolites of vitamin D_3 and D_2, although there is slightly less affinity for sterols with the vitamin D_2 side chain. The affinity of DBP for 25(OH)D is about 100-fold greater than for 1,25-dihydroxyvitamin D, and the affinity for vitamin D is considerably less [99]. The DBP concentration in plasma is considerably higher than the levels of most other hormone-binding proteins; its concentration is increased by oestrogens, with the DBP level at the end of pregnancy increasing to about twice the normal concentration [100]. The high affinity of 25(OH)D for DBP results in 25(OH)D having a volume of distribution within the body that is similar to that of DBP and equal to the plasma volume. The half-life of 25(OH)D in the circulation is about two to three weeks [101], and the 25-hydroxyvitamin D that is bound to DBP provides a sizeable reservoir of vitamin D within the circulation.

Because of its molecular weight, the VDR-bound 25(OH)D is readily filtered at the glomerulus; in the proximal tubule, megalin, a large-membrane lipoprotein that is present in the brush border of the proximal tubule, is largely responsible for the endocytic uptake of filtered VDR-bound 25(OH)D, numerous other peptide hormones, and other vitamin-binding proteins [102]. Through this action, megalin may be essential in making 25(OH)D available to the mitochondrial enzyme, 1α-Hydroxylase, for the subsequent synthesis of 1,25-dihydroxyvitamin D [103, 104]. Megalin is likely important in the conservation of both 25(OH)D and DBP [101]; this may explain the severe osteomalacia observed in animals devoid of megalin [2]. The DBP also has a high affinity for extracellular actin, and it may function as a "scavenger" for extracellular actin [105].

Assays of Vitamin D and Related Sterols

It is known that vitamin D and more than a score of its naturally occurring metabolites are found in plasma; the discussion of these compounds and their measurement is beyond the scope of this chapter; the interested reader is referred to other reviews on these topics [106, 107]. The assays of many of these are primarily of interest to investigators doing research into vitamin D metabolism. This chapter focuses on the measurements of diagnostic or therapeutic value in clinical medicine.

The measurement of vitamin D_3 or vitamin D_2 in plasma has no clinical value, but such determinations have been of value in clarifying certain aspects of vitamin D nutrition and synthesis. These sterols have very short circulating half-lives, and their levels are dependent on recent sunlight exposure and/or ingestion of the vitamin D.

25-Hydroxyvitamin D Levels

The measure of 25-hydroxyvitamin D is the most sensitive and specific method to detect vitamin D deficiency. As discussed later, most reports of vitamin D deficiency are largely based on the plasma levels of 25(OH)D [108]. The development of competitive protein binding assays for 25-hydroxyvitamin D has proven clinically useful, and such assays measure the combination of 25-hydroxyvitamin D_3 and 25-hydroxyvitamin D_2. For the separation of 25-hydroxyvitamin D_3 from 25-hydroxyvitamin D_2, high-pressure liquid chromatography (HPLC) is required; these determinations are primarily of value as a research tool, with details on methodology described elsewhere [107]. The measurements of 25(OH)D are of value to detect subclinical vitamin D deficiency or to confirm the diagnosis; also, such measurements may be useful to recognize vitamin D toxicity. Some of the conditions associated with high or low levels of 25(OH)D are shown in Table 8.3.

1,25-Dihydroxyvitamin D Levels

The measurements of plasma $1,25(OH)_2D$ levels can be useful in the clinical setting. Surprisingly, the plasma $1,25(OH)_2D$ levels are totally useless as indicators of nutritional vitamin D deficiency; indeed, the levels may be elevated as a consequence of secondary hyperparathyroidism and the highly efficient conversion of the markedly reduced body stores of 25(OH)D to $1,25(OH)_2D$ [110, 111]. In particular, the measurement of 1,25-dihydroxyvitamin D is useful to identify the cause of unexplained hypercalcaemia or hypercalciuria, particularly when serum PTH levels are normal or suppressed. Thus, serum 1,25-dihydroxyvitamin D levels may be useful to identify the presence of a granulomatous or lymphoproliferative disorder responsible for hypercalcaemia or hypercalciuria (see below).

The measurement of 1,25-dihydroxyvitamin D levels is of great value to identify patients with certain of the hereditary disorders causing hypocalcaemia and rickets: These include vitamin D-dependency rickets (VDDR) type I (pseudo vitamin D-deficient rickets), in which there is a lack or deficiency of 1α-Hydroxylase, and vitamin D-dependency rickets, type II, which represents an abnormality of the receptor of 1,25-dihydroxyvitamin D. The patients with VDDR type I exhibit hypocalcaemia, hypophosphataemia, and significant secondary hyperparathyroidism, and yet have subnormal or low normal levels of plasma 1,25-dihydroxyvitamin D [112]. The patients with VDDR type II also have hypocalcaemia and osteomalacia but exhibit strikingly elevated levels of 1,25-dihydroxyvitamin D; it has been suggested that this condition should be termed hereditary resistance to 1,25-dihydroxyvitamin D [113].

The interpretation of "normal" levels of 1,25-dihydroxyvitamin D is complex, in that the levels are normally modified by PTH levels and the levels of serum phosphorus and calcium. The finding of a value in the mid- or even low-normal range in association with a substantially elevated level of PTH in a patient with significant renal disease is quite abnormal since one would expect elevated levels of 1,25-dihydroxyvitamin D when the iPTH levels are high [114]. Unfortunately, nomograms, such as that used to evaluate serum renin levels in relation to the sodium intake, are not available to "correct" 1,25-dihydroxyvitamin D levels in relation to the levels of PTH, calcium, and/or inorganic phosphate. An earlier proposal to utilise the ratio of

Table 8.3. Changes in circulating concentrations of 25(OH)-vitamin D and 1,25(OH)$_2$-vitamin D in various clinical conditions

Condition	25(OH)D Level	1,25(OH)D Level	Comment[1]
Nutritional vitamin D deficiency	Low	High, normal or low	1,25(OH)$_2$D level low for the hypocalcaemia
Hypoparathyroidism	Normal*	Low normal or low	1,25(OH)$_2$D level low for the hypocalcaemia
Pseudohypoparathyroidism	Normal*	Normal or low	1,25(OH)$_2$D level low for the hypocalcaemia
Primary hyperparathyroidism	Normal*	High or high normal	1,25(OH)$_2$D level high for the hypercalcaemia
Tumour-induced osteomalacia	Normal*	Low normal or low	1,25(OH)$_2$D level low for the hypophosphataemia
X-linked hypophosphataemic rickets	Normal*	Low normal or low	1,25(OH)$_2$D level low for the hypophosphataemia
Vitamin D-dependency rickets, Type I	Normal*	Low or low normal	1,25(OH)$_2$D level very low for the hypocalcaemia and secondary hyperparathyroidism
Vitamin D-dependency rickets, Type II	Normal or low	Very high	
Chronic renal failure	Normal or low	Low or low normal	1,25(OH)$_2$D level low for the elevated PTH levels
Sarcoidosis or other granulomatous disease with hypercalcaemia	Normal*	High or high normal	iPTH levels depressed; 1,25(OH)$_2$D level high for hypercalcaemia
Pregnancy	High or normal	High	25(OH)D level reflects high DBP level; "free" 1,25(OH)$_2$D level high (?oestrogen effect)
Lactation	Normal*	High	? elevation of 1,25(OH)$_2$D level due to prolactin
Acromegaly	Normal*	High or normal	? acute rise of 1,25(OH)$_2$D level due to growth hormone
Vitamin D intoxication	High	Normal or low	iPTH levels depressed; 1,25(OH)$_2$D level may be high for hypercalcaemia

* Unless there is concomitant nutritional vitamin D deficiency.

[1] Since hypercalcaemia, per se, suppresses 1,25(OH)$_2$D levels and hypophosphataemia can elevate these values, while hyperphosphataemia reduces 1,25(OH)$_2$D levels, the serum levels of 1,25(OH)$_2$D must be evaluated in relation to the levels of serum calcium and phosphorus.

1,25-dihydroxyvitamin D/PTH levels to interpret 1,25-dihydroxyvitamin D levels in mild-to-moderate renal insufficiency [114] has never reached wide usage. The clinical conditions that affect the serum 1,25-dihydroxyvitamin D levels are listed in Table 8.3. Presently, there is no known clinical indication for the determination of plasma levels of 24,25-dihydroxyvitamin D; however, such measurements may have important research applications.

Clinical Disorders Related to Vitamin D or Affected by Vitamin D Treatment

Vitamin D Deficiency

Criteria for Vitamin D Deficiency

Vitamin D_3 and vitamin D_2 are in very wide use: they are added to certain foodstuffs in the US and the UK, and they are often given during pregnancy and to newborns and children. Despite the widespread use of vitamin D and the knowledge of its benefit to prevent or to cure vitamin D deficiency, various population studies indicate that vitamin D deficiency is common, even though overt rickets and/or osteomalacia are rare. As noted earlier, the plasma concentration of 25(OH)D provides a method for judging whether a person has adequate stores of vitamin D. However, the threshold concentration of plasma 25(OH)D that should be used to define vitamin D deficiency remains uncertain. Studies that evaluated serum PTH concentrations in relation to the concomitant levels of 25(OH)D revealed that elevated levels of intact PTH, providing evidence of secondary hyperparathyroidism, are very common in patients with 25(OH)D levels below 15 ng/ml (60 nmol/L) [115–118]. However, there is wide variation among individuals, and other population studies have shown a small but significant fraction of patients have elevated PTH levels when their 25(OH)D levels are between 15 and 30 ng/ml (60–129 nmol/L) [119, 120]. Others have considered the lowest serum 25(OH)D level found in association with normal levels of intact PTH. The threshold was estimated to be 27 ng/ml (110 nmol/L) in a large study in Boston [119]. Yet another study found that a plasma concentration of 25-hydroxyvitamin D greater than 22.5 mg/ml (90 nmol/L) was needed to prevent the rise of intact PTH levels during the wintertime [120]. Because of greater need for the substrate (25(OH)D) with the intake of a low calcium diet, greater vitamin D intake and the maintenance of higher levels of serum 25(OH)D would be required [23], while less vitamin D is needed when the dietary calcium intake is high, e.g., 800–1,000 mg/day [31].

Peacock et al. used the term vitamin D insufficiency to define a condition based on the serum level of 25-hydroxyvitamin D above which there was no further rise of the level of 1,25(OH)$_2$D after seven days of supplementation with 25(OH)D_3 [121]; patients were considered vitamin D sufficient if they demonstrated no increase in 1,25-dihydroxyvitamin D levels. This method requires the measurements of 1,25-dihydroxyvitamin D levels twice, separated by very specific treatment with 25(OH)D_3. It would seem that there would be greater precision, as well as less cost, with a single determination of an intact PTH level to recognise the presence of secondary hyperparathyroidism. The term vitamin D insufficiency is recommended to describe an intermittent level with less than optimal stores of vitamin D [122], while the term vitamin D deficiency is best reserved for severe deficiency, where histologic evidence of osteomalacia is likely.

From an evaluation of the various reference ranges for "normal levels" of plasma 25(OH)D, it seems apparent that many "normal" populations include individuals who should be considered to be vitamin D insufficient [119]. The frequency of lower than desired plasma levels of 25(OH)D has been high in several surveys [123, 124, 125]: In a study of 824 healthy elderly men and women, aged 70 to 76, and living in 19 towns in 12 European countries, the fraction of patients with 25(OH)D levels below 12 ng/ml (30 nmol/L) varied from 18% to 92% in the different towns; overall, the median value for plasma 25(OH)D was only 13 ng/ml (33 nmol/L) [123]. Of interest, the finding of low 25(OH)D levels was more common in southern European countries, such as Greece and Italy, than in the Scandinavian communities; this suggests that the clothing worn, the diet ingested, and other habits may be more important than the latitude in determining the 25(OH)D levels. In a Boston study of healthy ambulatory individuals aged 65 years and above, 33% of the men and 42% of women had 25(OH)D levels below 24 ng/ml (60 nmol/L) [124]. In Omaha, Nebraska, 23% of elderly women in nursing homes had 25(OH)D levels below 20 ng/ml (50 nmol/L), compared to 14% of elderly "free-living women" [125]. Also, the 25(OH)D levels in nursing home populations were as low or lower than in elderly free-living individuals despite slightly higher intake of vitamin D in the nursing home population [124, 125]. There is strong evidence, cited below, indicating that there are important clinical consequences for patients with suboptimal plasma levels of 25(OH)D.

Clinical Features of Vitamin D Deficiency

Overt clinical rickets or osteomalacia is usually only found when the 25(OH)D levels fall below 5 ng/ml (12.5 nmol/L). However, there is evidence of clinically significant features even in the absence of evidence of osteomalacia. Features of proximal myopathy [126] and acroparaesthesias [127], which improved or were reversed after replacement with vitamin D, were common among Arab women living in Denmark. Moreover, these symptoms occurred in many individuals with normal levels of both serum calcium and alkaline phosphatase activity. Population studies from both the Netherlands [128] and the Boston area [129] have shown a significant relationship between hip fractures and reduced serum levels of 25(OH)D. The patients admitted in Boston with hip fractures had lower levels of 25(OH)D and higher PTH levels in comparison both to other elderly patients without osteoporosis and to elderly patients admitted for elective hip replacement because of degenerative joint disease [129]. A very high fraction of elderly patients hospitalised with hip fracture in the Netherlands were found to have subnormal levels of 25(OH)D; moreover, there was evidence of both a lower intake of vitamin D and a reduced exposure to sunlight in those with hip fracture [128]. In a large, prospective European study, the daily administration of vitamin D_3, 800 IU, and calcium phosphate, which provided 1.2 g of elemental calcium, significantly reduced the incidence of hip fracture in comparison to patients who had received placebo [130]. A follow-up after a further 18 months of treatment continued to show substantial benefit compared to those receiving placebo [131]. This study and another from the UK [132] showed inverse relationships between femoral bone density and intact PTH levels. In a 12-month, placebo-controlled study in Boston, supplementation with vitamin D, 400 IU, partially prevented the bone loss observed during the winter in the placebo-treated patients [133]. The results of bone markers from serum of elderly patients with osteoporosis [122] indicate that the pathogenesis likely involves alterations that arise from secondary hyperparathyroidism rather than represent features of osteomalacia as the cause of reduced bone density and a tendency to fracture. It is further likely that myopathy of vitamin D deficiency and the associated weakness [127] may predispose to an increased risk of falling and thereby lead to increased risk of fracture.

Pending further studies, experts in this field recommend that serum 25(OH)D concentrations must be above 15 ng/ml (37.5 nmol/L) and ideally be above 25 ng/ml (62.5 nmol/L) throughout the year. A very large number of elderly people, particularly those institutionalised or homebound or living in the less sunny northern countries, are vitamin D-insufficient and may even be vitamin D-deficient [23, 31, 134, 135].

Vitamin D Deficiency and Rickets in Children

An extensive review of vitamin D deficiency in infants and children is beyond the scope of this chapter. However, vitamin D-deficiency rickets remains a significant problem, particularly in developing countries [136]. In both the UK and Europe, the problem is prevalent in immigrants from Africa, the Middle East, and India [122]. Vitamin D-deficiency rickets has been rare in the US; however, vitamin D deficiency is becoming more common with the increased prevalence of breast feeding, and it is more common in certain population groups, such as vegans [137, 138], children on macrobiotic diets [137], children who persist in breast feeding for prolonged periods [137, 139], and children of black and other mothers with increased melanin in the skin that leads to reduced synthesis of vitamin D_3 [140]. In Algeria, 24% of infants born to healthy middle-class mothers who had not received vitamin D supplementation had 25(OH)D levels below 12 ng/ml (30 nmol/L) and elevated PTH values [141]. In a placebo-controlled, prospective study, the maternal administration of either 100,000 IU of cholecalciferol or placebo at both the sixth and eighth month of gestation led to major benefits in the neonates of the "healthy" mothers who were assigned to the vitamin D_3 [142]. The differences in the infants from the vitamin D-supplemented mothers compared to those from mothers given placebo included: 1) a reduced incidence of neonatal hypocalcaemia to 8% compared to 48% in neonates from placebo-treated mothers; 2) increased body weights at birth, six months and at one year; 3) increased height at birth, six months, and at one year; and, 4) greater head circumference at birth and at six months. These differences existed even though all infants received vitamin D_3, 100,000 IU, every three months, from birth to one year of age. Such observations underscore major but subtle abnormalities that may arise from apparent "asymptomatic" vitamin D insufficiency during the maternal period.

Treatment and Prevention of Vitamin D Deficiency

Vitamin D replacement can be accomplished by the daily administration of 800 IU of ergocalciferol or by giving the equivalent total dose once a week or once monthly. Monitoring the therapy by following the 25(OH)D concentration is usually not necessary except in special cases, such as intestinal malabsorption. Higher doses of vitamin D should be used for patients with rickets or osteomalacia, for patients with malabsorption, and for patients taking drugs, such as certain antiepileptics, that activate the hepatic P450 enzyme system. The efficacy of vitamin D supplementation (800 IU/day) combined with calcium supplementation for the prevention of osteoporotic fractures has been demonstrated, with a 29% reduction compared to placebo

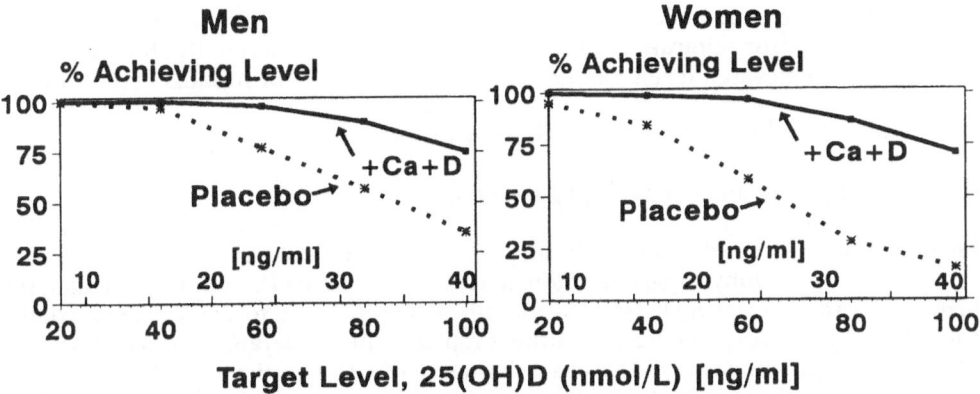

Figure 8.4. The percentages of elderly men and women whose plasma 25-hydroxyvitamin D levels achieved various target ranges. The dietary intake provided an average of 100 IU (2.5 mcg) vitamin D and 700 mg of calcium. Patients were randomly assigned to either "placebo" or "vitamin D₃" for one year. The placebo group received oral supplements providing 500 mg of elemental calcium and those assigned to vitamin D received the same calcium supplement and 700 IU (17.5 mcg) of vitamin D₃. The total vitamin D intake for the supplemented group was 800 IU/day or recommended intake for individuals of this age by the National Academy of Science. Over 10% of the men and nearly 14% of the women failed to reach the desired plasma level of 25(OH)D (80 nmol/L (32 ng/ml)). Adopted from Dawson-Hughes and Harris [143].

after 36 months in postmenopausal women [130, 131]. In elderly patients studied in the Boston area, there is some evidence that a vitamin D intake of 800 IU/day, the recommended allowance, may not be adequate to maintain the plasma levels of 25(OH)D in all elderly individuals at the desired values [143], as is shown in Figure 8.4.

The treatment of overt vitamin D deficiency with evidence of osteomalacia and severe secondary hyperparathyroidism requires larger doses. In Europe, vitamin D₂ is available in capsules containing 10,000 IU; however, in the US, the oral vitamin D dosage forms are either 400 IU, in various over-the-counter preparations, or a prescription preparation with 50,000 IU (1.25 mg) per capsule. A parenteral form of ergocalciferol, with 500,000 IU per ml, is available for intramuscular administration. It has been suggested that the administration of one single large intramuscular treatment may be curative for vitamin D deficiency [144], avoiding the problem with compliance, particularly in developing countries where either limited access to care and/or cultural differences or customs lead to poor compliance and treatment failure [145].

Use of Sterols Other than Vitamin D₂ or D₃ in Vitamin D Deficiency

The sterol 25(OH)D₃ or calcifediol has been available for more than 20 years; there may be a small pharmaceutical advantage in using calcifediol rather than calciferol or ergocalciferol to treat vitamin D deficiency, as the calcifediol is available in 20 mcg and 50 mcg capsules (Calderol$^{(R)}$, Organon); these doses may have less risk of producing hypercalcaemia with its daily administration compared to vitamin D₂, with only 50,000 IU capsules (1.25 mg) available; however, the added cost may not justify its use. For the prevention or treatment of vitamin D deficiency, there seems to be no place for the use of 1,25(OH)₂D₃, 1α-(OH)D₃, or any of the newer vitamin D analogues that have a 1α-hydroxyl group. When either vitamin D₂ or calcifediol is given to a patient with vitamin D deficiency and normal kidneys, there is a prompt increase of serum 1,25-dihydroxyvitamin D concentrations to values substantially above normal; moreover, such elevated levels can persist for several weeks or months [110, 136].

Conditions with Pharmacological Uses of Vitamin D

Hypoparathyroidism and Pseudohypoparathyroidism

There are several conditions, including idiopathic or post-surgical hypoparathyroidism and pseudohypoparathyroidism (resistance to PTH action) that can be managed using the 1α-hydroxylated vitamin D compounds to correct or partially correct the hypocalcaemia. Such treatment replaces the moderately reduced levels of endogenous 1,25 dihydroxyvitamin D_3 that arise due to the absence of PTH (or its ineffective action) combined with the presence of hyperphosphataemia. Vitamin D_2, itself, has been used, but there is the disadvantage of a very long half-life if toxicity should develop [146]. The use of 1,25-dihydroxyvitamin D or a 1α-hydroxylated vitamin D compound has the advantage of the relatively short durations of action of these sterols and the availability of more convenient dosage forms.

With these disorders, the vitamin D compounds should be given with caution. There is the lack of the PTH-stimulated renal tubular reabsorption of calcium; therefore, significant and sometimes marked hypercalciuria occurs when the serum calcium is raised into the normal range. Therefore, serum calcium should be maintained at a level slightly below the normal range in order to prevent hypercalciuria, nephrocalcinosis, hypercalcaemia and even progressive renal failure. The monitoring of serum calcium levels and urinary calcium excretion are required in this setting. The latter can be evaluated by measuring a fasting spot morning urine for the ratio of calcium/creatinine; the ratio (in mg/mg, calcium/creatinine) should be lower than 0.15 to 0.10; if the ratio exceeds 0.15 to 0.20, the vitamin D should be withheld or the dosage reduced.

Primary Osteoporosis and Glucocorticoid-Induced Osteoporosis

The 1α-hydroxylated vitamin D compounds have also been shown to be beneficial for the treatment of both primary osteoporosis and glucocorticoid-induced osteoporosis [147–150]. The incidence of vertebral fracture was significantly reduced by the administration of calcitriol (0.5 μg/day) to women with post-menopausal osteoporosis [151]. Alphacalcidol, usually in a dose of 0.5 to 1.0 μg/day, has also been extensively used in both Europe and Japan [152], with both a positive effect on BMD and reduced incidence of osteoporotic fractures [19, 152, 153, 154]. Presently, neither calcitriol nor alphacalcidol is approved for use in osteoporosis by the Food and Drug Administration (FDA) in the US. The magnitude of the benefit achieved in osteoporotic women during treatment with a bisphosphonate, both to increase the bone mineral density and to reduce the incidence of fractures, may be greater than those reported with the 1α-hydroxylated vitamin D compounds. However, no studies are available that directly compare a vitamin D sterol with any of the bisphosphonates.

X-Linked Hypophosphataemic Rickets or Osteomalacia

This condition represents the most frequent of the hypophosphataemic syndromes; it is inherited as an X-linked dominant trait with the mutant gene located in the distal part of the short arm of the X chromosome [155, 156]. The defective supply of inorganic phosphate in the extracellular fluid results in impaired mineralisation of bone, with the consequent findings of rickets at the growth plate and osteomalacia involving the cortical and endosteal surfaces of bone. The clinical expression of this disorder, which is fully expressed in the hemizygous male, consists of marked hypophosphataemia, bowing and other deformities of the lower limbs, and a stunted rate of growth. Hypophosphataemia is evident early after birth, but the deformities of the legs and the retarded growth become apparent only later when there is continued weight bearing. Because of the renal phosphate wasting, therapy was initially centred on the very aggressive replacement with inorganic phosphate (1–3 g elemental phosphorus per day, divided into four to five doses). In the past, large amounts of vitamin D, usually as ergocalciferol, were given to offset the hypocalcaemic actions of the phosphate administration and the consequent development of secondary hyperparathyroidism; however, these were often difficult to control. More recently, the use of calcitriol has allowed the more precise control of parathyroid hormone levels during treatment. The effect of calcitriol, both directly and indirectly, to reduce PTH secretion has permitted the maintenance of PTH levels within acceptable limits combined with adequate bone modelling and remodelling [157, 158, 159]. A major concern with this long-term therapy with calcitriol has been the gradual deterioration of renal function due to the development of interstitial nephrocalcinosis. Indeed, increased echogenicity of the renal pyramids has been noted with ultrasonography [160], and it has been shown histologically that the areas of echogenicity correspond to deposits of mineral that are comprised of calcium phosphate [161]. Whether the induction of such deposits is primarily related to the phosphate loading or to the long-term use of calcitriol and whether they are causally related to the reduction of renal function remain to be determined. Thus, long-term use of calcitriol combined with oral phosphate supplements, with frequent monitoring of the urinary calcium excretion to avoid episodes of hypercalciuria, is considered a reasonable approach to the control of the clinical expressions of X-linked hypophosphataemia. When hypercalciuria develops or the ratio of calcium/creatinine in the urine rises, the dose of calcitriol should be reduced.

Tumour-Induced Osteomalacia (Renal Phosphate Wasting)

The biochemical features of this syndrome include the presence of marked hypophosphataemia due to renal phosphate wasting, normal serum levels of 25(OH)D, and levels of serum $1,25(OH)_2D$ that are either slightly low or inappropriately normal in relation to the marked hypophosphataemia. Serum calcium levels are usually normal, and serum iPTH levels are normal or in the low–normal range. These metabolic abnormalities and in particular the marked hypophosphataemia lead to the development of osteomalacia. These patients present as adults with bone pain, muscular weakness, and, occasionally, recurrent fractures of long bones. The metabolic abnormalities completely disappear or markedly improve and there is healing of the osteomalacia after the surgical resection of a coexisting tumour. The tumours present in patients with tumour-induced osteomalacia have been of mesenchymal origin in the great majority of patients.

The treatment of choice for this disorder is the surgical resection of the tumour. When the tumour cannot be located, or there is a recurrence of an offending mesenchymal tumour, or it is impossible to resect the tumour completely, there may be the need for effective medical treatment for this metabolic disorder and its symptomatic muscle weakness and bone disease. The administration of calcitriol, in doses up to 1.5–3.0 μg/day, either alone or in combination with oral supplements of inorganic phosphate (providing elemental phosphorus, 2–4 g per day) has been useful to control the biochemical and osseous abnormalities of certain patients with tumour-induced osteomalacia [162, 163, 164].

Renal Failure and End-Stage Renal Disease (ESRD)

Oral and Intravenous Calcitriol

The "active" vitamin D compounds are the treatment of choice in conditions with structural, genetic, or functional deficiency of the activity of the renal 1α-Hydroxylase. Therefore, 1α-hydroxyvitamin D_3 (alphacalcidol) and 1,25 dihydroxy-vitamin D_3 (calcitriol) have been used for nearly 30 years to prevent and to treat the secondary hyperparathyroidism that develops in association with chronic renal failure. A decrease in the serum concentration of calcitriol, either the absolute level or a low value in relation to the substantially increased levels of serum parathyroid hormone is a major factor contributing to the development of secondary hyperparathyroidism as renal function decreases. Moreover, the failure of the kidneys and the consequently reduced generation of calcitriol, particularly in response to the elevated iPTH levels, contributes to the continued oversecretion of PTH by the parathyroid glands; this occurs directly through the loss of genomic control of the mRNA for prepro-PTH and indirectly through the reduced serum calcium levels and progressive hyperplasia of the parathyroid glands. With progressive loss of renal function, the decreased capacity of kidneys to excrete phosphorus leads to hyperphosphataemia. The latter further stimulates PTH production and reduces the levels of both calcium and calcitriol. Thus, calcitriol deficiency plays a major role in the development of secondary hyperparathyroidism in chronic renal failure and the consequent renal bone disease, osteitis fibrosa.

The recognition that calcitriol and other active vitamin D sterols can act directly on the parathyroid glands to reduce the synthesis of parathyroid hormone (PTH) has had a major impact on the therapeutic use of calcitriol and other vitamin D sterols to treat secondary hyperparathyroidism in patients with end-stage renal disease (ESRD). It is well known that calcitriol can reduce PTH secretion indirectly by raising serum calcium levels through the stimulation of intestinal calcium absorption [165]. The direct effect of calcitriol on the parathyroid cells occurs via its genomic action on the vitamin D receptor to inhibit the synthesis of mRNA for prepro-PTH, the initial step for PTH synthesis. In 1971, calcitriol was given for the first time to a patient with end-stage renal disease and secondary hyperparathyroidism [166]. An important shortcoming of treatment with calcitriol is the augmentation of the intestinal absorption of phosphate [167], although this effect is quantitatively smaller than its effect to enhance calcium absorption. When the kidneys function normally, they promptly excrete the excesses of calcium and phosphate that are absorbed. In the ESRD patient, this increased intestinal absorption commonly produces hypercalcaemia and aggravates the degree of hyperphosphataemia. An "ideal" goal would be to modify the treatment with vitamin D so that its action on the parathyroid glands is accentuated, while its stimulation of the intestinal absorption of calcium and phosphorus is minimised. The use of "pulse" or intermittent treatment with intravenous calcitriol, in doses of 0.5 mcg to 4.0 mcg given during haemodialysis two or three times weekly lowered serum PTH levels significantly while it had minimal effects on serum calcium and phosphorus [168]; these observations suggested that the action of "pulse" intravenous calcitriol on the parathyroid glands exceeded the actions on the gut. The initial "short-term" effects of intravenous calcitriol on PTH levels were confirmed [82, 169]. It was also suggested that intravenous "pulse" therapy might have major advantages over oral daily therapy as the trials with daily treatment rarely permitted the long-term administration of calcitriol in doses above 0.25 to 0.50 μg/day without producing hypercalcaemia [170, 171, 172].

Following the initial reports of the use of intravenous calcitriol, several studies reported substantial reductions of PTH levels following "pulse" oral therapy with calcitriol, 2.0 to 4.0 mcg, given either twice or thrice weekly [173–179]. Some investigators reported on intermittent therapy with calcitriol in selected patients with severe hyperparathyroidism [82, 178, 180, 181] and others included patients who had failed with daily oral therapy [82, 176, 182]. In some trials, the degree of hyperparathyroidism was only "mild" [168, 183, 184]. There are important theoretical differences between pulse treatment with calcitriol taken orally compared to that given intravenously: One relates to the blood levels achieved and the other to a potential selective "single pass" effect of oral calcitriol on the intestinal cells. The peak serum calcitriol level is much higher after intravenous treatment than oral therapy; however, this difference disappears in one to three hours and the serum levels are similarly increased above baseline over the next 24 to 48 hours with either treatment. In a study carried out in adolescent ESRD patients for 24 hours, the AUC was higher after intravenous than oral therapy [185]. Similar findings were reported by Ardissino et al. [186] in a study of older children and adolescents with a creatinine clearance that averaged approximately 21 ml/min; 10 patients were treated orally with calcitriol, and 10 intravenously, with doses of 1.5 mcg/M^2 body surface area. Despite an AUC that was approximately 35% higher, there was no difference in the stimulation of strontium absorption (used as a surrogate for calcium). Similar results were reported in a group of six adult ESRD patients, given single doses of 4.0 mcg. Yet another study done in adult patients who were followed for 48 hours found no difference in the AUC in a comparison of oral versus intravenous therapy. Overall, these observations suggest that there may be less than complete absorption of calcitriol or some modest degradation of the calcitriol as it is absorbed across the intestinal mucosa to enter the bloodstream. There are no data available comparing the effects of oral versus intravenous calcitriol on the intestinal absorption of calcium or phosphate which demonstrate any significant "first pass" effect of oral calcitriol on the gut before it reaches the bloodstream. There is agreement, however, that the plasma half-life (T$\frac{1}{2}$) of calcitriol, given intravenously, is substantially longer in patients with renal failure compared to the value observed in normal subjects [67, 68].

There have been six prospective studies that compared oral and intravenous or parenteral calcitriol; one is a crossover study [187] and the others compared two patient groups who received calcitriol by different routes [186, 188–191]. The crossover study was done in 10 patients with mild-to-moderate hyperparathyroidism [187]; half received intravenous calcitriol first and the others, oral calcitriol; after four months, there was crossover to the other modality. With both routes of therapy, serum iPTH fell significantly. Episodes of hypercalcaemia were common but not different with the two modalities: 11 episodes in 8 patients during intravenous therapy and 10 episodes in 7 patients during oral administration. There were no differences in serum phosphorus between the two treatment modalities.

The population studies randomly assigned patients to either oral or parenteral calcitriol. One study [191] compared oral and intraperitoneal calcitriol in paediatricage ESRD patients managed with peritoneal dialysis; there were no differences between the two groups with regard to the reduction of iPTH nor the incidence of hypercalcaemia or hyperphosphataemia, and the data were analysed together. Serum iPTH levels decreased significantly by eight months, and bone biopsies revealed significant improvement in most patients after 12 months; the latter occurred even though the iPTH levels had returned to a mean value at 12 months of treatment that did not differ from the pre-treatment value. Such observations suggest that calcitriol may have direct effects on bone, independent of the changes in serum PTH levels. The other three studies compared pulse oral and pulse intravenous calcitriol in adult ESRD patients [184, 188, 189]; they reported no advantage of one form of therapy over the other. One study noted improvement of both bone biopsy and PTH levels in 57% and 77% of the patients after intravenous and oral treatment, respectively (p = N.S.); however, unacceptable increments of serum Ca and the serum Ca X P product necessitated withholding therapy or a dose reduction in a high percentage of patients. Despite reductions of iPTH, one study found no reduction in parathyroid gland size as measured by either MRI or ultrasound [187]. In studies that evaluated the effect on skeletal effects of calcitriol, certain patients developed a low bone turnover condition, e.g., "adynamic" bone, on post-treatment bone biopsies; this occurred even though the iPTH levels were not suppressed to subnormal levels [79, 82, 190].

Thus, the data from these comparisons show little difference between oral and intravenous therapy. The number of patients with very severe hyperparathyroidism (e.g., iPTH levels > 1,200 pg/ml) was small, and it is possible that intravenous treatment may be more effective in certain patients.

Alphacalcidol (1α-Hydroxyvitamin D$_3$)

Intermittent intravenous doses of 1α-hydroxyvitamin D$_3$ has been shown to be effective in lowering PTH levels of ESRD patients [192, 193]; however, there have been no direct comparisons of oral versus intravenous treatment. One study showed effective suppression of iPTH levels with intravenous treatment in patients who had not responded well with previous daily oral treatment with alphacalcidol [194].

Potential Risks of Calcitriol and Alphacalcidol

Calcitriol and alphacalcidol have been efficacious for the control of PTH overproduction and for improved bone histology in patients with chronic renal failure and secondary hyperparathyroidism. The stimulatory effect of these compounds on intestinal calcium and phosphorus absorption, combined with the suppressive effects on bone turnover, can cause elevations in serum calcium, serum phosphorus, and the calcium–phosphorus product. The risk of hypercalcaemia and elevated calcium–phosphorus product is increased when the patients are utilising calcium-based phosphate binding agents [195, 196]. An increase of the calcium–phosphorus product increases the risk of valvular calcification among haemodialysis patients [197], with a greater risk of cardiac dysfunction and death. Furthermore, hyperphosphataemia, per se, has been identified as an independent risk factor for mortality in studies of large dialysis populations [198–201]. In a group of young dialysis patients who began dialysis in childhood or adolescence, studies of their coronary arteries were done utilising electron-beam computed tomography; the mean serum phosphorus concentration, the mean calcium–phosphorus product in serum, and the daily intake of calcium were higher among the patients who exhibited coronary artery calcification compared to those lacking such calcification [202].

New Vitamin D Sterols

With the recognition of the greatly exaggerated cardiovascular mortality among patients with kidney failure, and the possibility that treatment with the active vitamin D sterols might aggravate such risk, there is great interest in the discovery and development of new vitamin D analogues that might suppress PTH secretion but exert minimal hypercalcaemic and hyperphosphataemic actions. These analogues may offer greater selectivity and may potentially be safer and less toxic in comparison to calcitriol or alphacalcidol, although direct comparative studies are limited [203]. The first of these analogues of interest is 22-oxacalcitriol, as studies first demonstrated a reduced calcaemic effect in experimental animals in comparison to calcitriol [204–206]. This agent has undergone extensive clinical studies in Japan and application has been submitted in Japan for its clinical use. Paricalcitol (19-nor-1,25-hydroxyvitamin D_2) and doxercalciferol (1α-hydroxy-vitamin D_2) are two analogues that have been recently approved for use in the US for the treatment of secondary hyperparathyroidism. Falecalcitriol (hexafluorocalcitriol) is the fourth sterol; it has been studied in Japan, where an application for its use has been submitted. The structures of these four vitamin D sterols in comparison to calcitriol are shown in Figure 8.5.

Paricalcitol (19-nor-1,25-Dihydroxyvitamin D_2)

Paricalcitol has been evaluated and approved in its injectable form in the US. It has been shown in clinical trials to suppress the PTH levels substantially from pre-treatment levels in haemodialysis patients, although the final PTH levels did not differ between the placebo group and those receiving paricalcitol; there were small but statistically significant increments of serum levels of both calcium and phosphorus when compared to placebo [207]. In the reported trial, the number of patients with pre-treatment iPTH levels above 1,200 pg/ml was small. In general, the dose requirement for paricalcitol is threefold to fourfold higher than the requirement for intravenous calcitriol. To date, the clinical trials that compare intravenous paricalcitriol and intravenous calcitriol have not been reported; there are no data on changes in bone histology following its use.

Doxercalciferol (1α-hydroxyvitamin D_2)

Doxercalciferol is a prohormone that is converted in the liver to 1α,25-dihydroxyvitamin D_2, the active form. Doxercalciferol, when administered orally, was effective in decreasing intact PTH levels on haemodialysis patients with moderate-to-severe secondary hyperparathyroidism with a low incidence of hypercalcaemia and hyperphosphataemia [208–210]. The large trial using oral doxercalciferol provided good evidence that larger doses of this vitamin D sterol and a longer duration of therapy are required for adequate suppression of iPTH in patients with markedly elevated levels of iPTH (e.g., > 1,200 pg/ml), as shown in Figure 8.6.

An injectable form of doxercalciferol has also been approved in the US. The efficacy and safety of intravenous doxercalciferol were studied in haemodialysis patients with secondary hyperparathyroidism [211] and these results were compared

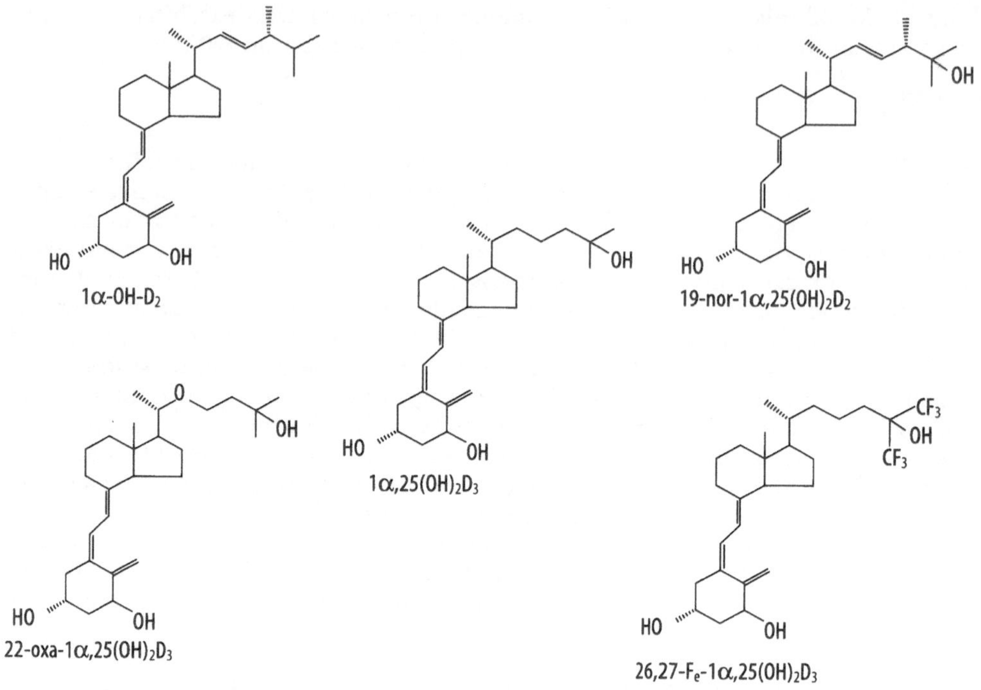

Figure 8.5. Chemical structure of calcitriol (1,25(OH)$_2$D$_3$) and the four newer vitamin D sterols described. Of these, 22-oxa-1,25-dihydroxyvitamin D$_3$ (22-oxa-1α,25(OH)$_2$D$_3$), 26,27-hexafluoro-1,25-dihydroxyvitamin D$_3$ (26,27-F$_6$-1α,25(OH)$_2$D$_3$), and calcitriol (1α,25(OH)$_2$D$_3$) have the vitamin D$_3$ side chain, while 1α-hydroxyvitamin D$_2$ (1α(OH)-D$_2$) and 19-nor-1,25-dihydroxyvitamin D$_2$ (19-nor-1α,25(OH)$_2$D$_2$) have the vitamin D$_2$ side chain. Reprinted with permission of the publishers from Coburn, JW and Salusky IB. Renal bone diseases. In: Bilezekian JP, Marcus R, Levine MA, editors. The parathyroids: Basic and clinical concepts, 2nd Edition. San Diego: Academic Press, 2001;635–61 [248].

with those of an earlier trial of intermittent oral doxercalciferol [210]. Of the 70 patients entering both trials, 64 completed both studies. The intravenous administration of doxercalciferol reduced intact PTH levels effectively and similarly when compared to the oral administration; however, a significantly lower incidence of hypercalcaemia and lower serum phosphorus levels were observed during treatment with the intravenous doxercalciferol compared to the oral form (Figure 8.7). These results suggest that therapy with intravenous doxercalciferol may be advantageous in dialysis patients prone to develop hypercalcaemia or hyper-phosphataemia [211].

22-Oxacalcitriol

The sterol, 22-oxacalcitriol, has shown promise in in vitro and in vivo studies. This compound was found to have 1/8th the affinity for the vitamin D receptor in the intestine and also 1/500th the affinity for the vitamin D binding protein [204, 212]. In rats, the suppressive effect on PTH secretion was noted with less hypercalcaemia in comparison to calcitriol [213]. The data on haemodialysis patients confirms this suppressive effect on PTH with only a modest effect on serum calcium levels [214, 215].

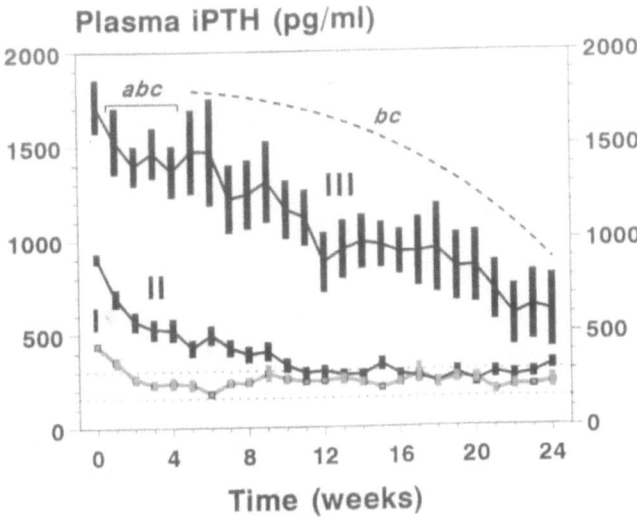

Figure 8.6.   Plasma immunoreactive iPTH levels, shown as mean ± SE, in 99 patients completing a 24-week trial with doxercalciferol with the dosage adjusted to achieve iPTH in the target range of 150 to 300 pg/ml (shown by the dotted lines). The patients are divided into three groups according to their entry (baseline) iPTH levels. Group I: iPTH 400 to 599 pg/ml, Group II: iPTH 600 to 1,200 pg/ml, and Group III: iPTH > 1,200 pg/ml. a, Group I differs from group II, b, Group I differs from Group III, c, Group II differs from Group III; each, P < 0.05. The doses of doxercalciferol given to Group III were significantly greater than that given to Groups I and II during much of treatment, and the doses in Group II exceeded that given to Group I (data not shown). Reprinted with permission from Frazão et al. [210] with permission of the publisher.

The medical community awaits published reports of recent clinical studies that document the safety and efficacy of this analogue.

Falecalcitriol (Hexafluoro-1,25-Dihydroxyvitamin D₃)

Falecalcitriol has been tested in haemodialysis patients with moderate hyperparathyroidism, and a highly significant decrease in PTH was observed compared to placebo. This treatment was associated with an increase in serum calcium levels of only 5% [216]. Another study involved a crossover trial in ESRD patients who had been stable while undergoing treatment with alphacalcidol with a serum intact PTH level between 200 and 800 pg/ml. After an eight-week observation period on their usual treatment with alphacalcidol, the patients were randomly divided into two groups. The first received falecalcitriol for 24 weeks and then were crossed over to alphacalcidol; the other group crossed over between the two drugs in the opposite sequence. The doses were adjusted throughout in an effort to maintain the serum calcium at the same level as it was when the patients entered the study. There was slightly better control of iPTH during treatment with falecalcitriol compared to alphacalcidol, with no differences in serum calcium levels; however, the serum phosphorus levels were lower during the treatment with falecalcitriol than with alphacalcidol [203]. This represents the only new vitamin D sterol with a comparison with either alphacalcidol or calcitriol, the "standards" of treatment of secondary hyperparathyroidism for the last 20 years. Such comparisons are needed in order to document the superiority or advantage of one compound over another.

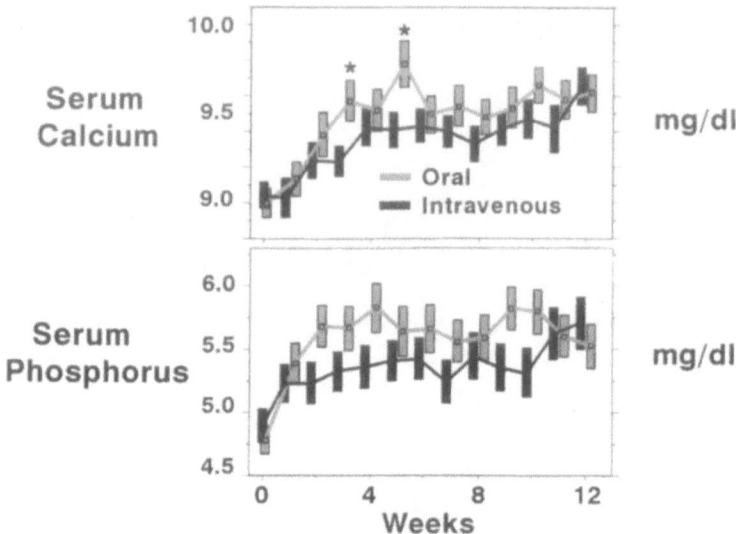

Figure 8.7. Measured levels of serum calcium (top) and phosphorus in 70 patients who completed sequential treatments with oral followed by intravenous doxercaliferol (1α(OH)-D₂) with the dosages of doxercalciferol adjusted to produce similar degrees of suppression of iPTH. Data are shown as mean ± SE. *P < 0.05 for comparison of the oral and intravenous trials. There was more variability of serum phosphorus levels; when the changes of serum phosphorus levels are shown as the percentage change from baseline, the values during the oral trial was significantly higher than during the intravenous at Weeks 1, 2, 4, 9 and 10. Data modified from Maung et al. [211] and shown with permission of the publishers.

Use of Vitamin D Sterols in Kidney Disease and Mild-to-Moderate Renal Insufficiency

There is considerable evidence, reviewed in detail elsewhere [217], that secondary hyperparathyroidism develops and is common in the early stages of progressive renal insufficiency; moreover, these patients can exhibit significant loss of bone mass [218]. When either calcitriol or alphacalcidol are given daily in modest doses to these patients, they can be effective and safe for the prevention of secondary hyperparathyroidism [217]. During such therapy, substantial improvement of the histomorphometric features of secondary hyperparathyroidism [219–222] and an increase of bone mineral density have been observed [223]. The results of six double-blind, placebo-controlled studies using calcitriol [219, 221, 223, 224] or alphacalcidol [222] are summarised in Table 8.4; the levels of renal function of these patients and the doses of the vitamin D sterols are noted.

Of concern is whether long-term treatment of such patients with an active vitamin D sterol might accelerate the progression of renal insufficiency. With doses of calcitriol that rarely exceeded 0.25 mcg/day and doses of alphacalcidol that were generally below 0.50 to 0.75 mcg/day, there was no evidence of worsening of renal function in comparison to the placebo-treated patients [219–224]. With daily doses of calcitriol of 0.5 mcg in patients with moderate renal insufficiency [225] and higher doses in patients treated for psoriasis [226], there was a significant reduction of the creatinine clearances both in the patients with renal disease [225] and in those with normal kidney function being treated for psoriasis [226]. In these trials, the glomerular fil-

Table 8.4.   Use of calcitriol or alphacalcidol in double-blind, placebo-controlled trials in patients with kidney disease and varying degrees of mild-to-moderate renal insufficiency[a]

Reference	Drug and average (μ) dose in mcg/day [range]:	Number of Patients	Duration of Therapy (months)	$C_{creat.}$ (ml/min)[b] [$S_{creat.}$ (mg/dl)]	Comments
Massry [219]	Calcitriol, μ0.46 [0.25 to 1.0 mcg/day]	50	12	C: 30 ± 12 P: 31 ± 15	With doses above 0.5 mcg/day, hypercalcaemia was common but resolved with dose reduction
Nordal and Dahl [220]	Calcitriol, μ0.42 [0.25 to 0.50 mcg/day]	30	8	[1.7 to 7.5]	Selected 30 consecutive patients having $S_{creat.}$ above 1.6 mg/dl and being stable for four months
Baker et al. [221]	Calcitriol, μ0.32 [0.25 to 0.50 mcg/day]	16	12	C: 35 ± 14 P: 45 ± 13	Serum phosphorus was reduced in calcitriol group in comparison to placebo
Przedlacki et al. [223]	Calcitriol, 0.25 mcg/day throughout	25	12	C: 22 ± 12 P: 31 ± 14	Increased BMD (both femoral neck and lumbar spine) in calcitriol group compared to placebo
Ritz et al. [224]	Calcitriol, 0.125 mcg/day throughout	45	12	[1.5 to 5.5]	Low dose chosen; there was no change in serum calcium, phosphorus or $S_{creat.}$ during calcitriol treatment
Hamdy et al. [222]	Alphacalcidol, [0.25 to 1.0 mcg/day]	176	24	A: 32 ± 11 P: 33 ± 12	Dose increased to raise serum Ca within the normal range; bone biopsies showed low frequency of developing adynamic (aplastic) bone after 48 months of therapy

[a] Modified from Coburn et al. [217].
[b] Results are shown as mean ± SD and the range []. Abbreviations: $C_{creat.}$, creatinine clearance; $S_{creat.}$, serum creatinine; C, calcitriol; A, alphacalcidol; P, placebo; BMD, bone mineral density.

tration rate, as measured by inulin clearance, the "gold standard", did not change [225, 226]. In the patients with renal disease [225], the reduction of creatinine clearance reversed totally within two weeks after the daily calcitriol therapy was stopped [225], providing further evidence for the functional nature of this effect on creatinine clearance. It is known that VDR exist in the renal tubules, and these effects on creatinine clearance could arise from an action of calcitriol to inhibit the tubular secretion of creatinine rather than representing an adverse effect on kidney function.

Several long-term, open-label therapeutic trials using calcitriol have also been reported [227–229]; they confirm the beneficial effects of calcitriol on bone reported in the placebo-controlled studies. One finding that differs significantly from observations in patients with ESRD has been the identification of significant histologic and histomorphometric abnormalities of bone with features of secondary hyperparathyroidism in these patients with "early renal failure" even though there were only modest elevations of iPTH levels [222, 228]. These findings differ significantly from observations in ESRD patients, in that the latter do not consistently exhibit histological features of secondary hyperparathyroidism unless the intact PTH levels are increased to at least three- to fourfold greater than the upper normal limit [230, 231].

Hereditary Disorders of Vitamin D Action

Vitamin D-Dependency Rickets, Type I

This genetic disorder, also termed "pseudovitamin D deficiency" is clinically recognised between six months and two years of life; it is characterised by profound hypocalcaemia and hypoplasia of tooth enamel [232, 233]. Hypotonia, proximal muscle weakness, and growth retardation are common; hypocalcaemic seizures are the initial manifestation in some patients. Radiographic features of rickets are present and can be profound. The syndrome responds to continuous treatment with large doses of vitamin D. Also, the syndrome responds to replacement with physiologic doses of calcitriol. Biochemically, the serum levels of 25(OH)D are normal (or high depending on earlier treatment), while the levels of 1,25-dihydroxyvitamin D are low but are usually detectible. This is an autosomal disorder resulting in defective activity of the renal 25(OH)D-1α-Hydroxylase that leads to insufficient synthesis of calcitriol [54]. The treatment of choice is calcitriol; however, because this treatment must be given throughout the lives of these individuals, they are prone to develop toxicity or side effects unless the treatment is adjusted appropriately.

Vitamin D-Dependency Rickets, Type II or Hereditary 1,25-Dihydroxyvitamin D-Resistant Rickets

This autosomal recessive condition is usually recognised by age two; it is characterised by rickets or osteomalacia of varying severity, hypocalcaemia and secondary hyperparathyroidism, either a lack of response or only partial response to physiological or pharmacological preparations of vitamin D, and the finding of greatly elevated serum levels of 1,25-dihydroxyvitamin D [55, 112]. About 70% of the afflicted patients have alopecia and often other skin abnormalities. In a few individuals the biochemical features do not become apparent until later – even as late as middle life. In patients with a delayed appearance, the hypocalcaemia is usually mild. Some patients show a partial correction of the hypocalcaemia following treatment with calciferol in large doses; because of such a partial response in the original patients, the term "vitamin D-dependency rickets, Type II" was used to describe the disorder [234]. Because most afflicted patients fail to respond to pharmacological doses of vitamin D or its metabolites, the term, "hereditary 1,25-dihydroxyvitamin D-resistant rickets" has been used to describe the disorder more accurately [55]. This disorder can arise from several different mutations that render the VDR ineffective [112]. The disorder is best defined by finding the abnormal receptor in cultures of dermal fibroblasts. These cells also fail to show the induction of 24-hydroxylase in response to the addition of 1,25-dihydroxyvitamin D.

The management of this disorder is not satisfactory; it has been shown that the skeletal abnormalities can be reversed by the daily infusion of calcium [78], and partial improvement was noted after treatment with large doses of oral calcium. Relapse is uniform when treatment is discontinued. It has been shown more recently that the initial healing produced by intravenous calcium infusions can later be maintained by subsequent treatment with large doses of oral calcium [235].

Psoriasis

Vitamin D sterols have been useful for the long-term treatment of patients with psoriasis. As noted above, keratinocytes and other skin cells are known target cells for the action of 1,25-dihydroxyvitamin D; also, 1,25-dihydroxyvitamin D has been shown to produce differentiation and maturation of cells from hyperplastic skin. These observations combined with the knowledge that sunlight exposure can produce improvement of psoriasis led to trials using the active vitamin D sterols for the management of psoriasis. Success was shown utilising the local application of calcitriol in an ointment in concentrations of 3 to 15 µg of calcitriol per g [236]. Subsequently, a trial lasting 6–36 months utilised the nocturnal oral administration of calcitriol, starting at a dose of 0.5 µg/day and with the dose increased at fortnightly intervals if serum and urinary calcium remained within acceptable limits. Some degree of improvement was noted in 88% of these patients. Serum calcium levels remained normal; urinary calcium excretion rose significantly but severe hypercalciuria was unusual [226]. Despite such favourable effects, oral calcitriol is not considered an indication for treatment of psoriasis by the Food and Drug Administration in the US, and a topical form of calcitriol has never been marketed.

For the management of psoriasis, there has been the introduction of calcipotriol, an analogue of vitamin D with a carbon ring in the side chain. This sterol has a very rapid plasma half-life, and its activity in producing hypercalcaemia and hypercalciuria in rats was 0.5% that of calcitriol. The topical application of calcipotriol (calcipotriene solution) produced marked improvement of psoriatic lesions in about two thirds of the patients treated in double-blind trials [237]. This compound has been approved for the treatment of psoriasis in the US by the FDA, and it has been used extensively for the management of psoriasis. Two patients who developed transient hypercalcaemia are described; these patients were applying doses far larger than recommended; moreover, they both had significantly reduced renal function [238]. Thus, the risk of hypercalcaemia of topical calcipotriol appears to be very low.

Vitamin D Intoxication (Including Granulomatous Disorders)

Vitamin D intoxication and various granulomatous and lymphoproliferative disorders that cause the excessive extrarenal generation of 1,25-dihydroxyvitamin D can produce hypercalciuria, hypercalcaemia, nephrocalcinosis and other soft tissue cal-cifications, and even renal insufficiency. These conditions usually come to the atten-tion of the clinician because of hypercalcaemia with its attendant symptoms, reviewed in detail in Chapter 3C. Hypercalciuria almost always precedes the appear-ance of hypercalcaemia for a considerable period of time; however, hypercalciuria is often not sought for as it is either asymptomatic or is associated with minor symp-toms that are overlooked. An exception occurs with pre-existing renal disease and reduced renal function, as hypercalciuria is very rare. When renal failure pre-exists or appears de novo due to nephrocalcinosis that develops during long-term vitamin D treatment, hypercalcaemia can appear abruptly and can be severe.

Vitamin D Intoxication

In the US, vitamin D was available without prescription in capsules containing 50,000 IU (1.25 mg) until the 1940s. During the 1930s and 1940s, vitamin D was given for a variety of disorders, including rheumatoid arthritis, hay fever, asthma, and other conditions. There has long been a folklore about the intake of a vitamins with the view that "if a little is good, a lot is better"; therefore, vitamin D intoxication was not uncommon [239]. Vitamin D has been totally abandoned for the uses noted above, and vitamin D toxicity is now rare. However, it still occurs accidentally when there is the faulty addition of vitamin D to foods, such as milk [240] or over-the-counter preparations [241]. It can also develop inadvertently during therapeutics, perhaps most often in patients who are prescribed vitamin D_2 or D_3 for the long-term man-agement of hypocalcaemia in patients with hypoparathyroidism or for the treatment of osteomalacia in patients with X-linked hypophosphataemic rickets. A major factor favouring the use of calcitriol or alphacalcidol in the management of such conditions is the fairly rapid recovery after the development of hypercalcaemia [146]. Vitamin D intoxication occurs in patients with advanced renal disease as a consequence of

Table 8.5.   Hypercalcaemic disorders associated with the extrarenal production of 1,25-dihydroxyvitamin D

Granulomatous Disorders:
Sarcoidosis
Tuberculosis
Coccidioidomycosis
Disseminated candidiasis
Leprosy
Histoplasmosis
Cryptococcoses
Eosinophilic granuloma
Wegener's granulomatosis
Silicon-induced granulomatosis
Berylliosis
Massive infantile fat necrosis
Lymphoproliferative Diseases:
Hodgkin's lymphoma
Non-Hodgkin's lymphoma
Lymphomatoid granulomatos

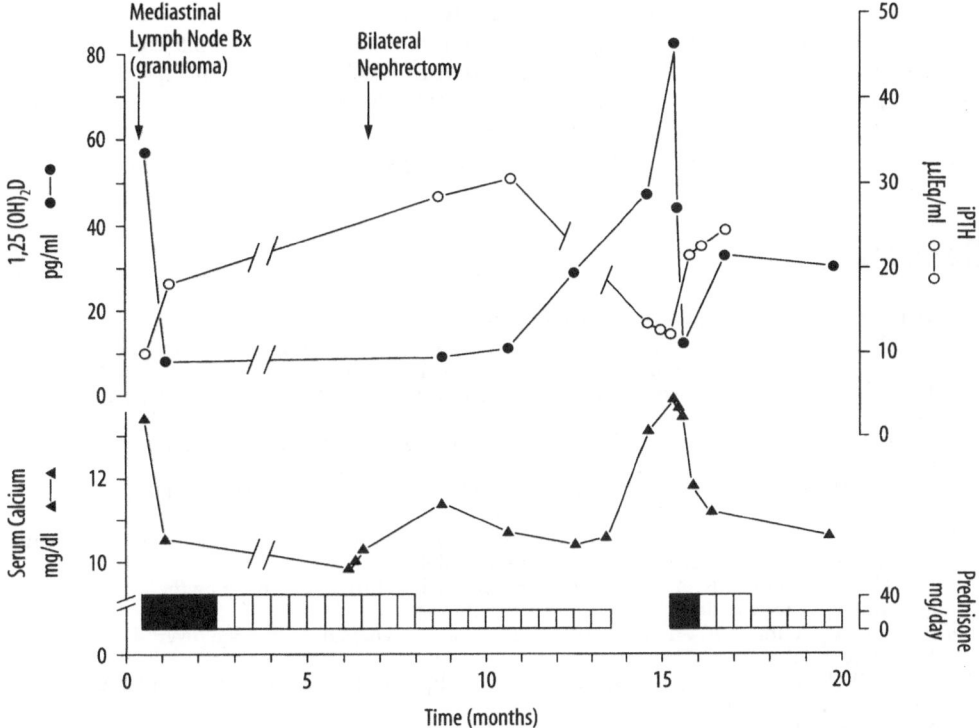

Figure 8.8. The relationship between changes in serum levels of 1,25(OH)₂D, immunoreactive PTH (iPTH), calcium and the doses of prednisone given to control hypercalcaemia in a patient with sarcoidosis and renal failure who had undergone bilateral nephrectomy. The solid portion of the bar represents the administration of daily therapy with prednisone and the open portion indicates alternate-day therapy. Bx denotes biopsy. Serum iPTH was measured using a mid-region assay. Reprinted from Barbour et al. [245] with the permission of the Massachusetts Medical Society.

treatment with calciferol, sometimes given in doses of 100,000 to 400,000 IU per day [242].

When vitamin D toxicity does develop, the serum levels of 25-hydroxyvitamin D are markedly elevated – always above 100 ng/ml and often above 200 ng/ml [243]. The serum levels of 1,25-hydroxyvitamin D are usually normal, and the toxicity is believed to arise as a consequence of the high levels of 25-hydroxyvitamin D. The management of vitamin D intoxication involves the withdrawal of the offending form of vitamin D combined with other measures to lower serum calcium levels.

Hypercalcaemia and Hypercalciuria from Endogenous Synthesis of 1,25-Dihydroxyvitamin D

Although first recognised in patients with sarcoidosis [244], it is now apparent that features of vitamin D toxicity can arise with several granulomatous or lymphoproliferative disorders due to the uncontrolled extrarenal synthesis of 1,25-dihydroxyvitamin D [37, 245]. Although the problem is not common, the number of disorders associated with such hypercalcaemia due to excessive generation of 1,25-dihydroxyvitamin D is numerous, and a partial list is shown in Table 8.5.

With sarcoidosis and certain of the granulomatous disorders, the administration of glucocorticoids leads to the rapid reduction of serum levels of 1,25-dihydroxyvitamin D with a concomitant fall in serum calcium level, as shown in Figure 8.8 [244, 245]. In patients who cannot tolerate glucocorticoids or in those who require excessive doses or prolonged treatment, ketoconazole, which leads to reduced steroid synthesis, has been useful for the management of hypercalcaemia [246, 247].

References

1. Holick MF. Vitamin D and the kidney. Kidney Int 1987;32:912–29.
2. Bouillon R. Vitamin D: From photosynthesis, metabolism, and action to clinical applications. In: DeGroot LJ, Jameson JL, editors. Endocrinology. Philadelphia: W.B. Saunders, 2001;1009–28.
3. Kumar R. Vitamin D. In: Bushinsky DA, editor. Renal osteodystrophy. Philadelphia: LippencottRaven Publishers, 1998;161–202.
4. Malluche HH, Mawad H, Koszewski NJ. Update on vitamin D and its newer analogues: Actions and rationale for treatment in chronic renal failure. Kidney Int 2002;(in press).
5. Feldman D, Glorieux FH, Pike JW. Vitamin D. San Diego: Academic Press, 1997;1–1285.
6. Mellanby E. An experimental investigation on rickets. Lancet 1919;1:407–12.
7. McCollum EV, Simmonds N, Becker JE, Shipley PG. An experimental demonstration of the existence of a vitamin which promotes calcium deposition. J Biol Chem 1922;53:293–8.
8. Goldblatt H, Soames KN. A study of rats on a normal diet irradiated daily by the mercury vapor quartz lamp or kept in darkness. Biochem J 1923;17:294–7.
9. Ponchon G, Kennan AL, DeLuca HF. "Activation" of vitamin D by the liver. J Clin Invest 1969;48:2032–7.
10. Fraser DR, Kodicek E. Unique biosynthesis by the kidney of a biologically active vitamin D metabolite. Nature 1970;228:764–6.
11. Nicolayson R. Studies on the mode of action of vitamin D. III. The influence of vitamin D on the absorption of calcium and phosphorus in the rat. Biochem J 1937;31:122–9.
12. Weinstein RS, Underwood JL, Hutson MS, DeLuca HF. Bone histomorphometry in vitamin D-deficient rats infused with calcium and phosphorus. Am J Physiol 1984;246:E499–E505.
13. Underwood JR, DeLuca H. Vitamin D is not necessary for bone growth and mineralization. Am J Physiol 1984;246:493–8.
14. Specker BL, Tsang RC, Hollis BW. Effect of race and diet on human-milk vitamin D and 25-hydroxyvitamin D. Am J Dis Child 1985;139:1134–7.
15. Lamberg-Allardt C, Karkkainen M, Seppanen R, et al. Low-serum 25-hydroxyvitamin D concentrations and secondary hyperparathyroidism in middle-aged white strict vegetarians. Am J Clin Nutr 1993;58:684–9.
16. Holick MF, Shao Q, Liu WW, Chen TC. The vitamin D content of fortified milk and infant formula. N Engl J Med 1992;326:1178–81.
17. Holick MF, MacLaughlin JA, Clark BM, Holick SA, Potts Jr JT, Anderson RR et al. Photosynthesis of previtamin D_3 in human skin and the physiologic consequences. Science 1980;210:203–5.
18. Holick MF, MacLaughlin JA, Doppelt SH. Factors that influence the cutaneous photosynthesis of previtamin D_3. Science 1981;211:590–3.
19. MacLaughlin J, Holick MF. Aging decreases the capacity of the human skin to produce vitamin D_3. J Clin Invest 1985;76:1536–8.
20. Matsuoko L, Ide L, Wortsman J, MacLaughlin JA, Holick MF. Sunscreens suppress cutaneous vitamin D_3 synthesis. J Clin Endocrinol Metab 1987;64:1165–8.
21. Holick MF. 1994 McCollum Award lecture: Vitamin D – New horizons for the twenty-first century. Am J Clin Nutr 1994;60:619–30.
22. Gannagé-Yared M-H, Chemali R, Yaacoub NE, Halaby G. Hypovitaminosis D in a sunny country: Relation to lifestyle and bone markers. J Bone Miner Res 2000;15:1856–62.
23. Chapuy MC, Preziosi P, Maamer M, Arnaud S, Galan P, Hercberg S et al. Prevalence of vitamin D insufficiency in an adult normal population. Osteoporosis Int 1997;7:439–43.
24. Gascon-Barre M. The vitamin D 25-hydroxylase. In: Feldman D, Glorieux FH, Pike JW, editors. Vitamin D. San Diego: Academic Press, 1997;41–85.
25. Mawer EB, Davies M. Bone disorders associated with gastrointestinal and hepatobiliary disease. In: Feldman D, Glorieux FH, Pike JW, editors. Vitamin D. San Diego: Academic Press, 1997;831–47.

26. Henry HL. The 25-hydroxyvitamin D 1α-Hydroxylase. In: Feldman D, Glorieux FH, Pike JW, editors. Vitamin D. San Diego: Academic Press, 1997;57–85.
27. Bell NH. Renal and nonrenal 25-hydroxyvitamin D-1α-hydroxylases and their clinical significance. J Bone Miner Res 1998;13:350–3.
28. Brown EM. Physiology and pathophysiology of the extracellular calcium-sensing receptor. Am J Med 1999;106:238–53.
29. Coburn JW, Maung HM. Calcimimetic agents and the calcium-sensing receptor. Curr Opin Nephrol Hypertens 2000;9:123–32.
30. Bikle DD, Gee E, Halloran B, Haddad JG. Free 1,25-dihydroxyvitamin D levels in serum from normal subjects, pregnant subjects, and subjects with liver disease. J Clin Invest 1984;74:1966–71.
31. Lips P, Graafmans WC, Ooms ME, Bezemer PD, Bouter LM. Vitamin D supplementation and fracture incidence in elderly persons. A randomized, placebo-controlled clinical trial. Ann Int Med 1996; 124:400–6.
32. Spanos E, Colston KW, Evans A, Galante LS, MacAuley SJ, MacIntyre I. Effect of prolactin on vitamin D metabolism. Nature 1976;5:163–7.
33. Spanos E, Barrett D, MacIntyre I, Pike JN, Safilian EC, Haussler MR. Effect of growth hormone on vitamin D metabolism. Nature (London) 1978;273:2420–33.
34. Adams ND, Garthwaite TL, Gray RW, Hagen TC, Lemann Jr J. The interrelationships among prolactin, 1,25-dihydroxyvitamin D, and parathyroid hormone in humans. J Clin Endocrinol Metab 1979; 49:628–30.
35. Bikle DD, Nemanic MK, Whitney JO, Ellas PW. Neonatal human foreskin keratinocytes produce 1,25-dihydroxyvitamin D_3. Biochemistry 1986;25:1545–8.
36. Bikle DD, Halloran BP, Riviere JE. Production of 1,25-dihydroxyvitamin D_3 by perfused pig skin. J Invest Dermatol 1994;102:796–8.
37. Adams JS. Extrarenal production of active vitamin D metabolites in human lymphoproliferative diseases. In: Feldman D, Glorieux FH, Pike JW, editors. Vitamin D. San Diego: Academic Press, 1977; 903–21.
38. Dusso AS, Lopez-Hilker S, Rapp N, Slatopolsky E. Extrarenal production of calcitriol in chronic renal failure. Kidney Int 1988;34:368–75.
39. Reddy GS, Tserng KY. Calcitroic acid, end product of renal metabolism of 1,25-dihydroxyvitamin D_3 through C-24 oxidation pathway. Biochemistry 1989;28:1763–9.
40. St-Arnaud R, Arabian A, Travers R, Barletta F, Raval-Pandya M, Chapin K, et al. Deficient mineralization of intramembranous bone in vitamin D-24-hydroxylase-ablated mice is due to elevated 1,25-dihydroxyvitamin D and not to the absence of 24,25-dihydroxyvitamin D. Endocrinol 2000; 141:2658–66.
41. St-Arnaud R. Novel findings about 24,25-dihydroxyvitamin D: an active metabolite? Curr Opin Nephrol Hypertens 1999;8:435–41.
42. St-Arnaud R, Glorieux FH. 24,25-dihydroxyvitamin D–Active metabolite or inactive catabolite? Endocrinol 1998;139:3371–3.
43. Henry HL, Norman AW. Vitamin D: Two dihydroxylated metabolites are required for normal chicken egg hatchability. Science 1978;201:835–7.
44. Seo EG, Einhorn TA, Norman AW. 24R,25-dihydroxyvitamin D_3: An essential vitamin D_3 metabolite for both normal bone integrity and healing of tibial fracture in chicks. Endocrinol 1997;138:3864–72.
45. Hodsman AB, Wong EGC, Sherrard DJ, Brickman AS, Lee DBN, Singer FR, et al. Preliminary trials with 24,25-dihydroxyvitamin D_3 in dialysis osteomalacia. Am J Med 1983;74:407–14.
46. Sherrard DJ, Ott SM, Andress DL, Coburn JW. Histologic response to 24,25-dihydroxyvitamin D in renal osteodystrophy. In: Norman AW, Schaefer K, Grigoleit HG, von Herrath D, Walter DE Gruyter, editors. Vitamin D: Chemical, biochemical and clinical endocrinology of calcium metabolism. Berlin, 1985;269–74.
47. Olgaard K, Finco D, Schwartz J, Arbelaez M, Teitelbaum S, Avioli L, et al. Effect of 24,25(OH)2D3 on PTH levels and bone histology in dogs with chronic uremia. Kidney Int 1984;26:791–7.
48. Ott SM, Maloney NA, Coburn JW, Alfrey AC, Sherrard DJ. The prevalence of bone aluminum deposition in renal osteodystrophy and its relation to the response to calcitriol therapy. N Engl J Med 1982;307:709–13.
49. Haussler MR, Whitfield GK, Haussler CA, Hsieh J-C, Thompson PD, Selznick SH, et al. The nuclear vitamin D receptor: Biological and molecular regulatory properties revealed. J Bone Miner Res 1998;13:325–49.
50. Crofts LA, Hancock MS, Morrison NA, et al. Multiple promoters direct the tissue-specific expression of novel N-terminal variant human vitamin D receptor gene transcripts. Proc Nat Acad Sci (USA) 1998;95:10529–34.

51. Wurtz J-M, Guillot B, Moras D. 3D model of the ligand binding domain of the vitamin D nuclear receptor based on the crystal structure of holo RAR-gamma. In: Norman AW, Bouillon R, Thomasset M, editors. Vitamin D: Chemistry, biology, and clinical applications of the steroid hormone. Riverside: University of California, 1997;165–72.
52. Xu L, Glass CK, Rosenfeld MG. Coactivator and corepressor complexes in nuclear receptor function. Curr Opin Genet Dev 1999;9:140–7.
53. Krishnan AV, Feldman D. Regulation of vitamin D receptor abundance. In: Feldman D, Glorieux FH, Pike JW, editors. Vitamin D. San Diego: Academic Press, 1997;179–200.
54. Marx SJ, Spiegel AM, Brown EM, Gardner DG, Downs Jr RW. A familial syndrome of decrease in sensitivity to 1,25-dihydroxyvitamin D. J Clin Endocrinol Metab 1978;47:1303–10.
55. Malloy PJ, Pike JW, Feldman D. Hereditary 1,25-dihydroxyvitamin D resistant rickets. In: Feldman D, Glorieux FH, Pike JW, editors. Vitamin D. San Diego: Academic Press, 1977;765–87.
56. Yoshizawa T, Handa Y, Uematsu Y, et al. Mice lacking the vitamin D receptor exhibit impaired bone formation, uterine hypoplasia and growth retardation after weaning. Nat Genet 1997;16:391–6.
57. Li YC, Amling M, Pirro AE, Preimel M, Meuse J, Baron R, et al. Normalization of mineral ion homeostasis by dietary means prevents hyperparathyroidism, rickets, and osteomalacia but not alopecia in vitamin D receptor-ablated mice. Endocrinol 1998;139:4391–6.
58. Walling MW. Intestinal Ca and phosphate absorption: Differential responses to vitamin D_3 metabolites. Am J Physiol 1977;233:E488–94.
59. Marcus CS, Lengemann FW. Absorption of Ca^{45} and Sr^{85} from solid and liquid food at various levels of the alimentary tract of the rat. J Nutr 1962;77:155–60.
60. Wasserman RH. Vitamin D and the intestinal absorption of calcium and phosphorus. In: Feldman D, Glorieux FH, Pike JW, editors. Vitamin D. San Diego: Academic Press, 1997;259–73.
61. Peerce BE. Identification of the intestinal Na-phosphate cotransporter. Am J Physiol 1989;256:G645–52.
62. Yaegci A, Werner A, Murer H, Biber J. Effect of rabbit duodenal mRNA on phosphate transport in Xenopus laevis oocytes: Dependence on 1,25-dihydroxyvitamin D_3. Pfleugers Arch 1992;422:211–16.
63. Lee DBN, Walling MW, Gafter U, Silis V, Coburn JW. Calcium and inorganic phosphate transport in rat colon: Dissociated response to 1,25-dihydroxyvitamin D_3. J Clin Invest 1980;65:1326–31.
64. Spielvogel AM, Farley RD, Norman AW. Studies on the mechanism of action of calciferol. Exp Cell Res 1972;74:359–66.
65. Goldstein DA, Horowitz RE, Petit S, Haldimann B, Massry SG. The duodenal mucosa in patients with renal failure: Response to 1,25(OH)$_2$D$_3$. Kidney Int 1981;19:324–31.
66. Omdahl J, May B. The 25-hydroxyvitamin D-24-hydroxylase. In: Feldman D, Glorieux FH, Pike JW, editors. Vitamin D. San Diego: Academic Press, 1997;69–85.
67. Hsu CH, Patel S, Buchsbaum BL. Calcitriol metabolism in patients with chronic renal failure. Am J Kidney Dis 1991;17:185–90.
68. Brandi L, Egfjord M, Olgaard K. Pharmacokinetics of 1,25(OH)$_2$D$_3$ and 1α(OH)D$_3$ in normal and uremic man. Nephrol Dial Transplant 2001;(in press).
69. Yamamoto M, Kawanobe Y, Takahashi H, et al. Vitamin D deficiency and renal calcium transport in the rat. J Clin Invest 1984;74:507–13.
70. Müller D, Hoenderop JGJ, van Os CH, Bindels RJM. The epithelial calcium channel, EcaC1: molecular details of a novel player in renal calcium handling. Nephrol Dial Transplant 2001;16:1329–35.
71. Silver J, Naveh-Many T, Mayer H, Schmelzer HJ, Popovtzer MM. Regulation by vitamin D metabolites of parathyroid hormone gene transcription in vivo in the rat. J Clin Invest 1986;78:1296–301.
72. Silver J, Naveh-Many T. Vitamin D and the parathyroid glands. In: Feldman D, Glorieux FH, Pike JW, editors. Vitamin D. San Diego: Academic Press, 1997;353–67.
73. Brown AJ, Zhong M, Finch J, Ritter C, McCracken R, Morrisey J, et al. Rat calcium sensing receptor is regulated by vitamin D but not by calcium. Am J Physiol 1996;270:F454–60.
74. Rogers KV, Dunn CK, Conklin RL, Hadfield S, Petty BA, Brown EA, et al. Calcium receptor messenger ribonucleic acid levels in the parathyroid glands and kidney of vitamin D-deficient rats are not regulated by plasma calcium or 1,25-dihydroxyvitamin D_3. Endocrinol 1995;136:499–504.
75. DeLuca HF. The vitamin D story: A collaborative effort of basic science and clinical medicine. Fed Proc Am Soc Exp Biol 1988;2:224–36.
76. Owen TA, Aronow MS, Barone LM, et al. Pleiotropic effects of vitamin D on osteoblast gene expression are related to the proliferative and differentiated state of the bone cell phenotype: Dependency upon basal levels of gene expression, duration of exposure, and bone matrix competency in normal rat osteoblast cultures. Endocrinol 1991;128:1496–504.
77. Matsumoto T, Igarashi C, Taksuchi Y, et al. Stimulation by 1,25-dihydroxyvitamin D_3 of in vitro mineralization induced by osteoblast-like MC3T3-E1 cells. Bone 1991;12:27–32.

78. Balsan S, Garabedian M, Larchet M, Groski AM, Cournot G, Tau C, et al. Long-term nocturnal calcium infusions can cure rickets and promote normal mineralization in hereditary resistance to 1,25-dihydroxyvitamin D. J Clin Invest 1986;77:1661-7.
79. Moscovici A, Gafter U, Popovtzer MM. Calcitriol pulse therapy arrests bone formation in dialysis patients with refractory secondary hyperparathyroidism (abstract). J Am Soc Nephrol 1994;5:854.
80. Goodman WG, Ramirez JA, Belin TR, Chon Y, Gales B, Segre GV, et al. Development of adynamic bone in patients with secondary hyperparathyroidism after intermittent calcitriol therapy. Kidney Int 1994;46:1160-6.
81. Kuizon BD, Goodman WG, Jüppner H, Boechat I, Nelson P, Gales B, et al. Diminished linear growth during intermittent calcitriol therapy in children undergoing CCPD. Kidney Int 1998;53:205-11.
82. Andress DL, Norris KC, Coburn JW, Slatopolsky EA, Sherrard DJ. Intravenous calcitriol in the treatment of refractory osteitis fibrosa of chronic renal failure. N Engl J Med 1989;321:274-9.
83. Li YC, Amling M, Pirro AE, et al. Targeted ablation of the vitamin D receptor: An animal model of vitamin D-dependent rickets type II with alopecia. Proc Nat Acad Sci (USA) 1997;94:9831-5.
84. Bikle DD, Pillai S. Vitamin D, calcium and epidermal differentiation. Physiol Rev 1993;14:3-19.
85. Segaert S, Garmyn M, Degreef H, Bouillon R. Retinoic acid modulates the antiproliferative effect of 1,25-dihydroxyvitamin D_3 in cultured human epidermal keratinocytes. J Invest Dermatol 1997; 109:46-54.
86. van Leeuwen JPTM, Pols HAP. Vitamin D: Anticancer and differentiation. In: Feldman D, Glorieux FH, Pike JW, editors. Vitamin D. San Diego: Academic Press, 1997;1089-105.
87. Zhao X-Y, Feldman D. Antiproliferative mechanisms of $1\alpha,25(OH)_2D_3$ in human prostate cancer cells. In: Norman AW, Bouillon R, Thomasset M, editors. Vitamin D endocrine system: Structural, biological, genetic and clinical aspects. Riverside: Printing and Reprographics, University of California, 2000;489-93.
88. Schwartz GG. Prostate cancer and vitamin D: From concept to clinic. A ten-year update. In: Norman AW, Bouillon R, Thomasset M, editors. Vitamin D endocrine system: Structural, biological, genetic and clinical aspects. Riverside: Printing and Reprographics, University of California, 2000;445-52.
89. Cross HS, Hofer H, Bareis P, Bises G, Posner GH, Peterlik M. Vitamin D compounds and colorectal cancer: A rationale for their use in prevention and therapy. In: Norman AW, Bouillon R, Thomasset M, editors. Vitamin D endocrine system: Structural, biological, genetic and clinical aspects. Riverside: Printing and Reprographics, University of California, 2000;495-502.
90. Colston KW, Berger U, Coombes RC. Possible role for vitamin D in controlling breast cancer cell proliferation. Lancet 1989;1:185-91.
91. Casteels K, Bouillon R, Waer M, Mathieu C. Immunomodulatory effects of 1,25-dihydroxyvitamin D_3. Curr Opin Nephrol Hypertens 1995;4:313-18.
92. Lemire J. The role of vitamin D_3 in immunosuppression: Lessons from autoimmunity and transplantation. In: Feldman D, Glorieux FH, Pike JW, editors. Vitamin D. San Diego: Academic Press, 1997;1167-81.
93. Mathieu C, van Etten E, Kato S, Verstuyf A, Laureys J, Dopovere J, et al. In vitro and in vivo analysis of the immune system of the VDR-KO mice. In: Norman AW, Bouillon R, Thomasset M, editors. Vitamin D endocrine system: Structural, biological, genetic and clinical aspects. Riverside: Printing and Reprographics, University of California, 2000;555-9.
94. Nemere I, Schwartz Z, Pedrozo H, et al. Identification of a membrane receptor for 1,25-dihydroxyvitamin D_3 which mediates rapid activation of protein kinase C. J Bone Miner Res 1998;13:1353-9.
95. Norman AW. Receptors for $1\alpha,25(OH)_2D_3$: Past, present, and future. J Bone Miner Res 1998;13:1360-9.
96. Norman AW, Bouillon R, Farach-Carson MC, et al. Demonstration that $1\beta,25$-dihydroxyvitamin D_3 is an antagonist of the nongenomic but not genomic biological responses and biological profile of the three A-ring diastereomers of $1\alpha,25$-dihydroxyvitamin D_3. J Biol Chem 1993;268:20022-30.
97. Song X, Bishop JE, Okamura WH, et al. Stimulation of phosphorylation of mitogen-activated protein kinase by $1\alpha,25$-dihydroxyvitamin D_3 in promyelocytic NB4 leukemia cells. Endocrinol 1998;139:457-68.
98. Cleve H, Constants J. The mutants of the vitamin-D-binding protein: More than 120 variants of the Gc\DBP system. Vox Sang 1988;54:215-25.
99. Cooke NE, Haddad JG. Vitamin D binding protein. In: Feldman D, Glorieux FH, Pike JW, editors. Vitamin D. San Diego: Academic Press, 1997;87-101.
100. Bouillon R, Van Assche FA, Van Baelen H, et al. Influence of the vitamin D-binding protein on the serum concentration of 1,25-dihydroxyvitamin D_3. J Clin Invest 1981;67:589-96.
101. Viccio D, Yergey A, Obrien K, et al. Quantitation and kinetics of 25-hydroxyvitamin D_3 by isotope dilution liquid chromatography/thermospray mass spectrometry. Biol Mass Spectrom 1993;22:53-8.
102. Christensen EI, Willnow TE. Essential role of megalin in renal proximal tubule for vitamin homeostasis. J Am Soc Nephrol 1999;10:2224-36.

103. Nykjaer A, Vorum H, Dragun D, Walther D, Jacobsen C, Melsen F, et al. Megalin, a member of the LDL receptor family, is essential for vitamin D homeostasis and bone formation. Cell 1999;96:507–15.
104. Brenza HL, DeLuca HF. Regulation of 25-hydroxyvitamin D₃ 1alpha-hydroxylase gene expression by parathyroid hormone and 1,25-dihydroxyvitamin D₃. Arch Biochem Biophys 2000;381:143–52.
105. Lee WM, Galbraith RM. The extracellular actin-scavenger system and actin toxicity. N Engl J Med 1992;326:1335–41.
106. Bouillon R, Okamura WH, Norman AW. Structure–function relationships in the vitamin D endocrine system. Endocr Rev 1995;16:200–57.
107. Hollis BW. Detection of vitamin D and its major metabolites. In: Feldman D, Glorieux FH, Pike JW, editors. Vitamin D. San Diego: Academic Press, 1977;587–606.
108. Bouillon R. Radiochemical assays for vitamin D metabolites: Technical possibilities and clinical applications. J Steroid Biochem 1983;19:921–7.
109. Hollis BW. Detection of vitamin D and its major metabolites. In: Feldman D, Glorieux FH, Pike JW, editors. Vitamin D. San Diego: Academic Press, 1997;587–606.
110. Papapoulos SE, Fraher LJ, Clemens TL, Gleed J, O'Riordan JLH. Metabolites of vitamin D in human vitamin-D deficiency: Effect of vitamin D₃ or 1,25-dihydroxy-cholecalciferol. Lancet 1980;2: 612–15.
111. Parfitt AM. Vitamin D and the pathogenesis of rickets and osteomalacia. In: Feldman D, Glorieux FH, Pike JW, editors. Vitamin D. San Diego: Academic Press, 1997;645–62.
112. Miller WL, Portale AA. Genetic disorders of vitamin D biosynthesis. Endocrinol Metab Clin North Am 1999;28:825–40.
113. Liberman UA, Marx SJ. Vitamin D-dependent rickets. In: Favus MJ, editor. Primer on the metabolic bone diseases and disorders of mineral metabolism. Philadelphia: Lippincott Williams & Wilkins, 1999;323–8.
114. Chesney RW, Hamstra AJ, Mazess RB, Rose P, DeLuca HF. Circulating vitamin D metabolite concentrations in childhood renal diseases. Kidney Int 1982;21:65–9.
115. Bouillon R, Auwerx JD, Lissens WD, Pelemans WK. Vitamin D status in the elderly, seasonal substrate deficiency causes 1,25-dihydroxycholecalciferol deficiency. Am J Clin Nutr 1987;45:755–63.
116. Ooms ME, Roos JC, Bezemer PD, Van Der Vijch WJF, Bouter LM, Lips P. Prevention of bone loss by vitamin D supplementation in elderly women: A randomized double blind trial. J Clin Endocrinol Metab 1995;80:1052–8.
117. Gloth FM. Vitamin D deficiency in homebound elderly persons. JAMA 1995;274:1683–6.
118. Schmidt-Gayk H, Bouillon R, Roth HJ. Measurement of vitamin D and its metabolites (calcidiol and calcitriol) and their clinical significance. Scand J Clin Lab Invest 1997;57:35–45.
119. Dawson-Hughes B, Harris SS, Dallal GE. Plasma calcidol, season, and serum parathyroid hormone concentrations in healthy elderly men and women. Am J Clin Nutr 1997;65:67–71.
120. Krall EA, Sahyoun N, Tannenbaum S, Dallal GE, Dawson-Hughes B. Effect of vitamin D intake on seasonal variations in parathyroid hormone secretion in postmenopausal women. N Engl J Med 1989;321:1777–83.
121. Peacock M, Selby PL, Francis RM, Brown WB, Hordon L. Vitamin D deficiency, insufficiency, sufficiency, and intoxication. What do they mean? In: Norman AW, Schaefer K, Grigoleit H-G, v.Herrath D, editors. Vitamin D: Chemical, biochemical, and clinical update. Berlin: de Gruyter, 1985;569–70.
122. Chapuy MC, Meunier PJ. Vitamin D insufficiency in adults and the elderly. In: Feldman D, Glorieux FH, Pike JW, editors. Vitamin D. San Diego: Academic Press, 1997;679–93.
123. van der Wielen RPJ, Löwik MRH, van den Berg H, de Groot LCPGM, Haller J, Moreiras O, et al. Serum vitamin D concentrations among elderly people in Europe. Lancet 1995;346:207–10.
124. Webb AR, Pilbeam C, Hanafin N, Holick MF. An evaluation of the relative contributions of exposure to sunlight and of diet to the circulating concentrations of 25-hydroxyvitamin D in an elderly nursing home population in Boston. N Engl J Med 1989;321:1777–83.
125. Kinyamu HK, Gallagher JC, Balhorn KE, Petranick KM, Rafferty KA. Serum vitamin D metabolites and calcium absorption in normal young and elderly free-living women and in women living in nursing homes. Am J Clin Nutr 1997;65:790–7.
126. Glerup H, Mikkelsen K, Poulsen L, Haas E, Overbeck S, Andersen H, et al. Hypovitaminosis D myopathy without biochemical signs of osteomalacic bone involvement. Calcif Tissue Int 2000;66:419–24.
127. Glerup H, Eriksen EF. Acroparesthesia – a typical finding in vitamin D deficiency. Br J Rheumatol 1999;39:482.
128. Lips P, van Ginkel FC, Jongen MJM, Rubertus F, Van der Vijgh WJF, Netelenbos JC. Determinants of vitamin D status in patients with hip fracture and in elderly control subjects. Am J Clin Nutr 1987;46:1005–10.
129. Leboff MS, Kohlmeier L, Hurwitz S, Franklin J, Wright J, Glowacki J. Occult vitamin D deficiency in postmenopausal US women with acute hip fracture. JAMA 1999;281:1505–11.

130. Chapuy MC, Arlot ME, Duboeuf F, Brun J, Crouzet B, Arnaud S, et al. Vitamin D₃ and calcium to prevent hip fractures in elderly women. N Engl J Med 1992;327:1637–42.
131. Chapuy MC, Arlot ME, Delmas PD, Meunier PJ. Effect of calcium and cholecalciferol treatment for three years on hip fractures in elderly women. Br Med J 1994;308:1081–2.
132. Khaw KT, Sneyd MJ, Compston J. Bone density, parathyroid hormone and 25-hydroxyvitamin D concentrations in middle aged women. Br Med J 1992;305:373–6.
133. Dawson-Hughes B, Dallal GE, Krall EA, Harris S, Sokoll LJ, Falconer G. Effect of vitamin D supplementation on wintertime and overall bone loss in healthy postmenopausal women. Ann Int Med 1991;115:505–12.
134. Thomas MK, Lloyd-Jones DM, Thadhani RI, Shaw AC, Deraska DJ, Kitch BT, et al. Hypovitaminosis D in medical inpatients. N Engl J Med 1998;338:777–783.
135. McKenna MJ. Differences in vitamin D status between countries in young adults and the elderly. Am J Med 1992;93:69–77.
136. Pettifor JM, Daniels ED. Vitamin D deficiency and nutritional rickets in children. In: Feldman D, Glorieux FH, Pike JW, editors. Vitamin D. San Diego: Academic Press, 1997;663–78.
137. Edidin DV, Levitsky LL, Shey W, Dumbuvic N, Campos A. Resurgence of nutritional rickets associated with breast-feeding and special dietary practices. Pediatrics 1980;65:232–5.
138. Dwyer JT, Dietz WH, Hass G, Suskind R. Risk of nutritional rickets among vegetarian children. Am J Dis Child 1979;133:134–40.
139. Rudolph M, Arulanantham K, Greenstein RM. Unsuspected nutritional rickets. Pediatrics 1980;66: 72–6.
140. Harris SS, Dawson-Hughes B. Seasonal changes in plasma 25-hydroxyvitamin D concentrations of young American black and white women. Am J Clin Nutr 1998;67:1232–6.
141. Zeghoud F, Vervel C, Guillozo H, Walrant-Debray O, Boutignon H, Garabédian M. Subclinical vitamin D deficiency in neonates: definition and response to vitamin D supplements. Am J Clin Nutr 1997; 65:771–8.
142. Zeghoud F, Ben-Mekhbi H, Garabedian M. Prevention of maternal vitamin D deficiency (one oral dose of 2.5 mg vitamin D₃ at the 6th and 8th month of pregnancy): Effects on neonatal calcium homeostasis and infantile growth. In: Norman AW, Bouillon R, Thomasset M, editors. Vitamin D endocrine system: Structural, biological, genetic and clinical aspects. Riverside: University of California Printing and Reprographics, 2000;839–42.
143. Dawson-Hughes B, Harris SS. Definition of optimal 25(OH)D status for bone. In: Norman AW, Bouillon R, Thomasset M, editors. Vitamin D endocrine system: Structural, biological, genetic and clinical aspects. Riverside: University of California Printing and Reprographics, 2000;909–15.
144. Shah BR, Finberg L. Single day therapy for nutritional vitamin D-deficiency rickets: A preferred method. J Pediatrics 1994;125:487–90.
145. Lubani MM, Al-Shab TS, Al-Saleh QA, Sharda DC, Quattawi SA, Ahmed SAH, et al. Vitamin D deficiency in Kuwait: The prevalence of a preventable disease. Ann Trop Paediatr 1989;3:134–9.
146. Kanis JA, Russell RGG. Rate of reversal of hypercalcaemia and hypercalciuria induced by vitamin D and its 1-alpha hydroxylated derivatives. Br Med J 1977;1:78–81.
147. Reid IR. Steroid osteoporosis. Calcif Tissue Int 1989;45:63–7.
148. Ringe JD. Active vitamin D metabolites in glucocorticoid-induced osteoporosis. Calcif Tissue Int 1997;60:124–7.
149. Eastell R, Reid DM, Compston J, Cooper C, Fogelman I, Francis RM, et al. A UK consensus group on the management of glucocorticoid-induced osteoporosis: An update. J Intern Med 1998;124:400–6.
150. Reid IR. Editorial: Glucocorticoid effects on bone. J Clin Endocrinol Metab 1998;83:1860–2.
151. Tilyard MW, Spears GFS, Thomson J, Dovey S. Treatment of postmenopausal osteoporosis with calcitriol or calcium. N Engl J Med 1992;326:357–62.
152. Orimo H, Shiraki M, Hayashi T, Nakamura T. Reduced occurrence of vertebral crush fractures in senile osteoporosis treated with 1α(OH)-vitamin D₃. Bone Miner 1987;3:47–52.
153. Lips P. Vitamin D deficiency and osteoporosis: The role of vitamin D deficiency and treatment with vitamin D an analogues in the prevention of osteoporosis-related fractures. Eur J Clin Invest 1996;26:436–42.
154. Nakamura T. Vitamin D for the treatment of osteoporosis. Osteoporosis Int 1997;7:S155–8.
155. Winters RW, Graham JB, WIlliams TF, McFalls VW, Burnett CH. A genetic study of familial hypophosphatemia and vitamin D-resistant rickets with a review of the literature. Medicine (Balt) 1958; 37:141–142.
156. Thakker RV, Read AP, Davies KE, Whyte WP, Weksberg R, Glorieux FH, et al. Bridging markers defining the map position of X-linked hypophosphatemic rickets. J Med Genet 1987;24:756–60.
157. Drezner MK, Lyles KW, Haussler MR, Harrelson JM. Evaluation of a role for 1,25-dihydroxyvitamin D in the pathogenesis and treatment of X-linked hypophosphatemic rickets and osteomalacia. J Clin Invest 1980;66:1020–32.

158. Harrell RM, Lyles KW, Harrelson JM, Friedman NE, Drezner MK. Healing of bone disease in X-linked hypophosphatemic rickets/osteomalacia. Induction and maintenance with phosphorus and calcitriol. J Clin Invest 1985;75:1858–68.
159. Bettinelli A, Bianchi ML, Mazazucchi E, Gandolini G, Appliani AC. Acute effects of calcitriol and phosphate salts on mineral metabolism in children with hypophosphatemic rickets. J Pediatrics 1991; 118:372–6.
160. Goodyer PR, Kronick JG, Jequier S, Reade TM, Scriver CR. Nephrocalcinosis and its relationship to treatment of hereditary rickets. J Pediatrics 1987;111:700–4.
161. Alon U, Donaldson DL, Hellerstein S, Warady BA, Harris DJ. Metabolic and histologic investigation of the nature of nephrocalcinosis in children with hypophosphatemic rickets and in the Hyp mouse. J Pediatrics 1992;120:899–905.
162. Drezner MK, Feinglos MN. Osteomalacia due to 1,25-dihydroxy-cholecalciferol deficiency. Association with a giant cell tumor of bone. J Clin Invest 1977;60:1046–53.
163. Lobaugh B, Burch WM, Jr., Drezner MK. Abnormalities of vitamin D metabolism and action in the vitamin D-resistant rachitic and osteomalacic diseases. In: Kumar R, editor. Vitamin D. Boston: Martinus Nijhoff, 1984;665–720.
164. Leicht E, Biro G, Langer H-J. Tumor induced osteomalacia: Pre- and post-operative biochemical findings. Horm Metab Res 1990;22:640–3.
165. Brickman AS, Coburn JW, Massry SG, Norman AW. 1,25-dihydroxyvitamin D_3 in normal man and patients with renal failure. Ann Int Med 1974;80:161–8.
166. Brickman AS, Coburn JW, Norman AW. Action of 1,25-dihydroxycholecalciferol, a potent kidney-produced metabolite of vitamin D_3 in uremic man. N Engl J Med 1972;287:891–5.
167. Brickman AS, Hartenbower DL, Norman AW, Coburn JW. Actions of 1α-hydroxy- and 1,25-dihydroxyvitamin D_3 on mineral metabolism in man. I. Effects on net absorption of phosphorus. Am J Clin Nutr 1977;30:1064–70.
168. Slatopolsky E, Weerts C, Thielan J, Horst RL, Harter H, Martin KJ. Marked suppression of secondary hyperparathyroidism by intravenous administration of 1,25-dihydroxycholecalciferol in uremic patients. J Clin Invest 1984;74:2136–43.
169. Dunlay R, Rodriguez M, Felsenfeld AJ, Llach F. Direct inhibitory effect of calcitriol on parathyroid function (sigmoidal curve) in dialysis. Kidney Int 1989;36:1093–8.
170. Coburn JW, DiDomenico NC, Bryce GF, Bassett LW, Shupien SA, Wong EGC, et al. Prospective, double-blind trial with calcitriol in the prophylaxis of bone disease in asymptomatic dialysis patients. In: Norman AW, Schaefer K, Grigoleit H-G, v.Herrath D, editors. Vitamin D: Chemical, biochemical, and clinical endocrinology of calcium metabolism. Berlin: deGruyter, 1982;833–4.
171. Baker LR, Muir JW, Sharman VL, Abrams SM, Greenwood RN, Cattell WR, et al. Controlled trial of calcitriol in hemodialysis patients. Clin Nephrol 1986;26:185–91.
172. Memmos DE, Eastwood JB, Talner LB, Gower PE, Curtis JR, Phillips ME, et al. Double-blind trial of oral 1,25-dihydroxy vitamin D_3 versus placebo in asymptomatic hyperparathyroidism in patients receiving maintenance haemodialysis. Br Med J 1981;282:1919–24.
173. Van der Merwe WM, Rodger RSC, Grant AC, Logue FC, Cowan RA, Beastall GH, et al. Low calcium dialysate and high-dose oral calcitriol in the treatment of secondary hyperparathyroidism in haemodialysis patients. Nephrol Dial Transplant 1990;5:874–7.
174. Fukagawa M, Kitaoka M, Yi H, Fukuda N, Matsumoto T, Ogata E, et al. Serial evaluation of parathyroid size by ultrasonography is another useful marker for the long-term prognosis of calcitriol pulse therapy in chronic dialysis patients. Nephron 1994;68:221–8.
175. Fukagawa M, Orazaki R, Takano K, Kaname S-Y, Ogata E, Kitoaka M, et al. Regression of parathyroid hyperplasia by calcitriol-pulse therapy in patients on long-term dialysis. N Engl J Med 1990;323: 421–2.
176. Tsukamoto Y, Nomura M, Takahashi Y, Takagi Y, Yoshida A, Nagaoka T, et al. The "oral 1,25-dihydroxyvitamin D_3 pulse therapy" in hemodialysis patients with severe secondary hyperparathyroidism. Nephron 1991;57:23–8.
177. Muramoto H, Haruki K, Yoshimura A, Mimo N, Oda K, Tofuku Y. Treatment of refractory hyperparathyroidism in patients on hemodialysis by intermittent oral administration of 1,25(OH)₂D₃. Nephron 1991;58:288–94.
178. Perez-Mijares R, Gomez-Fernandez P, Almaraz-Jimenez M, Ramos-Diaz M, Rivero-Bohorquez J. Treatment of severe secondary hyperparathyroidism with administration of calcium carbonate, intermittent high oral doses of 1,25-dihydroxyvitamin D_3 and dialysate with 3 mEq/1 calcium concentration. Am J Nephrol 1993;13:149–54.
179. Martin KJ, Ballal HS, Domoto DT, Blalock S, Weindel M. Pulse oral calcitriol for the treatment of hyperparathyroidism in patients on continuous ambulatory peritoneal dialysis: Preliminary observations. Am J Kidney Dis 1992;19:540–5.

180. Cannella G, Bonucci E, Rolla D, Ballanti P, Moriero E, De Grandi R, et al. Evidence of healing of secondary hyperparathyroidism in chronically hemodialyzed uremic patients treated with long-term intravenous calcitriol. Kidney Int 1994;46:1124–32.
181. Llach F, Hervas J, Cerezo S. The importance of dosing intravenous calcitriol in dialysis patients with severe hyperparathyroidism. Am J Kidney Dis 1995;26:845–51.
182. Malberti F, Surian M, Cosci P. Effect of chronic intravenous calcitriol on parathyroid function and set point of calcium in dialysis patients with refractory secondary hyperparathyroidism. Nephrol Dial Transplant 1992;7:822–8.
183. Herrmann P, Ritz E, Schmidt-Gayk H, Schäfer I, Geyer J, Nonnast-Daniel B, et al. Comparison of intermittent and continuous oral administration of calcitriol in dialysis patients: A randomized prospective trial. Nephron 1994;67:48–53.
184. Bechtel U, Mücke C, Feucht HE, Schiffl H, Sitter T, Held E. Limitations of pulse oral calcitriol therapy in continuous ambulatory peritoneal dialysis patients. Am J Kidney Dis 1995;25:291–6.
185. Salusky IB, Goodman WG, Horst R, Segre GV, Kim L, Norris KC, et al. Pharmacokinetics of calcitriol in CAPD/CCPD patients. Am J Kidney Dis 1990;16:126–32.
186. Ardissino G, Schmitt CP, Bianchi ML, Dacco V, Claris-Appiani A, Mehls O. European study group on vitamin D in children with renal failure. No difference in intestinal strontium absorption after oral or IV calcitriol in children with secondary hyperparathyroidism. Kidney Int 2000;58: 981–8.
187. Fischer ER, Harris DCH. Comparison of intermittent oral and intravenous calcitriol in hemodialysis patients with secondary hyperparathyroidism. Clin Nephrol 1993;40:216–20.
188. Quarles LD, Yohay DA, Carroll BA, Spritzer CE, Minda SA, Bartholomay D, et al. Prospective trial of pulse oral versus intravenous calcitriol treatment of hyperparathyroidism in ESRD. Kidney Int 1994;45:1710–21.
189. Levine BS, Song MM. Pharmacokinetics and efficacy of pulse oral versus intravenous calcitriol in hemodialysis patients. J Am Soc Nephrol 1996;7:488–96.
190. Faugere M-C, Friedler RM, Malluche HH. Efficacy and limitations of pulse IV and pulse oral 1,25 vit. D therapy in treatment of hyperparathyroidism in patients on long-term dialysis (abstract). J Am Soc Nephrol 1993;4:695.
191. Salusky IB, Ramirez JA, Belin T, Segre GV, Goodman WG. Pulse calcitriol therapy: A prospective randomized trial (abstract). J Am Soc Nephrol 1994;5:855.
192. Lind L, Wengle B, Wide L, Wrege U, Ljunghall S. Suppression of serum parathyroid hormone levels by intravenous alfacalcidol in uremic patients on maintenance hemodialysis: A pilot study. Nephron 1988;48:296–9.
193. Brandi L, Daugaard H, Tvedegaard E, Storm T, Olgaard K. Effect of intravenous 1-alpha-hydroxyvitamin D_3 on secondary hyperparathyroidism in chronic uremic patients on maintenance hemodialysis. Nephron 1989;53:194–200.
194. Moriniere P, Esper NE, Viron B, Judith D, Bourgeon B, Farquet C, et al. Improvement of severe secondary hyperparathyroidism in dialysis patients by intravenous $1\alpha(OH)$ vitamin D_3, oral $CaCO_3$ and low dialysate calcium. Kidney Int 1993;43 (Suppl 41):S121–4.
195. Emmett M, Sirmon MD, Kirkpatrick WG, Nolan CR, Schmitt GW, Cleveland MvB. Calcium acetate control of serum phosphorus in hemodialysis patients. Am J Kidney Dis 1991;17:544–50.
196. Sperschneider H, Gunther K, Marzoll I, Kirchner E, Stein G. Calcium carbonate ($CaCO_3$): an efficient and safe phosphate binder in haemodialysis patients? A 3-year study. Nephrol Dial Transplant 1993;8:530–4.
197. Ribeiro S, Ramos A, Brandao A, Rebelo JR, Guerra A, Resina C, et al. Cardiac valve calcification in haemodialysis patients: role of calcium-phosphate metabolism. Nephrol Dial Transplant 1998;13: 2037–40.
198. Block GA, Hulbert-Shearon TE, Levin NW, Port FK. Association of serum phosphorus and calcium X phosphate product with mortality risk in chronic hemodialysis patients: A national study. Am J Kidney Dis 1998;31:607–17.
199. Block GA, Port FK. Re-evaluation of risks associated with hyperphosphatemia and hyperparathyroidism in dialysis patients: Recommendations for a change in management. Am J Kidney Dis 2000;35:1226–37.
200. Amann K, Gross M-L, London GM, Ritz E. Hyperphosphatemia – a silent killer of patients with renal failure? Nephrol Dial Transplant 1999;14:2085–7.
201. Guerin AP, London GM, Marchais SJ, Metivier F. Arterial stiffening and vascular calcifications in end-stage renal disease. Nephrol Dial Transplant 2000;15:1014–21.
202. Goodman WG, Goldin J, Kuizon BD, Yoon C, Gales B, Sider D, et al. Coronary artery calcification in young adults with end-stage renal disease who are undergoing dialysis. N Engl J Med 2000; 342:1478–83.

203. Akiba T, Marumo F, Owada A, Kurihara S, Inoue A, Chida Y, et al. Controlled trial of falecalcitriol versus alfacalcidol in suppression of parathyroid hormone in hemodialysis patients with secondary hyperparathyroidism. Am J Kidney Dis 1998;32:238.

204. Abe J, Takita Y, Nakano T, Miyaura C, Suda T, Nishii Y. A synthetic analogue of vitamin D_3, 22-oxa-1Alpha,25-dihydroxyvitamin D_3, is a potent modulator of in vivo immunoregulating activity without inducing hypercalcaemia in mice. Endocrinol 1989;124:2645–7.

205. Brown AJ, Ritter CS, Finch JL, Morrissey J, Martin KJ, Murayama E, et al. The noncalcaemic analogue of vitamin D, 22-oxacalcitriol, suppresses parathyroid hormone synthesis and secretion. J Clin Invest 1989;84:728–32.

206. Brown AJ, Finch JL, Lopez-Hilker S, Dusso A, Ritter C, Pernalete N, et al. New active analogues of vitamin D with low calcaemic activity. Kidney Int 1990;29 (Suppl.29):S22–7.

207. Martin KJ, González EA, Gellens M, Hamm LL, Abboud H, Lindberg J. 19-Nor-1-α-25-Dihydroxyvitamin D_2 (Paricalcitol) safely and effectively reduces the levels of intact parathyroid hormone in patients on hemodialysis. J Am Soc Nephrol 1998;9:1427–32.

208. Tan Jr AU, Levine BS, Mazess RB, Kyllo DM, Bishop CW, Knutson JC, et al. Effective suppression of parathyroid hormone by 1α-hydroxyvitaminD$_2$ in hemodialysis patients with moderate to severe secondary hyperparathyroidism. Kidney Int 1997;51:317–23.

209. Frazão JM, Levine BS, Tan Jr AU, Mazess RB, Kyllo DM, Knutson JC, et al. Efficacy and safety of intermittent oral 1α (OH)-vitamin D_2 in suppressing 2° hyperparathyroidism in hemodialysis patients. Dial Transplant 1997;26:583–95.

210. Frazão JM, Elangovan L, Maung HM, Chesney RB, Acchiardo SR, Bower JD, et al. Intermittent doxercalciferol (1α-hydroxyvitamin D_2) therapy for secondary hyperparathyroidism: Results of a modified, double blinded, controlled study. Am J Kidney Dis 2000;36:550–61.

211. Maung HM, Elangovan L, Frazão JM, Bower JD, Kelley BJ, Acchiardo SR, et al. Efficacy and side-effects of intermittent intravenous and oral doxercalciferol (1α-hydroxyvitamin D_2) in dialysis patients with secondary hyperparathyroidism: A sequential comparison. Am J Kidney Dis 2001;37:532–43.

212. Dusso A, Gunawardhana S, Negrea L, Finch JL, Lopez-Hilker S, Mori T, et al. On the mechanisms for the selective action of vitamin D analogs. Endocrinol 1991;128:1687–92.

213. Finch JL, Brown AJ, Kubodera N, Nishii Y, Slatopolsky E. Differential effects of 1,25-(OH)$_2$D$_3$ and 22-oxacalcitriol on phosphate and calcium metabolism. Kidney Int 1993;43:561–6.

214. Akizawa T, Kurokawa K, Suzuki M, Akiba T, Nishizawa Y, Ohashi Y, et al. Suppression of PTH by 22-oxacalcitriol (OCT). A placebo controlled study in hemodialysis (HD) patients with secondary hyperparathyroidism (2 HPT) (abstract). J Am Soc Nephrol 1997;8:570A.

215. Akizawa T, Kurokawa K, Suzuki M, Akiba T, Nishizawa Y, Ohashi Y, et al. Suppressive effect of 22-oxacalcitriol (OCT) on secondary hyperparathyroidism (2 HPT) of hemodialysis (HD) patients. A double-blind comparison among four doses (abstract). J Am Soc Nephrol 1997;7:1810.

216. Morii H, Ogura Y, Koshikawa S, Mimura N, Suzuki M, Kurokawa K, et al. Efficacy and safety of oral falecalcitriol in reducing parathyroid hormone in hemodialysis patients with secondary hyperparathyroidism. J Bone Miner Metab 1998;16:34–43.

217. Coburn JW, Elangovan L. Prevention of metabolic bone disease in the pre-end-stage renal disease setting. J Am Soc Nephrol 1998;9:S71–7.

218. Rix M, Andreassen H, Eskildsen P, Langdahl B, Olgaard K. Bone mineral density and biochemical markers of bone turnover in patients with predialysis chronic renal failure. Kidney Int 1999;56: 1084–93.

219. Massry SG. Assessment of 1,25(OH$_2$D$_3$ in the correction and prevention of renal osteodystrophy in patients with mild to moderate renal failure. In: Norman AW, Schaefer K, Grigoleit H-G, v.Herrath D, editors. Vitamin D: A chemical, biochemical and clinical update. Berlin: de Gruyter, 1985;935–7.

220. Nordal KP, Dahl E. Low-dose calcitriol versus placebo in patients with predialysis chronic renal failure. J Clin Endocrinol Metab 1988;67:929–36.

221. Baker LRI, Abrams SML, Roe CJ, Faugere M-C, Fanti P, Subayti Y, et al. 1,25(OH)$_2$D$_3$ administration in moderate renal failure: A prospective double-blind trial. Kidney Int 1989;35:661–9.

222. Hamdy NAT, Kanis JA, Beneton MNC, Brown CB, Juttmann JR, Jordans JGM, et al. Effect of alfacalcidol on natural course of bone disease in mild to moderate renal failure. Br Med J 1995;310:358–63.

223. Przedlacki J, Manelius J, Huttunen K. Bone mineral density evaluated by dual-energy X-ray absorptiometry after one-year treatment with calcitriol started in the predialysis phase of chronic renal failure. Nephron 1995;69:433–7.

224. Ritz E, Küster S, Schmidt-Gayk H, Stein G, Scholz C, Kraatz G, et al. Low-dose calcitriol prevents the rise in 1,84 iPTH without affecting serum calcium and phosphate in patients with moderate renal failure (prospective placebo-controlled multicentre trial). Nephrol Dial Transplant 1995;10: 2228–34.

225. Bertoli M, Luisetto G, Ruffatti A, Urso M, Romagnoli G. Renal function during calcitriol therapy in chronic renal failure. Clin Nephrol 1990;33:98–102.
226. Perez A, Raab R, Chen TC, Turner AK, Holick MF. Safety and efficacy of oral calcitriol (1,25-dihydroxyvitamin D_3) for the treatment of psoriasis. Br J Derm 1996;134:1070–8.
227. Coen G, Mazzaferro S, Bonucci E, Ballanti P, Massimetti C, Donado G, et al. Treatment of secondary hyperparathyroidism of pre-dialysis chronic renal failure with low doses of $1,25(OH)_2D_3$: humoral and histomorphometric results. Miner Electrolyte Metab 1986;12:375–82.
228. Bianchi ML, Colantonio G, Campanini F, Rossi R, Valenti G, Ortolani S, et al. Calcitriol and calcium carbonate therapy in early chronic renal failure. Nephrol Dial Transplant 1994;9:1595–9.
229. Nordal KP, Dahl E, Halse J, Ajitamadal A, Flatmark A. Long-term low-dose calcitriol in predialysis chronic renal failure: Can it prevent hyperparathyroid bone disease? Nephrol Dial Transplant 1995;10:203–6.
230. Quarles LD, Lobaugh B, Murphy G. Intact parathyroid hormone overestimates the presence and severity of parathyroid-mediated osseous abnormalities in uremia. J Clin Endocrinol Metab 1992; 75:145–50.
231. Sherrard DJ, Hercz G, Pei Y, Maloney NA, Greenwood C, Manuel A, et al. The spectrum of bone disease in end-stage renal failure – An evolving disorder. Kidney Int 1993;43:435–6.
232. Scriver CR. Vitamin D dependency. Pediatrics 1970;45:361–3.
233. Coburn JW, Maung HM, Frazão JM, Elangovan L. New vitamin D analogs and calcimimetics. In: Drueke TB, Salusky IB, editors. Renal osteodystrophy. Oxford: Oxford University Press, 2002.
234. Brooks MG, Bell NH, Love L, Stern PH, Orfei E, Queener SF, et al. Vitamin D-dependency rickets type II. Resistance of target organs to 1,25-dihydroxyvitamin D. N Engl J Med 1978;298:996–9.
235. Weisman Y, Hochberg Z. Genetic rickets and osteomalacia. Curr Ther Endocrinol Metab 1994; 5:492–5.
236. Holick MF. Active vitamin D compounds and analogues: A new therapeutic era for dermatology in the 21st century. Mayo Clin Proc 1993;68:925–7.
237. Kragballe K, Gjertsein BT, De Hoop D, Karsmark T, van de Kerkhof PCM, Larkö O, et al. Double-blind, right–left comparison of calcipotriol and betamethasone valerate in treatment of psoriasis vulgaris. Lancet 1991;337:193–6.
238. Kragballe K. Psoriasis and other skin diseases. In: Feldman D, Glorieux FH, Pike JW, editors. Vitamin D. San Diego: Academic Press, 1997;1213–25.
239. Anning ST, Dawson J, Dolby DE, Ingram JT. The toxic effects of calciferol. Q J Med 1948;17:203–28.
240. Jacobus CH, Holick MF, Shao Q, Chen TC, Holm IA, Kolodny JM, et al. Hypervitaminosis D associated with drinking milk. N Engl J Med 1992;326:1173–7.
241. Koutkia P, Chen TC, Holick MF. Vitamin D intoxication associated with an over-the-counter supplement. N Engl J Med 2001;345:66–7.
242. Stanbury SW. The treatment of renal osteodystrophy. Ann Int Med 1966;65:1133–8.
243. Breslau NA, Zerwekh JE. Pharmacology of vitamin D preparations. In: Feldman D, Glorieux FH, Pike JW, editors. Vitamin D. San Diego: Academic Press, 1997;607–18.
244. Bell NH, Gill Jr JR, Barter FC. Abnormal calcium absorption in sarcoidosis: Evidence for increased sensitivity to vitamin D. Am J Med 1964;36:500–13.
245. Barbour GL, Coburn JW, Slatopolsky E, Norman AW, Horst RL. Hypercalcemia in an anephric patient with sarcoidosis, evidence for extrarenal generation of 1,25 dihydroxyvitamin D. N Engl J Med 1981;305:440–6.
246. Adams JS, Sharma OP, Diz MM, Endres DB. Ketoconazole decreases the serum 1,25-dihydroxyvitamin D and calcium concentration in sarcoidosis-associated hypercalcemia. J Clin Endocrinol Metab 1990;70:1090–5.
247. Bia MJ, Insogna K. Treatment of sarcoidosis-associated hypercalcemia with ketoconazole. Am J Kidney Dis 1991;18:702–5.
248. Coburn JW, Salusky IB. Renal bone diseases: clinical features, diagnosis and management. In: Bilezikian JP, Marcus R, Levine MA, editors. The Parathyroids. San Diego: Academic Press, 2001; 635–61.

Bone

Bone – Normal Structure and Physiology

M.-C. Monier-Faugere, B. P. Sawaya, M. C. Langub and H. H. Malluche

The macroscopic and microscopic characteristics of the organ "bone" are tightly related to its physiologic role in the body. The skeleton has a dual mechanical and metabolic function. The rigidity of the skeleton is responsible for maintenance of the human body configuration, protection of soft organs, and mobility through transmission of forces originated by muscle contraction. It is also the main reservoir for calcium, phosphate, magnesium, and bicarbonate; thus, the skeleton contributes to the minute-to-minute regulation of extracellular fluid and plays an important role in mineral homeostasis. There is a dynamic interaction between these two functions. For example, a prolonged demand for calcium will tax the mechanical strength of the skeleton by depleting its mineral content and affecting its hardness. Undermineralized bone is at risk of mechanical incompetence.

Normal Structure

The osseous framework comprises 206 distinct, differently shaped pieces, the organ bones. Each bone consists of bone tissue, haematopoietic marrow, and vasculature. Bones from the axial skeleton include the skull, spine, thorax, and pelvis, whereas bones from the extremities comprise the appendicular skeleton. Pathological processes and therapies can affect these skeletal sites in different ways. This is due to the relative amount of the two main structural types of bone tissues; that is, cortical and cancellous bone.

Cortical and Cancellous Bone

Cortical or compact bone can be distinguished macroscopically from cancellous or trabecular bone. Cortical bone is a dense tissue that contains less than 10% soft tissue. Cancellous or spongy bone is made up of trabecules shaped as plates or rods interspersed between bone marrow that represents more than 75% of the cancellous bone volume. Cortical bone forms the external layer of all bones but is found predominantly in the appendicular skeleton, particularly in diaphysis of long bones. Cancellous bone is found mainly in the axial skeleton, located between the cortices of smaller flat and short bones such as scapulae, vertebrae, and pelvis. It is also present in limited amounts in the juxta-articular extremities of appendicular skeleton. Cortical bone represents 80% of the skeletal mass and therefore supports most of the mechanical function [1]. Cancellous bone is only 20% of the skeletal mass but is metabolically four times more active per unit volume than cortical bone.

Thus, the metabolic function is equally distributed between cortical and cancellous bones [1].

Bone Envelopes and Bone Surfaces

At the microscopic level, each bone exhibits two envelopes. The periosteal envelope represents the outer layer of connective tissue that encloses both hard and soft tissue and separates bone from other organs. The inner or endosteal envelope surrounds all soft tissue within the bone (except osteocytes) and is the boundary between soft tissue, mainly bone marrow, and bone tissue. Within the endosteal envelope, three distinct but continuous surfaces are observed: the intracortical, including the Haversian and Volkmann canals, the endocortical, and the trabecular surfaces. Modelling and remodelling activities take place on bone surfaces. However, the levels of activity vary from one surface to another and can be affected differently by physiological or pathological events as well as by therapeutic agents.

The Bone Structural Unit (BSU)

Between the periosteal and endosteal envelopes, bone is further organised in several structural elements, the "bone structural units". The spatial arrangement of these smallest, individual units of bone and their cohesion are responsible for bone strength.

In cortical bone, BSU is represented by the osteon or Haversian system. Each osteon consists of a 200–250-µm wide cylinder running parallel to the long axis of cortical bone. The osteon centre is occupied by a 40–50-µm "Haversian" canal containing blood vessels, nerves, and connective tissue. The Haversian canals of adjacent osteons are linked transversally by Volkmann canals, creating an intracortical network, which is also connected with the periosteum and bone marrow. The central Haversian canal is surrounded by concentric layers of 20–30 osseous lamellae making the osteon wall approximately 70–100 µm thick. Osteons are densely packed together and separated only by interstitial lamellae, which are the remains of incompletely resorbed osteons.

In cancellous bone, BSUs are flat and can be envisioned as longitudinally cut and unfolded osteons. They appear as semilunar packets roughly parallel to the central axis of the trabecules. The trabecular surface corresponding to the open Haversian canal follows the shape of the trabecule and is in contact with bone marrow. Within trabecules, the BSUs are separated from each other by interstitial bone, which, as in cortical bone, represents the remainder of older incompletely resorbed packet.

Bone Matrix, Lamellar and Woven Bone

Each BSU consists of a specialised connective tissue; the osseous tissue made of an organic matrix upon which is deposited a mineralized protein matrix.

At the microscopic level, two different types of bone can be identified by the arrangement of collagen bundles.

In a normal mature skeleton, bone is of the lamellar type. In this bone, the orientation of collagen fibres alternates regularly from layer to layer. Each layer is approximately three µm thick with all collagen fibres deposited in the same direction. The deposited collagen exhibits an orderly lamellar pattern as circular layers alternate with longitudinal ones. The change in collagen fibre direction from layer to layer is responsible for the birefringence of bone under polarised light microscopy.

Contrasting with the regularity of lamellar bone, woven bone is composed of loose and randomly arranged collagen bundles. Woven bone is formed by irregular and unpolarised extrusion of protocollagen by osteoblasts. This matrix consists of an unordered, crisscross texture that lacks the birefringence typical of lamellar bone under polarised light. Woven bone is present in embryonic skeleton and in both cortical and cancellous bones during stages of rapid growth. After completion of bone growth, woven bone is replaced by lamellar bone in the normal skeleton. However, woven bone is also observed in certain pathologic conditions, such as Paget's disease of bone, fracture healing, osteogenesis imperfecta, and in primary or secondary hyperparathyroidism. In adults, woven bone is indicative of rapid, uncontrolled bone formation and high bone turnover that is attributable to either local or systemic factors.

Bone Modelling and Remodelling

During life, the skeleton is not static but undergoes numerous transformations. First, there is growth, also called bone modelling, when the overall shape and size of bones change. The two types of bone growth are longitudinal and appositional modelling. Longitudinal growth occurs by enchondral ossification; a process that takes place at the growth plate, when cartilage proliferates and progressively calcifies creating new trabeculae until the epiphyseal growth plate fuses. Appositional modelling, i.e., the growth of bone in width, proceeds by periosteal apposition of new bone and endosteal resorption of old bone.

After epiphyseal growth plates close, the adult skeleton continues to renew itself without noticeable changes in macroscopic shape. This is bone remodelling. The primary physiologic function of bone remodelling is to replace old bone with new bone. Old bone contains high mineral density and microfractures that decrease its mechanical properties. Approximately 3% of cortical bone and 25% of cancellous bone is renewed per year in the mature human skeleton [1]. Bone remodelling occurs in distinct locations on bone surfaces, the bone remodelling units (BRUs) [1]. BRUs require the involvement of a team of different cells, the bone multicellular units (BMUs). The remodelling of a "packet" or "quantum" of bone entails sequential events known as the remodelling cycle. This remodelling cycle includes activation, resorption, reversal, formation, and quiescence.

The signalling factors responsible for the initiation of the remodelling cycle are not well understood. Structural and/or biomechanical characteristics of old bone packets may play a role in this early phase, and signals may be relayed through the osteocyte-lining cell system to osteoblasts and bone marrow cells [1]. Numerous factors known to stimulate bone resorption are probably involved in the activation of bone remodelling that also requires interaction between hormones, cytokines, and growth factors and their receptors in bone cells and bone marrow cells. Activation of bone surface is accompanied by mobilisation of mononucleated osteoclast precursors, retraction of the lining cell layer, and exposure of bone matrix chemotactic substances such as osteocalcin, transforming growth factor β, and type I collagen.

The progressive fusion of mononucleated osteoclast precursors results in a team of one to four mature and active osteoclasts at each remodelling site. The team of osteoclasts then adheres to bone surface along a ring, i.e., the sealing zone, leaving an extracellular bone-resorbing compartment, and then bone resorption begins.

Osteoclasts dissolve bone mineral through acidification of the subosteoclastic compartment. This process involves proton pumps and carbonic anhydrase. Bone matrix is hydrolysed by proteolytic enzymes such as collagenase. Osteoclasts are motile cells and move along the erosion cavity. They are responsible for the rapid resorption (~7 days) of two-thirds of the final eroded cavity. Mononucleated cells resorb the other one-third at a slower rate (~36 days). These mononucleated resorbing cells consist of either unfused original osteoclast precursors or segmented osteoclasts. The rapid osteoclastic resorption leaves behind a rather rough surface that is transformed by the mononuclear cells into a smooth bone surface. In cortical bone, the resorption process takes place along the long axis of bone, and osteoclasts are observed in the cutting cone with depths reaching 100 μm [1]. In normal cancellous bone, osteoclasts erode bone parallel to the bone surface, forming a shallow cavity with a depth of 40–60 μm.

After resorption ceases, there is a transition period before bone formation occurs, i.e., the reversal phase. During this one-to-two-week phase, a layer of material with particular optical characteristics is deposited at the bottom of the resorption lacunae. On histologic section this "line" is positive for acid phosphatase staining and presents with a different birefringence under polarised or phase contrast light microscopy. This line (surface) serves as a "cement" or "glue" between old and new bone to be apposed and is referred to as the reversal or cement line. Only mononucleated cells are seen in front of the resorption lacunae during this phase and probably are responsible for the deposition of cement line. The exact origin of these cells is not known. Concurrent with the reversal phase, other events take place that are responsible for the coupling between resorption and formation. In the adult skeleton, bone is formed only on a previously eroded surface. This implies that signals are emitted to promote osteoblast proliferation and to direct osteoblast precursors to a precise location on the bone surface. The complex mechanisms responsible for this coupling phenomenon are not fully understood; however, several local agents such as chemotactic substances, growth factors, and cytokines are probably involved in this event.

After differentiation, osteoblasts form a layer in front of the reversal lacunae and deposit matrix protein, i.e., osteoid. Mineralization of the osteoid seam starts 5–15 days later when the osteoid seam is approximately 20 μm thick. Apposition of osteoid is rapid at first, then slows down progressively. The mineral apposition rate follows the same pattern. First, osteoid is mineralised at a rate of 1–2 μm/day and subsequently slows down but is still faster than matrix apposition at that time. When matrix apposition ceases, the remaining layers of osteoid are slowly mineralised. The first hydroxyapatite crystals are small and immature, and therefore calcium ions can chelate other substances present at high concentration in the bone microenvironment. These substances may include aluminium, fluoride, and others, in particular tetracycline hydrochloride. Under fluorescent light microscopy, deposits of tetracycline in new bone are spontaneously visible on unstained sections and appear as bright yellow bands (Figure 9.1). The technique of tetracycline double labelling uses this phenomenon to determine the rate of mineralization. However, when tetracycline is administered, it chelates reversibly to all exposed bone surfaces, i.e., in resorptive cavities and at the mineralization front. Therefore, it is necessary to perform a bone biopsy within a few days after the last administration of the antibiotic, usually four days. This ensures that the tetracycline chelated at the mineralisation front is protected from leaching out by a thin layer of newly mineralised bone. During bone formation, osteoblasts are first cuboidal and very active and thereafter flatten, and

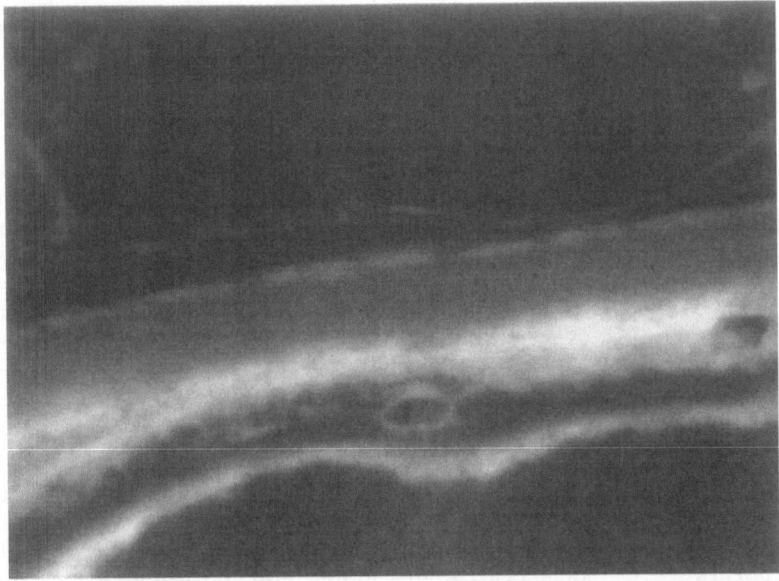

Figure 9.1. Double tetracycline labelling. Unstained section viewed under fluorescent light microscopy. Original magnification: ×125.

become lining cells when the osteoid seam is completely mineralised. Moreover, a certain fraction of osteoblasts are embedded in bone matrix and become osteocytes. Also, some osteoblasts may locally undergo apoptosis (programmed cell death) [2]. The total duration of the bone formative phase in normal skeleton is approximately three months.

A phase of quiescence follows. During the beginning of this phase (three to six months), the newly formed "young" bone packet will mature by increasing its mineral density. New bone is separated from bone marrow by the layer of lining cells and a thin collagenous membrane. In normal adult bone, the majority (80%) of bone surface is in a quiescent stage.

Three types of bone remodelling have been described: random, selective, and redundant. With random remodelling, any packet of bone has the same probability to be remodelled, regardless of its age. Selective remodelling involves the remodelling of the older, more fragile bone due to fatigue fracture, local trauma, or osteocyte death. Redundant remodelling renews the youngest bone packets. In normal adult bone, remodelling is mainly random, and the overall age of bone has been calculated to be 20 years in cortical and four years in cancellous bone [1]. The age of bone and its mechanical properties can be influenced by pathological processes that can preferentially induce a specific mode of remodelling [1].

Bone Balance and Bone Turnover

In normal young adults, coupling between the amount of bone resorbed by osteo-clasts and the amount of bone formed by osteoblasts results in bone balance. Un-coupling between formation and resorption will result either in negative or positive bone balance at the remodelling site. In states of negative bone balance, the amount of bone resorbed is disproportionately higher than bone formed; conversely, more bone is deposited than resorbed when bone balance is positive.

If bone remodelling represents the cellular-based events that occur at a specific site of the bone surface, bone turnover represents the rate at which the skeleton is renewed. This depends on the activation rate and the extent and distribution of the remodelling sites among the various bone envelopes. If bone turnover is high, the minute changes observed at the bone remodelling sites are amplified. For example, in case of negative bone balance, bone loss is greater in individuals with high bone turnover, and reduction of turnover will result in slower bone loss.

Bone Cells

Osteoclasts (Figure 9.2)

Under light microscopy, osteoclasts are large multinucleated cells located in resorption lacunae in the vicinity of mineralised bone. Osteoclasts represent the main cells in the breakdown of bone matrix and bone mineral. They vary in size from 20 to 100 μm in diameter and usually display projections and lobes that give them an irregular appearance. Osteoclasts are highly mobile and go through cycles of resorption and rest. Thus, it is not surprising that these cells vary in histological appearance, depending on the stage of the cycle at which they are observed. In normal skeleton, osteoclasts are somewhat larger than macrophages and may have from two to five nuclei. In pathological states, osteoclasts are large with up to 100 nuclei. The nuclei are found in the centre of the cell, they are characteristically round or oval and

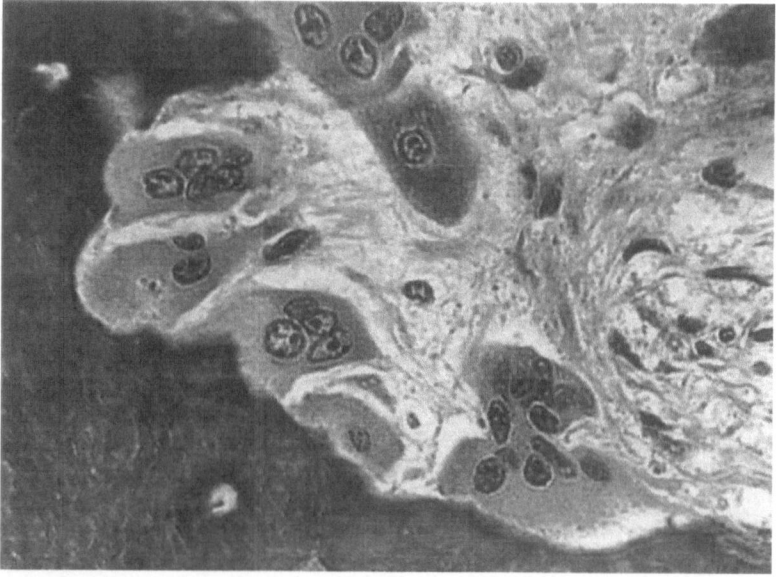

Figure 9.2. Resorption cavity with numerous multinucleated osteoclasts. Masson–Goldner trichrome stain. Light microscopy. Original magnification: ×500.

usually contain one or two prominent nucleoli. Osteoclasts may appear mononucleated depending on the plane of the sections; however, serial sections may show other nuclei. These cells exhibit a characteristic pink staining with the modified Masson–Goldner trichrome stain due to abundance of mitochondria, lysosomes, and ribosomes in their cytoplasm. Their foamy appearance reflects the presence of numerous endocytic and lysosomal vacuoles. The ruffled border, the ultrastructural trademark of osteoclasts, consists of numerous foldings of the apical membrane and is occasionally identified at high magnification on thin histologic bone sections. However, in the majority of bone samples, the ruffled border is not recognisable.

Osteoblast Lineage Cells

Osteoblasts are mononucleated cells responsible for production of bone matrix and are involved in its mineralisation (Figure 9.3). Under light microscopy, mature osteoblasts in normal bone form a monolayer of cells in front of the bone formation site. They are cuboidal, polarized cells measuring between 15 and 25 μm in diameter. Their round nuclei are located at the basal pole of the cells (away from bone) and contain one or more nucleoli. The cytoplasm is strongly basophilic due to the large amount of endoplasmic reticulum and Golgi apparatus responsible for active production of type-I collagen and other substances found in the bone matrix. Osteoblasts at a bone formation site give the appearance of an epithelium-like organisation. They exhibit gap junctions that ensure their cohesion and provide cell-to-cell communication. During active bone formation, osteoblast cells are plump. Towards completion of the new bone packet, the osteoblast shape flattens and its cytoplasm loses its basophilic characteristics, and its nuclei become elongated. The

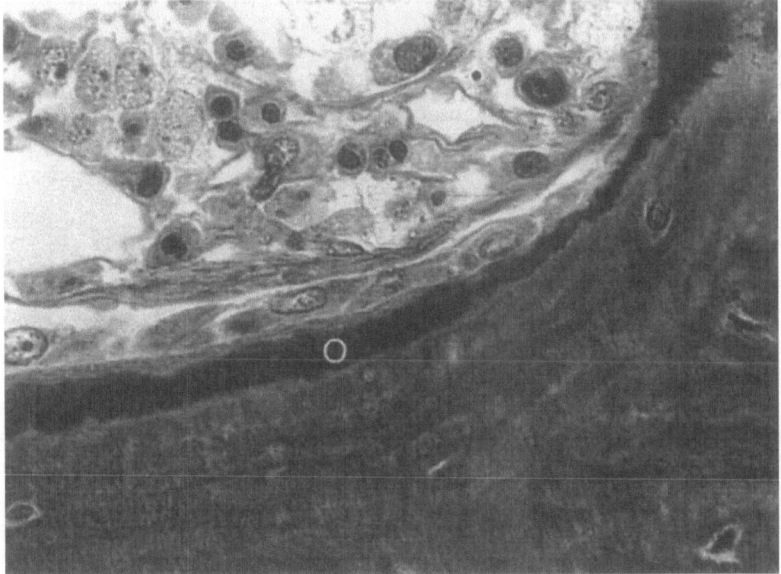

Figure 9.3. Osteoblasts in front of an osteoid seam (O). Masson–Goldner trichrome stain. Light microscopy. Original magnification: ×500.

majority of these resting osteoblasts become bone-lining cells, observed above the quiescent surfaces.

Also, approximately 10% of osteoblasts become embedded in the bone matrix before its mineralisation and become stellar cellular complexes with a central nucleus and numerous long cellular processes, the osteocytes. Gap junctions have also been described between the peripheral processes of osteocytes, osteoblasts, and bone-lining cells from the same bone area. When matrix mineralises, osteocytes are found in osteocytic lacunae, the border of which has been found to be calcified by osteocytes. Long cytoplasmic processes are seen in canaliculae. Newly embedded osteocytes conserve the cellular characteristics of osteoblasts, and the oldest ones in deeper bone areas lose signs of protein synthesis and accumulate glycogen. The periosteocytic space between osteocytes and mineralised bone contains the bone extracellular fluid (1–1.5 l) and the bone-lining cells–osteocytes network plays an important role in the minute-to-minute calcium homeostasis as the osteocytes are rapidly exposed to changes in circulating factors. Due to their deep location in mineralised bone, osteocytes are also considered sensors for fatigue damage and microfractures and thus may represent an important contributor to the regulation of bone remodelling and bone turnover.

Bone Marrow Cells

Besides bone cells, other cells are present in the bone microenvironment. There is ample evidence that the bone marrow cells play a role in the local regulation of bone remodelling and bone turnover. Cells from the monocyte–macrophage and lymphoid lineages produce various substances such as cytokines and growth factors that directly or indirectly act on bone cell recruitment and activity [3]. Moreover, macrophages have the capability to produce calcitriol. Mast cells, which produce heparin, a proven stimulator of bone resorption, may also be involved in the local regulation of bone.

In bone biopsies, it is common to observe that states of high bone turnover are often associated with various degrees of bone marrow cell hyperplasia, whereas low turnover states of bone are accompanied by variable hypoplasia of haematopoietic cells [4].

Normal Bone Physiology

Role of Bone in Calcium and Phosphate Metabolism

Bone has an important role in mineral homeostasis. Bone possesses two fundamental properties that greatly facilitate this function. First, there is the enormous capacity of the apatite crystals and the calcium phosphate salts to adsorb bone-seeking elements such as calcium, phosphorus, magnesium, aluminium, and zinc. Second, the large skeletal surface between the osteocyte-lining cell network and the extracellular fluid compartment facilitates ion exchange. Besides facilitating mineral movement in and out of bone, these highly complex anatomical arrangements contribute to the metabolic and electrical coupling of bone cells.

Bone is the largest primary calcium reservoir of the body and plays a major role in calcium homeostasis. Extracellular calcium concentration is determined by the rate of calcium entry into the extracellular fluid and the rate of calcium loss from it.

Calcium enters the extracellular space via three routes: intestinal absorption, which is the major contributor to the available calcium pool; calcium release from bone; and renal tubular reabsorption. Calcium loss occurs by means of gastrointestinal digestive juices, bone uptake during mineralisation, and urinary loss.

There is a daily flux of approximately 110 nmoles of calcium into and from the bone [5]. Only 10% of this calcium is exchanged through remodelling surfaces [5]; the rest is transferred across the quiescent surfaces of bone and bone-lining cells. The exact cellular mechanism(s) that influence calcium fluxes across this bone membrane are poorly understood. However, there is considerable evidence to suggest that parathyroid hormone and calcitriol, individually or synergistically, play an important role in governing calcium translocation across bone surfaces, independently of their role in bone remodelling [5].

The Role of Bone in Acid-Base Homeostasis

It has long been recognised that bone mineral contributes to the buffering mechanisms in acute and chronic acidosis [6]. Both low pH and low bicarbonate concentration independently influence calcium flux from bone [6]. Current evidence indicates that short-term acidosis mainly influences the physicochemical solution equilibrium, while long-term acidosis affects calcium efflux via the activation of bone resorption mechanism [6].

Factors Affecting Bone Metabolism

Bone cells are regulated by a complex interplay between systemic hormonal signals and local factors [7]. Bone cells are influenced by systemic factors and various circulating blood cells, particularly leukocytes that reach bone via capillary circulation. Bone cells are also in close proximity to local cells such as endothelial cells, chondrocytes and stromal (haematopoietic) cells that are capable of responding to circulating substances as well as secreting their own growth-regulating factors. It is only for simplification and practicality that one can categorically separate circulating factors from local factors.

Circulating Factors

Parathyroid Hormone (PTH)

The major role of PTH, an 84-amino-acid polypeptide, is to tightly regulate extracellular calcium concentration and keep it within narrow limits. This function is achieved by exerting important direct actions on bone and kidney and indirect effects on the intestine. Receptors for PTH have recently been cloned and characterised in cells in the bone microenvironment. PTH interacts with these specific membrane-bound receptors to initiate a cascade of intracellular events that involve protein G activation with resultant cAMP generation, phosphatidylinositol and calcium transport activation [8].

PTH maintains extracellular calcium concentration by facilitating calcium fluxes from and to the skeleton. Calcium movements from bone proceed in at least two phases. First, there is a rapid mobilisation from the lining cells and osteocytes as evidenced by the prompt structural response of these cells to injected PTH as well as the rapid release of radio-labelled calcium from bone surfaces following PTH administration. Second, there is a slow (hours) response that is dependent on mineral release by osteoclasts during bone resorption. One of the most recognised effects of PTH on bone is the enhancement of osteoclast activity and numbers [9]. This effect is, in fact, dependent on osteoblast activation since, in vitro, osteoblasts are required for PTH control of bone resorption and osteoblast release factor(s) that stimulate bone osteoclastic resorption. However, most recently, Langub et al. were able to show PTH/PTHrP receptors in osteoclasts of bone biopsies from normal human individuals and patients with renal failure [10].

In vivo, PTH increases not only osteoclastic resorption but also osteoblastic anabolic activity. In vitro, however, PTH inhibits osteoblast activity [11]. This apparent discrepancy can be related to differences in dosages or to the intermittent and pulsatile secretion of PTH in vivo, as opposed to the continuous effect in vitro [11]. Under physiological conditions, these synchronised anabolic and resorptive actions of PTH contribute to the maintenance of skeletal balance, which can be markedly disturbed in hyperparathyroidism.

Another important renal action of PTH is its influence on the production of $1,25(OH)_2D$, the active metabolite of vitamin D. PTH directly activates $25(OH)D-1\alpha$-hydroxylase found in proximal tubular cells leading to an increase in $1,25(OH)_2D$ production. Finally, PTH also inhibits proximal bicarbonate reabsorption. This effect is usually minor at physiological concentrations of the hormone or is overridden by other homeostatic mechanisms. However, it is not unusual to see systemic acidosis in hyperparathyroidism secondary to urinary bicarbonate losses.

Vitamin D

Vitamin D molecules are fat-soluble steroids known to have a protective effect against rickets. It is only in recent years that modern molecular techniques have uncovered remarkable findings that have substantially augmented the role of these hormones. It is now recognised that the most biologically active vitamin D metabolite, calcitriol $(1,25(OH)_2D_3)$ binds to a cytoplasmic vitamin D receptor (VDR). The vitamin D–VDR complex subsequently binds to nuclear DNA and alters a variety of transcriptional genes [12]. VDRs have been found not only in the conventional target organs for vitamin D such as the intestine, the kidney and bone, but also in a number of diverse tissues: parathyroid glands, pituitary gland, ovaries, skin, hair follicles, stomach, pancreas, thymus, breast, peripheral leukocytes, cardiac and skeletal muscles, tumour cell lines, and others. The physiological and clinical implications of these newly discovered functions of vitamin D are exciting and hold tremendous potential.

It is well known that vitamin D deficiency leads to rickets or osteomalacia. Therefore, the role of vitamin D in bone mineralisation has been suspected. Evidence from rat studies suggests that vitamin D indirectly affects mineralisation through maintaining normal serum calcium and phosphorus levels. However, it was demonstrated in dogs, which have skeletons more akin to humans, that vitamin D is required in addition to normal calcium and phosphorus concentrations for adequate mineralization [13]. In vitro, vitamin D has a direct resorptive effect on bone. It also stimulates calcium mobilisation from the bone fluid compartment to the extracellular space. In vivo, however, the direct resorptive effect of vitamin D is shown only in hypocalcaemic conditions or states of vitamin D deficiency. There is also significant evidence to suggest that the effect of PTH on bone is facilitated by vitamin D.

Recently, diverse effects of vitamin D have been described. Vitamin D directly inhibits PTH production, PTH secretion, and parathyroid cell proliferation. Vitamin D also inhibits the proliferation of cultured melanoma cells, fibroblasts, and keratinocytes. There is evidence to suggest that vitamin D enhances cell differentiation and inhibits cell growth. Finally, by affecting lymphocytes, vitamin D may play a role as an immunoregulatory hormone. These effects of vitamin D are the result of its role in regulating a large number of genes upwards or downwards [14].

Calcitonin

Calcitonin, a 32-amino-acid peptide, is secreted by the parafollicular cells of the thyroid gland. Its main action is to inhibit osteoclast resorption activity via a cAMP-mediated mechanism. The antiresorptive effect is quite rapid and evident at the physiological concentration of the hormone. The hypocalcaemic effect of calcitonin is probably related to its antiresorptive activity and usually is not evident unless a state of high bone turnover is present. In other words, calcitonin does not induce hypocalcaemia in normal subjects. Many factors affect calcitonin secretion, the most important of which is hypercalcaemia. Other factors include gastrin, cholecystokinin, and probably oestrogen and calcitriol. Calcitonin gene-related peptide (CGRP) is another peptide encoded by the calcitonin gene. Like calcitonin, this hormone has a similar action on osteoclasts, but it might also have a PTH-like effect on osteoblasts. The physiological role of this peptide in bone remodelling may be related more to its local abundance at the bone level rather than as a circulating hormone.

Other Circulating Factors

Other hormonal factors such as thyroid hormone, oestrogen, testosterone, and glucocorticoid have an impact on bone metabolism [14]. Also, vitamins A, C, and K all play a role in maintaining normal bone metabolism. Hypovitaminosis A may lead to inhibition of bone resorption and enhancement of bone formation. Vitamin C is required for the hydroxylation of proline and lysine, an essential step in the synthesis of bone matrix collagen. Finally, vitamin K is required for the synthesis of many proteins including osteocalcin, a protein that may play a role in bone mineralisation and calcium homeostasis.

Local Factors

The coupling between osteoblasts/stromal cells and osteoclasts to bring about the highly coordinated process of bone remodelling, i.e., bone formation and bone resorption, is dependent upon the cross talk between these bone cells through elaboration of local signals (factors) [15]. Within the bone microenvironment, resorption-stimulating cytokines such as IL-1β, IL-6, IL-11, IL-17, TNF-α, M-CSF, and RANK-ligand (RANKL) or resorption-inhibiting cytokines such as IL-13, INF-γ, TGF-β1, and osteoprotegerin (OPG) impact the differentiation and/or activation of osteoclast progenitor cells and mature osteoclasts [16–20]. These local factors produced in osteoblasts, osteoclasts, osteocytes, and marrow stromal cells have been shown at the synthetic and secretion levels to be regulated by PTH and calcitriol [8, 15]. Beginning with the discovery of OPG, a potent osteoclastogenesis-inhibitory factor, rapid progress was made to the isolation of RANKL, a ligand expressed on osteoblasts/stromal cells, that binds to receptor activator of NF-κB (RANK), a receptor on haemopoietic osteoclast precursor and mature osteoclasts. The binding of

RANKL by RANK initiates a signalling and gene expression cascade that results in differentiation and maturation of osteoclast precursor cells to active osteoclasts capable of resorbing bone. OPG is a decoy receptor that binds RANKL and blocks its interaction with RANK, thus inhibiting osteoclast development. For example, striking new advances have been made in understanding the molecular mechanisms that govern the cross talk between osteoblasts/stromal cells and haemopoietic osteoclast precursor cells that leads to osteoclastogenesis. Many of the calciotropic hormones and cytokines, including PTH, calcitriol, prostaglandin E2 and interleukin-11, appear to stimulate osteoclastogenesis through the dual action of inhibiting production of OPG while stimulating production of RANKL [21, 22]. Thus, local factors serve critical roles in transducing integral information, which guarantees the coupling between osteoblasts/stromal cells and osteoclasts to promote bone remodelling. Clearly, dysregulation of local factors may underlie many if not all of the pathological bone abnormalities.

References

1. Parfitt A. The physiologic and clinical significance of bone histomorphometric data. In: Recker RR, editor. Bone histomorphometry: techniques and interpretation. Boca Raton, FL: CRC Press, 1983; 132–223.
2. Jilka RL, Weinstein RS, Bellido T, Parfitt AM, Manolagas SC. Osteoblast programmed cell death (apoptosis): modulation by growth factors and cytokines. J Bone Miner Res 1998;13:793–802.
3. Canalis E. Regulation of bone remodeling. In: Favus MJ, editor. Primer on the metabolic bone diseases and disorders of mineral metabolism. New York: Raven Press, 1993.
4. Malluche HH, Faugere MC. Atlas of mineralized bone histology. New York: Karger, 1986.
5. Parfitt AM. Plasma calcium control at quiescent bone surfaces: a new approach to the homeostatic function of bone lining cells. Bone 1989;10:87–8.
6. Bushinsky DA. Hydrogen ions. In: Bushinsky DA, editor. Renal osteodystrophy. Philadelphia, PA: Lippincott-Raven, 1998.
7. Hruska KA, Teitelbaum SL. Renal osteodystrophy. N Engl J Med 1995;333:166–74.
8. Potts JJ, Juppner H. Parathyroid hormone and parathyroid hormone-related peptide in calcium homeostasis, bone metabolism, and bone development: the proteins, their genes, and receptors. In: Avioli L, Krane M, editorss. Metabolic bone disease. New York: Academic Press, 1997.
9. Chambers TJ. The cellular basis of bone resorption. Clin Orthop Rel Res 1980;151:283–93.
10. Langub M, Monier-Faugere M, Qi Q, Koszewski N, Malluche HH. PTH/PTHrP receptor expression in biopsy specimens from individuals with normal and elevated serum PTH. J Bone Min Metab 2001;16(3):448–56.
11. Heersche JMM, Aubin JE. Regulation of cellular activity of bone-forming cells. In: Hall BK, editor. Bone. Vol. 1. The osteoblast and osteocyte. Boca Raton, FL: CRC Press, 1992.
12. Haussler MB, Donaldson CA, Kelly MA, et al. Functions and mechanism of action of the 1,25-dihydroxyvitamin D_3 receptor. In: Norman AW, Schaefer K, Grigoleit HG, von Herrath D, editors. Vitamin D: Chemical, biochemical and clinical opdate. Berlin, Germany: Walter de Gruyter, 1985.
13. Malluche HH, Matthews C, Faugere MC, Fanti P, Endres DB, Friedler RM. 1,25-dihydroxyvitamin D maintains bone cell activity, and parathyroid hormone modulates bone cell number in dogs. Endocrinology 1986;119:1298–304.
14. Krieger NS. Bone formation, resorption, and turnover. In: Bushinsky DA, editor. Renal osteodystrophy. Phildelphia, PA: Lippincott-Raven, 1998.
15. Suda T, Takahashi N, Udagawa N, Jimi E, Gillespie MT, Martin TJ. Modulation of osteoclast differentiation and function by the new members of the tumor necrosis factor receptor and ligand families. Endocr Rev 1999;20:345–57.
16. Spelsberg TC, Subramaniam M, Riggs BL, Khosla S. The actions and interactions of sex steroids and growth factors/cytokines on the skeleton. Mol Endocrinol 1999;13:819–28.
17. Hofbauer LC, Khosla S, Dunstan CR, Lacey DL, Boyle WJ, Riggs BL. The roles of osteoprotegerin and osteoprotegerin ligand in the paracrine regulation of bone resorption. J Bone Miner Res 2000;15:2–12.
18. Nakashima T, Kobayashi Y, Yamasaki S, Kawakami A, Eguchi K, Sasaki H et al. Protein expression and functional difference of membrane-bound and soluble receptor activator of NF-kappaB ligand:

modulation of the expression by osteotropic factors and cytokines. Biochem Biophys Res Commun 2000;275:768–75.

19. Manolagas SC. Birth and death of bone cells: basic regulatory mechanisms and implications for the pathogenesis and treatment of osteoporosis. Endocr Rev 2000;21:115–37.

20. McSheehy PMJ, Chambers TJ. Osteoblastic cells mediate osteoclastic responsiveness to parathyroid hormone. Endocrinology 1986;118:824–8.

21. Udagawa N, Takahashi N, Yasuda H, Mizuno A, Itoh K, Ueno Y et al. Osteoprotegerin produced by osteoblasts is an important regulator in osteoclast development and function. Endocrinology 2000; 141:3478–84.

22. Aubin JE, Bonnelye E. Osteoprotegerin and its ligand: A new paradigm for regulation of osteoclasto-genesis and bone resorption. Medscape Womens Health 2000;5:5.

9B

Primary Osteoporosis

H. Morii, T. Miki, M. Ito, Y. Uchida, Y. Takaishi, K. Nakatsuka and H. Tahara

Definition and Classification

Osteoporosis is defined as a condition characterised by low bone mass, microarchitectural deterioration of bone tissue, leading to enhanced bone fragility and a consequent increase in fracture risk [1]. Since the first consensus development conference in Hong Kong in 1993 there have been no changes of the definition until 2001.

Such pathological states are induced in various situations. As age-related osteoporosis, Khosla et al. [2] proposed type 1 and 2 osteoporosis. Type 1 osteoporosis denotes postmenopausal and type 2 senile osteoporosis [2]. Although osteoporosis in men occurs after the age of 70 [2], there are distinct gender differences in cortical bone structure [3]. There are differences in bone remodelling during puberty and in ageing men in whom the periosteal expansion compensates for intracortical and endosteal loss of bone, and the strength is well maintained in contrast to women in whom insufficient periosteal formation and/or excessive endosteal resorption may lead to regression of strength [3].

Thus, primary osteoporosis can be classified in the following way:

Postmenopausal osteoporosis,

Age-related osteoporosis in females,

Age-related osteoporosis in males.

In secondary osteoporosis, the following disorders may be important:

Glucocorticoid use associated,

Rheumatoid arthritis,

Hyperthyroidism,

Diabetes mellitus,

Hypogonadism,

Immobilisation.

Epidemiology

The interim report of the WHO Task Force for Osteoporosis [5] reported that hip fracture rates increase exponentially with age; by age 80, a Caucasian woman has about a 3% annual risk of hip fracture. The incidence of fracture risk is three times more in women compared with elderly men. Fracture risk is calculated as one-third in black people and one-half in Asians and Hispanics compared with Caucasians. The WHO Study Group reported in 1994 that about 1.7 million hip fractures occurred worldwide in 1990 and that age-adjusted hip fracture incidence rates by sex are higher in Scandinavia than in North America or Oceania and lower in the countries of southern Europe [5]. Lau compared the age-adjusted rate of hip fracture in females and males in Asia in comparison with Caucasians and showed that the rate was higher in northern European countries and that the female to male ratio was 1 : 1 in Chinese and the Bantu, while it was 3 : 1 in Caucasians [6]. While Japan has its own diagnostic criteria in which osteopenia is defined as having a BMD between 70% and 80% (−1.5 and −2.5 SD) of young adult mean and osteoporosis a BMD less than 70% (−2.5 SD), it is estimated that 7.83 million female and 2.26 million male patients (total 10.09 million) (female : male ratio of 3.46) [7].

The incidence of vertebral fracture was compared between residents of Rochester, Minnesota and Beijing, China [8]. The incidence was higher in Rochester in most of the age groups except for the 70–79 group than in Beijing [8]. The incidence of vertebral fracture and the increasing rate with ageing were dependent on the year of birth during the period 1880–1939 in Japan. Although the incidence was higher in males with earlier period of the year of birth than in later period, there was no tendency of the increase with aging in each birth-year group. On the other hand, the incidence increased with ageing in each birth-year group in females. The cause was ascribed to the very low level of calcium intake in this period [9]. Daily calcium intake was estimated at 368 mg in Japan by the Food and Agricultural Organization/World Health Organization (FAO/WHO) [10]. The level was close to that in India, at 347 mg, while the highest level was 1,329 mg in Finland [10].

With regard to hip fracture, Khosla et al. [2] showed that the incidence of hip fracture increases exponentially throughout life in both men and women. Garraway et al. showed data for the incidence of limb fractures, comparing males and females from childhood to over 85 years of age. Although the incidence was much higher in male subjects with ages younger than 25 years of age than in female ones with corresponding ages, there was a reversal of the incidence between males and females at the ages 45 to 54 years of age. There was a sharp increase of incidence in females until later in life, while that in males increased after the ages of 65–74 years [11]. The prevalence of hip fracture was estimated at 53,000 in 1987, 76,600 in 1992 and 92,400 in 1997 among the Japanese population of 120 million. [12]. There was an increase of age-adjusted rate of hip fracture in Hong Kong almost twice as much from 1865–1967 to 1985 [6]. A cross-national study of hip fracture incidence was carried out in five geographic areas – Beijing, Budapest, Hong Kong, Porto Alegre, and Reykjavik during the years 1990–1992. Cases of hip fracture among women and men of age 20 years and older were identified using hospital discharge data in conjunction with medical records, operating room logs, and radiology logs. The estimated incidence rates varied widely, with Beijing reporting the lowest rates (age-adjusted rate per 100,000 population for men aged 20 years and older = 45.4; women = 39.6) and Reykjavik the highest rates (men = 141.3; women = 274.1). Rates were higher for women than for men in every area except Beijing [13]. The incidence was highest in every age group in Iceland. Although the incidence was lower in Asian countries and Budapest compared with Brazil and Iceland, it was higher in Hong Kong among other Asian countries.

Risk Factors

It was recommended to provide a diet that maintains normal body weight through-out life and with a calcium intake of some 1,000 mg/day from late childhood to middle life [4]. A physically active lifestyle, maintenance of eugonadism, avoidance of smoking and of high alcohol intake, minimisation of glucocorticoid use and pro-motion of vitamin D supplementation were also recommended [4].

Risk factors for hip fractures were analysed in 14 centres from Portugal, Spain, France, Italy, Greece and Turkey over a one-year period (Mediterranean Osteoporosis Study: MEDOS). Significant risk factors were low body mass index (BMI), short fertile period, low physical activity, lack of exposure to sunlight, low milk consumption, no consumption of tea and a low IQ [14]. The same system of questionnaire was applied to a Japanese population [15]. Significant risk factors were drinking more than three cups of coffee, living in rural areas in the past, sleep disturbance, stroke with hemiplegia and sleeping in a Western-type bed. Factors reducing risk were large BMI, moderate alcohol intake and eating fish [15].

When the two studies are compared, there are at least two factors common to two regions, Europe and Japan: large BMI and calcium intake (Table 9.1). All kinds of nutrients contribute to large BMI but protein intake is thought to be the most important. Ovariectomised rats were fed a low-protein diet and showed low BMD of the femur and genetically analbuminaemic Nagsse rats also showed low BMD in the proximal tibia region [16]. Bonjour reported that protein repletion after hip fracture was associated with a more favourable outcome, including shorter hospital rehabilitation. The protein-supplemented patients had significant greater gains in serum prealbumin, IGF-1 and IgM and a decrease in dexypyridinoline. In the protein-supplemented patients, the decrease in proximal femur BMD observed at one year in the placebo group was attenuated [17].

There is uncertainty about the role of calcium deficiency in the pathogenesis of osteoporosis. There is no evidence that calcium deficiency depresses bone formation

Table 9.1. Factors reducing hip fracture in Europe and Japan [14, 15]

Europe (MEDOS)	Japan (modified MEDOS)
large BMI	large BMI
active physical life	low coffee consumption
more sun exposure	high fish intake
more milk intake	moderate alcohol intake (less than 27 g/day)
more tea consumption	using matt in bedtime
long fertile period	
high mental score	

but overwhelming evidence that it increases bone resorption. The commonest type of osteoporosis which follows menopause in women could be due to calcium deficiency [18]. Alactasia, milk allergy and habit are other causes of calcium deficiency [18]. Reid performed a randomised controlled study by adding 1g of elemental calcium to dietary calcium of 640–700 mg in postmenopausal women for four years. There was a sustained reduction in the rate of loss of total body BMD in the calcium group and bone loss was significantly less in the calcium-treated group in years two through four. In the lumbar spine, bone loss was reduced in the calcium group in the first year but not subsequently. There was a significant treatment effect at this site over the whole four-year period. In the proximal femur, the benefit of calcium treatment tended to be greater in the first year and was significant over the four-year study period in the femoral neck and the trochanter [19].

Case Report

KH, a 47-year-old male with calcium deficiency and vertebral fracture.

Diagnosis: idiopathic osteoporosis, hyperuricaemia, hypertriglyceridaemia.

Chief complaint: back pain due to a second lumbar spine fracture.

Family history: nothing in particular.

Previous history: at 44 years of age, suffered a right rib fracture during a massage.

His body weight was 70 kg at age 20, 80 kg at 35 and 65 kg at 40. In addition, he suffered from obesity during his middle school years (12–15 years of age).

Present illness: Two months before he was admitted to the Second Department of Internal Medicine, Osaka City University Hospital, Osaka, he noticed an intense back pain and fracture noise while testing the strength of his back muscles. He was referred to a physician who found a fracture of the second lumbar vertebra. A corset was applied and analgesic drugs were prescribed.

Physical findings at admission: Height: 164.6 cm, weight: 70.1 kg with BMI of 25.87. Blood pressure was 118/64 mmHg, heart rate 76/min with regular beat. No anaemia and jaundice were seen in conjunctiva. No goitre and lymphadenopathy were palpated in the neck. No abnormal findings were noted in the abdomen. There was a pain in the back at the height of the second lumbar vertebra. Neurological findings were negative.

Laboratory: Urine: glucose (–), protein (–), ketone body (–) and haematuria (–).

Stool: no blood, CBC: WBC 9,600, RBC 465 × 10^4, Hb 14.5 g/dl, Ht 41.9%, Platelet 24.8 × 10^4, ESR 9 mm/1h, 24 mm/2 h, normal coagulation test, lymphocyte subset: OKT4 = 54.1%, OKT8 = 23.0%, CRP = 0.4 mg/dl, RAHA (–), TPHA (–), Anti-DNA Ab (–), IgG 1,090 mg/dl, IgA 227 mg/dl, IgM 62 mg/dl, total protein 6.8 g/dl, Alb 4.3 g/dl, alkaline phosphatase 248 (type II & III) (80–230), BUN 14 mg/dl, creatinine 1.1 mg/dl, uric acid 6.9 mg/dl, Na 143 mEq/l, K 3.9 mEq/l, Cl 109 mEq/l, GFR 92 ml/min, Chol 205 mg/dl, FBS 83 mg/dl, HbA1c 5.3%, AFP 2.4 ng/ml (<10), CEA 9.8 ng/ml (<5), CA19-9 5 U/ml (<37), β 2microglobulin 1.67 mg/l, free T4 1.6 ng/dl, TSH 1.08 μIU/ml, cortisol 5.5 μg/dl (10–15), urinary 17KS 7.0 mg/day (4.6–16.4), 170HCS 3.1 mg/day (2.9–11.6), 17 OH progesterone 0.4 ng/ml (0.5–2.9), oestradiol 20 pg/ml (20–60), progesterone 0.2 ng/ml

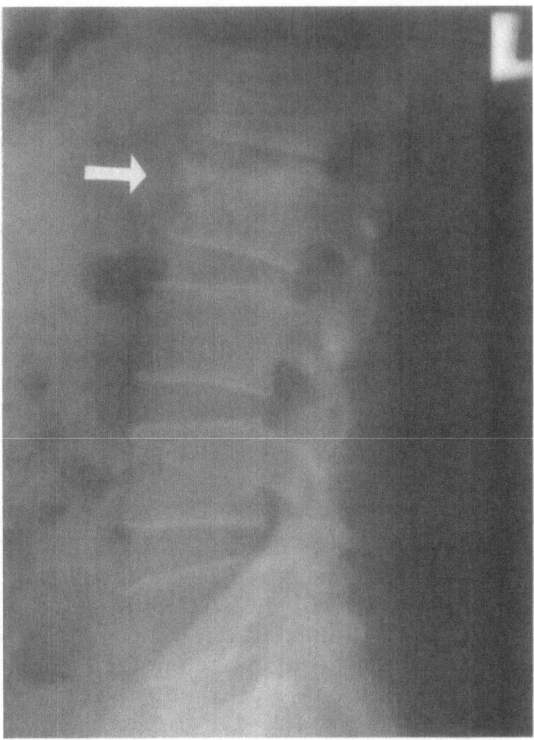

Figure 9.4.   Lateral radiography of lumbar spine with wedge-shaped fracture in L2 in a 47-year-old male.

(<0.7), aldosterone 13.2 ng/dl (2–13), urinary aldosterone 2.0 µg/day (2–13), ACTH 18 pg/ml (9–52), testosterone 5.7 ng/dl (5.46 ± 0.75), DHEA 4.5 ng/ml (1.7–6.0), 5, HIAA 2.5 mg/day (0.6–4.1), urinary epinephrine 6.3 µg/day (1–23), urinary nore- pinephrine 48.3 µg day (29–120), dopamine 720 µg/day (100–1000), urinary VMA 3.6 mg/day (1.4–4.9), non-functioning tumour in the left adrenal gland from CT imaging, Ca 4.7 mEq/l, P 3.7 mg/dl, ionised Ca 1.27 mmol/l, middle molecule PTH 189 pg/ml (74–273), intact PTH 7.4 pg/ml (10–65), calcitonin 26 pg/ml (30.9–120.1), urinary Ca/creatinine 0.09 mg/mg (<0.3), 1,25(OH)$_2$D 40.2 pg/ml (15.8–50.8), L2-4 BMD 0.723 g/cm^2 (Hologic QDR 2000, no reference data established but female YAM 1.011 ± 0.119), ECG within normal range. Lateral radiography of the lumbar spine showed wedge-shaped fracture of L$_2$ (Figure 9.4).

Course of Illness

Although vertebral BMD showed a low level, the parameters of endocrine function and calcium metabolism showed normal levels. A rather low level of PTH was sup- posed to be due to the intake of small amounts of alphacalcidol after rib fracture. History taking was repeated and it became clear that the patient smoked 20 ciga- rettes/day after the age of 20 and 40 cigarettes/day after 42 years of age, that he was engaged in combative sports such as wrestling, sumo and judo during his years of junior and senior school but never did any kind of exercise during college life. He had taken taxis to work every day for 10 years. Regarding his diet, he never ate seaweed, raw fish, food in sweetened vinegar, shrimp, chicken, soy bean cake, bean sprouts, spinach, vegetables – except for lettuce – mushrooms, apples, cheese and dairy products. He never drank milk because of diarrhoea since his senior high school years.

It was suspected that his calcium intake had been between 100 and 200 mg/day for many years in addition to the insufficient exercise and smoking. He was advised to eat calcium- rich foods and to take appropriate exercise.

Pathogenesis

Role of Vitamin D

Riggs classified two types of osteoporosis: type 1, which is postmenopausal osteoporosis and type 2, which is age-related osteoporosis [2]. The mechanism common to both types of osteoporosis is that the 1,25-dihydroxyvitamin D level is low, although the mechanisms are different between the two. In type 1 osteoporosis, the effects of 1,25-dihydroxyvitamin D on bone, intestine and kidney are reduced after menopause, resulting in increased bone loss, reduced PTH secretion and reduced 1,25-dihydroxyvitamin D levels. In type 2 osteoporosis, ageing results in decreased bone formation, intestinal resistance to 1,25-dihydroxyvitamin D, and decreased production of 1,2-dihydroxyvitamin D due to decreased activity of 25-hydroxyvitamin D 1, α hydroxylase. After this proposal many studies became concerned with this issue. Ott summarised them, showing that the levels of this sterol in women with osteoporosis were compared to similarly aged women without fractures. Although Ott claimed that the results were inconsistent – and the largest studies have found no significant difference – some of the results showed higher levels of 1,25-dihydroxyvitamin D in patients with osteoporosis compared with normal women [20].

The total circulating 1,25-dihydroxyvitamin D increases after oestrogen therapy in postmenopausal women, but there is disagreement about the free levels, because oestrogen can increase the D binding protein. Although there is a view that oestrogen deficiency plays a key role in the pathogenesis of postmenopausal osteoporosis [21], definite conclusions have not yet been reached.

Calcium and vitamin D supply is insufficient in some cohorts of the elderly. From such observations the idea of hypovitaminosis D and vitamin D insufficiency were proposed [22] (Table 9.2). The term hypovitaminosis D originated in the earlier stages of the study; for example, Albright used the term to designate a cause of osteomalacia [23]. To describe a lack of vitamin D, he proposed a simple lack of vitamin D, resistance to vitamin D and steatorrhoea [23].

One of the biggest achievements during the twentieth century in the field of endocrinology will be the discovery of mechanisms of activation of vitamin D [24]. Vitamin D is hydroxylated in the liver at the 25 position and again in the kidney in the 1 position to form 1,25-dihydroxyvitamin D. 1,25-dihydroxyvitamin D is the most active of the metabolites in promoting intestinal absorption of calcium, in modulating bone resorption and formation, in suppressing PTH secretion, in accelerating

Table 9.2. Definition of vitamin D status based on 25(OH)D concentration
(McKenna & Freaney [22])

Vitamin D Status	Serum 25(OH)D	
	nmol/l	ng/ml
Desirable	>100	>40
Hypovitaminosis D	<100	<40
Vitamin D Insufficiency	<50	<20
Vitamin D Deficiency	<25	<1

tubular reabsorption of calcium and others. The Development Committee of the
National Osteoporosis Foundation of the USA issued the special publication of the
results of a three-year project to systematically review all available evidence about
assessing and reducing the risk of osteoporotic fractures [25]. The Committee said
that vitamin D by itself is probably inactive, contrasting with the by now well-
accepted opinion that vitamin D itself is biologically inactive [24]. It was also said
that 25-hydroxyvitamin D has one-tenth to one-hundredth of the biological activity
of 1,25-dihydroxyvitamin D, depending on the assay endpoint used [25]. If differences in the level of 25-hydroxyvitamin D between hypovitaminosis D and vitamin
D insufficiency and that between vitamin D insufficiency and vitamin D deficiency
make the difference in bone pathology, how does such a small difference in 25-
hydroxyvitamin D levels produce osteoporosis and not osteomalacia? Chapuy et al.
reported that serum intact PTH held a stable plateau level at 36 pg/ml as long as
serum 25-hydroxyvitamin D values were higher than 31 ng/ml, but increased when
the serum 25-hydroxyvitamin D value fell below this level [26] (Figure 9.5). While
PTH secretion is suppressed by 1,25-dihydroxyvitamin D, there may be other
mechanisms than direct action of decreased 25-hydroxyvitamin D.

The most important evidence to support this idea was from Chapuy et al., who
showed that elderly women who were administered 800 IU of vitamin D_3 for 18 months
showed significant less incidence of fracture [27]. Other evidence is that the European
population is at high risk for vitamin D insufficiency because the continent is located
at high latitude, the nutritional supply of vitamin D is low in most countries and fortification of food is done mostly in Northern Europe [28] and that the incidence of hip
fracture is higher in Northern Europe than in Mediterranean countries [29]. Regarding the correlation between BMD and 25-hydroxyvitamin D, it was significant at a 25-
hydroxyvitamin D concentration of less than 30 nmol/l [30]. However, this level of
25-hydroxyvitamin D falls in the range of vitamin D deficiency according to the
classification of McKenna [22]. This is the range of osteomalacia.

The answer to this question has not been solved. Parfitt [31] suggested that the
problem will be microenvironmental activation of vitamin D in bone tissue. Although
the secretion of 1,25-dihydroxyvitamin D from the kidney was shown to be under
the control of many factors [32], the regulation of extrarenal production of 1,25-
dihydroxyvitamin D has not been fully elucidated until now. From the clinical standpoint, it may be important to elucidate the microenvironmental vitamin D status in
patients with osteoporosis with hypovitaminosis D or vitamin D insufficiency.

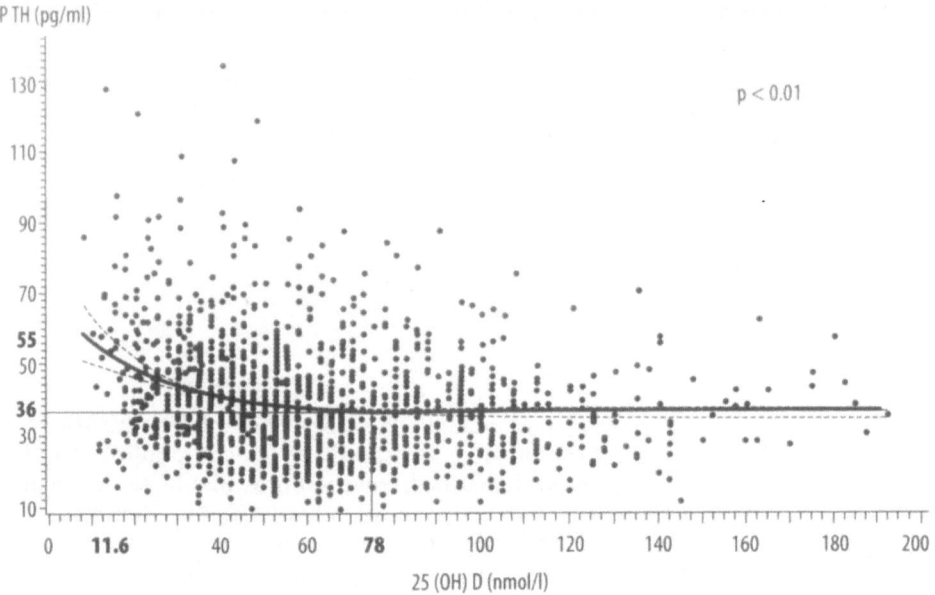

Figure 9.5. Relationship between serum intact PTH and 25,hydroxyvitamin D in 1,569 individuals in 20 French cities (765 men, aged 45–65, 804 women, aged 35–60) (26). Cited with permission from the author and the editor-in-chief.

Significance of Active Vitamin D as a Remedy for Osteoporosis

There have been controversial results regarding the effects of vitamin D analogues in the treatment of osteoporosis. Regarding the clinical efficacy of these active vitamin D preparations, Nakamura overviewed some clinical studies, showing that 700–800 mg of calcium supplementation may be needed to have obvious effects in decreasing vertebral fractures [36] (Figure 9.6).

However, the following questions need to be answered.

1. How was the vitamin D state in each of the studies related to the efficacy of calcitriol or alphacalcidol?
2. It should be investigated whether or not there is a basis for how vitamin D insufficiency causes biological effects.
3. It should be investigated whether calcium supplementation is beneficial in treatment with plain vitamin D.
4. Are there differences in the biological effects between plain vitamin D and calcitriol or alphacalcidol?

Regarding point 1, there have been no such clinical studies to compare the effects of calcitriol or alphacalcidol in patients with different 25-hydroxyvitamin D levels. Harris et al. [32] studied wintertime vitamin D and PTH status of an elderly population in Boston. Twenty-one per cent of black subjects and 11% of whites had very low concentrations of 25-hydroxyvitamin D (<25 μmol/L) and 75% of blacks and 35% of whites showed 25-hydroxyvitamin D concentrations less than 50 μmol/L. Low 25-

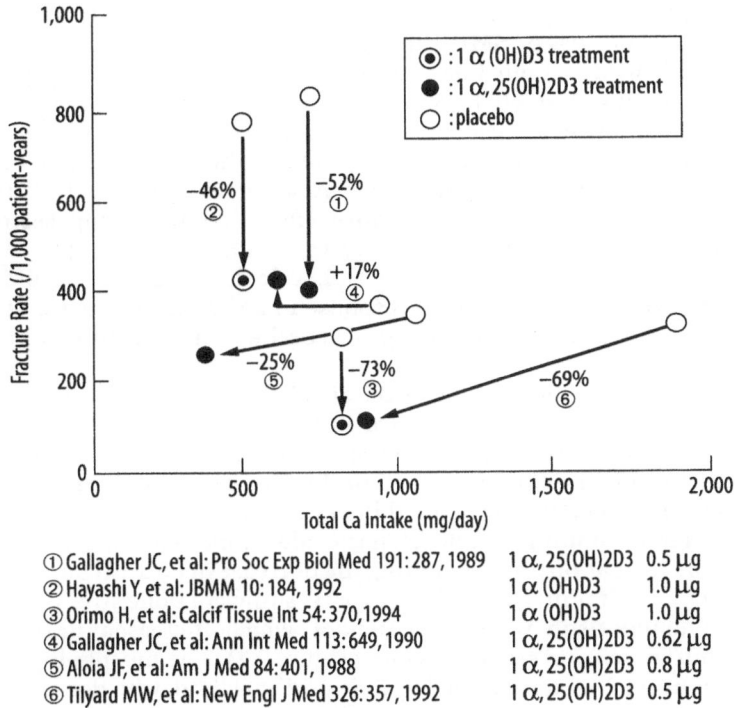

① Gallagher JC, et al: Pro Soc Exp Biol Med 191:287, 1989   1 α, 25(OH)2D3 0.5 μg
② Hayashi Y, et al: JBMM 10:184, 1992   1 α (OH)D3   1.0 μg
③ Orimo H, et al: Calcif Tissue Int 54:370, 1994   1 α (OH)D3   1.0 μg
④ Gallagher JC, et al: Ann Int Med 113:649, 1990   1 α, 25(OH)2D3 0.62 μg
⑤ Aloia JF, et al: Am J Med 84:401, 1988   1 α, 25(OH)2D3 0.8 μg
⑥ Tilyard MW, et al: New Engl J Med 326:357, 1992   1 α, 25(OH)2D3 0.5 μg

Figure 9.6. The vertebral fracture incidences were plotted against the basal levels of daily calcium intake. Open circles indicate data of controls and closed ones those of active vitamin D-treated groups. Arrows represent connect controls and active vitamin D groups (36). Reproduced by permission of the publishers, Springer-Verlag GmbH & Co, KG.

hydroxyvitamin D concentrations were associated with increased PTH and reduced serum calcium. The vitamin D intake was closely related to serum 25-hydroxy-vitamin D levels. From this study it seems likely that serum 25-hydroxyvitamin D and hence PTH levels have a close relationship with vitamin D intake. However, there was no discussion of whether 25-hydroxyvitamin D itself has a biological activity, nor has the optimum 25-hydroxyvitamin D level been defined. However, there was a suggestion that racial difference may be affect the optimum concentration of 25-hydroxyvitamin D.

Regarding point 2, Harris et al. [32] are affirmative, but the mechanism of action has not yet been clarified. In a Swedish population living at home, serum 25-hydroxy-vitamin D and 1,25-dihydroxyvitamin D levels were below the reference ranges in 4% and 5% of the subjects, respectively. When serum levels of 25-hydroxyvitamin D were lower than 30 ng/ml, the serum intact PTH began to increase from a level of 43 pg/ml [34]. Multiple regression analysis demonstrated femoral neck BMD to be significantly and positively associated with higher BMI, male gender, no history of fragility fracture and 25-hydroxyvitamin D [34]. Another study indicates that there is a high prevalence of vitamin D insufficiency in the elderly active community with established vertebral osteoporosis, which leads to increased risk of further osteo-porotic fractures in comparison with vitamin D-sufficient subjects [35].

Regarding calcium supplementation, Peacock et al. [35] showed that a supplement of 750 mg/day prevents loss of BMD and reduces femoral neck expansion, secondary hyperparathyroidism and high bone turnover. A supplement of 15 μg/day of 25-hydroxyvitamin D_3 was less effective and because its effects are seen only at low calcium intake levels, its beneficial effect is to reverse calcium insufficiency.

As to the last point, there have been many studies showing the effect of calcitriol or alphacalcidol in reducing fractures, which were reviewed by Nakamura [36] who proposed that most marked preventive effects on fracturing seemed to be obtained with daily calcium intakes at the levels of 400–800 mg (Figure 9.3). Nakamura stated that the effect of calcitriol or alphacalcidol in increasing BMD is dose dependent. However, the magnitude of response in BMD falls in the range of 2–3%. On the other hand, the effect of reducing vertebral and hip fracture is pronounced [37, 38].

In summary of this section on the effect of calcitriol and alphacalcidol, it is suggested that these compounds are effective in reducing the incidence of fractures. However, it has not been established whether they may differ in the mechanism of action from that of plain vitamin D. In clinical settings, such comparisons have not been made, but Masaki et al. [39] showed in experimental studies in ovariectomised rats that both plain vitamin D_3 and alphacalcidol inhibited bone resorption and maintained bone formation, thus inhibiting the decrease of cortical bone to the same extent, but that plain vitamin D_3 could not express sufficient bone strength by increasing dosage in comparison with alphacalcidol, which increased BMD and bone strength without increasing serum calcium (Figure 9.7). The morphological basis of this effect of alphacalcidol was studied by Shiraishi et al. [40], who demonstrated that

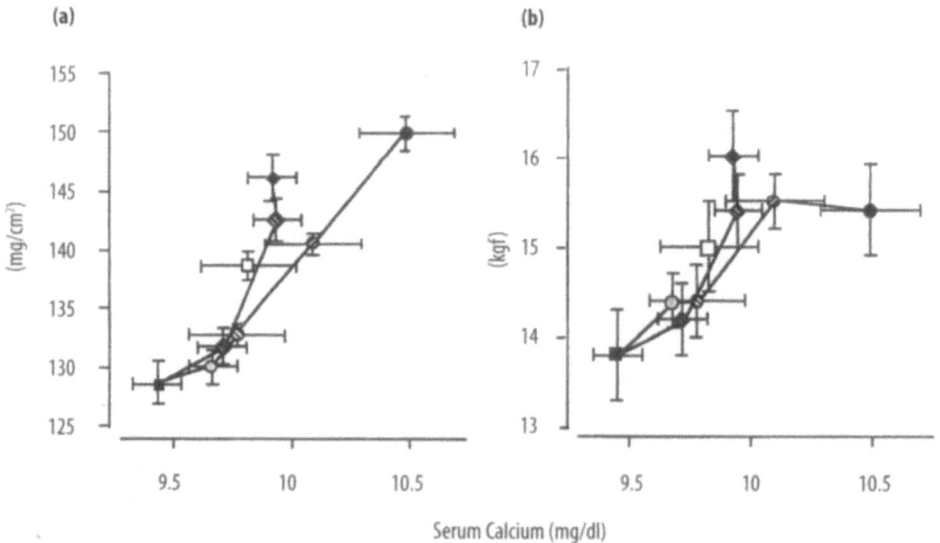

Figure 9.7. Effects of alfacalcidol and plain vitamin D_3 on femoral BMD (a) and strength (b): Relationship to serum Ca levels. Ovariectomised (OVX) rats were treated with alfacalcidol (◈ : 0.025, ◈ : 0.05, ◆ : 0.1 μg/kg) or vitamin D_3 (○ : 50, ◈ : 100, ◗ : 200, ● : 400 μg/kg) orally five times a week for 3 months. □ : Sham-operated rats, ■ : OVX rats. Each value represents mean ±S.E.

(a)   (b)   (c)

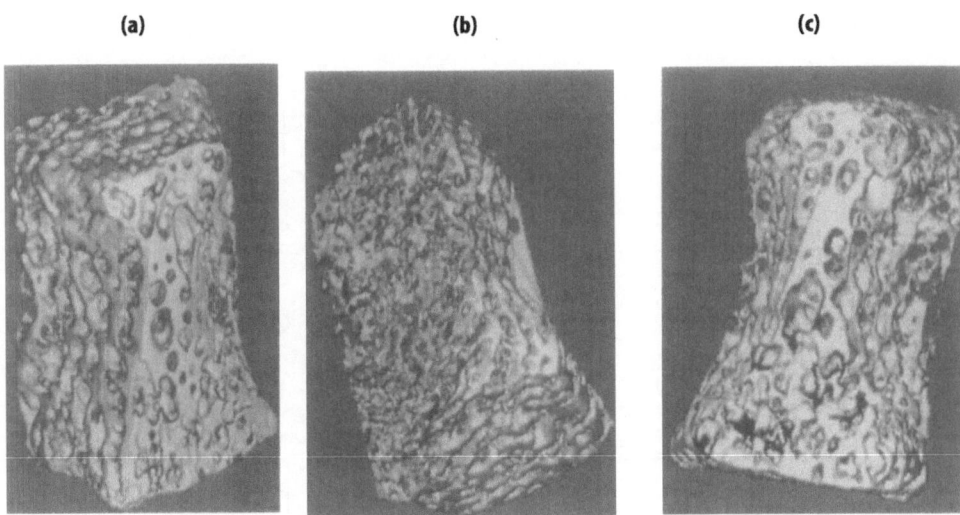

Figure 9.8. Three-dimensional trabecular microarchitectural images of rat 5th lumbar vertical body (L5) using micro-computerized tomography. **a** Sham-operated rat **b** OVX-control rat and **c** OVX/alfacalcidol 0.1 μg/kg treated rat at 3 months of observation.

trabecular microarchitecture was maintained when investigated using computed tomography (Figure 9.8). Thus, a vitamin D analogue and not plain vitamin D may have direct effects on bone by improving strength and microstructure in ovariectomised animals.

ED-71, another analogue of 1,25-dihydroxyvitamin D, has been under development in Japan (Figure 9.9d) and it was demonstrated that this compound has a potent action in increasing bone mass, increasing bone strength and improving microarchitectural structure in ovariectomised rats. A randomised controlled clinical trial was conducted in 108 Japanese osteoporotic patients who were administered 0.25, 0.5, 0.75 or 1.0 μg of ED-71 for six months. The percentages of patients that showed an increase in L2-4 BMD over 3% increased dose-dependently. Twenty-three per cent of patients showed serum 25-hydroxyvitamin D levels below 20 ngml but the effect of ED-71 on L2-4 BMD was not affected by the level of serum 25-hydroxyvitamin D [41]. The effect was thought to be a preferential one on bone.

Mortality and Morbidity

Mortality and morbidity associated with hip and vertebral fractures were well reviewed by Cooper and Melton [42]. Hip fracture was found to lead to a reduction in the survival of 10–20% of patients, and the majority of deaths occur within the first six months after fracture. Regarding vertebral fracture, it was described that overall survival among patients was 61% compared to an expected survival for those of like age and sex of 76%. Compared with women without prevalent vertebral deformities, women with prevalent deformities had higher risk of mortality and hospitalisation [43].

Seventy-six (8.9%) patients who were 90 years of age or older had significantly longer mean hospital lengths of stay than younger individuals, and were more likely

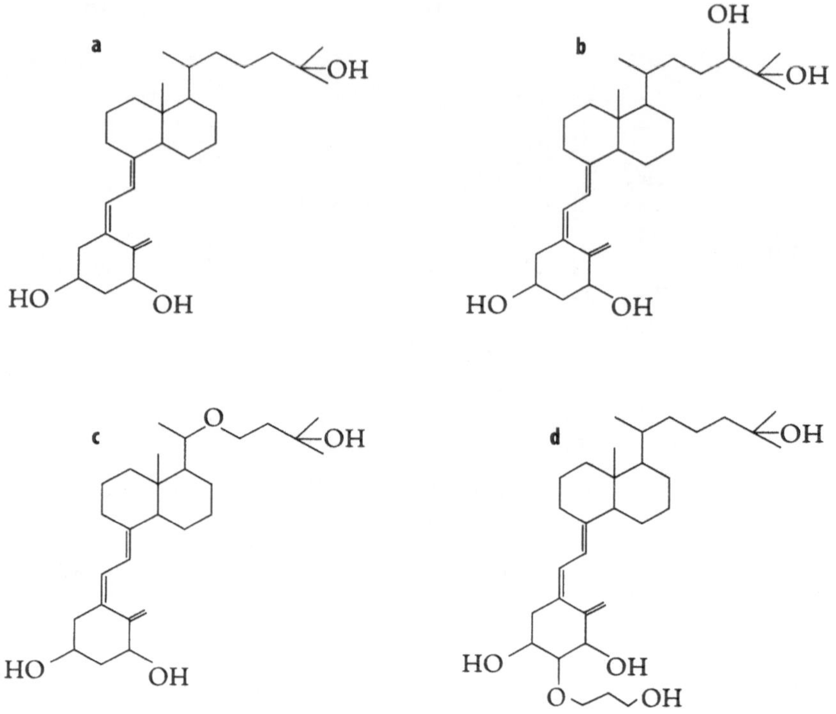

Figure 9.9. 1,25 dihydroxycholecalciferol and its analogue **a** 1,25 dihydroxycholecalciferol, **b** 24, 25 dihydroxycholecalciferol, **c** maxacalcitol (OCT), **d** ED-71.

to die during the hospital stay and within one year of surgery. Younger patients had a higher standard mortality ratio than did patients off 90 years of age and older [44]. The all cause mortality rate was 24% at 12 months. As the American population ages, hip fractures will substantially affect the utilisation of hospital resources. Venous thrombosis, delirium, nutrition and urinary tract management are important in the care of these patients [45].

Forty-two per cent of Australian women of 50 years of age and over sustain at least one fracture in their remaining lifetime. The proportion of women expected to sustain their first fracture increased from 1.9% of the population less than 55 years of age up to 49.1% of women over 89 years of age [46]. Davison [47] compared three treatments for displaced intracasular fractures of the hip in 280 patients of ages 65–79 years. The mean patient survival was significantly higher in the group undergoing reduction and internal fixation (79 months) compared with that with a cemented Thompson hemiarthroplasty or a cemented Monk bipolar hemiarthroplasty (61 and 68 months, respectively) [47].

Predictors of mortality and morbidity are not known, however, a biochemical marker could have such a role. Browner et al. [48] reported that there were 154 women who suffered fractures, including 34 with wrist fractures and 28 with hip fractures. Serum osteoprotegerin (OPG) levels increased with age with a mean of 0.23 ± 0.12 ng/ml. The OPG level was slightly higher in women with hypertension and about

30% greater in those with diabetes. OPG levels were also greater in women who were current users of hormone replacement therapy. Greater serum OPG levels were associated with increased all-cause and cardiovascular mortality and with fatal stroke. There was a significant association between OPG levels and subsequent fractures, but not wrist fractures [48].

Comorbidity

The association of osteoporosis with other abnormalities have been described. Abnormalities include cardiovascular disease, cancer, psychiatric issues and others.

Cardiovascular Disease: Atherosclerosis

Kado et al. [49] studied the correlation between the rate of bone loss at heel and hip and the cause of mortality in 9,704 elderly white women. Bone loss was an independent predictor of mortality even after adjustment for several potential confounders. In adjusted analyses, each SD increase of bone loss was associated with a 1.3-fold increased risk of dying from coronary heart disease and a 1.2-fold increased risk of dying from atherosclerosis. Rate of bone loss at the total hip was associated with a 1.6-fold increased mortality because of pulmonary causes. In the discussion, the association of the rate of bone loss and coronary heart disease was ascribed to oestrogen deficiency.

The correlation between BMD and stroke was studied by Jorgensen et al. [50], who demonstrated that femoral neck BMD in female stroke patients was 8% lower than in the control subjects. Women with BMD values in the lowest quartile had a higher risk of stroke than women with BMD values in the highest quartile. It was concluded that low BMD may predict stroke in women.

The correlation between high blood pressure and bone mineral loss was studied by Bulpitt et al. [51], who showed that initial bone mineral density, weight and weight change, smoking and regular use of hormone replacement therapy, the rate of bone loss increased at the femoral neck with blood pressure at baseline. For diastolic blood pressure, there was an association with bone loss in women younger than 75.

The mechanical influence of vertebral fracture appears as a trauma to the abdominal aorta [52]. Eight cases of blunt abdominal aortic disruption (BAAD) were identified from 11,645 trauma admissions, and six with concurrent thoracolumbar spine fractures. Three injury types were classified: 1. haemodynamically unstable (uncontained full-thickness laceration), 2. stable symptomatic (intimal dissection with occlusion), and 3. stable asymptomatic (contained full-thickness laceration or intimal dissection without occlusion).

Case Report

TK, an 80-year-old man.

Diagnosis: Osteoporosis, hypertension, atherosclerosis of the aorta, acute renal failure.

Present Illness: He has been suffering from hypertension for almost five years. He played golf until 60 years of age, but quit because of lumbago and gonalgia.

Physical Findings: Body length 159.9 cm, body weight 61.9 kg, BMI 24.2, BP 186/ 80 mmHg.

No abnormal findings in the chest, but there was an abnormal pulsation in the abdomen. Neurological examination showed absent deep tendon reflex but no other abnormal findings were noted.

Laboratory Examination: RBC 471 × 104, WBC 8990, platelet 25.4 × 10⁴, creatinine 1.0 mg/dl, cholesterol 232 mg/dl, albumin 4.6 g/dl, BS 116 mg/dl.

Bone and Calcium Metabolism: Serum Ca 4.7 mEq/l, P 3.0 mg/dl, bone alkaline phosphatase (BAP) 44.5 U/l (13.0–33.9), urine Ca/Cr 0.024 mg/mg (<0.3), deoxypyridinoline/Cr 7.2 μmol/mmol (2.1–5.4).

Bone X-ray and BMD: No deformity of thoracic and lumbar spine with decreased density of vertebral bodies (Figure 9.10a), BMD L2-4 (Hologic QDR 2000) 0.664 g/cm² T = −2.97, radius distal 1/3 0.684 g/cm² T = −4.82.

Course of Illness: From BMD he was diagnosed as having osteoporosis, characterised by high turnover metabolism of bone. Although urinary excretion of calcium was rather low, bone resorption marker urinary deoxypyridinoline excretion and serum bone formation marker BAP were increased in the absence of apparent fracture findings. Calcium intake may have been insufficient, especially since the previous year when his wife died.

Intermittent cyclic treatment with etidronate was initiated. Etidronate was orally administered at a dose of 400 mg/day for 14 days and this prescription was repeated with an off-period of 10 weeks. There was decreased excretion of deoxypyridinoline to 4.0 nmmol/mmol creatinine in four months and serum BAP to 26.5 U/l in eight months with an increase of L2-4 BMD four months later to 0.774 g/cm² (an increase of 11.5%). The increased level persisted for the next five months.

Eight months after the initiation of the treatment, a sudden backache appeared and a radiography of the lateral view of the spine showed a compression fracture at the level of Th 12 (Figure 9.10b). Blood pressure showed a remarkable increase to over 230 mmHg systolic and 130 mmHG diastolic. He had decreased appetite, general malaise and headache. He was admitted to the National Cardiovascular Center and was found to be suffering from renal failure with creatinine levels of up to 5.6 mg/dl, hyperpotassaemia, metabolic acidosis, and anaemia.

Case Summary: Since high-turnover osteoporosis was the main feature of his disease, etidronate may have been the best choice of treatment. There was a remarkable increase of BMD even after four months of treatment. The analgesic effect of etidronate was reported after four weeks of treatment from the clinical trials [53]. Freedom from pain may have revived his daily activity and this may have been the cause of the vertebral fracture, although there was some improvement in BMD value.

Accelerated hypertension may have been induced by the vertebral fracture. It was reported that accelerated hypertension was preceded by essential hypertension in 47 out of 121 cases with accelerated hypertension [54]. Hypertensive vascular crisis, which was investigated by this group, occurred secondary to chronic occlusion of the renal artery [55]. Ageing itself makes the large artery stiffen and there is an age-related increased impedance to ejection, a greater systolic load, a lower coronary perusion pressure and an increased pulse wave velocity. The entire beta-sympathetic system responds with a resultant decrease in the vasodilating response [56]. Magnetic resonance imaging at the National Cardiovascular Center indicated stenosis in abdominal aorta, carotid and femoral arteries.

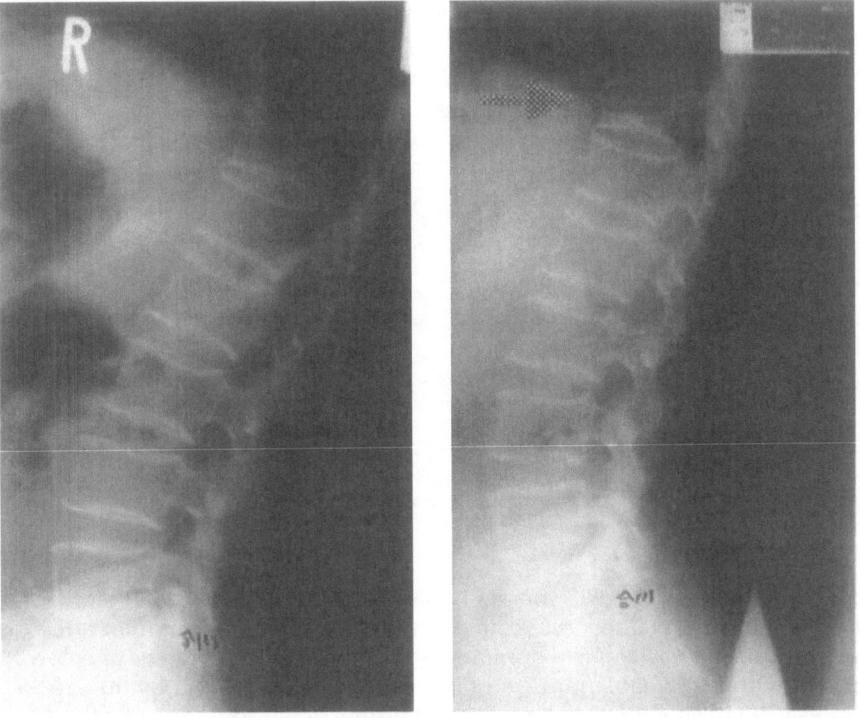

a   b

Figure 9.10.   Lateral radiography of lumbar spine before (a) and after (b arrow) fracture of Th12 in an 80-year-old man with osteoporosis. After the fracture, accelerated hypertension with acute renal failure occurred.

Each standard deviation (SD) increase of the rate of bone loss from the average value was associated with 1.2 relative hazard due to atherosclerosis [49]. Correlations between the skeletal and vascular system were shown by Shioi [57], who demonstrated that vascular cells of mesenchymal origin differentiate into osteoblastic cells in atherosclerotic cells. Many factors related to bone remodelling are involved in the process of calcification of vascular smooth muscle cells [57] (Table 9.3, originally prepared from articles by Shioi [56] and by Demer [58]. Kiel et al. [59] showed that BMD of the proximal femur was negatively correlated with coronary calcification as estimated by electron beam computed tomography in 157 women of 35–77 years of age. Osteoporosis and atherosclerosis are closely related disorders, and the pathogenesis of these two abnormalities should be investigated further from such a viewpoint.

Cancer

Colon Cancer

Epidemiological studies on calcium, vitamin D and colon cancer are inconsistent, although experimental data have shown a definite preventive effect of vitamin D on colon cancer. In the United States, data were collected from northern California, Utah

Table 9.3.   Comparison of factors in calcification between bone and artery (prepared from references 57 and 58)

	Bone	Artery
cells involved in calcification	osteoblast	vescular cell (smooth muscle cell, pericyte-like cell)
matrix vesicle	+	+
morphological features	lamellar/woven bone	bone-like cell
factors in extracellular matrix	osteoid collagen 1 osteocalcin osteopontin matrix gla protein sialoprotein osteoprotegerin	tissue of atherosclerosis collagen 1 osteocalcin osteopontin matrix gla protein osteonectin osteoprotegerin
response to Ca regulating hormone	+	+
significance of calcification	storage of calcium supporting tissue	possible stabilisation of plaque

and Minnesota involving 1,993 incident colon cancer cases and 2,410 controls [60]. Dietary calcium was inversely associated with colon cancer risk in men and women, but no significant correlation was demonstrated between the incidence of colon cancer and dietary vitamin D consumption or sunshine exposure [60]. Colon cancer incidence rates have been shown to be inversely proportional to intake of calcium. Most cases of colon cancer may be prevented with regular intake of calcium in the range of 1,800 mg per day, in a dietary context that includes 800 IU per day of vitamin D_3 (in women, an intake of approximately 1,000 mg of calcium per 1,000 kcal of energy with 800 IU of vitamin D-fortified milk). Vitamin D may also be obtained from oily fish [61]. An epidemiological study was performed in the Faroe Islands in the North Atlantic during the period 1979–1993 [62]. The incidence of colorectal cancer was the lowest among countries in north-western Europe and North America. Nutrients rich in calcium and vitamin D from fish were supposed protective against colorectal cancer.

The effects of vitamin D analogues were studied in experimental animals [63, 64]. 22-oxacalcitriol (OCT) or maxacalcitol is a new analogue of 1,25-dihydroxyvitamin D_3, and has one oxygen atom at position 22, replacing the carbon atom in the molecule, and has been shown to have a potent PTH suppressive effect [65] (Figure 9.9c). Otoshi et al. showed that OCT has a suppressive effect on carcinoma of the small intestine and there was a tendency to the same effect on carcinoma of the large intestine [63] (Figure 9.11a).

Taniyama et al. [64] also showed that 24R,25-dihydroxyvitamin D_3 has a chemopreventive effect on colon carcinogenesis in rats. 24R,25-dihydroxyvitamin D_3 (Figure 9.6b) is synthesised in the kidney from 25-dihydroxyvitamin D_3. The biological activity of 24R,25-dihydroxyvitamin D_3 as to intestinal absorption of calcium is much less compared to 1,25-dihydroxyvitamin D_3 but still has a preventive action on carcinoma, which is induced by five kinds of carcinogens with other kinds of carcinomas (Figure 9.10b).

Vitamin D insufficiency or deficiency is closely related to pathogenesis of osteoporosis and colon cancer. However, more detailed studies on the correlation between osteoporosis and colon cancer will be needed before definite conclusions can be reached.

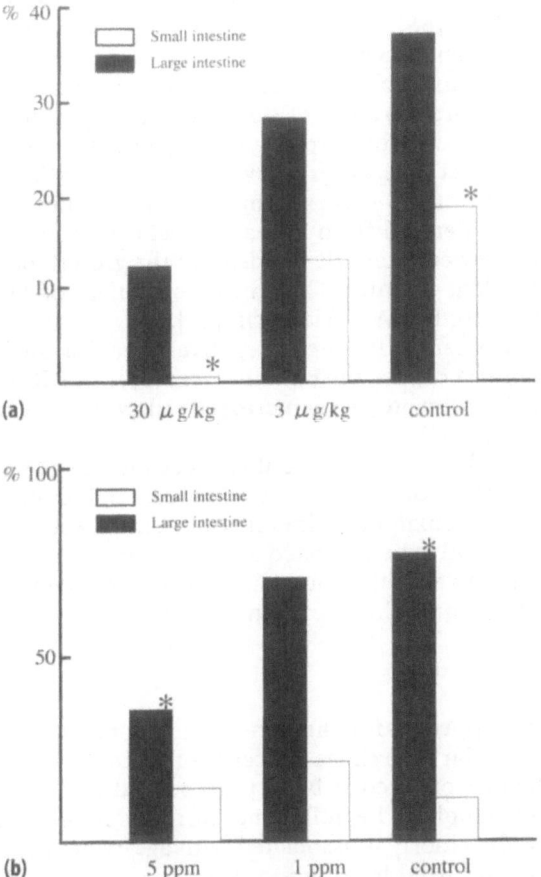

Figure 9.11.   **a** Male F344 rats were given 22-oxacalcitriol (OCT) for six weeks after being treated with five kinds of car-
cinogen for four weeks, followed by two weeks of intermittent period. The incidence of carcinoma in the small intestine (0%)
was significantly less compared to the vehicle group with administration of 30 μg/kg of OCT (18%) ($p < 0.05$). The incidence
of carcinoma in the large intestine showed a tendency to decrease in rats treated with 30 μg/kg of OCT (Otoshi et al. [63]). **b**
Male F344 rats were given 24R,25-dihydroxyvitamin D_3 mixed in their diet after being treated with N,N'-dimethyohydrazine
with four other carcinogens. Thirty weeks after the initiation of the study, the incidence of adenocarcinoma of the large intestine in the group to which the diet containing 5 PPM of 24R,25-dihydroxyvitamin D_3 was given was significantly less than that
in the vehicle group (Taniyama et al. [64]).

Breast Cancer

There are some correlations proposed regarding breast cancer and osteoporosis:

1. BMD and incidence of breast cancer,

2. Vitamin D intake and breast cancer,

3. Selective oestrogen receptor modulator in the treatment of osteoporosis.

In a prospective cohort study, 131 incident breast cancer women were identified from
among 8,203 postmenopausal women in the state of Washington, USA. An increase

in breast cancer risk was noted among women with higher BMD, independent of age, geographic area, and BMI [66]. Barger-Lux et al. [67] stated that long-term calcium restriction and/or insufficient vitamin D may promote the development of bone fragility, high blood pressure, colon cancer, and breast cancer in susceptible individuals. OCT inhibited the proliferation of both ER (oestrogen receptor) positive and ER-negative breast cancer cells in vitro in a time- and dose-dependent manner [66]. The antiproliferative effect was observed with a concentration as low as 10–11 M OCT. The in vivo effect of OCT was examined in athymic mice implanted with ER-negative MX-1 tumours derived from human breast carcinoma. Intratumour transplantation three times a week remarkably delayed the growth of MX-1 tumours in a time- and dose-dependent manner. The antitumour effect of 1 μg/kg BW OCT was greater than that of 500 μg/kgBW adriamycin [68].

In studies in 241 women following a negative screening mammogram and 181 women with known breast cancer, the vitamin D receptor (VDR) gene polymorphism Bsm 1 was significantly associated with increased risk with odds ratio bb vs BB genotype 2.32 (95% CI, 1.23–4.39) (69).

It was demonstrated that tamoxifen could reduce breast cancer incidence by 50%. Nevertheless, only half the women who develop breast cancer have risk factors other than age. Tamoxifen and raloxifene maintain bone density, and raloxifene is now used to prevent osteoporosis and is being tested as a preventive for coronary heart disease and breast cancer. If the twentieth century was the era of chemotherapy, the twenty-first century will be the era of chemoprevention [70].

Periodontitis

Tooth health is very important for the general health of youngsters as well as for elderly individuals. Alveolar pyorrhoea is a cause of loss of alveolar bones but another aspect is whether the loss of alveolar bone is associated with the process of general bone loss in elderly people. The following working hypotheses will be issued. Alveolar bone loss in the elderly is associated with age-related osteoporosis, and can be treated with medicines used for osteoporosis.

A study from the US indicated that the mean alveolar bone loss was significantly correlated with BMD of the trochanter, Ward's triangle and total regions of the femur and that the mean clinical attachment loss showed a tendency of a relation to BMD consistently at all regions of the skeleton [71]. Payne et al. [72] showed that osteoporotic women exhibited a higher frequency of alveolar bone height loss and crestal and subcrestal density loss relative to women with normal BMD. Oestrogen deficiency was associated with increased frequency of alveolar bone crestal density loss in osteoporotic women and in the overall study population.

One of the main causes of alveolar bone loss is periodontitis, which should be treated with antimicrobial agents and other dental procedures. However, cytokines such as interleukins are supposed to accelerate bone resorption. Bisphosphonates are potent inhibitors of bone resorption and have been utilised for the treatment of several bone diseases including osteoporosis [73]. Alendronate was effective in preventing alveolar bone loss when evaluated by subtraction radiography [74]. Takaishi et al. [75] treated four patients with etidronate who had periodontitis and resultant alveolar pyorrhoea. Etidronate was administered orally at a daily dose of 200 mg for two weeks followed by 10 weeks of off-period.

The gross appearance of gingival mobility of the teeth, depth of periodontal pockets (Table 9.4) and X-ray findings of alveolar bones improved markedly.

Table 9.4. Effect of treatment including oral administration of Etidronate on the course of alveolar pyorrhoea in four cases as evaluated by mobility score and measurement of the depth of periodontal pockets (Takaishi et al. [75])

Case	month of treatment	Mobility Score*** Maxilla Total No of teeth	Score 0	1	2	3	Mandibula No of teeth	Score 0	1	2	3	Depth of periodontal pockets (mm) Maxilla	Mandibula
1	0	14	0	8	5	1	12	0	6	6	0	5.1±1.0	5.1±0.9
	12	14	9	4	0	1	12	12	0	0	0	3.1±0.6*	3.1±0.5*
	31	14	10	3	0	1	12	12	0	0	0	2.4±0.6*	2.1±0.3*
2	0	12	3	3	6	0	14	8	6	0	0	5.4±0.8	4.7±0.5
	18	12	6	5	1	0	14	11	3	0	0	3.3±0.6*	2.9±0.6*
	22	12	11	0	1	0	14	14	0	0	0	3.1±1.1*	2.1±0.3*
3	0	13	11	2	0	0	12	0	12	0	0	3.5±0.6	3.5±0.7
	17	13	13	0	0	0	12	12	0	0	0	2.6±0.6*	2.6±0.5*
	25	13	13	0	0	0	11	11	0	0	0	2.1±0.3*	2.4±0.5*
4	0	13	6	5	2	0	12	9	3	0	0	3.2±0.6	2.8±0.4
	6	13	11	2	0	0	11	10	1	0	0	2.2±0.4*	2.5±0.5**
	12	13	13	0	0	0	11	11	0	0	0	2.2±0.4*	2.6±0.5*
	21	13	13	0	0	0	11	11	0	0	0	2.2±0.6*	2.0±0.0*
	30	13	13	0	0	0	11	11	0	0	0	2.2±0.6*	2.0±0.0*

* $p < 0.005$; ** $p < 0.025$; (Mean ± SD)
Significance of difference was shown at time indicated in comparison with data of zero month of treatment.
*** First degree mobility is defined as movement of the top of a tooth of 0.2–1.0 mm in the horizontal direction, second degree as movement of more than 1.0 mm in the horizontal direction and third degree as movement in both the horizontal and vertical direction.
This table was reproduced with the permission of the author and Cambridge Medical Publications.

Alveolar bone loss in elderly women occurs as a result of inflammation of periodontal tissues and oestrogen deficiency- induced processes and thus there is a rational basis for treatment with bisphosphonates.

Depression

Depression has been shown to be associated with diabetes, cardiovascular disease, immunological abnormalities, multiple sclerosis, cancer, osteoporosis and ageing [76]. A study in Portugal showed that women with osteoporosis have significantly higher levels of depressive symptoms and a corresponding higher prevalence of depression, independent of other factors strongly associated with osteoporosis, such as age or BMI [77]. The Quality of Life Questionnaire of the European Foundation for Osteoporosis (QUALEFFO) was performed in 159 patients aged 55–80 years with clinical osteoporosis. The analysis of QUALEFFO indicated that five domains (pain, physical activity, social function, general health perception and mental function) were significantly predictive of vertebral fractures [78]. In a survey in the USA there was no difference in mean BMD of the hip and lumbar spine in women with depression compared to those without depression. Women with depression were more likely to experience subsequent falls than women without depression. Women with depression had a 40% increased rate of non-vertebral fracture compared with women without depression [79]. Azuma et al. [80] showed that the self-rating depression scale (SDS) was significantly higher in patients with vertebral fractures than in those without fractures (Table 9.5).

Thus, there are controversial opinions regarding the correlation between BMD and depression but fracture rates were consistently higher in patients with fractures than in those without depression. The cause–result relationship has not been established but either one of fragility fracture and depression can be cause for another event.

Table 9.5. Comparison of QOL between osteoporotic women with fractures and non-fractures (Azuma et al. [80])

	Patients with Fractures	Patients without Fractures	Significance
LSIA 1	12.1 ± 4.1 (48)	12.5 ± 3.3 (141)	NS
SES 2	35.8 ± 5.3 (56)	36.0 ± 6.4 (146)	NS
STAI 3	38.9 ± 8.1 (49)	40.3 ± 9.3 (144)	NS
SDS 4	41.5 ± 8.0 (34)	37.8 ± 8.3 (83)	<0.05

Numbers in parentheses indicate number of subjects studies.   (Mean ± SD)
1. LSIA   Life satisfaction index A (Neugarten BL J Geront 1961, 16:134–143.
2. SES   Self-esteem scale (Rosenberg M. Society and Adolescent Self-Image, Princeton U Press, 1965).
3. STAI   The state-trait anxiety inventory (Spielberger CD, Manual for the State-Trait Anxiety Inventory, Consult Psychologist Press, 1970).
4. SDS   Self-rating depression scale (Zung WWK. Arch Gen Psychiat 1965, 12:63–75).

References

1. Consensus Development Conference on Osteoporosis. Hong Kong, April 1–2, 1993. Am J Med. 1993;95(5A):1S–78S.
2. Khosla S, Riggs BL, Melton III LJ. Clinical spectrum. In: Riggs BL, Melton III LJ, editors. Osteoporosis: Etiology, diagnosis, and management 2nd ed. Philadelphia: Lippincott-Raven, 1995.
3. Martin BR. Aging and changes in cortical mass and structure. In: Orwoll ES, editor. Osteoporosis in Men. San Diego: Academic Press, 1999;Ch 7.
4. Genant HK, Cooper C, Poor G, Reid I, Ehrlich G, Kanis J et al. Interim report and recommendations of the World Health Organization Task Force for Osteoporosis. Osteoporosis Int 1999;10(4):259–64.
5. WHO Study Group. Assessment of fracture risk and its application to screening for postmenopausal osteoporosis. Geneva: World Health Organization, 1994.
6. Lau EMC. The epidemiology of osteoporosis in Asia. In: Lau EDC, editor. Osteoporosis in Asia. Singapore: World Scientific, 1997.
7. Yamamoto I. Estimation of osteoporosis population in Japan. Osteoporosis Jpn 1999;9:1183–64.
8. Ling XU, Cummings SR, Mingwel Q, Xihe Z, Xioashu C, Nevitt M et al. Vertebral fractures in Beijing, China: The Beijing Osteoporosis Project. J Bone Miner Res 2000;15:2019–25.

9. Fujiwara S. Ross PD. Epidemiological studies of osteoporosis in Japan. In: Lau EMC, editor. Osteoporosis in Asia. Singapore: World Scientific, 1997;21–29.
10. Nordin BEC. Nutritional considerations. In: Nordin BEC, editor. Calcium, phosphate and magnesium metabolism. Edinburgh: Churchill Livingsone, 1976;1–35.
11. Garraway WM, Stauffer RN, Kurland LT, O'Fallon WM. Limb fractures in a defined community. 1. Frequency and distribution. Mayo Clinic Proc 1979;54:701–7.
12. Orimo H et al. Survey of hip fracture in Japan. Japan Med Archiv 1999;3916:46–49.
13. Schwartz AV, Kelsey JL, Maggi S, Tuttleman M, Ho SC, Jonsson PV et al. International variation in the incidence of hip fractures: cross-national project on osteoporosis for the World Health Organization Program for Research on Aging. Osteoporos Int 1999;9:242–53.
14. Johnell O, Gullberg B, Kanis JA, Allander E, Elffors L, Dequeker J et al. Risk factors for hip fracture in European women: the MEDOS Study. Mediterranean Osteoporosis Study. J Bone Miner Res 1995;10:1802–15.
15. Suzuki T, Yoshida H, Hashimoto T, Yoshimura N, Fujiwara S, Fukunaga M et al. Case-control study of risk factors for hip fractures in the Japanese elderly by a Mediterranean Osteoporosis Study (MEDOS) questionnaire. Bone 1997;21:461–7.
16. Morii H, Shioi A, Inaba M, Goto H, Kawagishi T, Nakatsuka K et al. Significance of albumin in the pathogenesis of osteoporosis: Bone changes in genetically analbuminaemic rats and rats fed a low albumin diet. Osteoporos Int 1997;7 Suppl 3:S30–35.
17. Schurch MA, Rizzoli R, Vadas L, Slosman DO, Bonjour JP. Protein supplements in the elderly with a recent fracture increase serum IGF-1, decrease urinary deoxypyridinoline, and prevent proximal femur bone loss. J Bone Miner Res 1996;11(suppl 1) S139.
18. Nordin BEC. The calcium controversy. Osteoporos Int 1997;7(suppl 3):S17–23.
19. Reid IR, Ames RW, Evans MC, Gamble GD, Sharpe SJ. Long-term effects of calcium supplementation on bone loss and fractures in postmenopausal women: a randomized controlled trial. Am J Med 1995;98:331–3.
20. Ott SM. Calcium and vitamin D in the pathogenesis of osteoporosis. In: Marcus R, editor. Osteoporosis. Oxford, UK: Blackwell Scientific Publications, 1994;237–92.
21. Caniggia A, Lore F, di Cairano G, Nuti R. Main endocrine modulators of vitamin D hydroxylases in human pathophysiology. J Steroid Biochem 1987;27(4–6):815–24.
22. McKenna MJ, Freaney R. Secondary hyperparathyroidism in the elderly: Means to defining hypovitaminosis D. Osteoporos Int 1998;8(Suppl 2):S3–6.
23. Albrigth F, Reifenstein J. The parathyroid glands and metabolic bone disease. Baltimore: Williams & Wikins, 1948.
24. DeLuca HF. 1,25-dihydroxyvitamin D_3 in the pathogenesis and treatment of osteoporosis. Osteoporos Int 1997;7(Suppl 3):S24–9.
25. National Osteoporosis Foundation. Osteoporosis: Review of the evidence for prevention, diagnosis, and treatment and cost-effectiveness analysis. Osteoporos Int 1998;8(Suppl 4).
26. Chapuy MC, Preziosi P, Maamer M, Arnaud S, Galan P, Hereberg S et al. Prevalence of vitamin D insufficiency in an adult normal population. Osteoporos Int 1997;7:439–43.
27. Chapuy MC, Arlot ME, Duboeuf F, Brun J, Crouzet B, Arnaud S et al. Vitamin D_3 and calcium to prevent hip fractures in elderly women. N Engl J Med 1992;327:1637–42.
28. Scharla SH. Prevalence of subclinical vitamin D deficiency in different European countries. Osteoporos Int 1998;8(Suppl 2):S7–12.
29. Meunier P. Osteoporosis: Diagnosis and management. London: Martin Dunitz, 1998.
30. Ringe JD. Vitamin D deficiency and osteopathies. Osteoporos Int 1998;8(Suppl 2):S35–9.
31. Parfitt AM. Osteomalacia and related disorders. In: Avioli LV, Krane SM editors. Metabolic Bone Disease and Clinically Related Disorders, IIIrd ed. Academic Press, San Diego 1998;327–86.
32. Harris S, Soteriadew E, Coolidge JAS, Mudgal S, Dawson-Hughes B. Vitamin D insufficiency and hyperparathyroidism in low income, multiracial, elderly population. J Clin Endocrinol Metab 2000;85:4125–30.
33. Melin AL, Wilske J, Ringertz, Saaf M. Vitamin D status, parathyroid function and femoral bone density in an elderly Swedish population living at home. Aging (Milano) 1999;11:200–7.
34. Sahota O, Masud T, San P, Hosking DJ. Vitamin D insufficiency increases bone turnover markers and enhances bone loss at the hip in patients with established vertebral osteoporosis. Clin Endocrinol (Oxf) 1999;51:217–21.
35. Peacock M, Liu G, Carey M, McClintock R, Ambrosius W, Hui S et al. Effect of calcium or 25 OH vitamin D_3 dietary supp. Clin Endocrinol METAB 2000;85:3009–10.
36. Nakamura T. The importance of genetic and nutritional factors in responses to vitamin D and its analogs in osteoporotic patients. Calcif Tissue Int 1997;60:119–23.

37. Tilyard M, Spears GF, Thomson J, Davey S. Treatment of postmenopausal osteoporosis with calcitriol or calcium. N Engl J Med 1992;326:357–62.
38. Tanizawa T, Imura K, Ishii Y, Nishida S, Takano Y, Mashiba T et al. Treatment with active vitamin D metabolites and concurrent treatments in the prevention of hip fractures: A retrospective study. Osteoporos Int 1999;9:163–70.
39. Masaki T, Shiraishi A, Higashi S, Uchida Y, Saito M, Ogata E et al. Superiority of alphacalcidol over plain vitamin D: Failure to increase bone strength at cortical site by the treatment with plain vitamin D but no alphacalcidol might be related to the enhanced porosity. J Bone Miner Res 2000;15(Suppl 1)M449;S563.
40. Shiraishi A, Higashi S, Masaki T, Uchida Y, Saito M, Ito M et al. Improvement of bone strength by alpha-calcidol treatment: Relationship to trabecular microarchitecture and antiresorptive activity. J Bone Miner Res 2000;15(Suppl 1)SA449;S315.
41. Matsumoto T, Miki T, Sugimoto T, Teshima R, Kato Y, Okamoto S et al. A new active vitamin D analog, ED-71, increases bone mass with preferential effects on bone in osteoporotic patients. J Bone Miner Res 2001;(Suppl 1) to be presented in October, 2001.
42. Cooper C, Melton III LJ. Magnitude and impact of osteoporosis and fractures. In: Marcus R, Feldman D, Kelsey J, editors. Osteoporosis. San Diego: Academic Press, 1996.
43. Ensrud KE, Thompson DE, Cauley JA, Nevitt MC, Kado DM, Hochberg MC et al. Prevalent vertebral deformities predict mortality and hospitalization in older women with low bone mass, Fracture Intervention Trial Research Group. J Am Geriar Soc 2000;48:241–9.
44. Shah MR, Aharonoff GB, Wolinsky P, Zuckerman JD, Koval KJ. Outcome after hip fracture in individuals ninety years of age and older. J Orthop Trauma 2001;15:34–9.
45. Huddleston JM, Whitford KJ. Medical care of elderly patients with hip fractures. Mayo Clin Proc 2001;76:295–8 (referred to Medline).
46. Doherty DA, Sanders KM, Kotowicz MA, Prince RL. Lifetime and five-year age-specific risks of first and subsequent osteoporotic fractures in postmenopausal women. Osteoporos Int 2001;12:16–23.
47. Davison JN, Caider SJ, Anderson GH, Ward G, Japper C, Harper WM et al. Treatment for displaced intracapsular fracture of the proximal femur. A prospective, randomized trial in patients aged 65 to 79 years. J Bone Joint Surg Br 2001;83:206–12 (referred to Medline).
48. Browner WS, Lui L-Y, Cummings SR. Association of serum osteoprotegerin levels with diabetes, bone density, fractures, and mortality in elderly women. J Clin Endocrinol Metab 2001;86:631–7.
49. Kado DM, Browner WS, Blackwell T, Gore R, Cummings SR. Role of bone loss is associated with mortality in older women: A prospective study. J Bone Miner Res 2000;15:1974–80.
50. Jorgensen L, Engstad T, Jacobsen BK. Bone mineral density in acute stroke patients: Low bone mineral density may predict first stroke in women. Stroke 2001;32:47–51.
51. Bulpitt C, Fletcher A, Beckett N, Coope J, Gil-Extremera B, Forette F et al. Hypertension in the very elderly trial (HYVET) protocol for the main trial. Drugs Aging 2001;18:151–64.
52. Inaba K, Kirkpatrick AW, Finkelstein J, Murphy J, Brenneman FD, Boulanger BR et al. Blunt abdominal aortic trauma in association with thoracolumbar spine fractures. Injury 2001;32:201–7.
53. Fujita T, Orimo H, Inoue T, Kaneda K, Sakurai M, Morita R et al. Double-blind, multicenter comparative study with alphacalcidol of etidronate disodium (EHDP) in involutional osteoporosis. Clin Eval 1993;21:261–302.
54. Scarpelli PT, Livi R, Caselli GM, Di Maria L, Teghini L, Montemurro V et al. Accelerated (malignant) hypertension: a study of 121 cases between 1974 and 1996. J Nephrol 1997;10:207–15.
55. Scarpelli PT, Livi R, Caselli GM, Chiari G, Gallo M, Teghini L et al. Hypertensive vascular crisis secondary to chronic total renal artery occlusion. J Nephrol 1998;11:325–9.
56. Weisfeldt M. Aging, changes in the cardiovascular system, and response to stress. Am J Hypertens 1998;11:41S.
57. Shioi A. Vascular calcification. This monograph.
58. Demer LL. A skeleton in the atherosclerotic closet. Circulation 1995;92:2029–31.
59. Kiel DP, Hannan MT, Cupples LA, Wilson PWF, Levy W, Clouse ME et al. Low bone mineral density (BMD) is associated with coronary artery calcification. J Bone Miner Res 2000;15(Suppl 1):S160, 1087.
60. Kampman E, Stattery ML, Caan B, Potter JD. Calcium, vitamin D, sunshine exposure, dairy products and colon cancer risk (United States). Cancer Causes Control 2000;11:459–66.
61. Garland CF, Garland FC, Gorham ED. Calcium and vitamin D. Their potential roles in colon and breast cancer prevention. Ann NY Acad Sci 1999;889:107–19.
62. Dalberg J, Jacobsen O, Nielsen NH, Steig BA, Storm HH. Colorectal cancer in the Faroe Islands – a setting for the study of the role of diet. J Epidemiol Biostat 1999;4:31–6.
63. Otoshi T, Iwata H, Kitano M, Nishizaawa Y, Morii H, Yano Y et al. Inhibition of intestinal tumor development in rat multi-organ carcinogenesis and aberrant crypt foci in rat colon carcinogenesis by 22 oxa-calcitriol, a synthetic analogue of 1α, 25-dihydroxyvitamin D_3. Carcinogenesis 1995;16:2091–7.

64. Taniyama T, Wanibuchi H, Salim EI, Yano Y, Otani S, Nishizawa Y et al. Chemopreventive effect of 24R,25-dihydroxyvitamin D_3 in N,N'-dimethyohydrazine-induced rat colon carcinogenesis. Carcinogenesis 2000;21:173–8.
65. Sato K, Tominaga Y, Ichikawa F, Uchida K, Takezawa J, Tsuchiya T et al. In vitro suppression of parathyroid hormone secretion by 22-oxa-calcitriol in human parathyroid hyperplasia due to uraemia. Nephrology 1998;4:177–82.
66. Buist DS, LaCroix AZ, Barlow WE, White E, Weiss NS. Bone mineral density and breast cancer risk in postmenopausal women. J Clin Epidemiol 2001;54:417–22.
67. Barger-Lux MJ, Heaney RP. The role of calcium intake in preventing bone fragility, hypertension, and certain cancers. J Nutr 1994;124:1406S–11S.
68. Abe J, Nakano T, Nishii Y, Matsumoto T, Ogata E, Ikeda K. A novel vitamin D analog, 22-oxa-1α,25dihydroxyvitamin D_3 inhibits the growth of human breast cancer in vitro and in vivo without causing hypercalcemia. Endocrinology 1991;129:832–7.
69. Bretherton-Watt D, Given-Wilson R, Mansi JL, Thomas V, Carter N, Colston KW. Vitamin D receptor gene polymorphisms are associated with breast cancer risk in a UK Caucasian population. Br J Cancer 2001;85:171–5.
70. Jordan S. Progress in the prevention of breast cancer: incept to reality. J Steroid Biochem Mol Biol 2000;74:269–77.
71. Tezal M, Wactawski-Wende J, Grossi SG, Ho AW, Dunford R, Genco RJ. The relationship between bone mineral density and periodontitis in postmenopausal women. J Periodontol 2000;71:1492–8.
72. Payne JB, Reinhardt RA, Nummikoski PV, Patil KD. Longitudinal alveolar bone loss in postmenopausal osteoporotic/osteopenic women. Osteoporos Int 1999;10:34–40.
73. Fleisch H. Bisphosphonates in bone disease. 4th ed. San Diego: Academic Press, 2000.
74. Jeffcoat MK, Reddy MS. Alveolar bone loss and osteoporosis: evidence for a common mode of therapy using the bisphosphonate alendronate. In: Davidovitch Z, Norton LA, editors. Biological mechanisms of tooth movement and cranofacial adaptation. Proc Second Int Conf at Tara Ferncroft Conference Resort, Danvers, MA, 1995;Oct 19–22:365.
75. Takaishi Y, Miki T, Nishizawa Y, Morii H. Clinical effect of etidronate on alveolar pyorrhea associated with chronic marginal periodontitis. J Int Med Res 2001;29:355–65.
76. Horrobin DF, Bennett CN. Depression and bipolar disorder: relationships to impaired fatty acid and phospholipid metabolism and to diabetes, cardiovascular disease, immunological abnormalities, cancer, ageing and osteoporosis. Possible candidate genes. Prostaglandins Leukot Essent Fatty Acids 1999;60:217–34.
77. Coelho R, Silva C, Maia A, Prata J, Barros H. Bone mineral density and depression: a community study in women. J Psychosom Res 1999;46:29–35.
78. Lips P, Cooper C, Agnusdei D, Caulin F, Egger P, Jonell O et al. Quality of life in patients with vertebral fractures: validation of the Quality of Life Questionnaire of the European Foundation for Osteoporosis (QUALEFFO) Working Party for Quality of Life of the European Foundation for Osteoporosis. Osteoporos Int 1999;10:150–60.
79. Whooley MA, Kip KE, Cauley JA, Ensrud KE, Nevitt MC, Browner WS. Depression, falls, and risk of fracture in older women. Study of Osteoporotic Fractures Research Group. Arch Intern Med 1999;159:484–90.
80. Azuma M, Shirota K, Yasumori Y, Tabata K, Maeda Y, Yoshimura Y et al. Quality of life form mental state in patients with osteoporosis. Osteoporosis Jpn 2000;8:269–71 (in Japanese).

Secondary Osteoporosis

M. Inaba and E. Ishimura

Osteoporosis, the most common metabolic bone disease, is by definition a systemic skeletal disease characterised by low bone mass and microarchitectual deterioration of bone tissue, with resultant increase in bone fragility and susceptibility to fracture. While primary osteoporosis is a condition of reduced bone mass appearing in postmenopausal women (postmenopausal osteoporosis) and in elderly individuals (senile osteoporosis), secondary osteoporosis is a condition of reduced bone mass resulting from a variety of specific and well-defined disorders, such as thyrotoxicosis, glucocorticoid use, and immobilisation (Table 9.6).

To understand the mechanism of development of osteoporosis, the intimate involvement in it of bone remodelling, a continuous process of bone resorption and formation throughout life, should be borne in mind. In the remodelling cycle, the amount of bone replaced by formation is not always sufficient to replace the amount previously removed during the resorption phase. The cause of this imbalance can be an increase in the amount of bone resorbed, or a decrease in the amount formed, or both. Even if the magnitude of this imbalance is small during each remodelling cycle, bone mass is lost as remodelling cycles proceed. Therefore, both the activation frequency, birth rate of new remodelling units, and net imbalance during each remodelling cycle are major factors determining the rate of bone loss. Furthermore, the concept of "remodelling space" should be recalled. This term is applied to the temporarily missing bone space which has not yet been replaced by bone formation after resorption during a remodelling cycle. Remodelling space expands as bone turnover increases. In contrast, it contracts when bone turnover slows. Therefore, enhanced bone turnover may reduce bone mass in part by expanding remodelling space.

These mechanisms together explain why acceleration of bone turnover increases bone loss and why inhibition of it reduces bone loss. A variety of diseases enhancing bone turnover, mainly by stimulating bone resorption, correspondingly cause the development of secondary osteoporosis. In contrast, the principal mechanism by which glucocorticoid excess and diabetes mellitus cause secondary osteoporosis is their profound inhibitory effect on bone formation.

Thyrotoxicosis

Thyroid hormone exerts its effects on osteoblasts via nuclear receptors and thereby stimulates osteoclastic bone resorption [1-3]. Both osteoblast and osteoclast activities are increased, with predominance of the latter activity, and hyperthyroidism is thus one of the major causes of secondary osteoporosis [4]. Although the presence

347

Table 9.6.   Aetiology of Secondary Osteoporosis

Endocrine/Metabolic Diseases
   Hyperparathyroidism (primary, secondary)
   Thyrotoxicosis
   Cushing's syndrome
   Hypogonadism
   Diabetes mellitus
   Pregnancy
   Anorexia nervosa

Inflammatory Diseases
   Rheumatoid Arthritis
   Ankylosing spondylitis

Functional
   Immobilisation/weightlessness
   Chronic obstructive lung diseases
   Post-gastrectomy
   Hepatic disease (particularly primary biliary cirrhosis)
   Alcohol abuse
   Following organ transplantation

Haematopoietic
   Multiple myeloma
   Lymphoma/Leukaemia
   Mastocytosis

Congenital
   Osteogenesis imperfecta
   Menkes' syndrome
   Ehlers–Danlos syndrome
   Homocysteinuria
   Marfan's syndrome

Drugs
   Corticosteroid
   Thyroxine
   Anticonvulsant (barbiturates, phenytoin)
   Anticoagulants (heparin, coumarin)
   Antimetaolites (methotrexate, cyclosporine)

of thyroid hormone receptors has been demonstrated in osteocytes and osteoclasts, the physiological significance of this has not been thus far elucidated [5]. It has been clearly demonstrated that hyperthyroid patients exhibit increases in both osteoclastic and osteoblastic activity with a predominance of bone resorption, resulting in a decrease in bone mineral density (BMD) [6]. Furthermore, normalisation of thyroid function was found to be associated with an increase in lumbar spine BMD, which was preceded by significant attenuation of increased bone turnover [7]. Although there is an increase in the number and length of osteoid borders in hyperthyroid patients, the increased osteoid is always associated with bone-forming surfaces with normal calcification. As shown in Figure 9.12, the increased entry of calcium from bone into blood due to increased bone resorption suppresses PTH secretion from the parathyroid gland and thus the PTH-induced vitamin D activation in the kidneys. The resultant negative calcium balance due both to decreased intestinal absorption and increased urinary excretion may further accelerate bone loss in hyperthyroidism

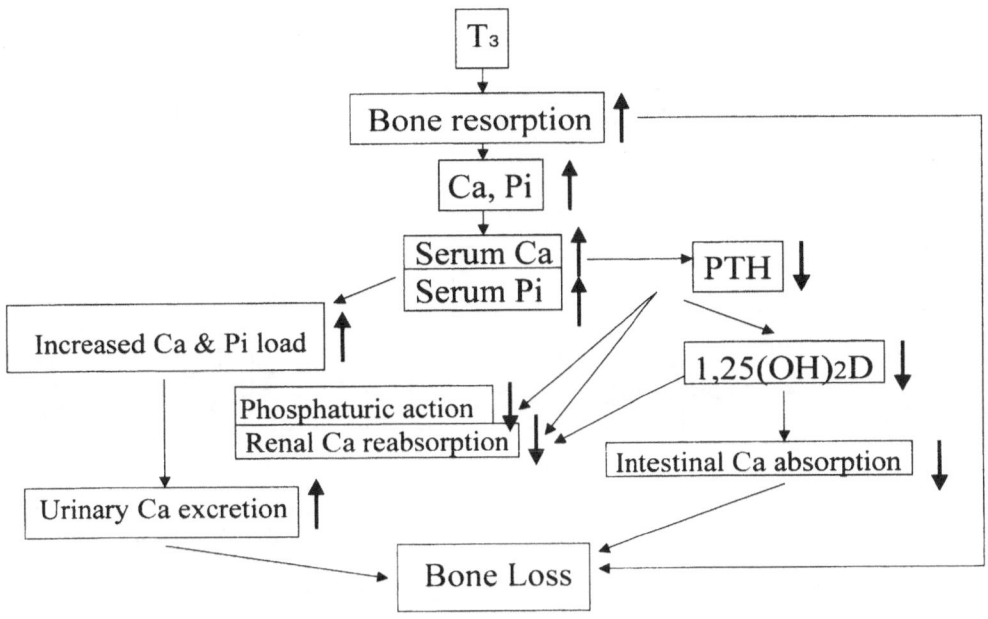

Figure 9.12. Mechanism of bone loss in patients with thyrotoxicosis.

[8]. Although calcium load in the blood is increased, since the capacity of the kidneys to excrete calcium into urine is usually enormous under the condition of suppressed PTH, clinically significant hypercalcaemia in patients with this condition has seldom been reported [9]. However, development of severe hypercalcaemia was reported in a rare case in which renal function was so impaired that increased calcium load could not be excreted into urine [10].

Bone loss is dependent on the duration of the hyperthyroid state, and the response of bone to thyroid hormone is augmented after menopause [11]. Consensus has not been reached concerning the prevalence of low bone mass in hyperthyroid patients, particularly those older than 50 years. Since elderly hyperthyroid patients often exhibit the vague symptoms such as depression and body weight loss with relatively few physical signs, the diagnosis may be missed for a long period of time. Therefore, hyperthyroidism should be recalled as a cause of osteoporosis in elderly individuals. Osteoporosis preferentially occurs in bone areas enriched in cortical bone such as the femur.

Controversy exists whether antithyroid drugs (ATD) can completely normalise bone metabolism in Graves' disease (GD) patients and whether the reduction in bone mineral density is restored to completely normal levels [8, 9, 12, 13]. Of GD patients treated with ATD, some exhibited persistent suppression of thyroid-stimulating hormone (TSH) long after their serum levels of free triiodothyronine (FT$_3$) and free thyroxine (FT$_4$) were normalised; these patients are in a so-called subclinical hyperthyroid state [10–14]. We recently reported that GD patients whose serum TSH has been suppressed even after ATD treatment still retain increased bone turnover state (Figure 9.13), and are thus at high risk for accelerated bone loss [15], indicating that even a minimal hyperthyroid state sustained on a chronic basis can decrease cortical bone mass. We suggest the possibility that TSH receptor antibody may directly stimulate bone metabolism, since 1) a strong correlation exists between TSH receptor

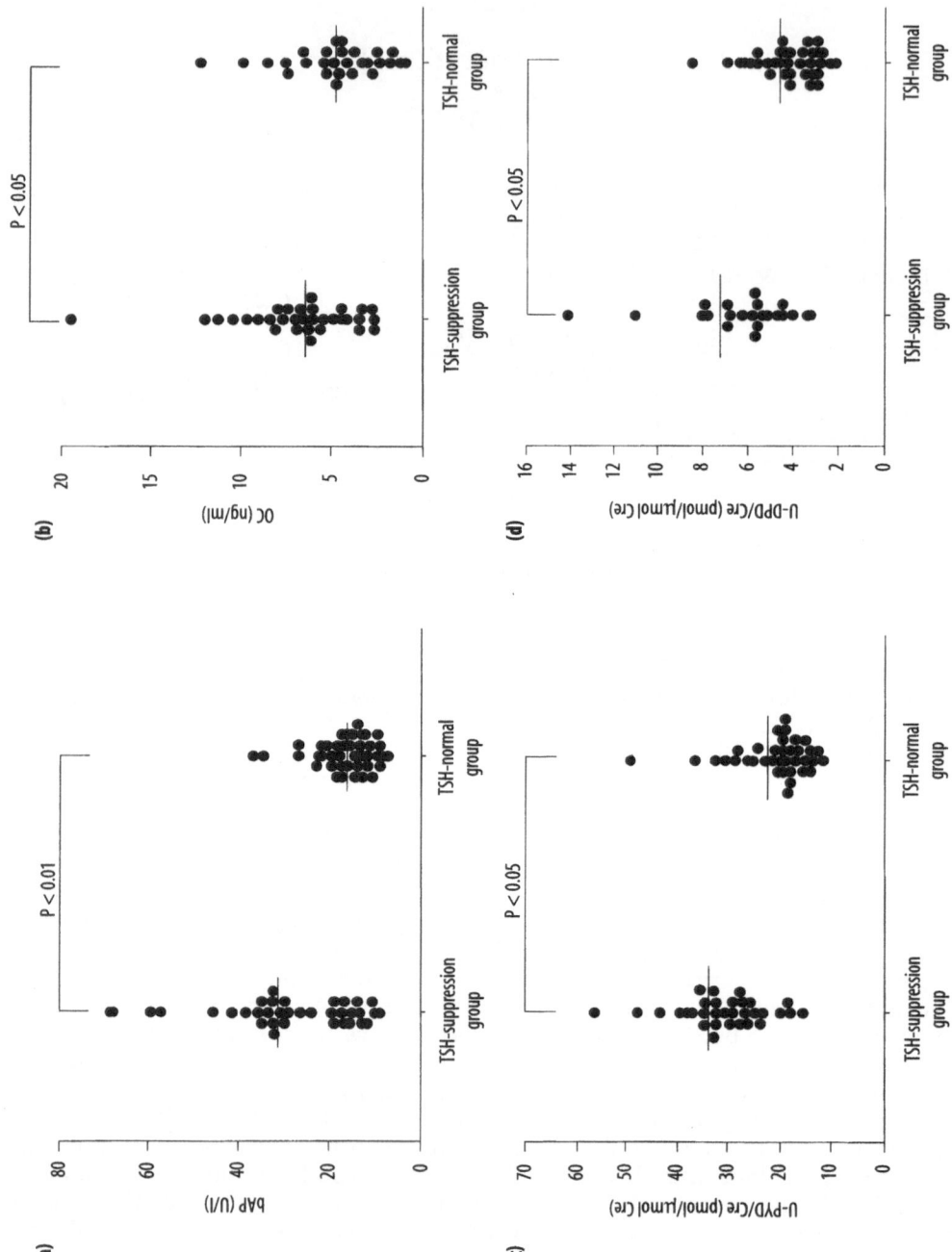

antibody titre and bone markers including bone-specific alkaline phosphatase and urinary excretions of pyridinoline and deoxypyridinoline, and 2) TSH receptors are present in osteoblasts [16]. To normalise bone turnover in these patients and thus to protect against accelerated bone loss, bone-antiresorptive agents such as bisphosphonates and oestrogen may be useful since these agents were demonstrated to effectively protect bone loss in hyperthyroid patients [17, 18]. However, when ATD normalises serum TSH in GD patients, bone markers fall to within a normal range, suggesting that bone metabolism and the risk of osteoporosis in these patients may be restored to completely normal levels. This phenomenon is similar to that in patients receiving thyroid hormone. Patients with serum TSH normalised and suppressed exhibited no and high risk of increased bone loss, respectively [19, 20].

In summary, patients whose serum TSH is normalised are not likely to need additional treatment for osteoporosis. In contrast, those with suppressed serum TSH should be treated by block replacement therapy (large doses of ATD plus L-thyroxine) to normalise serum TSH or by additional treatment for osteoporosis, mainly with the use of bone antiresorptive drugs such as bisphosphonate or oestrogen derivatives.

Glucocorticoid-Induced Osteoporosis

Although glucocorticoids (GC) are effective for control of inflammatory diseases and thus widely used, osteoporosis is one of the major problems associated with their long-term use [21, 22]. The incidence of osteoporosis-related fractures in patients receiving GC for more than six months is reported to be between 30% and 60% [23], indicating that GC-induced osteoporosis is probably the most important causative disease among secondary osteoporosis. Although the mechanism by which GC induces osteoporosis has not been completely elucidated because of its complexity, the main mechanism appears to be its profound inhibitory effects on osteoblasts (Figure 9.14). GC directly inhibits osteoblast replication and differentiation at supraphysiological concentrations [24]. On the other hand, GC enhances bone turnover overall by stimulating bone resorption. Increased bone resorption is due mainly to the development of secondary hyperparathyroidism resulting from both GC-induced increase in urinary loss of calcium [25] and decreased intestinal calcium absorption probably independent of vitamin D metabolism [26]. Furthermore, GC inhibits

---

Figure 9.13.   Bone parameters in TSH-normal and TSH-suppressed Graves' patients. Sixty-seven patients with Graves' disease (52 female and 15 male) had been maintained euthyroid (2.6 pg/ml < FT3 < 4.3 pg/ml and 0.8 ng/dl < FT4 < 1.8 ng/dl) for at least 10 months under treatment with less than 5 mg/day thiamazole or 50 mg/day propylthiouracil. Patients were divided into two groups on the basis of their serum TSH level; TSH-normal group (10/32, M/F) who showed normal FT3 and FT4 with serum TSH within normal range (0.4 μIU/ml ≦ TSH < 4.0 μIU/ml) and TSH-suppression group (5/20, M/F) whose serum TSH remained suppressed (<0.4 μIU/ml) for at least six months even after their serum FT3 and FT4 had been normalised. The former group included three postmenopausal women aged 55–56 years and the latter no postmenopausal women. No patient had a past history of hepatic or renal disorders, alcoholism, or other major medical condition, or had taken any medicine which might affect calcium metabolism. Serum bone-specific alkaline phosphatase (bAP) (A) and urinary excretions of pyridinoline (U-PYD) (C) and deoxypyridinoline (U-DPD) (D), but not serum-intact osteocalcin (OC) (B), were significantly increased in the TSH-suppression group (n = 25) compared to the TSH-normal group (n = 42). Horizontal lines indicate the mean value for each group. The differences in the means of these bone parameters between TSH-suppression group and TSH-normal group were statistically significant (bAP; p < 0.01, OC; p < 0.05, U-PYD; p < 0.05, U-DPD; p < 0.05).

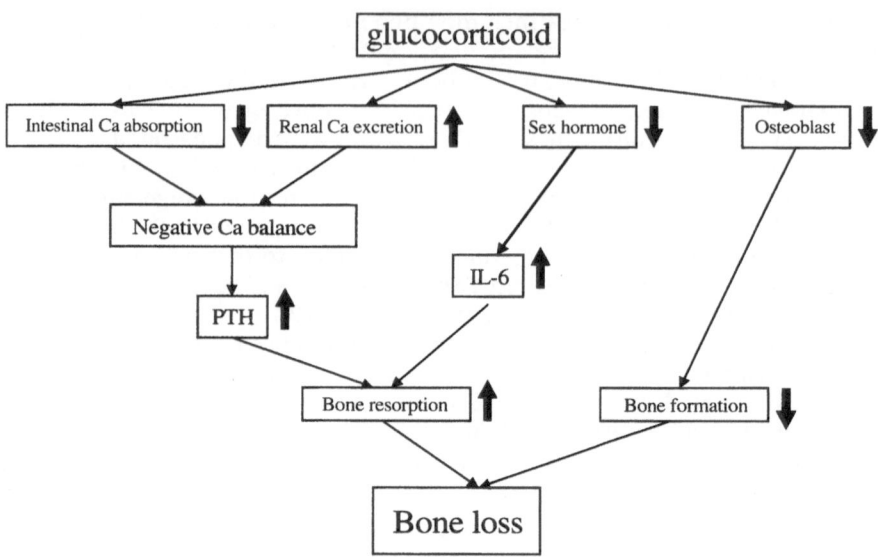

Figure 9.14. Mechanism for the development of glucocorticoid-induced osteoporosis.

secretion of sex hormone from the adrenal gland and gonads by inhibiting adreno-corticotropic hormone and gonadotropin secretion from the pituitary gland, thus enhancing bone resorption. As a result of these abnormalities, since bone formation is not increased sufficiently to compensate for increased bone resorption, uncoupling between bone formation and resorption is responsible for enhanced bone loss. The induction by GC of bone loss is dependent on several factors including age, sex and BMI of the patients, as well as the duration, type and dose of GC treatment. GC has a greater effect on trabecular bone than on cortical bone, since metabolism in the latter bone is more rapid than in the former. Therefore, GC-induced fractures occur preferentially in vertebrae, ribs, and the ends of appendicular bones. Bone loss is highest during the first three months of GC treatment [27], and the loss during the first year of GC therapy was as great as 20% [28]. As the dose of GC increases, spinal BMD decreases in association with reduction in serum OC levels [29]. The reduction in serum OC correlates well with that in serum testosterone in male patients. Although serum levels of alkaline phosphatase and procollagen extension peptides were also reported to decrease upon high-dose GC treatment, it is likely that serum OC is the best marker for predicting GC-induced impairment of bone metabolism. Serum markers of bone resorption are reported to be either increased or normal, possibly depending on the underlying diseases for which GC treatment is needed.

It should initially be kept in mind that patients who will be taking GC for more than two months at either high or low dose are at risk for GC-induced fracture. Therefore, it is extremely important to prevent GC-induced bone loss to the extent possible in those who are expected to take GC (Table 9.7). Since the degree to which GC-induced bone loss depends on the dose and duration of GC treatment, the most obvious way to prevent GC-induced osteoporosis is to withdraw glucocorticoid use, although this is not possible in most instances. A second choice is use of the minimal

Table 9.7. Prevention and Therapy for Glucocorticoid-Induced Osteoporosis

Glucocorticoid Treatment
  Lowest effective dose or discontinue if feasible
  Derivatives with shorter half-life

Food
  Sufficient intake of calcium/vitamin D
  Restricted intake of NaCl

Exercise
  Weight-bearing exercise
  Isometric exercise

Drugs
  Thiazide (to reduce calcium loss into urine when urinary Ca > 4 mg/kg/d)
  Sex steroid replacement (for those in hypogonadal state)
  Active vitamin D derivatives
  Calcitonin
  Bisphosphonate (Alendronate, Risedronate, etc.)

effective dose and short-acting GC or topical use of GC. Furthermore, a new derivative of GC, deflazacort, is recommended for use because it is less harmful to bone, although its biological potency as a GC has not been confirmed [30]. Development of secondary hyperparathyroidism may be attenuated by increasing calcium absorption in the intestine with sufficient intake of calcium and vitamin D from food or by reducing calcium loss into urine with salt restriction +/− thiazide. Exercise is helpful in maintaining bone mineral density by increasing mechanical stress to bone. Drug therapy is needed to prevent bone loss in patients receiving GC. Due to the inhibitory effect of GC on calcium absorption in the intestine and renal tubules, supplementation with vitamin D and calcium is reasonable. Active vitamin D derivatives effectively protect against the development of GC-induced osteoporosis [31]. Gonadal hormone replacement may be useful in those whose gonadal hormone levels are extremely low due to excess GC administration. A recent study demonstrated that bisphosphonate is superior to active vitamin D derivatives in the prevention and treatment of GC-induced osteoporosis [32]. Among various bisphosphonates, etidronate, alendronate, pamidronate and risedronate have been shown to effectively prevent GC-induced bone loss [33]. Alendronate and risedronate have been reported to significantly reduce the rate of GC-induced fracture. For assessment of drug therapy, bone density should be measured every six months, particularly during the first two years of treatment.

Rheumatoid Arthritis

Rheumatoid arthritis (RA), the hallmark of which is the presence of an intense inflammation of the synovium, is a systemic autoimmune inflammatory disease that causes secondary osteoporosis para-articularly as well as systemically (Figure 9.15). The incidence of hip fracture is increased approximately twofold in patients with RA and the risk of vertebral fracture is likely to be increased. RA is associated with three types of bone loss: 1) local subchondral and joint margin bone erosion, 2) paraarticular osteoporosis adjacent to inflamed joints, and 3) generalised osteoporosis

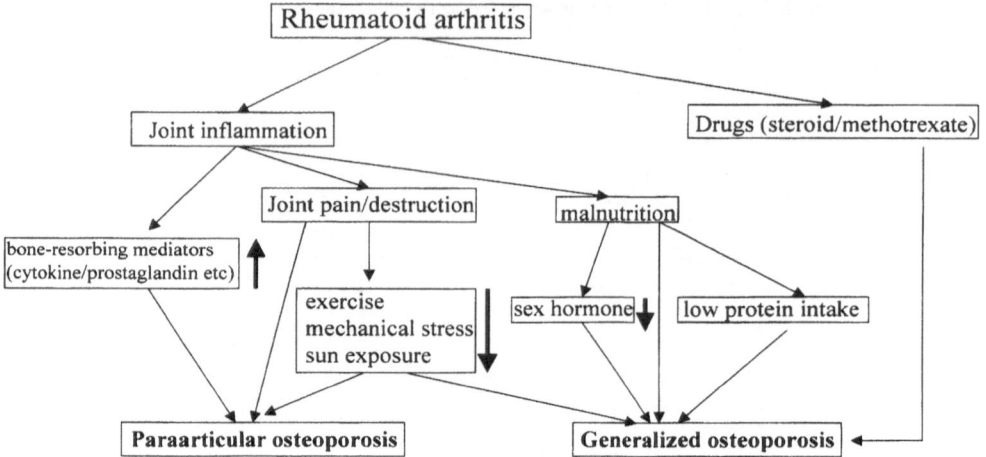

Figure 9.15.   Mechanism for the development of osteoporosis in patients with rheumatoid arthritis.

[34]. Para-articular osteoporosis is the radiographic hallmark of early RA; it typically appears in the area near active synovitis and usually precedes the appearance of focal erosion.

Para-articular osteoporosis is associated with frequent osteoclasts and increased osteoid and resorptive surfaces, indicating local increase of bone turnover. Decreased motion and local atrophy of muscle around the diseased joints may be additional factors contributing to para-articular osteoporosis. The level of deoxypyridinoline (DPD), a product of degradation of collagen type I, is significantly increased in synovial fluid obtained from RA patients compared to that from patients with osteoarthritis [35], suggesting that bone resorption may be enhanced in para-articular bone. Furthermore, DPD in synovial fluid is significantly and positively correlated with synovial interleukins 1α and 6, both of which are bone-resorption-stimulating cytokines. In addition to these cytokines, tumour-necrosis factor-α, prostaglandins, and 1,25-dihydorxyvitamin D have been implicated in the development of para-articular osteoporosis [36, 37].

RA is also associated with generalised osteoporosis [38]. Biochemical markers of bone metabolism are significantly increased in RA patients and correlate strongly with changes in bone mineral density [39], suggesting that enhanced bone resorption is a major mechanism in the development of generalised osteoporosis. In contrast, histomorphometric study showed that generalised osteoporosis in RA even without glucocorticoid treatment is related to a decrease in bone formation rather than an increase in bone formation. Since it has recently been reported that tumour necrosis factor-α (TNF-α) has a suppressive effect on bone formation and that serum TNF-α level is increased in RA patients, TNF-α might have an important role in the suppression of osteoblast dysfunction in RA patients. A study using monozygotic twins revealed that RA is an independent risk for osteoporosis because of a significant reduction in BMD in one twin with RA either at lumbar spine, femoral neck, or total body respectively, as compared to the twin without RA [40]. The available data suggest that the factors that are major determinants of bone loss in RA are age, menopausal status, impairment of activity of daily living and RA activity. Although consensus has not been reached, recent findings reveal reduced levels of the major adrenal androgen precursor to oestrogen, dehydroepiandrosterone sulphate (DHEAS), in postmenopausal women with RA, possibly due to chronic illness and alteration in the hypothalamic–pituitary–adrenal axis and suppression of the immune system. Since these sex steroids have bone-protective effects, reduction of these hormones may contribute to accelerated bone loss. In addition, drugs used against RA such as glucocorticoid and methotrexate [41], and malnutrition should have profoundly harmful effects on bone metabolism. A large-scale prospective study indicated that RA patients lost a significant amount of bone mass during the early phases of RA, mostly due to increased bone resorption.

Diabetes Mellitus

Diabetes mellitus is associated with various disorders of calcium metabolism, such as impairment of calcium absorption [42], and increased loss of calcium into urine [43], which can be followed by the development of secondary hyperparathyroidism and the resultant bone loss [44]. However, diabetic osteopenia is characterised by low bone turnover resulting from osteoblast deficiency (Figure 9.16). Therefore, it should be emphasised that bone loss in diabetics is responsible for impaired bone formation that does not compensate for increased bone resorption. Deficiency of insulin and the resultant insulin-like growth factor has been proposed to be the most important factor in impaired bone formation [45–47]. However, osteoblast dysfunction results either from insulin deficiency [48] or from sustained high glucose levels [49–53]. High

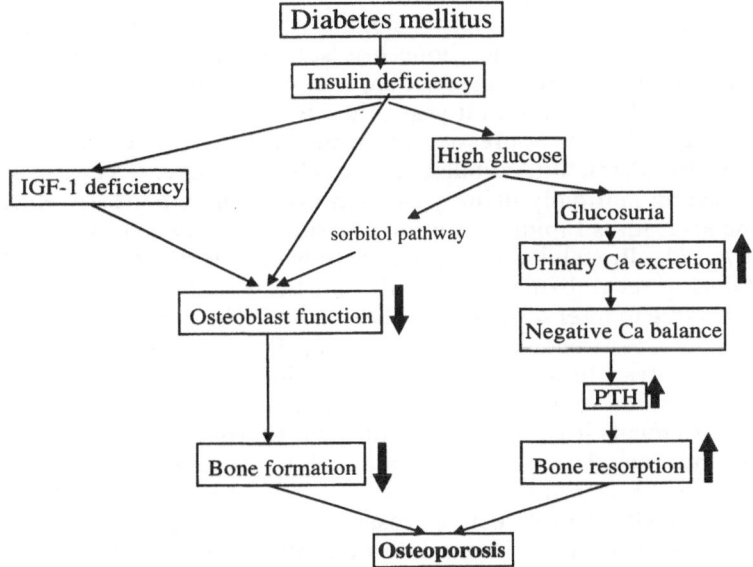

Figure 9.16.   Mechanism for the development of osteoporosis in patients with diabetes mellitus.

glucose level impairs osteoblast function both in vivo [52] and in vitro [49, 50], al least in part through accumulation of intracellular sorbitol [51]. Additional factors in diabetic osteopenia are secondary hyperparathyroidism resulting from nephropathy, impaired blood perfusion to bone due to macroangiopathy, decreased mechanical stress to bone due to neuropathy and myopathy. Low bone turnover resulting from osteoblast dysfunction may retard bone accumulation during growth and may slow age-related bone loss after peak bone mineral density is obtained [54]. Although retardation of age-related bone loss may result in higher bone mineral density, it may increase bone fragility independent of bone mineral density by promoting accumulation of fatigue microdamage. A recently performed prospective study clearly indicated that type 2 diabetes is a risk factor for hip, proximal humerus, and foot fractures in older women despite significantly higher BMD in these bones, although elevated risk was not found for vertebral fracture [55]. These findings suggest that prevention of the development of fracture may be needed in the treatment of diabetes. Since diabetic osteopenia is characterised by impairment of osteoblast dysfunction, drugs to stimulate osteoblast function should be prescribed. Among such drugs, a retrospective study reported that HMG-CoA reductase inhibitors increased BMD of the femur in male patients with type 2 diabetes mellitus during 14–15 months of treatment [56], although contradictory findings were also reported [57].

Immobilisation

Evidence has accumulated to indicate that bone mass changes in amount and distribution to accommodate the mechanical stress imposed on it [58, 59]. Although the exact mechanism is not yet elucidated, osteocytes have been proposed to be the major bone cells sensing mechanical stress. As mechanical stress imposed on bone is increased, bone mass increases. The relationship between mechanical stress and bone mass is curvilinear, with a much steeper slope at very low levels of mechanical stress. Thus, immobilisation may induce bone loss so rapid and great in degree that it becomes a major clinical problem. Sudden, complete immobilisation can induce as much as 40% bone loss in the first year. Although serum calcium is in most cases maintained normal due to suppression of serum PTH and the resultant suppression of serum $1,25(OH)_2D$, calcium mobilisation from bone is increased by increased bone resorption. Fasting and daily urinary calcium excretions are grossly elevated. Trabecular bone loses mass rapidly, resulting in a substantial deficit by six months that subsequently gradually stabilises. A previous study measured lumbar spine BMD longitudinally during recumbency in 34 patients who remained at bed rest, suffering from back pain due to intervertebral disc hernia [60]. The patients lost lumbar BMD by as much as 0.9% per week in the initial period. Histomorphometric study demonstrated that, in contrast to increased bone resorption, bone formation is profoundly reduced, as measured both with static parameters and in tetracycline-related dynamic measurements [61]. A previous study has shown increased secretion of cortisol from the adrenal gland. Therefore, a possible relationship with an increase in endogenous cortisol secretion should be recalled since histological findings are similar to those of GC-induced osteoporosis.

Prevention and treatment of immobilisation osteoporosis remains an enormous challenge. In healthy subjects at bed rest, four hours of daily ambulation is sufficient to correct negative calcium balance. Although pharmacological approaches have been unsuccessful, bisphosphonates have met with some success.

Liver Disease

Liver diseases known to induce secondary osteoporosis are the chronic cholestatic diseases, the most important of which are primary biliary cirrhosis, chronic active hepatitis and alcoholic cirrhosis. The mechanism of bone loss in chronic liver diseases is impairment of bone-forming activity, which may be related in part to inadequate ingestion of vitamin D, protein and calcium, and increased morbidity of associated disorders. Among liver diseases, primary biliary cirrhosis is strongly associated with osteoporomalacia resulting from severe intestinal malabsorption of calcium and phosphate due to disturbance of the enterohepatic circulation of vitamin D metabolites. Another factor that should be mentioned is high levels of corticosteroid, gonadal insufficiency, and possibly alcoholism. Furthermore, impaired production of serum vitamin D-binding protein might be involved in the development of osteoporosis when liver function is severely impaired.

Gastrointestinal Diseases

Osteoporosis is known to occur several years after partial gastrectomy and intestinal bypass surgery and is associated with inflammatory bowel diseases. The Billroth II procedure appears to cause bone disease more than the Billroth I procedure. Although the exact mechanism is yet to be known, it may be related to malabsorption of vitamin D and calcium, resulting in the development of secondary hyperparathyroidism. Post-gastrectomy patients exhibit mild degrees of fat malabsorption. Fat malabsorption reduces calcium absorption by stimulating the formation of calcium complexes and by inhibiting vitamin D absorption. Calcium absorption is also impaired by reduction of acid output from the gastric epithelium. Spinal osteoporosis was found in 50% of patients following partial gastrectomy, in contrast to 22% of age-matched controls.

Drugs

Gonadotropin-releasing hormone and luteinising hormone-releasing antagonists or agonists are increasingly being used for the treatment of premenstrual syndrome, polycystic ovary syndrome, endometriosis, breast feeding and prostatic cancer. Patients undergoing this kind of therapy lose bone due to suppression of gonadal steroids, since gonadal steroids prevent bone loss by suppressing bone turnover. Furthermore, anticonvulsant therapy is known to be associated with skeletal abnormalities. Anticonvulsant bone disease was initially considered a form of osteomalacia, since these drugs accelerate hepatic inactivation of vitamin D metabolites. However, it is now evident that the characteristic lesion in most patients is not osteomalacia but high-turnover osteoporosis, resulting principally from the development of secondary hyperparathyroidism by anticonvulsant-mediated inhibition of intestinal calcium absorption.

References

1. Britto JM, Fenton AJ, Holloway WR, Nicholson GC. Osteoblasts mediate thyroid hormone stimulation of osteoclastic bone resorption. Endocrinology 1992;134:327–31.

2. Kim CH, Kim HK, Shong YK, Lee KU, Kim GS. Thyroid hormone stimulates basal and interleukin (IL)-1-induced IL-6 production in human bone marrow stromal cells: a possible mediator of thyroid hormone-induced bone loss. J Endocrinol 1999;160:97-102.

3. Allain TJ, McGregor AM. Thyroid hormones and bone. J Endocrinol 1993;139:9-18.

4. Riggs BL, Melton III LJ. Involutional osteoporosis. N Engl J Med 1986;314:1676-86.

5. Abu EO, Horner A, Teti A, Chatterjee VK, Compston JE. The localization of thyroid hormone receptor mRNA in human bone. Thyroid 2000;10:287-93.

6. Jódar E, Muñoz-Torres M, Escobar-Jiménez F, Quesada-Charneco M, Luna del Castillo JD. Bone loss in hyperthyroid patients and in former hyperthyroid patients controlled on medical therapy: influence of aetiology and menopause. Clin Endocrinol (Oxf) 1997;47:279-85.

7. Nagasaka S, Sugimoto H, Nakamura T, Kusaka I, Fujisawa G, Sakuma N et al. Antithyroid therapy improves bony manifestations and bone metabolic markers in patients with Graves' thyrotoxicosis. Clin Endocrinol (Oxf). 1997;47:215-21.

8. Shafer RB, Gregory DH. Calcium malabsorption in hyperthyroidism. Gastroenterology 1972;63:235-9.

9. Epstein FH, Freedman LR, Levitin H. Hypercalcemia, nephrocalcinosis and reversible renal insufficiency associated with hyperthyroidism. N Engl J Med 1958;259:782-8.

10. Inaba M, Hamada N, Ito K, Mimura T, Ohno M, Yamakawa J et al. A case report on disequiribrium hypercalcemia in hyperthyroidism. Endocrinol Jpn 1982;29:389-93.

11. Jódar E, Muñoz-Torres M, Escobar-Jiménez F, Quesada-Charneco M, Luna del Castillo JD. Bone loss in hyperthyroid patients and in former hyperthyroid patients controlled on medical therapy: influence of aetiology and menopause. Clin Endocrinol (Oxf).1997;47:279-85.

12. Langdahl BL, Loft AGR, Eriksen EF, Mosekilde L, Charles P. Bone mass, bone turnover, body composition, and calcium homeostasis in former hyperthyroid patients treated by combined medical therapy. Thyroid 1996;6:161-8.

13. Muddle AH, Houben AJ, Nieuwenhuijzen Kruseman AC. Bone metabolism during anti-thyroid drug treatment of endogenous subclinical hyperthyroidism. Clin Endocrinol (Oxf). 1994;41:421-4.

14. Kasagi K, Takeuchi R, Misaki T, Kousaka T, Miyamoto S, Iida Y et al. Subclinical Graves' disease as a cause of subnormal TSH levels in euthyroid subjects. J Endocrinol Invest 1997;20:183-8.

15. Kumeda Y, Inaba M, Tahara H, Kurioka Y, Ishikawa T, Morii H et al. Persistent increase in bone turnover in Graves' patients with subclinical hyperthyroidism. J Clin Endocrinol Metab 2000;85:4157-61.

16. Inoue M, Tawata M, Yokomori N, Endo T, Onaya T. Expression of thyrotropin receptor on clonal osteoblast-like rat osteosarcoma cells. Thyroid 1998;8:1059-64.

17. Lupoli G, Nuzzo V, Di Carlo C, Affinito P, Vollery M, Vitale G et al. Effects of alendronate on bone loss in pre- and postmenopausal hyperthyroid women treated with methimazole. Gynecol Endocrinol Oct 1996;10(5):343-8.

18. Kung AW, Ng F. A rat model of thyroid hormone-induced bone loss: effect of antiresorptive agents on regional bone density and osteocalcin gene expression. Thyroid 1994;4(1):93-8.

19. Faber J, Galloe AM. Changes in bone mass during prolonged subclinical hyperthyroidism due to L-thyroxine treatment: a meta-analysis. Eur J Endocrinol 1994;130:350-6.

20. Duncan WE, Chung A, Solomon B, Wartofsky L. Influence of clinical characteristics and parameters associated with thyroid hormone therapy on the bone mineral density of women treated with thyroid hormone. Thyroid 1994;4:183-90.

21. Reid IR. Glucocorticoid effects on bone. J Clin Endocrinol Metab 1998;83:1860-2.

22. Lukert BP, Raisz LG. Glucocorticoid-induced osteoporosis: pathogenesis and management. Ann Intern Med 1990;112:352-64.

23. Adinoff AD, Hollister JR. Steroid-induced fractures and bone loss in patients with asthma. N Engl J Med 1983;309:265-8.

24. Lukert B, Mador A, Raisz LG, Kream BE. The role of DNA synthesis in the responses of fetal rat calvariae to cortisol. J Bone Miner Res 1991;6:158-66.

25. Suzuiki Y, Ichikawa Y, Saito E, Homma M. Importance of increased urinary calcium excretion in the development of secondary hyperparathyroidism of patients under glucocorticoid therapy. Metabolism 1983;32:151-6.

26. Lukert BP, Stanbury SW, Mawer EB. Vitamin D and intestinal transport of calcium: effects of prednisolone. Endocrinology 1973;93:718-22.

27. Rickers H, Deding A, Christiansen C, Rodbro P, Naestoft J. Coritcosteroid-induced osteopenia and vitamin D metabolism: effect of vitamin D2, calcium, phosphate, and sodium fluoride administration. Clin Endocrinol 1982;16:409-15.

28. Gennari C, Imbimbo B, Montaganani M, Bernini M, Nardi P, Avioli LV. Effects of prednisolone and deflazacort on mineral metabolism and parathyroid hormone activity in humans. Calcif Tissue Int 1984;36:245-52.

29. Pearce G, Tabensky DA, Delmas PD, Baker HW, Seeman E. Corticosteroid-induced bone loss in men. J Clin Endocrinol Metab 1998;83:801–6

30. Krossgaard MR et al. Changes in bone mass during low dose coricosteroid treatment in patients with polymyalgia rheumatica: a double blind, prospective comparison between prednisolone and deflazacort. Ann Rheum Dis 1996;55:143–6.

31. Ringe JD. Active vitamin D metabolites in glucocorticoid-induced osteoporosis. Calcif Tissue Int. 1997;60:124–7

32. Papapoulos SE. Bisphosphonates. In: Rosen CJ, Glowacki J, Bilezikian JP, editors. The aging skeleton. San Diego, London: Academic Press, 1999;541–9.

33. Saag KG, Emkey R, Schnitzer TJ, Brown JP, Hawkins F, Goemaere F et al. Alendronate for the prevention and treatment of glucocorticoid-induced osteoporosis. N Engl J Med 1998;339:292–9.

34. Deodhar AA, Woolf AD. Bone mass measurement and bone metabolism in rheumatoid arthritis: A review. Br J Rheumatol 1996;35:309–22.

35. Furumitsu Y, Inaba M, Yukioka K, Yukioka M, Kumeda Y, Azuma Y et al. Levels of serum and synovial fluid pyridinium crosslinks in patients with rheumatoid arthritis. J Rheumatol 2000;27:64–70.

36. Chu CQ, Field M, Allard S, Abney E, Feldmann M, Maini RN. Detection of cytokines at the cartilage/pannus junction in patients with rheumatoid arthritis; implications for the role of cytokines in cartilage destruction and repair. Br J Rheumatol 1992;32:653–61.

37. Inaba M, Yukioka K, Furumitsu Y, Murano M, Goto H, Nishizawa Y et al. Positive correlation between levels of IL-1 or IL-2 and 1,25(OH)$_2$D/25-OH-D ratio in synovial fluid of patients with rheumatoid arthritis. Life Sci 1997;61:977–85.

38. Sambrook PN, Reeve J. Bone disease in rheumatoid arthritis. Clin Sci 1988;74:225–30.

39. Cortet B, Flipo RM, Pigny P, Duquesnoy B, Boersma A, Marchandise X et al. Is bone turnover a determinant of bone mass in rheumatoid arthritis? J Rheumatol 1998;25:1251–3.

40. Sambrook PN, Spector TD, Seeman E, Bellamy N, Buchanan RR, Duffy DL et al. Osteoporosis in rheumatoid arthritis: A monozygotic co-twin control study. Arthritis Rheum 1995;38:806–9.

41. Buckley LM, Leib ES, Cartularo KS, Vacek PM, Cooper SM. Effect of low dose methotrexate on the bone mineral density of patients with rheumatoid arthritis. J Rheumatol 1997;24:1489–94.

42. Schneider LE, Schedl HP. Diabetes and intestinal calcium absorption in the rat. Am J Physiol 1972;223:1319–23.

43. Leman Jr J, Lennon EJ, Piering WR, Prien Jr EL, Ricinati ES. Evidence that glucose ingestion inhibits net renal tubular reabsorption of calcium and magnesium in man. J Lab Clin Med 1970;75:578–85.

44. Imura H, Seino Y, Nakagawa S, Goto Y, Kosaka K, Sakamoto N et al. Diabetic osteopenia in Japanese: a geographic study. J Jpn Diabetes Soc 1987;30:9924–9.

45. Klein M, Frost HM. The numbers of bone resorpion and formation in rib. Henry Ford Hos Med Bull 1964;12:527–36.

46. Rico H, Hernandez ER, Cabranes JA, Gomez-Castresana F. Suggestion of a deficient osteoblastic function in diabetes mellitus: the possible cause of osteopenia in diabetics. Calcif Tis Int 1989;45:71–3.

47. Ishida H, Seino Y, Taminato T, Usami M, Takeshita N, Seino Y et al. Circulating levels and bone contents of bone γ-carboxyglutamic acid-containing protein are decreased in streptozotocin-induced diabetes: possible marker of diabetic osteopenia. Diabetes 1988;37:702–6.

48. Wettenhall REH, Schwqarz PL, Bornstein J. Actions of insulin and growth hormone on collagen and chondroitin sulfate synthesis in bone organ cultures. Diabetes 1969;18:280–4.

49. Inaba M, Terada M, Koyama H, Yoshida O, Ishimura E, Kawagishi T et al. Influence of high glucose on 1,25-dihydroxyvitamin D3-induced effect on human osteoblast-like MG-63 cells J Bone Miner Res 1995;10:1050–60.

50. Terada M, Inaba M, Yano Y, Hasuma T, Nishizawa Y, Morii H et al. Growth-inhibitory effect of a high glucose concentration on osteoblast-like cells. Bone 1998;22:17–23.

51. Inaba M, Terada M, Nishizawa Y, Shioi A, Ishimura E, Otani S et al. Protective effect of an aldose reductase inhibitor against bone loss in galactose-fed rats: possible involvement of the polyol pathway in bone metabolism. Metabolism 1999;48:904–9.

52. Inaba M, Nishizawa Y, Mita K, Kumeda Y, Emoto M, Kawagishi T et al. Poor glycemic control impairs the response of biochemical parameters of bone formation and resorption to exogenous 1,25-dihydroxyvitamin D$_3$ in patients with type 2 diabetes. Osteoporos Int 1999;9:525–31.

53. Inaba M, Nishizawa Y, Shioi A, Morii H. Importance of sustained high glucose condition in the development of diabetic osteopenia: Possible involvement of the polyol pathway. Osteoporosis Int 1997;7(suppl. 3):S209–12.

54. Krakauer JC, McKenna MJ, Buderer NF, Rao DS, Whitehouse FW, Parfitt AM. Bone loss and bone turnover in diabetes. Diabetes 1995;44(7):775–82.

55. Scwartz AV, Sellmeyer DE, Ensrud KE, Cauley JA, Tabor HK, Schreiner PJ et al. Older women with diabetes have an increased risk of fracture: a prospective study. J Clin Endocrinol Metab 2001;86:32–8.

56. Chung YS, Lee MD, Lee SK, Kim HM, Fitzpatrick LA. HMG-CoA reductase inhibitors increase BMD in type 2 diabetes mellitus patients. J Clin Endocrinol Metab. 2000;85:1137–42.
57. Wada Y, Nakamura Y, Koshiyama H. Lack of positive correlation between statin use and bone mineral density in Japanese subjects with type 2 diabetes. Arch Intern Med 2000;160:2865.
58. Frost HM. The mechanostat: a proposed pathogenic mechanism of osteoporosis and the bone mass effects of mechanical and non-mechanical agents. Bone Miner 1987;2:73–85.
59. Carter DR. Mechanical loading history and skeletal biology. J Biomech 1987;20:1095–9.
60. Krolner B, Toft B. Vertebral bone loss: an unheeded side effect of therapeutic bed rest. Clin Sci 1983;64:537–40
61. Minaire P, Neunier P, Edouard C, Bernard J, Courpron P, Bourret J. Quantitative histological data on disuse osteoporosis. Calcif Tissue Int 1974;17:57–73.

Rickets and Osteomalacia

Y. Yamanaka and Y. Seino

Rickets is a disorder of mineralisation of the bone matrix, or osteoid in growing bones, involving both the growth plate and newly formed trabecular and cortical bone. Osteomalacia also is a deficient bone matrix mineralisation, but it occurs after the growth is completed and involves only the bone and not the growth plate. The clinical and radiological spectrum of rickets is highly variable, depending on the age, the aetiology (Table 9.8), and the duration and severity of the mineralisation defect.

Nutritional Rickets and Osteomalacia

Nutritional rickets and osteomalacia were major problems, particularly in cities in temperate zones, before 1920. Intakes of Vitamin D, calcium, or phosphate substantially below required doses may result in rickets or osteomalacia.

Vitamin D Deficiency

Inadequate sunlight exposure, avoidance of vitamin D-supplemented foods, prematurity, and rapid growth with a consequent need for adequate skeletal calcium and phosphorus may contribute to nutritional vitamin D deficiency. The main natural sources of vitamin D in foods are fish livers. Human milk and cow's milk are inadequate sources. In the United States, vitamin D is added routinely to cow's milk and many other foods are similarly fortified [1]. Breast-fed infants who do not receive vitamin D supplementation run the risk of vitamin D deficiency. The avoidance of vitamin D supplementation remains a public health problem in many countries.

Pathogenesis

$1,25(OH)_2D$ increases calcium and phosphate absorption from the small intestine. During vitamin D deficiency, intestinal calcium and phosphate absorption are reduced, causing hypocalcaemia. The hypocalcaemia stimulates parathyroid hormone (PTH) secretion from the parathyroid glands. PTH promotes bone resorption and increases calcium and phosphate available to the blood. In addition, PTH acts on the kidney to promote tubular calcium resorption and increases phosphate excretion. PTH also stimulates the renal conversion of $25(OH)D$ to $1,25(OH)_2D$.

Table 9.8. Aetiology of Rickets

Abnormalities of vitamin D
Nutritional deficiency
Premature infants
Fat malabsorption
Severe liver disease
Long-term anticonvulsant treatment
Metabolic disturbance
1α-Hydroxylase deficiency (VDDR-I)
Abnormal receptor for $1,25(OH)_2D_3$
Deficiency of calcium
Hypophosphataemic rickets
Phosphate deficiency
X-linked hypophosphataemic rickets (XLH)
Autosomal dominant hypophosphataemic rickets (ADHR)
Oncogenic
Renal osteodystrophy
Renal tubular disorders
Distal renal tubular acidosis
Fanconi syndrome

Clinical and Laboratory Features

The clinical features of vitamin D deficiency include hypotonia, muscle weakness, bone pain, bone deformity, and fracture, depending principally on the age of onset. The most rapidly growing bones demonstrate the most striking abnormalities. In the first year of life, the most rapidly growing bones are in the cranium, ribs, and wrists. Vitamin D deficiency at this time leads to widened cranial sutures, frontal bossing, parietal flattening, bulging of costochondral junctions, and enlargement of the wrists. The rib cage may be so deformed that it contributes to an increased incidence of pneumonia [2]. There is delayed eruption of permanent dentition, and enamel defects can occur. Tetany is unusual because hypocalcaemia is mild and its progress is slow. Deformities of the back, including kyphosis and lordosis, along with limb bowing, can contribute to a waddling gait. There is also an increased susceptibility to fractures. Weight bearing produces a limb-bowing deformity, after age one. In children, mineralisation defects result in abnormalities of diaphyses, metaphyses, and epiphyses. Vitamin D deficiency seen in the adult causes less severe clinical features. In the mature skeleton, less than 5% of the calcium is newly deposited per year. Thus, a mineralisation defect in adults must be present for several years to be manifested. The characteristic symptom is bone pain when weight or pressure is added to the affected bones.

The earliest and most common sites of radiographic change due to vitamin D deficiency are the long bones (Figure 9.17). In children, the failure of cartilage mineralisation is manifested by delayed opacification of the epiphyses and widening of growth plates. In vitamin D deficiency, the usually straight ends of the mineralising metaphyses are frayed or irregular. In the diaphyses, the cortex is thin, the periosteum may be fuzzy, and bone trabeculae are sparse and coarse. Pseudofractures (looser zones) are uncommon in children. Secondary hyperparathyroidism may

Figure 9.17. Wrist demonstrating typical findings of rickets in the radius and ulna. Note widening, irregularity, and cupping of the metaphyses. The failure of cartilage mineralisation is usually observed.

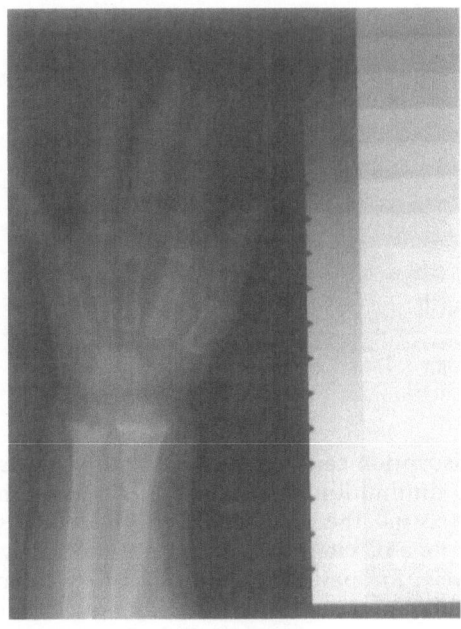

cause sub-periosteal erosions, but bone cysts are rare in children. The characteristic radiographic feature of adult osteomalacia is the pseudofracture localisation, often symmetrically, at the concave sides of the shafts of long bones, as well as the scapulae, ribs, and pubic rami.

Deficient intestinal calcium adsorption resulting from vitamin D deficiency leads to mild hypocalcaemia (Table 9.9). This in turn induces secondary hyperparathyroidism. The features of secondary hyperparathyroidism include an increased serum PTH level, a decreased renal threshold for phosphate excretion and hypophosphataemia, and an increased serum alkaline phosphatase (ALP) level. The serum calcium level is maintained within the lower portion of the normal range because of secondary hyperparathyroidism. The reduced filtered load and the enhanced action of PTH on the renal transport of calcium lead to low urinary calcium. Serum level of 25(OH)D are low whereas 1,25(OH)$_2$D levels are variable. Assessment of the total serum 25(OH)D level is a useful indicator of the state of vitamin D nutrition.

Treatment and Prevention

In vitamin D deficiency, owing to avoidance of vitamin D-supplemented foods, use of vitamin D$_2$ (usually 1,500–5,000 IU/day orally [3] or 10,000–50,000 IU/month intramuscularly [3]) is recommended. Particularly, in vitamin D deficiency resulting from fat malabsorption, administration of a more polar compound, such as 25(OH)D, 20–30 µg/day, or 1,25(OH)$_2$D, 0.15–0.5 µg/day, may be more efficacious. In unsupplemented, breast-fed infants, vitamin D$_2$ doses for prevention of vitamin D-deficiency rickets have ranged from 10,000 to 50,000 IU per month orally [3], The dose and frequency of administration must be adjusted depending on the serum levels of calcium, phosphorus, ALP, and 25(OH)D. The response to ergocalciferol begins within days, but if an active metabolite such as calcitriol is given, a change in intestinal calcium

Table 9.9.   Laboratory Data in Rickets

Type	Serum levels						Gene defect
	Calcium	Phosphorus	ALP	25(OH)D	1,25(OH)$_2$D	PTH	
Vitamin D deficiency	↓	↓	↑	↓	↑↑–↓	↑	–
VDDR-I	↓	↓	↑	→	↓	↑	1α-hydroxylase
VDDR-II	↓	↓	↑	→	↑↑	↑↑	VDR
XLH	→	↓	↑	→	→–↓	→	PHEX

→, normal; ↓, low; ↑, high.

absorption can be observed within hours. Successful initiation of therapy is proven by diminution of secondary hyperparathyroidism; serum PTH and urinary cAMP levels fall, the serum phosphorus level rises, and the serum ALP level falls. Complications of vitamin D$_2$ therapy include hypercalcaemia and hypercalciuria. Alternatively, for patients with rickets or osteomalacia resulting from lack of adequate sunlight, either increased exposure to sunlight or ultraviolet lamp treatment will effect a satisfactory cure [1]. For prevention of vitamin D deficiency not due to malabsorption or prematurity, consumption of fortified milk, dietary supplementation of 400 IU, or daily exposure to adequate sunlight is recommended.

Calcium Deficiency

Decreased calcium intake or intestinal absorption contributes to rickets. Rickets can develop during rapid adolescent growth in children who consume a diet low in calcium. In South Africa, children who consumed no milk or dairy products and whose daily calcium intake was estimated to be only 125 mg, whereas phosphate intake is adequate, demonstrated biochemical and radiographic evidence of rickets [4]. In this case, calcium supplementation led to both biochemical and histological improvement [5].

Among extremely premature infants, the growing bones get insufficient calcium and phosphate from intestinal absorption. In this case, both undeveloped intestinal responsivity to 1,25(OH)$_2$D and an imbalance between skeletal needs and dietary supply are underlined reasons for calcium deficiency.

The pathogenesis of calcium-deficiency rickets is similar to that of vitamin D-deficiency rickets in that hypocalcaemia induces secondary hyperparathyroidism. Parathyroid hormone enhances the renal conversion of 25(OH)D to 1,25(OH)$_2$D to increase intestinal calcium and phosphate absorption and also increases bone resorption. Normal serum 25(OH)D levels may exclude the possibility of vitamin D deficiency. Clinical features of calcium-deficiency rickets are similar to those described for vitamin D deficiency.

For patients with rickets resulting from lack of dietary calcium, use of calcium (700 mg/day, orally) is recommended. Special premature baby formulas now contain 75–150 mg calcium/dl. Approximately 200 mg/kg/day is recommended for daily oral calcium intake [6].

Phosphate Deficiency

Phosphate is uniformly distributed in foods and, therefore, a nutritional deficiency of phosphate cannot be usually induced. Severe phosphate deficiency can develop with consumption of antacid phosphate chelators such as aluminium hydroxide or when receiving dialysis [7]. Nutritional hypophosphataemic rickets has also been reported in a premature infant who was breast-fed without calcium or phosphate supplements [8].

The pathogenesis of phosphate-deficiency rickets is different from that of vitamin D and calcium deficiency in that neither secondary hyperparathyroidism nor vitamin D deficiency exist. Phosphate deficiency causes an increase in renal 25(OH)D1 α-hydroxylase that is independent of parathyroid function and enhances intestinal actions of 1,25(OH)$_2$D. If serum phosphorus is insufficient to support mineralisation of osteoid, osteomalacia develops, even though a high serum 1,25(OH)$_2$D level is present. Continuation of this state for several months will cause muscle weakness, bone pain, and occasionally elevation of serum ALP level.

In breast-fed infants with phosphate deficiency, supplementation of phosphate (25 mg/kg/day) is recommended [3]. For patients undergoing long-term anti-ulcer therapy, alternation to non-phosphate chelators such as histamine-receptor antagonists should be considered.

Vitamin D-Dependent Rickets

Vitamin D-dependent rickets (VDDR) is a rare entity resulting from a primary defect of vitamin D metabolism. There are two different types (type I and type II), both familial with autosomal recessive transmission. They present as a severe hypocalcaemic vitamin D-deficiency rickets that fails to respond to treatment with vitamin D and persists despite an almost normal serum 25(OH)D level.

Vitamin D-Dependent Rickets Type I

Affected persons develop early-onset vitamin D deficiency with hypercalcaemia, secondary hyperparathyroidism, and hypophosphataemia. Typically, rickets is diagnosed between the ages of 4 and 12 months and is unresponsive to amounts of ergocalciferol or calcifediol that are effective in nutritional rickets. Complete remission can be obtained but is dependent on continuous therapy with high doses of vitamin D. Serum levels of 25(OH)D are normal or markedly elevated in patients treated with high doses of vitamin D or 25(OH)D (Table 9.9). Serum levels of 1,25(OH)$_2$D were very low in several studies of this disorder. Although massive doses of vitamin D or 25(OH)D$_3$ are required to maintain remission, 0.25–1.0 μg/day of 1,25(OH)$_2$D are sufficient to obtain the same effect. Taken together, these observation imply that patients with VDDR-I have a hereditary defect in the renal tubular 25(OH)D-1α-hydroxylase.

Takeyama et al. first isolated a candidate 1α-hydroxylase (CYP1 α or Cyp1 α) cDNA using a vitamin D receptor (VDR) knockout mouse model [9]. The absence of functional VDR in the knockout mouse results in the loss of feedback control of renal 1,25(OH)$_2$D production, and leads to constitutive over-expression of 1α-hydroxylase. The mouse 1α-hydroxylase protein was shown to be homologous to other members of the cytochrome P450 family. Subsequent cloning of a human 1α-hydroxylase cDNA was achieved [10]. The first description of a mutation in the 1 α -hydroxylase gene associated with defective 1α-hydroxylase activity was carried out using mRNA isolated using cultured keratinocytes from a VDDR-I patient [10]. The absence of 1α-hydroxylase activity in these cells was associated with deletion/frameshift mutations at codons 211 or 231, indicating that the patient was a compound heterozygote for two null mutations. Subsequent to this study, several other reports have been published which have documented families with mutations in the 1α-hydroxylase gene [11, 12].

Vitamin D-Dependent Rickets Type II

Brooks et al. [13] described a patient with osteomalacia who exhibited hypocalcaemia, hypophosphataemia, and secondary hyperparathyroidism. Interestingly, the patient had markedly increased serum levels of 1,25(OH)$_2$D. They suggested that the rickets was caused by impaired responsiveness of target organs to 1,25(OH)$_2$D. They termed this disorder vitamin D-dependent rickets type II (VDDR-II) to distinguish it from a closely related syndrome known as vitamin D-dependent rickets type I (VDDR-I). Subsequently, Marx et al. [14] reported similar findings in two children, and the authors again suggested that the disease was caused by end-organ resistance to 1,25(OH)$_2$D. Since these initial studies, a number of different terms have been used to describe this syndrome. The disorder also has been called pseudo-vitamin D-deficiency rickets type II (PDDR-II), hereditary hypocalcaemic vitamin D-resistant rickets and hereditary vitamin D-resistant rickets (HVDRR).

Clinical and Laboratory Features

The major clinical findings in patients with VDDR-II, hypocalcaemia and rickets, are due to defective intestinal calcium absorption leading to impaired mineralisation of newly forming bone and preosseous cartilage. The rickets is often severe and is usually exhibited within months of birth. Patients suffer from bone pain, muscle weakness, hypotonia, and occasionally convulsions from hypocalcaemia. Children are often growth retarded, and in some cases they develop severe dental caries or exhibit hypoplasia of the teeth [15–19]. Some infants have died from pneumonia caused by poor respiratory movement due to severe rickets of the rib cage [16, 18, 20]. Many children with VDDR-II have sparse body hair, and some have total scalp and body alopecia (Figure 9.18), including eyebrows and in some cases eyelashes. Hair loss may be evident at birth or it occurs during the first few months of life. An analysis of VDDR-II patients shows that there is some correlation between the severity of rickets and the presence of alopecia [21]. Patients with alopecia generally have more severe resistance to calcitriol than those without alopecia. In families with a prior history of the disease, the absence of scalp hair in newborns provides initial diagnostic evidence for VDDR-II.

The abnormalities include low levels of calcium and phosphorus and elevated serum ALP (Table 9.9). The hypocalcaemia leads to secondary hyperparathyroidism with elevated PTH levels and hypophosphataemia. The 25(OH)D levels are normal and, importantly, the 1,25(OH)$_2$D levels are elevated. This clinical feature distinguishes VDDR-II from VDDR-I (1α-hydroxylase deficiency) since the serum 1,25(OH)$_2$D levels in the latter syndrome are low. In the cases in which it has been measured, 24,25(OH)$_2$D levels have been normal or low [16, 18, 22–28]. Unlike

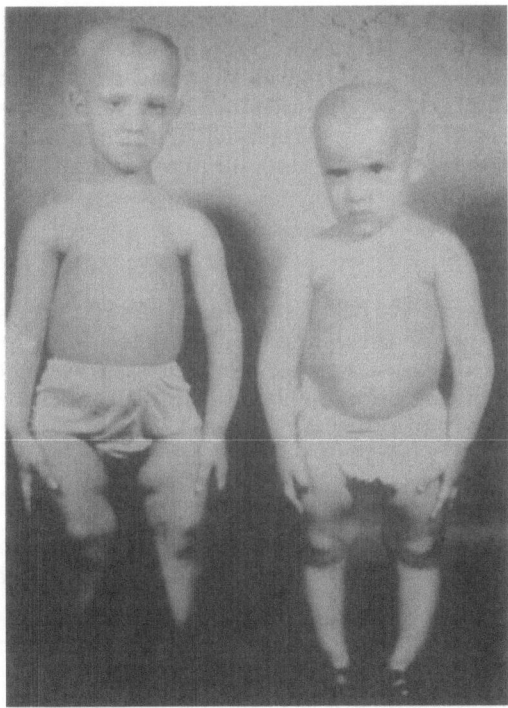

Figure 9.18.   Sisters with alopecia and rickets from vitamin D-dependent rickets type II. Reprinted with permission [15].

patients with 1α-hydroxylase deficiency, most VDDR-II individuals are resistant to supraphysiological doses of all forms of vitamin D therapy.

VDDR-II follows an autosomal recessive pattern of inheritance. The recessive nature of the disease is evident from the parents, who are heterozygous for the genetic trait but show a normal phenotype with no symptoms of the disease and normal bone development. In many, if not all, cases, parental consanguinity is associated with the disease. Males and females are equally affected and often a family has several affected children.

Pathogenesis

Among the many biological processes attributed to vitamin D, maintenance of calcium and bone homeostasis is most apparent. $1,25(OH)_2D$ is essential for promoting calcium and phosphate transport across the small intestine and into the circulation, which is necessary for the normal mineralisation of bone. Approximately 50% of the total intestinal calcium absorption is attributed to $1,25(OH)_2D$ action, while the remaining 50% is due to passive absorption [29]. It is now well established that the biological actions of $1,25(OH)_2D$ are mediated by the VDR, a nuclear transcription factor that regulates gene expression in $1,25(OH)_2D$-responsive cells. Since vitamin D regulates the translocation of calcium and phosphate, interference with the $1,25(OH)_2D$ action pathway causes decreased mineral transport and hypocalcaemia. The hypocalcaemia, in turn, results in secondary hyperparathyroidism, which induces hypophosphataemia. The calcium and phosphate deficiencies interfere with normal bone mineralisation, leading to rickets in children and osteomalacia in adults. In VDDR-II, $1,25(OH)_2D$ target organs, such as the intestine, are resistant to hormone action, and therefore the intestine is less efficient in promoting calcium and phosphate absorption into the circulation. The vitamin D resistance is due to mutations in the VDR that render the receptor non-functional or less functional than the wild-type VDR.

Mutations in the VDR Gene

Mutations in the VDR gene have been identified in almost every patient with this disorder (Figure 9.19). Four of these mutations resulted in a nonsense change that caused a stop codon predicting a truncated VDR that lacks hormone-binding or both hormone- and DNA-binding domains. Mis-sense VDR mutations were observed in cells derived from almost all other patients or kindreds examined. The functional characterisation of the patient's VDR reflected the localisation of the point mutation, as presented below:

1. Mutations localised to the DNA-binding, N-terminal, and zinc-finger region.
2. Mutations localised to the C-terminal hormone-binding domain confirming defects in ligand binding.
3. Mutations in a subregion of the C-terminal domain that affect heterodimerisation of the VDR and RXR.

VDDR-II Mouse Model

Recently, VDR gene knockout mouse models have been created using targeted gene disruption strategies. Yoshizawa et al. [30] deleted the first zinc finger module by ablating exon 2, while Li et al. [31] deleted the second zinc finger module by ablating exon 3. In each VDR($-/-$) null mouse, the animals displayed the classic features of VDDR-II. The mice exhibited hypocalcaemia and hypophosphataemia and developed rickets and secondary hyperparathyroidism after weaning. Serum levels of $1,25(OH)_2D$ were elevated and $24,25(OH)_2D$ levels were low. Alopecia, which

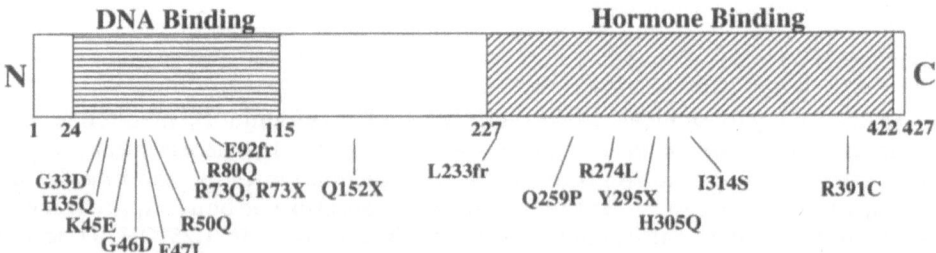

Figure 9.19. Schematic presentation of the homozygous mutations in the vitamin D receptor in patients with vitamin D-dependent rickets type II (VDDR-II). Known natural point mutations in human patients with VDDR-II are indicated by the following abbreviations. Fr indicates a frame-shift mutation, leading to premature termination, whereas X refers to a premature stop codon. Mutants (from positions 33–80) are defective in DNA binding, and all X and Fr mutants cannot bind hormone. Other point mutations in the C-terminus (from positions 259–391) are defective in hormone binding.

developed progressively after birth, was also manifested in these animals. Kinuta et al. [32] reported VDR(−/−) null mice that also showed gonadal insufficiencies. Uterine hypoplasia and impaired folliculogenesis were observed in the female, and decreased sperm count and decreased motility with histological abnormality of the testis were observed in the male. Both activity and expression of aromatase were reduced in these organs. Calcium supplementation increased aromatase activity and partially corrected the hypogonadism. This report indicated that vitamin D is essential for full gonadal function in both sexes. Thus, the VDR knockout mouse model provides a valuable tool with which to examine the effects of the VDR null mutation in various target tissues not readily available for study in humans.

Treatment

Because VDDR-II is caused by a generalised resistance to $1,25(OH)_2D$, to overcome the resistance to vitamin D, a number of treatment therapies using active vitamin D metabolites and calcium have been attempted. Affected persons show widely varied responses. The presence or absence of alopecia is one simple predictor of potential for response to therapy [21]. Patients without alopecia responded to 20–200 µg/day of 25(OH)D and 17–20 µg/day of $1,25(OH)_2D$ [14]. In general, VDDR-II patients with alopecia appear to be more resistant to treatment with vitamin D metabolites. Among patients with alopecia, approximately half have not responded to the highest doses of calciferols available; half have shown satisfactory calcaemic response, but the dose requirement is typically tenfold higher than in patients without alopecia. Basically, the most mildly affected patients can be treated with calciferols (vitamin D_3, vitamin D_2, 25(OH)D),whereas more severely affected patients may respond only to extremely high doses of analogues ($1,25(OH)_2D$, $1\alpha(OH)D$) that do not require 1α-hydroxylation. Based on the gene defects, patients who have mutations that result in a totally unresponsive VDR, such as those due to premature stop signals or those in the DNA-binding domain that cause abnormal DNA binding, are unresponsive even to pharmacological doses of calcitriol. On the other hand, in patients with mis-sense mutations in the ligand-binding domain that result in a fully translated VDR, high doses of calcitriol are more likely to be effective. In cases that fail to respond to 1,25 $(OH)_2D$, intensive calcium therapy is indicated.

High-dose oral calcium therapy was given to a patient who has the Gly46Asp mutation and had failed to respond to calciferols [33, 34]. The patient received 3–4 g of elemental calcium orally per day and showed clinical improvement during four months of therapy. The most rapid way to deliver sufficient amounts of calcium into the bloodstream is by intravenous infusion. Several investigators have demonstrated beneficial effects of intravenous calcium infusion in VDDR-II children [19, 35–37]. Balsan et al. [35] successfully used long-term intravenous calcium infusions in a child with VDDR-II who failed prior treatment with massive doses of vitamin D derivatives and/or oral calcium [18]. High doses of calcium were infused intravenously during the nocturnal hours over a nine-month period. Relief of bone pain was observed within the first two weeks of therapy. Subsequently, the child gained both weight and height. Eventually, the serum calcium normalised, the secondary hyperparathyroidism was reversed, and the rickets was ultimately cured. When the intravenous infusions were discontinued, however, the disorder recurred. In another study, two patients treated with calcium infusion demonstrated a decrease in serum ALP and an increase in their serum calcium and phosphorus over a one-year period [36]. After the improvement of rickets, high-dose oral calcium therapy has been shown to be effective in maintaining normal serum calcium levels in some cases [37]. In cases that fail to respond to high-dose calcitriol, this two-step protocol may be reasonable. With high doses of calcium intravenously or orally, the serum phosphorus level may decrease and, therefore, oral phosphate should be supplemented.

Hypophosphataemic Rickets/Osteomalacia

Hypophosphataemia has a decrease in net renal tubular phosphate reabsorption as a major underlying abnormality and is associated with rickets and osteomalacia. X-linked hypophosphataemic rickets (XLH) is described here as well as an acquired disorder, oncogenic hypophosphataemic osteomalacia (OHO), because of the phenotypic similarities between them.

Clinical Features

XLH is an X-linked dominant disorder with a prevalence of ~1 : 20,000 and is characterised by 1) hypophosphataemia, 2) renal phosphate leak as expressed as a lowered transfer maximum of phosphate per unit volume of glomerular filtrate (TMPO4/GFR), 3) inappropriate vitamin D metabolism in the presence of low serum phosphorus (low to normal serum 1,25(OH)$_2$D), 4) high ALP, and 5) skeletal defects (Table 9.9). However, severity of the phenotype varies considerably. Indeed, affected members of the same family may have markedly different phenotypes, and some individuals have only minimal symptoms.

OHO shares many clinical, biochemical, and physiological features with XLH [38, 39]. Patients with OHO have biochemical abnormalities that include hypophosphataemia, reduced renal tubular phosphate reabsorption, low to normal 1,25(OH)$_2$D levels (despite the low serum phosphorus levels), low to normal serum calcium levels, normal PTH and PTH-related peptide levels, and no other renal reabsorptive defects [40–44]. The patients have osteomalacia or rickets on radiological or histological examination of bone [42]. The tumours are mainly of mesenchymal origin, although a number of different tumour types have been reported. On histological examination, the tumours have included haemangiopericytomas, mesenchymal tumours with vascular elements and giant cells, epidermal naevi, osteosarcomas, chondroblastomas, histiocytomas, neuroblastomas, neurilemomas, paragangliomas, oat cell tumours of the lung, and prostate adenocarcinoma [45, 46]. Evidence indicates that changes in renal phosphate handling, vitamin D metabolism, and skeletal mineralisation are caused by factor(s) secreted by OHO tumours [39–41, 47–49]. The resection of OHO tumours results in the disappearance of disease symptoms and bone healing, and is a key observation that supports this.

The Rickets Gene PHEX

The rickets gene, PHEX (phosphate regulating gene with endopeptidase activity on the X chromosome) [50, 51], is a zinc metalloendopeptidase, which includes neprilysin (NEP), endothelin-converting enzymes (ECE1 and ECE2), and the Kell antigen. The predicted amino acid sequence of PHEX suggests that its protein structure closely resembles that proposed for NEP, ECE1 and ECE2, and Kell, which are type II integral membrane proteins. There is evidence for a short NH2-terminal cytoplasmic domain, followed by a hydrophobic region, which is likely to be a transmembrane domain, and then a large extracellular domain. This latter domain contains several highly conserved sequence motifs thought to be involved in zinc binding and substrate catalysis. Human PHEX and its cDNA have been characterised fully [50, 52], and extensive mutation analysis has confirmed that the defect in XLH can be attributed to primary mutations in the PHEX gene [52–55]. Two murine homologues (HYP and Gy), for X-linked rickets have been characterised [56, 57], and deletions in the murine Phex gene were discovered at the 3' and 5' ends, respectively [58, 59]. A defined pattern of PHEX/Phex expression has emerged after screening a large range of tissue types. Mouse bone [59, 60], mouse teeth [61], human foetal bone [59, 60], human lung [59, 62], and adult human ovary [62] express PHEX/Phex mRNA. The highest relative levels of expression were measured in bones [59] and teeth [61]. No PHEX expression has yet been recorded in kidney [59, 61, 62], while PHEX expression occurs in OHO tumours [60]. Although little is known about the Kell blood group protein, ECE1, ECE2, and NEP function as ectoenzymes. Thus, it is likely that PHEX functions to either activate or degrade a peptide hormone.

PHEX normally degrades a phosphate transport inhibitor (phosphatonin), which normally acts in the kidney to increase phosphate excretion. Excessive amounts of this inhibitor circulate as a result of inadequate proteolysis, and cause hypophosphataemia and low $1,25(OH)_2D$ levels. In both XLH and OHO, it appears as if the biochemical phenotype is caused by excessive amounts of a circulating factor. It may be hypothesised that OHO is a disease in which large amounts of phosphatonin are generated by the tumour and overwhelm endogenous PHEX activity. On the other hand, XLH is due to an excess of circulating phosphatonin on account of reduced or absent phosphatonin-specific protease activity.

Pathogenesis

Based on research carried out on murine models of rickets, several distinct molecular processes are defective and include: 1) Renal phosphate wasting (renal Na-dependent phosphate co-transport), 2) vitamin D metabolism, and 3) osteoblast function and skeletal mineralisation. All the abnormal changes described are now known to be due to a primary defect in the Phex gene [50, 52–55, 58, 59].

Tumour osteomalacia has very similar clinical and biochemical features to XLH, and it is likely that the tumour may secrete a key factor regulated or processed by PHEX.

Renal Phosphate Wasting

Studies on the Hyp mouse demonstrated that the renal phosphate wasting resulted from decreased sodium-dependent phosphate transport in the brush border membrane of the renal proximal tubule [63]. Subsequent studies have shown that, although the low-affinity/high-capacity transport mechanism is intact, there is a defect in the high-affinity/low-capacity transporter [64]. Indeed, the maximal transport rate (V_{max}) is about one-half of normal with no change in affinity for phosphate, a finding consistent with a decreased transporter number in the Hyp mouse. This high-affinity/low-capacity sodium-dependent phosphate co-transporter (Npt-2) mRNA and protein were decreased by ~50% in the Hyp mouse [65, 66]. Studies on the Gy mouse also demonstrate a reduction in V_{max} of the high-affinity/low-capacity transport system [67], and a decrease in Npt-2 mRNA and protein was also observed by Tenenhouse et al. [68]. Conversely, Collins and Ghishan [69] have reported that

Hyp and Gy mice differ in this regard, since they found normal levels of Npt-2 mRNA in Gy mice, with decreased levels of Npt-2 protein. The Phex gene may be involved in regulation of NPT-2 expression.

Vitamin D Metabolism

In addition to the phosphate-wasting defect seen in XLH, defects in vitamin D metabolism were investigated. Treatment of XLH patients with phosphate alone does not result in resolution of the osteomalacia [70]. A combination of phosphate and high-dose calcitriol is required [71, 72]. In XLH and OHO, low serum phosphorus does not initiate an increase in calcitriol and the levels are either inappropriately low to normal (XLH), or severely reduced (OHO). Studies on the Hyp mouse confirm and expand the observations seen in humans [73, 74]. Drezner et al. [73] investigated the activity of renal 1α-hydroxylase in Hyp, normal, and phosphate-depleted mice. Although normal mice on phosphate-depleted diets had profoundly increased 1α-hydroxylase activity compared with normal mice on the control diet, 1α-hydroxylase activity in Hyp mice was lower than phosphate-depleted controls, despite similar serum phosphorus levels. Moreover, 24-hydroxylase is increased in Hyp mice compared with controls, indicating that catabolism of calcitriol is probably increased [75, 76]. Interestingly, when Hyp mice are placed on phosphate-restricted diets, their calcitriol levels paradoxically decrease [74]. This abnormal regulatory response of at least two key enzymes, renal 1α-hydroxylase and 24-hydroxylase, are responsible for the abnormal modulation of serum calcitriol [77, 78].

Osteoblast Function and Skeletal Mineralisation

Since bone from Hyp mice does not mineralise normally, osteoblasts from the Hyp mouse have been investigated. Intramuscular transplantation of normal and XLH mice osteoblasts or periostea have provided experimental evidence for osteoblast defects [79–81]. When normal cells were transplanted into Hyp mice, mineralisation was impaired. However, when Hyp cells were transplanted into normal mice, reduction but not normalisation of the defect was observed. These observations supported the hypothesis that there is a primary osteoblast defect in the Hyp mouse.

Treatment

Treatment of XLH

The current standard therapy continues to be a combination of phosphate and $1,25(OH)_2D$ or 1α-hydroxyvitamin D3 [71, 82–84]. In general, therapy is initiated during childhood and is continued at least until growth is completed. The decision as to whether to continue therapy in adulthood must be individualised. The objectives of therapy in children are to correct or prevent deformity from rickets, improve growth rate, and lessen bone pain. In adults, indications for treatment are controversial. Pseudofractures are common in moderately to severely affected adult XLH patients. Since pseudofractures are often painful and generally respond well to treatment, patients with pseudofractures should be treated.

It is best to start at a low dose of calcitriol + phosphate and gradually increase the doses over several months. The high-dose phase was maintained for up to one year. During this phase, 1–2 g of phosphate is administered in four divided doses. Also, the calcitriol dose may be as high as 50 ng/kg/day in two divided doses, but not more than 3.0 μg/day. Many patients do not require such a high dose of calcitriol. During this high-dose phase, serum calcium, phosphorus, and creatinine levels, as well as urine calcium and creatinine levels should be monitored on a monthly basis. The doses of calcitriol and phosphate are adjusted based on the laboratory data. Hypercalciuria may be the first sign of healing of osteomalacia. The high-dose phase may lead to normalisation of bone histology [72]. After approximately one year on the high-dose phase, patients are switched to a long-term maintenance phase with 10–20 ng/kg/day of calcitriol and no change in the dose of phosphate. During the maintenance phase, serum and urine biochemistries should be monitored at least every three to four months. Administration of phosphate alone will lead to secondary hyperparathyroidism, and administration of calcitriol alone leads to hypercalciuria and hypercalcaemia. Therefore, anytime it becomes necessary to interrupt phosphate administration, it is necessary to discontinue calcitriol as well.

The most common and potentially serious complication of therapy is the development of nephrocalcinosis. Hypercalciuria and hypercalcaemia may contribute to this problem. Long-term use of calcitriol associated with supplemental phosphate, with careful monitoring of urinary calcium excretion to avoid episodes of hypercalciuria, should be considered. When hypercalciuria develops, adjustment of the calcitriol dosage is required.

Miyamura et al. [85] reported the studies on bone marrow transplantation (BMT) from wild-type mice to Hyp mice. Transplantation of normal BM cells improved abnormal bone mineral metabolism in Hyp mice; increased serum phosphorus level and decreased serum ALP level. This result may provide evidence for the clinical application of BMT in XLH patients.

Treatment of OHO

The first and foremost treatment of OHO is complete resection of the induced tumour. The difficulty is often that of finding the tumour. If no tumour can be found, then the alternative is treatment with combined phosphate and calcitriol, as in treatment for XLH. This therapy may heal the osteomalacia, but must be continued until the tumour is removed.

New Insights into Hypophosphataemic Rickets

In addition to XLH, autosomal dominant hypophosphataemic rickets (ADHR), another inherited disorder that results in isolated phosphate wasting, has been described. Although it is less common than XLH, its prevalence is unknown. Up to 80% of familial cases of phosphate wasting have PHEX mutations, but far fewer patients with sporadic forms of phosphate wasting (less than 50%) have PHEX mutations [52, 53, 55, 86]. Thus, many sporadic cases may have ADHR, instead of XLH. ADHR is characterised by low serum phosphorus levels, rickets, osteomalacia, lower-extremity deformities, short stature, bone pain and dental abscesses [87]. Recent positional cloning approaches using DNA from index patients of four families that had male-to-male transmission and clinical features compatible with ADHR identified the ADHR gene. Mis-sense mutations in FGF23, a gene encoding a new member of the fibroblast growth factor (FGF) family, gives rise to ADHR [88]. FGF23 may have an important role in human phosphate homeostasis.

References

1. Stamp TC. Factors in human vitamin D nutrition and in the production and cure of classical rickets. Proc Nutr Soc 1975;34:119–30.
2. Muhe L, Lulseged S, Mason KE, Simoes EA. Case-control study of the role of nutritional rickets in the risk of developing pneumonia in Ethiopian children. Lancet 1997;349:1801–4.
3. Klein GL. Nutritional rickets and osteomalacia. In: Favus MJ, editor. Primer on the metabolic bone diseases and disorders of mineral metabolism. Philadelphia: Lippincott Williams & Wilkins, 1999;315–9.
4. Pettifor JM, Ross FP, Travers R, Glorieux FH. Dietary calcium deficiency: a syndrome associated with bone deformities and elevated serum 1,25-dihydroxyvitamin D concentrations. Metab Bone Dis Rel Res 1981;2:301–6.
5. Marie PJ, Pettifor JM, Ross FP, Glorieux FH. Histological osteomalacia due to dietary calcium deficiency in children. N Engl J Med 1982;307:584–8.
6. Greer FR, Steichen JJ, Tsang RC. Effects of increased calcium, phosphorus, and vitamin D intake on bone mineralization in very low-birth-weight infants fed formulas with Polycose and medium-chain triglycerides. J Pediatr 1982;100:951–5.
7. Carmichael KA, Fallon MD, Dalinka M, Kaplan FS, Axel L, Haddad JG. Osteomalacia and osteitis fibrosa in a man ingesting aluminum hydroxide antacid. Am J Med 1984;76:1137–43.
8. Rowe JC, Wood DH, Rowe DW, Raisz LG. Nutritional hypophosphatemic rickets in a premature infant fed breast milk. N Engl J Med 1979;300:293–6.
9. Takeyama K, Kitanaka S, Sato T, Kobori M, Yanagisawa J, Kato S. 25-Hydroxyvitamin D3 1alpha-hydroxylase and vitamin D synthesis. Science 1977;277:1827–30.
10. Fu GK, Lin D, Zhang MY, Bikle DD, Shackleton CH, Miller WL et al., Cloning of human 25-hydroxyvitamin D-1 alpha-hydroxylase and mutations causing vitamin D-dependent rickets type 1. Mol Endocrinol 1997;11:1961–70.
11. Kitanaka S, Takeyama K, Murayama A, Sato T, Okumura K, Nogami M et al., Inactivating mutations in the 25-hydroxyvitamin D3 1alpha-hydroxylase gene in patients with pseudovitamin D-deficiency rickets. N Engl J Med 1998;338:653–61.
12. Wang JT, Lin CJ, Burridge SM, Fu GK, Labuda M, Portale AA et al., Genetics of vitamin D 1alpha-hydroxylase deficiency in 17 families. Am J Hum Genet 1998;63:1694–702.
13. Brooks MH, Bell NH, Love L, Stern PH, Orfei E, Queener SF et al., Vitamin-D-dependent rickets type II. Resistance of target organs to 1,25-dihydroxyvitamin D. N Engl J Med 1978;298:996–9.
14. Marx SJ, Spiegel AM, Brown EM, Gardner DG, Downs Jr RW, Attie M et al., A familial syndrome of decrease in sensitivity to 1,25-dihydroxyvitamin D. J Clin Endocrinol Metab 1978;47:1303–10.
15. Rosen JF, Fleischman AR, Finberg L, Hamstra A and DeLuca HF. Rickets with alopecia: an inborn error of vitamin D metabolism. J Pediatr 1979;94:729–35.
16. Liberman UA, Samuel R, Halabe A, Kauli R, Edelstein S, Weisman Y et al., End-organ resistance to 1,25-dihydroxycholecalciferol. Lancet 1980;1:504–6.
17. Sockalosky JJ, Ulstrom RA, DeLuca HF, Brown DM. Vitamin D-resistant rickets: end-organ unresponsiveness to 1,25(OH)$_2$D$_3$. J Pediatr 1980;96:701–3.
18. Balsan S, Garabedian M, Liberman UA, Eil C, Bourdeau A, Guillozo H et al., Rickets and alopecia with resistance to 1,25-dihydroxyvitamin D: two different clinical courses with two different cellular defects. J Clin Endocrinol Metab 1983;57:803–11.
19. Bliziotes M, Yergey AL, Nanes MS, Muenzer J, Begley MG, Vieira NE et al., Absent intestinal response to calciferols in hereditary resistance to 1,25-dihydroxyvitamin D: documentation and effective therapy with high-dose intravenous calcium infusions. J Clin Endocrinol Metab 1988;66:294–300.
20. Fraher LJ, Karmali R, Hinde FR, Hendy GN, Jani H, Nicholson L et al., Vitamin D-dependent rickets type II: extreme end-organ resistance to 1,25-dihydroxy vitamin D3 in a patient without alopecia. Eur J Pediatr 1986;145:389–95.
21. Marx SJ, Bliziotes MM, Nanes M. Analysis of the relation between alopecia and resistance to 1,25-dihydroxyvitamin D. Clin Endocrinol (Oxf) 1986;25:373–81.
22. Beer S, Tieder M, Kohelet D, Liberman OA, Vure E, Bar-Joseph G et al., Vitamin D-resistant rickets with alopecia: a form of end-organ resistance to 1,25 dihydroxy vitamin D. Clin Endocrinol (Oxf) 1981;14:395–402.
23. Feldman D, Chen T, Cone C, Hirst M, Shani S, Benderli A et al., Vitamin D resistant rickets with alopecia: cultured skin fibroblasts exhibit defective cytoplasmic receptors and unresponsiveness to 1,25(OH)$_2$D$_3$. J Clin Endocrinol Metab 1982;55:1020–2.
24. Hochberg Z, Benderli A, Levy J, Vardi P, Weisman Y, Chen T et al., 1,25-Dihydroxyvitamin D resistance, rickets, and alopecia. Am J Med 1984;77:805–11.

25. Hirst MA, Hochman HI, Feldman D. Vitamin D resistance and alopecia: a kindred with normal 1,25-dihydroxyvitamin D binding, but decreased receptor affinity for deoxyribonucleic acid. J Clin Endocrinol Metab 1985;60:490–5.
26. Takeda E, Kuroda Y, Saijo T, Naito E, Kobashi H, Yokota I et al., 1 alpha-hydroxyvitamin D3 treatment of three patients with 1,25-dihydroxyvitamin D-receptor-defect rickets and alopecia. Pediatrics 1987;80:97–101.
27. Yagi H, Ozono K, Miyake H, Nagashima K, Kuroume T, Pike JW. A new point mutation in the deoxyribonucleic acid-binding domain of the vitamin D receptor in a kindred with hereditary 1,25-dihydroxyvitamin D-resistant rickets. J Clin Endocrinol Metab 1993;76:509–12.
28. Malloy PJ, Weisman Y, Feldman D. Hereditary 1 alpha,25-dihydroxyvitamin D-resistant rickets resulting from a mutation in the vitamin D receptor deoxyribonucleic acid-binding domain. J Clin Endocrinol Metab 1994;78:313–6.
29. Heaney RP, Barger-Lux MJ, Dowell MS, Chen TC, Holick MF. Calcium absorptive effects of vitamin D and its major metabolites. J Clin Endocrinol Metab 1997;82:4111–6.
30. Yoshizawa T, Handa Y, Uematsu Y, Takeda S, Sekine K, Yoshihara Y et al., Mice lacking the vitamin D receptor exhibit impaired bone formation, uterine hypoplasia and growth retardation after weaning. Nat Genet 1997;16:391–6.
31. Li YC, Pirro AE, Amling M, Delling G, Baron R, Bronson R et al., Targeted ablation of the vitamin D receptor: an animal model of vitamin D-dependent rickets type II with alopecia. Proc Natl Acad Sci U S A 1997;94:9831–5.
32. Kinuta K, Tanaka H, Moriwake T, Aya K, Kato S, Seino Y. Vitamin D is an important factor in estrogen biosynthesis of both female and male gonads. Endocrinology 2000;141:1317–24.
33. Sakati N, Woodhouse NJ, Niles N, Harfi H, de Grange DA, Marx S. Hereditary resistance to 1,25-dihydroxyvitamin D: clinical and radiological improvement during high-dose oral calcium therapy. Horm Res 1986;24:280–7
34. Lin NU, Malloy PJ, Sakati N, al-Ashwal A, Feldman D. A novel mutation in the deoxyribonucleic acid-binding domain of the vitamin D receptor causes hereditary 1,25-dihydroxyvitamin D-resistant rickets. J Clin Endocrinol Metab 1996;81:2564–9.
35. Balsan S, Garabedian M, Larchet M, Gorski AM, Cournot G, Tau C et al., Long-term nocturnal calcium infusions can cure rickets and promote normal mineralization in hereditary resistance to 1,25-dihydroxyvitamin D. J Clin Invest 1986;77:1661–7.
36. Weisman Y, Bab I, Gazit D, Spirer Z, Jaffe M, Hochberg Z. Long-term intracaval calcium infusion therapy in end-organ resistance to 1,25-dihydroxyvitamin D. Am J Med 1987;83:984–90.
37. Hochberg Z, Tiosano D, Even L. Calcium therapy for calcitriol-resistant rickets. J Pediatr 1992;121:803–8.
38. Rowe PS. The role of the PHEX gene (PEX) in families with X-linked hypophosphataemic rickets. Curr Opin Nephrol Hypertens 1988;7:367–76.
39. Nelson AE, Robinson BG, Mason RS. Oncogenic osteomalacia: is there a new phosphate-regulating hormone? Clin Endocrinol (Oxf) 1997;47:635–42.
40. Aschinberg LC, Solomon LM, Zeis PM, Justice P, Rosenthal IM. Vitamin D-resistant rickets associated with epidermal nevus syndrome: demonstration of a phosphaturic substance in the dermal lesions. J Pediatr 1977;91:56–60.
41. Cai Q, Hodgson SF, Kao PC, Lennon VA, Klee GG, Zinsmiester AR et al., Brief report: inhibition of renal phosphate transport by a tumor product in a patient with oncogenic osteomalacia. N Engl J Med 1994;330:1645–9.
42. Fukumoto Y, Tarui S, Tsukiyama K, Ichihara K, Moriwaki K, Nonaka K et al., Tumor-induced vitamin D-resistant hypophosphatemic osteomalacia associated with proximal renal tubular dysfunction and 1,25-dihydroxyvitamin D deficiency. J Clin Endocrinol Metab 1979;49:873–8.
43. Harrison HE. Oncogenous rickets: possible elaboration by a tumor of a humoral substance inhibiting tubular reabsorption of phosphate. Pediatrics 1973;52:432–4.
44. Ryan EA, Reiss E. Oncogenous osteomalacia. Review of the world literature of 42 cases and report of two new cases. Am J Med 1984;77:501–12.
45. Stone MD, Quincey C, Hosking DJ. A neuroendocrine cause of oncogenic osteomalacia. J Pathol 1992;167:181–5.
46. Weidner N, Bar RS, Weiss D, Strottmann MP. Neoplastic pathology of oncogenic osteomalacia/rickets. Cancer 1985;55:1691–705.
47. Miyauchi A, Fukase M, Tsutsumi M, Fujita T. Hemangiopericytoma-induced osteomalacia: tumor transplantation in nude mice causes hypophosphatemia and tumor extracts inhibit renal 25-hydroxyvitamin D 1-hydroxylase activity. J Clin Endocrinol Metab 1988;67:46–53.
48. Kumar R, Haugen JD, Wieben ED, Londowski JM, Cai Q. Inhibitors of renal epithelial phosphate transport in tumor-induced osteomalacia and uremia. Proc Assoc Am Physicians 1995;107:296–305.

49. Nelson AE, Namkung HJ, Patava J, Wilkinson MR, Chang AC, Reddel RR et al., Characteristics of tumor cell bioactivity in oncogenic osteomalacia. Mol Cell Endocrinol 1996;124:17–23.
50. Hyp Consortium. A gene (PEX) with homologies to endopeptidases is mutated in patients with X-linked hypophosphatemic rickets. The HYP Consortium. Nat Genet 1995;11:130–6.
51. Rowe PS, Goulding JN, Francis F, Oudet C, Econs MJ, Hanauer A et al., The gene for X-linked hypophosphataemic rickets maps to a 200–300 kb region in Xp22.1, and is located on a single YAC containing a putative vitamin D response element (VDRE). Hum Genet 1996;97:345–52.
52. Rowe PS, Oudet CL, Francis F, Sinding C, Pannetier S, Econs MJ et al., Distribution of mutations in the PEX gene in families with X-linked hypophosphataemic rickets (HYP). Hum Mol Genet 1997;6:539–49.
53. Holm IA, Huang X, Kunkel LM. Mutational analysis of the PEX gene in patients with X-linked hypophosphatemic rickets. Am J Hum Genet 1997;60:790–7.
54. Econs MJ, Friedman NE, Rowe PS, Speer MC, Francis F, Strom TM et al., A PHEX gene mutation is responsible for adult-onset vitamin D-resistant hypophosphatemic osteomalacia: evidence that the disorder is not a distinct entity from X-linked hypophosphatemic rickets. J Clin Endocrinol Metab 1998;83:3459–62.
55. Dixon PH, Christie PT, Wooding C, Trump D, Grieff M, Holm I et al., Mutational analysis of PHEX gene in X-linked hypophosphatemia. J Clin Endocrinol Metab 1998;83:3615–23.
56. Eicher EM, Southard JL, Scriver CR, Glorieux FH. Hypophosphatemia: mouse model for human familial hypophosphatemic (vitamin D-resistant) rickets. Proc Natl Acad Sci USA 1976;73:4667–71.
57. Lyon MF, Scriver CR, Baker LR, Tenenhouse HS, Kronick J, Mandla S. The Gy mutation: another cause of X-linked hypophosphatemia in mouse. Proc Natl Acad Sci USA 1986;83:4899–903.
58. Strom TM, Francis F, Lorenz B, Boddrich A, Econs MJ, Lehrach H et al., Pex gene deletions in Gy and Hyp mice provide mouse models for X-linked hypophosphatemia. Hum Mol Genet 1997;6:165–71.
59. Beck L, Soumounou Y, Martel J, Krishnamurthy G, Gauthier C, Goodyer CG et al., Pex/PEX tissue distribution and evidence for a deletion in the 3' region of the Pex gene in X-linked hypophosphatemic mice. J Clin Invest 1997;99:1200–9.
60. Lipman ML, Panda D, Bennett HP, Henderson JE, Shane E, Shen Y et al., Cloning of human PEX cDNA. Expression, subcellular localization, and endopeptidase activity. J Biol Chem 1998;273:13729–37.
61. Ruchon AF, Marcinkiewicz M, Siegfried G, Tenenhouse HS, DesGroseillers L, Crine P et al., Pex mRNA is localized in developing mouse osteoblasts and odontoblasts. J Histochem Cytochem 1998;46:459–68.
62. Grieff M, Mumm S, Waeltz P, Mazzarella R, Whyte MP, Thakker RV et al., Expression and cloning of the human X-linked hypophosphatemia gene cDNA. Biochem Biophys Res Commun 1997;231:635–9.
63. Tenenhouse HS, Scriver CR, McInnes RR, Glorieux FH. Renal handling of phosphate in vivo and in vitro by the X-linked hypophosphatemic male mouse: evidence for a defect in the brush border membrane. Kidney Int 1978;14:236–44.
64. Tenenhouse HS, Klugerman AH, Neal JL. Effect of phosphonoformic acid, dietary phosphate and the Hyp mutation on kinetically distinct phosphate transport processes in mouse kidney. Biochim Biophys Acta 1989;984:207–13.
65. Collins JF, Scheving LA, Ghishan FK Decreased transcription of the sodium–phosphate transporter gene in the hypophosphatemic mouse. Am J Physiol 1995;269:F439–48.
66. Tenenhouse HS, Werner A, Biber J, Ma S, Martel J, Roy S et al., Renal Na(+)-phosphate cotransport in murine X-linked hypophosphatemic rickets. Molecular characterization. J Clin Invest 1994;93:671–6.
67. Tenenhouse HS, Meyer Jr RA, Mandla S, Meyer MH, Gray RW. Renal phosphate transport and vitamin D metabolism in X-linked hypophosphatemic Gy mice: responses to phosphate deprivation. Endocrinology 1992;131:51–6.
68. Tenenhouse HS, Beck L. Renal Na(+)-phosphate cotransporter gene expression in X-linked Hyp and Gy mice. Kidney Int 1996;49:1027–32.
69. Collins JF, Ghishan FK. The molecular defect in the renal sodium-phosphate transporter expression pathway of Gyro (Gy) mice is distinct from that of hypophosphatemic (Hyp) mice. Faseb J 1996;10:751–9.
70. Lyles KW, Harrelson JM, Drezner MK. The efficacy of vitamin D2 and oral phosphorus therapy in X-linked hypophosphatemic rickets and osteomalacia. J Clin Endocrinol Metab 1982;54:307–15.
71. Glorieux FH, Marie PJ, Pettifor JM, Delvin EE. Bone response to phosphate salts, ergocalciferol, and calcitriol in hypophosphatemic vitamin D-resistant rickets. N Engl J Med 1980;303:1023–31.
72. Harrell RM, Lyles KW, Harrelson JM, Friedman NE, Drezner MK. Healing of bone disease in X-linked hypophosphatemic rickets/osteomalacia. Induction and maintenance with phosphorus and calcitriol. J Clin Invest 1985;75:1858–68.
73. Lobaugh B, Drezner MK. Abnormal regulation of renal 25-hydroxyvitamin D-1 alpha-hydroxylase activity in the X-linked hypophosphatemic mouse. J Clin Invest 1983;71:400–3.
74. Meyer Jr RA, Gray RW, Meyer MH. Abnormal vitamin D metabolism in the X-linked hypophosphatemic mouse. Endocrinology 1980;107:1577–81.

75. Cunningham J, Gomes H, Seino Y, Chase LR. Abnormal 24-hydroxylation of 25-hydroxyvitamin D in the X-linked hypophosphatemic mouse. Endocrinology 1983;112:633–8.

76. Tenenhouse HS, Yip A, Jones G. Increased renal catabolism of 1,25-dihydroxyvitamin D3 in murine X-linked hypophosphatemic rickets. J Clin Invest 1988;81:461–5.

77. Yamaoka K, Seino Y, Satomura K, Tanaka Y, Yabuuchi H, Haussler MR. Abnormal relationship between serum phosphate concentration and renal 25-hydroxycholecalciferol-1-alpha-hydroxylase activity in X-linked hypophosphatemic mice. Miner Electrolyte Metab 1986;12:194–8

78. Roy S, Tenenhouse HS. Transcriptional regulation and renal localization of 1,25-dihydroxyvitamin D3-24-hydroxylase gene expression: effects of the Hyp mutation and 1,25-dihydroxyvitamin D3. Endocrinology 1996;137:2938–46.

79. Ecarot B, Glorieux FH, Desbarats M, Travers R, Labelle L. Defective bone formation by Hyp mouse bone cells transplanted into normal mice: evidence in favor of an intrinsic osteoblast defect. J Bone Miner Res 1992;7:215–20.

80. Ecarot B, Glorieux FH, Desbarats M, Travers R, Labelle L. Effect of 1,25-dihydroxyvitamin D3 treatment on bone formation by transplanted cells from normal and X-linked hypophosphatemic mice. J Bone Miner Res 1995;10:424–31.

81. Tanaka H, Seino Y, Shima M, Yamaoka K, Yabuuchi H, Yoshikawa H et al., Effect of phosphorus supplementation on bone formation induced by osteosarcoma-derived bone-inducing substance in X-linked hypophosphatemic mice. Bone Miner 1988;4:237–46.

82. Seino Y, Shimotsuji T, Ishii T, Ishida M, Ikehara C, Yamaoka K et al., Treatment of hypophosphataemic vitamin D-resistant rickets with massive doses of 1 alpha-hydroxy-vitamin D3 during childhood. Arch Dis Child 1980;55:49–53.

83. Glorieux FH, Scriver CR, Reade TM, Goldman H, Roseborough A. Use of phosphate and vitamin D to prevent dwarfism and rickets in X-linked hypophosphatemia. N Engl J Med 1972;287:481–7.

84. Costa T, Marie PJ, Scriver CR, Cole DE, Reade TM, Nogrady B. X-linked hypophosphatemia: effect of calcitriol on renal handling of phosphate, serum phosphate, and bone mineralization. J Clin Endocrinol Metab 1981;52:463–72.

85. Miyamura T, Tanaka H, Inoue M, Ichinose Y, Seino Y. The effects of bone marrow transplantation on X-linked hypophosphatemic mice. J Bone Miner Res 2000;15:1451–8.

86. Francis F, Strom TM, Hennig S, Boddrich A, Lorenz B, Brandau O et al., Genomic organization of the human PEX gene mutated in X-linked dominant hypophosphatemic rickets. Genome Res 1997; 7:573–85.

87. Econs MJ, McEnery PT. Autosomal dominant hypophosphatemic rickets/osteomalacia: clinical characterization of a novel renal phosphate-wasting disorder. J Clin Endocrinol Metab 1997;82:674–81.

88. The ADHR Consortium. Autosomal dominant hypophosphataemic rickets is associated with mutations in FGF23. Nat Genet 2000;26:345–8.

Bone Disease in Chronic Renal Failure

T. Akizawa and E. Kinugasa

Changes in mineral metabolism associated with progressive renal failure bring about disturbances of bone metabolism. Metabolic bone disease induced by chronic renal failure has been called renal bone disease or renal osteodystrophy. Some therapeutic interventions to this abnormal mineral metabolism also contribute to the development of bone disorder.

Renal Osteodystrophy

There are three types of typical renal bone disease; i.e., osteitis fibrosa, osteomalacia and adynamic bone disease; and patients who showed symptoms of both osteitis fibrosa and osteomalacia have been diagnosed as having mixed uraemic osteodystrophy [1].

Osteitis Fibrosa

Bone turnover is markedly increased in osteitis fibrosa with a great number of osteoclasts and an increased number and activity of osteoblasts (Figure 9.20). It also shows bone fibrosis, marked resorptive surfaces and distinct tetracycline double staining. These indicate accelerated bone resorption and bone formation. However, bone mineral decreases with greater bone resorption than bone formation in the long-term course of osteitis fibrosa. Secondary hyperparathyroidism leads to osteitis fibrosa.

Osteomalacia

Osteomalacia is characterised by a remarkable increase in osteoid volume (Figure 9.21) with little erosive surface and tetracycline staining. Bone mass decreases with the calcification defect of osteoid. The major pathogenesis is a calcitriol deficiency in children and an accumulation of trace metals (aluminium in major cases and iron in rare cases) at the calcification front of bone in adult patients. Histological examination detects aluminium in bone by aurintricarboxylic acid staining (aluminium staining: Figure 9.22).

Adynamic Bone Disease

Adynamic bone disease or aplastic bone disease is distinct from osteitis fibrosa, showing defective bone formation and bone resorption without increase in osteoid

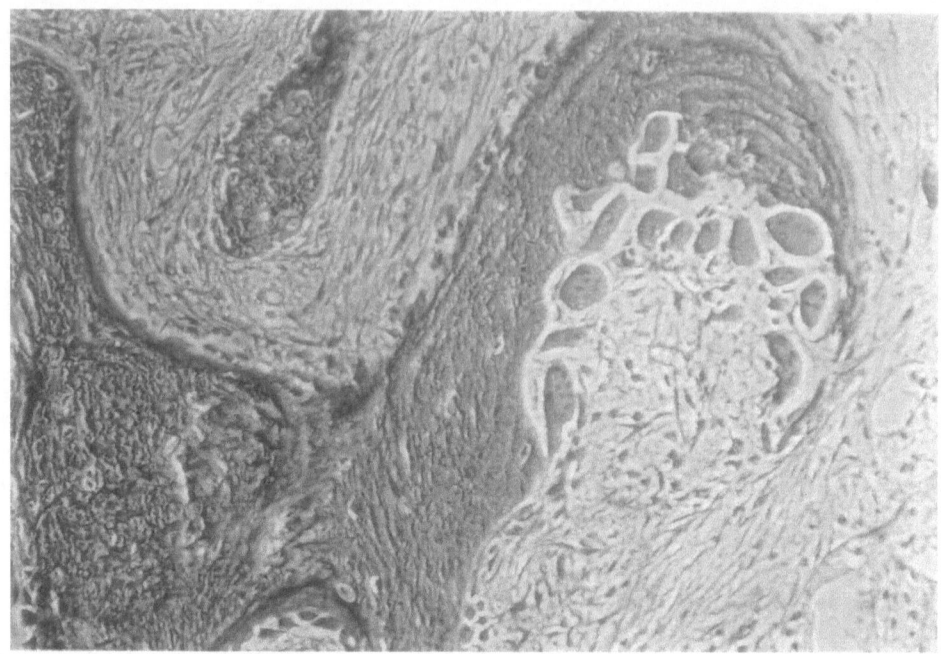

Figure 9.20.   Histological finding of osteitis fibrosa.

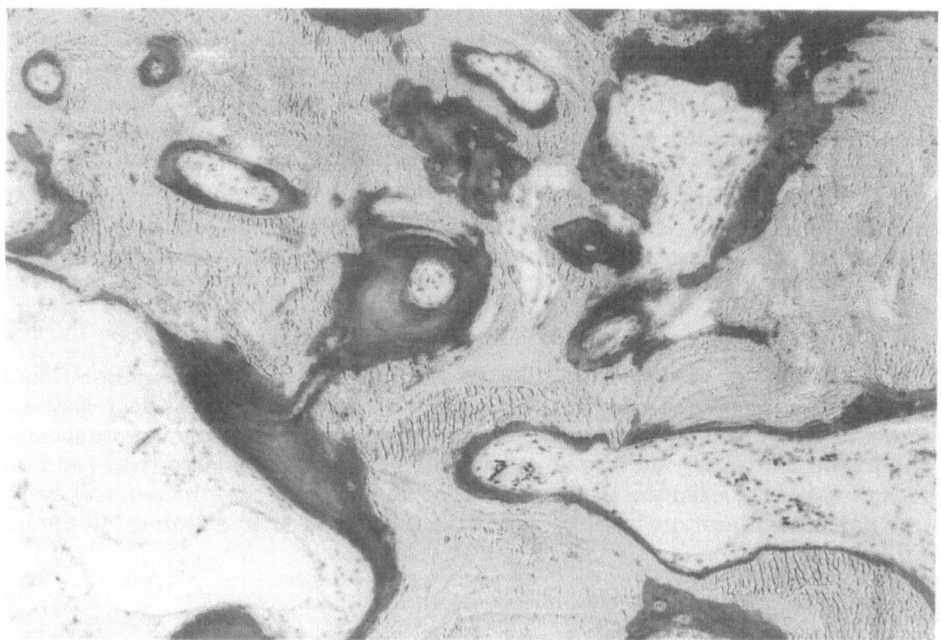

Figure 9.21.   Histological finding of osteomalacia.

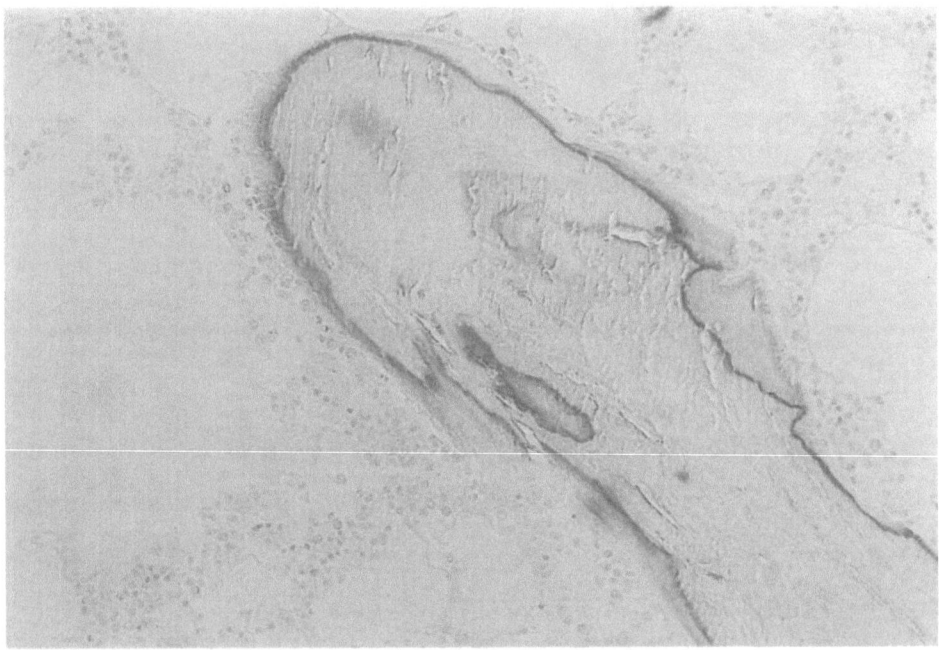

Figure 9.22. Aluminium staining of aluminium bone disease.

or fibrosis (Figure 9.23). It has been reported that a higher than normal level of parathyroid hormone (PTH) is necessary for uraemic patients to maintain their bone turnover at a normal level [2]. Adynamic bone disease is frequently associated with relative PTH deficiency in which normal or even supernormal PTH is not sufficient for the normal bone metabolism in uraemia [3]. This relative hypoparathyroidism in uraemia is now speculated to be a major pathogenesis of adynamic bone disease.

Secondary Hyperparathyroidism

Parathyroid hyperplasia is another characteristic feature of secondary hyperparathyroidism in uraemia. Tominaga et al. classified parathyroid hyperplasia into two groups; i.e., nodular hyperplasia and diffuse hyperplasia, using surgically removed parathyroids [4]. They showed that a larger gland (nodular hyperplasia) has a greater growth potential than a smaller gland (diffuse hyperplasia). They also reported that the recurrence rate of hyperparathyroidism in forearm autograft is correlated with the growth potential of the gland, and a recurrence was more frequent in nodular hyperplasia [5]. Parathyroid cells show monoclonal proliferation [6] in this large nodular hyperplasia, in contrast to polyclonal proliferation observed in diffuse hyperplasia. A genetic abnormality, distinct from primary hyperparathyroidism, may contribute to these neoplastic transforming processes in nodular hyperplasia [7]. However, the precise mechanism has not been fully elucidated.

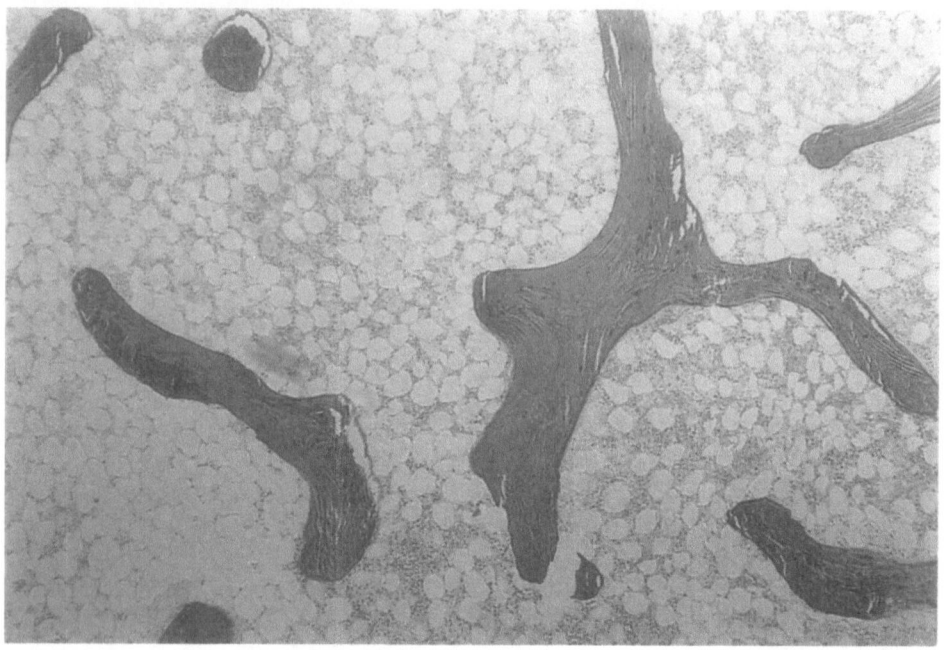

Figure 9.23. Histological finding of adynamic bone disease

Pathogenesis

Hypocalcaemia, resulting from the accumulation of phosphorus and the reduced production of calcitriol due to the deterioration of renal function, has been regarded as the classical cause of secondary hyperparathyroidism. Calcitriol deficiency also stimulates PTH secretion by reducing its direct inhibitory effect on PTH secretion. Furthermore, calcitriol deficiency stimulates hypertrophy of the parathyroid gland. In these processes, it is reported that the reduced induction of p21,cyclin/cyclin-dependent kinase inhibitor and increased induction of transforming growth factor α (TGF-α) may function as an autocrine signal to stimulate parathyroid hyperplasia [8]. TGF-α induction involves activation of mitogen-activated protein kinase cascades and binding to the epidermal growth factor receptor. Inhibition of p21 stimulates the cycle of cell division. Furthermore, recent progress in cellular and molecular biology has had a great impact on understanding the pathogenesis of secondary hyperparathyroidism (Table 9.10).

Resistance of Parathyroid Cells to Physiological Levels of Calcitriol

The resistance leads to hypersecretion and excessive synthesis of PTH, and proliferation of parathyroid cells. This resistance is partially explained by the decreased density of vitamin D receptors in parathyroid glands. Low calcitriol levels may be an important factor for the reduction of calcitriol receptors, because large dose supplementations of calcitriol can up-regulate the receptor number and correct the

Table 9.10.   Pathogenesis of secondary hyperparathyroidism

1. Phosphorus retention
   Hypocalcaemia, inhibition of vitamin D-1α-hydroxylase activity, direct effect

2. Calcitriol deficiency
   Hypocalcaemia, decrease in VDR, reduction of direct inhibitory effects on PTH production and parathyroid
     proliferation

3. Abnormality in VDR and post VDR
   Decreased number of VDR, inhibition of VDR and nuclear chromatin interaction

4. Abnormality in CaR
   Decreased number of CaR

5. Skeletal resistance to PTH
   Reference in Table 9.14

6. Genetic background
   Polymorphism of VDR, CaR, etc.

7. Others
   Mutation of parathyroid cell, VDR or CaR, inhibition of parathyroid cell apoptosis, etc.

VDR: vitamin D receptor, CaR: Ca-sensing receptor.

resistance to calcitriol. Clinically, parathyroid cells in nodular hyperplasia have lower calcitriol receptor density than cells in diffuse hyperplasia [9]. Also, some uraemic toxins or uraemic degradation products inhibit the internalisation of vitamin D and vitamin D receptor complexes into the nucleus and its interaction with nuclear chromatin [10]. These factors interfere with the transcriptional regulation of PTH by calcitriol, and calcitriol is unable to exert a sufficient biological action on target cells.

Decreased Sensitivity to Extracellular Calcium Concentration

Changes in extracellular calcium are detected by calcium-sensing receptors in parathyroid cells, and the receptor regulates the PTH secretion. The calcium-sensing function of parathyroid cells has been evaluated by the relation between serum intact PTH and calcium ion levels during acute changes in extracellular calcium (sigmoidal curve) [11]. Uraemic patients undergoing dialysis show decreased sensitivity to extracellular calcium as evidenced by the greater concentration of extracellular calcium that is required to suppress intact PTH in their sigmoidal curves [12, 13]. The resistance was reported to vary depending on the underlying renal disease of uraemia or duration of dialysis history. This resistance is also more remarkable in patients with secondary hyperparathyroidism than those with low-turnover aluminium bone disease or adynamic bone disease.

With the recent cloning of the calcium sensing receptor gene, the abnormal-calcium-sensing mechanism has been analysed at the molecular level [14]. In the enlarged parathyroid gland of uraemic patients, the decreased number of calcium-sensing receptors has been reported, in particular within cells in nodular hyperplasia [15, 16]. The decrease in calcium-sensing receptors was detected in association with the proliferation of parathyroid cells in uraemic rats developed by hyperphosphorus feeding [17]. These results suggest a strong relation between parathyroid hyperplasia and the down-regulation of the calcium-sensing receptor.

The Direct Effects of Phosphorus

Phosphorus accumulation has been a classical pathogenesis of secondary hyper-parathyroidism, and the mechanism has been attributed to the decreasing effect on serum calcium and calcitriol production in the remnant kidney. Recently, several reports have pointed out the direct effect of phosphorus on PTH synthesis and secretion. In vitro studies using hyperplastic parathyroid tissue from uraemic rats or patients with renal failure found that high phosphorus concentrations in the medium increased prepro-PTH mRNA expression and PTH secretion in primary culture systems [18–20]. This effect was more remarkable in diffuse hyperplasia glands than that in nodular hyperplasia. In vivo studies of uraemic rats and humans also strongly support the direct stimulating action of phosphorus on PTH secretion that is independent of the plasma calcium and calcitriol concentrations [21, 22]. Several mechanisms have been postulated to affect the direct action of phosphorus. Reduced arachidonic acid production by parathyroid tissue under high phosphorus concentrations may be an explanation, because high arachidonic acid suppresses PTH production [23]. But the precise mechanisms responsible for the direct effect of phosphorus on PTH production and parathyroid proliferation are not well understood.

Other Factors

The influence of polymorphism of vitamin D receptors or Ca-sensing receptors on secondary hyperparathyroidism has been reported [24–26]. Also, the possible contribution of point mutation within those related hormone receptors, or somatic mutation of the proto-oncogene suppressive gene to the pathogenesis of secondary hyperparathyroidism has not been completely excluded [27]. With regard to the effect of apoptosis, several recent papers do not support the decrease in apoptosis of parathyroid cells in secondary hyperparathyroidism [28, 29].

Skeletal resistance to the calcaemic action of PTH is also an important aetiological factor of secondary hyperparathyroidism. Recently, many new mechanisms have been elucidated and these are described below.

Treatment

Phosphorus Management

Phosphorus control is a fundamental treatment for and prevention of secondary hyperparathyroidism. Dietary phosphorus restriction by a low-protein diet lowers plasma phosphorus concentration, but it is not an appropriate way for dialysis patients because of the risk of malnutrition. Removal of the amount of phosphorus by dialysis is strictly limited by the frequency and duration of dialysis. As a result, oral phosphorus binders are almost always required for the management of hyper-phosphataemia in dialysis patients.

Among the phosphorus binders listed in Table 9.11, antacid containing aluminium was contraindicated in 1992 in Japan, because of aluminium intoxication due to the gradual accumulation of this substance in brain and bone, which brought about dialysis dementia and vitamin D-resistant osteomalacia (aluminium bone disease), respectively. It also led to microcytic anaemia that was resistant to recombinant human erythropoietin therapy.

Table 9.11. Agents for the treatment of hyperphosphataemia

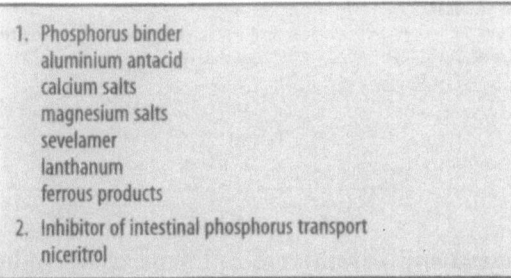

1. Phosphorus binder
   aluminium antacid
   calcium salts
   magnesium salts
   sevelamer
   lanthanum
   ferrous products
2. Inhibitor of intestinal phosphorus transport
   niceritrol

Calcium salts, the major alternative phosphorus binder to aluminium, has the disadvantage of inducing hypercalcaemia in patients taking vitamin D preparations. Hypercalcaemic patients should take limited doses of calcium salts as well as vitamin D preparations. As a result, phosphorus and PTH levels could not be well controlled, and greater phosphorus and calcium × phosphorus products promoted the development of vascular calcification and atherosclerosis [30]. These factors were reported to determine the mortality risk of dialysis patients [31, 32].

Magnesium salts have not been accepted in Japan, because of the risk of hypermagnesaemia and the inconvenience of using dialysates with low magnesium concentrations that are not commercially available.

These problems have now been solved by the introduction of the cationic polymer agent sevelamer (RenaGel), which is non-absorbable and contains neither calcium nor aluminium [33]. It adsorbs negatively charged phosphorus ions by cationic force, and removes phosphorus in faeces. It also has a lowering action on total and LDL cholesterol concentration, and an elevating action on HDL cholesterol [34]. By the application of sevelamer, serum phosphorus is effectively and safely managed in dialysis patients, and the effect of sevelamer on reducing the hospitalisation risk was also reported [35].

Niceritrol (nicotinate preparation) was found to have a suppressive action on phosphorus transport in the intestine, and to lower the plasma phosphorus levels in dialysis patients. But it has been reported in Japan that chronic use of niceritrol on dialysis patients may lead to anaemia or a decrease in platelet counts.

Calcitriol and Alphacalcidol

Since the first clinical trials of calcitriol and alphacalcidol, secondary hyperparathyroidism was treated by daily oral administration of these drugs. However, hypercalcaemia and hyperphosphataemia limited the dosage and the therapeutic effect. Slatopolsky first reported that intermittent high doses of intravenous calcitriol could suppress PTH secretion effectively without causing significant hypercalcaemia, probably due to the transient but supraphysiological high peak concentration of calcitriol in the circulation [36]. Since this first report in 1984, the effect of intravenous calcitriol pulse therapy on secondary hyperparathyroidism has been well established. In Japan, where calcitriol for intravenous administration was not commercially available, oral calcitriol pulse therapy (four to six µg, one or two times each week) has been applied, and it showed an effect comparable with intravenous pulse therapy [37, 38]. The mechanism of the effects of calcitriol pulse therapy has been explained by the intermittent pharmacological level of calcitriol in the circulation, which directly

Table 9.12. New vitamin D analogues for the treatment of secondary hyperparathyroidism

1. Paricalcitol (19-nor-1-alpha, 25-dihydroxyvitamin D_2)
2. Doxercalciferol (1-alpha-hydroxyvitamin D_2)
3. Maxacalcitol (22-oxacalcitriol)
4. Falecalcitriol (26,26,26,27,27,27-hexafluoro-calcitriol)

suppressed PTH secretion and parathyroid cell hyperplasia induced by lack of calcitriol, and up-regulated the decreased number of calcitriol receptors in parathyroid cells [39]. But hypercalcaemia and hyperphosphataemia are the most frequent complications that require at least a transient reduction in the dose of calcitriol or calcium salts for phosphorus binder.

New Calcitriol Analogue

New calcitriol analogues with less calcaemic action than calcitriol have been developed to circumvent the above side effects of calcitriol (Table 9.12 and Figure 9.24).

Paricalcitol (19-nor-1-alpha, 25-dihydroxyvitamin D_2) is an intravenous vitamin D_2 sterol, and showed an effect on PTH suppression comparable with calcitriol. Hypercalcaemic episodes prior to attaining target PTH levels were less frequent than those by calcitriol [40, 41]. The decreased calcaemic action of paricalcitol is partially explained by an acquired, postreceptor resistance of the intestine and bone to chronic treatment with the paricalcitol [42, 43]. Paricalcitol is commercially available in the United States.

Doxercalciferol (1-alpha-hydroxyvitamin D_2) is also a vitamin D_2 sterol, which is effective in suppressing PTH secretion using oral and intravenous administration

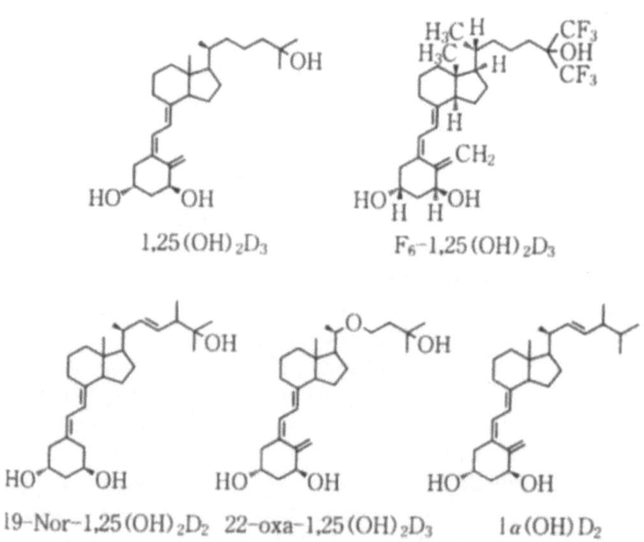

Figure 9.24. Chemical structure of newly developed vitamin D analogues.

[44]. However, it is not clear whether the onset of hypercalcaemic and/or hyper-phosphataemic episodes by this drug is less frequent than those by calcitriol. Doxercalciferol has been approved in the United States.

Maxacalcitol (22-oxacalcitriol: OCT) is a newly developed calcitriol analogue in Japan. OCT has shown a less calcaemic action and a strong suppressive effect on PTH in uraemic rats and dogs [45, 46]. In uraemic patients with secondary hyper-parathyroidism, OCT dose dependently suppressed PTH secretion and also increased serum calcium levels. But over 60% of patients achieved a decrease in intact PTH levels of greater than 30% from base line in long-term treatment of OCT for up to one year without non-physiological increases in mean serum calcium [47]. The low calcaemic action of OCT was explained by the difference in its affinity to vitamin D receptors (1/8 of calcitriol) and that to the vitamin D binding protein (1/500 of cal-citriol), the short plasma half-life, and the difference in the required cofactors at the binding to vitamin D receptors from calcitriol [48–50]. OCT is commercially available in Japan.

Falecalcitriol (26,26,26,27,27,27-hexafluoro-calcitriol) is a calcitriol analogue for oral intake developed in Japan. It dose-dependently suppresses PTH secretion and increases serum calcium levels, but it was reported that daily oral administration of falecalcitriol was more effective on PTH suppression than that of alphacalcidol, with a comparable calcaemic action [51, 52]. Falecalcitriol is now commercially available in Japan.

Calcium-sensing Receptor Agonists

Drugs that activate calcium-sensing receptors as calcium were named calcimimetics. They activate calcium-sensing receptors in parathyroid cells and suppress PTH secretion with negligible effects on serum calcium levels. Calcimimetics were also reported to have a suppressive effect on parathyroid proliferation despite the hypocalcaemia induced by the marked reduction of PTH secretion, and they ameliorated osteitis fibrosa in uraemic rats [53, 54]. Preliminary clinical studies of the calcimimetic R-568 produced a dose-dependent reduction in the plasma PTH concentration and a decrease in serum calcium in patients on regular haemodialysis with mild to severe hyperparathyroidism [55, 56]. A transient increase in serum calcitonin levels, possibly resulting from the activation of thyroid calcium-sensing receptors, was also observed. However, long-term clinical studies of R-568 were abandoned because several patients lacked its metabolic enzyme in their liver. A clinical trial is taking place in the USA on a new calcimimetic that is metabolised by other hepatic enzymes.

Percutaneous Direct Injection Therapy into Parathyroid Gland

With the improvement of ultrasonography technologies, the size and blood flow of the parathyroid gland can be precisely measured. Using this technique, percutaneous ethanol injection therapy (PEIT) into the parathyroid gland has become a more prac-tical procedure. In particular, repeated ethanol injection therapy was reported to be an effective alternative to surgical parathyroidectomy [57]. However, this procedure requires an experienced physician as well as a very sensitive scanner because it is sometimes difficult to make an exact identification of the parathyroid gland [58]. It may also cause a diplegia of recurrent laryngeal nerves especially by the concurrent injection of ethanol on both sides. Furthermore, fibrosis of parathyroid glands and

Table 9.13. Parathyroid intervention therapies

1. Percutaneous ethanol injection therapy (PEIT)
2. Percutaneous calcitriol injection therapy (PCIT)
3. Percutaneous calcitriol analogue injection therapy
4. Percutaneous calcimimetic injection therapy
5. Surgical parathyroidectomy

possibly their surrounding tissue after ethanol injection may bring some technical difficulties for subsequent surgical parathyroidectomy.

In order to avoid these complications, an intravenous preparation of calcitriol was applied for percutaneous direct injection therapy. Repeated injection of calcitriol into the parathyroid gland was reported to be effective way to suppress PTH secretion without the risk of recurrent nerve paralysis [59]. Direct injection of a calcitriol analogue or calcimimetic into the parathyroid gland has also been reported in animal models of secondary hyperparathyroidism [60]. Both agents showed strong suppressive effects on PTH secretion. These results suggest that percutaneous direct injection therapy may become a powerful tool for the treatment of secondary hyperparathyroidism. However, careful attention should be paid to the possibility that a parathyroid with monoclonal cell proliferation may have escaped from some inhibitory mechanism controlling cell growth and direct injection into the parathyroid may stimulate the invasion of such cells to surrounding tissues.

Surgical Parathyroidectomy

Despite recent advances in the non-surgical treatment of secondary hyperparathyroidism, severe cases of hyperparathyroidism still require surgical parathyroidectomy. However, the risk of persistent and/or recurrence of hyperparathyroidism or development of adynamic bone disease resulting from the inadequate secretion of PTH after parathyroidectomy is a new problem of surgical parathyroidectomy [61]. To resolve the former problem, total parathyroidectomy with auto-transplantation is widely selected, because this approach enables easy removal of recurrent hyperplastic glands from the auto-transplanted site.

Considering the development of the above direct therapies of parathyroid glands for the treatment of secondary hyperparathyroidism, these techniques have been classified into a new concept, i.e., parathyroid intervention therapy (Table 9.13).

Osteomalacia

It had been assumed that vitamin D deficiency was an important factor for the pathogenesis of osteomalacia. This was true for uraemic children in whom the epiphyseal line had not disappeared. Calcitriol supplementation brought about a considerable effect on their bone lesions. However, in most adult cases, aluminium intoxication resulted from intestinal absorption of aluminium-containing phosphorus binders and/or inflow via dialysis membranes from contaminated dialysate appeared to be a major cause (aluminium bone disease). Aluminium accumulation is accelerated in the bone in particular in the case of low PTH concentrations such as after surgical parathyroidectomy. By avoiding aluminium-containing phosphorus binders and using the complete water treatment system for dialysate with reverse osmosis, the prevalence of osteomalacia has fallen dramatically.

For the precise diagnosis of aluminium bone disease, bone histological examination by aurintricarboxylic acid staining is necessary. But it is difficult to diagnose by bone histology in most patients, and the change of serum aluminium concentration by deferoxamine mesylate (DFO) infusion has been used as an alternative diagnostic procedure (DFO test). Using the DFO test, the rise in plasma aluminium concentration after an acute intravenous infusion of DFO (5–50 mg/kg) was evaluated. Low intact PTH levels also have diagnostic value for aluminium accumulation. Stimulated increase in plasma aluminium of more than 150–200 µg/L and a low intact PTH level have been reported to give the most specific value for aluminium bone disease.

For the treatment of aluminium bone disease, the effect of long-term DFO therapy over one year has been reported. However, DFO has a number of side effects such as mucormycosis, ocular and auditory neurotoxicity, gastrointestinal disturbances, hypotension, anaphylaxis and liver dysfunction. Therefore, treatment once per week with a low-dose (5 mg/kg or 500 mg) DFO and the efficient removal of DFO–aluminium complex by haemodialysis or haemodiafiltration with high-flux or high-performance dialysis membranes are recommended. Furthermore, patients treated with DFO should be referred to ophthalmology and otorhinolaryngology for regular examinations before and during the treatment.

Adynamic Bone Disease

Since first reported, adynamic bone disease, originally called aplastic bone, with little or no aluminium accumulation has been observed in 15–60% of bone specimens obtained from dialysis and pre-dialysis patients [62]. The prevalence is increasing. Adynamic bone disease is histologically characterised by the defect of osteoid and fibrosis with markedly reduced bone resorption and formation: that with aluminium deposition is classified as aluminium bone disease.

A higher than normal level of PTH is necessary to maintain normal bone turnover in uraemic patients [2]. Adynamic bone disease is frequently associated with relatively low PTH secretion, which has been referred to as relative hypoparathyroidism. The incidence and prevalence of relative hypoparathyroidism has also been increasing [63].

Pathogenesis

Adynamic bone disease is classified into two types: with aluminium accumulation and without aluminium accumulation. Even in patients with minimum aluminium exposure, iron deficiency and up-regulation of transferrin receptors have been reported to increase aluminium uptake in the osteoblast and parathyroid, and suppress bone turnover and PTH production. These situations frequently resulted from the use of recombinant human erythropoietin [64]. In cases without aluminium deposition, several risk factors have been pointed out: diabetes, older age, short dialysis history, intake of vitamin D derivatives and/or calcium-containing phosphorus binders, chronic ambulatory peritoneal dialysis (CAPD), male under CAPD treatment, use of high-calcium dialysate, and hypermagnesaemia [65]. PTH secretion is often suppressed in diabetes, and the lack of insulin decreases bone metabolism. Low bone metabolism enhances hypercalcaemia, which in turn further suppresses PTH secretion and bone remodelling. Hyperglycaemia and advanced glycation end products are suspected to be responsible for the low secretion of PTH. It has also been reported that adynamic bone disease developed in patients treated with intermittent calcitriol in spite of the persistent hyperparathyroidism [66]. These findings suggest that calcitriol may bring about adynamic bone disease not only by reducing PTH but also by directly suppressing bone turnover.

Malnutrition has been reported to be a contributing factor for the development of adynamic bone disease. Positive correlations between intact PTH and serum albumin, creatinine, pre-albumin and cholesterol were pointed out in haemodialysis or CAPD patients. In malnutritional patients, decreased protein intake may lead to hypophosphataemia, and immobilisation may cause hypercalcaemia. Patients with adynamic bone disease easily become hypercalcaemic due to the reduced buffering capacity of bone for calcium. These abnormalities further reduce the production and secretion of PTH.

Outcome

Short duration of dialysis history is one of the characteristic backgrounds of patients with adynamic bone disease. This may be a reflection of poor survival of these patients. Several reports observed a greater mortality risk in patients with low PTH [67]. Considering these results, adynamic bone disease and relative hypoparathyroidism seem to reflect, at least in part, a malnutritional state. Atherosclerosis and ectopic vascular calcification by persistent hypercalcaemia, resulting from the reduced buffering capacity of bone, have been suspected to be another major risk factor of mortality.

Bone fracture is another important index of prognosis. Several recent papers found the increased risk of vertebral fracture in patients with adynamic bone disease and/or low PTH values [68]. These results suggest that close attention should be paid to the prognosis of bone as well as life itself in patients with adynamic bone disease and relative hypoparathyroidism.

Mechanism of Skeletal Resistance to PTH

Skeletal resistance to PTH has been a classical pathogenesis of secondary hyperparathyroidism, and is now one of the major causes of adynamic bone disease. It is widely accepted that two or three times the physiological level of PTH is required for the normal bone turnover in uraemic patients. A mechanism has been reported for the explanation of this skeletal resistance to PTH. Recent studies have further clarified the new mechanisms (Table 9.14). Consensus has been reached as to the decreased number of PTH/PTH-related peptide (PTHrP) receptors (PTH 1 receptor) in the growth plate cartilage. Down-regulation of the PTH 1 receptor was recently reported in uraemic bone, and in particular in adynamic bone disease [69]. Although this abnormality can clearly explain the skeletal resistance to PTH, it remains controversial whether the PTH 1 receptor is really reduced in osteoblasts of uraemic patients. PTH exerts its action on osteoblasts via the induction of osteoclast differentiation factor (ODF or RANKL). Osteoclastogenesis inhibitory factor (OCIF), also named osteoprotegerin, is a decoy receptor of ODF. In uraemia, serum concentrations of ODF are reported to be increased significantly, and the level reaches five times that of normal renal function. At that concentration, ODF inhibits about 50% of osteoclastogenesis in vitro. This result suggests that accumulated ODF inhibits

Table 9.14. Pathogenesis of skeletal resistance to PTH in uraemia

1. Classically reported genesis
   Deficiency of calcitriol, phosphorus retention, high serum PTH, acidosis, etc.

2. Recently reported genesis
   Down-regulation of PTH 1 receptor in bone
   Accumulation of PTH fragment (7,84-PTH)
   Accumulation of osteoprotegerin
   Direct effect of calcitriol
   Deficiency of BMP-7
   Abnormality of IGFBP

BMP-7: bone morphogenetic protein-7, IGFBP: insulin-like growth factor binding protein.

osteoclastogenesis in uraemia, and the inhibition may play a responsible role for the skeletal resistance to PTH in uraemia.

Recently, a new PTH assay that detects only whole molecules of PTH (1–84) has been developed. Since conventional intact PTH assay has recognised the 1–84 PTH and its fragment of 7–84 PTH, plasma PTH concentrations measured by the new assay are much less than those by intact PTH. It has been reported that 7–84 PTH competitively inhibits the action of 1–84 PTH, and about 40% of PTH stored in hyperplastic parathyroid of uraemic patients is 7–84 PTH [70]. Thus, in uraemia, 7–84 PTH inhibits the action of 1–84 PTH, and the inhibition may lead to an overestimation of PTH concentration.

Bone morphogenetic protein 7 (BMP 7), also named osteogenetic protein-1, is mainly produced in the kidney, and strongly activates osteoblastic differentiation. Reduced production of BMP 7 in uraemia was supposed to be responsible for low bone formation.

Insulin-like growth factor I (IGF-I) and IGF-II levels were reported to increase in uraemic patients. Recently, a marked increase in IGF binding protein-4 (IGFBP-4) and a decrease in IGFBP-5 were observed in uraemic plasma. Since the IGF family regulates bone formation, the abnormal metabolism of the IGF family may contribute to reduced bone formation in uraemia.

Prevention and Treatment

Hypercalcaemia is an important factor to suppress PTH and bone turnover. Therefore, active vitamin D sterols and calcium-containing phosphorus binders should be used with care. Withdrawal of vitamin D sterols stimulates PTH secretion in short-term studies, but the effects should be determined in long-term studies.

Calcitriol directly suppresses bone turnover. On this point, maxacalcitol showed less suppressive effects on bone turnover in vitro and in animal models of uraemia.

Favourable effects on adynamic bone disease have also been reported in the attempt to stimulate PTH secretion by low-calcium dialysate in CAPD patients.

Several authors have found, in patients with adynamic bone disease, an increase in PTH and bone metabolic markers by vitamin K supplementation. Vitamin K is essential for the proper carboxylation of bone matrix proteins, including osteocalcin, and low vitamin K activity is frequently observed in patients on regular haemodialysis. Therefore, vitamin K deficiency might have some role for the development of adynamic bone disease.

With respect to new therapeutic agents, calcilytics, which stimulate PTH secretion by sending hypocalcaemic signals on calcium-sensing receptors, appear to be very effective for the treatment of adynamic bone disease. Preliminary results suggest that calcilytics increased bone formation as well as PTH administration.

Conclusion

Among bone diseases associated with chronic renal failure, osteomalacia can be now effectively treated by active vitamin D sterols and the prevention of aluminium inflow into the human body. For secondary hyperparathyroidism and osteitis fibrosa, powerful therapeutic tools such as new phosphorus binders, active vitamin D analogues, calcimimetics and several kinds of parathyroid interventions have been developed. For adynamic bone disease, it is very important to elucidate the mechanisms of skeletal resistance to PTH in uraemia. Keeping PTH at a relatively higher level than normal for the treatment of adynamic bone disease leads to parathyroid hyperplasia, which eventually brings about resistance to medical therapy for PTH oversecretion.

References

1. Coburn JW. Renal osteodystrophy. Kidney Int. 1980;17:677–23.
2. Quarles LD, Lobaugh B, Murphy G. Intact parathyroid hormone overestimates the presence and severity of parathyroid-mediated osseous abnormalities in uremia. J Clin Endocrinol Metab 1992;75: 145–50.
3. Hercz G, Pei Y, Greenwood C, Manuel A, Saiphoo C, Goodman WG et al. Aplastic osteodystrophy without aluminum: the role of "suppressed" parathyroid function. Kidney Int 1993;44:860–66.
4. Tominaga Y, Sato K, Tanaka Y, Numano M, Uchida K, Takagi H. Histopathology and pathophysiology of secondary hyperparathyroidism due to chronic renal failure. Clin Nephrol 1995;44 Suppl 1: S42–7.
5. Tominaga Y, Takagi H. Molecular genetics of hyperparathyroid disease. Curr Opin Nephrol Hypertens 1996;5:336–41.
6. Arnold A, Brown MF, Urena P, Gaz RD, Sarfati E, Drueke TB. Monoclonality of parathyroid tumors in chronic renal failure and in primary parathyroid hyperplasia. J Clin Invest 1995;95:2047–53.
7. Tominaga Y, Tsuzuki T, Uchida K, Haba T, Otsuka S, Ichimori T et al. Expression of PRAD1/cyclin D1, retinoblastoma gene products, and Ki67 in parathyroid hyperplasia caused by chronic renal failure versus primary adenoma. Kidney Int 1999;55:1375–83.
8. Dusso AS, Pavlopoulos T, Naumovich L, Lu Y, Finch J, Brown AJ et al. p21WAF1 and transforming growth factor-alpha mediate dietary phosphate regulation of parathyroid cell growth. Kidney Int 2001;59:855–65.
9. Fukuda N, Tanaka H, Tominaga Y, Fukagawa M, Kurokawa K, Seino Y. Decreased 1,25-dihydroxy-vitamin D3 receptor density is associated with a more severe form of parathyroid hyperplasia in chronic uremic patients. J Clin Invest 1993;92:1436–43.
10. Patel SR, Koenig RJ, Hsu CH. Effect of Schiff base formation on the function of the calcitriol receptor. Kidney Int 1996;50:1539–45.
11. Kinugasa E, Akizawa T, Koshikawa S. Parathyroid Function in end-stage renal failure. J Bone Mineral Met 1993;11 (suppl):53–8.
12. Goodman WG, Veldhuis JD, Belin TR, Van Herle AJ, Juppner H, Salusky IB. Calcium sensing by parathyroid glands in secondary hyperparathyroidism. J Clin Endocrinol Metab 1998;83:2765–72.
13. Akizawa T, Fukagawa M. Modulation of parathyroid cell function by calcium ion in health and uremia. Am J Med Sci 1999;317:358–62.
14. Brown AJ, Zhong M, Finch J, Ritter C, McCracken R, Morrissey J et al. Rat calcium-sensing receptor is regulated by vitamin D but not by calcium. Am J Physiol 1996;270:F454–60.

15. Kifor O, Moore Jr FD, Wang P, Goldstein M, Vassilev P, Kifor I et al. Reduced immunostaining for the extracellular Ca^{2+}-sensing receptor in primary and uremic secondary hyperparathyroidism. J Clin Endocrinol Metab 1996;81:1598–606.

16. Gogusev J, Duchambon P, Hory B, Giovannini M, Goureau Y, Sarfati E et al. Depressed expression of calcium receptor in parathyroid gland tissue of patients with hyperparathyroidism. Kidney Int 1997;51:328–36.

17. Brown AJ, Ritter CS, Finch JL, Slatopolsky EA. Decreased calcium-sensing receptor expression in hyperplastic parathyroid glands of uremic rats: role of dietary phosphate. Kidney Int 1999;55:1284–92.

18. Almaden Y, Canalejo A, Hernandez A, Ballesteros E, Garcia-Navarro S, Torres et al. Direct effect of phosphorus on PTH secretion from whole rat parathyroid glands in vitro. J Bone Miner Res 1996;11:970–6.

19. Slatopolsky E, Finch J, Denda M, Ritter C, Zhong M, Dusso A et al. Phosphorus restriction prevents parathyroid gland growth. High phosphorus directly stimulates PTH secretion in vitro. J Clin Invest 1996;97:2534–40.

20. Almaden Y, Hernandez A, Torregrosa V, Canalejo A, Sabate L, Fernandez Cruz L et al. High phosphate level directly stimulates parathyroid hormone secretion and synthesis by human parathyroid tissue in vitro. J Am Soc Nephrol 1998;9:1845–52.

21. Hernandez A, Concepcion MT, Rodriguez M, Salido E, Torres A. High phosphorus diet increases pre-proPTH mRNA independent of calcium and calcitriol in normal rats. Kidney Int 1996;50:1872–8.

22. de Francisco AL, Cobo MA, Setien MA, Rodrigo E, Fresnedo GF, Unzueta MT et al. Effect of serum phosphate on parathyroid hormone secretion during hemodialysis. Kidney Int 1998;54:2140–5.

23. Almaden Y, Canalejo A, Ballesteros E, Anon G, Rodriguez M. Effect of high extracellular phosphate concentration on arachidonic acid production by parathyroid tissue in vitro. J Am Soc Nephrol 2000;11:1712–8.

24. Yokoyama K, Shigematsu T, Tsukada T, Ogura Y, Takemoto F, Hara S et al. Apa I polymorphism in the vitamin D receptor gene may affect the parathyroid response in Japanese with end-stage renal disease. Kidney Int 1998;53:454–8.

25. Nagaba Y, Heishi M, Tazawa H, Tsukamoto Y, Kobayashi Y. Vitamin D receptor gene polymorphisms affect secondary hyperparathyroidism in hemodialyzed patients. Am J Kidney Dis 1998;32:464–9.

26. Yano S, Sugimoto T, Kanzawa M, Tsukamoto T, Hattori T, Hattori S et al. Association of polymorphic alleles of the calcium-sensing receptor gene with parathyroid hormone secretion in hemodialysis patients. Nephron 2000;85:317–23.

27. Falchetti A, Bale AE, Amorosi A, Bordi C, Cicchi P, Bandini S et al. Progression of uremic hyperparathyroidism involves allelic loss on chromosome 11. J Clin Endocrinol Metab 1993;76:139–44.

28. Canalejo A, Almaden Y, Torregrosa V, Gomez-Villamandos JC, Ramos B, Campistol JM et al. The in vitro effect of calcitriol on parathyroid cell proliferation and apoptosis. J Am Soc Nephrol 2000;11:1865–72.

29. Zhang P, Duchambon P, Gogusev J, Nabarra B, Sarfati E, Bourdeau A et al. Apoptosis in parathyroid hyperplasia of patients with primary or secondary uremic hyperparathyroidism. Kidney Int 2000;57:437–45.

30. Goodman WG, Goldin J, Kuizon BD, Yoon C, Gales B, Sider D et al. Coronary artery calcification in young adults with end-stage renal disease who are undergoing dialysis. N Engl J Med 2000; 342:1478–83.

31. Block GA, Hulbert-Shearon TE, Levin NW, Port FK. Association of serum phosphorus and calcium × phosphate product with mortality risk in chronic hemodialysis patients: a national study. Am J Kidney Dis 1998;31:607–17.

32. Block GA, Port FK. Re-evaluation of risks associated with hyperphosphatemia and hyperparathyroidism in dialysis patients: recommendations for a change in management. Am J Kidney Dis 2000; 35:1226–37.

33. Chertow GM, Burke SK, Dillon MA, Slatopolsky E. Long-term effects of sevelamer hydrochloride on the calcium × phosphate product and lipid profile of haemodialysis patients. Nephrol Dial Transplant 1999;14:2907–14.

34. Slatopolsky EA, Burke SK, Dillon MA. Renagel, a nonabsorbed calcium- and aluminum-free phosphate binder, lowers serum phosphorus and parathyroid hormone. The RenaGel Study Group. Kidney Int 1999;55:299–307.

35. Collins AJ, St Peter WL, Dalleska FW, Ebben JP, Ma JZ. Hospitalization risks between Renagel phosphate binder treated and non-Renagel treated patients. Clin Nephrol 2000;54:334–41.

36. Slatopolsky E, Weerts C, Thielan J, Horst R, Harter H, Martin KJ. Marked suppression of secondary hyperparathyroidism by intravenous administration of 1,25-dihydroxy-cholecalciferol in uremic patients. J Clin Invest 1984;74:2136–43.

37. Tsukamoto Y, Nomura M, Takahashi Y, Takagi Y, Yoshida A, Nagaoka T et al. The oral 1,25-dihydroxy-vitamin D3 pulse therapy in hemodialysis patients with severe secondary hyperparathyroidism. Nephron 1991;57:23–8.

38. Liou HH, Chiang SS, Huang TP, Shieh SD, Akmal M. Comparative effect of oral or intravenous calcitriol on secondary hyperparathyroidism in chronic hemodialysis patients. Miner Electrolyte Metab 1994;20:97–102.

39. Akizawa T, Fukagawa M, Koshikawa S, Kurokawa K. Recent progress in management of secondary hyperparathyroidism of chronic renal failure. Curr Opin Nephrol Hypertens 1993;2:558–65.

40. Martin KJ, Gonzalez EA, Gellens M, Hamm LL, Abboud H, Lindberg J. 19-Nor-1-alpha-25-dihydroxy-vitamin D2 (Paricalcitol) safely and effectively reduces the levels of intact parathyroid hormone in patients on hemodialysis. J Am Soc Nephrol 1998;9:1427–32.

41. Llach F, Keshav G, Goldblat MV, Lindberg JS, Sadler R, Delmez J et al. Suppression of parathyroid hormone secretion in hemodialysis patients by a novel vitamin D analog: 19-nor-1,25-dihydroxyvita-min D2. Am J Kidney Dis 1998;32(Suppl 2):S48–54.

42. Brown AJ, Finch J, Takahashi F, Slatopolsky E. Calcemic activity of 19-Nor-1,25(OH)2D2 decreases with duration of treatment. J Am Soc Nephrol 2000;11:2088–94.

43. Holliday LS, Gluck SL, Slatopolsky E, Brown AJ. 1,25-Dihydroxy-19-nor-vitamin D2, a vitamin D analog with reduced bone resorbing activity in vitro. J Am Soc Nephrol 2000;11:1857–64.

44. Tan Jr AU, Levine BS, Mazess RB, Kyllo DM, Bishop CW, Knutson JC et al. Effective suppression of parathyroid hormone by 1 alpha-hydroxy-vitamin D2 in hemodialysis patients with moderate to severe secondary hyperparathyroidism. Kidney Int 1997;51:317–23.

45. Fukagawa M, Kaname S, Igarashi T, Ogata E, Kurokawa K. Regulation of parathyroid hormone synthesis in chronic renal failure in rats. Kidney Int 1991;39:874–81.

46. Kubrusly M, Gagne ER, Urena P, Hanrotel C, Chabanis S, Lacour B et al. Effect of 22-oxa-calcitriol on calcium metabolism in rats with severe secondary hyperparathyroidism. Kidney Int 1993;44:551–6.

47. Akizawa T, Suzuki M, Akiba T, Nishizawa Y, Kurokawa K. Clinical effects of maxacalcitol on secondary hyperparathyroidism of uremic patients. Am J Kidney Dis 2001;38(Suppl 1):S147–51.

48. Denda M, Finch JL, Brown AJ, Nishii Y, Kubodera N, Slatopolsky E. 1,25-Dihydroxyvitamin D3 and 22-oxacalcitriol prevent the decrease in vitamin D receptor content in the parathyroid glands of uremic rats. Kidney Int 1996;50:34–9.

49. Ichikawa F, Hirata M, Endo K, Katsumata K, Ohkawa H, Kubodera N et al. Attenuated up-regulation of vitamin D-dependent calcium-binding proteins by 22-oxa-1, 25-dihydroxyvitamin D3 in uremic rats. A possible mechanism for less-calcemic action. Nephrology 1998;4:391–5.

50. Takeyama K, Masuhiro Y, Fuse H, Endoh H, Murayama A, Kitanaka S et al. Selective interaction of vitamin D receptor with transcriptional coactivators by a vitamin D analog. Mol Cell Biol 1999;19: 1049–55.

51. Morii H, Ogura Y, Koshikawa S, Mimura N, Suzuki M, Kurokawa K et al. Efficacy and safety of oral falecalcitriol in reducing parathyroid hormone in hemodialysis patients with secondary hyper-parathyroidism. J Bone Mineral Metab 1998;16:34–43.

52. Akiba T, Marumo F, Owada A, Kurihara S, Inoue A, Chida Y et al. Controlled trial of falecalcitriol versus alfacalcidol in suppression of parathyroid hormone in hemodialysis patients with secondary hyper-parathyroidism. Am J Kidney Dis 1998;32:238–46.

53. Wada M, Nagano N, Furuya Y, Chin J, Nemeth EF, Fox J. Calcimimetic NPS R-568 prevents parathyroid hyperplasia in rats with severe secondary hyperparathyroidism. Kidney Int 2000;57:50–8.

54. Wada M, Ishii H, Furuya Y, Fox J, Nemeth EF, Nagano N. NPS R-568 halts or reverses osteitis fibrosa in uremic rats. Kidney Int 1998;53:448–53.

55. Antonsen JE, Sherrard DJ, Andress DL. A calcimimetic agent acutely suppresses parathyroid hormone levels in patients with chronic renal failure. Rapid communication. Kidney Int 1998;53:223–7.

56. Goodman WG, Frazao JM, Goodkin DA, Turner SA, Liu W, Coburn JW. A calcimimetic agent lowers plasma parathyroid hormone levels in patients with secondary hyperparathyroidism. Kidney Int 2000;58:436–45.

57. Kitaoka M, Fukagawa M, Ogata E, Kurokawa K. Reduction of functioning parathyroid cell mass by ethanol injection in chronic dialysis patients. Kidney Int 1994;46:1110–17.

58. Fukagawa M, Kitaoka M, Tominaga Y, Akizawa T, Kurokawa K. Selective percutaneous ethanol injection therapy (PEIT) of the parathyroid in chronic dialysis patients – the Japanese strategy. Nephrol Dial Transplant 1999;14:2574–7.

59. Kitaoka M, Fukagawa M, Kurokawa K. Direct injection of calcitriol into parathyroid hyperplasia in chronic dialysis patients with severe parathyroid hyperfunction. Nephrology 1995;1:563–8.

60. Kakuta T, Fukagawa M, Fujisaki T, Hida M, Suzuki H, Sakai H et al. Prognosis of parathyroid function after successful percutaneous ethanol injection therapy guided by color Doppler flow mapping in chronic dialysis patients. Am J Kidney Dis 1999;33:1091–9.

61. Fukagawa M, Tominaga Y, Kitaoka M, Kakuta T, Kurokawa K. Medical and surgical aspects of parathyroidectomy. Kidney Int 1999;73(Suppl):S65–9.
62. Torres A, Lorenzo V, Hernandez D, Rodriguez JC, Concepcion MT, Rodriguez AP et al. Bone disease in predialysis, hemodialysis, and CAPD patients: evidence of a better bone response to PTH. Kidney Int 1995;47:1434–42.
63. Akizawa T, Kinugasa E, Akiba T, Tsukamoto Y, Kurokawa K. Incidence and clinical characteristics of hypoparathyroidism in dialysis patients. Kidney Int 1997;62(Suppl):S72–4.
64. Smans KA, D'Haese PC, Van Landeghem GF, Andries LJ, Lamberts LV, Hendy GN et al. Transferrin-mediated uptake of aluminium by human parathyroid cells results in reduced parathyroid hormone secretion. Nephrol Dial Transplant 2000;15:1328–36.
65. Fukagawa M, Kurokawa K. Is aplastic osteodystrophy a disease of malnutrition? Curr Opin Nephrol Hypertens 2000;9:363–7.
66. Goodman WG, Ramirez JA, Belin TR, Chon Y, Gales B, Segre GV et al. Development of adynamic bone in patients with secondary hyperparathyroidism after intermittent calcitriol therapy. Kidney Int 1994;46:1160–6.
67. Avram MM, Sreedhara R, Avram DK, Muchnick RA, Fein P. Enrollment parathyroid hormone level is a new marker of survival in hemodialysis and peritoneal dialysis therapy for uremia. Am J Kidney Dis 1996;28:924–30.
68. Atsumi K, Kushida K, Yamazaki K, Shimizu S, Ohmura A, Inoue T. Risk factors for vertebral fractures in renal osteodystrophy. Am J Kidney Dis 1999;33:287–93.
69. Picton ML, Moore PR, Mawer EB, Houghton D, Freemont AJ, Hutchison AJ et al. Down-regulation of human osteoblast PTH/PTHrP receptor mRNA in end-stage renal failure. Kidney Int 2000;58:1440–9.
70. Slatopolsky E, Finch J, Clay P, Martin D, Sicard G, Singer G et al. A novel mechanism for skeletal resistance in uremia. Kidney Int 2000;58:753–61.

10

Organ Diseases Associated with Deranged Calcium Metabolism

10A

Calcium and Ageing

T. Fujita

Introduction

Calcium (Ca), the most abundant inorganic element and the fifth most abundant of all elements in the human body, is characterised by its unique distribution (Table 10.1). More than 99% of Ca in the human body is found in the bone, representing the first and the largest Ca compartment of the body. Ca is found at a concentration around $10\,M$ in this pool. Much less Ca is found in the second compartment, blood and extracellular fluid, at a strictly maintained concentration of $10^{-3}\,M$, only $1/10^{-4}$ the level in the bone. Blood Ca concentration is the most strictly guarded and maintained biological constant in the living organism. The borderline between bone and blood is a layer of lining cells derived from osteoblasts. Inside this layer is found the bone extracellular fluid, with components somewhere between blood and bone, serving as a kind of buffering transitional zone between the two compartments with a large difference in concentration (Figure 10.1).

The third Ca compartment with the lowest concentration is the cytoplasmic microcompartment. Bordered by the plasma membrane, a lipid bilayer surrounding each cell, cytosolic free Ca concentration ($[Ca]i$) is maintained at $10^{-7}\,M$, against the ever-constant Ca level in the surrounding extracellular fluid, $10^{-3}\,M$, a concentration difference of $10^4\,M$. This is strangely similar to the difference between the first and second Ca compartment, the bone and extracellular fluid. This common intercompartmental Ca concentration difference may be called the Ca gap. The second Ca gap across the plasma membrane is essential for the role of Ca as the messenger for cellular function. The unique nature of the Ca gap is realised when one compares the intra- and extracellular Ca concentration with that of other inorganic ions. The concentration differences for Mg and K with higher intracellular than extracellular concentration and that of sodium with higher extracellular concentration, never exceeding $10–10^2$ times, are all far lower than the 10^4 times difference of Ca.

Table 10.1. Calcium compartments in young and old age

Compartments	Young	Old
Bone Calcium	Maintained	Decreased
Blood Calcium	Maintained	Maintained
Cytosolic Calcium	Maintained	Increased

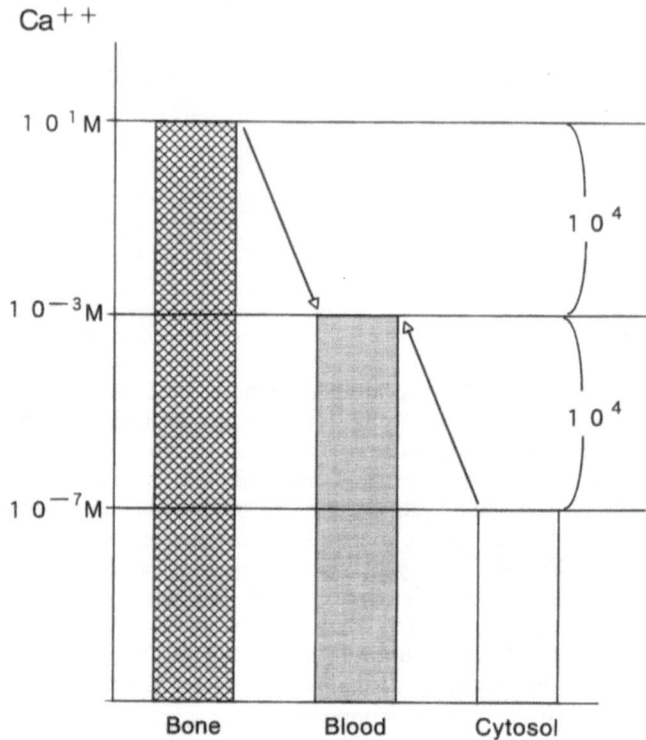

Figure 10.1. Concentrations of ionised calcium in three compartments, bone, blood and cytoplasm, in human body. The intercompartmental calcium concentration difference, "Calcium Gap", is approximately 10^4 higher than that of any other inorganic ions.

Ageing is associated with abnormalities of calcium (Ca) metabolism and its distribution among compartments, making such intercompartmental concentration difference less and less distinct. The Ca comes out from the first compartment, bone, with advancing age, leaving osteoporosis behind. Osteoporosis is a universal phenomenon associated with ageing, regardless of race, gender and environment. Ca deficiency plays a key role in the development of osteoporosis. Even the rapid postmenopausal bone loss, which helped the establishment of the gold standard concept of postmenopausal osteoporosis, may depend on Ca deficiency in addition to œstrogen deficiency, in view of the decrease of Ca absorption on œstrogen removal, returns to normal after œstrogen replacement and the effect of Ca supplementation enforcing the œstrogen effect in the treatment of postmenopausal osteoporosis. Along with ageing and decrease of bone mass, calcification of the blood vessels, especially the aorta, becomes evident. [Ca]i also shows a steady increase with advance in age, apart from transient increases associated with each phase of the cell function. A steady and permanent rise of the basal [Ca]i may compromise the amplitude of Ca transients and consequently interfere with cell function itself. The age-bound deterioration of

Table 10.2. Characteristics of Ageing

1. Calcium decompartmentalisation with skeletal decalcification and ectopic calcification. Less distinct calcium concentration difference between compartments.
2. Slower and compromised response to physical, chemical and biological stimuli.
3. Less efficient adjustment to environmental changes.
4. Variable, inconstant, hidden and prolonged manifestations of diseases.

Table 10.3. Causes of Ca Deficiency in Ageing. Decreased intake, intestinal malabsorption and increased loss are the three major factors for Ca deficiency in ageing

I.	Decreased Intake
II.	Intestinal Calcium Malabsorption
	1. Decreased solar ultraviolet ray exposure and vitamin D biosynthesis in the skin
	2. Decreased renal 1,25(OH) vitamin D biosynthesis
	3. Oestrogen and androgen deficiency
	4. Growth hormone deficiency
	5. Ageing of the intestine
III.	Increased Ca Loss
	1. Increased urinary Ca
	2. Increased faecal Ca
	3. Increased sweat Ca

cell function may find some explanation in the decreased concentration difference between intra- and extracellular compartments.

Life is said to be a miracle, hardly explained by ordinary physicochemical principles. The Intricate mechanism of life is like a state with a low entropy according to a thermodynamic concept. The active, healthy, but frail state of life with low entropy, high tension and high potential energy like the top of the mountain is always exposed to dangers of decline and dissipation into an inactive state of high entropy, low tension and low potential energy state like the foot of a mountain, whenever the complex protective mechanisms fail in accidents, diseases and ageing. In terms of Ca physiology, the low entropy state of active and healthy life depends on the huge concentration difference between the three Ca compartments; bone, blood and cytosol. As the differences among these compartments are lost, the entropy of life rises and ageing and diseases advance.

In terms of Ca physiology, the low entropy, high tension or high potential energy state depends on the large concentration gradients between the three Ca compartments. Ageing is definded as a gradual rise of entropy in the form of gradual loss of intercompartmental Ca concentration differences (Table 10.2).

Age-bound Ca deficiency is the background for decompartmentalisation (Table 10.3). Even a slight fall of serum Ca prompts parathyroid hormone (PTH) secretion, causing secondary hyperparathyroidism of ageing. Age-bound increase of PTH in blood, especially in those with osteoporosis, has been repeatedly demonstrated since the pioneer report by Fujita et al. [1]. Persitent excess PTH plays a key role in the ageing phenomenon both in the cellular and whole body level. The only way to alleviate and delay secondary hyperparathyroidism of ageing and to prevent the Ca paradox is to provide adequate supplementation with good absorbability along with the active form of vitamin D, bypassing the deteriorating renal function for biosynthesis of $1,25(OH)_2$ vitamin D.

The Calcium Paradox

The strict compartmentalisation of the human body as to Ca makes it necessary to view the human body as a joint complex of three pools; bone, blood and cytoplasm, each behaving according to its own rules with mutual interactions. Ca deficiency or excess occurs in each compartment separately. Calcium deficiency in the whole organism is primarily reflected in the bone compartment because of the overwhelming majority of Ca contained in this compartment. Blood Ca is maintained constant regardless of the Ca economy of the body. Unlike Fe, with a high concentration in the blood on excessive intake and low concentration on deficient intake because of the absence of sufficient storage space within the body, Ca concentration in the blood is strictly guarded and kept constant by the concerted effort of calcium regulating hormones, most importantly parahtyroid hormone (PTH), and enormous storage space, the bone. Slight hypocalcaemia on Ca deficiency is immediately restored to normal by an increase of PTH secretion and slight hypercalcaemia caused by excess Ca intake is again normalised by a decrease of PTH secretion. Increased calcitonin secretion also participates only when bone resorption is increased as in tumour-induced hypercalcaemia caused by pathologically increased bone resorption. Hypercalcaemia or hypocalcaemia therefore indicate abnormalities of controlling factors, mainly PTH and $1,25(OH)_2$ vitamin D, and not the state of body Ca economy, deficit or excessive Ca intake or absorption. PTH also exerts a strange action to increase the entrance of Ca from the extracellular space into the cytosol of not only the cells traditionally known as targets for PTH such as osteoblasts and renal epithelial cells, but also cardiomyocytes, vascular wall cells, pancreatic endocrine cells, brain cells, hepatocytes and blood cells. Ca deposition also occurs in the interstitial connective tissue with physiologically very low Ca content such as cartilage and blood vessels on PTH excess regardless of the serum Ca level. Ca deficiency and PTH hyper-secretion therefore increases [Ca]i and soft tissue Ca contents. Thus, the Ca deficiency of the whole body and skeletal compartment causes Ca excess in the cytoplasmic compartment and interstitial tissues usually almost free of Ca, giving the impression of a uniform and non-specific distribution of Ca within the body instead of specific, uneven and selective distribution with distinct compartmentalisation. Such unique behaviour of Ca with apparent excess prompted by or associated with deficiency may appear paradoxical in the absence of accurate knowledge on the three distinct Ca pools, their mutual interaction and the mechanism to control it [2].

Suggestions of contrasting behaviours of bone and soft tissue Ca have been made for a long time based on clinical experience and radiographic observations. Osteoporosis frequently coexists with calcification of the aorta and other blood vessels (Figure 10.2). Chronic renal failure with a longstanding Ca deficiency because of a decrease of $1,25(OH)_2$ vitamin D biosynthesis in the kidney and consequent intestinal Ca malabsorption causes typical secondary hyperparathyroidism. Vascular calci-

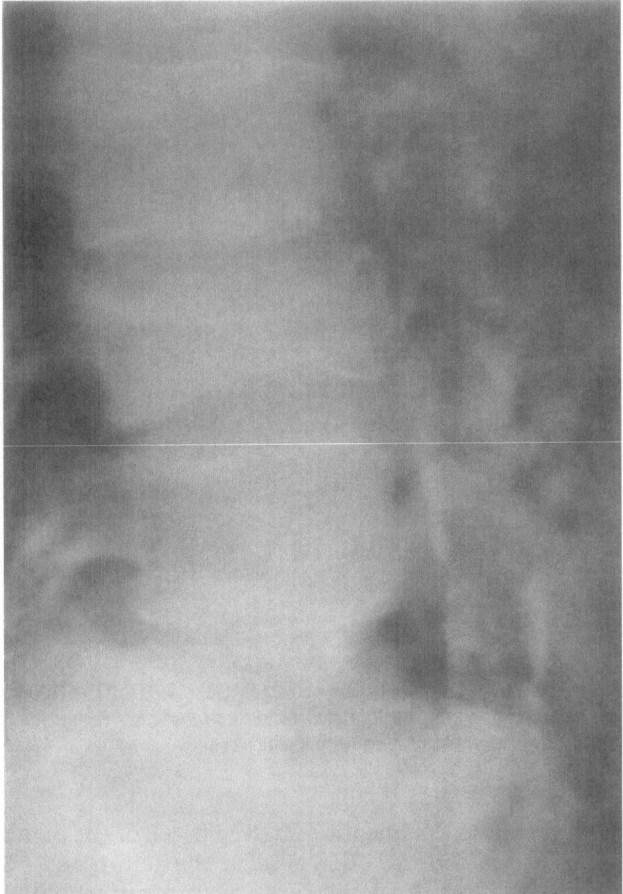

Figure 10.2.   Coexistence of compression fracture due to osteoporosis, and aortic calcification due to atherosclerosis in a patient, exhibiting so-called "Calcium Shift"; loss of calcium from bone and its deposition in the blood vessel, or "Calcium Paradox"; calcium excess in soft tissue in the presence of calcium deficiency in the bone.

fication, especially the Mönckeberg type medial sclerosis, periarticular calcification and renal osteodystrophy manifested by skeletal decalcification and occasional hyperostosis characterise the disturbance of Ca distribution with decompartmental-isation in chronic renal failure. Since ageing is also associated with a mild and gradual fall of kidney function and renal $1,25(OH)_2$ vitamin D biosynthesis and intestinal Ca malabsorption, secondary hyperparathyroidism is also expected to occur in ageing, though in milder forms than the typical ones seen in chronic renal failure. Ageing may therefore be regarded as a mild and slowly progressive chronic renal failure with all its complications. Skeletal decalcification accompanied by soft tissue calcification, sometimes called "Ca shift", thus characterise both ageing and chronic renal failure.

Ageing is associated with progressive Ca deficiency, secondary hyperparathy-roidism and Ca paradox diseases in addition to simple Ca deficiency of the skeleton or osteoporosis. Increased Ca content of the vascular system causes hypertension and

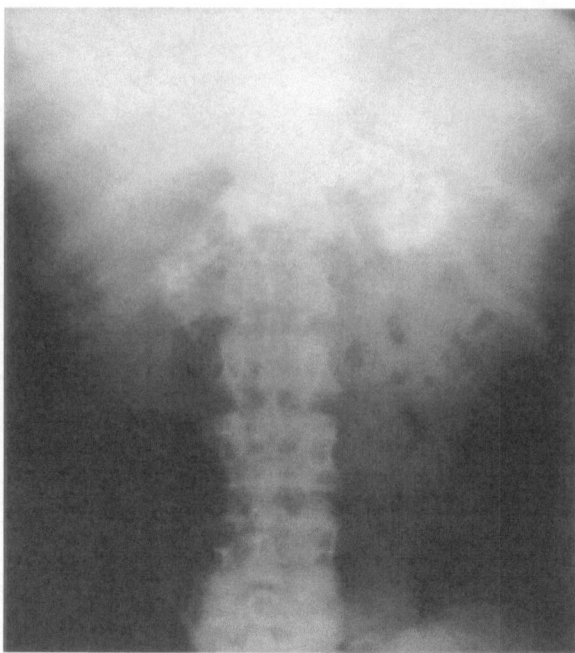

Figure 10.3.   Pancreatic calcification in a case of primary hyperparathyroidism with sustained hypercalcaemia and diabetes mellitus. Chronic excess of PTH caused a soft tissue calcification and diabetes mellitus. Although this is an unusual example, calcium accumulation in the pancreas induced a disturbance of insulin secretion.

arteriosclerosis. Ca inflow into the insulin-secreting β-cell of the pancreas causes diabetes mellitus and its complications. To symbolise the action of PTH increasing Ca entry into the pancreas, pancreatic calcification seen in a patient with primary hyperparathyroidism is presented (Figure 10.3). Pancreatic calcification in primary hyperparathyroidism is not as common as renal calcification or nephrolithiasis, but chronic pancreatitis and pancreatic stones are frequently seen in this disease, probably on account of the increased Ca content of the pancreatic juice. Increase of Ca in the brain and nerves causes neurodegenerative diseases such as Alzheimer's disease, amyotrophic lateral sclerosis and multiple sclerosis. Sustained increases of cytoplasmic free Ca may impair energy expenditure, and initiate cytoskeletal degradation, prompting cell death, apoptosis, cell cycle activation and malignant proliferation [3]. Increased urinary Ca excretion in response to Ca deficiency and PTH excess with an increase of bone resorption may cause nephrolithiasis in contrast to the one-time belief of high Ca intake facilitating kidney stone formation. Ca deposition in cartilage in response to Ca deficiency, secondary hyperparathyroidism and Ca release from bone, cause cartilage degeneration, hardening, wearing out, and loss of cartilage, leading to a direct contact between bones and, finally, degenerative joint disease.

Calcium and Blood Vessels

The role of Ca deficiency in essential hypertension was first pointed out in epidemiological studies [4] and repeatedly confirmed. The favourable effects of Ca supplementation on hypertension were also documented, but the mechanism of the action of Ca on blood vessels still remains to be fully explained. Calcium supplementation does not always decrease blood pressure, but seems to be more effective in Ca-deficient subjects with low renin, low plasma ionised Ca, high $1,25(OH)_2$ vitamin D and high PTH. Ca also stimulates secretion of calcitonin and calcitonin gene-related peptide (CGRP), which lowers blood pressure though vasorelaxation.

Unlike the clinical and epidemiological setting, in which multiple complex factors of daily lifestyle and variable environmental factors make the picture quite complicated, and the effect of nutritional intervention rather vague and equivocal, experimental animal models of hypertension have provided a much more clear-cut picture. Hypertension in spontaneously hypertensive rat (SHR) was found to depend on deficient dietary Ca intake [5]. In experimental models of hypertension such as spontaneously hypertensive rat (SHR), Lyon hypertensive rat and Dahl salt sensitive hypertensive rat, Ca supplementation prevented the occurrence of hypertension. In SHR, renal $1,25(OH)_2$ vitamin D biosynthesis is decreased, intestinal Ca absorption decreased, plasma PTH is raised, and urinary Ca excretion increased due to secondary hyperparathyroidism, which explains the dramatically increased bone turnover and osteopenia in this model [6]. At the same time, the Ca contents of the platelets and blood vessels are also increased. Parathyroidectomy disrupts the sequence of events and inhibits the rise of blood pressure. Dietary supplementation of alpha-lipoic acid capable of increasing intracellular thiol compound, binding excess endogenous aldehyde and normalising Ca channels and [Ca]i normalised blood pressure in SHR, suggesting the importance of PTH-induced rise of [Ca]i in hypertension [7]. Rise of [Ca]i in vascular smooth muscle cells causes vasoconstriction and in platelets causes augmentation of the response to ADP and epinephrine. Boood pressure was related to a thrombin-evoked rise of [Ca]i in platelets.

Progressive aortic calcification is associated with bone loss after menopause in a population-based longitudinal study (Figure 10.4) [8]. Mitral annular calcification, detected by means of high-speed angiographic tomography, occuring with advancing age may predict coronary calcification and atherosclorosis. Despite the male predisposition of atherosclerotic cardiovascular morbidity over females, females paradoxically showed higher incidence of mitral annular calcification thicker than 5 mm, 6% of 238 compared to 9% of 284 males [9]. This may also be related to the more frequent occurrence of osteoporosis in females. Left ventricular hypertrophy with fibrosis, the major complication of hypertension, is mediated by a rise of [Ca]i and an increase of collagen type I and II induced by angiotensin II and inhibited by angiotensin II receptor antagonist. Both of the two major therapeutic agents for hypertension, Ca channel antagonist and angiotensin II receptor antagonist, widely used in the treatment of hypertension, thus inhibit the rise of [Ca]i and disrupt the sequence of events following Ca deficiency. In order to evaluate the relationship between Ca accumulation among bone, cartilage, artery and vein, the relative contents of Ca were analysed in human bones, arteries, veins and cartilages in 27 subjects (17 men and 10 women who died between 40 and 98 years of age). Ca released from bone appears to accumulate in the artery and cartilage [10]. In addition to Ca deficiency and secondary hyperparathyroidism, PTHrP secretion was also reported to be abnormal in spontaneous hypertensive rats [11]. Foetal plasma PTHrP, but not total plasma Ca concentration, was lower in SHR than in WKY controls. Placental and amniotic PTHrP were also lower in SHR than In WKY controls, and no rise towards term occurred in SHR, unlike WKY. Such reduced PTHrP secretion in SHR may be related to the growth disturbance and developmental inhibition in SHR.

Ageing and Ca Paradox

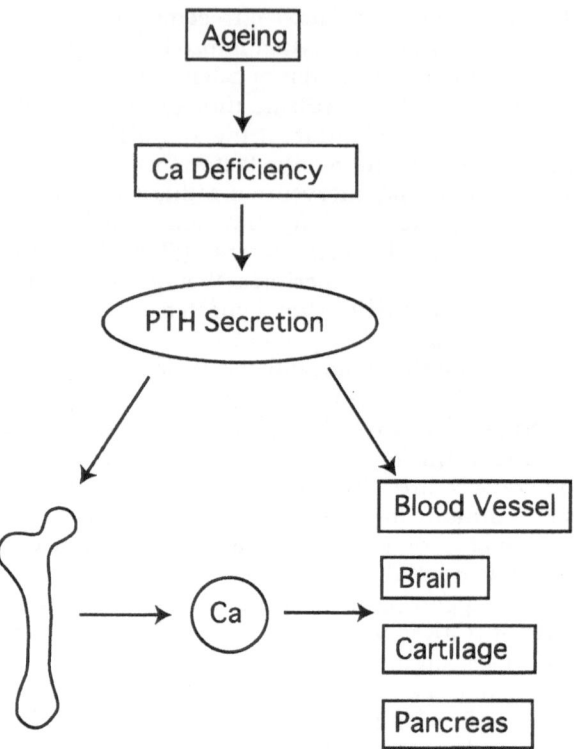

Figure 10.4.   Ageing and calcium paradox. Ageing is accompanied by calcium deficiency. Even a slight fall of serum calcium prompts extra-secretion of parathyroid hormone (PTH), in addition to the normal level for the maintenance of the physiological function of the organism. PTH increases bone resorption and augments calcium release from bone. At the same time, PTH increases calcium deposition into the interstitial tissues of blood vessels, brain, cartilage and pancreas, as well as the entrance into the cytosolic compartment.

Water hardness with high Ca and Mg content was pointed out to be inversely related to death rate due to cardiovascular disease, especially atherosclerosis. An inverse correlation was found between coronary arterial calcification and serum $1,25(OH)_2$ vitamin D levels. Shear stress increases [Ca]i in the endothelial cells with the release of vasorelaxing nitrous oxide (NO). NO deficiency, induced by treatment with N-nitroso-L-arginine methyl ester (L-NAME) for eight weeks to inhibit NO synthetase, also causes hypertension [12]. A high-calcium diet attenuated the development of hypertension by antagonising the impairment of endothelium-independent arterial relaxation. This anti-hypertensive effect of Ca is thought to be accomplished by enhanced hyperpolarisation and increased sensitivity to nitric acid in arterial smooth muscle, and decreased vascular production of superoxide and vasoconstrictor prostanoids.

Low-density lipoprotein (LDL) with a high affinity to the vascular wall, representing an important risk factor for atherosclerosis, increases [Ca]i in vascular smooth muscle cells. Oxidised LDL causes a more sustained increase. Lipid oxidation is one of the major pathways of pathological changes of the cardiovascular system associated with ageing, including hypertension and arteriosclerosis. Rise of [Ca]i increase lipid peroxidation, and lipid peroxide injures the plasma membrane and further increases [Ca]i, creating a vicious cycle. Lipid peroxide, free radicals or reactive oxygen species are thus one of the factors causing cytoplasmic Ca overload [13]. The deleterious effects of reactive species consists in producing changes of subcellular organelles and inducing intracellular Ca overload. Antioxidant therapy proved to be effective on hypertension, atherosclerosis and cardiomyopathy, probably through inhibiting the production of lipid peroxide. Vascular calcification occurs like bone calcification under the participation of matrix vesicles, bone morphogenic protein, osteopontin, osteocalcin and type I collagen. Ca deposits, formerly regarded as an end-stage phenomenon in the degenerated atherosclerotic lesion, were shown to occur quite early in the development of atherosclerosis under a close association with lipid [14]. Minimally modified low-density lipoprotein (MM-LDL), an LDL oxidation product responsible for vascular injury and advance of atherosclerosis, inhibited differentiation and mineralisation of osteoblastic cell line MC3T3-E1 [15]. However, the effects on osteoblastic cell culture cannot explain the whole complex process of the development of arteriosclerosis. Cellular ionic components change with age. [Ca]i, increasing with age, is already high in young hypertensives and diabetics, without a further rise with age [16].

Ca binds lipids in the intestinal lumen, forming hardly absorbable soap and inhibiting their absorption. High Ca intake thus lowers plasma cholesterol and triglyceride level [17–19]. Since excess lipid intake and deposition is the prerequisite for lipid peroxidation, sufficient Ca intake should be useful in inhibiting lipid peroxidation and all its consequences.

Diabetes Mellitus

Diabetes mellitus, a manifestation of absolute or relative insulin secretion, is fundamentally a problem of intracellular Ca overload of insulin-secreting pancreatic β-cells and all the target cells for insulin. Ca deficiency leading to secondary hyperparathyroidism facilitates Ca entrance into these cells with subsequent functional derangement. Fasting [Ca]i in erythrocytes was higher in older than in younger subjects, with a highly significant correlation between age and [Ca]i. No significant difference was found between younger subjects with diabetes and older normal subjects, suggesting a premature occurrence of the age-bound rise of [Ca]i in young diabetic patients. In type II diabetes mellitus, regular oscillation of insulin secretion is lost even before the onset of hyperglycaemia, suggesting the absence of the basic control mechanism underlying insulin secretion in diabetes mellitus. In experimental type II diabetes mellitus, a persistent elevation of [Ca]i in the insulin-secreting β-cells was noted. On stimulation with glusose, the rise of [Ca]i seen in normal islets was absent or reduced. Persistently higher [Ca]i may have contributed to the restriction of a further rise when needed, compromising the insulin response depending on it [20, 21]. Increased [Ca]i in β-cells may also cause apoptosis, leading to a decrease in the number of β-cells [22].

Increased [Ca]i in target cells such as vascular and nerve tissue may explain the phenomenon of insulin resistance and complications such as angiopathy and neuropathy [23]. High [Ca]i in the endothelial cells may contribute to thickening of basement membranes as seen in diabetic microangiopathy. In addition to Ca deficiency and secondary hyperparathyroidism, hyperglycaemia causing non-enzymatic glycosulation of Ca^{2+} ATP-ase and increase of protein kinase C activity may also contribute to elevation of [Ca]i [24]. PTH excess in vivo decreases the insulin effect on glucose utilisation, probably through the increase of [Ca]i [25]. Diabetes mellitus is generally associated with an increase of [Ca]i and hydrogen ions [26].

Alzheimer's Disease and Other Neurodegenerative Diseases

One of the major causes of brain cell death on exposure to hypoxia is an abnormal rise of [Ca]i. Cerebral vascular disease causes brain cell death and functional defects through this mechanism [27]. Reducing Ca overload may alleviate ischaemic brain changes [28]. Alzheimer's disease, a selective degenerative disease of unknown cause, mainly affecting the hippocampal formation indispensable for memory, is also associated with neuronal cell death with the characteristic neuropathological picture such as senile plaque and neurofibrillary changes and biochemical changes of amyloid β-protein deposition [29]. Although the neuropathological similarities between hereditary Alzheimer's disease of early onset and the much more widespread syndrome of memory disturbance and progressive deterioration of intelligence and personality seen in more advanced age with features common to the ageing phenomena are quite impressive, multiple environmental and lifestyle factors, including nutritional intake, may have modified the clinical picture. Serum Ca was reported to be lower in elderly subjects and in patients with Alzheimer's disease, associated with a rise of serum PTH [30]. Ca uptake by brain cells and fibroblasts was reported to decrease with age, along with a fall of buffering capacity and rise of [Ca]i. Age-dependent changes were also seen in the density of voltage-operated Ca channels, especially in the hippocampal cells. Secreted forms of beta APP, APPss, are released in response to electrical activity and can modulate neuronal response to glutamate. Neural injury increases beta APP, and APPss can protect neurons against injury by stabilising the intracellular Ca^{2+} concentration. An alternate beta APP processing pathway liberates intact beta amyloid, which forms aggregates that disrupt Ca^{2+} homeostasis and render neurons vulnerable to metabolic and toxic injury. Calcium deficiency and secondary hyperparathyroidism reduce energy availability due to hypoxia and increase oxidative stress: this may favour intracellular accumulation of Ca, which destabilises beta-amyloid. Although this amyloid cascade theory assuming the main role of amyloid protein offers a predominant explanation of the pathogenesis of Alzheimer's disease, amyloid plaque may only represent the end results or by-products of the disease process. In any case, amyloid β increases [Ca]i, creating a vicious cycle.

PTH causes a gradual increase of [Cai]i by activation of dihydropyridine-sensitive Ca channels via specific PTH receptors in the brain [31]. Primary cultures of hippocampal cells from 18-day foetal rat [Ca]i were measured by fura-2 fluorometry. PTH 1-84 and 1-34 raised [Ca]i but PTHrP 1-34 did not. Rise of [Ca]i in brain cells occurs in brain trauma, ischaemia, viral infection and ageing. Glutamate, an excitatory amino acid essential for physiological brain function in the introduction of Ca, may be toxic on excessive and long-term exposure in raising [Ca]i permanently. The PTH effect was inhibited by nifedipine, an L-type Ca channel antagonist, and augmented by BAY K 8644, an agonist. Exposure of cultured hippocampal slices to PTH resulted in a gradual increase of [Ca]i, especially in the apical CA1 subfield. On addition of 1 nM PTH 1-34 to the culture medium, significant toxicity was noted after three days, but this was inhibited by an L-type Ca channel antagonist [32]. Heated algal ingredient (HAI) from Cystophyllum fusiforme, an active component of the Active Absorbable Algal Calcium (AAACa), showed a potent inhibitory effect on Ca entry into hippocampal slices, suggesting its usefulness as a therapeutic agent for Alzheimer's disease [33]. As in ageing, Alzheimer's disease is associated with a disturbed Ca homeostasis of the neuronal cells [34]. Increased apoptosis of nerve cells may be involved in Alzheimer's disease. Presenilin-1 and -2, with certain homology to a Ca channel protein abundantly found in neuronal cells (especially in endoplasmic reticulum), may be related to apoptosis possibly through an increase of [Ca]i. Œstrogen, which inhibits bone resorption and Ca release from bone directly or through an increase of intestinal Ca absorption and inhibition of PTH secretion, was reported to be protective in the development of Alzheimer's disease.

In Alzheimer's disease, the evening fall of serum Ca is more pronounced than in normal subjects, coinciding with the "sundowning" syndrome, evening restlessness and abnormal behaviours with the rise of serum PTH. Administration of AAACa, efficiently absorbed from the intestine in the morning, effectively prevented the evening fall of serum Ca and some of the abnormal behaviours [35]. Sufficient Ca supplementation may prevent calcium deficiency, secondary hyperparathyroidism and inhibit the advance of Alzheimer's disease, in addition to alleviating the sundowning syndrome.

Other degenerative diseases of the nervous system may also be related to Ca deficiency and secondary hyperparathyroidism. In multiple sclerosis, with relapses and recurrences in its long course of progressive deterioration, the frequency of recurrences was reduced by increased Ca intake. Dementia – Parkinsonian complex was highly prevalent among the inhabitants of the southern part of Guam with its volcanic soil and extremely low concentration of Ca in drinking water, but was practically non-existent in the northern part, where there was adequate Ca content in the water. Ca deficiency was thought to be responsible for this disease, which is not prevalent any more. The Kii Peninsula in the southern part of central Japan was also associated with another degenerative disease of the central nervous system, amyotropic lateral sclerosis. Ca deficiency with secondary hyperparathyroidism, low bone mineral density and aortic calcification was also prevalent in this mountainous part of the Kii Peninsula compared to the nearby coastal area where Ca intake was adequate and there was no unusually high prevalence of amyotropic lateral sclerosis [36, 37]. Along with an improvement of the economic situation and increase of Ca intake to an adequate level, amyotrophic lateral sclerosis has become as rare a disease as in the rest of Japan. Simultaneous occurrence and disappearance of a high prevalence of osteoporosis, aortic calcification and amyotrophic lateral sclerosis provides direct support for the theory of the Ca paradox, Ca deficiency in the bone and Ca excess in the blood vessels and brain occurring in this setup. Suicide occurred much more frequently in the mountain area with low Ca intake than in the coastal area with adequate Ca intake. This may be related to panic disorder caused by Ca deficiency and frequent fall of serum Ca and rise of PTH secretion. Many children watching a provocative TV program developed convulsive seizures in 1998. The fall of blood ionised Ca was reproduced in young student volunteers watching the same program and this was successfully prevented by taking 600 mg AAACa three hours before watching TV [38]. The Ca paradox may be extended over functional disorders in addition to degenerative diseases of the nervous system.

Malignancy

Injection of Ca in ova facilitates cell division and proliferation. Increase of [Ca]i is one of the prerequisites for the action of protein kinase C (C stands for calcium), which is a universal regulator of the signal transduction of the cell, especially proliferation and malignant transformation, mediated by a series of oncogenes, cytokines and growth factors. Ca transients and [Ca]i oscillations are involved in driving cell cycles. Ca regulates cross-linking of membrane immunoglobulin of lymphocytes and driving them from G_0 to G_1, with an apparent dissociation between [Ca]i and cell proliferation [39]. Growth factors such as epidermal growth factor and platelet-derived growth factor raised [Ca]i. Protein kinase C activation and Ca mobilisation are essential for cell proliferation and mitogenic response of macrophage-depleted lymphocytes [40]. Normal keratinocytes exposed to increased [Ca]i exhibited terminal differentiation characteristics. Late passage HPV-18 immobilised keratinocytes resistant to the terminal differentiation signal showed progressively higher levels of [Ca]i in the immortalised cells with passage in cultures when compared to primary keratinocytes. Proliferating tumour cells transplanted and growing in nude mice showed high [Ca]i.

Low Ca and vitamin D intake were reported to be a risk factor for carcinoma of the colon and rectum, and Ca supplementation alleviated the risk [41, 42]. The cumulative survival rate of patients after surgery due to colorectal carcinoma is significantly higher in a Ca chemoprevention group supplemented with Ca. Adenomatous polyp recurrences after polypectomy were much lower (12.9%) in the chemoprevention group than in the controls (55%). Calcium may combine with bile acids and other lipids with potential toxicity and tumour-inducing effects on the intestinal mucosa to prevent the occurrence of cancer. The Ca paradox, cytoplasmic Ca overload secondary to Ca and vitamin D deficiency, may facilitate cell proliferation and possibly cancerogenesis. High Ca intake also inhibited the development of gastric cancer. A Ca-deficient diet enhanced gastric carcinogenesis by N-methyl-N'nitro-N-nitrosoguanidine in Wistar rats [43]. A Western-style diet containing high fat and phosphate and low Ca and vitamin D caused colonic epithelial cell hyperproliferation and hyperplasia leading to neoplasia. Ca in the diet inhibited this process. Structural and functional integrity of plasma membranes is necessary to avoid neoplastic development. Plasma membrane injury by anoxia, lipid peroxidation and other causes may open the way to Ca influx by cytolol. This is the most important step in cell proliferation and the development of cancer. Calcium-dependent endonuclease may then cause DNA damage followed by repair and neoplastic proliferation.

Nephrolithiasis

Apart from urate or cysteine stones which usually do not contain Ca, most of the kidney stones do contain Ca, and high Ca intake was thought to facilitate kidney stone formation. Idiopathic hypercalcaemia with increased Ca excretion in urine was frequently associated with kidney stones. Kidney stones and nephrolithiasis are a major manifestation of primary hyperparathyroidism with increased bone resorption and high urinary excretion of Ca. PTH augments bone resorption and releases a large amount of Ca into the bloodstream, causing hypercalcaemia and hypercalciuria, despite the effect of PTH itself diminishing urinary Ca excretion. Most of the Ca utilised for kidney stone formation probably comes from the bone, and not from the food taken. Deficient dietary Ca intake causes PTH hypersecretion leading to secondary hyperparathyroidism, resulting in a release of Ca from bone and increased Ca originating from bone into urine. Sufficient Ca intake would suppress PTH secretion, decrease bone resorption, decrease urinary Ca excretion and prevent nephrolithiasis.

Curhan et al. [44, 45] demonstrated a lower incidence of kidney stones in subjects with high Ca intake in a prospective study on 45,619 subjects between 40 and 75 years of age without previous history of kidney stones. The group in the highest quartile of Ca intake showed an incidence of kidney stone formation of 56% of that in the lowest quartile of Ca intake. An effect of increased dietary Ca intake to decrease kidney stone formation was suggested, but no effect of Ca supplementation in the form of calcium carbonate was established. As another explanation for the preventive effect of high Ca intake on kidney stone formation, binding with oxalate, a frequent constituent of Ca-containing kidney stones, in the intestinal lumen to prevent its absorption and excretion into urine has been suggested. Ohgitani and Fujita [46] reported a decrease in urinary excretion of oxalate, Ca × oxalate product and Ca oxalate crystals on administration of AAACa, but not calcium carbonate. In addition to being more absorbable, AAACa is probably more reactive, readily binding with oxalate and possibly with fat or other undesirable anions in the food to exert a more efficient detoxifying action. The increase of kidney stone formation with low dietary Ca intake and its prevention by increasing food Ca intake may also be called an example of the Ca paradox, though it is still misunderstood by many.

Degenerative Joint Disease

Both osteoporosis and degenerative joint diseases, such as spondylosis deformans and osteoarthritis of various joints, increase with age, representing major causes of musculoskeletal pain in advanced age. In contrast to osteoporosis, characterised by a decrease in bone mineral density, degenerative joint diseases are usually associated with an increase in bone mineral density. Such diverse findings have led some investigators to believe that osteoporosis and degenerative joint diseases, especially spondylosis deformans, represent changes in two opposite directions, contrasting, incompatible and mutually exclusive [47]. Spondylosis deformans mainly occurs in males, especially in muscular, heavy-set ones, who are used to bearing weight or labouring, whereas osteoporosis occurs in females, especiially slender and underweight "petite" ones.

Questions, however, have been raised against this simple traditional concept. Osteoarthritis of the knees is quite common in postmenopausal women, even those with osteoporosis. Women accustomed to carrying heavy loads and to manual labour are especially prone to spondylosis deformans along with osteoporosis. Compression of the vertebral body with deformity may also prompt the occurrence of secondary spondylotic changes. Oestrogen is effective in the prevention and treatment for both, suggesting a common aetiology. Women who were currently using oral oestrogen had a significantly reduced risk of osteoarthritis of the knee with an OR of 0.62 (0.49–0.86) against those who were not using oestrogen. The longer the use of oestrogen, the more pronounced was the reduction of the risk [48]. Ipriflavone, one of the synthetic phyto-oestrogens used for the treatment of osteoporosis because of its action inhibiting bone resorption, was also reported to be effective in preventing osteoarthritis. Upper extremity bone mass was lower in subjects with osteoarthritis of the knee, suggesting a parallel relationship between osteoarthritis and osteoporosis instead of an inverse relationship [49]. Once osteoarthritis or spondylosis develops, osteophyte and localised sclerosis may increase the apparent bone mineral density. In portions other than those involved in ectopic calcifications, bone mineral density was reported to be lower and fracture risk higher in patients with degenerative joint disease. Direct measurement of the bone mineral density of the 4th lumbar vertebra during postmortem examination revealed lower density in those with osteoarthritis. The question as to the parallel or inverse relationship between osteoporosis and degenerative joint disease has not yet been settled and the opinions of experts are still divided.

Interleukin 1 (IL-1), one of the cytokines involved in osteoarthritis with an increase in the synovial fluid and probably responsible for the inflammation, increased bone resorption, collagen degradation and pain, is released in response to oestrogen deficiency, calcium deficiency and secondary hyperparathyroidism. Neovascularisation of the cartilage common in osteoarthritis may also be due to the action of IL-1, mediated by NO synthesised in response to mechanical stimuli and a rise of [Ca]i. IL-6, another bone-resorbing cytokine released in osteoporosis, and oestrogen and Ca deficiency, was also increased in the synovial fluid in osteoarthritis. Pseudogout or chondrocalcification causing gout-like joint pain occurs in association with PTH excess, in primary and secondary hyperparathyroidism. In the cartilage of patients with osteoarthritis, PTHrP, a foetal form of PTH playing an important role in the development of cartilage during foetal development, is expressed in increased quantities, possibly in an attempt to protect the cartilage from calcification and degeneration.

Fujita suggested a calcium paradox mechanism for degenerative joint disease [50]. Calcium deficiency stimulates secretion of PTH, which augments bone resorption and release of Ca from bone and its deposition into cartilage. Ca-laden cartilage loses its elasticity, hardens and wears out during constant and unavoidable physical stimuli on the joint. Physical strain causes a rise of [Ca]i in chondrocytes and calcium deficiency and secondary hyperparathyroidism aggravates it. Excessive physical strain alone may cause chondrocyte degeneration and calcium deposition, leading to similar loss of cartilage. Direct contact between bones would ensue in consequence. Unlike non-viable structures, bones do not simply wear out on repeated physical strain, but react to become hyperplastic and more mineralised by producing osteophytes and hyperostosis (Figure 10.5).

Summary

Ageing, a universal, unavoidable, progressive and deleterious phenomenon, is characterised by a gradual decompartmentalisation of Ca, and loss of the distinct and pronounced concentration gradient between bone, blood and cytoplasm. Blood Ca, by far the most strictly controlled biological constant, must be maintained even at the cost of bone Ca, to keep the heart and brain running. Ca deficiency causes a slight temporary decrease of serum Ca, which prompts an increase of PTH secretion and Ca loss from bone by means of PTH, which increases bone resorption, leading to osteoporosis. PTH at the same time increases interstitial tissue Ca in otherwise Ca-free tissue such as blood vessels and brain, and [Ca]i causing a decrease in the concentration difference between the three Ca compartments. The rise of [Ca]i and consequent loss of the extra- and intracellular Ca concentration gradient leads to an impairment of cell function, apoptosis, necrosis and malignant transformation.

Ca excess in cytoplasm and interstitial tissue otherwise free of Ca prompted by Ca deficiency in the whole body and bone may appear paradoxic. This "Ca Paradox" characterises the ageing process, which is a gradual change from distinction to ambiguity, from order to chaos and from a highly developed, low entropy state to a primitive, high entropy state. There is also decompartmentalisation by the attenuation

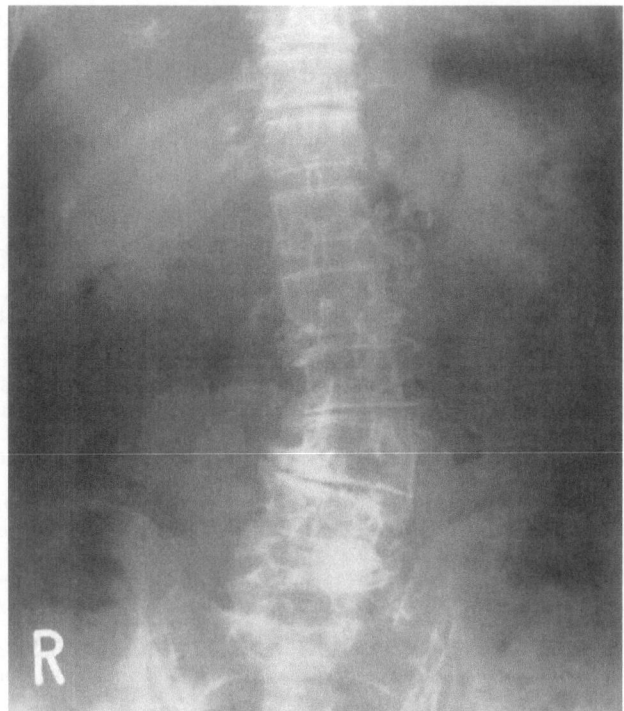

Figure 10.5. X-ray picture of an individual with typical calcium paradox. Loss of calcium from bone in osteoporosis and calcium deposition into blood vessel and cartilage, in atherosclerotic vascular calcification and spondylosis deformans.

of the concentration gradient between the three compartments; bone, blood and cytoplasm.

Prevention and delay of the Ca paradox, accomplished by supplying enough Ca and promoting its absorption to eliminate Ca deficiency, would eventually delay and antagonise the ageing phenomenon. This would improve the quality of life and provide people with a new and fruitful meaning. Sufficient supply of Ca with good absorbability is the most effective means to inhibit the development and progress of

Table 10.4.   Calcium Deficiency-Associated Diseases

I.	Skeletal Calcium Deficiency – Osteoporosis
II.	Calcium Paradox Diseases – Calcium Excess in Soft Tissue and Cytosol in the Presence of Calcium Deficiency
	1. Hypertension
	2. Arteriosclerosis
	3. Diabetes Mellitus
	4. Neurodegenerative Diseases Especially Alzheimer's Disease
	5. Malignancy
	6. Nephrolithiasis
	7. Degenerative Joint Diseases

age-bound or lifestyle-dependent diseases such as hypertension, arteriosclerosis, diabetes mellitus, Alzheimer's disease, malignancy and degenerative joint diseases, thereby providing a more healthy and fruitful life (Table 10.4).

References

1. Fujita T, Orimo H, Ohata M, Okano K, Yoshikawa M, Inoue T, Itami Y. Radioimmunoassay of serum parathyroid hormone in postmenopausal osteoporosis. Endocrinol Japon 1972;19:571–7.
2. Fujita T, Palmieri GMA. Calcium paradox disease: Calcium deficiency prompting secondary hyperparathyroidism and cellular calcium overload. J Bone Miner Metab 2000;18:109–25.
3. Nicotera P, Orrenius S. The role of calcium in apoptosis. Cell Calcium 1998;23:173–80.
4. McCarron D, Morris CD, Cole C. Dietary calcium in essential hypertension. Science 1982;217:267–9.
5. Schleiffer R, Pernot F, Berthelot A, Gairard A. Low calcium diet enhances development of hypertension in the spontaneous hypertensive rat. Clin Exper Hypertens Ther Pract 1984;A6:783–93.
6. Wright GL, DeMoss D. Evidence for dramatically increased bone turnover in spontaneously hypertensive rat. Metabol 2000;49:1130–3.
7. Vasdev S, Ford CA, Parai S, Lomgerich L, Gadag V. Dietary alphlipoic acid supplementation lowers blood pressure in spontaneously hypertensive rats. J Hypertens 2000;18:567–73.
8. Hak AE, Pols HA, van Hemert AM, Hofman A, Witteman JC. Progression of aortic calcification is associated with metacarpal bone loss during menopause. A population-based longitudinal study. Atheroscleros Thrombos Vasc Biol 2000;20:1926–31.
9. Tenenbaum A, Fisman EZ, Pines A, Shemesh J, Shapira I, Alder A et al. Gender paradox in cardiac calcium deposits in middle-aged and elderly patients: Mitral annular and coronary artery calcification interrelationship. Maturitus 2000;36:35–42.
10. Tohno Y, Tohno S, Minami T, Moriwake Y, Utsumi M, Nishiwake F et al. A possible balance of calcium accumulation among bone, cartilage, artery and vein in single human individuals. Biol Trace Element Res 1998;63:105–11.
11. Wlodek ME, Westcott KT, Ho PW, Serruto A, Di-Nicolantonio R, Farrugia W et al. Reduced fetal, placental and amniotic fluid PTHrP in the growth-restricted spontaneously hypertensive rat. Am J Physiol 2000;279:1231–8.
12. Jolma P, Kalliovalkama J, Tolvanen JP, Koobi P, Kahonen M, Hutri-Kahonen N et al. High calcium diet enhances vasorelaxation in nitric oxide-deficient hypertension. Am J Physiol 2000;279:H1036–43.
13. Dhalla NS, Temsah RM, Netticadan T. Role of oxidative stress in cardiovascular diseases. J Hypertension 2000;18:655–73.
14. Hirsch D, Azoury R, Sarig S, Kruth HS. Colocalization of cholesterol and hydroxyapatite in human atherosclerotic lesions. Calcif Tissue Int 1993;52:94–8.
15. Parhami F, Morrow AD, Balucan J, Leitinger N, Watson AD, Tintut Y et al. Lipid oxidation products have opposite effects on calcifying vascular cell and bone cell differentiation. A possible explanation for the paradox of arterial calcification in osteoporotic patients. 1997.
16. Barbagallo M, Gupta RK, Dominguez LJ, Resnick LJ. Cellular ionic alterations with age; relation to hypertension and diabetes. J Am Geriat Soc 2000;48:1111–6.
17. Yocowitz H, Fleischman AI, Bierenbaum ML. Effects of oral calcium upon serum lipids in man. Brit Med J 1965;1:1352–4.
18. Nazir DJ, Mishkel MA. The effect of calcium on plasma lipids and bile acid and fecal fat excretion in normolipidemic subjects. Clin Chim Acta 1975;62:117–23.
19. Karanja N, Morris CD, Rufolo P, Snyder G, Illinworth DR, McCarron DA. Impact of increasing calcium in the diet on nutrient consumption, plasma lipids and lipoproteins in humans. Am J Clin Nutr 1994;59:900–7.
20. Wang L, Bhattacharjee A, Fu J, Li M. Abnormally expressed low voltage activated calcium channels in β-cells from NOD mice and a related clonal cell line. Diabetes 1996;45:1678–83.
21. Islam S. Calcium and diabetes. In: Pochet R, editor. Calcium. Dordrecht Kluwer Academic Publishers, 2000;401–3.
22. Efanova IB, Zaitsev SV, Zhivotovsky B, Kohler M, Efendie S, Orrenius S et al. Glucose and tolbutamide induce apoptosis in pancreatic β-cells. J Biol Chem 1998;273:33501–7.
23. Massry SG, Smogorzewski M. Role of elevated cytosolic calcium in the pathogenesis of complications in diabetes mellitus. Miner Electrol Metab 1997;23:253–60.
24. Gonzalez-Flech FL, Castello PR, Gagliardino JJ, Rossi JP. Molecular characterization of the glycated plasma membrane calcium pump. J Membr Biol 1999;171:25–34.

25. Saxe AW, Gibson G, Gingerich RL, Levy J. Parathyroid hormone decreases in vivo insulin effect on glucose utilization. Calcif Tissue Int 1995;57:127–32.
26. Okorodudu AO, Adegboyega PA, Scholz CI. Intracellular calcium and hydrogen ion in diabetes mellitus. Ann Clin Lab Sci 1995;25:394–401.
27. Choi DW. Calcium-mediated neurotoxicity: Relationship to specific channel types and role in ischemic damage. Trends Neurosc 1998;7:369–71.
28. Zipfel GJ, Lee JM, Choi DW. Reducing calcium overload in the ischemic brain. N Engl J Med 1999;341:1543–54.
29. Mattson MP, Barger SW, Cheng B, Lieberberg I, Smith-Swintosky VL, Rydel RE. β-amyloid precursor protein metabolites and loss of neuronal Ca^{2+} homeostasis in Alzheimer's disease. Trends Neurosci 1993;16:409–14.
30. Disterhoft JF, Moyer Jr JR. The calcium rationale in aging and Alzheimer's disease. Evidence from an animal model of normal aging. Ann N Y Acad Sci 1994;744:382–406.
31. Hirasawa T, Nakamura T, Morita M, Ezawa I, Miyakawa H, Kudo Y. Activation of dihydropyridine-sensitive Ca^{2+} channels in rat hippocampal neurons in culture by parathyroid hormone. Neurosci Letters 1998;256:139–42.
32. Hirasawa T, Nakamura T, Mizushima A, Morita M, Ezawa I, Miyakawa H et al. Adverse effect of an active fragment of parathyroid hormone on rat hippocampal organotypic cultures. Bri J Pharmacol 2000;129:21–8.
33. Fujita F, Fujii Y, Goto B, Miyauchi A, Takagi Y, Kobayashi S. Increase of intestinal calcium absorption and bone mineral density by heated algal ingredient (HAI) in rats. J Bone Miner Metab 2000;18:165–9.
34. Kirishuk S, Verkhratsky A. Calcium homeostasis in aged neurons. Life Sci 1996;59:451–9.
35. Fujita T, Fujii Y, Ohgitani S, Iwamoto B, Miki K, Takahashi Y. Exaggerated evening fall of serum calcium as a possible factor in sundowning syndrome in Alzheimer's disease, Fifth International Congress on Alzheimer's Disease and related Disorders. Abstract 681 1996.
36. Fujita T, Okamoto Y, Tomita T, Sakagami Y, Ota K, Ohata M. Calcium metabolism in aging inhabitants of mountain versus seacoast communities in the Kii Peninsula. J Am Geriatr Soc 1997;24:254–8.
37. Fujita T, Okamoto Y, Sakagami Y, Ota K, Ohata M. Bone changes and aortic calcification in aging inhabitants of mountain versus seacoast communities in the Kii Peninsula. J Am Geriatr Soc 1984;32:124–8.
38. Fujita T, Ohgitani S, Nomura M. Fall of blood ionized calcium on watching a provocative TV program and its prevention by active absorbable algal calcium (AAA Ca) J Bone Miner Metab 1999;17:131–6.
39. Livroz C, Grillo-Courbalin C, Labaume S, Miglieriner R, Broust J-C. Cross-linking of membrane IgM on B CLL cells: Dissociation between intracellular free Ca^{2+} and cell proliferation. Europ J Immunol 1988;18:1811–7.
40. Kakibuchi K, Takai Y, Nishizuka Y. Protein kinase C and calcium is associated with progression of HPV-18 immobilized human peripheral lymphocytes. J Biol Chem 1985;260:1366–9.
41. Garrett LR, Coder DM, McDougall JK. Increased intracellular calcium is associated with progression of HPV-18 immobilized human keratinocytes to tumorigenicity. Cell Calcium 1991;12:343–9.
42. Garland C, Shekelle RB, Barrett-Connor E, Criqui MH, Rosoff AH, Paul O. Dietary vitamin D and calcium and risk of colorectal cancer: a 19 year prospective study in men. Lancet 1992;1:307–9.
43. Tatsuta M, Ishii H, Baba M, Uehara H, Nakaizumi A, Taniguchi H. Enhancing effects of calcium-deficient diet on gastric carcinogenesis by N-methyl-N'-nitro-N-nitrosoguanidine in Wistar rats. Japan J Cancer Res 1993;84:945–50.
44. Curhan CC, Willet WC, Rimm EB, Stampfer MJ. A prospective study of dietary calcium and other nutrients and the risk of symptomatic kidney stones. N Engl J Med 1993;329:508–9.
45. Curhan CC, Willet WC, Speizer EF, Spiegelman D, Stampfer MJ. Comparison of dietary calcium with supplemental calcium and other nutrients as factors affecting the risk for kidney stones in women. Ann Intern Med 1997;126:553–5.
46. Ohgitani S, Fujita T. Heated oyster shell with algal ingredient (AAA Ca) decreases urinary oxalate excretion. J Bone Miner Metab 2000;18:283–6.
47. Dequeker J. Inverse relationship between osteoporosis and osteoarthritis. J Rheumatol 1996;35:813–20.
48. Nevitt MC, Cummings SL, Lane NE, Hochberg MC, Scott JC, Pressman AR et al. Association of estrogen replacement therapy with the risk of osteoarthritis of the hip in elderly white women. Arch Intern Med 1996;156:2073–80.
49. Hochberg MC, Lethbridge-Cejku M, Scott Jr WW, Plato CC, Tobin JD. Upper extremity bone mass and osteoarthritis of the knee: data from the Baltimore Longitudinal Study of Aging. J Bone Miner Res 1995;10:432–8.
50. Fujita T. Degenerative joint disease: An example of calcium paradox. J Bone Miner Metab 1998; 16:195–205.

10B

Hypertension

N. Hasebe and K. Kikuchi

Hypertension is the most common disease-specific reason for which Japanese visit a physician. It is currently among the leading causes of morbidity and mortality in the world and is expected to have an even greater impact on the health of the public as more of the world becomes developed [1].

A considerable body of evidence suggests a relationship between abnormalities of calcium metabolism and hypertension [2]. Calcium intake has been linked to blood pressure regulation for many years. This possibility is evident in the many studies that have found reductions of blood pressure with dietary Ca^{2+} supplementation in experimental models of hypertension. The linkage has been made at the cellular level and, more recently, through numerous observational and intervention studies. The majority of the observational studies have found an inverse relationship between blood pressure and calcium or magnesium or both.

Since elevated peripheral resistance is the hallmark of hypertension, it follows that aberrations of Ca^{2+} metabolism at either the cellular or systemic level may be involved in the pathogenesis of hypertension. Investigation of factors regulating cellular Ca^{2+} ($[Ca^{2+}]_i$) function may be of seminal importance in understanding the mechanisms underlying hypertension. Regulation of $[Ca^{2+}]_i$ is especially important in vascular smooth muscle (VSM) tissue, as $[Ca^{2+}]_i$ is essential in the generation and maintenance of active tension. As reviewed in this chapter, there is evidence that $[Ca^{2+}]_i$ is abnormally high in some cell types in human hypertension. Most measurements in people have been made in accessible cells such as platelets, erythrocytes, and lymphocytes, and these data have been extrapolated to what may be occurring in vascular tissue in hypertension. The alterations of Ca^{2+} metabolism found in human hypertension are closely approximated in the model of experimental genetic hypertension. The most commonly studied model of genetic hypertension is the spontaneously hypertensive rat (SHR), of which the normotensive control strain is the Wistar-Kyoto (WKY) rat. Although the SHR is an imperfect model of human essential hypertension, it remains the most important experimental model for research in the mechanism of hypertension including abnormalities in cellular Ca^{2+} metabolism.

Calcium antagonists have become the most popular class of agents used in the treatment of hypertension, particularly in Japan [3]. They are effective in hypertensive patients of all ages and races and with any types of vascular complications.

This chapter summarises the relationship between calcium and hypertension.

Hypertension and Calcium Balance

Associations were reported between decreased hypertension or heart disease and water hardness, a characteristic of water dependent on its calcium and magnesium content. There have been many reports on observational studies of calcium and blood pressure. McCaron and co-workers reported an association between calcium and blood pressure, a lower calcium intake in hypertensives than normotensives. A large number of observation studies have been reported in which dietary calcium intake was related inversely to blood pressure. The studies have generally been designed as different types in different racial or ethnic groups. Thus, direct comparison across studies is difficult. McCarron et al. [4] found higher intakes of calcium were associated with lower systolic blood pressure and lower absolute risk hypertension. Some authors have suggested a non-linear relationship between calcium and blood pressure, which might help to explain the conflicting results seen in the observational studies. Others have suggested possible effect modifiers, such as dietary sodium, potassium, or vitamin D. In the Nurses' Health Study, a prospective study of dietary nutrients and blood pressure, the risk of developing hypertension over a four-year period was investigated [5]. More than 58,000 women completed a food frequency questionnaire and were followed for incidence of self-reported hypertension. During those four years 3,275 incident cases of hypertension were reported. After controlling for age, body mass index, and alcohol consumption, both calcium and magnesium intake were inversely associated with the risk of hypertension. In another prospective study, an inverse relationship was reported between the blood pressure in infants and prenatal maternal calcium intakes [6].

It has been argued that a systemic Ca^{2+} deficiency resulting from biochemical alterations in transmembrane Ca^{2+} transport in multiple systems may play a causal role in hypertension. Although the ultimate lesions responsible for altered Ca^{2+} metabolism in the SHR are not known, it has been proposed that there is a defect in some aspect of cellular Ca^{2+} handling that is common to all cell types including vascular smooth muscle, renal tubular cells, and intestinal epithelial cells. According to this hypothesis, both the relative Ca^{2+}-deficient state and the hypertensive condition would arise from this same pancellular defect in Ca^{2+} handling. The candidates include defective Ca^{2+} binding proteins such as calmodulin, altered Ca^{2+} pump activity, and/or sodium pump activity.

An impact of dietary Ca^{2+} on calmodulin may explain these outcomes. Using duodenal enterocytes, Roullet et al. [7] have observed that calmodulin levels are lower in the SHR than in the WKY, and that increasing dietary Ca^{2+} eliminates that difference, and can actually up-regulate calmodulin levels. Correction of a defect in calmodulin activity by Ca^{2+} could provide a mechanism whereby a multitude of molecular and cellular processes might be modified. Calmodulin plays a pivotal role in intracellular Ca^{2+} regulation and could be responsible for diet-induced variations in Ca^{2+}-ATPase [8], an enzyme of critical importance to extrusion of Ca^{2+} from the cell and therefore of vasorelaxation.

Urinary Ca^{2+} excretion is increased in the mature SHR although there is controversy as to whether the hypercalciuria reflects primary intestinal hyperabsorption or a primary renal leak. In contrast to the evidence, which supports the idea of a primary renal Ca^{2+} leak, Lau et al. [9] propose that the hypercalciuria occurs because of primary intestinal hyperabsorption of Ca^{2+}. The distinction between renal and absorptive hypercalciuria in the mature SHR is important because of what it implies about the overall Ca^{2+} status of the hypertensive animal; i.e., a relative deficit or surfeit of Ca^{2+} compared to the normotensive control animal. Finally, several studies have found evidence for renal hypercalciuria in human essential hypertension, which is independent of urinary sodium excretion or filtered load of Ca^{2+} [10].

Unfortunately, knowledge of Ca^{2+} absorption alone does not reveal the overall Ca^{2+} status of the organism because intestinal hyperabsorption would be predicted for both renal and absorptive calciurias in their pure forms. Assuming that all of these studies purporting to measure whole-animal Ca^{2+} absorption are valid, a very confusing and inconsistent picture results. Given the inherent difficulty in performing accurate balance studies in rats, it is tempting to look at other measures of cumulative Ca^{2+} balance.

In the male SHR, bone Ca^{2+} is normal in early life, but falls below WKY as the animals reach late adolescence and early adulthood. Reduced bone Ca^{2+} and mineralisation in the male SHR suggests reduced Ca^{2+} retention over time mediated by decreased Ca^{2+} absorption or increased excretion. In the male SHR, a small but significant decrease in serum-ionised Ca^{2+} has been reported in animals between 8 and 45 weeks of age by several investigators. Decreased serum-ionised Ca^{2+} has also been found in the uninephrectomised deoxycorticosterone acetate (DOCA)-salt hypertensive rat relative to its sham-operated control. Similarly, low serum-ionised Ca^{2+} has been reported in human essential hypertension and in some patients with primary aldosteronism. Resnick et al. demonstrated that the range of plasma renin activity in essential hypertension shows a positive correlation with the serum-ionised calcium level (Figure 10.6) [11].

Calcium Intervention Studies in Hypertension

The increased ability of vessels to relax following Ca^{2+} supplementation is an important finding. Previous studies of isolated vessels from Ca^{2+}-supplemented animals concentrated on maximal contractility as the measure of vascular function. Either to change or even enhance contraction following Ca^{2+} supplementation was typically found. However, in examining relaxation, an entirely different picture emerges. Here, it can more clearly be seen that Ca^{2+} supplementation has an impact on the vasculature commensurate with an in vivo finding of lowered blood pressure. Experimental support for this hypothesis comes from many studies that have found that oral Ca^{2+} supplementation lowers blood pressure in the SHR and other experimental models of hypertension.

In humans, approximately two-thirds of Ca^{2+} supplementation studies have found a reduction in blood pressure. The net mean changes in systolic and diastolic blood pressure at the end of the study period are shown in Figure 10.7 [12]. Many studies have small samples and report variable responses to calcium supplementation, and the conflicting findings from these reports have been attributed to inadequate sample size to detect modest blood pressure effects. Cutler and Brittain [13] performed a meta-analysis of trials to overcome the problem of numerous underpowered studies. They reported a small decrease in systolic pressure of 1.8 mm Hg when studies were combined. Another characteristic of many smaller studies is the lack of blood pressure decrease in normotensive persons but a decrease in hypertensive persons. When the effect size is small, it is more difficult to find statistically significant effects with lower pressures. The Trials of Hypertension Prevention study (TOHP) measured the efficacy of non-pharmacological intervention for individuals with high–normal dias-

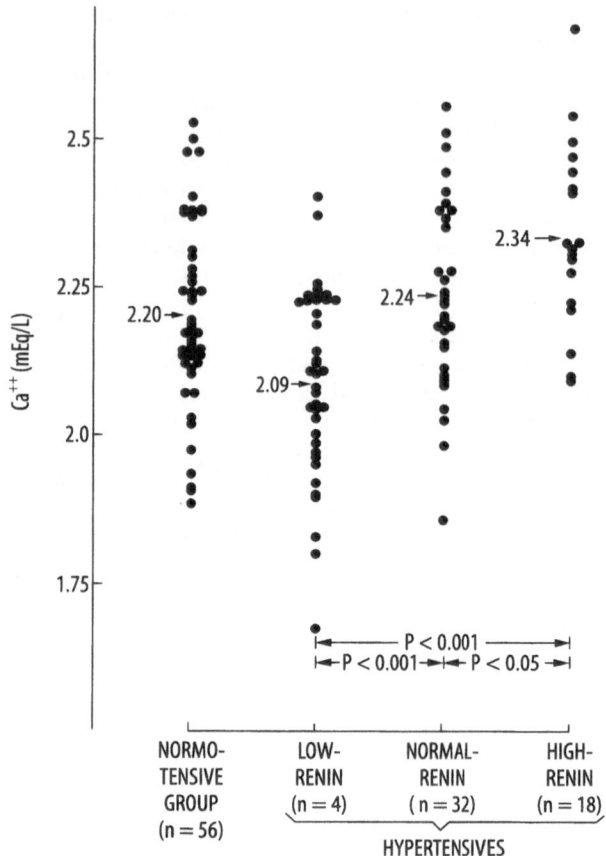

Figure 10.6.   Direct relation of serum levels of ionised calcium to plasma renin activity in patients with essential hypertension. (From Resnick LM et al. Divalent cations in essential hypertension. Relations between serum ionised calcium, magnesium,
and plasma renin activity. N Engl J Med 1983;309:888–91).

tolic blood pressure (80–89 mm Hg) [14]. The study investigated the effect of weight
reduction, sodium reduction, stress management, usual care, calcium supplementation, magnesium supplementation, potassium, fish oil, or placebo on blood pressure.
Some 237 subjects were given 1 g calcium per day. Compared with those on placebo,
the calcium-supplementation group had no significant change in either systolic or
diastolic blood pressure. This carefully performed study measured blood pressure
using blinded observers and followed the participants for six months. The absence
of a significant effect on blood pressure provides evidence that calcium supplementation has little or no efficacy in lowering blood pressure of individuals with high
normal pressures, a group for whom non-pharmacological intervention is desirable.
    Overall, calcium supplementation is not always effective for treatment of hyper-
tension. Only a subgroup of persons given increased calcium responded with a drop
in blood pressure, and these individuals have been classified as calcium-responsive.
There is no means of determining those who are calcium responders except by testing
the blood pressure response to increased intake. Morris and McCarron [15] reported

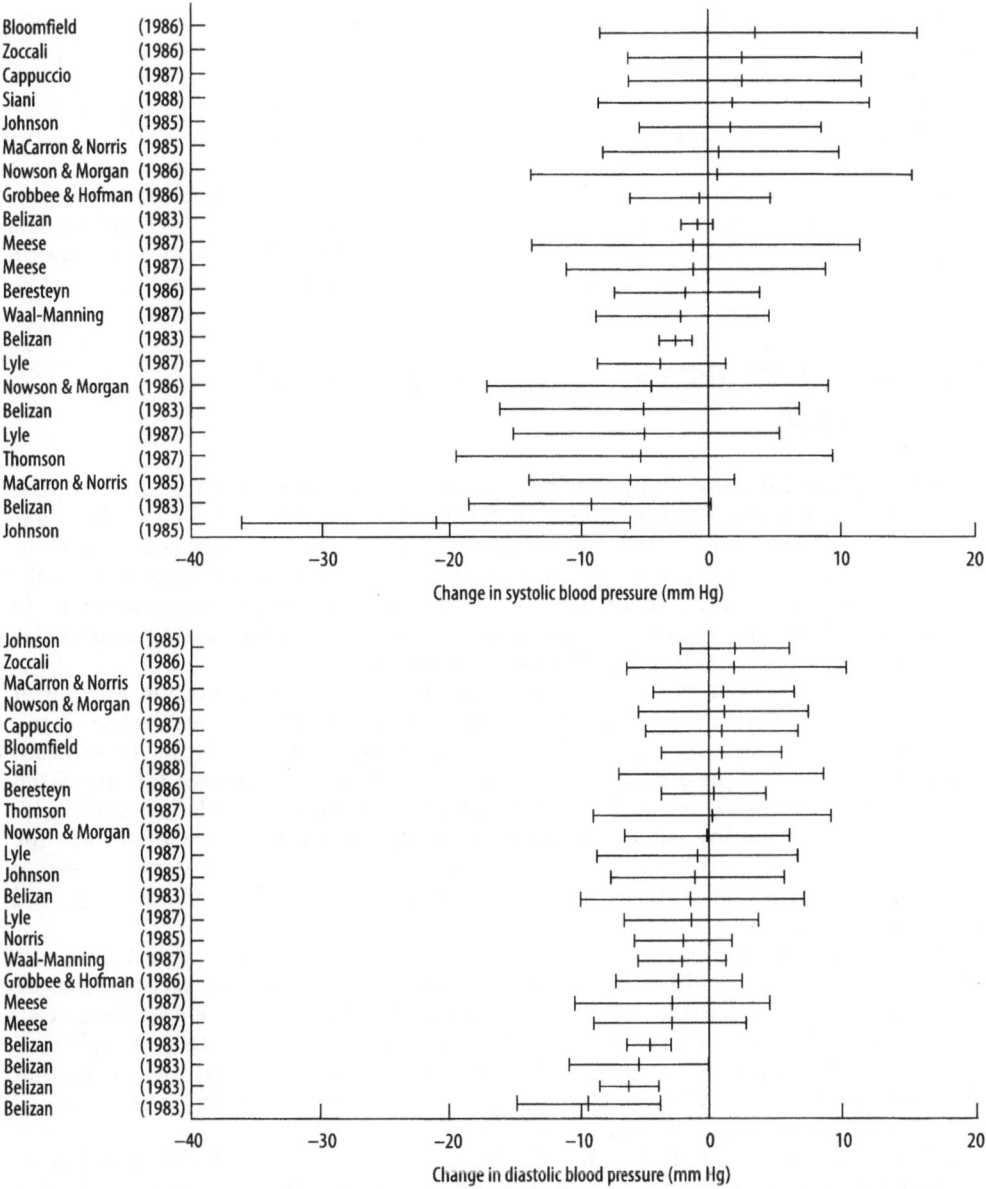

Figure 10.7.   Effect of calcium supplements on systolic (upper panel), and diastolic blood pressure (lower panel) (mean and 90% confidence interval). The figure is, in general, based on the difference in blood pressure change from baseline to the final measurement in the study. (Modified from Grobbee DE, Waal-Manning HJ. The role of calcium supplementation in the treatment of hypertension. Current Evidence. Drugs 1990;39:7−18.)

on a study of calcium supplementation in individuals aged 50 to 80 years with mildly elevated blood pressure. At 12 weeks there was no difference in the mean blood pressures between those on placebo and those on 1 g of calcium carbonate. However, 42 of the 103 study subjects had a decrease in blood pressure of at least 5 mm Hg and were continued on calcium supplementation. Their pressures remained lower through the 48-week study. After completion of 48 weeks of calcium supplementation, 12 of these subjects were given placebos for 12 weeks. Systolic blood pressures increased significantly. The authors suggest that although general use of calcium in older individuals may not be warranted, there are individuals who respond to calcium supplementation and whose lowered blood pressure can be maintained for at least a year.

Effective Mechanism of Calcium Supplementation in Hypertension

Several explanations for responsiveness have been offered. Perhaps the simplest explanation is a direct calcium effect on sodium balance. Increased calcium intake acts to increase sodium excretion in the urine, and thereby decreases pressure. Considerable attention has been directed toward increased calcium or magnesium intake and the resultant changes in hormonal levels of calcium-regulating hormones in rennin-angiotension, and in the interaction between the two hormonal systems; these alterations have been explored by Resnick and co-workers.

To date, other convincing evidence suggests that Ca^{2+} supplementation may lower blood pressure through a modulatory action on sympathetic nervous system activity. There are consistent reports of altered sympathetic nervous system activity associated with variations in dietary Ca^{2+}. Increased dietary Ca^{2+} reverses the increased blood pressure induced by sodium chloride while reducing circulating catecholamines. Lower circulating catecholamines are likely due to increased hypothalamic norepinephrine (NE) levels and a pursuant reduction of sympathetic nervous system outflow. Restricting dietary Ca^{2+} has the opposite effect in both normotensive and hypertensive rats.

In addition to altered sympathetic nervous system outflow, blood pressure responses to NE are modified by dietary Ca^{2+}. Pressor responses are reported to be potentiated by restricted Ca^{2+} diets and dampened by high Ca^{2+} diets. Hatton et al. [16] reported diminished pressor responses to exogenous NE in SHR on high-Ca^{2+} diets. The reduced pressor response did not occur to angiotensin II, suggesting that it was not due to a generalised change in vascular responsiveness. Doris et al., Baksi, and Peuler have reported similar results with regard to NE.

It has been proposed that urinary calcium wasting is an underlying defect in calcium-sensitive hypertension. A normal or increased calcium intake offsets this hypercalciuria and moderates the hormonal imbalance. This mechanism implies that the blood pressure changes result from the hormonal alterations rather than from a direct calcium effect. The hormonal changes could also change blood pressure through an effect on sodium metabolism. Some findings support this mechanism. Persons whose blood pressure responds to increased calcium intake are reported to have higher levels of parathyroid hormone, which may directly or indirectly influence blood pressure.

The cellular effects of calcium on vascular tone have been responsible for much of the interest in calcium intake and blood pressure and seem to provide a physiological explanation and biological plausibility for the epidemiological observations. It is difficult to directly link calcium intake and cellular events except through hormonal changes. This does not imply that cellular calcium flux and levels of cytosolic calcium ion are not important in determining vascular tone or in regulating blood pressure. Rather, considerably more evidence is needed to support a proximate physiological mechanism.

Abnormalities in Cellular Ca^{2+} Metabolism in Hypertension

While a variety of alterations in cellular and systemic Ca^{2+} handling have been described in experimental hypertension, the causal link remains speculative at this time. In essential hypertension there is evidence for an array of alterations of the Ca^{2+} regulatory system, including increased cell $[Ca^{2+}]_i$, increased Ca^{2+} uptake and decreased removal of $[Ca^{2+}]_i$ [17]. Increased $[Ca^{2+}]_i$ is directly related via increases in myosin light chain phosphorylation to an increase in active tension and therefore could directly contribute to the development of hypertension. In this regard a number of laboratories have reported elevated platelet $[Ca^{2+}]_i$ and a positive relationship with elevated blood pressure in hypertensives. Indeed, a positive correlation between blood pressure and platelet $[Ca^{2+}]_i$ has also been reported in normotensive individuals. Also, antihypertensive therapy has been reported to reduce platelet $[Ca^{2+}]_i$. These observations suggest that basal platelet $[Ca^{2+}]_i$ levels are generally higher in hypertensives but may be affected dramatically by antihypertensive therapy.

A number of possibilities exist to explain the higher $[Ca^{2+}]_i$ observed in essential hypertension. For example, individuals with hypertension have been reported to display reduced binding of Ca^{2+} to erythrocyte membranes as compared to normotensives. A reduction in Ca^{2+} binding to inner cell membranes could indirectly cause an increase in $[Ca^{2+}]_i$, as it results in reduced potassium conductance, membrane depolarisation, and activation of voltage-operated Ca^{2+} channels.

There is also evidence that cell membrane Ca^{2+}-ATPase may function in a reduced state in hypertension, either due to an intrinsic defect in the pump itself or inadequate activation. The V_{max} for Ca^{2+}-ATPase in the platelet membranes is lower in essential hypertensives [18]. Since the level and distribution of calmodulin is normal in hypertensive individuals, the observed reduction in the V_{max} may reflect 1) impaired calmodulin binding kinetics, or 2) an intrinsic abnormality in the expression of the pump, leading to reduced calmodulin activation of Ca^{2+}-ATPase.

Theoretically, significant increases in $[Na^+]_i$ would result in inhibition of Na^+/Ca^{2+} exchange, increasing $[Ca^{2+}]_i$. The $[Na^+]_i$, in turn, is predominantly regulated by the activity of plasma membrane Na^+, K^+-ATPase. Endogenous regulators of the Na^+, K^+-ATPase include ouabain and perhaps other circulating factors. In this regard, elevations in endogenous Na^+ K^+-ATPase inhibitors have been reported in essential hypertension, and abnormalities of the Na^+–Ca^{2+} exchanger per se have been inferred in experimental hypertension.

There is evidence of Ca^{2+}-sensitive deficits in each of these cellular components in the SHR. Porsti et al. found that potassium-induced relaxation of mesenteric arterial rings was augmented by Ca^{2+} supplementation in SHR. Since ouabain was able to prevent the relaxation, the difference between diet groups was attributed to increased Na-K-ATPase activity. This outcome is consistent with the notion of a reciprocal relationship between intracellular Ca^{2+} and Na-K-ATPase activity; as intracellular Ca^{2+} declines, Na-K-ATPase activity increases.

Activity of the Na^+–H^+ exchanger has been reported to be elevated in hypertensive states. Increases in activity of the Na^+–H^+ exchanger may be secondary to increases in $[Na^+]_i$ and $[Ca^{2+}]_i$ or may reflect a primary abnormality in membrane transport mechanisms in hypertension. Another disturbance in cellular metabolism that has been observed in hypertension is that of increased platelet cAMP, an abnormality that predisposes to exaggerated agonist-evoked $[Ca^{2+}]_i$ responses. Thus, there are a number of cellular metabolic and transport abnormalities that may characterise the disease known as essential hypertension.

Abnormalities in Ca^{2+} Responsiveness in Hypertension

The vascular smooth muscle cell has available a large transcellular gradient of Ca^{2+} that can be used for translating extracellular signals into functional activity. It is imperative that the smooth muscle cells maintain this gradient for normal function. It is not surprising, therefore, that redundant systems exist for achieving this end; i.e., Ca^{2+} uptake by intracellular organelles, and Ca^{2+} extrusion across the cell membrane via both the Ca^{2+} pump and the Na^+–Ca^{2+} exchange process. When considering a disorder such as hypertension, which is characterised in part by altered vascular reactivity, it is logical to propose that dysfunction of one or some of these Ca^{2+}-regulatory mechanisms may result in altered cell activation. For example, there could be enhanced contractile activity as a result of sustained activation of a Ca^{2+} channel or vascular tone may be sustained because of an impairment in vasorelaxation caused by a depressed rate of Ca^{2+} extrusion. This could occur secondary to defective Ca^{2+}/calmodulin stimulation of Ca^{2+} ATPase, for example. Furthermore, Ca^{2+}-dependent regulation of second messenger generation may be altered such that there could be altered growth patterns in the cell secondary to altered growth factor signal processing. Thus, investigators have sought to determine whether Ca^{2+} metabolism is altered in vascular smooth muscle of hypertensive animals, and whether these alterations are translated into dysfunctional contractile or proliferative responses.

The extracellular Ca^{2+} appears to enter the endothelial cell through a Ca^{2+}-permeable, non-selective cation channel activated by a receptor, a second messenger such as IP_4, or depletion of Ca^{2+} stores in the endoplasmic reticulum. Although there do not appear to be any voltage-dependent Ca^{2+} channels on endothelial cells, Ca^{2+} influx is modulated by membrane potential. Ca^{2+} influx is facilitated by a potassium current that serves to hyperpolarise the cell, thus increasing the electrochemical gradient for Ca^{2+} which ultimately facilitates extracellular Ca^{2+} entry into the cell.

An equally pertinent question is whether the defects that have been described to date, i.e., attenuated ATP-supported ^{45}Ca uptake by subcellular fractions, increase basal and agonist-induced influx, and perhaps modestly elevate levels of free intracellular Ca^{2+}, are manifest in altered contractile function.

It has been consistently observed that active tension generation, force per unit length of vessel, is elevated in resistance arteries of the SHR. However, this difference is not present if active wall stress, tension normalised to cross-sectional area of the vessel wall, is calculated instead. The apparent difference in tension development by the vessel from the hypertensive animal is therefore the result of the presence of structural changes and does not reflect a difference in the intrinsic ability of the muscle cell to contract. This finding alone indicates that the contractile proteins are likely to be intact and that excitation–contraction coupling mechanisms, including changes in free intracellular Ca^{2+}, function normally in the SHR.

In summary, there appears to be only minor differences in the intrinsic contractile properties of resistance arteries of the SHR and WKY. There may be more pronounced differences in vasodilation either as a consequence of endothelial influences or aberrations in Ca^{2+} extrusion, making it more difficult for the vessel to relax. Additional data need to be gathered to assess the extent and nature of the differences in vasorelaxation between experimental models and their controls.

Another possibility to be considered is that whereas only minor differences in intrinsic contractile ability of vessels of the SHR and WKY are detectable in vitro, the vessel may demonstrate altered reactivity in vivo secondary to neural and/or humoral influences. Nerve activity could be catecholaminergic or peptidergic, whereas humoral factors could be either platelet-associated growth factors or hormones responsible for Na^+ or Ca^{2+} metabolism of the whole animal.

Perturbations of extracellular Ca^{2+} have significant ramifications for the synthesis and release of EDRF precisely because extracellular Ca^{2+} levels determine free cytosolic Ca^{2+} levels [19]. Synthesis of EDRF from L-arginine by NO synthase varies with free cytosolic Ca^{2+} levels because it is a Ca^{2+}/calmodulin-dependent process. Likewise, release of EDRF depends upon free cytosolic Ca^{2+} levels. The importance of extracellular Ca^{2+} is particularly apparent in the context of flow-dependent vasodilation where there is continuous modulation of Ca^{2+} influx by mechanical stimulation of the blood vessel.

The stabilising effect of extracellular Ca^{2+} on the vascular smooth muscle cell may be reinforced by release of EDRF. As outlined above, the synthesis and release of EDRF is directly dependent upon extracellular Ca^{2+} levels. Consequently, optimal release of EDRF will depend, to some extent, on maintaining adequate levels of extracellular Ca^{2+}. Assuming that serum-ionised Ca^{2+} levels are depressed in SHR, as would be indicated by elevated PTH levels, it would be predicted that EDRF release would be reduced in SHR.

Calcium-Regulatory Hormones in Hypertension

Hypertension Associated with Hyperparathyroidism

Of the calcitrophic hormones, parathyroid hormone (PTH) and the biologically active vitamin D metabolites appear to be the most important in the regulation of blood pressure. In target tissues, PTH binds to a membrane receptor and exerts many of its metabolic actions by activation of adenylate cyclase, catalysing formation of cAMP from ATP. The cAMP binds to a specific cAMP-binding protein coupled to protein kinase A (PKA).

Hypertension is often present in disorders in which hyperparathyroidism (hPTH) (with or without hypercalcaemia) is present. The occurrence of hypertension in hPTH was approximately twice that expected for sex-, age-, and race-matched populations. The current consensus is that the pathogenesis of hypertension associated with hPTH is multifactored and remains poorly understood. Chronic infusion of PTH in normal individuals resulted in a reversible rise in blood pressure to hypertensive levels. In contrast, acute infusion of PTH, or its related analogue parathyroid hormone-related protein PTHrp, consistently resulted in vasodilation and a reduction in blood pressure. PTH and PTHrp appear to share the same membrane receptors on VSMC. PTHrp is also present in significant amounts in VSMC, and vasoactive peptides such as angiotensin II, endothelin, and norepinephrine can modulate its synthesis. These results suggest that PTHrp may act in a pericranial fashion to mediate vasoconstrictor activity. Conversely, PTH can decrease angiotensin II and vasopressin-induced intracellular Ca^{2+} transients. It has been suggested that conditions in which there is a chronic elevation of PTH and PTHrp may be associated with desensitisation to their vasodilatory responses and relatively unopposed vasoconstriction responses [20].

A close relationship between serum Ca^{2+} concentration and blood pressure has been observed in both individual patients as well as groups of individuals with end-stage renal disease treated with haemodialysis. In this regard, dialysis with a high Ca^{2+} concentration in the dialysate can reduce or block the occurrence or severity of postdialysis hypotension. Furthermore, patients with end-stage renal disease have an augmented pressor response to acute Ca^{2+} infusion. A linear relationship has been observed between levels of intact PTH and mean arterial pressure. PTH levels were also highly correlated with levels of platelet $[Ca^{2+}]_i$. Treatment with alphacalcidol lowered plasma PTH, platelet $[Ca^{2+}]_i$ and blood pressure. Calcitriol or alphacalcidol had previously been shown to decrease PTH secretion and blood pressure in patients with chronic renal disease.

Individuals with various PTH-resistance syndromes have an increased prevalence of hypertension. Patients with these disorders develop secondary hyperparathyroidism usually in the presence of hypocalcaemia. The mechanism of the hypertension is poorly defined although studies do suggest altered sympathetic nervous system activity, especially abnormal dopaminergic regulatory mechanisms, and altered circadian rhythms of blood pressure.

Elevated serum PTH levels have been reported in humans with essential hypertension, or gestational hypertension in those at genetic risk for hypertension, and individuals with primary aldosteronism. Furthermore, PTH has been found to be significantly correlated with both blood pressure in hypertensive individuals and renal failure patients and with left ventricular weight in untreated hypertensive subjects. Serum PTH has been found either to be elevated in the SHR or not different from the WKY, apparently as a function of the specific assay used. When elevated, it has been proposed to be a secondary response to the low serum-ionised Ca^{2+}. Although serum PTH may be elevated in the SHR in response to a decreased serum-ionised Ca^{2+}, end-organ responsiveness to the hormone may be abnormal. Reduced renal responsiveness to PTH is suggested by findings of hypercalciuria, decreased basal and stimulated $1,25(OH)_2$ vitamin D_3 production, and low-to-normal urinary and nephrogenous cyclic AMP, all in the face of apparently elevated immunoreactive PTH. Moreover, infusion of PTH caused a smaller increment in serum-ionised Ca^{2+} in 13-week-old male SHR than in WKY. The diminished response in the SHR may be due to down-regulation of renal PTH receptors as a consequence of chronically elevated circulating PTH. DiPette et al. found that PTH receptor levels in the SHR were only 60% of that observed in WKY tissue. Thus, it would appear that failure of the SHR to increase serum-ionised Ca^{2+} adequately results in chronic elevation of circulating PTH and a subsequent down-regulation of renal PTH receptors.

Vitamin D Metabolism and Hypertension

Vitamin D and its metabolites are steroid hormones that are biochemically related to cholesterol. Human vitamin D is derived from two sources: 1) photoconversion of 7-dehydrocholesterol to vitamin D_3 (cholecalciferol) in the epidermis under the influence of ultraviolet radiation, and 2) dietary sources containing vitamin D_2 (ergocalciferol). The active D metabolite, $1,25(OH)_2D_3$, binds to a receptor in the target cells and causes transcription of the specific genes that code for proteins including a Ca^{2+}-binding protein called calbindin-D, involved in the intracellular transport of Ca^{2+}.

$1,25(OH)_2D_3$ has been suggested to have a hypertensive effect. In this regard, DNA-binding receptors for D_3 have recently been demonstrated in VSMC. Walters et al. have observed that nanomolar quantities of D_3 cause a threefold increase in cardiomyocyte Ca^{2+} uptake. Indeed, this effect on $[Ca^{2+}]_i$ may account for the effect of D_3 to increase vascular tone.

Vitamin D regulation appears to be abnormal in the SHR and may partially explain disordered Ca^{2+} homeostasis in the strain. The decrease serum concentration of $1,25(OH)_2$ vitamin D_3 in the SHR may be explained entirely by decreased production because metabolic clearance of the hormone is the same in the SHR and WKY. PTH may exert an effect on blood pressure by stimulating the synthesis and release of $1,25(OH)_2$ vitamin D_3. Despite evidence of direct effects on VSM, observations of elevated blood pressure as a consequence of administration of $1,25(OH)_2$ vitamin D3 have been inconsistent.

Parathyroid Hypertensive Factor (PHF) and Calcitonin Gene-Related Peptide (CGRP)

This putative hypertensive factor, produced by the parathyroid glands but not related structurally or biochemically to PTH, may contribute to the development of high blood pressure in some segments of the hypertensive population. PHF has been reported to be increased in low-renin essential hypertension with salt sensitivity. This characteristic presumably is related to the fact that PHF can decrease urinary sodium excretion. Furthermore, PHF has been reported to increase cell $[Ca^{2+}]_i$ and is correlated in experimental deoxycorticosterone acetate (DOCA)-salt hypertension and in essential hypertension with increased cell $[Ca^{2+}]_i$.

The importance of CGRP actions in regulating the cardiovascular system is indicated by the fact that peripheral administration of CGRP causes vasodilation and positive chronotropic and inotropic effects. The release of CGRP from vascular and cardiac nerve fibres may play a functional role in the neurogenic regulation of VSM tone and cardiac contractility.

Calcium and Insulin

The recently recognised relationship between insulin resistance, hyperinsulinaemia, and high blood pressure may be mediated through alterations in calcium metabolism. Many conditions such as obesity, diabetes mellitus, and essential hypertension have common abnormalities in insulin function and calcium metabolism. Insulin is now recognised as a risk factor for hypertension, glucose intolerance, dyslipidaemia, and atherosclerosis. The contrasting effects of insulin to raise sympathetic activity yet reduce vascular resistance suggest that insulin can have properties of both a vasoconstrictor and a vasorelaxant.

Recent research has linked both hypertension and diabetes to common defects in calcium metabolism. In humans, induced hyperinsulinaemia leads to exaggerated urinary calcium excretion and diabetic subjects have hypercalcuria, decreased duodenal calcium transport, and low serum ionised calcium. Studies have found elevated cytosolic calcium in platelets from subjects with Type 2 diabetes mellitus. Interestingly, several of the defects in renal and gastrointestinal calcium metabolism and cytosolic calcium reported in diabetics are also found in patients with essential hypertension. Several mechanisms may explain the link between abnormal calcium homeostasis and hypertension in diabetes. Higher levels of free 1,25-dihydroxyvitamin D reported in diabetics could increase calcium uptake in both cardiac tissue and vascular smooth muscle. Insulin may also directly alter the vascular smooth muscle calcium signal leading to change in vascular tone. Although the role of insulin as a regulator of calcium metabolism remains controversial, several effects of insulin have been reported on calcium mobilisation and transport. Insulin has direct effects on calcium metabolism in skeletal tissue, cardiac muscle cells, adipocytes and kidney tissue. These studies suggest that insulin alters calcium transport in several tissues including vascular smooth muscle. Whether insulin alone modulates levels of cytosolic calcium is less certain.

When insulin is administered in vivo or added to an in vitro preparation, most studies find acute vasodilation. Insulin added to vascular preparations attenuates vasoconstriction induced by norepinephrine in resistance beds such as the isolated perfused rat tail. Yagi et al. performed extensive studies with rabbit femoral arteries and veins showing that 30-minute incubations with insulin dose-dependently inhibited norepinephrine- and angiotensin II-induced contractions.

Other reports have described potential hypertensive effects of insulin. Yanagisawa-Miwa et al. observed an enhancing effect of insulin on contraction to a thromboxane A_2 analogue in porcine coronary artery. In alloxan- and streptozotocin-induced diabetic animals, insulin reverses diabetes-induced vascular contractility and long-term insulin therapy (8–12 weeks), and has been shown to increase vascular contractility to high potassium. Numerous studies have shown that both Type 1 and Type 2 diabetic patients, whether normotensive or hypertensive, with or without complications, have enhanced blood pressure responses to administration of angiotensin II and norepinephrine. The cause of this enhanced vascular reactivity in diabetes mellitus is unknown, but could relate to abnormal vascular calcium metabolism.

Kahn et al. [21] examined the effect of insulin in individual vascular smooth muscle cells using photomicroscopy to measure contraction. Preincubation with insulin attenuated angiotensin II- and serotonin-induced contractions in a dose-dependent manner. Serotonin-induced intracellular calcium transients were inhibited by insulin and this effect was blocked by verapamil and ouabain, indicating an effect of insulin on calcium channels and the Na–K pump. As insulin has been shown to increase the activity of the Na^+–K^+ pump, it was proposed that insulin stimulation of the Na^+–K^+ pump leads to hyperpolarisation of the cell and reduced calcium influx through the voltage-operated calcium channels.

Insulin also regulates cell calcium levels by altering the activity of the Ca-ATPase membrane pump. Levy et al. found that insulin increased Ca-ATPase activity in a variety of tissues. However, in both human and experimental diabetes, there is a generalised reduction in Ca-ATPase activity that might indicate a resistance to insulin's normal effect on this pump. Reduced calcium efflux through this pathway would lead to calcium accumulation in vascular smooth muscle cells and provide an explanation for the high tissue levels of calcium, increased vascular reactivity and high incidence of hypertension in diabetes mellitus. Insulin also stimulates the Na^+/H^+ antiporter transport system and this transport pathway has been shown to be elevated in diabetes and hypertension and linked to abnormal calcium metabolism.

Calcium Channel Blocker in Hypertension

In the last two decades, calcium channel blockers (CCB) have emerged as popular compounds for controlling hypertension. Although the three major types of CCB, verapamil, diltiazem and dihydropiridines (DHP) act on different sites of the voltage-operated calcium channels, they all block the mediator function of Ca^{2+} in excitation–contraction coupling of heart and vascular smooth muscle, lowering Ca^{2+}-dependent vascular tone and spasm. They differ in both their site and modes of action (Table 10.5). The short-acting formulations of diltiazem and nifedipine were approved for both chronic stable and vasospastic angina in the US but not for hypertension; subsequently, the long-acting formulations of all three and the short-acting verapamil were approved for both angina and hypertension. In contrast, verapamil is not approved for hypertension in Japan. Nevertheless, CCB is an antihypertensive and anti-anginal drug that has been most frequently used in Japan [3], and has become the most popular choice of antihypertensive therapy in the US and most of the world.

In the randomised, controlled trials of monotherapy with representatives of all five major classes of antihypertensive drugs, the CCBs were the most effective and best tolerated. A CCB but not an alpha beta-blocker nor an ACE inhibitor completely normalised the haemodynamic abnormalities of hypertensive patients [22]. In addition to correcting the haemodynamic alterations of hypertension, CCB may provide numerous special ancillary effects, i.e., regression of left ventricular hypertrophy, anti-arrhythmic and anti-atherosclerotic effects, renal vasodilation and natriuresis. Moreover, CCB has attracted attention because of the lack of metabolic effects, the impairment of glucose metabolism, and because lipid profile and insulin resistance are only rarely encountered. There is also interest in their potential, especially in the agents that have anti-oxidative actions, to prevent complication of oxidative organ damage in hypertension. Sugawara et al. compared antioxidant effects of nine calcium antagonists on rat myocardial membrane lipid peroxidation with a non-enzymatic active oxygen-generating system (DHF/FeCl3-ADP) [23]. The order of antioxidant potency of these agents was nilvadipine > nisoldipine > felodipine > nicardipine > verapamil > benidipine (Figure 10.8). The antioxidant properties of

Table 10.5. Pharmacological effects of calcium antagonists

	Dihydropyridine	Diltiazem	Verapamil
Vascular Effects			
Vascular Selectivity	(+++)	(+)	0
Peripheral Vasodilation	(++)	(+)	(+)
Brain	(++)	(+)	(+0)
Coronary	(+++)	(++)	(++)
Renal	(++)	(+)	(+)
Cardiac Effects			
Nodal Conduction	0	(−)	(−−)
Myocardial Contractility	(−0)	(−)	(−−−)
Heart Rate	(+0)	(−)	(−)

(+): increase, (−): decrease, 0: no change.

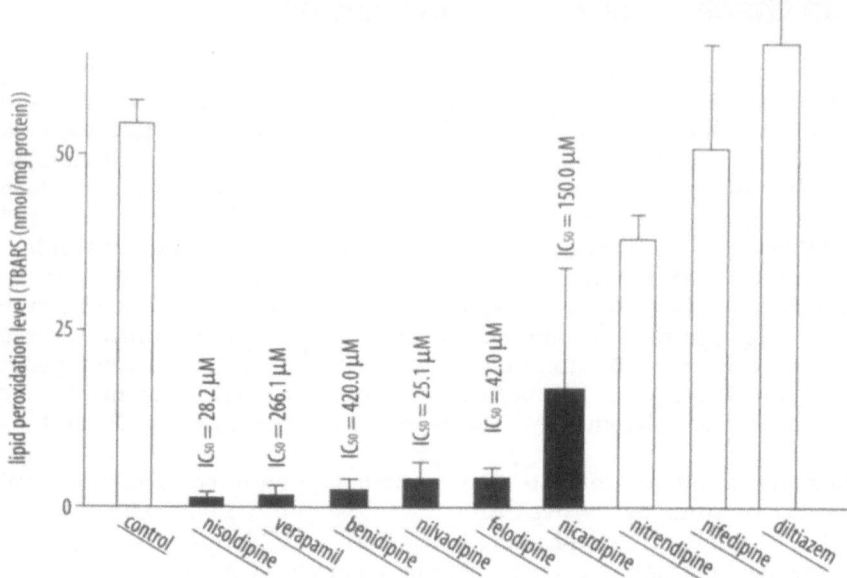

Figure 10.8. Comparative antioxidant effects of calcium antagonists on lipid peroxidation in rat myocardial membranes. (From Sugawara H et al. Antioxidant effects of calcium antagonists on rat myocardial membrane lipid peroxidation. Hypertens Res 1996;19:223–8.)

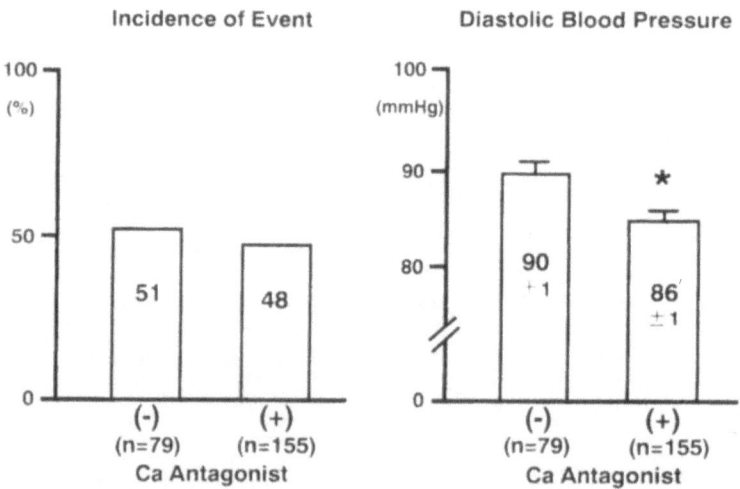

Figure 10.9. Effects of calcium antagonist on the incidence of cardiovascular events and the levels of diastolic blood pressure in hypertensive patients with angina pectoris.

some CCBs may be beneficial clinically in protecting against cellular damage caused by lipid peroxidation in hypertension.

CCB is effective in hypertensive patients of all ages and races and in hypertensive diabetics. Therapy based on the DHP nitrendipine provided even greater protection to elderly patients with isolated systolic hypertension in the Syst-Eur trial than did chlorthalidone in the SHEP trial, particularly in those in the two trials with diabetes accompanying the hypertension. Therefore, CCB is recommended for treatment of elderly patients with isolated systolic hypertension. DHP was also the foundation of therapy in other large trials that found significant reduction in cardiovascular events. CCB is recommended for treatment of hypertensive patients with angina pectoris and peripheral vascular disease. Non-DHP CCB is recommended for hypertensive patients with atrial tachycardia, atrial fibrillation and migraine. In contrast, non-DHP CCB is contraindicated for patients with second- or third-degree heart block.

Recently, there was debate on the potential hazard of calcium antagonists in the treatment of ischaemic heart disease and hypertension [24]. We investigated the relationship among the use of CCB, the treated levels of blood pressure, and the incidence of cardiovascular events in 234 hypertensive patients with angina pectoris [25]. More than two-thirds of patients were prescribed CCB, nearly 90% of which were long-acting. The incidence of events was not significantly high at 48% in the patients treated with CCB compared to 51% in the patients treated without CCB. In contrast, the treated levels of diastolic blood pressure were significantly lower in patients treated with CCB (Figure 10.9), suggesting that CCB is potent in lowering blood pressure; however, it does not seem to induce cardiovascular events in hypertensive patients with angina pectoris. A recent study of overviews of randomised trials demonstrated strong evidence of beneficial effects of CCB in the treatment of hypertension [26].

References

1. Murray CJ, Lope AD. Evidence-based health policy – lessons from the Global Burden of Disease Study. Science 1996;274:740–3.
2. Sowers JR, Standley PR, Tuck ML, Ram JL. Calcium and calcium-regulatory hormones in hypertension. In: Laragh JH, Brenner BM, editors. Hypertension: Pathophysiology, diagnosis, and management, 2nd edn. New York: Raven Press, 1995;1155–68.
3. Saruta T. Current status of calcium antagonist in Japan. Am J Cardiol 1998;82:32R–4R.
4. McCarron JA, Morris CD, Henry HJ, Stanton JL. Blood pressure and nutrient intake in the United States. Science 1984;224:1392–8.
5. Witteman JCM, Willett WC, Stampfer MJ, Colditz GA, Sacks FM, Speizer FE, et al. A prospective study of nutritional factors and hypertension among US women. Circulation 1989;80:1320–7.
6. McGarvey S, Zinner SH, Willett WC, Rosner B. Maternal prenatal dietary potassium, calcium, magnesium and infant blood pressure. Hypertension 1991;17:218–24.
7. Roullet CM, Poullet J-B, Duchambon P, et al. Abnormal intestinal regulation of calbindin-D9K and calmodulin by dietary calcium in genetic hypertension. Am J Physio 1991;261:F474–80.
8. Drueke TB. Mechanisms of action of calcium absorption: factors that influence bioavailability. In: Langford H, Levine B, Ellenbogen L, editors. Nutritional factors in hypertension. New York: Alan R Liss, 1990;155–73.
9. Lau K, Chen S, Eby B. Evidence for an intestinal mechanism in hypercalciuria of the spontaneously hypertensive rat. Am J Physiol 1984;247:E625–33.
10. Hatton DC, Young EW, Bukoski RD, McCarron DA. Calcium metabolism in experimental genetic hypertension. In: Laragh JH, Brenner BM, editors. Hypertension: Pathophysiology, diagnosis, and management, 2nd edn. New York: Raven Press, 1995;1193–211.
11. Resnick LM, Laragh JH, Sealey JE, Alderman MH. Divalent cations in essential hypertension. Relations between serum ionized calcium, magnesium, and plasma renin activity. N Engl J Med 1983;309:888–91.

12. Grobbee DE, Waal-Manning HJ. The role of calcium supplementation in the treatment of hypertension. Current Evidence Drugs 1990;39:7–18.
13. Cutler JA, Brittain E. Calcium and blood pressure: an epidemiologic perspective. Am J Hypertens 1990;3:137S–46S.
14. The Trials of Hypertension Prevention Collaborative Research Group. The effects of nonpharmacologic interventions on blood pressure of persons with high normal levels. Results of the Trials of Hypertension Prevention. Phase I. JAMA 1992;267:1213–20.
15. Morris CD, McCarron DA. Effect of calcium supplementation in an older population with mildly increased blood pressure. Am J Hypertens 1992;5:230–7.
16. Hatton DC, Scrogin KE, Metz JA, MaCarron DA. Dietary calcium alters blood pressure reactivity in spontaneously hypertensive rats. Hypertension 1989;13:622–9.
17. Ram JL, Standley PR, Sowers JR. Calcium function in vascular smooth muscle and its relationship to hypertension. In: Epstein M, editor. Calcium antagonists and the kidney. St Louis: Hanley & Belfus, 1993;29–48.
18. Takaya J, Lasker N, Bamforth R, Gutkin M, Byrd LH, Aviv A. Kinetics of Ca^{2+}-ATPase activation in planet membranes of essential hypertensives and normotensives. Am J Physiol 1990;258:C988–94.
19. Lopez-Jaramiilo, Gonzalez MC, Palmer RMJ, Moncada S. The crucial role of physiological Ca^{2+} concentrations in the production of endothelial nitric oxide and the control of vascular tone. Br J Pharmacol 1990;101:489–93.
20. Brickman AS, Nyby M, von Hunger K, Eggena P, Tuck ML. Parathyroid hormone, platelet calcium and blood pressure in normal subjects. Hypertension 1991;18:176–82.
21. Kahn AM, Seidel CL, Allen JC, O'Neil RG, Shelat H, Song T. Insulin reduces contraction and intracellular calcium concentration in vascular smooth muscle. Hypertension 1993;22:735–42.
22. Ting CT, Chen JW, Chang MS, Yin FCP. Arterial hemodynamics in human hypertension: effects of the calcium channel antagonist nifedipine. Hypertension 1995;25:1326–32.
23. Sugawara H, Tobise K, Kenjiro K. Antioxidant effects of calcium antagonists on rat myocardial membrane lipid peroxidation. Hypertens Res 1996;19:223–8.
24. Psaty BM, Heckbert SR, Koepsell TD, Siscovick DS, Raghunathan TE, Weiss NS, et al. The risk of myocardial infarction associated with antihypertensive drug therapies. JAMA 1995;274:620–5.
25. Hasebe N, Kido S, Ido A, Kikuchi K. Use of calcium antagonist and low diastolic pressure do not predict cardiac events in hypertensive patients with angina pectoris. Circulation 1997;96:1–762.
26. Blood Pressure Lowering Treatment Trialist's Collaboration. Effects of ACE inhibitors, calcium antagonists, and other blood-pressure-lowering drugs: results of prospectively designed overviews of randomised trials. Lancet 2000;355:1955–64.

Lipid Disorder

T. Shoji and Y. Nishizawa

Introduction

Epidemiological studies in a cohort of 2,185 male subjects identified a higher serum calcium concentration within the normal range as an independent prospective risk factor for myocardial infarction during 18 years of follow-up [1]. This was confirmed by a recent cross-sectional study of 12,865 men and 14,293 women showing that serum calcium was higher in men with a history of myocardial infarction than in those without [2]. Since calcium has diverse effects on cellular functions, extracellular and intracellular abnormalities in calcium may affect the cardiovascular risk factor profile. This chapter summarises the findings regarding the possible relationship between calcium and lipid disorder (Table 10.6). First, we briefly describe human lipoprotein metabolism in normal conditions, then changes in lipid metabolism in chronic renal failure, as a model for the study of the effects of calcium and related factors on lipid disorder (Table 10.7). This chapter also deals with possible lipid abnormalities in primary hyperparathyroidism and other topics.

Normal Lipoprotein Metabolism

Plasma Lipoproteins

According to density, plasma lipoproteins are classified into five major factions, namely: chylomicron (d < 0.96 g/mL), very low-density lipoprotein (VLDL, d = 0.96–1.006 g/mL), intermediate-density lipoprotein (IDL, d = 1.006–1.019 g/mL), low-density lipoprotein (LDL, d = 1.019–1.063 g/mL), and high-density lipoprotein (HDL, d = 1.063–1.21 g/mL). These lipoproteins contain not only lipids (free and esterified forms of cholesterol, triglycerides, and phospholipids), but also protein moieties called apolipoproteins (apo for short). Plasma lipoproteins are metabolised in the following three pathways.

Exogenous Lipid Pathway

Dietary lipids are absorbed in the small intestine, and secreted into lymph, and appear in the circulation as triglyceride-rich chylomicrons. Chylomicrons are metabolised into smaller chylomicron remnants by hydrolysis of triglycerides by

Table 10.6.   Studies examining changes in plasma lipid levels in disease conditions associated with calcium abnormalities

1968	Plasma triglycerides are elevated in patients with chronic renal failure. (Bagdade JD)
1973	Abnormally low serum cholesterol in patients with primary hyperparathyroidism is increased following parathyroidectomy. (DeMoor P)
1977	HDL cholesterol is decreased in patients with chronic renal failure. (Bagdade JD)
1977	Serum cholesterol and triglycerides are reduced in primary hyperparathyroidism, and these lipids are normalised after parathyroidectomy. (Christensson T)
1983	No specific lipid abnormality is found in primary hyperparathyroidism, and no change is observed following parathyroidectomy. (Vaziri ND)
1986	Serum triglycerides are increased in primary hyperparathyroidism and it is normalised by parathyroidectomy. (Lacour B)
1991	HMG-CoA reductase inhibitors do not inhibit vitamin D biosynthesis. (Dobs AS)
1992	Plasma 25(OH)D$_3$ level correlated positively with HDL cholesterol and apo A-I levels in healthy people. (Auwerx J)
1992	Bisphosphonates reduces cholesterol biosynthesis by inhibiting squalene synthase. (Amin D)
1997	A phosphate binder, RenaGel, decreased serum cholesterol in normal volunteers. (Burke SK)
1997	Total and LDL cholesterol decreased following RenaGel in haemodialysis patients. (Chertow GM)
1997	Long-term (three years) vitamin D$_3$ supplementation increased LDL cholesterol and decreased HDL cholesterol in postmenopausal hormone replacement therapy. (Heikkinen AM)
1997	Nifedipine had no effect on total cholesterol or triglycerides in young hypertensive patients. (Masuo K)
1998	Oral vitamin D$_3$ pulse therapy for secondary hyperparathyroidism increased HDL cholesterol and apo A-I in haemodialysis patients. Intact PTH correlated inversely with HDL levels. (Lim PS)
1998	Patients with asymptomatic primary hyperparathyroidism had higher cholesterol and triglycerides. (Lundgren E)
1998	A dihydropyridine calcium channel blocker, felodipine, increased HDL cholesterol without affecting plasma total cholesterol and triglycerides in hypertensive patients. (Reuter MK)
1998	PTX for primary hyperparathyroidism was followed by reduction of triglycerides only in male patients. (Valdemarsson S)
1999	Long-term treatment with RenaGel decreased LDL cholesterol by 30% and increased HDL cholesterol by 18% in 192 adult patients with ESRD. (Chertow GM)
1999	PTX for secondary hyperparathyroidism did not change plasma cholesterol level in haemodialysis patients. (Khajehdehi P)
1999	Cholesterol reduction by simvastatin was greater in patients treated with diltiazem. (Yeo KR)
2000	Supplementation of 1–2 g/day of elemental calcium did not change total or HDL cholesterol in 193 subjects. (Bostick RM)
2001	Calcium supplementation of chocolate reduced LDL cholesterol by 15% without changing HDL cholesterol. (Shahkhalili Y)

lipoprotein lipase (LPL) bound to the endothelial surface. Apo C-II on the lipoprotein promotes LPL action, whereas apo C-III inhibits it. The tissue utilises the released free fatty acids as a source of energy. Chylomicron remnants are atherogenic but are rapidly cleared from the blood by the liver via receptors for remnants. The uptake of chylomicron remnants is mediated by apo E bound to the lipoprotein, and it is inhibited by apo C-III. Lipoproteins in this pathway contain one molecule of apo B48 per particle.

Table 10.7.   Studies examining mechanisms for the effects of calcium and related factors on lipoprotein metabolism

1977	Patients with chronic renal failure have normal lipoprotein lipase but suppressed hepatic lipase activity. (Mordasini RN)
1983	Intracellular calcium is essential for the long-term activation of lipoprotein lipase. (Vydelingum N)
1987	PTH promotes lipolysis in human adipose tissue. (Taniguchi A)
1988	Calcitonin administration lowers serum cholesterol and triglycerides in Watanabe hereditable hyperlipidaemic rabbits and high-fat-fed rats by suppressing hepatic lipid synthesis through a calcium–calmodulin-dependent pathway. (Nishizawa Y)
1989	Depletion of extracellular calcium by EGTA suppresses lipoprotein lipase activity in adipocytes. (Soma MR)
1990	Dogs with chronic renal failure show secondary hyperparathyroidism, lowered lipoprotein lipase activity and increased plasma triglycerides. These abnormalities are restored following parathyroidectomy. (Akmal M)
1989	Active form vitamin D_3 increases lipid uptake via the scavenger receptor pathway in human monocytes/macrophages. (Roullet J-B)
1992	Suppressed hepatic lipase activity in haemodialysis patients has independent association with lowered ionised calcium and increased PTH levels. (Shoji T)
1993	Verapamil prevents chronic renal failure-induced abnormalities in lipid metabolism. (Akmal M)
1995	$1,25(OH)_2D_3$ suppresses expression of the scavenger receptor and uptake of acetylated LDL in THP-1 macrophages. (Suematsu Y)
1997	AE0047, a dihydropyridine-type calcium antagonist, decreased plasma triglycerides and increased HDL cholesterol in obese Zucker rats. This compound inhibited chylomicron secretion by human intestinal cell line Caco-2, and enhanced uptake of VLDL by human hepatoma cell line HepG2. (Hayashi K)
1997	Nifedipine suppressed cholesterol accumulation in macrophages without changing cholesterol efflux. This calcium channel blocker suppressed LDL oxidation in vitro and by macrophages. (Lesnik P)
1997	Verapamil decreased apoB (chylomicron) secretion by Caco-2 cells without changing microsomal triglycerides transfer protein mRNA and activity. (Mathur SN)
1997	Monatepil, a calcium antagonist, enhanced expression of LDL receptor mRNA and activity in human fibroblasts. (Matsunaga A)
1997	A calcium antagonist, monatepil, decreased plasma VLDL and total cholesterol, hepatic ACAT activity and cholesteryl ester content in cholesterol-fed Japanese monkeys. (Sumiya T)
1998	VLDL secretion reached maximum in rat hepatocytes cultured in a medium containing calcium 0.8–2.4 mmol/L. Calcium depletion reduced lipoprotein secretion and increased cellular lipid content, whereas calcium loading suppressed both lipid secretion and cellular lipids. (Bjornsson OG)
1998	A calcium channel blocker, AE0047, inhibited LDL oxidation induced by copper ion and degradation of oxidised LDL by macrophages. (Hayashi K)
1998	A calcium channel blocker, monatepil, accelerated the clearance of LDL from circulation and increased bile acid excretion through an LDL receptor-mediated pathway. This compound tended to inhibit intestinal absorption of cholesterol. (Ikeno A)
1998	A calcium channel blocker, isradipine, lowered retention of LDL in carotid and femoral arteries of hypertensive patients. (Kritz H)
1999	Cholesterol efflux to free apo A-I from murine macrophage RAW264 cells involves calcium-dependent endocytotic pathway, followed by recycling and the subsequent release of the nascent lipoprotein particle from the cells. (Takahashi Y)
2000	Nifedipine inhibited uptake of acetylated LDL by endothelial cells. (Berkels R)
2000	Nilvadipine treatment reduced the ratio of 7-keto cholestadien to cholesterol, an index of cholesterol oxidation of LDL, in hypertensive patients in vivo. (Inouye M)
2000	Calcium and lipoprotein lipase synergistically enhanced the binding and uptake of native and oxidised LDL in mouse peritoneal macrophages. (Wang X)

Endogenous Lipid Pathway

The liver can convert free fatty acids and carbohydrate into triglycerides and secrete them into the circulation. Lipids derived from the liver are secreted as triglyceride-rich VLDL. Like chylomicron, VLDL is metabolised into VLDL remnants, or IDL, by the function of LPL. IDL is atherogenic. Some IDL particles are taken up by the hepatic LDL receptors but most of the IDL is further metabolised into LDL by hepatic triglyceride lipase (HTGL, or HL). LDL carries cholesterol for peripheral cells requiring cholesterol for cell membranes and sterol hormones. Approximately one-third of plasma LDL is taken up by peripheral tissues and two-thirds by the liver via the LDL receptors. Lipoproteins in this pathway contain one molecule of apo B100 per particle.

Reverse Cholesterol Transport

Excessive cholesterol accumulation leads to tissue damage; for example, atherosclerosis. Since peripheral cells do not have cholesterol 7-hydroxylase – the only enzyme degrading cholesterol into bile acids – excess cholesterol in peripheral tissue needs to be transported back to the liver. Free apo A-I appears to serve as the initial acceptor for cellular cholesterol, and forms discoidal HDL. This immature form of HDL shows the prebeta mobility on agarose gel electrophoresis, in contrast to the alpha mobility of mature HDL. Free apo A-I and discoidal HDL are thought to be generated through lipolysis of triglyceride-rich lipoproteins by LPL and by direct secretion from hepatic and intestinal cells. Discoidal HDL, or nascent HDL, incorporates cholesterol into the lipoprotein core through esterification by lecithin–cholesterol acyltransferase (LCAT), and becomes larger spherical HDL called HDL_3 and HDL_2 showing the alpha electrophoretic mobility. Esterified cholesterol in HDL can reach the liver via at least two pathways. First, HDL cholesteryl esters are selectively taken up via the HDL receptor, SR-B1. Second, HDL cholesteryl esters are transferred to VLDL in exchange with VLDL triglycerides, metabolised in the VLDL–IDL–LDL pathway, and finally taken up via the hepatic LDL receptor. The lipid exchange between HDL and VLDL (and other apo B-containing lipoproteins) is mediated by cholesteryl ester transfer protein (CETP).

Calcium, Phosphate, PTH and Lipoprotein Abnormalities in Chronic Renal Failure

Changes in Lipoprotein Metabolism in Chronic Renal Failure

Patients with chronic renal failure have various abnormalities in plasma lipids and lipoproteins [3]. Uraemic dyslipidaemia is characterised by the increased plasma triglycerides and lowered HDL cholesterol. Hypertriglyceridaemia in renal failure is due to increased triglyceride-rich lipoproteins, VLDL and IDL [4]. Cholesterol-rich LDL is usually in the normal range or even reduced in renal failure. Hyperchylomicronaemia is rarely seen.

In vivo kinetic studies directly demonstrated that catabolism of VLDL was impaired [5] and also that the conversion of IDL to LDL was severely impaired [6]. These changes indicate that functions of LPL and HTGL, respectively, are both suppressed in chronic renal failure. The fractional catabolic rate of LDL was shown to be reduced in patients with renal failure [7], suggesting that the LDL-receptor activity is also down-regulated in this condition.

LPL is produced by adipose and muscle cells, and anchored on heparan sulphate proteoglycans at endothelial surfaces, where LPL exerts its functions. To assess the amount of functional LPL in vivo, heparin is injected intravenously to release the endothelium-bound LPL into the circulation, and post-heparin plasma LPL activity is measured. In spite of the defective catabolism of VLDL in uraemia, LPL activity in post-heparin plasma is reduced in a few but not all studies [8]. LPL protein mass in post-heparin plasma is also comparable with the reference level [9]. In contrast, the plasma level of apo C-III, an LPL inhibitor bound to VLDL, is increased in patients with renal failure [10–12]. Therefore, it is reasonable to conclude that the impaired catabolism of VLDL in renal failure is due mainly to "LPL resistance" of the lipoprotein: reduction of functional LPL may have some additional contributions in some cases.

Lowered HDL in renal failure is explained by hypertriglyceridaemia and lowered activity of lecithin–cholesterol acyltransferase (LCAT) [3]. Since the precursors of mature HDL are generated as a by-product of LPL-mediated degradation of chylomicrons and VLDL, impaired LPL function in vivo results in increased plasma triglycerides on one hand, and lowered HDL on the other. In addition, hypertriglyceridaemia promotes lipid exchange between VLDL and HDL, leading to triglyceride enrichment and cholesterol depletion of HDL. Partial LCAT deficiency in renal failure inhibits maturation of HDL, and could cause a low cholesterol content of HDL per particle [9].

Relationship Between Calcium and Lipid Metabolism in Chronic Renal Failure

In chronic renal failure, serum calcium is lowered, whereas phosphate and PTH levels are increased, and 1,25-hydroxyvitamin D_3 level is low. These changes in calcium homeostasis may account for, at least partly, the alterations in lipoprotein metabolism in renal failure of experimental animals and humans.

Akmal et al. [13] showed that elevated PTH in uraemia was responsible for hypertriglyceridaemia in experimental renal failure. They utilised a dog model of chronic renal failure showing secondary hyperparathyroidism, lowered post-heparin LPL activity, and hypertriglyceridaemia. These abnormalities were restored following PTX and calcium supplementation, suggesting the direct role of excess PTH in down-regulation of LPL and hypertriglyceridaemia. Vaziri et al. [14] showed a decrease in LPL mRNA in five-sixths-nephrectomised rats and restoration of it after PTX. These studies suggest the direct role of excess PTH in dyslipidaemia.

Akmal et al. [15] further explored the mechanisms for PTH-mediated alterations in lipid metabolism using five-sixths-nephrectomised rats. Chronic renal failure rats displayed hypertriglyceridaemia, fat intolerance, reduced post-heparin plasma lipoprotein and hepatic lipase activities, decreased hepatic lipase in liver homogenate, and elevated calcium content in liver and epididymal fat. Treatment of the CRF rats with verapamil prevented all these derangements in lipid metabolism. These effects of verapamil were similar to those produced by PTX of CRF rats. These data are consistent with the hypothesis that chronic excess of PTH increases the calcium burden of liver and adipose tissue and consequently impairs the synthesis and/or release of lipoprotein and hepatic lipases. Reduced availability of these enzymes results in impaired peripheral removal of triglycerides, leading to hyper-triglyceridaemia.

Vaziri et al. [16] examined the effects of chronic renal failure and PTH on HDL metabolism using a rat model. They determined hepatic apo A-I and an HDL receptor (SR-B1) mRNA abundance, and HDL receptor protein mass in five-sixths-nephrectomised rats (CRF), parathyroidectomised CRF rats (CRF-PTX) and sham-operated controls. The CRF group exhibited normal hepatic HDL receptor mRNA and HDL receptor protein abundance coupled with reduced hepatic apo A-I mRNA. Hepatic apo A-I mRNA, HDL receptor mRNA and protein abundance were not affected by PTX. These results suggest that CRF results in the down-regulation of hepatic apo A-I gene expression, whereas CRF does not affect HDL receptor mRNA or protein expression in this model. Also, the results suggest no contribution of PTH on either apo A-I or HDL receptor expression in CRF rats.

Accumulating evidence suggests that changes in calcium, phosphate and PTH in patients with CRF have influence on lipoprotein disorder. We revealed the correlation between altered calcium homeostasis and lipoprotein-regulating enzymes in patients with ESRD [17]. In that study, post-heparin plasma LPL activity and LPL protein concentration were both comparable to the values of normal subjects, whereas both HTGL activity and protein concentration in post-heparin plasma were severely lowered in the patient group. LCAT activity was also significantly reduced in the patients. Multiple regression analysis in the subjects showed that a low HTGL level was independently associated with age, female gender, a low ionised calcium concentration and increased PTH level, and that a low LCAT activity was independently associated with age and a low ionised calcium concentration. The reduction in HTGL and LCAT explain increased IDL [18] and abnormal sub-fraction distribution of HDL [9] in ESRD.

It is expected that the lipoprotein abnormalities may be corrected following successful PTH reduction in CRF patients with secondary hyperparathyroidism. So far, there is only limited information regarding this possibility. Lacour et al. [19] examined the changes in lipids following PTX in 34 haemodialysis patients. After surgical correction of secondary hyperparathyroidism, serum triglycerides and increase of HDL cholesterol occurred in short- and long-term follow-up. In contrast, in a recent report by Khajehdehi et al. [20], decrease in serum cholesterol was not changed following PTX in 15 patients with secondary hyperparathyroidism. According to Lim et al. [21], reduction of PTH by an oral vitamin D_3 pulse therapy was followed by a significance increase in HDL cholesterol and apo A-I. Also, they showed significant correlation of intact PTH level with HDL cholesterol and apo A-I, There is no data whether correction of secondary hyperparathyroidism results in reduction of IDL, which is difficult to detect by measuring serum total cholesterol and triglycerides [4].

A calcium-free, aluminium-free phosphate binder RenaGel lowered phosphate levels in normal volunteers [22] and in patients with end-stage renal disease (ESRD) [23]. In a placebo-controlled, double-blind trial with RenaGel in 36 maintenance haemodialysis patients over an eight-week period, reduction of phosphate was accompanied by a significant decreased in total cholesterol and LDL cholesterol [23]. Long-term treatment with this phosphate binder decreased LDL cholesterol by 30% and increased HDL cholesterol by 18% in 192 adult patients with ESRD [24]. It is unknown whether the observed changes in lipid levels were directly related to serum phosphate levels, since pre-treatment LDL cholesterol in ESRD is lower than normal even in the presence of hyperphosphataemia. It is possible that RenaGel worked as a bile acid sequestrant in the intestine, inhibiting the enterohepatic circulation of bile acids, reducing cholesterol pool in hepatocytes, and finally up-regulating hepatic LDL receptor expression.

Mak [25] showed evidence that 1,25-dihydroxyvitamin D_3 corrects carbohydrate and lipid metabolism in uraemic patients in the absence of PTH suppression. Eight patients received intravenous 1,25-dihydroxyvitami D_3 therapy for four weeks; although serum PTH level did not change during the period, normalisation was observed in glucose tolerance, insulin sensitivity during euglycemic clamp studies, insulin secretion, and plasma triglyceride levels. Plasma total and HDL cholesterol levels did not change following the 1,25-dihydroxyvitamin D_3 treatment.

Navarro et al. [26] failed to observe a significant role of PTH in dyslipidaemia in renal failure. There was no correlation between serum lipid parameters and PTH in 34 haemodialysis patients. Also, lipid profile did not change after PTX in seven patients with severe hyperparathyroidism.

Calcium, Lipoprotein Disorder and Atherosclerosis in Chronic Renal Failure

Cardiovascular mortality is more than 10 times higher in ESRD patients than that in the general population [27, 28], suggesting that atherosclerosis is advanced in ESRD. Atherosclerosis has the major features of thickening and stiffening of arterial walls. Patients with ESRD have increased intima-media thickness (IMT) of carotid and femoral arteries [29, 30] and increased stiffness of the aorta [31]. Non-HDL cholesterol (total cholesterol minus HDL cholesterol) was an independent factor associated with arterial wall thickness [30] and stiffness [31] in patients with ESRD. Within non-HDL lipoproteins (VLDL, IDL, LDL and Lp(a)), IDL was shown as the lipoprotein that was most closely associated with aortic atherosclerosis [31]. IDL level in haemodialysis patients is two-to-three times higher than that of healthy subjects [4]. Raised IDL is independently associated with lowered HTGL [18]. The low HTGL is independently associated with hypocalcaemia and hyperparathyroidism [9]. Taken together, these studies suggest the sequence of events that renal failure results in hypocalcaemia and secondary hyperparathyroidism, which adversely affect lipoprotein-regulating enzymes, especially HTGL, which directly increases IDL, accounting for advanced atherosclerosis in ESRD. In fact, Kawagishi et al. [32] reported that IMT of carotid and femoral arteries had independent positive associations with serum phosphate and PTH levels in haemodialysis patients.

Lipid Disorder in Primary Hyperparathyroidism

In primary hyperparathyroidism, PTH and serum calcium levels are increased whereas phosphate is lowered. These changes may affect lipid and lipoprotein metabolism. De Moor reported that serum total cholesterol was decreased in 75 patients with primary hyperparathyroidism, and that PTX was followed by an increase in cholesterol [33]. In a study by Christensson et al. [34], both serum total cholesterol and triglycerides were lowered in 42 patients with primary hyperparathyroidism, which were normalised following PTX. Vaziri et al. [35] found no significant alterations in serum lipid levels in 24 cases with this disease, and no change was observed in VLDL,

LDL or HDL fractions after PTX. In contrast, Lacour et al. [19] reported that 86 patients with primary hyperparathyroidism had increased triglyceride levels, which were normalised following PTX. A recent study by Lundgren et al. [36] in 102 asymptomatic patients with primary hyperparathyroidism showed that serum triglycerides, VLDL cholesterol and VLDL triglycerides were higher in the patients. Valdemarsson et al. [37] did not find changes in cholesterol or triglycerides in a total of 117 patients following PTX, although male patients had significantly lower triglyceride levels at follow-up. These studies do not provide evidence for a consistent tendency of alterations in cholesterol or triglyceride levels in primary hyperparathyroidism. However, no study has examined whether that primary hyperparathyroidism, as is the case with secondary hyperparathyroidism [18], affects IDL levels by modulating hepatic lipase activity.

Effects of Calcium, Calcium-Modulating Hormones, and Calcium Channel Blockers on Lipid Metabolism in Other Conditions

Extracellular Calcium Level and Calcium Supplementation

In experimental conditions, the extracellular calcium level could alter synthesis and secretion of lipids by hepatocytes. Bjornsson et al. [38] reported that in rat hepatocytes in culture, depleting the cells of calcium by incubating in calcium-free medium or by treating cells with a Ca^{2+}-ATPase inhibitor – thapsigargin – suppressed lipoprotein secretion and increased cellular triglyceride content. Calcium loading by a high-calcium medium (2.4 mmol/L) suppressed both lipid secretion and cellular lipid level, suggesting that overall lipid synthesis was decreased.

Extracellular calcium appears to be important in cholesterol efflux from cells. Takahashi and Smith [39] reported that apo A-I-mediated efflux of cellular cholesterol depended on extracellular calcium in mouse macrophage RAW264 cells. In the macrophage model system, free apo A-I was bound to coated pits, taken up by the cells, and resecreted back into the medium with cholesterol from cells. Their data suggest that cholesterol efflux to apo A-I involves a calcium-dependent endocytic pathway, followed by secretion of nascent HDL particles from macrophages.

In contrast, calcium supplementation does not have a consistent effect on serum lipid levels. Bostic et al. [40] performed a randomised, double-blind, placebo-controlled clinical trial to examine whether calcium has beneficial effects on total cholesterol, HDL cholesterol and blood pressure. One hundred and ninety-three men and women, aged 30 to 74 years, were treated with a supplement of 1.0–2.0 g per day of elemental calcium over four months. However, no significant change was observed in total cholesterol, HDL cholesterol or blood pressure. In contrast, Shahkhalili et al. [41] made an interesting observation in a randomised, double-blind, crossover study to examine the effect of calcium supplementation of chocolate on plasma lipids in 10 men. The subjects were fed control diets containing 98–101 g of chocolate per day with or without 0.9 g/day of calcium supplementation for two periods of two weeks each. Calcium supplementation of chocolate increased faecal fat from 4.4 g to 8.4 g per day, and reduced the absorption of cocoa butter by 13%. Also, LDL cholesterol was significantly reduced by 15% but HDL cholesterol was not.

Calcium-Modulating Hormones

PTH

In an early study by Lacour et al. [42], the effects of hyperparathyroidism were examined in rat models. They induced an endogenous hyperparathyroid state by a calcium-poor diet. This type of hyperparathyroidism was associated with a significant increase in serum cholesterol and triglyceride concentrations, a significant decrease in the serum clearance rate of intravenously infused intralipid and a significant increase in hepatic tissue triglyceride content as compared to the values obtained in control animals. They induced an exogenous type of hyperparathyroidism by parathyroid extract injections, which were accompanied by a slight (insignificant) increase in serum cholesterol and triglyceride concentrations as well as a decrease in post-heparin plasma lipolytic activity (PHLA). In the aparathyroid state, a significant decrease in serum cholesterol and triglyceride concentrations and a significant increase in PHLA were observed as compared to control rats. They also demonstrated that the endogenous hyperparathyroid state was not associated with increased intestinal lipid absorption or increased hepatic triglyceride secretion. These results indicate that hyperparathyroidism induces an increase in serum cholesterol and triglyceride concentrations by impairing peripheral removal of triglyceride-rich lipoproteins in rats.

Arnadottir and Nilsson-Ehle [43] examined the possibility that PTH is an inhibitor of lipoprotein lipase. Addition of intact PTH molecules or fragments of PTH (39–84) had no significant effect on LPL activities in vitro. The authors concluded that PTH in not a direct inhibitor of LPL activity in vitro.

Calcitonin

There is only limited information on the effects of calcitonin on lipid metabolism. Nishizawa et al. [44] found that calcitonin suppresses lipogenesis by the liver and lowers plasma cholesterol and triglycerides. They showed the effects of calcitonin on lipid in three kinds of rats, one strain of rabbit, and a primary culture of rat hepatocytes. In a short-term experiment, calcitonin decreased serum cholesterol and triglycerides after injection in rats on either an ordinary or high-fat diet. In a long-term experiment, calcitonin decreased the serum cholesterol and triglycerides in uraemic rats, hypothalamic obese rats, and Watanabe-heritable hyperlipidaemic rabbits. In cultured hepatocytes, calcitonin reduced the incorporation of ^{14}C-acetate into cholesterol and triglycerides in a dose-dependent way. Treatment with W7, a calmodulin inhibitor, overcame the decrease caused by calcitonin in serum lipids in rats and in the synthesis of triglycerides from acetate or palmitate in the hepatocytes, but did not alter the intracellular cAMP level or incorporation of ^{32}P-Pi into Pi in the cells. These results suggest that calcitonin lowers serum lipid levels and lipogenesis in hepatocytes in a calcium/calmodulin-dependent way.

Vitamin D₃

De Novellis et al. [45] examined the effect of vitamin D deficiency on lipid levels in rats. Rats exposed to a vitamin D-deficient diet showed a decreased ionised calcium concentration, whereas no change was observed in cholesterol or triglycerides.

Heikkinen et al. [46] found adverse effects on serum lipids in long-term vitamin D_3 supplementation during postmenopausal hormone replacement therapy. They randomised 464 women into four treatment groups: 1) HRT (sequential combination of 2 mg oestradiol valerate and 1 mg cyproterone acetate), 2) Vit D_3 (vitamin D_3, 300 IU/day), 3) HRT+Vit D_3 (both as above), 4) placebo (calcium lactate 500 mg/day). After three years of treatment, serum concentrations of low-density lipoprotein (LDL) cholesterol decreased in the HRT group (−10.1%) and the HRT+Vit D_3 group (−5.9%), increased in the Vit D_3 group (+4.1%) but remained unchanged in the placebo group. HDL cholesterol decreased in the Vit D_3 group (−5.2%), HRT+Vit D_3 group (−3.7%), and the placebo group (−4.5%) but did not change significantly in the HRT group. The HDL/LDL ratio increased in the HRT group (+10.5%) and decreased in the Vit D_3 group (−10.5%), whereas no changes occurred in the other two groups. In addition, serum triglycerides increased similarly in all groups. These results suggest that long-term vitamin D_3 supplementation may have unfavourable effects on lipid profile in postmenopausal women.

Calcium Channel Blockers

Mathur et al. [47] found that apo B secretion from human intestinal cell line Caco-2 was inhibited by treatment with verapamil.

Treatment with a dihydropyridine-type calcium antagonist AE0047 decreased plasma triglyceride and increased HDL cholesterol, whereas plasma total and LDL cholesterol did not change in obese Zucker rats [48]. This calcium channel blocker was able to suppress basolateral secretion of apo B by human intestinal cell line Caco-2. Uptake of VLDL by human hepatoma cell line HepG2 was enhanced by AE0047. These results indicate that triglyceride reduction by calcium channel blockade may be mediated by suppressed intestinal chylomicron secretion and enhanced hepatic VLDL uptake.

Masuo et al. [49] examined the long-term effect of nifedipine on plasma lipids in hypertensive patients. Although treatment with nifedipine retard in 78 hypertensive patients had an ameliorative effect on glucose metabolism, it did not improve total cholesterol or triglycerides.

The effect of felodipine ER, a dihydropyridine calcium antagonist, on serum lipids was analysed in 106 hypertensive patients during a 24-week period. Total cholesterol and triglyceride did not change, whereas HDL cholesterol increased to a significant but modest extent from 1.30 to 1.33 mmol/L [50].

Calcium channel blockade may inhibit hepatic acyl-CoA:cholesterol acyltransferase (ACAT) and decrease plasma lipid levels. Sumiya et al. [51] examined the effect of monatepil, a calcium channel blocker with alpha-1 adrenoceptor blocking properties, on hepatic ACAT activity in Japanese monkeys. Both ACAT activity and esterified cholesterol content in the livers of monkeys fed on a cholesterol-rich diet for six months significantly increased about 7- and 16-fold, respectively, as compared with those in monkeys fed on a standard diet. Oral administration of monatepil inhibited the increases of ACAT activity and esterified cholesterol content by 51% and 71%, respectively. In in vitro experiments, this agent inhibited ACAT activity in a concentration-dependent manner, whereas it did not affect HMG-CoA reductase activity. Hepatic ACAT activity correlated to hepatic esterified cholesterol content, to plasma very-low-density lipoprotein (VLDL) level, and to plasma total cholesterol concentration. These results suggest that the ACAT-inhibiting effect of this calcium channel blocker plays an important role in the reduction of plasma lipids.

Ikeno et al. [52] performed in vivo kinetic studies of LDL labelled with ^{3}H-cholesterol and followed the metabolism of LDL in the circulation and cholesterol metabolites into bile. Monatepil significantly accelerated the clearance of radioactivity from the blood after intravenous injection of LDL, increasing biliary excretion of ^{3}H-bile acids without modifying bile acid composition. Furthermore, monatepil tended to inhibit the absorption of orally administered ^{3}H-cholesterol from the gastrointestinal tract in these rabbits. In Watanabe heritable hyperlipidaemic (WHHL) rabbits, an animal model of hepatic LDL receptor deficiency, monatepil did not suppress the increase in plasma lipids. These results suggest that monatepil enhanced the clearance of plasma LDL by up-regulation of hepatic LDL receptors, and resulted in accelerated bile acid secretion by the liver. Monatepil is also shown to enhance mRNA expression of the LDL receptor in human fibroblasts [53].

Kritz et al. [54] examined the effect of calcium channel blockade by isradipine on retention of ^{123}I-labeled LDL in human carotid and femoral arteries in 12 hypertensive patients. Following the drug; plasma levels of total, HDL, and LDL cholesterol did not change significantly. The types of entry kinetics reflecting vascular surface lining did not change, while the LDL retention 20 hours after tracer application was depressed by up to 23.5%. These results indicate a decreased LDL retention in the arterial wall of hypertensive patients induced by isradipine, although it is unknown whether the effect was a direct action of the calcium channel blocker or was due to reduced blood pressure.

Lesnik et al. examined the effects of nifedipine and a beta blocker, atenolol, on LDL oxidation in vitro and cholesterol metabolism in macrophages. Atenolol had no effect on the copper-mediated in vitro oxidation of LDL, whereas nifedipine showed a significant dose-dependent inhibition of LDL oxidation. The anti-oxidative ability of nifedipine was also significant in a system using macrophage-mediated LDL oxidation. However, neither nifedipine nor atenolol protected mouse J774 macrophages from oxidised LDL-mediated foam cell formation. Cholesterol efflux from preloaded human macrophages was not affected by the addition of either nifedipine or atenolol. These results suggest that calcium blockade with nifedipine exerts potentially anti-atherogenic properties by reducing oxidative modification of LDL. Anti-oxidative properties of calcium channel blockers were also reported for AE0047 [55].

The in vivo anti-oxidative action of a calcium channel blocker, nilvadipine, was evaluated by measuring the ratio of 7-keto cholestadien to cholesterol in plasma LDL, a marker of lipid peroxidation in vivo. Inouye et al. [56] measured the ratio of 7-keto cholestadien to cholesterol in 15 healthy subjects and 15 hypertensive patients. The ratio was 0.2% in the healthy controls, whereas it was 6.5% in the patients. Treatment with nilvadipine for four weeks reduced the oxidation level to 3.8%, suggesting that nilvadipine inhibited LDL oxidation in vivo.

Calcium channel blockade may be protective against uptake of modified LDL by cells. Berkels et al. [57] reported that nifedipine inhibited the uptake of acetylated LDL into endothelial cells via an NO- but presumably not by a cGMP-mediated process. Although they concluded that it may possibly contribute to the anti-atherogenic action of this drug, the role of endothelial uptake of acetylated LDL in atherogenesis is not fully understood.

Conclusion

As reviewed above, calcium and related factors affect various aspects of lipid metabolism. Both depletion and excess of calcium and vitamin D_3 are associated with unfavourable changes of lipid metabolism in the circulation and in the cells. Calcium homeostasis appears to be of importance not only in bone mineral metabolism but also in lipids and lipoproteins, and also in cardiovascular disease.

References

1. Lind L, Skarfors E, Berglund L, Lithell H, Ljunghall S. Serum calcium: a new, independent, prospective risk factor for myocardial infarction in middle-aged men followed for 18 years. J Clin Epidemiol 1997;50:967–73.
2. Jorde R, Sundsfjord J, Fitzgerald P, Bonaa KH. Serum calcium and cardiovascular risk factors and diseases: the Tromso study. Hypertension 1999;34:484–90.
3. Attman PO, Samuelsson O, Alaupovic P. Lipoprotein metabolism and renal failure. Am J Kidney Dis 1993;21:573–92.
4. Shoji T, Nishizawa Y, Kawagishi T, Tanaka M, Kawasaki K, Tabata T et al. Atherogenic lipoprotein changes in the absence of hyperlipidemia in patients with chronic renal failure treated by hemodialysis. Atherosclerosis 1997;131:229–36.
5. Savdie E, Gibson JC, Crawford GA, Simons LA, Mahony JF. Impaired plasma triglyceride clearance as a feature of both uremic and post-transplant triglyceridemia. Kidney Int 1980;18:774–82.
6. Chan PC, Persaud J, Varghese Z, Kingstone D, Baillod RA, Moorhead JF. Apolipoprotein B turnover in dialysis patients: its relationship to pathogenesis of hyperlipidemia. Clin Nephrol 1989;31:88–95.
7. Horkko S, Huttunen K, Kesaniemi YA. Decreased clearance of low-density lipoprotein in uremic patients under dialysis treatment. Kidney Int 1995;47:1732–40.
8. Mordasini R, Frey F, Flury W, Klose G, Greten H. Selective deficiency of hepatic triglyceride lipase in uremic patients. N Engl J Med 1977;297:1362–6.
9. Shoji T, Nishizawa Y, Nishitani H, Yamakawa M, Morii H. Impaired metabolism of high density lipoprotein in uremic patients. Kidney Int 1992;41:1653–61.
10. Shoji T, Nishizawa Y, Nishitani H, Yamakawa M, Morii H. Roles of hypoalbuminemia and lipoprotein lipase on hyperlipoproteinemia in continuous ambulatory peritoneal dialysis. Metabolism 1991;40:1002–8.
11. Nishizawa Y, Shoji T, Nishitani H, Yamakawa M, Konishi T, Kawasaki K et al. Hypertriglyceridemia and lowered apolipoprotein C-II/C-III ratio in uremia: effect of a fibric acid, clinofibrate. Kidney Int 1993;44:1352–9.
12. Samuelsson O, Attman PO, Knight-Gibson C, Kron B, Larsson R, Mulec H et al. Lipoprotein abnormalities without hyperlipidaemia in moderate renal insufficiency. Nephrol Dial Transplant 1994;9:1580–5.
13. Akmal M, Kasim SE, Soliman AR, Massry SG. Excess parathyroid hormone adversely affects lipid metabolism in chronic renal failure. Kidney Int 1990;37:854–8.
14. Vaziri ND, Liang K. Down-regulation of tissue lipoprotein lipase expression in experimental chronic renal failure. Kidney Int 1996;50:1928–35.
15. Akmal M, Perkins S, Kasim SE, Oh HY, Smogorzewski M, Massry SG. Verapamil prevents chronic renal failure-induced abnormalities in lipid metabolism. Am J Kidney Dis 1993;22:158–63.
16. Vaziri ND, Deng G, Liang K. Hepatic HDL receptor, SR-B1 and Apo A-I expression in chronic renal failure. Nephrol Dial Transplant 1999;14:1462–6.
17. Shoji T, Nishizawa Y, Koyama H, Hagiwara S, Aratani H, Sasao K et al. High-density lipoprotein metabolism during a very-low-calorie diet. Am J Clin Nutr 1992;56:297S–8S.
18. Shoji T, Nishizawa Y, Kawagishi T, Emoto M, Morii H. Secondary hyperparathyroidism, decreased hepatic triglyceride lipase, elevated intermediate density lipoprotein and atherosclerosis in hemodialysis patients. Nephron 1998;78:121–2.
19. Lacour B, Roullet JB, Liagre AM, Jorgetti V, Beyne P, Dubost C et al. Serum lipoprotein disturbances in primary and secondary hyperparathyroidism and effects of parathyroidectomy. Am J Kidney Dis 1986;8:422–9.
20. Khajehdehi P, Ali M, Al-Gebory F, Henry G, Bastani B. The effects of parathyroidectomy on nutritional and biochemical status of hemodialysis patients with severe secondary hyperparathyroidism. J Ren Nutr 1999;9:186–91.
21. Lim PS, Hung TS, Yeh CH, Yu MH. Effects of treatment of secondary hyperparathyroidism on the lipid profile in patients on hemodialysis. Blood Purif 1998;16:22–9.
22. Burke SK, Slatopolsky EA, Goldberg DI. RenaGel, a novel calcium- and aluminium-free phosphate binder, inhibits phosphate absorption in normal volunteers. Nephrol Dial Transplant 1997;12:1640–4.

23. Chertow GM, Burke SK, Lazarus JM, Stenzel KH, Wombolt D, Goldberg D et al. Poly[allylamine hydrochloride] (RenaGel): a noncalcemic phosphate binder for the treatment of hyperphosphatemia in chronic renal failure. Am J Kidney Dis 1997;29:66–71.

24. Chertow GM, Burke SK, Dillon MA, Slatopolsky E. Long-term effects of sevelamer hydrochloride on the calcium x phosphate product and lipid profile of haemodialysis patients. Nephrol Dial Transplant 1999;14:2907–14.

25. Mak RH. 1,25-Dihydroxyvitamin D3 corrects insulin and lipid abnormalities in uremia. Kidney Int 1998;53:1353–7.

26. Navarro JF, Teruel JL, Lasuncion MA, Mora-Fernandez C, Ortuno J. Relationship between serum parathyroid hormone levels and lipid profile in hemodialysis patients. Evolution of lipid parameters after parathyroidectomy. Clin Nephrol 1998;49:303–7.

27. Lindner A, Charra B, Sherrard DJ, Scribner BH. Accelerated atherosclerosis in prolonged maintenance hemodialysis. N Engl J Med 1974;290:697–701.

28. Foley RN, Parfrey PS, Sarnak MJ. Clinical epidemiology of cardiovascular disease in chronic renal disease. Am J Kidney Dis 1998;32:S112–9.

29. Kawagishi T, Nishizawa Y, Konishi T, Kawasaki K, Emoto M, Shoji T et al. High-resolution B-mode ultrasonography in evaluation of atherosclerosis in uremia. Kidney Int 1995;48:820–6.

30. Shoji T, Kawagishi T, Emoto M, Maekawa K, Taniwaki H, Kanda H et al. Additive impacts of diabetes and renal failure on carotid atherosclerosis. Atherosclerosis 2000;153:257–8.

31. Shoji T, Nishizawa Y, Kawagishi T, Kawasaki K, Taniwaki H, Tabata T et al. Intermediate-density lipoprotein as an independent risk factor for aortic atherosclerosis in hemodialysis patients. J Am Soc Nephrol 1998;9:1277–84.

32. Kawagishi T, Nishizawa Y, Konishi T, Kawasaki K, Emoto M, Shoji T et al. High-resolution B-mode ultrasonography in evaluation of atherosclerosis in uremia. Kidney Int 1995;48:820–6.

33. De Moor P, Creyttens G, Bouillon R, Joossens J. Results obtained in 75 patients operated upon for hyperparathyroidism: low cholesterol levels in overt primary hyperparathyroidism. Ann Endocrinol 1973;34:616–620.

34. Christensson T, Einarsson K. Serum lipids before and after parathyroidectomy in patients with primary hyperparathyroidism. Clin Chim Acta 1977;78:411–5.

35. Vaziri ND, Wellikson L, Gwinup G, Byrne C. Lipid fractions in primary hyperparathyroidism before and after surgical cure. Acta Endocrinol (Copenh) 1983;102:539–42.

36. Lundgren E, Ljunghall S, Akerstrom G, Hetta J, Mallmin H, Rastad J. Case-control study on symptoms and signs of asymptomatic primary hyperparathyroidism. Surgery 1998;124:980–5; discussion 985–6.

37. Valdemarsson S, Lindblom P, Bergenfelz A. Metabolic abnormalities related to cardiovascular risk in primary hyperparathyroidism: effects of surgical treatment. J Intern Med 1998;244:241–9.

38. Bjornsson OG, Bourgeois CS, Gibbons GF. Varying very-low-density lipoprotein secretion of rat hepatocytes by altering cellular levels of calcium and the activity of protein kinase C. Eur J Clin Invest 1998;28:720–9.

39. Takahashi Y, Smith JD. Cholesterol efflux to apolipoprotein AI involves endocytosis and resecretion in a calcium-dependent pathway. Proc Natl Acad Sci USA 1999;96:11358–63.

40. Bostick RM, Fosdick L, Grandits GA, Grambsch P, Gross M, Louis TA. Effect of calcium supplementation on serum cholesterol and blood pressure. A randomized, double-blind, placebo-controlled, clinical trial. Arch Fam Med 2000;9:31–8.

41. Shahkhalili Y, Murset C, Meirim I, Duruz E, Guinchard S, Cavadini C et al. Calcium supplementation of chocolate: effect on cocoa butter digestibility and blood lipids in humans. Am J Clin Nutr 2001;73:246–52.

42. Lacour B, Basile C, Drueke T, Funck-Brentano JL. Parathyroid function and lipid metabolism in the rat. Miner Electrolyte Metab 1982;7:157–65.

43. Arnadottir M, Nilsson-Ehle P. Parathyroid hormone is not an inhibitor of lipoprotein lipase activity. Nephrol Dial Transplant 1994;9:1586–9.

44. Nishizawa Y, Okui Y, Inaba M, Okuno S, Yukioka K, Miki T et al. Calcium/calmodulin-mediated action of calcitonin on lipid metabolism in rats. J Clin Invest 1988;82:1165–72.

45. De Novellis V, Loffreda A, Vitagliano S, Stella L, Lampa E, Filippelli W et al. Effects of dietary vitamin D deficiency on the cardiovascular system. Res Commun Chem Pathol Pharmacol 1994;83:125–44.

46. Heikkinen AM, Tuppurainen MT, Niskanen L, Komulainen M, Penttila I, Saarikoski S. Long-term vitamin D3 supplementation may have adverse effects on serum lipids during postmenopausal hormone replacement therapy. Eur J Endocrinol 1997;137:495–502.

47. Mathur SN, Born E, Murthy S, Field FJ. Microsomal triglyceride transfer protein in CaCo-2 cells: characterization and regulation. J Lipid Res 1997;38:61–7.

48. Hayashi K, Gohda M, Matzno S, Kubo Y, Kido H, Yamauchi T et al. Possible mechanism of action of AE0047, a calcium antagonist, on triglyceride metabolism. J Pharmacol Exp Ther 1997;282:882–90.

49. Masuo K, Mikami H, Ogihara T, Tuck ML. Metabolic effects of long-term treatments with nifedipine-retard and captopril in young hypertensive patients. Am J Hypertens 1997;10:600–10.
50. Reuter MK, Lorenz H, Verho P, Smith N, Degen A, Verho M. Effects of felodipine ER, a dihydropyridine calcium antagonist, on blood pressure and serum lipids. Curr Med Res Opin 1998;14:97–103.
51. Sumiya T, Ikeno A, Kato H, Fujitani B, Masuda Y, Hosoki K et al. Inhibitory effect of monatepil maleate on acyl-CoA:cholesterol acyltransferase activity in the liver of cholesterol-fed Japanese monkeys. Am J Hypertens 1997;10:779–85.
52. Ikeno A, Sumiya T, Minato H, Fujitani B, Masuda Y, Hosoki K et al. Effects of monatepil maleate, a new Ca^{2+} channel antagonist with alpha1-adrenoceptor antagonistic activity, on cholesterol absorption and catabolism in high cholesterol diet-fed rabbits. Jpn J Pharmacol 1998;78:303–12.
53. Matsunaga A, Inoue T, Koga T, Mori K, Kugi M, Sasaki J et al. Effects of monatepil, a novel calcium antagonist with alpha 1-adrenergic blocking activity, on the low-density lipoprotein receptor in human skin fibroblasts. Cardiovasc Drugs Ther 1997;11:747–50.
54. Kritz H, Sinzinger H, Fitscha P, O'Grady J. Isradipine lowers human arterial low-density lipoprotein retention in vivo. Prostaglandins Leukot Essent Fatty Acids 1998;59:305–12.
55. Hayashi K, Imada T, Yamauchi T, Kido H, Shinyama H, Matzno S et al. Possible mechanism for the anti-atherosclerotic action of the calcium channel blocker AE0047 in cholesterol-fed rabbits. Clin Exp Pharmacol Physiol 1998;25:17–25.
56. Inouye M, Mio T, Sumino K. Nilvadipine protects low-density lipoprotein cholesterol from in vivo oxidation in hypertensive patients with risk factors for atherosclerosis. Eur J Clin Pharmacol 2000; 56:35–41.
57. Berkels R, Hass U, Klaus W. The calcium antagonist nifedipine inhibits the uptake of acetylated LDL into endothelial cells. Naunyn Schmiedebergs Arch Pharmacol 2000;362:91–5.

10D

Atherosclerosis

Y. Nishizawa and H. Koyama

Atherosclerosis is a cause of cardiovascular disease, one of the most common causes of death. Since high plasma concentrations of cholesterol are one of the principal risk factors for atherosclerosis, the process of atherogenesis has been considered to mainly consist of the accumulation of lipids within the arterial wall. Recently, in broad outline, atherosclerosis can be considered to be a form of chronic inflammation resulting from the interaction between modified lipoproteins, monocyte-derived macrophages, T cells, and the normal cellular elements of the arterial wall [1]. This inflammatory process can ultimately lead to the development of complex lesions, or plaques, which are frequently associated with calcium accumulation (calcification). Importantly, these complexed lesions are sometimes complicated with plaque rupture and thrombosis, resulting in the acute clinical complications of myocardial infarction and stroke. Calcium deposition in atherosclerosis lesions is known to occur just after fatty streak formation. Microscopic analysis has shown small aggregates of crystalline calcium among the lipid particles of lipid cores in lesions of young adults [2]. Deposition of calcium is found in greater amounts in elderly and more advanced lesions. Atherosclerotic calcification is now known to be an organised, regulated process similar to bone formation that occurs only when other aspects of atherosclerosis are also present [3]. Osteocalcin, osteopontin and its mRNA and bone morphogenetic protein-2a, known to be involved in bone mineralisation and osteoblastic differentiation, have been identified in calcified atherosclerotic lesions.

Calcium and Atherosclerosis: Calcium Channel Blockers Retard Animal Models of Atherosclerosis

Calcium plays a pivotal role in numerous biological processes, ranging from a structure stabilising effect (i.e., bone) to signalling transduction. These physiological events include vasoconstriction, thrombosis, cell growth and cell motility, all of which are important in the pathogenesis of atherosclerosis. Calcium-mediated signal transduction invariably involves the interaction of ionised calcium with specific proteins, resulting in conformational changes in specific proteins, which lead to a cascade of protein–protein interactions. Although calcium ions are pivotal for these events, excessive accumulation inside the cell or failure of the intracellular calcium homeostasis results in death or necrosis of the cells. Based on the rationale that inhibiting calcium ion entry would reduce the calcium overloading and hence cell death or necrosis, the calcium channel blockers were developed as therapeutic agents. These agents are now widely used in clinical fields in the treatment of patients in vascular

Table 10.8. Suppressive effects of calcium channel blockers on progression of atherosclerosis

Dihydropyridine-based compounds		
Amlodipine	cholesterol-fed rabbit	[77]
	Atherogenic diet-fed monkey	[6]
Isradipine	cholesterol-fed rabbit	[8]
Nifedipine	cholesterol-fed rabbit	[5, 78]
Nicardipine	cholesterol-fed rabbit	[78]
Nilvadipine	cuff-induced rabbit carotid thickness	[79]
Nisoldipine	cholesterol-fed rabbit	[9]
Phenylalkylamine-based compounds		
Verapamil	cholesterol-fed rabbit	[80]
Anipamil	Watanabe hereditary hyperlipidaemic rabbit	[81]
Benzothiazepine-based compounds		
Diltiazem	cholesterol-fed rabbit	[82]

diseases including hypertension or angina pectoris. Currently, the possibility has been tested of whether the reagents may be effective in slowing the progression of atherosclerosis [4].

Evidence has supported the hypothesis that organic calcium channel blockers suppress the progression of atherosclerosis in animal models (Table 10.8). Henry and Bentley [5] first reported that nifedipine reduces lesion formation in cholesterol-fed rabbits at a dose below that used to lower blood pressure. Several studies confirmed this finding in a cholesterol-fed rabbit model and extended the effect to other calcium channel blockers (Table 10.8). It appears that the effects of the reagents are not associated with significant changes in serum lipid profiles, and are accompanied by reduced aortic calcium contents. Using amlodipine, Kramsch [6] also showed the anti-atherogenic effect of calcium channel blocker in monkey atherosclerosis models. In marked contrast to control animals receiving an atherogenic diet that developed lesions by 100%, almost 50% of the animals receiving amlodipine developed no lesions at all.

Pharmacological Effects of Calcium Channel Blockers on Restenosis

Reperfusion of the occluded coronary artery by percutaneous transluminal coronary angioplasty (PTCA) during the early phase of acute myocardial infarction has been successful in reducing infarction size and preserving left ventricular function of the heart. However, the major limitation of the intervention is restenosis following angioplasty, which was shown to occur in 10–60% of patients after PTCA. Pharmacological interventions have been tested in several trials to prevent restenosis. Drugs such as heparin, platelet-suppressing drugs (i.e., dipyridamole, ticlopidine, prostacyclin, thromboxane receptor blockers), and anti-coagulants have all failed to improve the long-term outcome of PTCA therapy. Cholesterol-lowering interventions, which have been shown to successfully reduce the occurrence of reinfarction and angina

Table 10.9. The effect of calcium channel blockers on restenosis following coronary angioplasty

Drugs	N	Restenosis rate (%) Blocker placebo (95% CI)		Odds ratio	Ref
Diltiazem	92	15	22	0.65 (0.22–1.88)	[83]
Nifedipine	241	28	30	0.93 (0.50–1.71)	[84]
Verapamil	196	47	62	0.56 (0.30–1.02)	[85]
Diltiazem	201	36	36	1.02 (0.48–2.15)	[86]
Diltiazem	189	21	38	0.44 (0.22–0.86)	[87]
Various	919	31	39	0.68 (0.49–0.94)	[88]

pectoris in large multicentre trials, have also failed to reduce the loss of minimal luminal diameter after PTCA.

For calcium channel blockers, five trials were performed in angioplasty patients (Table 10.9 and reviewed in [7]). The results of the individual trials were unconvincing, largely because of small sample sizes. Hillegas et al. [88] undertook a meta-analysis of the combined results. In total, more than 900 patients were enrolled in these trials and angiographic follow-up was performed on average in 82% of the patients. Restenosis was angiographically evident in 22–62% of placebo-treated patients, most commonly observed in the verapamil trial, which included high-risk patients. On average, 39% of placebo-treated patients had restenosis compared with 31% of patients treated with calcium channel blockers (odds ratio of 0.68 with a 95% CI of the combined data of the five trials 0.49–0.94, p = 0.03). Thus, the combined findings of the five trials suggested calcium-channel blockers may be effective on the reduction of restenosis following therapeutic angioplasty.

Pathogenesis of Atherosclerosis: Calcium Channel Blockers Inhibit Multiple Pathways

Even though numerous studies have reported the anti-atherogenic effects of calcium channel blockers in experimental animal models of the disease, the exact mechanism underlying these effects is not necessarily clear. In addition to the effect on retardation of atherosclerosis progression, isladipine [8] and nisoldipine [9] preserve endothelium-dependent vascular relaxation in a cholesterol-fed rabbit model. It is well recognized that endothelial dysfunction is the first step in progression of atherosclerosis [1]. Flow-mediated vasodilation, an endothelium-dependent response [10] based on shear stress-induced release of an endothelium-derived relaxing factor [11], is impaired in the systemic arteries of asymptomatic subjects with hypercholesterolaemia, healthy smokers and patients with coronary artery disease [12, 13], while nitroglycerin-induced, endothelium-independent dilation, primarily reflecting vascular smooth muscle function, is preserved. Other possible causes of endothelial dysfunction include hypertension and diabetes mellitus; genetic alteration; elevated plasma homocysteine concentrations; infectious micro-organisms such as herpes viruses or Chlamydia pneumoniae [1].

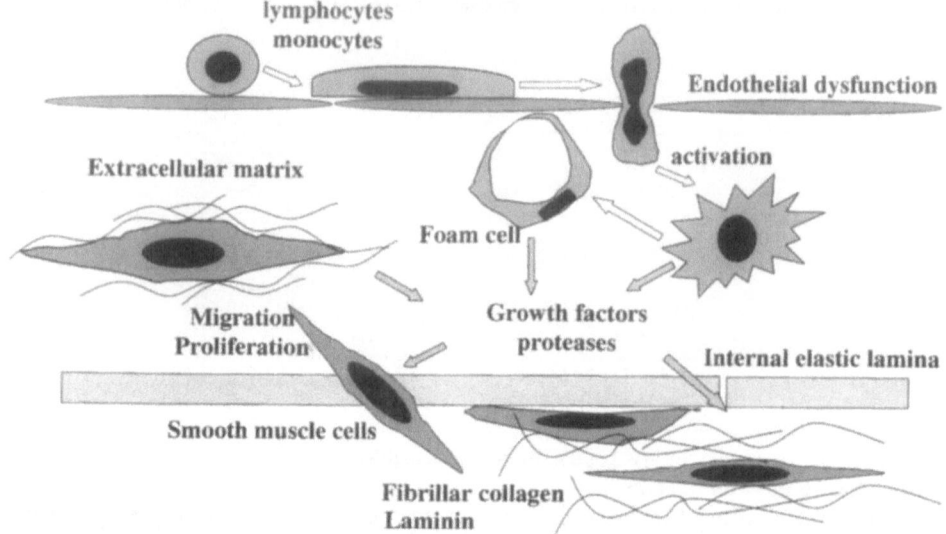

Figure 10.10. Progression of atherosclerosis.

According to the Response-to-Injury Hypothesis [1], the functionally impaired endothelium can initiate the cascade of responses, which eventually leads to progression of atherosclerosis (Figure 10.10). Initially, endothelium increases in adhesiveness with respect to leukocytes or platelets, increases in its permeability, has procoagulant instead of anticoagulant properties, and forms vasoactive molecules, cytokines and growth factors. These inflammatory responses stimulate migration and proliferation of smooth muscle cells to form an intermediate lesion. If these responses continue, they can thicken the arterial wall with compensatory dilatation (remodelling). As for the inflammatory cells, the response is mediated by monocyte-derived macrophages and specific subtypes of T lymphocytes at every stage of atherogenesis. Continued inflammation results in increased numbers of macrophages and lymphocytes, both of which emigrate from the blood and multiply within the lesion. These cells can release proteases, cytokines, chemokines, and growth factors, which can induce further cycles of accumulation of inflammatory cells, migration and proliferation of smooth muscle cells, and accumulation of extracellular matrices. These cycles eventually lead to the formation of a necrotic core, and enlargement and restructuring of the lesion that is known as advanced complicated lesion.

Calcium channel blockers are known to inhibit multiple points in vitro in the above-mentioned pathways during progression of atherosclerosis (Table 10.10). First, calcium channel blockers appear to inhibit lipid accumulation and foam cell formation in subendothelial tissues. A calcium antagonist was shown to reduce U937 monocytic cell movement through the endothelial cell barrier, although it increased binding and uptake of low-density lipoproteins by monocyte and smooth muscle cells [14]. They promote cholesterol efflux from macrophages [15] and human aortic tissue [16]. Calcium antagonists are also shown to inhibit lipid peroxidation in a cell-free system in vitro [17]. Second, the specific role of calcium in cellular proliferation has been suggested in smooth muscle cells [18]. Calcium antagonists were shown to inhibit proliferation [19] and migration [20] of vascular smooth muscle cells. Third,

Table 10.10. Putative mechanisms of calcium channel blockers to inhibit progression of atherosclerosis

Modulation of lipid metabolism and macrophage function	
Inhibit lipid peroxidation	[17]
Inhibit monocyte transmigration through endothelial cells	[14]
Promote cholesterol efflux from macrophages	[15]
Promote cholesterol efflux from human aorta	[16]
Modulation of smooth muscle cell (SMC) function	
Inhibit SMC proliferation (in vitro)	[19]
Inhibit SMC proliferation (in vivo)	[18]
Inhibit SMC migration	[20]
Inhibition of platelet aggregation	[21, 22]

calcium antagonists may also inhibit several platelet functions [21, 22], including Ca^{2+}-dependent processes of adhesion and aggregation and the release of platelet factors, which may participate in atherosclerosis and restenosis following injury.

Bisphosphonates: Potential Drugs to Retard Atherosclerosis

Bisphosphonates are drugs used for the treatment of bone resorption and hyper-calcaemia associated with osteolytic metastatic bone tumours, Paget's disease and hyperparathyroidism. They are currently approved also for the treatment of osteoporosis. Bisphosphonates are known to effectively inactivate bone-resorbing osteoclasts. Bisphosphonates (etidronate, pamidronate and clodronate) have also been found to inhibit the development of experimental atherosclerosis in several animal species without altering serum cholesterol levels [23–26]. Etidronate was also effective in reducing pre-existing atherosclerosis in rabbits and swine [27, 28]. Recently, Koshiyama [29] showed that etidronate retards progression of intimal–medial thickness in carotid artery in type 2 diabetic patients.

At present, it is not fully understood how bisphosphonates retard atherogenesis. Bisphosphonates, especially liposome-encapsulated clodronate, inhibit proinflammatory cytokine secretion from activated macrophages in vitro, and eliminate macrophages from the spleen, liver and lymph nodes. Thus, bisphosphonates may inhibit inflammatory processes, which is now regarded as an important mechanism for progression of atherosclerosis. Bisphosphonates are also known to markedly accumulate in the aorta of healthy and atherosclerotic rabbits, even though they are practically insoluble in lipids [30]. Thus, relatively high drug concentrations in arteries can interact with cells and other structures in the arterial wall, especially with subendothelial atheromatous lesions. In addition, liposome-encapsulated bisphosphonates were recently shown to inhibit the activity of pharocytic cells in internalising and degrading atherogenic modified low-density lipoproteins [31].

Calcium Homeostasis and Atherosclerosis

Some studies have suggested that serum calcium levels may be related to cardiovascular events. Patients with primary hyperparathyroidism are at increased risk for death, particularly from ischaemic heart disease [32, 33], which may be associated with hypercalcaemia [34, 35]. Moreover, necropsy analyses of patients with chronic hypercalcaemia revealed accelerated deposition of calcium not only in myocardial fibres but also in the media and intima of coronary arteries [34]. In rabbits fed on an atherogenic diet, antihypercalcaemic–hyperphosphataemic agents appear to normalise serum calcium levels, but not serum phosphorus or cholesterol levels, and inhibit calcium/phosphorus deposition in aortic tissue in addition to massive fibrous, fatty aortic plaques [36]. In contrast, in an experimental study in young goats, supplemental calcium with normal amounts of vitamin D reduced serum cholesterol and total lipid deposition in the aorta [37]. However, a high intake of vitamin D accompanied by excessive intake of calcium appeared to accelerate the development of atherosclerosis.

The mechanisms involved in the relationship between serum calcium and the atherosclerotic process are not fully understood and require further clarification. The serum calcium concentration, adjusted with serum albumin levels, or calcium influx through voltage-independent calcium channels of red cells, appears to correlate with plasma high-density lipoprotein levels [38, 39]. Total serum calcium has also been positively correlated with systolic and diastolic blood pressure in some studies [40, 41] but not in others [42]. Recently, it was reported that hypercalcaemia was associated with impairment of endothelium-independent vasodilation in patients with primary hyperparathyroidism [43]. They also reported that the altered vasoreactivity in these patients was reversible following parathyroidectomy [44]. There is evidence that vascular smooth muscle cells [45], as well as cardiac myocytes [46], are targets for parathyroid hormone. PTH increases receptor-mediated $(Ca^{2+})i$ of cardiac myocytes, which is due to augmented entry of calcium into myocytes as well as to mobilization of calcium from the sarcoplasmatic reticulum by a calcium-induced calcium release mechanism [46]. As a result of this calcium overload, myocytes may get damaged [47]. Parathyroid hormone is also known to be a vasodilator [48]. A functional endothelium is not required for this relaxing effect [49] but increases in intracellular cyclic AMP concentrations were temporally and quantitatively correlated with PTH-induced smooth muscle relaxation [45]. Calcitonin, another calcium-regulating hormone secreted from C cells in the thyroid, suppresses serum cholesterol and triglycerides through a calcium/calmodulin-mediated mechanism [50], and may also be involved in progression of atherosclerosis. The role of vitamin D in the progression of atherosclerosis is not clear at present. The active vitamin D metabolite, 1,25-dihydroxyvitamin D, which is serially 25 and 1α-hydroxylated in the liver and kidney, respectively, appears to have diverse functions in vascular cells. Active vitamin D_3 suppresses proliferation [51, 52] and promotes prostacyclin production in vascular smooth muscle cells in vitro [53], which may be protective against progression of atherosclerosis. In contrast, we showed that active vitamin D_3 induces foam cell formation from THP-1 monocytic cells through up-regulation of a scavenger receptor [54]. In the rat, pharmacological amounts of vitamin D were reported to induce proliferation of smooth muscle cells without endothelial damage [55]. In a calcification model system in vitro, we showed that active vitamin D_3 stimulates calcification of vascular smooth muscle cells [56].

There is very little information on the association between serum phosphorus levels and the severity of atherosclerosis. Low serum calcium and high serum phosphorus levels are often seen in patients with chronic renal failure. The prevalence of coronary artery disease is increased in patients with chronic renal insufficiency, including those being treated with dialysis. Although hypertension is common in patients with renal failure, the excess risk of coronary artery disease in these patients does not appear to be fully explained by conventional risk factors. We examined the levels of atherosclerosis by arterial ultrasound in patients undergoing haemodialysis and found that serum phosphorus levels, in addition to aging and smoking, were independently associated with carotid atherosclerosis [57]. Thus, it is possible that calcium homeostasis is involved in the progression of atherosclerosis, particularly in this predisposition.

Calcium and Coronary Artery Atherosclerosis

Coronary calcium deposits have been widely regarded to result from a passive process of encrustation or adsorption of minerals onto advanced atherosclerotic lesions. Nevertheless, many investigators recognised that non-invasive imaging of coronary calcium might be useful in identifying patients with unsuspected coronary artery disease. Although calcification is found more frequently in advanced lesions, it may also occur in small amounts in earlier lesions. In an in vitro investigation, Mautner et al. [58] examined 4,298 sections of 3-mm thickness of coronary arteries obtained from >50 heart specimens. Calcific deposits measured at histomorphometry were seen most frequently in the proximal three to four centimetres of the left anterior descending and circumflex coronary arteries and in the proximal right coronary artery with a second, less striking peak, of calcium more distally in the right coronary artery. The bifurcation of the left coronary artery, the orifice of the first diagonal branch, and the proximal right coronary artery have been described as predirection sites of lesion formation.

Imaging Methods

Realization of this finding has generated the need to determine clinically useful threshold levels of coronary calcium content to make appropriate management decisions. Coronary artery calcification is potentially detectable in vivo by the following methods: plain-film roentgenography; coronary arteriography; fluoroscopy; cine-fluorography; conventional, helical-, electron beam-computed tomography (EBCT); intravascular ultrasound; magnetic resonance imaging; and transthoracic and trans-esophageal echocardiography. Fluoroscopy, electron beam-, and helical-computed tomography can identify calcific deposits; EBCT and, to a lesser extent, double-helical CT have the enhanced capability to localize coronary calcification and detect smaller and less dense calcific deposits.

Fluoroscopy has frequently been used to detect calcification in coronary arteries. Sensitivity in detecting significant stenoses in coronary arteries (greater than 50% diameter obstruction) ranged from 40–79% with specificity ranging from 52% to 95% [59]. In a study of 613 asymptomatic subjects who underwent coronary arteriography because of one or more abnormal screening tests [60], coronary artery calcification had a 66.3% sensitivity and a 77.6% specificity in detecting significant coronary artery stenoses (greater than 50%-diameter narrowing). Although fluoroscopy detects moderate to large calcifications, its ability to identify small calcific deposits is low.

Helical CT has considerably faster scan times than conventional CT. Imaging by helical CT is reported to have sensitivity of 91% and a specificity of 52% detecting angiographically significant coronary obstructive diseases [61]. Even though there is an argument that even at accelerated scan times, especially with single helical CT,

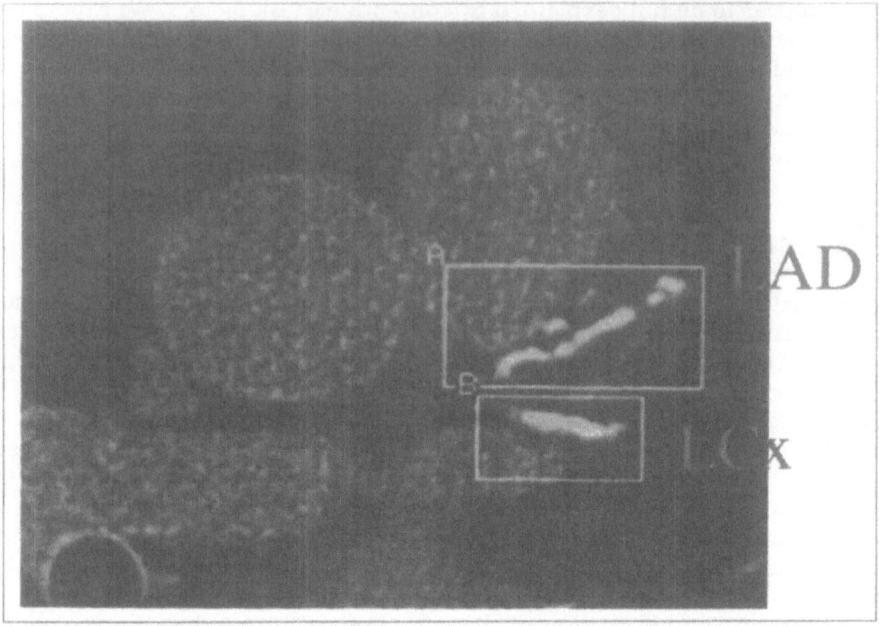

Figure 10.11. Detection of coronary calcium with electron beam computed tomography (EBCT). Typical image of calcification in left coronary artery detected by EBCT. Dr. Masayuki Hosoi, Osaka City General Hospital, provided the image.

cardiac deposits are blurred due to cardiac movement, and small calcifications may not be seen, helical CT remains superior to fluoroscopy and conventional CT in detecting calcification. Double helical CT appears to be more sensitive than single helical scanners in detection of coronary calcification.

EBCT uses an electron gun and a stationary tungsten "target" rather than a standard X-ray tube to generate X-rays, permitting very rapid scanning times. For purposes of detecting coronary calcium, EBCT images are obtained at 100 ms with a scan slice thickness of 3 mm. Thirty to forty adjacent axial scans are obtained. The scans are triggered by electrocardiac signals to minimise the effect of cardiac motion. The rapid image acquisition time virtually eliminates motion artifacts related to cardiac motion. EBCT is considered the non-invasive standard method for delineating and quantifying coronary calcium [62]. Among the methods described above, only EBCT can quantitate the amount or volume of calcium (Figure 10.11). EBCT is attractive due to the ease of measurement of coronary calcium compared with other techniques. Considerable previous evidence suggests that the extent of coronary calcification is strongly related to plaque burden [63]. Moreover, in symptomatic patients, EBCT appears to be moderately useful in predicting the severity of coronary artery diseases with sensitivities and specificities comparable with that of exercise testing [62]. Comparison of calcium detection by EBCT with angiographic findings of significant stenosis (more than 50% narrowing) on coronary arteries reveals sensitivities ranging from 85–100% and specificities of 41–76%. Thus, the absence of calcific deposits on an EBCT scan implies the absence of significantly angiographic coronary narrowing; however, it does not imply the absence of atherosclerosis, including unstable plaque. Similarly, calcification may frequently be seen in the absence of significant angiographic narrowing and before there has been sufficient plaque build-up to narrow the vessel. Adult females tend to have low scores or negative scans before menopause. The greater the amount of calcification, the greater the likelihood of obstructive disease, but there is no one-to-one relation, and the findings are not site specific.

Event Prediction with EBCT Coronary Calcium Scores

Some prospective studies are now available, which examined the usefulness of quantitation of coronary calcium by EBCT for prediction of cardiac events. From the St. Francis Medical Center and the South Bay Heart Watch Program [64, 65], the prognostic value of EBCT was reported using an array of combined end points, including cardiac death, acute coronary syndrome and coronary revascularization, which turned out to be controversial. Arad et al. [65] showed that in 1,172 asymptomatic subjects with an average follow-up of 3.6 years, 39 coronary events occurred: three coronary deaths, 15 nonfatal myocardial infarctions and 21 coronary artery revascularisation procedures. In that study, the mean coronary artery calcium score was 764 ± 935 among subjects with events, compared with 135 ± 432 among those without events (p < 0.0001). For the prediction of all coronary events and of nonfatal myocardial infarction and deaths, the areas under the receiver–operator characteristics curve were 0.84 and 0.86, respectively, and a coronary calcium score larger than 160 was associated with odds ratios of 15.8 and 22.2, respectively. Although using three different endpoints for cardiac events can be controversial, EBCT of the coronary arteries in asymptomatic adults appears to predict coronary events.

In contrast, Detrano et al. [64] failed to show any benefit to using coronary calcium score to predict cardiovascular events. They used 1,196 asymptomatic high-coronary-risk subjects with 41 months follow up, and applied the Framingham model and their data-derived risk model to determine the three-year likelihood of a coronary event. Using EBCT, 818 subjects had detectable coronary calcium. Seventeen coronary deaths (1.4%) and 29 nonfatal myocardial infarctions (2.4%) occurred. They showed that the receiver–operator characteristic curve areas calculated from the Framingham model, their data-derived risk model, and the calcium score were 0.69 ± 0.05, 0.68 ± 0.05, and 0.64 ± 0.05, respectively (n.s.). When the calcium score was included as a variable in the data-derived model, the receiver–operator characteristics area did not change significantly (0.68+/−0.05 to 0.71+/−0.04; P = NS). The same group [66] recently reported the modest usefulness of EBCT determination of coronary artery calcium (CAC) in 926 asymptomatic persons with a follow-up questionnaire two to four years (mean 3.3) after scanning, inquiring about myocardial infarction, stroke, and revascularization. They showed that persons with scores at or above the median had a relative risk of 4.5 (p < 0.01) for new events. Thus, in an asymptomatic population, assessment of coronary calcium could be added to other risk predictors of future cardiovascular events. However, the current finding may not be adequate to support the use of EBCT to screen asymptomatic populations, and more prospective findings need to be accumulated.

Coronary Calcium and Plaque Rupture

Compelling evidence has implicated plaque rupture and subsequent thrombosis as the pathophysiologic event precipitating the clinical syndromes of unstable angina, myocardial infarction, and sudden death. Plaque volume or mass appears to have little bearing on the likelihood of rupture. Plaque structure and composition appears to be a major determinant of the risk of plaque rupture; ruptured plaque contains high lipid and macrophages, less collagen, smooth muscle cells and calcium [67]. Calcium deposits may tend to impart stability to an atherosclerotic lesion and decrease the probability of plaque rupture [68, 69]. In contrast, other studies claimed that the presence of calcified plaques implies the likely association of lipid-rich and possible unstable plaque. Demer et al. [70] argued that the presence of a soft plaque, with a point of weakness induced by an inflammation adjacent to an area of calcification, predisposes the plaque to rupture because of the presence of a tissue interface of differing physical properties that is subjected to the pulsatile changes in arterial pressure.

Calcium Deposition in an Aortic Valve as a Window for Systemic Atherosclerosis

Experimentally induced systemic arterial atherosclerosis is associated with the deposition of fatty plaques on the aortic surface of the aortic valve cusps. In a necropsy study of persons aged >65 years, it was shown that 100% of those with aortic valve calcium or mitral annulus calcification had calcific deposits in one coronary artery [71]. These findings suggest that coronary atherosclerosis, aortic valve calcium, and mitral annulus calcification in the elderly have a similar etiology. Boon et al. [72] reported that age, hypertension, diabetes mellitus, and hypercholesterolaemia were significantly associated with aortic valve calcium and mitral annulus calcification, suggesting they should be considered as manifestations of generalised atherosclerosis. Several studies showed a highly significant association between mitral annulus calcification and general atherosclerosis [73, 74].

The association between aortic valve calcium and stroke is still controversial. Boon et al. [75] prospectively analysed the occurrence of stroke, stroke subtypes, and concomitant cardiovascular risk factors in 300 patients with echocardiographic evidence of aortic valve calcium, 515 patients with aortic valve calcium and stenosis, and 562 control subjects. They concluded that aortic valve calcium with or without stenosis was not a risk factor for stroke. Otto et al. [76] reported on 5,621 males and females, in 29% of whom aortic valve calcium was detected by echocardiography, who were enrolled in a population-based prospective study. The risk of death from any cause and from cardiovascular causes was assessed after a mean of five years. They showed that aortic valve calcium is common in the elderly and is associated with increased risk of cardiovascular death, myocardial infarction, and stroke.

References

1. Ross R. Atherosclerosis – an inflammatory disease. N Engl J Med 1999;340:115–26.
2. Stary HC, Chandler AB, Dinsmore RE, Fuster V, Glagov S, Insull Jr W, et al. A definition of advanced types of atherosclerotic lesions and a histological classification of atherosclerosis. A report from the Committee on Vascular Lesions of the Council on Arteriosclerosis, American Heart Association. Circulation 1995;92:1355–74.
3. Schinke T, McKee MD, Kiviranta R, Karsenty G. Molecular determinants of arterial calcification. Ann Med 1998;30:538–41.
4. Nayler WG. Review of preclinical data of calcium channel blockers and atherosclerosis. J Cardiovasc Pharmacol 1999;33:S7–11.

5. Henry PD, Bentley KI. Suppression of atherogenesis in cholesterol-fed rabbits treated with nifedipine. J Clin Invest 1981;68:1366–9.
6. Kramsch DM, Sharma RC. Limits of lipid-lowering therapy: the benefits of amlodipine as an anti-atherosclerotic agent. J Hum Hypertens 1995;9 Suppl 1:S3–9.
7. Thaulow E. Pharmacologic effects of calcium channel blockers on restenosis. J Cardiovasc Pharmacol 1999;33:S12–16.
8. Habib JB, Bossaller C, Wells S, Williams C, Morrisett JD, Henry PD. Preservation of endothelium-dependent vascular relaxation in cholesterol-fed rabbits by treatment with the calcium blocker PN 200110. Circ Res 1986;58:305–9.
9. Kappagoda CT, Thomson AB, Senaratne MP. Effect of nisoldipine on atherosclerosis in the cholesterol fed rabbit: endothelium-dependent relaxation and aortic cholesterol content. Cardiovasc Res 1991;25:270–82.
10. Rubanyi GM, Romero JC, Vanhoutte PM. Flow-induced release of endothelium-derived relaxing factor. Am J Physiol 1986;250:H1145–9.
11. Joannides R, Haefeli WE, Linder L, Richard V, Bakkali EH, Thuillez C, et al. Nitric oxide is responsible for flow-dependent dilatation of human peripheral conduit arteries in vivo. Circulation 1995;91:1314–19.
12. Celermajer DS, Sorensen KE, Gooch VM, Spiegelhalter DJ, Miller OI, Sullivan ID, et al. Non-invasive detection of endothelial dysfunction in children and adults at risk of atherosclerosis. Lancet 1992;340:1111–15.
13. Celermajer DS, Sorensen KE, Georgakopoulos D, Bull C, Thomas O, Robinson J, et al. Cigarette smoking is associated with dose-related and potentially reversible impairment of endothelium-dependent dilation in healthy young adults. Circulation 1993;88:2149–55.
14. Alexander JJ, Miguel R, Piotrowski JJ. The effect of nifedipine on lipid and monocyte infiltration of the subendothelial space. J Vasc Surg 1993;17:841–7; discussion 847–8.
15. Schmitz G, Robenek H, Beuck M, Krause R, Schurek A, Niemann R. Ca^{++} antagonists and ACAT inhibitors promote cholesterol efflux from macrophages by different mechanisms. I. Characterization of cellular lipid metabolism. Arteriosclerosis 1988;8:46–56.
16. Etingin OR, Hajjar DP. Calcium channel blockers enhance cholesteryl ester hydrolysis and decrease total cholesterol accumulation in human aortic tissue. Circ Res 1990;66:185–90.
17. Ondrias K, Misik V, Gergel D, Stasko A. Lipid peroxidation of phosphatidylcholine liposomes depressed by the calcium channel blockers nifedipine and verapamil and by the antiarrhythmic–antihypoxic drug stobadine. Biochim Biophys Acta 1989;1003:238–45.
18. Jackson CL, Bush RC, Bowyer DE. Inhibitory effect of calcium antagonists on balloon catheter-induced arterial smooth muscle cell proliferation and lesion size. Atherosclerosis 1988;69:115–22.
19. Nilsson J, Sjolund M, Palmberg L, Von Euler AM, Jonzon B, Thyberg J. The calcium antagonist nifedipine inhibits arterial smooth muscle cell proliferation. Atherosclerosis 1985;58:109–22.
20. Nomoto A, Mutoh S, Hagihara H, Yamaguchi I. Smooth muscle cell migration induced by inflammatory cell products and its inhibition by a potent calcium antagonist, nilvadipine. Atherosclerosis 1988;72:213–19.
21. Pales J, Palacios-Araus L, Lopez A, Gual A. Effects of dihydropyridines and inorganic calcium blockers on aggregation and on intracellular free calcium in platelets. Biochim Biophys Acta 1991;1064:169–74.
22. Zucker ML, Budd SE, Dollar LE, Chernoff SB, Altman R. Effect of diltiazem and low-dose aspirin on platelet aggregation and ATP release induced by paired agonists. Thromb Haemost 1993;70:332–5.
23. Rosenblum IY, Flora L, Eisenstein R. The effect of disodium ethane-1-hydroxy-1,1-diphosphonate (EHDP) on a rabbit model of athero-arteriosclerosis. Atherosclerosis 1975;22:411–24.
24. Kramsch DM, Chan CT. The effect of agents interfering with soft tissue calcification and cell proliferation on calcific fibrous-fatty plaques in rabbits. Circ Res 1978;42:562–71.
25. Kramsch DM, Aspen AJ, Rozler LJ. Atherosclerosis: Prevention by agents not affecting abnormal levels of blood lipids. Science 1981;213:1511–12.
26. Ylitalo R, Oksala O, Yla-Herttuala S, Ylitalo P. Effects of clodronate (dichloromethylene bisphosphonate) on the development of experimental atherosclerosis in rabbits. J Lab Clin Med 1994;123:769–76.
27. Hollander W, Paddock J, Nagraj S, Colombo M, Kirkpatrick B. Effects of anticalcifying and antifibrobrotic drugs on pre-established atherosclerosis in the rabbit. Atherosclerosis 1979;33:111–23.
28. Daoud AS, Frank AS, Jarmolych J, Fritz KE. The effect of ethane-1-hydroxy-1,1-diphosphonate (EHDP) on necrosis of atherosclerotic lesions. Atherosclerosis 1987;67:41–8.
29. Koshiyama H, Nakamura Y, Tanaka S, Minamikawa J. Decrease in carotid intima-media thickness after 1-year therapy with etidronate for osteopenia associated with type 2 diabetes. J Clin Endocrinol Metab 2000;85:2793–6.

30. Ylitalo R, Monkkonen J, Urtti A, Ylitalo P. Accumulation of bisphosphonates in the aorta and some other tissues of healthy and atherosclerotic rabbits. J Lab Clin Med 1996;127:200–6.
31. Ylitalo R, Monkkonen J, Yla-Herttuala S. Effects of liposome-encapsulated bisphosphonates on acety-lated LDL metabolism, lipid accumulation and viability of phagocyting cells. Life Sci 1998;62:413–22.
32. Palmer M, Adami HO, Bergstrom R, Akerstrom G, Ljunghall S. Mortality after surgery for primary hyperparathyroidism: a follow-up of 441 patients operated on from 1956 to 1979. Surgery 1987; 102:1–7.
33. Niederle B, Roka R, Woloszczuk W, Klaushofer K, Kovarik J, Schernthaner G. Successful parathy-roidectomy in primary hyperparathyroidism: a clinical follow-up study of 212 consecutive patients. Surgery 1987;102:903–9.
34. Roberts WC, Waller BF. Effect of chronic hypercalcemia on the heart. An analysis of 18 necropsy patients. Am J Med 1981;71:371–84.
35. Palmer M, Adami HO, Bergstrom R, Jakobsson S, Akerstrom G, Ljunghall S. Survival and renal func-tion in untreated hypercalcaemia. Population-based cohort study with 14 years of follow-up. Lancet 1987;1:59–62.
36. Chan CT, Wells H, Kramsch DM. Suppression of calcific fibrous-fatty plaque formation in rabbits by agents not affecting elevated serum cholesterol levels. The effect of thiophene compounds. Circ Res 1978;43:115–25.
37. Hines TG, Jacobson NL, Beitz DC, Littledike ET. Dietary calcium and vitamin D: risk factors in the development of atherosclerosis in young goats. J Nutr 1985;115:167–78.
38. Stimpel M, Neyses L, Locher R, Knorr M, Vetter W. High density lipoproteins – modulators of the calcium channel? J Hypertens 1985;Suppl 3:S49–51.
39. De Bacquer D, De Henauw S, De Backer G, Kornitzer M. Epidemiological evidence for an association between serum calcium and serum lipids. Atherosclerosis 1994;108:193–200.
40. Klatsky AL, Friedman GD, Armstrong MA. The relationships between alcoholic beverage use and other traits to blood pressure: a new Kaiser Permanente study. Circulation 1986;73:628–36.
41. Staessen J, Sartor F, Roels H, Bulpitt CJ, Claeys F, Ducoffre G, et al. The association between blood pres-sure, calcium and other divalent cations: a population study. J Hum Hypertens 1991;5:485–94.
42. Cooper RS, Shamsi N. Ionized serum calcium in black hypertensives: absence of a relationship with blood pressure. J Clin Hypertens 1987;3:514–19.
43. Neunteufl T, Katzenschlager R, Abela C, Kostner K, Niederle B, Weidinger F, et al. Impairment of endothelium-independent vasodilation in patients with hypercalcemia. Cardiovasc Res 1988;40: 396–401.
44. Neunteufl T, Heher S, Prager G, Katzenschlager R, Abela C, Niederle B, et al. Effects of successful parathyroidectomy on altered arterial reactivity in patients with hypercalcaemia: results of a 3-year follow-up study. Clin Endocrinol (Oxf) 2000;53:229–33.
45. Nickols GA. Actions of parathyroid hormone in the cardiovascular system. Blood Vessels 1987;24:120–4.
46. Smogorzewski M, Zayed M, Zhang YB, Roe J, Massry SG. Parathyroid hormone increases cytosolic calcium concentration in adult rat cardiac myocytes. Am J Physiol 1993;264:H1998–2006.
47. Bogin E, Massry SG, Harary I. Effect of parathyroid hormone on rat heart cells. J Clin Invest 1981;67:1215–27.
48. Saglikes Y, Massry SG, Iseki K, Nadler JL, Campese VM. Effect of PTH on blood pressure and response to vasoconstrictor agonists. Am J Physiol 1985;248:F674–81.
49. Nickols GA, Nana AD, Nickols MA, DiPette DJ, Asimakis GK. Hypotension and cardiac stimulation due to the parathyroid hormone-related protein, humoral hypercalcemia of malignancy factor. Endocrinology 1989;125:834–41.
50. Nishizawa Y, Okui Y, Inaba M, Okuno S, Yukioka K, Miki T, et al. Calcium/calmodulin-mediated action of calcitonin on lipid metabolism in rats. J Clin Invest 1988;82:1165–72.
51. Carthy EP, Yamashita W, Hsu A, Ooi BS. 1,25-Dihydroxyvitamin D3 and rat vascular smooth muscle cell growth. Hypertension 1989;13:954–9.
52. Mitsuhashi T, Morris Jr RC, Ives HE. 1,25-dihydroxyvitamin D3 modulates growth of vascular smooth muscle cells. J Clin Invest 1991;87:1889–95.
53. Wakasugi M, Noguchi T, Inoue M, Kazama Y, Tawata M, Kanemaru Y, et al. Vitamin D3 stimulates the production of prostacyclin by vascular smooth muscle cells. Prostaglandins 1991;42:127–36.
54. Suematsu Y, Nishizawa Y, Shioi A, Hino M, Tahara H, Inaba M, et al. Effect of 1,25-dihydroxyvitamin D3 on induction of scavenger receptor and differentiation of 12-O-tetradecanoylphorbol-13-acetate-treated THP-1 human monocyte-like cells. J Cell Physiol 1995;165:547–55.
55. Mohtai M, Yamamoto T. Smooth muscle cell proliferation in the rat coronary artery induced by vitamin D. Atherosclerosis 1987;63:193–202.

56. Jono S, Nishizawa Y, Shioi A, Morii H. 1,25-Dihydroxyvitamin D3 increases in vitro vascular calcification by modulating secretion of endogenous parathyroid hormone-related peptide. Circulation 1998;98:1302–6.
57. Kawagishi T, Nishizawa Y, Konishi T, Kawasaki K, Emoto M, Shoji T, et al. High-resolution B-mode ultrasonography in evaluation of atherosclerosis in uremia. Kidney Int 1995;48:820–6.
58. Mautner GC, Mautner SL, Froehlich J, Feuerstein IM, Proschan MA, Roberts WC, et al. Coronary artery calcification: assessment with electron beam CT and histomorphometric correlation. Radiology 1994;192:619–23.
59. Detrano R, Froelicher V. A logical approach to screening for coronary artery disease. Ann Intern Med 1987;106:846–52.
60. Loecker TH, Schwartz RS, Cotta CW, Hickman Jr JR. Fluoroscopic coronary artery calcification and associated coronary disease in asymptomatic young men. J Am Coll Cardiol 1992;19:1167–72.
61. Shemesh J, Apter S, Rozenman J, Lusky A, Rath S, Itzchak ••, et al. Calcification of coronary arteries: detection and quantification with double-helix CT. Radiology 1995;197:779–83.
62. Wexler L, Brundage B, Crouse J, Detrano R, Fuster V, Maddahi J, et al. Coronary artery calcification: pathophysiology, epidemiology, imaging methods, and clinical implications. A statement for health professionals from the American Heart Association. Writing Group. Circulation 1996;94:1175–92.
63. Rumberger JA, Simons DB, Fitzpatrick LA, Sheedy PF, Schwartz RS. Coronary artery calcium area by electron-beam computed tomography and coronary atherosclerotic plaque area. A histopathologic correlative study. Circulation 1995;92:2157–62.
64. Detrano RC, Wong ND, Doherty TM, Shavelle RM, Tang W, Ginzton LE, et al. Coronary calcium does not accurately predict near-term future coronary events in high-risk adults. Circulation 1999;99:2633–8.
65. Arad Y, Spadaro LA, Goodman K, Newstein D, Guerci AD. Prediction of coronary events with electron beam computed tomography. J Am Coll Cardiol 2000;36:1253–60.
66. Wong ND, Hsu JC, Detrano RC, Diamond G, Eisenberg H, Gardin JM. Coronary artery calcium evaluation by electron beam computed tomography and its relation to new cardiovascular events. Am J Cardiol 2000;86:495–8.
67. Libby P. Molecular bases of the acute coronary syndromes. Circulation 1995;91:2844–50.
68. Cheng GC, Loree HM, Kamm RD, Fishbein MC, Lee RT. Distribution of circumferential stress in ruptured and stable atherosclerotic lesions. A structural analysis with histopathological correlation. Circulation 1993;87:1179–87.
69. Doherty TM, Detrano RC. Coronary arterial calcification as an active process: a new perspective on an old problem. Calcif Tissue Int 1994;54:224–30.
70. Demer LL, Watson KE, Bostrom K. Mechanism of calcification in atherosclerosis. Trends Cardiovasc Med 1994;4:45–49.
71. Roberts WC. Morphologic features of the normal and abnormal mitral valve. Am J Cardiol 1983;51:1005–28.
72. Boon A, Cheriex E, Lodder J, Kessels F. Cardiac valve calcification: characteristics of patients with calcification of the mitral annulus or aortic valve. Heart 1997;78:472–4.
73. Adler Y, Herz I, Vaturi M, Fusman R, Shohat-Zabarski R, Fink N, et al. Mitral annular calcium detected by transthoracic echocardiography is a marker for high prevalence and severity of coronary artery disease in patients undergoing coronary angiography. Am J Cardiol 1998;82:1183–6.
74. Adler Y, Koren A, Fink N, Tanne D, Fusman R, Assali A, et al. Association between mitral annulus calcification and carotid atherosclerotic disease. Stroke 1998;29:1833–7.
75. Boon A, Lodder J, Cheriex E, Kessels F. Risk of stroke in a cohort of 815 patients with calcification of the aortic valve with or without stenosis. Stroke 1996;27:847–51.
76. Otto CM, Lind BK, Kitzman DW, Gersh BJ, Siscovick DS. Association of aortic-valve sclerosis with cardiovascular mortality and morbidity in the elderly. N Engl J Med 1999;341:142–7.
77. Nayler WG. The antiatherogenic effects of amlodipine: promise of preclinical data. J Hum Hypertens 1992;6 Suppl 1:S19–23.
78. Willis AL, Nagel B, Churchill V, Whyte MA, Smith DL, Mahmud I, et al. Antiatherosclerotic effects of nicardipine and nifedipine in cholesterol-fed rabbits. Arteriosclerosis 1985;5:250–5.
79. Nomoto A, Hirosumi J, Sekiguchi C, Mutoh S, Yamaguchi I, Aoki H. Antiatherogenic activity of FR34235 (Nilvadipine), a new potent calcium antagonist. Effect on cuff-induced intimal thickening of rabbit carotid artery. Atherosclerosis 1987;64:255–61.
80. Blumlein SL, Sievers R, Kidd P, Parmley WW. Mechanism of protection from atherosclerosis by verapamil in the cholesterol-fed rabbit. Am J Cardiol 1984;54:884–9.
81. Hansen BF, Mortensen A, Hansen JF, Frandsen H. (-)-anipamil retards atherosclerosis in Watanabe heritable hyperlipidemic rabbits. J Cardiovasc Pharmacol 1995;26:485–9.

82. Sugano M, Nakashima Y, Matsushima T, Takahara K, Takasugi M, Kuroiwa A, et al. Suppression of atherosclerosis in cholesterol-fed rabbits by diltiazem injection. Arteriosclerosis 1986;6:237–41.
83. Corcos T, David PR, Val PG, Renkin J, Dangoisse V, Rapold HG, et al. Failure of diltiazem to prevent restenosis after percutaneous transluminal coronary angioplasty. Am Heart J 1985;109:926–31.
84. Whitworth HB, Roubin GS, Hollman J, Meier B, Leimgruber PP, Douglas Jr JS, et al. Effect of nifedipine on recurrent stenosis after percutaneous transluminal coronary angioplasty. J Am Coll Cardiol 1986;8:1271–6.
85. Hoberg E, Schwartz F, Schmig A. Prevention of restenosis by verapamil in the Verapamil Angioplasty Study (VAS) [abstract]. Circulation 1990;82 (Suppl 3):III–428.
86. O'Keefe Jr JH, Giorgi LV, Hartzler GO, Good TH, Ligon RW, Webb DL, et al. Effects of diltiazem on complications and restenosis after coronary angioplasty. Am J Cardiol 1991;67:373–6.
87. Unverdorben M, Kunkel B, Leucht M, Bachman K. Reduction of restenosis after PTCA by diltiazem [abstract]. Circulation 1992;86:1–53.
88. Hillegass WB, Ohman EM, Leimberger JD, Califf RM. A meta-analysis of randomized trials of calcium antagonists to reduce restenosis after coronary angioplasty. Am J Cardiol 1994;73:835–9.

Central Nervous System

M. Smogorzewski

Introduction

Calcium ion (Ca^{2+}) is an omnipresent intracellular messenger controlling multiple cell functions such as growth, differentiation, membrane permeability and exocytosis, synaptic activity and gene regulation. In neuronal cells, resting levels of intracellular calcium ($[Ca^{2+}]i$) are around 50–200 nM, 10^4 times lower than extracellular Ca^{2+}, which allows them to achieve a high ratio of signal to resting/background $[Ca^{2+}]i$ when $[Ca^{2+}]i$ is suddenly increased. Thus, relatively small, transient and localised increases in $[Ca^{2+}]i$ can induce a physiological response by activation of enzymes, change in membrane channels activity, neurotransmitter release modulation of synaptic transmission, programmed cell death through apoptosis and alteration of gene expression. Under normal conditions, the delicate balance between Ca^{2+} influx, Ca^{2+} buffering, intracellular Ca^{2+} storage and Ca^{2+} efflux is maintained, preserving wide options for Ca^{2+} signalling. However, it is currently established that excessive Ca^{2+} entry or inadequate buffering mechanisms can lead to acute overactivation of neurons or chronic neurotoxicity. The precise mechanisms by which neurotoxicity occurs are still not completely understood, despite the research efforts of the past 30 years.

This chapter will briefly discuss the role of calcium in the physiological functions of the brain and address issues of calcium neurotoxicity in uraemia, brain degeneration and Alzheimer's disease, ischaemia of the brain and AIDS dementia complex.

Role of Calcium in Normal Brain Function

Brain function requires several types of cellular elements integrated by physiological and biochemical processes. The neuron is the principal player in intracellular and intercellular information transfer or signalling. This highly polarised (differentiated) cell is typically composed from the cell body, axon and multiple dendrites. The neuron's body contains major cytoplasmic components and nucleus, and generates a single axon, which usually extends over long distances and transfers information between the cell body and effector cells such as other neurons or muscles. The majority of axons are surrounded by an insulating myelin sheath, which facilitates electrical impulse conduction. In addition, multiple dendrites emanate from the cell body, giving rise to a dense network of processes known as dendritic trees. Dendrites extend over the surface of the body and they primarily receive information from outside of the cell but also, depending on neuron specificity, may transmit electrical

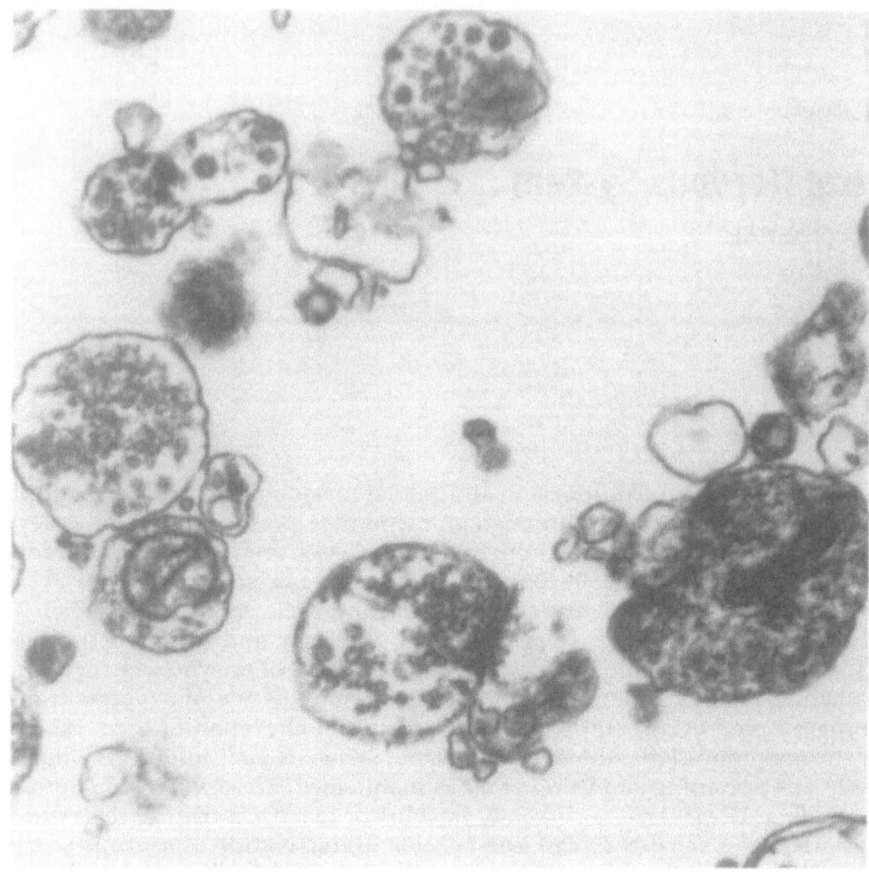

Figure 10.12. An electron micrograph of a section of the synaptosomal pellet (×21,000). Notice well-preserved membranes, multitude of vesicles and mitochondria presence in synaptosomes.

signals. The axon terminal area (synaptic terminal) is responsible for transmission of excitatory or inhibitory inputs between neurons and between neuron and other specialised tissue (e.g., muscle). Synaptic transmission is either chemical or electrical in nature by releasing specific neurotransmitters (chemical) or through gap junction (electrical). These two modes of neuronal connection exist side by side in the brain. Chemical synapses occur in areas where two neurons come within 20–50 nm of each other. Each synapse contains an axon ending, which forms presynaptic elements and postsynaptic elements, where transmitter receptors are localised at specialised regions of axons, dendrites or cell bodies.

Presynaptic nerve terminals are filled with vesicles containing either single or multiple neurotransmitter substances [1]. Detached synapses, called synaptosomes, can be generated from brain nerve terminals and varicosities by application of shear forces (Figure 10.12). Synaptosomes undergo active metabolism utilising oxygen and glucose, extrude Na$^+$, accumulate K$^+$, maintain normal membrane potential and, during depolarisation, release neurotransmitters in a Ca^{2+}-dependent manner. They provide an excellent preparation [2, 3, 4] to study in vitro ion transport and neurotransmitter metabolism in various conditions including chronic renal failure without complications encountered with synapses in in vivo conditions, and are used in in vitro studies as a surrogate of synapse.

Neuronal function is supported and maintained by several classes of glial cells such as astrocytes and oligodendrocytes in the brain or Schwann cells in the peripheral nervous system. Glia maintain extracellular milieu, act as scaffolding for neuronal migration and axon outgrowth, and participate in metabolism of the neurotransmitters and scavenging of cellular debris.

Sources of Intracellular Calcium

The concentration of resting intracellular free Ca^{2+} in most neurons is between 50 and 200 nM. At the same time, extracellular calcium concentration is 10^4 higher (1–1.25 mM). As mentioned before, Ca^{2+} homeostasis is achieved by Ca^{2+} influx, Ca^{2+} intracellular buffering and Ca^{2+} movement into storage, and Ca^{2+} efflux (Figure 10.13).

The best-characterised ways of neuronal calcium entry are voltage-dependent (gated) calcium channels (VDCC). These plasma membrane proteins are pores that allow selective movement of calcium from extracellular fluids into cytoplasm. A single opening of a Ca^{2+} channel can allow the flow of many thousands of Ca^{2+} ions resulting in a rise in $[Ca^{2+}]i$. [5]. Various subtypes of VDCC were found in brain tissue and, based on their voltage- and time-dependence and pharmacological properties, are classified as T-, L-, N-, P-, Q- and R-type. L-type channels are high-voltage activated and strongly sensitive to low concentrations of dihyropyridines (DHP) such as nifedipine. These channels are involved in neurotransmitter release and Ca^{2+} signal regulation of gene expression. T-type channels are less sensitive to DHP and show slow deactivation following a sudden repolarisation. They are mainly distributed in dendrites. N-type Ca^{2+} channels are insensitive to DHP, but are high-voltage activated, similar to L-type channels, and are primarily found in presynaptic nerve terminals, dendrites and cell bodies. L-type and N-type channels are involved in neurotransmitter release and Ca^{2+} signal regulation of gene expression.

Neurotransmitter receptors also serve as the site of calcium influx into neurons. They can operate through G-protein-regulated Ca channels or allow calcium entry through neurotransmitter receptors. The best-known neurotransmitter Ca-permeable ion channels are receptors for acetylcholine and glutamate. Glutamate, a major excitatory neurotransmitter, in combination with depolarisation of the membrane, can activate NMDA (N-methyl-D-aspartate) receptors, which in the open state are highly permeable to Ca^{2+}. The NMDA receptor channels are particularly suitable to enhance excitability because they allow Ca^{2+} entry into neurons, which activate calcium-dependent K^+ channels. There is strong evidence that NMDA receptors, which enhance calcium entry into neurons, are critical to neuronal plasticity, and for the glutamate toxicity during brain hypoxia [6, 7].

Entry of calcium into neurons can be modified by calcium-sensing receptors (CaSR) which are found to be expressed in bovine [8] and rat [9] brains, with the highest levels of gene expression in the hypothalamus and corpus striatum. The function of CaSR in the brain is not well understood, but it is possible that CaSR may detect local changes in Ca^{2+} in the synaptic cleft after neuronal depolarisation and may modify neuronal function accordingly.

Intracellular membrane stores can also release Ca^{2+} in response to many stimuli (such as neurotransmitters and hormones), which act through specific G-protein-linked receptors [10, 11]. These receptors, after interaction with neurotransmitters,

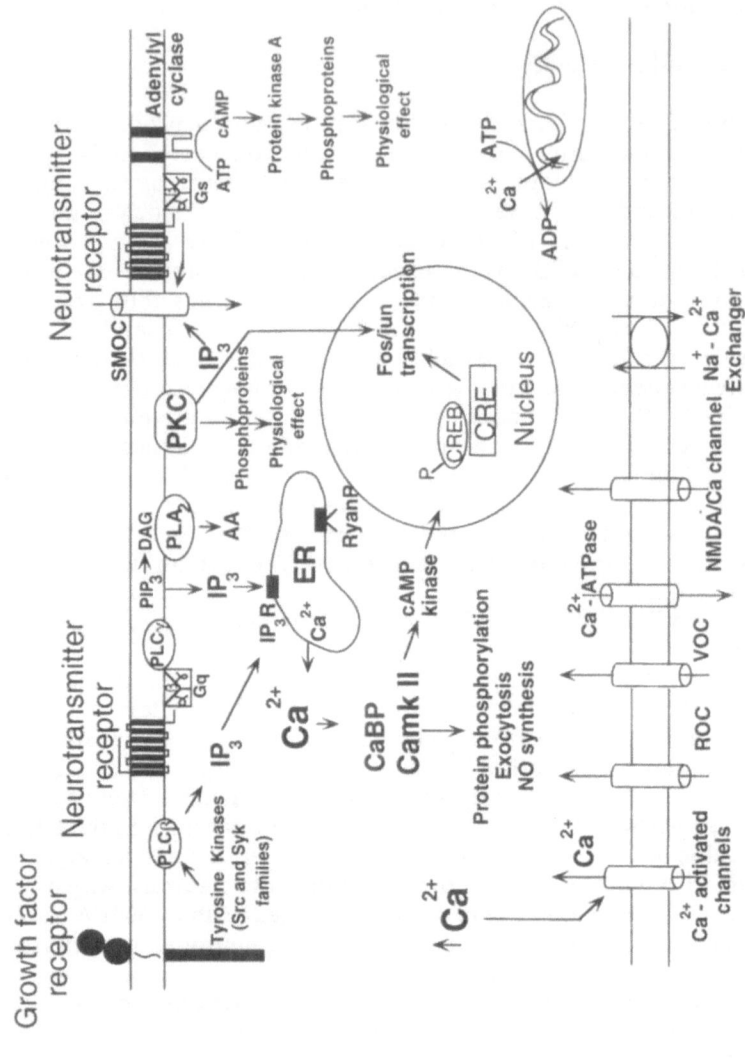

Figure 10.13. Selected, relatively well-defined signal transduction pathways in neuron. Abbreviations: AA = arachidonic acid; CaBP = calcium binding protein; Camk II = calmodulin kinase II; cAMP = cyclic-adenosine monophosphate; CRE = cAMP response element; CREB = cAMP response element binding protein; DAG = diacylglycerol; ER = endoplasmic reticulum; IP_3 = inositol 1,4,5-triphosphate; NMDA = N-methyl-D-aspartate; NO = nitric oxide; PKA = protein kinase A; PKC = protein kinase C; PLA = phospholipase A; PLC = phospholipase C; ROC = receptor-operated channel; Ryan = Ryanodine; SMOC = second messenger-operated channel; VOC = voltage-operated channel.

undergo conformational changes resulting in the formation of Gα and G$\beta\gamma$ subunits. The Gα subunits exchange guanosine 5'-diphosphate (GDP) for guanosine 5"-triphosphate (GTP) and activate phospholipase C-β1 which hydrolyses the phosphoinositol 4,5-bisphosphate to give diacylglycerol (DAG) and inositol (1,4,5)-trisphosphate (IP$_3$). The latter binds to IP$_3$ receptors present at ER membranes and mobilises calcium from intracellular stores, whereas DAG activates protein kinase C.

Another mechanism by which Ca^{2+} is released from intracellular stores is by Ca^{2+}-mediated Ca^{2+} release. It is regulated by ryanodine receptors (RyanR) present in intracellular membranes. RyanRs in the brain are activated by Ca^{2+} entry through dihydropiridine-sensitive Ca^{2+} channels [10, 111]. The RyanR-activated channels open at [Ca^{2+}]i concentrations between 100 nM and 1 μM, but are inhibited at [Ca^{2+}]i concentrations above 10 μM.

Intracellular Calcium Binding and Membrane Sequestration

Several intracellular proteins that bind Ca^{2+} (calcium-binding protein; CaBP) can temporarily decrease intracellular calcium concentrations, and at the same time, these CaBPs may undergo conformational changes and mediate the activation of intracellular enzymes such as calcium-dependent protein kinases, adenylate cycles and others [12]. The best-understood CaBP is calmodulin, a small (16.7 kDa) protein. Calcium binding induces large conformational changes in calmodulin, exposing a hydrophobic domain that can interact with a variety of effector proteins. One of those proteins is calcium/calmodulin (Ca^{2+}/Cam)-dependent protein kinase (Ca^{2+}/Cam kinase II). This enzyme reaches very high concentrations in the brain since it comprises approximately 1% of total brain proteins. It becomes activated upon calcium–calmodulin complex binding to Ca^{2+}/Cam kinase II and is able to transfer phosphate groups from ATP to different proteins such as synapsin I. Phosphorylation of synapsin I, one of the synaptic vesicle proteins, reduces the attachment of synaptic vesicles to actin at nerve terminals and facilitates release of neurotransmitters from synaptic vesicles. Interestingly, Ca^{2+}/Cam kinase II is able to phosphorylate itself and stay active for a long time, a finding that led to the hypothesis that this enzyme works as a "calcium switch" in the expression of long-lasting effects of calcium on the neuron [13].

Mitochondria act as a sink for cytosolic calcium. An ATP-dependent calcium pump moves Ca^{2+} from cytosol into the mitochondrial matrix where calcium is stored as calcium–phosphate complex. However, a slow leak of Ca^{2+} into cytosol also occurs. Movement of Ca^{2+} into the endoplasmic reticulum (ER) is controlled by Ca^{2+}-ATPase, which pumps Ca^{2+} against a steep concentration gradient. High calcium concentration in ER is necessary for the folding and processing of newly synthesised proteins and for depletion of ER by exposure of neuronal cells to tapsigargin (irreversible inhibitor of ER Ca^{2+}-ATPase), triggering the process of apoptosis.

Calcium Extrusion from Neurons

The low resting [Ca^{2+}]i concentration is maintained by membrane-associated calcium-transporting proteins which move Ca^{2+} against its concentration gradient into three major places: extracellular compartment, endoplasmic reticulum (ER) or mitochondria (Figure 10.13). Two major transporting proteins include the Ca^{2+}-ATPase and Na$^+$/Ca^{2+} exchanger.

The Ca^{2+}-ATPase pump uses ATP energy to extrude Ca^{2+} from the cell or to move Ca^{2+} into endoplasmic reticulum [14]. Its high calcium affinity allows for the fine-tuning of the calcium in the cytosol.

The Na^+/Ca^{2+} exchanger couples the transport of Ca^{2+} to the co-transport of Na^+. The exchanger uses the inwardly directed electrochemical gradient of Na^+ (produced by the Na^+-K^+-ATPase pump) to extrude Ca^{2+} from the cell [15]. The direction of exchange depends on the Na^+ gradient, the Ca^{2+} gradient and the membrane depolarisation. Under physiological conditions in polarised neurons at $-80\,mV$, the Ca^{2+} extrusion prevails. ATP stimulates the exchanger most likely by increasing the sarcolemmal level of phosphatidyl inositol-4-5-bisphosphate [16].

Calcium Transients and Neurotransmitter Release in Synapses

Neurotransmitter release is controlled by Ca^{2+} influx into the presynaptic body [17–19]. This very localised, high-concentration spike of $[Ca^{2+}]i$ of short duration induces the movement of neurotransmitters from presynaptic nerve terminals into the synaptic cleft [20, 21, 22]. The molecular events that accompany neurotransmitter release are complex and not fully understood. The short delay (less than $200\,\mu sec$) between Ca^{2+} signal and neurotransmitter release requires fast interaction between the vesicles docking at the active zone of the plasma membrane, fusion proteins and Ca^{2+} [23]. Synaptootagmins I and II are the strongest candidates to serve as Ca^{2+} sensors in the process of fast release of neurotransmitters [24].

Role of $[Ca^{2+}]i$ in Learning and Memory

Learning can be considered a process by which animals' behavioural responses are altered as a result of environmental input. Although the process of learning is extremely complex and difficult to study, especially in humans, it appears that short-term memory does not require ongoing protein synthesis, while long-term memory does [25]. Changes in behaviour are encoded as change in the properties of a large number of individual neurons and synapses. This phenomenon is called neural plasticity. Neural plasticity can occur as a consequence of spike activities due to the accumulation of Ca^{2+} entering through Ca^{2+} channels and subsequent change in either their excitability or transmitter release. Activation of synapse, depletion of releasable pool of transmitters or desensitisation of postsynaptic receptors can also be part of neural plasticity.

Increases in the levels of intracellular Ca^{2+} can be a potent activator of nuclear factor kappa B (NF-κB) complex that induces the expression of several genes, including one necessary for memory formation, promotion of neuron survival and synaptic plasticity [26].

Long-term memory formation in the sea snail, Aplysia californica, can be blocked by nuclear injection of synthetic cAMP response element (CRE) oligonucleotide, which prevents activation of cAMP-inducible genes [27]. Thus, regulation of gene expression by cAMP is critical for learning. It is known that cAMP-dependent protein kinase phosphorylates a transcription factor called cAMP response element binding protein (CREB). Phosphorylated CREB binds to CRE and turns on transcription of specific genes. CREB activity can also be induced by elevated $[Ca^{2+}]i$ [28].

Studies with brain hippocampal slices – another model of learning known as "long-term potentiation" (LTP) – show also that Ca^{2+} entry into the neuron through NMDA receptor channels and subsequent activation of Ca^{2+}/Cam kinase II is necessary for this learning process to occur [29].

Role of Calcium in Neuron Death

A continuous, unopposed increase in the concentration of intracellular Ca^{2+} can lead to disintegration of the cell (necrosis) through the activity of calcium-sensitive, protein-digesting enzyme [30]. Calcium is also involved in the active process of programmed cell death called apoptosis. Apoptosis is morphologically and pathologically distinct from necrosis and involves compaction of the cell body, nuclear DNA fragmentation and formation of surface blebs [31]. This programmed cell death process primarily occurs during brain development, but an episode of transient ischaemia, brain trauma or epilepsy can also induce the apoptotic process and produce delayed neuronal death [30]. The major factors that allow Ca^{2+} accumulation in the neuron have been identified as glutamate and anoxia. NMDA receptor activation by glutamate facilitates excessive influx of Ca^{2+} and activation of calcium-dependent processes of neurotoxicity. The VDCC influx of Ca^{2+} seems to be less harmful in this context. Mitochondria are seen as contributors to apoptotic cell death. Prevention in the rise of Ca^{2+} in mitochondria significantly reduced neurotoxicity of glutamate [7]. Since the loss of neurons occurs slowly in neurodegenerative disorders, controlling $[Ca^{2+}]i$ levels may be an effective way to prevent neuronal death. On the other hand, since $[Ca^{2+}]i$ is so important in the normal function of the cells, it could be difficult to prevent apoptosis alone without disrupting other survival pathways.

Calcium and Specific Brain Disorders

Calcium and the Brain in Renal Failure

Neurological complications are commonly encountered in patients with advanced chronic renal failure. They include neurobehavioural abnormalities, peripheral neuropathy and autonomic nervous system dysfunction [32–34]. The severity of signs and symptoms is usually quite variable in individual patients and only partially corrected by dialysis, nutritional and other therapies such as erythropoetin. In addition, dialysis treatment itself has been associated with at least two other central nervous system disorders such as disequilibrium syndrome and dialysis dementia.

The mechanisms underlying the central nervous system derangements are multifactorial and only partially characterised. Biochemical studies of the brain in animals with renal failure show normal brain content of several ions such as sodium, potassium, magnesium, chloride and bicarbonate, as well as water. In dogs with CRF, intracellular pH of the brain was normal despite the presence of metabolic acidosis [35]. This steady state finding does not necessarily reflect the functional integrity of the neurons since despite normal sodium content, there is an increase in veratridine-stimulated Na^+ uptake in uraemia, primarily due to altered Na^+-K^+-ATPase pump activity [36].

Brain osmolality in CRF is increased mainly due to the increased urea concentration of the brain, and due to the increased production of idiogenic osmoles [37]. Brain tissue of rats with acute renal failure showed an increase of creatinine phosphate, ATP and glucose, with a corresponding decrease in creatinine, ADP, AMP and lactate, suggesting that the brain displays less active metabolism and utilises less ATP in uraemia [38]. Similar studies are not available in CRF, but we found that in rats with CRF of 21-days duration, the ATP content of brain synaptosomes was 35% lower than in normal animals [39].

A consistent biochemical finding in the brain of humans and animals with acute or chronic renal failure is an elevation of total brain calcium content [34]. This derangement appears to be mediated by the state of secondary hyperparathyroidism of the renal failure since parathyroidectomy (PTX) of dogs with renal failure prevented the accumulation of calcium in the brain [38]. The pathophysiological importance of calcium accumulation in the brain was provided in the report of Guisado et al. [40] who showed that the electroencephalogram (EEG) with the characteristic uraemic changes becomes normal in PTX dogs with acute uraemia. Also, Cogan et al. [41] and Goldstein [42] reported that the changes in EEG in patients with CRF were either improved or reversed after PTX or after the suppression of the parathyroid gland activity by treatment with $1,25(OH)_2D_3$. Finally, Akmal et al. [43] found that PTX-CRF dogs had normal EEG and almost normal calcium content.

The increase in total calcium content of the brain in uraemia is accompanied by an increased basal level of $[Ca^{2+}]i$ [44]. Indeed, resting levels of cytosolic calcium of brain synaptosomes in rats with CRF of 21-days duration were significantly higher ($437 \pm 9\,nM$) as compared to normal rats ($345 \pm 9\,nM$). This rise in $[Ca^{2+}]i$ was prevented either by parathyroidectomy prior to induction of CRF or by treatment of CRF rats with the calcium channel blocker, verapamil. Brain synaptosomes obtained from normal rats treated with PTH (1–84) for 21 days show increased basal levels of $[Ca^{2+}]i$ with the value being $428 \pm 5.6\,nM$. These findings suggest that endogenous as well as exogenous excess of PTH stimulates entry of calcium into synaptosomes through calcium channels inhibitable by verapamil [45].

Support for PTH action on the brain is provided by the studies using molecular techniques. The high-affinity binding sites for PTH were demonstrated in the brain membranes [46]. The presence of mRNA for PTH/PTHrP [47], mRNA for PTH2 receptor [48], and cDNA for PTH (1–84) were also shown in rat brains [49], especially in the hypothalamus. PTH stimulates cAMP production by cultured murine brain [50] and generation of IP_3 in rat brain synaptosomes [51]. Hiasawa et al. [52] show that PTH (1–34) gradually increases $[Ca^{2+}]i$ in rat hippocampal slices via activation of DHP-sensitive Ca^{2+} channels and the hormone had a marked toxic effect on brain tissue after three days of exposure.

The source of PTH in the brain is still unclear. Autocrine production and paracrine action is one possibility. Some indirect evidence also suggests the penetration of PTH into the brain through the blood–brain barrier [53] when blood levels of PTH are high or the blood–brain barrier defective.

The effect of PTH on calcium transport by brain synaptosomes was investigated in acutely uraemic rats [54, 55] and in rats with normal renal function [56]. It was found that acute uraemia with excess PTH is associated with an increase in calcium uptake by synaptasomes, and this effect is mediated via the Na^+–Ca^{2+} exchanger and via the ATP-dependent calcium transport mechanism [54]. PTX of acutely uraemic rats prevented the accumulation of calcium in brain synaptosomes [55], and administration of PTH to rats with normal renal function augmented calcium transport into the synaptosomes [55]. This acute effect of PTH on calcium influx in brain synaptosomes is not cyclic AMP-dependent [56], but may be regulated by inositol-1, 4,5-triphosphate [57].

Despite augmented entry of Ca^{2+} into synaptosomes, the basal levels of $[Ca^{2+}]i$ should not increase since cells have powerful pumps that extrude excess Ca^{2+} and thus prevent Ca^{2+} accumulation [14]. These pumps, as discussed above, are directly or indirectly involved in $[Ca^{2+}]i$ homeostasis which includes Ca^{2+}-ATPase, Na^+-K^+-ATPase, and Na^+/Ca^{2+} exchanger.

The finding in our studies, which showed that basal levels of $[Ca^{2+}]i$ of brain synaptosomes are elevated in CRF, implies that the processes which mediate Ca^{2+} exit out of the brain synaptosomes are impaired. Indeed, the V_{max} for Ca^{2+} of synaptosomal Ca^{2+}-ATPase is significantly lower in CRF rats [58], as well as activity of Na^+-K^+-ATPase [39]. The molecular basis for the decreased activity of these pumps in brain tissue is not clear. These changes could be due to a reduction in the number of the enzyme units per cell, in the activity of each unit, or both. The lower levels of ATP content of the synaptosomes in CRF would also impair the activity of Ca^{2+}-ATPase and Na^+-K^+-ATPase, since ATP is required for normal function of these enzymes [59]. Furthermore, Fraser and Sarnacki [55] reported that the calcium uptake by the Na^+–Ca^{2+} exchanger of synaptosomes of uraemic rats is augmented and this abnormality was corrected by PTX.

The initial event is PTH-induced augmentation of calcium entry into synaptosomes that would impair mitochondrial production of ATP. The subsequent decrease in the ATP content of cells would impair the activities of Ca^{2+}-ATPase and Na^+-K^+-ATPase since ATP is required for normal function of these enzymes. The reduction in the activity of Na^+-K^+-ATPase and Ca^{2+}-ATPase enzymes leads to a decrease in calcium extrusion out of synaptosomes. Such an effect in the face of continued calcium entry through both voltage-dependent calcium channels as well as Na^+–Ca^{2+} exchanger would result in calcium accumulation in synaptosomes and to an increase in their basal levels of $[Ca^{2+}]i$. This elevation in $[Ca^{2+}]i$ would further inhibit mitochondrial oxidation and ATP production. Thus, a vicious circle develops until a new steady state is achieved with a low ATP content, reduced activity of both Ca^{2+}-ATPase and Na^+-K^+-ATPase, and elevated $[Ca^{2+}]i$. It is evident that these changes could be prevented by abolishing the state of excess PTH or by blocking the PTH-induced calcium entry into synaptosomes. This is supported by studies demonstrating that prior PTX of CRF rats or their treatment with verapamil prevents the metabolic derangements in brain synaptosomes [39, 44, 60].

The sequence of events described above does not continue but stabilises at a steady state. This may be due to down-regulation of the PTH-PTHrP receptors in CRF [61, 62], an adaptive process that would protect the cells from further action of the high blood levels of PTH and consequent continued accumulation of calcium that may eventually led to cell death.

The calcium signal, which is defined as the ratio of $[Ca^{2+}]i$ triggered by action potential to the basal levels of $[Ca^{2+}]i$, is important for the aggregation of synaptic vesicles, the release of neurotransmitters from presynaptic terminals into synaptic space, and the successful transfer of appropriate information [22]. The increase in basal levels of $[Ca^{2+}]i$ may decrease the calcium signal and impair the release of at least one neurotransmitter such as norepinephrine.

Cytosolic Calcium and Metabolism of Norepinephrine (NE) and Acetylcholine (ACh) of Brain Synaptosomes

It is obvious that significant calcium derangements in neurons may modify neurotransmitter metabolism. We studied these issues in synaptosomes from CRF rats and from normal rats treated with PTH-(1–84) for 21 days (both approaches increase $[Ca^{2+}]i$ in brain synaptosomes) and found that NE content, uptake, and release are significantly reduced in these animals [39]. Normalisation of synaptosomal $[Ca^{2+}]i$ in CRF rats by their treatment with verapamil [60] or by PTX prevented these derangements. Also, synaptosomal activity of the key enzymes involved in NE production (tyrosine hydroxylase) and degradation (monoamine oxidase) [63] are impaired in CRF, and these derangements are also due to the state of secondary hyperparathyroidism [64]. Our finding of reduced NE content in brain synaptosomes is in agreement with a previous report showing a decrease in NE content in the brain of rats with chronic renal failure [65].

Since the NE content of brain synaptosomes is reduced in CRF, one must conclude that the decrement in NE production is greater than in NE degradation. It is well established that changes in concentration or metabolism of cerebral catecholamines can modify both somatic and behavioural patterns. Thus, the abnormalities in NE metabolism not only play a role in the pathogenesis of hypertension in uraemia but also in behavioural dysfunction of uraemia.

The metabolism of acetylcholine (ACh) in brain synaptosomes is also impaired in CRF and this abnormality is again at least partially mediated by the rise in the $[Ca^{2+}]i$ of the synaptosomes. Ni et al. [66] studied brain synaptosomes from rats with excess PTH which, in the presence or absence of CRF, displayed increased ACh content, decreased choline content, reduced choline kinase activity, and increased ACh release. PTX of the CRF rats or their treatment with verapamil prevented these abnormalities.

Available data indicate that the cholinergic system plays a major role in both behaviour and motor function [67]. It has been shown that increased cholinergic activity is associated with arousal and may therefore lead to insomnia. Parkinson-like tumour and dyskinetic movements have been produced by centrally active cholinergic agents [67]. Finally, topical application of ACh to the cerebral cortex increased cerebral excitability and caused convulsions [68]. Therefore, one may propose that the tremor, insomnia, and seizures frequently encountered in CRF are at least partly due to the increase in ACh content and release by synaptosomes.

Cytosolic Calcium, PTH and Phospholipids of Brain Synaptosomes

Phospholipids are the components of the cell membrane and play an important role in maintaining membrane fluidity and cell function. Changes in lipid composition of cell membranes may affect the activity of proteins and enzymes present in such membranes [69].

CRF in rats and the chronic treatment with PTH-(1–84) on animals with normal renal function was associated with significant decrements in synaptosomal total phospholipids, phosphatidylinostol (PI), phosphatidylserine (PS), and phosphatidylethanolamine (PE). PTX of the CRF rats [70] or their treatment with verapamil [71] (Figure 10.14) prevented the changes in the phospholipids of brain synaptosomes. These observations support the notion that derangements in synaptosomal phospholipids are also mediated by the PTH-induced elevation in $[Ca^{2+}]i$ of brain synaptosomes.

The changes in the content of various phospholipids of brain synaptosomes may affect their function in many ways. First, it is known that PS is important for the aggregation of synaptosomal vesicles, a critical step in the fusion of these vesicles with the synaptosomal membrane and the subsequent release of their contents [69]. A reduction in synaptosomal PS may, therefore, affect neurotransmitter release. Indeed, norepinephrine (NE) release by brain synaptosomes is reduced in rats with CRF and in animals treated with PTH [37]. Second, it is possible that the changes in synaptosomal phospholipids are associated with decreased activity of Na^+-K^+-

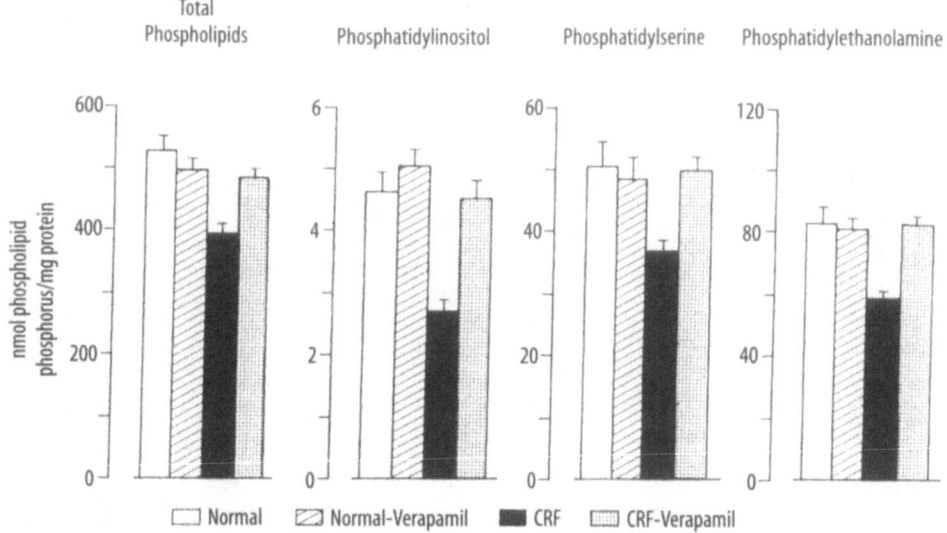

Figure 10.14. Total phospholipid, phosphatidylinositol, phosphatidylserine, and phosphatidylethanolamine contents of brain synaptosomes in the normal (n = 8), normal–verapamil treated (n = 6), chronic renal failure (n = 6) and chronic renal failure–verapamil treated (n = 6) animals. Each column represents mean values and the brackets denote LSE. The values in CRF groups are significantly lower $p < 0.01$ versus all other groups. (Reprinted with permission [71].)

ATPase, an enzyme that affects NE uptake by brain synaptosomes [72]. Third, PI metabolism is critical for the responses induced by agonist–receptor interaction [73]. Therefore, a reduction in synaptosomes PI may render them less responsive to many agonists, and this affects the function. Finally, alteration in phospholipids of brain synaptosomes may render them more permeable to calcium and allow an increase in calcium influx.

Studies in patients undergoing chronic dialysis treatment found mild to moderate cognitive performance abnormalities on measures of memory and learning, attention and concentration, and tasks requiring mental manipulation [74, 75]. Treatment with erythropoetin and correction of anaemia only partially improved cognitive and electrophysiological function in dialysis patients, but there was no change in the latency of event-related potential P3 [75], which is characteristically prolonged in dialysis patients. The P3 latency potential is a good indicator of subtle impairments of brain function in early dementia. It is possible that the inadequate or excessive Ca^{2+} signals in uraemia are involved in the deficiency of learning, memory and cognitive function.

Phosphate Depletion and Cytosolic Calcium

Phosphate depletion is associated with multiple organ dysfunction including nervous system abnormalities (PD) [76]. Decreased availability of energy-rich phosphate compounds such as ATP was found in red cells, leukocytes, pancreatic islets, the kidney, brain and heart [77].

We postulated that PD could be another state of calcium derangement. Indeed, PD causes marked alterations in the metabolism of $[Ca^{2+}]i$ in pancreatic islets and leuko-

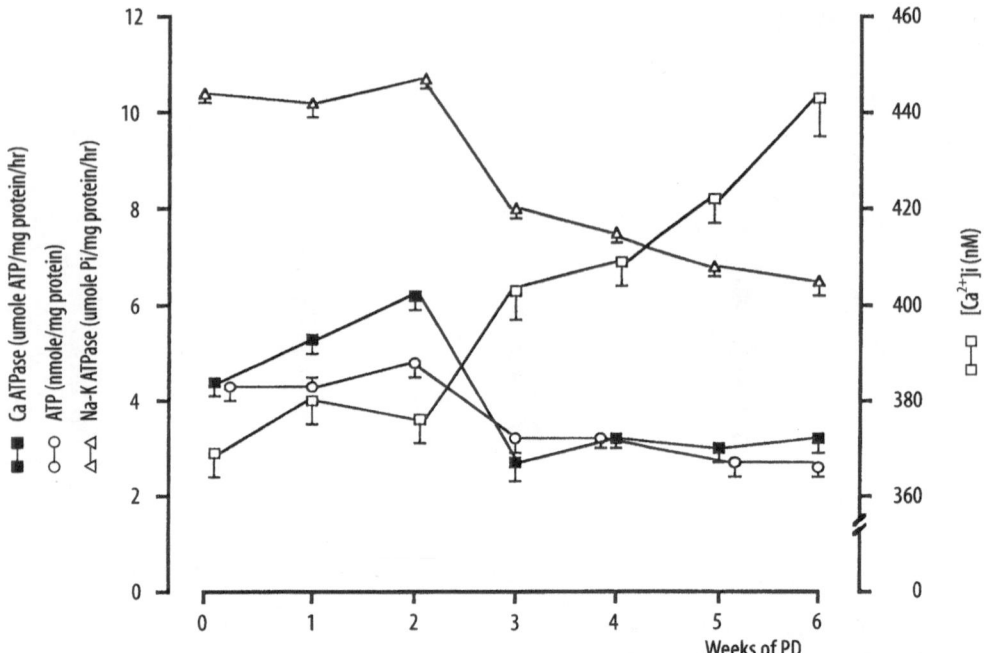

Figure 10.15. The chronological relationship between the ATP content, Ca^{2+}-ATPase and Na^+-K^+-ATPase activity and $[Ca^{2+}]i$ in brain synaptosomes studied during the evolution of phosphate depletion in rats over a period of six weeks. (Reprinted with permission [79].)

cytes [74]. Brain synaptosomes from rats with chronic PD (six weeks) display a decrease in ATP content, a marked rise in resting levels of $[Ca^{2+}]i$, and a significant impairment in the activity of Na^+-K^+-ATPase and Ca^{2+}-ATPase of their membrane [78]. This occurs despite suppressed blood levels of PTH in PD. Verapamil, by blocking the enhanced calcium entry, prevents the rise in $[Ca^{2+}]i$ and normalised synaptosomal $[Ca^{2+}]i$ (361 ± 8.5 nM) in PD-V animals. In verapamil-treated, paired-weight rats (PW-V), synaptosomal $[Ca^{2+}]i$ (359 ± 8.3 nM) was not affected. The time relationship between the changes in ATP, Na^+-K^+-ATPase and Ca^{2+}-ATPase is presented in Figure 10.15 [79].

Total phospholipid content as well as content of individual fractions of PI, PS, and PE of brain synaptosomes are significantly reduced in PD rats, and treatment of these rats with verapamil mitigated these changes but did not normalise the synaptosomal contents of total phospholipids, PS, and PE [80]. It suggests that these derangements are due to reduced availability of phosphorus and to the rise in $[Ca^{2+}]i$.

The content of total phospholipid, PI, PS, and PE of the synaptosomes is determined by the rate of their synthesis and degradation. Our studies did not evaluate the effect of PD on these two parameters. However, it is obvious that inorganic phosphorus is essential for adequate synthesis of these compounds. The reduced availability of inorganic phosphorus in PD would interfere with synthesis of these phospholipids, resulting in a decrease in their synaptosomal content. Such an effect should not be limited to the phospholipids of brain synaptosomes but should affect the phospholipid contents of other organs as well. Indeed, PD is associated with reduced PI, PS, and PE content of myocardium [82] and the brush border membrane of their proximal tubules [83]. The predominant pathway responsible for the synthesis of PS in animals is the base exchange reaction with PE [84]. Thus, it is not surprising to find that the synaptosomal content of both PE and PS is reduced in PD animals.

Calcium and Degenerative Brain Disorders

Dysregulation of calcium homeostasis has been suggested as a cause of cell dysfunction associated with ageing and degenerative brain disorders such as Alzheimer's disease (AD) [85, 86]. This disease is clinically characterised by progressive cognitive decline, and pathological hallmarks include neuronal loss in the hippocampus, cerebral cortex, nucleus basalis and locus ceruleus, and neurofibrillary tangles and neuritic plaques containing β-amyloid. Despite the lack of a comprehensive theory on the pathogenesis of AD, it is clear that calcium dysregulation plays a prominent role in this disease [87].

The β-amyloid can increase neuronal $[Ca^{2+}]i$ by incorporation into lipid bilayers and formation of calcium channels that flux Ca^{2+}; it can also enhance Ca^{2+} entry by activation of NMDA receptor/Ca^{2+} channel and generation of free radicals [88, 89]. It is of interest that neuronal cells can produce β-amyloid intracellularly. Intracellular β-amyloid acts similarly to ionophore depleting ER calcium store [90]. On the other hand, brain tissue obtained at autopsy from patients with Alzheimer's disease has revealed increased NF-κB activity in cells showing degeneration. This activation of NF-κB most likely has a neuroprotective effect associated with β-amyloid deposition and increased levels of intracellular calcium [26]. Remaining neurons obtained postmortem from AD hippocampus show increased L-type calcium channel density [91], which may place AD neurons at risk for higher Ca^{2+} influx. The use of calcium channel blockers to ameliorate calcium neurotoxicity was evaluated in animal models of degenerative brain disease and humans with AD. Despite some encouraging early studies in animals [92] and humans [93], the use of these agents was, in general, disappointing and at the present time have limited value in the treatment of degenerative brain diseases [88].

Calcium and Brain Ischaemia

Brain ischaemia, either acute (stroke) due to sudden disruption of blood supply to part of the brain, or chronic as a result of disturbed microvascular blood flow, reduces the supply of oxygen and glucose to the brain cells. Cellular ATP levels plummet rapidly following the onset of acute ischaemia, which impairs the activity of ATPase to remove Na^+ and Ca^{2+} from the cell, resulting in membrane depolarisation. Depolarisation, together with excessive accumulation of extracellular glutamate and activation of NMDA receptors, result in calcium influx into neurons [94]. Mitochondrial dysfunction further disrupts intracellular calcium homeostasis [7]. Production of free radicals in brain ischaemia is activated by Ca^{2+} as well as activation of enzymes such as nitric oxide synthase and cycloxygenase [94]. Release of Ca^{2+} from endoplasmic reticulum also plays a role in cell injury. Datrolone, which blocks calcium release through Ryan receptors, has a protective effect against ischaemia [90].

Calcium Dysregulation in HIV Infection

Brain diseases, both primary and secondary, are found in up to 90% of patients with HIV infection. Primary brain diseases are caused by direct effects of HIV on the brain, and secondary diseases are a consequence of immunodeficiency states and include opportunistic infections [96]. Primary vital infection may eventually lead to dysfunction in cognition, movement and sensation, which is called AIDS Dementia Complex. The process is initiated by HIV infection of a small number of brain macrophages and microglia that initiates the variety of neurotoxins including TNF-α, nitric oxide, superoxide anion, glutamate-like agonist, interleukin-1β, etc. Some of these cytokines and HIV-1 envelop protein gp-120 and can activate voltage-dependent Ca^{2+} channels and NMDA receptor-operated channels, which permit Ca^{2+} influx and initiate events leading to degeneration of the neuron by apoptosis or necrosis [96, 97]. Specific antagonists of the NMDA receptor block gp-120-mediated neurotoxicity, both in vitro and in animal models, supporting the hypothesis that Ca^{2+} plays an important role in the pathogenesis of brain damage in HIV infection. In addition, HIV-1-related protein (Tat) can release intracellular calcium from IP_3-sensitive intracellular pools [98] and inhibition of this pathway prevented Tat-induced neuronal cell death.

Despite significant progress in the area of HIV therapy, HIV infection of the CNS is very difficult to suppress. Thus, the development of additional therapies targeting specific receptors involved in calcium signalling, such as NMDA or IP_3 receptors, may be of help [97, 98].

References

1. Kelly RB. An introduction to the nerve terminal. In: Ballen HJ, editor. Neurotransmitter release. New York: Oxford University Press, 1999;1–33.
2. Whittaker VP. The isolation and characterization of acetylcholine-containing particles from brain. Biochem J 1959;72:694–706.
3. Booth RFG, Clark JB. A rapid method for the preparation of relatively pure metabolically competent synaptosomes from rat brain. Biochem J 1978;176:365–70.
4. Whittaker VP. Thirty years of synaptosomes research. J Neurocytology 1993;22:735–42.
5. Tsien RW, Wheeler DB. Voltage-gated calcium channels. In: Carafoli E, Klee C, editors. Calcium as a cellular regulator. New York: Oxford University Press, 1999;171–200.
6. Choi DW. Calcium-mediated neurotoxicity: relationship to specific channel types and role in ischemic damage. TINS 1988;11:465–9.
7. Stout AK, Raphael HM, Kanterewicz BI, Klann E, Reynolds IJ. Glutamate-induced neuron death requires mitochondrial calcium uptake. Nature Neuroscience 1998;1:366–73.
8. Brown EM, Gamba G, Riccardi D, Lombardi D, Butters R, Kifor O, et al. Cloning and characterization of an extracellular Ca^{2+}-sensing receptor from bovine parathyroid. Nature 1993;366:575–89.
9. Ruat M, Molliver ME, Snowman AM, Snyder SH. Calcium sensing receptor: Molecular cloning in rat and localization to nerve. Proc Natl Acad Sci USA 1995;92:3161–5.
10. Berridge MJ. Inositol trisphosphate and calcium signaling. Nature 1993;361:315–25.
11. Ghosh A, Greenberg ME. Calcium signaling in neurons: molecular mechanisms and cellular consequences. Science 1995;268:239–47.
12. Kawasaki H, Kretsinger RH. Calcium-binding proteins.I:EF-hands. Protein Profile 1994;1:342–91.
13. Lisman JE, Goldring MA. Feasibility of long-term storage of graded information by Ca^{2+}/calmodulin-dependent protein kinase molecules of postsynaptic density. Proc Natl Acad Sci USA 1988;85:5320–4.
14. Carafoli E, Stauffer T. The plasma membrane calcium pump: functional domains, regulation of the activity, and tissue specificity of isoform expression. J Neurobiol 1994;25:312–24.
15. Philipson KD, Nicoli DA. Sodium–calcium exchange. Curr Opin Cell Biol 1992;4:678–83.
16. Hilgemann DW, Ball R. Regulation of cardiac Na^+, Ca^{2+} exchanger and K^+-ATP potassium channels by PIP_2. Science 1996;273:956–60.

17. Dodge Jr FA, Rahamimoff R. Cooperative action of Ca ions in transmitter release at the neuromuscular junction. J Physiol (Lond) 1967;138:434–44.
18. Katz B, Miledi R. The effect of calcium on acetylcholine release from motor nerve terminals. Proc R Soc Lond B Biol Sci 1965;161:496–503.
19. Silinsky EM. The biophysical pharmacology of calcium-dependent acetylcholine secretion. Pharmacol Rev 1985;37:81–132.
20. Miledi R. Transmitter release induced by injection of calcium ions into nerve terminals. Proc R Soc Lond (Biol) 1973;183:421–5.
21. Llinas R, Nicholson C. Calcium role in depolarisation–secretion coupling: an aequorin study in squid giant synapse. Proc Natl Acad Sci USA 1975;72:187–90.
22. Fossier P, Tauc L, Baux G. Calcium transients and neurotransmitter release at an identified synapse. Trends Neurosci1999;22:161–6.
23. Südhof TC. The synaptic vesicle cycle: a cascade of protein–protein interaction. Nature 1995; 375:645–53.
24. Kell RB. Synaptotagmin is just a calcium sensor. Curr Biol 1995;5:257–9.
25. Agranoff BW, Uhler MD. Learning and memory. In: Siegel GJ, editor. Basic neurochemistry: molecular, cellular, and medical aspects. New York: Raven Press, 1994;1025–43.
26. Mattson MP, Camandola S. NF-κB in neuronal plasticity and neurodegenerative disorders. J Clin Invest 2001;107:247–54.
27. Baily CH, Kandel ER. Structural changes accompanying memory storage. Ann Rev Physiol 1993;55:397–426.
28. McFadden SM, Greenberg ME. Membrane depolarization and calcium-induced c-fos transcription via phosphorylation of transcription factor CREB. Neuron 1990;4:571–82.
29. Larkman AU, Jack JB. Synaptic plasticity: hippocampal LTP. Curr Opin Neurobiol 1995;5:324–34.
30. Sattler R, Tymianski M. Calcium and cellular death. In: Verkhratsky A, Toescu EC, editors. Integrative aspects of calcium signaling. New York: Plenum Press, 1998;267–90.
31. Ojcius DM, Zychlinsky A, Zheng LM, Young D. Ionophore-induced apoptosis: role of DNA fragmentation and calcium fluxes. Exp Cell Res 1991;197:43–9.
32. Tyler HR. Neurologic disorders in renal failure. Am J Med 1968;44:734–48.
33. Nielsen VA. The peripheral nerve function in chronic renal failure. Acta Med Scand 1974;195:155–62.
34. Fraser CL. Neurologic manifestations of the uremic state in metabolic brain dysfunction. In: Arieff AI, Griggs RG, editors. Systemic disorders. Boston Toronto London: Little Brown & Company, 1992; 139–66.
35. Mahoney LA, Arieff IA. Central and peripheral nervous system effects of chronic renal failure. Kidney Int 1983;24:170–7.
36. Fraser CL, Sarnacki P, Arieff AI. Abnormal sodium transport in synaptosomes from brain in uremic rats. J Clin Invest 1985;74:2014–23.
37. Van den Noort S, Eckel RE, Bvine K. Brain metabolism in uremic and adenosine-infused rats. J Clin Invest 1968;47:2133–42.
38. Arieff AI, Massry SG. Calcium metabolism of brain in acute renal failure. J Clin Invest 1974;53:387–92.
39. Smogorzewski M, Campese VM, Massry SG. Abnormal norepinephrine uptake and release in brain synaptosomes in chronic renal failure. Kidney Int 1989;36:458–65.
40. Guisado R, Arieff AI, Massry SG. Changes in the electroencephalogram in acute uremia. J Clin Invest 1975;55:738–45.
41. Cogan MG, Covey CM, Arieff AI, Wisniesky A, Clark OH, Lazarowitz V, et al. Central nervous system manifestations of hyperparathyroidism. Am J Med 1978;65:963–71.
42. Massry SG. The relationship between the abnormalities in electroencephalogram and blood levels of PTH in dialysis patients. J Clin Endocrinol Metab 1980;51:130–4.
43. Akmal M, Goldstein DA, Multani S, Massry SG. Role of uremia, brain calcium and parathyroid hormone on change in electroencephalogram in chronic renal failure. Am J Physiol 1984;246:F575–9.
44. Smogorzewski M, Koureta P, Fadda GZ, Perna AF, Massry SG. Chronic parathyroid hormone excess in vivo increases resting levels of cytosolic calcium in brain synaptosomes: Studies in the presence and absence of chronic renal failure. J Am Soc Nephrol 1991;1162–8.
45. Massry SG, Smogorzewski M. The mechanisms responsible for the PTH-induced rise in cytosolic calcium in various cells – one not uniform. Miner Electrolyte Metab 1995;21:13–28.
46. Harvey S, Hayer S. Parathyroid hormone binding sites in the brain. Peptides 1993;14:1187–91.
47. Urena P, Kong X-F, Abou-Samra A-B, Juppner H, Kronenber HM, Pott Jr JT, et al. Parathyroid hormone (PTH)/PTH-related peptide receptor messenger ribonucleic acids are widely distributed in rat tissue. Endocrinology 1993;133:617–23.
48. Usdin TB, Bonner TI, Harata G, Mezey E. Distribution of parathyroid hormone-2 receptor messenger ribonucleic acid in rat. Endpcrinology 1996;137:4285–97.

49. Nutley MT, Parimi SA, Harvey S. Sequence analysis of hypothalamic parathyroid hormone messenger ribonucleic acid. Endocrinology 1995;136:5600-7.
50. Loffler F, Van Calker D, Hamprecht B. Parathyrin and calcitonin stimulate cyclic AMP accumulation in cultured murine brain cells. EMBO J 1982;1:297-302.
51. Smogorzewski M. Parathyroid hormone stimulates the generation of inositol 1,4,5-triphosphate in brain synaptosomes. Am J Kidney Dis 1995;26:814-17.
52. Hirasawa T, Nakamura T, Mizushima A, Morita M, Ezawa I, Miyakawa H, et al. Adverse effects of active fragment of parathyroid hormone on rat hippocampal organotypic cultures. Brit J Pharmacol 2000;129:21-8.
53. Joburn C, Hetta J, Niklasson F, Rastad J, Wide L, Agren H, et al. Cerebrospinal fluid, calcium, parathyroid hormone and monoamine and purine metabolites and blood-brain barrier function in primary parathyroidism. Psychoneroendocrinology 1991;16:311-22.
54. Fraser CL, Sarnacki P, Arieff AI. Calcium transport abnormality in uremic rat brain synaptosomes. J Clin Invest 1985;76:1789-95.
55. Frasier CL, Sarnacki P. Parathyroid hormone mediates changes in calcium transport in uremic rat brain synaptosomes. Am J Physiol 1988;254:F837-47.
56. Fraser CL, Sarnacki P, Pudyar A. Evidence that parathyroid hormone-mediated calcium transport in rat brain synaptosomes is independent of cyclic adenosine monophosphate. J Clin Invest 1988;81:928-88.
57. Fraser CL, Sarnacki P. Inositol 1,4,5-trisphosphate may regulate rat brain $Ca^{++}i$ by inhibiting membrane-bound Na^{+}-Ca^{++} exchanger. J Clin Invest 1990;86:2169-73.
58. Hajjar SM, Smogorzewski M, Zayed MA, Fadda GZ, Massry SG. Effect of chronic renal failure on Ca^{2+}-ATPase of brain synaptosomes. J Am Soc Nephrol 1991;2:1115-21.
59. Blaustein MP, Hodgkin AC. The effect of cyanide on the efflux of calcium from squid axon. J Physiol 1969;200:497-527.
60. Smogorzewski M, Islam A, Minasain R, Soliman AR, Massry SG. Verapamil corrects abnormalities in norepinephrine metabolism of brain synaptosomes in CRF. Am J Physiol 1990;258:F1036-41.
61. Tian J, Smogorzewski M, Kedes L, Massry SG. PTH-PTHrP receptors mRNA is down-regulated in chronic renal failure. Am J Nephrol 1994;14:41-6.
62. Urena P, Kubrusly M, Mannstadt M, Kurby M, Tan M-MTT, Silve C, et al. The renal PTH/PTHrP receptor is down-regulated in rats with chronic renal failure. Kidney Int 1994;45:605-11.
63. Bradford HF. Chemical neurobiology. New York: W.H. Freeman, 1986.
64. Islam A, Smogorzewski M, Zayed MA, Massry SG. Effect of chronic renal failure with and without secondary hyperparathyroidism on the activities of synaptosomal tyrosine hydroxylase and monoamine oxidase. Nephron 1992;61:32-6.
65. Ali F, Tayeh O, Attallah A. Plasma and brain catecholamines in experimental uremia: Acute and chronic studies. Life Sci 1985;37:1757-64.
66. Ni Z, Smogorzewski M, Massry SG. Derangements in acetylcholine metabolism in brain synaptosomes in chronic renal failure. Kidney Int 1993;44:630-7.
67. DeFeudis FV. Central cholinergic system and behaviour. London: Academic Press, 1974.
68. Lalley PM, Rossi GV, Baker WW. Analysis of local cholinergic tremor mechanisms following selective neurochemical lesions. Exp Neurol 1970;27:258-75.
69. Yeagle PL. The membranes of cells. Orlando, FL: Academic Press, 1987.
70. Islam A, Smogorzewski M, Massry SG. Effect of chronic renal failure and parathyroid hormone on phospholipid content of brain synaptosomes. Am J Physiol 1989;256:F705-10.
71. Islam A, Smogorzewski M, Massry SG. Effect of verapamil on CRF-induced abnormalities in phospholipid contents of brain synaptosomes. Proc Soc Exp Biol Med 1990;194:16-20.
72. Bogdanski DF. Mechanisms of transport for the uptake and release of biogenic amines in nerve endings. Adv Exp Med Biol 1976;69:291-305.
73. Berridge MJ. Cell signaling through phospholipid metabolism. J Cell Sci 1986;4(suppl):137-53.
74. Ratner DP, Adams KM, Levin NW, Rourke BP. Effects of hemodialysis on cognitive and sensorimotor functioning of the adult chronic hemodialysis patient. J Behav Med 1983;6:291-311.
75. Marsh JT, Brown WS, Wolcott D, Carr CR, Harper R, Schweitzer SV, et al. HuEPO treatment improves brain and cognitive function of anemic dialysis patients. Kidney Int 1991;39:155-63.
76. Berner YN, Shike M. Consequences of phosphate imbalance. Ann Rev Nutr 1988;8:121-48.
77. Massry SG, Fadda GZ, Perna AF, Kiersztejn M, Smogorzewski M. Mechanism of organ dysfunction in phosphate depletion: critical role for rise in cytosolic calcium. Miner Electrolyte Metab 1992;18:133-40.
78. Massry SG, Hajjar SM, Koureta P, Fadda GZ, Smogorzewski M. Phosphate depletion increases cytosolic calcium of brain synaptosomes. Am J Physiol 1991;260:F12-18.

79. Rios T, Smogorzewski M, Ni Z, Levi E, Massry SG. Sequence of appearance of the metabolic derangements in rat brain synaptosomes during phosphate depletion. Nephron 1994;67:54–8.
80. Smogorzewski M, Islam A, Koureta P, Fadda GZ, Massry SG. Reduced phospholipid contents of brain synaptosomes in phosphate depletion. Am J Physiol 1991;261:E742–7.
81. Smogorzewski M, Isla A, Koureta P, Massry SG. Abnormal norepinephrine metabolism in rat brain synaptosomes in phosphate depletion. Am J Nephrol 1993;13:43–52.
82. Brautbar N, Tabernero-Romo J, Coats JC, Massry SG. Impaired myocardial lipid metabolism in phosphate depletion. Kidney Int 1984;26:18–23.
83. Levi M, Jameson DM, Van Der Meer BW. Role of BBM lipid composition and fluidity in impaired renal Pi transport in aged rat. Am J Physiol 1989;256:F85–94.
84. Molitoris BA, Alfrey AC, Harris RA, Simon FR. Renal apical membrane cholesterol and fluidity in regulation of phosphate transport. Am J Physiol 1985;24:F12–19.
85. Hartman H, Eckert A, Muller W. Disturbances of the neuronal calcium homeostasis in the aging nervous system. Life Science 1994;55:2011–18.
86. Landfield P, Thibault O, Mazzanti M, Porter N, Kerr D. Mechanisms of neuronal death in brain aging and Alzheimer's disease; role of endocrine-mediated calcium dyshomeostasis. J Neurobiol 1992;23:1247–60.
87. Missiaen L, Robberecht W, van de Bosch L, Callewaert G, et al. Abnormal intracellular Ca^{2+} homeostasis and disease. Cell Calcium 2000;28:1–21.
88. Pascale A, Etcheberrigaray R. Calcium alteration in Alzheimer's disease: pathophysiology, models and therapeutic opportunities. Pharmacological Res 1999;2:81–8.
89. Cotter RL, Burke WJ, Thomas VS, Potter J, Zheng J, Gendelman HE. Insight into the neurodegenerative process of Alzheimer's disease: role of mononuclear phagocyte-associated inflammation and neurotoxicity. J Leukoc Biol 1999;65:416–27.
90. Paschen W, Doutheil J. Disturbances of the functioning of endoplasmic reticulum: A key mechanism underlying neuronal cell injury? J Cerebral Blood Flow and Met 1999;19:1–18.
91. Coon AL, Wallace DR, Mactutus CF, Booze RM. L-type calcium channels in the hippocampus and cerebellum of Alzheimer's disease brain tissue. Neurobiol Aging 1999;20:597–603.
92. De Jonge MC, Traber J. Nimodipine: cognition, aging and degeneration. Clin Neuropharmacol 1993;16:S25–30.
93. Tollefson GD. Short-term effects of the calcium channel blocker nimodipine (Bay-e-9736) in the management of primary degenerative dementia. Biol Psychiatry 1990;27:1133–42.
94. Dirnagl U, Iadecola C, Moskowita MA. Pathobiology of ischemic stroke: an integrated view. Trends Neurosci 1999;22:391–7.
95. Mattson MP, Culmose C, Zai FY. Apoptotic and antiapoptotic mechanisms in stroke. Cell Tissue Res 2000;301:173–87.
96. Pazos-Trillo G, Everall IP. From human immunodeficiency virus (HIV) infection of the brain to dementia. Genitourinary Med 1997;73:343–7.
97. Lipton SA. Neuronal injury associated with HIV-1: Approaches to treatment. Ann Rev Pharmacol Toxicol 1998;38:159–77.
98. Haughey NJ, Holden CP, Nath A, Geiger JD. Involvement of inositol 1,4,5-triphosphate-regulated stores of intracellular calcium and calcium dysregulation and neuronal cell death caused by HIV-1 protein Tat. J Neurochemistry 1999;73:1363–74.

10F

Vascular Calcification

A. Shioi

Introduction

Vascular calcification is a pathological calcification process termed "dystrophic calcification", which may occur despite normal serum levels of calcium and phosphate. Dystrophic calcification is often found in areas of necrosis or injury including atherosclerotic plaque, aging or damaged cardiac valves, or neoplasia [1]. It has many similarities to physiologic skeletal mineralisation. Virchow noted that vascular calcification was similar to bone and described calcified coronary artery lesions as "an ossification, and not a mere calcification", based on the observation of fully formed lamellar bone including trabeculae, lacunae, and areas resembling marrow [2, 3]. Recent evidence demonstrates that vascular calcification such as atherosclerotic plaques have many features of bone. These areas of mineralisation in cardiovascular tissues are composed of both hydroxyapatite and organic matrix including type I collagen and non-collagenous proteins regulating mineralisation (NCPs) [4, 5]. Both physiologic and pathologic calcifications involve collagen-associated crystal deposition and initiation of mineralisation within matrix vesicles [6]. This has led to the concept that vascular calcification is an active, regulated process rather than a passive accumulation of mineral [3]. Indeed, several groups have demonstrated that subpopulations of vascular smooth muscle cells behave, in vitro, like bone-forming osteoblasts [7–10].

On the other hand, it has long been known that calcification follows necrosis. Of the tissue components, the most vulnerable are the cells. Nevertheless, little attention has been paid to the role of cell death in calcification. Recent advances on the mechanism of cell death indicate that the loss of ion regulation in cell death may play the primary role in calcification. Concomitant increases in Ca^{2+} and inorganic phosphate (Pi) in blebs (and matrix vesicles) formed by apoptotic and/or necrotic cells may be the primary mechanism of calcification. In addition, membranous cellular degradation products (CDP) resulting from cell disintegration in toto frequently serve as the nidus of calcification. Apoptotic and/or necrotic cell death play an important role in both physiological and pathological calcification [11].

There are several forms of dystrophic calcification found in cardiovascular tissues: atherosclerotic calcification, Mönckeberg's calcification and cardiac valve calcification. This chapter will describe several forms of vascular calcification and discuss its pathogenic mechanism, especially focuss on osteogenesis and apoptotic cell death.

Pathology of Vascular Calcification

Atherosclerotic (Intimal) Calcification

Various degrees of mineralisation are found in atherosclerotic lesions, which advance to type IV (lesions with lipid core) and beyond [12]. With refined microscopic methods, the lesions of younger adults have revealed small aggregates of crystalline calcium among the lipid particles of lipid cores. Calcium deposits are found more frequently and in greater amounts in elderly individuals and more advanced lesions.

The first evidence of calcification is seen in the organelles (probably mitochondria) of intimal smooth muscle cells that remain embedded among the accumulated particles of lipid cores. Calcium ions presumably enter the cells at an increased rate through the damaged cell membrane. When the cells die and disintegrate, mineralised organelles become part of the extracellular accumulation. The extracellular membranous vesicles and particles of the lipid core may also serve as sites of calcium deposition. Extracellular particles of calcium progressively increase in size and fuse into aggregates. Large clumps of mineral may dominate the core of a lesion by replacing the extracellular lipid at the base or throughout the lesion core. In late stage atherosclerotic plaques, trabecular bone including marrow elements (osseous metaplasia) may also be seen (Figure 10.16). Multinucleated giant cells are observed

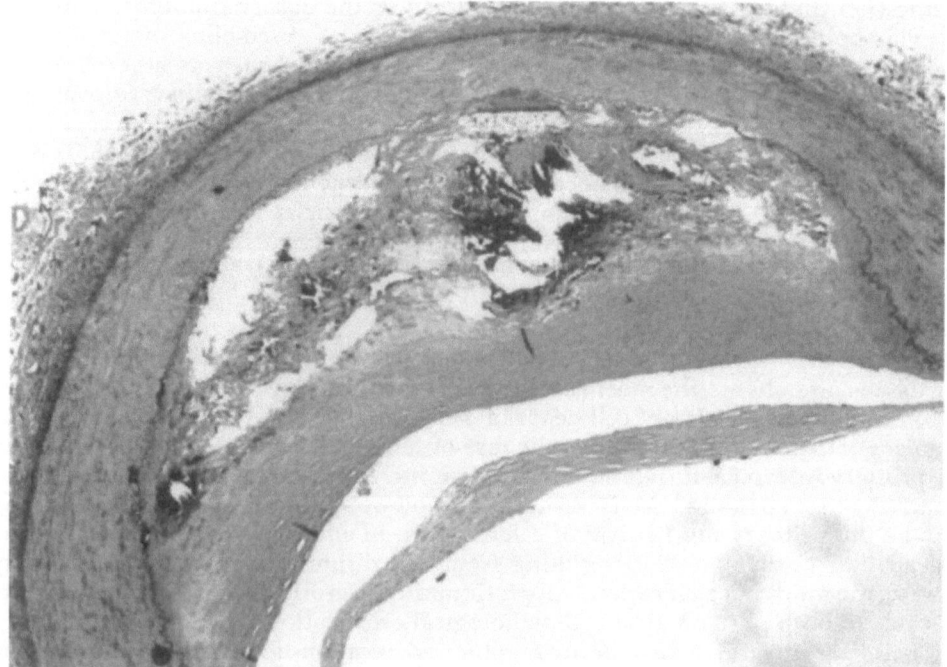

Figure 10.16.   Calcified lesion in an atherosclerotic plaque. This photomicrograph represents a section of femoral artery obtained at autopsy (haematoxylin and eosin stain). Osseous metaplasia including marrow elements are present at the base of the plaque.

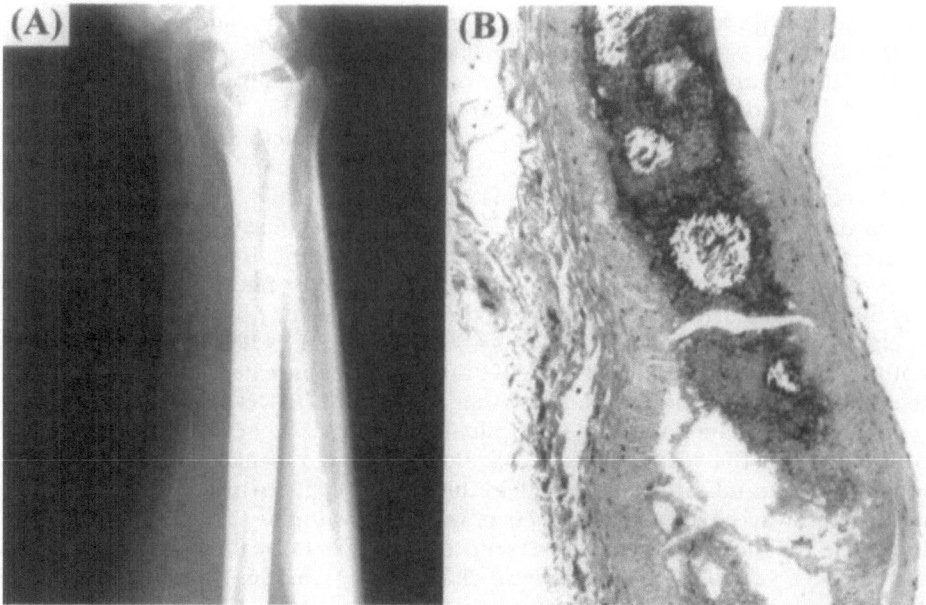

Figure 10.17. Mönckeberg's medial sclerosis in a patient with end-stage renal disease. (A) The roentogenogram of the left forearm before surgery shows diffuse medial calcification of radial artery. (B) The photomicrograph represents a section of radial artery obtained at the surgery of arteriovenous fistula for haemodialysis (haematoxylin and eosin stain).

in close apposition to the mineralised bone, suggesting that osteoclastic bone resorption may occur in ectopic bone tissue.

Non-collagenous proteins regulating skeletal mineralisation such as osteopontin (OPN), matrix gla protein (MGP), osteocalcin (OC), and osteonectin (ON) are demonstrated in the atherosclerotic calcified lesions [5, 13, 14, 15]. However, the functions of these NCPs in atherosclerotic calcification remain largely unknown. Additionally, in vivo models of atherosclerotic calcification have not been established, although various models of atherosclerosis have been developed.

Mönckeberg's Medial Sclerosis (Medial Calcification)

Mönckeberg's medial calcification is an age-related degenerative process in which the media of large and medium-sized muscular arteries calcify, and it is a fundamentally different process from occlusive atherosclerosis and has little or no clinical significance. The vessels most commonly affected are the femoral, tibial, radial, ulnar, and uterine arteries. This calcified lesion may coexist with atherosclerosis in the same vessels. Affected vessels, which are hard when palpated and are demonstrable on roentogenograms, generally are dilated and show transverse ridges of medial calcification under the intima (Figure 10.17). Sometimes, medial calcification exhibits osseous metaplasia containing marrow elements. Mönckeberg's medial calcification is frequently seen in diabetic patients with neuropathy or uraemic patients undergoing haemodialysis [16, 17].

In Mönckeberg's medial calcific lesions, matrix proteins regulating mineralisation such as matrix gla protein (MGP), osteocalcin (OC), alkaline phosphatase (ALP), and type II collagen are expressed [18]. Since these matrix proteins are exclusively elaborated by osteoblasts and/or chondrocytes, osteogenic and/or chondrogenic processes may be involved in the development of medial calcification. Animal models of medial calcification have also been developed. MGP-deficient mice develop diffuse arterial calcification of medial tissue [19]. Therefore, MGP may function as an inhibitor of calcification in the arteries. Furthermore, osteoprotegerin (OPG) knockout mice show early-onset osteoporosis and medial calcification of aorta and renal arteries [20]. OPG is a soluble member of the tumour necrosis factor receptor family that inhibits osteoclastogenesis. In the calcified lesions of OPG-deficient mice, osteoclast-like multinucleated giant cells are observed [21]. OPG ligand (OPGL) and receptor activator of NF-kappaB (RANK) are expressed in calcified lesions of OPG-deficient mice. These findings indicate that OPG/OPGL/RANK signalling pathways may play an important role in medial calcification process. Additionally, such findings may explain the coincidence of osteoporosis and vascular calcification. Mice lacking Smad6, an inhibitory molecule in the TGF-β (transforming growth factor-β) signalling pathway, have various defects of the cardiovascular system including bone formation in the media of the outflow tracts of the heart. Bone formation is restricted to areas where vascular smooth muscle cells express Smad6. Mutant mice exhibit cartilaginous metaplasia and trabeculated bone structures, sometimes containing marrow elements, suggesting a process similar to endochondral bone formation mediated by BMPs (bone morphogenetic proteins) [22].

Mechanisms of Vascular Calcification

In Vitro Models of Vascular Calcification

As described in the previous section, vascular wall cells of mesenchymal origin including smooth muscle cells may play an important role in the development of both types of vascular calcification. Several groups have developed in vitro models of vascular calcification (Table 10.11) [7–10, 23–25]. Bovine vascular smooth muscle cells calcify their extracellular matrix in the presence of β-glycerophosphate (β-GP). Bovine microvascular pericytes and pericyte-like cells present in vascular walls form nodules, which are subsequently mineralised. Human vascular smooth muscle cells

Table 10.11.   In Vitro Models of Vascular Calcification

Cells	Origin	Patterns of Calcification	Ref.
Calcifying vascular cells (pericyte-like cells)	Bovine aortic media	Spontaneous/Nodule formation	[7]
Vascular smooth muscle cells	Bovine aortic media	Addition of β-GP/Diffuse	[8, 23]
Microvascular pericytes	Bovine retina	Spontaneous/Nodule formation	[10, 24]
Vascular smooth muscle cells	Human aortic media	Spontaneous/Nodule formation	[9]
Vascular smooth muscle cells	Human aortic media	Addition of Pi/Diffuse	[25]

β-GP, beta-glycerophosphate 10 mM; Pi, inorganic phosphate 0.4–2 mM.

also have the calcifying capacity, and inorganic phosphate induces extensive calcification [25]. In these models of vascular calcification, the cells express phenotypic markers of osteoblastic differentiation such as Core-binding factor-α subunit 1 (Cbfa1), OPN, OC, and ALP. These findings suggest that vascular cells of mesenchymal origin preserve the osteogenic potential.

Origin of Calcifying Cells in Vascular Walls

Since primary culture of smooth muscle cells derived from the aorta may include microvascular pericytes of vasa vasorum (feeding vessels of the aortic wall), it is likely that pericyte-like vascular cells found in primary culture of smooth muscle cells may be derived from pericytes of vasa vasorum. Actually, pericytes can function as a supplementary source of osteoblasts in periosteal osteogenesis [26]. In angiogenesis in vitro, vascular smooth muscle cells may acquire pericyte-like phenotypes when co-cultured with endothelial cells [27]. In embryonic vascular development, pericytes and smooth muscle cells may be differentiated from mesenchymal progenitor cells. However, precise relationships between pericytes and smooth muscle cells in vascular development largely remain unknown. Anyway, both types of vascular cells may exhibit interchangeable phenotypes under certain pathological conditions.

In endochondral bone formation, vascular endothelial growth factor (VEGF)-mediated vascular invasion is an essential signal that regulates growth plate morphogenesis and triggers cartilage remodelling [28]. Angiogenesis may recruit progenitors of bone-forming cells and promote their differentiation into osteoblasts. In the development of atherosclerotic plaques, angiogenesis has been frequently found and may stimulate remodelling of vascular wall structures. In this process, vascular invasion into atherosclerotic plaques may recruit the mesenchymal cells that can differentiate into bone-forming cells. Subsequently, these cells may induce calcification and form trabecular bone-like structures in atherosclerotic plaque lesions (Figure 10.18).

Osteoblastic Differentiation of Vascular Cells

Since osseous metaplasia has often been observed in the progression of vascular calcified lesions, it is an intriguing hypothesis that subpopulations of vascular cells may differentiate into bone-forming osteoblasts in the process of vascular calcification. Actually, bovine vascular smooth muscle cells or pericyte-like vascular cells express some of the phenotypic markers of osteoblastic differentiation such as ALP, OPN, and OC [29, 30]. Microvascular pericytes or pericyte-like vascular cells form nodules, which are subsequently calcified as seen in osteoblast cultures [7, 10].

Osteoblastic differentiation is a multi-step process, proceeding through defined stages of maturation from a committed progenitor cell of mesenchymal origin capable of proliferation to a post-proliferative osteoblast expressing bone phenotypic markers. However, the molecular basis of osteoblast-specific gene expression and differentiation remains unclear. Recently, a key regulatory transcription factor in osteoblastic differentiation, Cbfa1, has been identified [31]. Cbfa1-deficient mice show a total lack of bone and retention of the partially calcified cartilaginous skeleton [32]. In humans, mutations of this gene cause cleidocranial dysplasia, an autosomal dominant skeletal disorder. Moreover, over-expression of the Cbfa1 gene in non-osteoblastic cells induces expression of the principal osteoblast-specific genes, such as $\alpha 1$(I) procollagen, OPN, bone sialoprotein (BSP), and OC. Therefore, Cbfa1

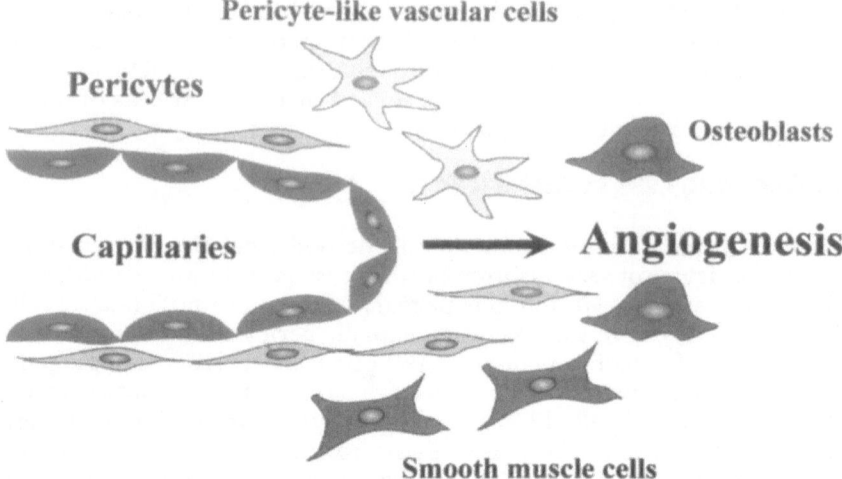

Figure 10.18.   Possible origin of bone-forming vascular cells. In the development of atherosclerotic plaque lesions, angiogenesis may stimulate remodelling of arterial wall structures. In this process, vascular invasion into atherosclerotic plaques may recruit the mesenchymal cells that can differentiate into bone-forming cells.

may be one of the "master genes" for osteoblastic differentiation. The expression of the Cbfa1 gene has been demonstrated in various vascular cells including bovine vascular smooth muscle cells (BVSMCs), pericyte-like vascular cells, and microvascular pericytes. Furthermore, dexamethasone, a potent synthetic glucocorticoid, increases the expression of the Cbfa1 gene in BVSMCs and enhances in vitro calcification [29]. These findings indicate that Cbfa1 may function as an inducer that differentiates vascular cells into osteoblastic cells.

Roles of Non-Collagenous Proteins Regulating Mineralisation

Non-collagenous proteins regulating mineralisation (NCPs) are frequently found in atherosclerotic and medial calcification (Table 10.12) [1]. Like bone tissues, these areas of mineralisation are composed of both hydroxyapatite and organic matrix, including type I collagen and NCPs. Since NCPs are important for matrix organisation and regulation of mineralisation in bone, it may also be involved in the development of vascular calcification.

Table 10.12.   Expression of Non-Collagenous Proteins in Vascular Calcification

Lesions	Non-collagenous matrix protein
Atherosclerotic calcification	OPN, MGP, OC
Mönckeberg's medial sclerosis	OPN, MGP, OC, BSP, ALP, ON
Valvular calcification	OC, OPN, ON

OPN, osteopontin; MGP, matrix gla protein; OC, osteocalcin; ON, osteonectin; BSP, bone sialoprotein; ALP, alkaline phosphatase.

Osteopontin (OPN) (also known as bone sialoprotein I, 2ar, Spp-I, an Eta-1) is a multifunctional, highly phosphorylated glycoprotein containing the RGD (Arg-Gly-Asp) sequence which is a motif common to many matrix proteins and serves as a recognition site for diverse integrin receptors. OPN is widely expressed in skeletal and non-skeletal tissues: bone, kidney, brain, smooth muscle cells, activated T cells, macrophages, and many others. OPN has been implicated in biomineralisation owing to its ability to bind Ca^{2+} at high capacity and low affinity. OPN has been localised at the mineralisation front, the lamina limitans, and cement lines. It has been demonstrated that OPN functions as an inhibitor of vascular calcification by using an in vitro model and that this inhibitory function is dependent upon phosphorylation of its molecule [23, 33]. Additionally, OPN is chemotactic for vascular smooth muscle cells, endothelial cells, and macrophages in vitro.

Matrix gla protein (MGP) is a γ-carboxylated mineral-binding extracellular matrix protein (ECM). MGP is abundantly expressed in vascular smooth muscle cells and chondrocytes. As demonstrated by a classical knockout experiment, MGP acts in vivo as an inhibitor of mineralisation in arteries and cartilage. Mice deficient in MGP are normal at birth, but develop severe calcification of all their arteries within weeks [19]. This calcification is lethal with a 100% penetrance in different genetic backgrounds, and all MGP-deficient mice die at around eight weeks of age, mostly due to rupture of the aorta. Histologically, the arterial calcification in the MGP-deficient mice appears in the media and, once initiated, advances rapidly along elastin fibres. There is no indication of atherosclerotic lesions or ectopic bone formation in these mice. MGP expression in bovine vascular smooth muscle cells (VSMCs) in vitro decreases along with the progression of VSMC calcification, and inhibition of mineralisation by bisphosphonates such as etidronate and alendronate restores its expression [34].

Humans lacking functional MGP have a disease called Kuetel syndrome [35]. These patients show calcification of cartilage, but their arteries are not affected. This finding suggests that human arteries do not only rely on the presence of MGP in their ECM, but that other proteins acting in a similar fashion participate in the inhibition of mineralisation. Eight polymorphic sites have been identified in the coding region and 5'-flanking sequences of the human MGP gene. Among these polymorphisms, the A:-7 or Ala 83 alleles may confer an increased risk of plaque calcification and myocardial infarction [36]. Additionally, serum levels of MGP have been found to be significantly increased in patients with severe atherosclerosis [37]. Therefore, MGP may play an important role in the progression of vascular calcification.

Osteocalcin (OC) is another γ-carboxylated mineral-binding ECM protein. OC is the most abundant gla protein in human bone. Localised mainly to mineralised tissues, its temporal and spatial orientation suggests that it plays a role in tissue mineralisation. Although its function is still unknown, OC has been localised to cement lines and outer lamellae in bone. In human vascular smooth muscle cells, inorganic phosphate induces the expression of OC in association with the progression of in vitro calcification [25]. In atherosclerotic plaques, OC is also localised only in calcified areas. Interestingly, the bones of OC-deficient mice have an increase in bone mass and functional quality, suggesting that OC may normally function to limit bone formation without impairing bone resorption [38]. However, how OC functions in the development of vascular calcification still remains unclear.

Osteonectin (ON) or SPARC (secreted protein acidic and rich in cysteine, also termed BM40, and 43 K protein) is a collagen-binding glycoprotein widely distributed in human tissues undergoing cell proliferation, migration, and developmental remodelling. The ON molecule contains several different structural features, the most notable of which is the two EF-hand high-affinity calcium-binding sites. ON has been implicated in regulating progression of the cell through the cell cycle, cell shape, cell-interactions, binding to metal ions, binding to growth factors, and modulating enzyme activities. ON-deficient mice have decreased numbers of osteoblasts and osteoclasts, indicating decreased bone turnover as a result of decreased basic multi-cellular units of remodelling [39]. Compared with control animals, the decrease in bone formation exceeds the decrease in osteoclast number, causing a progressive decline in trabecular bone volume in the ON-null mice and leading to low-turnover osteopenia. Therefore, ON may be critical in the support of remodelling and main-tenance of bone mass. Although ON has been detected in atherosclerotic plaques, its expression decreases with development and progression of atherosclerosis. Its role in atherosclerotic calcification is not clear.

Roles of Sodium-Dependent Phosphate Co-Transport in Vascular Cells

Inorganic phosphate (Pi) is essential for formation and mineralisation of bone. A positive relationship exists between the plasma Pi level and the rate of bone growth and/or mineralisation. Observations in cultured osteogenic cells and in the matrix vesicles they produce suggest that Pi transport plays an important role in the initia-tion of extracellular matrix calcification, both in vitro and in vivo. Type III sodium-dependent phosphate co-transporter (Pit-1) has been identified in osteogenic cells. Since β-GP accelerates BVSMC calcification, we hypothesise that Pi released by the action of alkaline phosphatase may regulate vascular calcification through Pi trans-port. Pi directly increases in vitro calcification by bovine and human vascular smooth muscle cells [25, 40]. These cells express Pit-1 mRNA and a functional sodium-dependent phosphate co-transport system. Furthermore, phosphonoformic acid (foscarnet) and arsenate are inhibitors of sodium-dependent phosphate co-transporter, in vitro calcification as well as Pi transport. In addition, Pi transport directly regulates gene expression such as Cbfa1, OC, and OPN in bovine and human vascular smooth muscle cells or murine osteoblastic cells [25, 41]. Finally, Pi directly induces apoptosis of osteoblast-like cells in culture and this effect is also dependent on Pi transport [42]. These findings suggest that sodium-dependent Pi co-transport may play a crucial role in vascular calcification (Figure 10.19).

Regulation by Hormones, Growth Factors, and Cytokines

Vascular calcification often occurs in women with osteoporosis. Moreover, there is an inverse relationship between the degree of vascular calcification and bone mineral content. As mentioned earlier, osteoprotegerin (OPG)-deficient mice show early-onset osteoporosis and medial calcification of aorta and renal arteries [20]. These findings indicate that calcium homeostasis is important in vascular calcification as well as in osteoporosis. Therefore, various hormones, growth factors, and cytokines that regulate skeletal metabolism may modulate vascular calcification (Table 10.13). We have demonstrated that parathyroid hormone (PTH) and PTH-related peptide (PTHrP) inhibit in vitro calcification through depression of ALP activity and that PTHrP elaborated by BVSMC functions as an endogenous inhibitor of vascular calcification, suggesting that vascular smooth muscle cells are equipped with an autocrine and/or paracrine system that regulates calcium metabolism [43]. $1\alpha,25$-dihydroxyvitamin D_3 ($1,25(OH)_2D_3$), an active metabolite of vitamin D, can induce vascular calcification in both humans and experimental animals. In the BVSMC

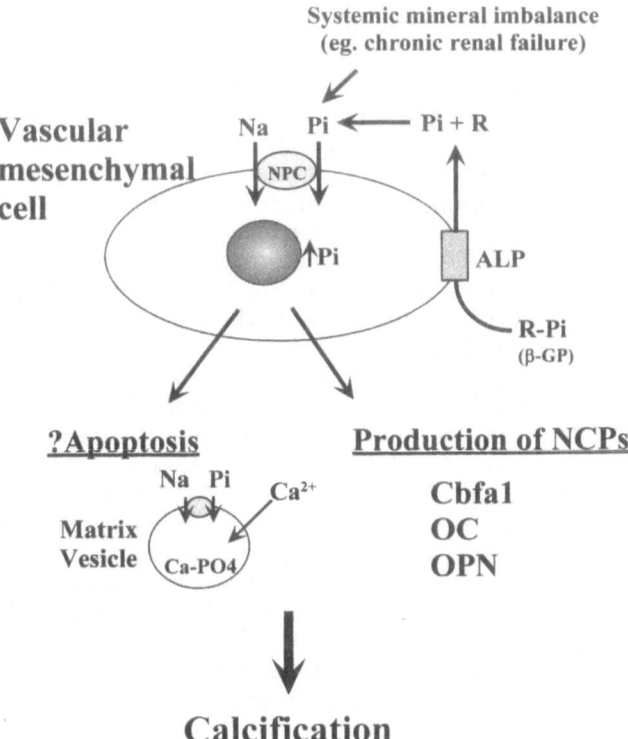

Figure 10.19. Possible mechanism of calcification by vascular mesenchymal cells. Sodium-dependent phosphate co-transport activity (probably type III co-transporter) is present in the plasma membrane of vascular mesenchymal cells and in their derived matrix vesicles. Alkaline phosphatase expressed on the cell surface may release inorganic phosphate (Pi) from phosphorylated proteins. Pi transport activity in matrix vesicles may play an important role in vascular wall calcification. Apoptosis may promote matrix vesicle formation. NCP, sodium-dependent phosphate co-transporter; ALP, alkaline phosphatase; OC, osteocalcin; Cbfa1, core binding factor-a subunit 1; OPN, osteopontin.

Table 10.13.   Factors Regulating Vascular Calcification

Regulators	In Vitro Calcification
Hormones	
Parathyroid hormone (PTH)	Inhibition
Parathyroid hormone-related peptide (PTHrP)	Inhibition
1α,25-Dihydroxyvitamin D	Stimulation
17β-Oestradiol	Stimulation
Glucocorticoid (dexamethasone)	Stimulation
Growth Factors/Cytokines	
Transforming growth factor-β1 (TGF-β1)	Stimulation
Tumour necrosis factor-α (TNF-α)	Stimulation
Oxidised Lipids	
25-Hydroxycholesterol	Stimulation
Others	
Advanced glycation end-product (AGE)	Stimulation

model of vascular calcification, 1,25(OH)$_2$D$_3$ increases mineralisation and ALP activity, while it decreases secretion of PTHrP and its gene expression [44]. Furthermore, exogenous PTHrP antagonises the stimulatory effect of 1,25(OH)$_2$D$_3$ on the calcium deposition. Therefore, 1,25(OH)$_2$D$_3$ exerts a stimulatory effect on vascular calcification by impacting the calcium-regulating system of VSMC.

Oestrogen deficiency is one of the major causes of postmenopausal osteoporosis and may also promote the progression of atherosclerosis in women. 17β-oestradiol, an active form of oestrogen, exerts various effects on vascular smooth muscle cells. While many in vitro and in vivo studies demonstrate the anti-atherogenic effects of oestrogen, 17β-oestradiol promotes osteoblastic differentiation of pericyte-like vascular cells and their in vitro calcification [45]. Therefore, oestrogen may play a regulatory role in vascular calcification.

Dexamethasone (Dex), a potent, synthetic glucocorticoid, is well known to promote osteoblastic differentiation of osteoprogenitor cells. Dex increases calcium deposition in BVSMC culture and the expression of phenotypic markers of osteoblasts such as Cbfa1 and ALP [29]. Therefore, Dex may enhance vascular calcification by promoting osteoblastic differentiation of vascular cells.

Various cytokines and growth factors are elaborated in atherosclerotic lesions and may be implicated in progression of the lesions. Transforming growth factor-β1 (TGF-β1) and tumour necrosis factor-α (TNF-α) have been demonstrated to increase in vitro calcification by pericyte-like vascular cells [7, 46]. Oxidised lipids including oxysterol are also localised in atherosclerotic plaques and may be involved in the development of the lesions. 25-Hydroxycholesterol, which is a major component of oxidised LDL, stimulates calcified nodule formation by pericyte-like vascular cells [7]. Therefore, cytokines, growth factors, and oxidised lipids may be involved in the progression of atherosclerotic calcification.

Roles of Inflammatory Cells in Vascular Calcification

Inflammatory cells such as macrophages and T lymphocytes are frequently found in advanced atherosclerotic lesions, particularly fibro-fatty or fibrous plaques in which calcification usually initiates. Calcium deposits are usually observed at the periphery of the lipid core. It has been pointed out that macrophages are the predominant cell type associated with different stages of calcification in atherosclerotic plaques [47]. T lymphocytes and macrophage infiltrates are also associated with calcification in human native and porcine bioprosthetic valves [48]. It has been suggested that production of non-collagenous proteins and cytokines by these cells may contribute to vascular calcification.

We hypothesised that an interaction between macrophages and vascular smooth muscle cells (VSMCs) may be implicated in the phenotypic changes of VSMCs, resulting in atherosclerotic calcification. One of the possible mechanisms responsible for this interaction is 1,25(OH)$_2$D production by macrophages, because monocytes and tissue macrophages synthesise 1,25(OH)$_2$D after stimulation with interferon-γ (IFN-γ). IFN-γ, one of the major cytokines elaborated by T lymphocytes, is implicated in the development of atherosclerosis and strongly stimulates production of 1,25(OH)$_2$D in macrophages through inducing expression of the vitamin D activating enzyme, 25-hydroxyvitamin D 1α-hydroxylase [40]. We have demonstrated using the co-culture of human vascular smooth muscle cells (HVSMCs) with THP-1 cells or human peripheral blood monocytes in the presence of 1,25(OH)$_2$D$_3$ and IFN-γ that ALP is strongly induced in HVSMCs and that in vitro calcification is brought about in HVSMCs with high expression of ALP in the presence of β-GP (Shioi et al. submitted, 2001). These findings suggest that interactions among smooth muscle cells, macrophages, and T lymphocytes in atherosclerotic lesions may contribute to the development of atherosclerotic calcification.

Interestingly, multinucleated giant cells resembling osteoclasts are sometimes found in calcified vascular lesions such as osseous metaplasia in the intima and medial calcified lesions [49]. We found that these osteoclast-like cells are present in medial calcified lesions in patients with end-stage renal disease (Figure 10.20). As mentioned above, OPG ligand and RNAK are expressed in calcified lesions of OPG-deficient mice [21]. Since these molecules are involved in osteoclastogenesis, osteoclasts may play a role in progression of vascular calcified lesions.

Roles of Apoptotic Cell Death in Vascular Calcification

Apoptotic cell death has frequently been observed in atherosclerotic plaques by the TUNEL technique, which detects DNA fragmentation in tissue sections [50, 51]. As described above, the first evidence of calcification is seen in the organelles (probably mitochondria) of intimal smooth muscle cells that remain embedded among the accumulated particles of lipid cores from the morphological studies of atherosclerotic calcification. The extracellular membranous vesicles and particles of the lipid core may also serve as sites of calcium deposition. Additionally, we detected apoptotic cell death of medial smooth muscle cells in medial calcified lesions in patients with end-stage renal disease by the TUNEL technique (Shioi et al. submitted, 2001). These observations suggest that apoptotic cell death of smooth muscle cells and the subsequent membranous vesicle formation may be implicated in the initiation of vascular calcification. Although apoptosis-inducing factors in atherosclerotic plaques still remain to be clarified, oxidised low-density lipoprotein, oxysterols, and nitric oxide have to be considered.

Interestingly, inorganic phosphate (Pi) induces apoptosis of osteoblast-like cells in culture [42]. Apoptotic cell death of osteoblasts is inhibited by treating the cells with phosphonoformic acid, an inhibitor of the plasma membrane, sodium-dependent Pi co-transporter, suggesting that induction of apoptosis by Pi is dependent on Pi transport. In vascular smooth muscle cells, Pi may induce apoptosis as well as in vitro calcification by a Pit-1-dependent mechanism. Actually, apoptotic cell death has been observed in the cultures of vascular smooth muscle cells that subsequently form calcified nodules [52]. The inhibition of apoptosis by caspase inhibitor decreases in vitro calcification, while stimulation of apoptosis by anti-Fas antibody increases calcium deposition. Therefore, apoptotic cell death of vascular smooth muscle cells followed by membranous vesicle formation may be the initial event in developing vascular calcification.

Integrated Hypothesis for Pathogenesis of Vascular Calcification

In the previous section, various factors contributing to the development of vascular calcification have been discussed. The most important point is that vascular cells in mesenchymal origin differentiate into osteoblastic cells in atherosclerotic lesions. These osteoblast-like cells elaborate matrix proteins and mineralise their

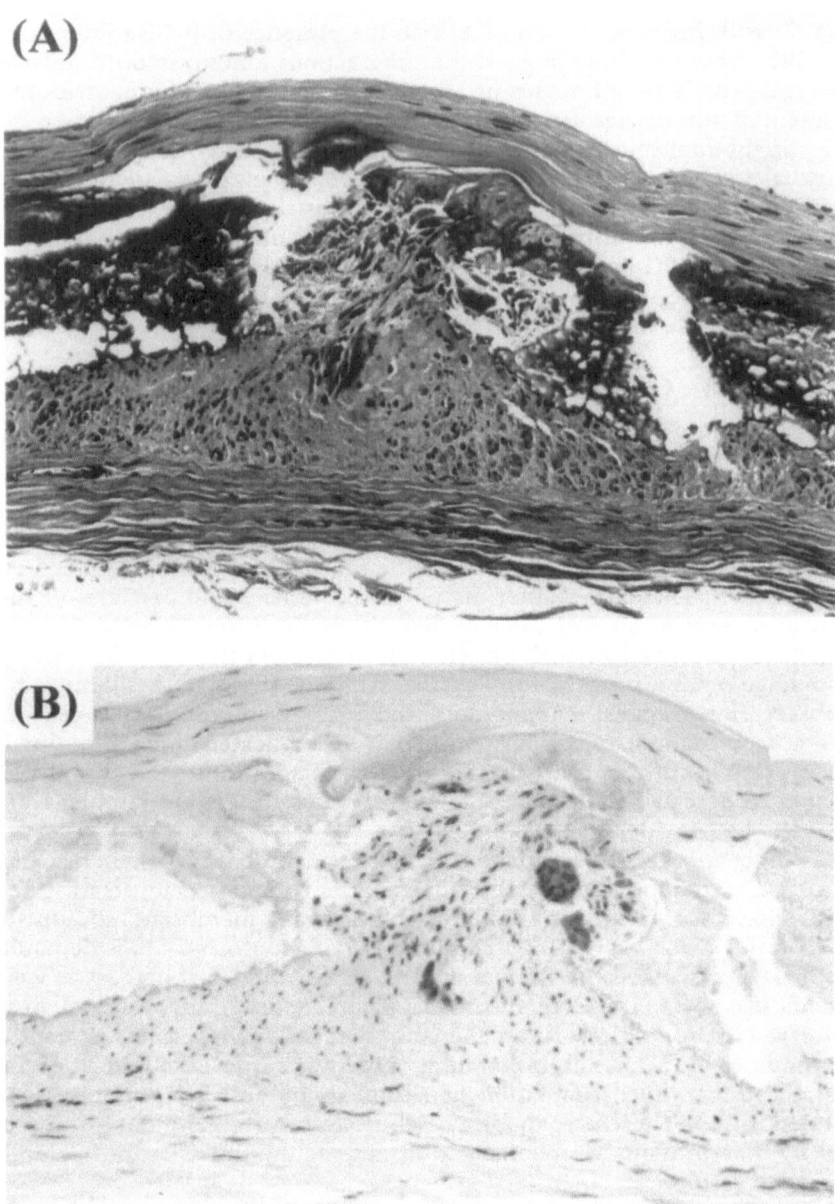

Figure 10.20.   Multinucleated giant cells in calcified lesions. Multinucleated cells are CD68-positive and may be derived from monocyte/macrophage-lineage cells. (A) The photomicrograph represents a section of human radial artery obtained at the surgery of arteriovenous fistula for haemodialysis (haematoxylin and eosin stain). (B) The photomicrograph represents a serial section of the same lesion presented in (A) and demonstrates that multinucleated giant cells are CD68-positive (immuno-histochemistry for CD68).

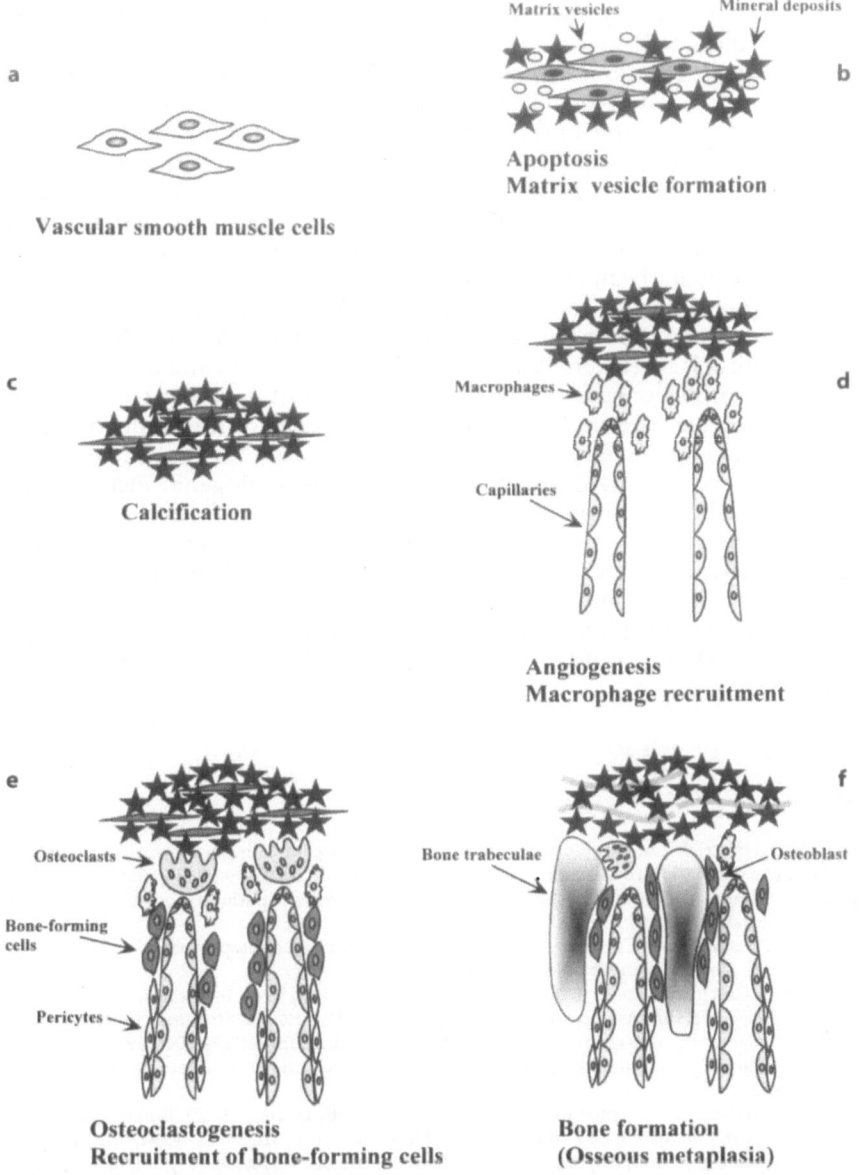

Figure 10.21. Integrated hypothesis for pathogenesis of vascular calcification. Vascular calcification may simulate the process of endochondral ossification. In arteries, vascular smooth muscle cells in the plaque or the media may undergo apoptosis, followed by matrix vesicle formation and calcification. Angiogenesis may be induced into the calcified lesions by angiogenic factors including VEGF. Osteoclast-like multinucleated giant cells are formed around calcified lesions and resorb calcified matrix. In association with vascular invasion, mesenchymal cells including microvascular pericytes migrate and differentiate into bone-forming osteoblastic cells. Finally, trabecular bone-like structures are formed.

extracellular matrix. Finally, these osteogenic reactions may lead to trabecular bone formation in arterial walls. However, the relationship between apoptotic cell death of SMC and osteogenic reactions remains to be clarified. Furthermore, the role of osteoclast-like multinucleated giant cells in calcified lesions of the arteries is not known.

To clarify these issues, it is important to compare vascular calcification with endochondral ossification (Figure 10.21). Endochondral ossification is the process whereby cartilage is replaced by bone, and it is essential for the development of long bones. Blood vessel invasion of cartilage, which is normally avascular, is the critical first step in this process. Endochondral ossification occurs as the chondrocytes of the growth plate mature into a final hypertrophic state. These hypertrophic chondrocytes secrete vascular endothelial growth factor (VEGF), which induces vascular invasion into the growth plate [28]. These cells then undergo apoptosis and are subsequently calcified. Chondroclasts (osteoclasts) aid in the resorption of the calcified cartilage, paving the way for osteoblasts to migrate in and deposit bone matrix.

In arteries, vascular smooth muscle cells in the plaque or the media may undergo apoptosis, followed by calcification. Angiogenesis (vascular invasion) may be induced into the plaques or the medial apoptotic lesions by angiogenic factors including VEGF. Osteoclast-like multinucleated giant cells are formed around calcified lesions and resorb calcified matrix. In association with vascular invasion, mesenchymal cells including microvascular pericytes migrate and differentiate into bone-forming osteoblastic cells [26]. Finally, trabecular bone-like structures are formed. This is a working hypothesis to explain how apoptosis of SMC is related to osteogenic reaction in vascular lesions. In order to prove this hypothesis, several studies should be performed.

References

1. Donley GE, Fitzpatrick LA. Noncollagenous matrix proteins controlling mineralization: possible role in pathologic calcification of vascular tissue. Trens Cardiovasc Med 1998;8:199–206.
2. Virchow R. Cellular pathology: as based upon physiological and pathological histology. New York: Dover, 1863 (unabridged reprinting, 1971).
3. Doherty TM, Detrano RC. Coronary arterial calcification as an active process: a new perspective on an old problem. Calcif Tissue Int 1994;54:224–30.
4. Anderson HC. Calcific diseases: a concept. Arch Pathol Lab Med 1983;107:341–8.
5. Fitzpatrick LA, Severson A, Edwards WD, Ingram RT. Diffuse calcification in human coronary arteries. Association of osteopontin with atherosclerosis. J Clin Invest 1994;94:1597–604.
6. Kim K. Calcification of matrix vesicles in human aortic valve and aortic media. Fed Proc 1976;35:156–62.
7. Watson KE, Bostrom K, Ravindranath R, Lam T, Norton B, Demer LL. TGF-beta 1 and 25-hydroxycholesterol stimulate osteoblast-like vascular cells to calcify. J Clin Invest 1994;93:2106–13.
8. Shioi A, Nishizawa Y, Jono S, Koyama H, Hosoi M, Morii H. Beta-glycerophosphate accelerates calcification in cultured bovine vascular smooth muscle cells. Arterioscler Thromb Vasc Biol 1995;15:2003–9.
9. Proudfoot D, Skepper JN, Shanahan CM, Weissberg PL. Calcification of human vascular cells in vitro is correlated with high levels of matrix Gla protein and low levels of osteopontin expression. Arterioscler Thromb Vasc Biol 1998;18:379–88.
10. Doherty MJ, Ashton BA, Walsh S, Beresford JN, Grant ME, Canfield AE. Vascular pericytes express osteogenic potential in vitro and in vivo. J Bone Miner Res 1998;13:828–38.
11. Kim KM. Apoptosis and calcification. Scanning Microsc 1995;9:1137–75.
12. Stary HC. Atlas of atherosclerosis: progression and regression. New York London: Parthenon Publishing, 1999.
13. Giachelli CM, Bae N, Almeida M, Denhardt DT, Alpers CE, Schwartz SM. Osteopontin is elevated during neointima formation in rat arteries and is a novel component of human atherosclerotic plaques. J Clin Invest 1993;92:1686–96.

14. Ikeda T, Shirasawa T, Esaki Y, Yoshiki S, Hirokawa K. Osteopontin mRNA is expressed by smooth muscle-derived foam cells in human atherosclerotic lesions of the aorta. J Clin Invest 1993;92: 2814–20.
15. Shanahan CM, Cary NR, Metcalfe JC, Weissberg PL. High expression of genes for calcification-regulating proteins in human atherosclerotic plaques. J Clin Invest 1994;93:2393–402.
16. Edmonds ME, Morrison N, Laws JW, Watkins PJ. Medial arterial calcification and diabetic neuropathy. Br Med J 1982;284:928–30.
17. Ejerblad S, Ericsson JLE, Eriksson I. Arterial lesions of the radial artery in uremic patients. Acta Chir Scand 1979;145:415–28.
18. Shanahan CM, Cary NR, Salisbury JR, Proudfoot D, Weissberg PL, Edmonds ME. Medial localization of mineralization-regulating proteins in association with Mönckeberg's sclerosis: evidence for smooth muscle cell-mediated vascular calcification. Circulation 1999;100:2168–76.
19. Luo G, Ducy P, McKee MD, Pinero GJ, Loyer E, Behringer RR, et al. Spontaneous calcification of arteries and cartilage in mice lacking matrix GLA protein. Nature 1997;386:78–81.
20. Bucay N, Sarosi I, Dunstan CR, Morony S, Tarpley J, Capparelli C, et al. Osteoprotegerin-deficient mice develop early onset osteoporosis and arterial calcification. Genes Dev 1998;12:1260–8.
21. Min H, Morony S, Sarosi I, Dunstan CR, Capparelli C, Scully S, et al. Osteoprotegerin reverses osteoporosis by inhibiting endosteal osteoclasts and prevents vascular calcification by blocking a process resembling osteoclastogenesis. J Exp Med 2000;192:463–74.
22. Galvin KM, Donovan MJ, Lynch CA, Meyer RI, Paul RJ, Lorenz JN, et al. A role for smad6 in development and homeostasis of the cardiovascular system. Nat Genet 2000;24:171–4.
23. Wada T, McKee MD, Steitz S, Giachelli CM. Calcification of vascular smooth muscle cell cultures: inhibition by osteopontin. Circ Res 1999;84:166–78.
24. Yamagishi S, Fujimori H, Yonekura H, Tanaka N, Yamamoto H. Advanced glycation end-products accelerate calcification in microvascular pericytes. Biochem Biophys Res Commun 1999;258:353–7.
25. Jono S, McKee MD, Murry CE, Shioi A, Nishizawa Y, Mori K, et al. Phosphate regulation of vascular smooth muscle cell calcification. Circ Res 2000;87:e10–17.
26. Diaz-Flores L, Gutierrez R, Lopez-Alonso A, Gonzalez R, Varela H. Pericytes as a supplementary source of osteoblasts in periosteal osteogenesis. Clin Orthop 1992;275:280–6.
27. Nicosia RF, Villaschi S. Rat aortic smooth muscle cells become pericytes during angiogenesis in vitro. Lab Invest 1995;73:658–66.
28. Gerber HP, Vu TH, Ryan AM, Kowalski J, Werb Z, Ferrara N. VEGF couples hypertrophic cartilage remodeling, ossification and angiogenesis during endochondral bone formation. Nat Med 1999;5:623–8.
29. Mori K, Shioi A, Jono S, Nishizawa Y, Morii H. Dexamethasone enhances In vitro vascular calcification by promoting osteoblastic differentiation of vascular smooth muscle cells. Arterioscler Thromb Vasc Biol 1999;19:2112–18.
30. Tintut Y, Parhami F, Bostrom K, Jackson SM, Demer LL. cAMP stimulates osteoblast-like differentiation of calcifying vascular cells. Potential signaling pathway for vascular calcification. J Biol Chem 1998;273:7547–53.
31. Ducy P, Zhang R, Geoffroy V, Ridall AL, Karsenty G. Osf2/Cbfa1: a transcriptional activator of osteoblast differentiation. Cell 1997;89:747–54.
32. Komori T, Yagi H, Nomura S, Yamaguchi A, Sasaki K, Deguchi K, et al. Targeted disruption of Cbfa1 results in a complete lack of bone formation owing to maturational arrest of osteoblasts. Cell 1997;89:755–64.
33. Jono S, Peinado C, Giachelli CM. Phosphorylation of osteopontin is required for inhibition of vascular smooth muscle cell calcification. J Biol Chem 2000;275:20197–203.
34. Mori K, Shioi A, Jono S, Nishizawa Y, Morii H. Expression of matrix Gla protein (MGP) in an in vitro model of vascular calcification. FEBS Lett 1998;433:19–22.
35. Munroe PB, Olgunturk RO, Fryns JP, Van Maldergem L, Ziereisen F, Yuksel B, et al. Mutations in the gene encoding the human matrix Gla protein cause Keutel syndrome. Nat Genet 1999;21:142–4.
36. Herrmann SM, Whatling C, Brand E, Nicaud V, Gariepy J, Simon A, et al. Polymorphisms of the human matrix gla protein (MGP) gene, vascular calcification, and myocardial infarction. Arterioscler Thromb Vasc Biol 2000;20:2386–93.
37. Braam LA, Dissel P, Gijsbers BL, Spronk HM, Hamulyak K, Soute BA, et al. Assay for human matrix gla protein in serum: potential applications in the cardiovascular field. Arterioscler Thromb Vasc Biol 2000;20(5):1257–61.
38. Ducy P, Desbois C, Boyce B, Pinero G, Story B, Dunstan C, et al. Increased bone formation in osteocalcin-deficient mice. Nature 1996;382:448–52.
39. Delany AM, Amling M, Priemel M, Howe C, Baron R, Canalis E. Osteopenia and decreased bone formation in osteonectin-deficient mice. J Clin Invest 2000;105:915–23.

40. Shioi A, Mori K, Jono S, Wakikawa T, Hiura Y, Koyama H, et al. Mechanism of atherosclerotic calcification. Z Kardiol 2000;89(Suppl 2):75–9.
41. Beck Jr GR, Zerler B, Moran E. Phosphate is a specific signal for induction of osteopontin gene expression. Proc Natl Acad Sci USA 2000;97:8352–7.
42. Meleti Z, Shapiro IM, Adams CS. Inorganic phosphate induces apoptosis of osteoblast-like cells in culture. Bone 2000;27:359–66.
43. Jono S, Nishizawa Y, Shioi A, Morii H. Parathyroid hormone-related peptide as a local regulator of vascular calcification. Its inhibitory action on in vitro calcification by bovine vascular smooth muscle cells. Arterioscler Thromb Vasc Biol 1997;17:1135–42.
44. Jono S, Nishizawa Y, Shioi A, Morii H. 1,25-Dihydroxyvitamin D3 increases in vitro vascular calcification by modulating secretion of endogenous parathyroid hormone-related peptide. Circulation 1998; 98:1302–6.
45. Balica M, Bostrom K, Shin V, Tillisch K, Demer LL. Calcifying subpopulation of bovine aortic smooth muscle cells is responsive to 17 beta-estradiol. Circulation 1997;95:1954–60.
46. Tintut Y, Patel J, Parhami F, Demer LL. Tumor necrosis factor-alpha promotes in vitro calcification of vascular cells via the cAMP pathway. Circulation 2000;102:2636–42.
47. Jeziorska M, McCollum C, Woolley DE. Calcification in atherosclerotic plaque of human carotid arteries: associations with mast cells and macrophages. J Pathol 1998;185:10–17.
48. Srivatsa SS, Harrity PJ, Maercklein PB, Kleppe L, Veinot J, Edwards WD, et al. Increased cellular expression of matrix proteins that regulate mineralization is associated with calcification of native human and porcine xenograft bioprosthetic heart valves. J Clin Invest 1997;99:996–1009.
49. Jeziorska M, McCollum C, Woolley DE. Observations on bone formation and remodeling in advanced atherosclerotic lesions of human carotid arteries. Virchow Arch 1998;433:559–65.
50. Bjoerkerud S, Bjoerkurud B. Apoptosis is abundant in human atherosclerotic lesions, especially in inflammatory cells (macrophages and T cells), and may contribute to the accumulation of gruel and plaque instability. Am J Pathol 1996;149:367–80.
51. Kockx MM, Herman AG. Apoptosis in atherogenesis: implications for plaque destabilization. European Heart Journal 1998;19 (Supplement G):G23–8.
52. Proudfoot D, Skepper JN, Hegyi L, Bennett MR, Shanahan CM, Weissberg PL. Apoptosis regulates human vascular calcification In vitro: evidence for initiation of vascular calcification by apoptotic bodies. Circ Res 2000;87:1055–62.

Index